Alphabetical Table of the Elements

Element	Symbol	Atomic Number	Atomic Mass	Element	Symbol	Atomic Number	Atomic Mass	Element	Symbol	Atomic Number	Atomic Mass
Actinium	Ac	89	(227)	Holmium	Ho	67	164.9	Rhenium	Re	75	186.2
Aluminum	Al	13	26.98	Hydrogen	H	1	1.008	Rhodium	Rh	45	102.9
Americium	Am	95	(243)	Indium	In	49	114.8	Rubidium	Rb	37	85.47
Antimony	Sb	51	121.8	Iodine	I	53	126.9	Ruthenium	Ru	44	101.1
Argon	Ar	18	39.95	Iridium	Ir	77	192.2	Samarium	Sm	62	150.4
Arsenic	As	33	74.92	Iron	Fe	26	55.85	Scandium	Sc	21	44.96
Astatine	At	85	(210)	Krypton	Kr	36	83.80	Selenium	Se	34	78.96
Barium	Ba	56	137.3	Lanthanum	La	57	138.9	Silicon	Si	14	28.09
Berkelium	Bk	97	(247)	Lawrencium	Lr	103	(260)	Silver	Ag	47	107.9
Beryllium	Be	4	9.012	Lead	Pb	82	207.2	Sodium	Na	11	22.99
Bismuth	Bi	83	209.0	Lithium	Li	3	6.941	Strontium	Sr	38	87.62
Boron	B	5	10.81	Lutetium	Lu	71	175.0	Sulfur	S	16	32.06
Bromine	Br	35	79.90	Magnesium	Mg	12	24.31	Tantalum	Ta	73	180.9
Cadmium	Cd	48	112.4	Manganese	Mn	25	54.94	Technetium	Tc	43	(98)
Calcium	Ca	20	40.08	Mendelevium	Md	101	(258)	Tellurium	Te	52	127.6
Californium	Cf	98	(251)	Mercury	Hg	80	200.6	Terbium	Tb	65	158.9
Carbon	C	6	12.01	Molybdenum	Mo	42	95.94	Thallium	Tl	81	204.4
Cerium	Ce	58	140.1	Neodymium	Nd	60	144.2	Thorium	Th	90	232.0
Cesium	Cs	55	132.9	Neon	Ne	10	20.18	Thulium	Tm	69	168.9
Chlorine	Cl	17	35.45	Neptunium	Np	93	(237)	Tin	Sn	50	118.7
Chromium	Cr	24	52.00	Nickel	Ni	28	58.69	Titanium	Ti	22	47.88
Cobalt	Co	27	58.93	Niobium	Nb	41	92.91	Tungsten	W	74	183.9
Copper	Cu	29	63.55	Nitrogen	N	7	14.01	Unnilennium	Une	109	(267)
Curium	Cm	96	(247)	Nobelium	No	102	(259)	Unnilhexium	Unh	106	(263)
Dysprosium	Dy	66	162.5	Osmium	Os	76	190.2	Unniloctium	Uno	108	[illegible]
Einsteinium	Es	99	(252)	Oxygen	O	8	16.00	Unnilpentium	Unp	105	[illegible]
Erbium	Er	68	167.3	Palladium	Pd	46	106.4	Unnilquadium	Unq	104	[illegible]
Europium	Eu	63	152.0	Phosphorus	P	15	30.97	Unnilseptium	Uns	107	[illegible]
Fermium	Fm	100	(257)	Platinum	Pt	78	195.1	Uranium	U	92	
Fluorine	F	9	19.00	Plutonium	Pu	94	(244)	Vanadium	V	23	
Francium	Fr	87	(223)	Polonium	Po	84	(209)	Xenon	Xe	54	
Gadolinium	Gd	64	157.3	Potassium	K	19	39.10	Ytterbium	Yb	70	
Gallium	Ga	31	69.72	Praseodymium	Pr	59	140.9	Yttrium	Y	39	
Germanium	Ge	32	72.59	Promethium	Pm	61	(145)	Zinc	Zn	30	
Gold	Au	79	197.0	Protactinium	Pa	91	(231)	Zirconium	Zr	40	
Hafnium	Hf	72	178.5	Radium	Ra	88	226				
Helium	He	2	4.003	Radon	Rn	86	(222)				

Essentials of Chemistry

Extended Edition

Essentials of Chemistry
Extended Edition

William Rife

Professor, Chemistry Department
California Polytechnic State University

Saunders College Publishing
A Harcourt Brace Jovanovich College Publisher

Fort Worth Philadelphia San Diego New York Orlando Austin San Antonio Toronto Montreal London Sydney Tokyo

Text Typeface: Caledonia
Compositor: York Graphic Services
Acquisitions Editor: John Vondeling
Developmental Editors: Kate Pachuta, Jennifer L. Bortel
Managing Editor: Carol Field
Project Editor: Nancy Lubars
Copy Editing: General Graphic Services, Martha Colgan
Manager of Art and Design: Carol Bleistine
Art Director: Christine Schueler
Art Coordinator: Caroline McGowan
Photo Research Editor: Dena Digilio-Betz
Text Designer: Tracy Baldwin
Cover Designer: Lawrence R. Didona
Text Artwork: Rolin Graphics
Layout Artist: Tracy Baldwin, Bill Boehm
Director of EDP: Tim Frelick
Production Manager: Charlene Squibb
Senior Marketing Manager: Marjorie Waldron

Cover Credit: © Jay Freis/Image Bank

Printed in the United States of America

ESSENTIALS OF CHEMISTRY

0-03-030353-2

Library of Congress Catalog Card Number: 92-060345

2345 071 987654321

(Diane Schiumo) Fundamental Photographs, New York

Preface

An increasing number of students who are required to take a college or university course in general chemistry face that requirement with anxiety because they do not have the preparation in knowledge or skills to meet it. The purpose of this book is to prepare these students to take a course in general chemistry confidently and enjoyably by giving them a thorough understanding of the most fundamental principles of chemistry: the atomic theory, periodicity, bonding and interparticle forces, chemical notation and nomenclature, chemical calculations, and the nature of chemical reactions in aqueous solutions.

A student who sees chemistry as isolated scraps of information to be memorized will be defeated by it; to be learned effectively, chemistry must make sense. For that reason, I have emphasized the framework of reasoning within which the facts of chemistry can be understood.

Chemists know that chemistry is central between physics and biology and that it is crucial for modern civilized life, but students need help to develop these perspectives. As a framework for their learning and as a motivation, I have stressed in the text information about the unique importance of chemistry as a science and as an essential part of modern technology.

The motto of my university is "Learn by Doing," and this book uses that method. The text includes many exercises and problems to test students' comprehension and to help them build confidence by applying their newly learned concepts successfully. This book is an expanded version of my earlier book, *Essentials of Chemistry*. At the request of colleagues, this version contains chapters on nuclear chemistry, organic chemistry, and biochemistry.

I would of course be delighted to have suggestions from colleagues or students on how the book can be improved.

William Rife
Chemistry Department
California Polytechnic State University
San Luis Obispo, California 93407
July, 1992

Fundamental Photographs, New York

Features of the Book

Clarity

Emphasis on Reasoning Students who see chemistry as a collection of isolated facts and arbitrary rules to be memorized do poorly in introductory chemistry, while students who see the subject as a coherent way of thinking do well. For that reason, there is an emphasis throughout this book on the framework of reasoning—the interplay among observation, experiment, theory, and application—that makes chemistry a coherent body of knowledge. The Periodic Table is especially emphasized as a resource, always available to the student, that summarizes many important facts and embodies many fundamental theories.

Emphasis on Active Learning Chemistry must be learned actively: its concepts must be applied to be understood, and its vocabulary—both words and symbols—must be practiced to be mastered. Throughout this book concepts and their applications are presented together, rather than separately, to give the concepts immediate operational meaning. Many exercises and problems, described below, require the student to practice the application of chemical concepts.

Emphasis on Relevance In this book chemistry is presented, not as a fixed collection of facts and theories detached from human experience, but as an expanding body of knowledge growing out of centuries of collective human effort, a body of knowledge with profound implications for our understanding of the world and with enormous capacities for changing it. At the end of each chapter a short essay called *Inside Chemistry* describes an important event in the creation or application of chemistry. Throughout the book, shorter essays called *Chemistry Insight* present interesting information related to the surrounding text, and marginal notes remind the student of important ideas or provide additional details on complex concepts without breaking the flow of the text.

Section Titles and Chapter Previews Several features of the book focus the student's attention on the most important information to be learned. Each chapter is divided into sections, and the title of each section is a sentence that states the most important concept in the section. On the opening page of each chapter a list of these section titles gives a preview of the important concepts in the chapter.

Chapter Summaries and Glossary A summary at the end of each chapter reviews all of the important concepts in the chapter in a flashcard format that encourages self testing and facilitates frequent review. The glossary at the end of the book gives quick access to definitions of all important terms and refers the reader to appropriate sections of the book for further information.

Examples, Exercises, and Problems Throughout the book many examples demonstrate applications of concepts in solving problems. Many exercises in each chapter and many problems at the end of each chapter provide thorough practice and self testing on newly learned concepts. At the end of each chapter, consecutive odd- and even-numbered problems are matched pairs; answers to exercises and odd-numbered problems are in Appendix 1.

Illustrations Throughout the book many drawings and photographs illustrate important concepts and encourage the student's interest. Each pedagogical figure has a complete, self-contained legend, so the figure can be studied without hunting for explanations in the text.

Friendliness

Many students begin introductory chemistry with the expectation that the subject will be cold, drab, and irrelevant. I have tried to overcome that expectation.

The tone of the text is friendly but not condescending. Each chapter is divided into several relatively short sections, and each section presents a concept, building on the sections before it. Because of this format, a student who works successfully through the book is rewarded with a growing sense of achievement, and a student who becomes confused can easily backtrack to find the source of confusion.

Adaptability

Instructors of introductory chemistry teach with a wide variety of styles to a wide variety of students in a wide variety of time frames. This book can be adapted to meet many of their needs and preferences.

Chapters or sections of the book can be treated briefly or omitted to accommodate a short time frame, (say, a quarter-length course) and these adaptations can be made in such a way that the aspects of chemistry that the instructor values most are emphasized. For example, Chapter 10, The Periodic Table as a Guide to Formulas and Reactions, can be used extensively by instructors who emphasize descriptive chemistry, or all or part of it can be assigned for enrichment reading, or it can be omitted, without breaking the continuity of reasoning in the text. Because each section heading throughout the book is a sentence that states the main concept in the section, instructors will find it easy to use the Table of Contents, which includes these section headings, to design the reading assignments for their own courses.

In presenting certain topics in introductory chemistry, many instructors have strong preferences of method and style; examples of such topics are scientific notation, electron configurations, and the gas laws. In presenting these topics, I have emphasized the approaches that I think work best, but I have briefly described the alternative approaches and alerted the student that his instructor may prefer one of the alternatives. Teaching is an art, and the function of a textbook is to enable, as fully as possible, each instructor to practice the art in his or her own way.

Ancillaries

This text is accompanied by an extensive set of support materials.

Study Guide The Study Guide provides review and study aids to further enhance the students' understanding of the text. It includes an overview of each chapter, a chapter outline, learning objectives, and review questions and answers.

Laboratory Manual The laboratory manual, *Introduction to Chemical Principles: A Laboratory Approach, 4th edition,* by veteran authors Susan Weiner of West Valley College and Edward I. Peters, provides 33 experiments that illustrate principles of introductory chemistry, including common laboratory operations and the collection and analysis of experimental data. The new fourth edition emphasizes laboratory safety and includes a new experiment using household products. The laboratory manual is supplemented with an instructor's manual.

Student Solutions Manual The *Student Solutions Manual* by Jan William Simek of California Polytechnic State University provides solutions to all of the exercises and the odd-numbered problems in the text, and it includes notes to the student that emphasize important principles of problem solving.

Instructor's Manual The *Instructor's Manual* by Jan William Simek of California Polytechnic State University provides teaching goals, alternate sequences of topics, answers to even-numbered problems in the text, and a short bibliography. The manual also provides a concordance to the Shakhashiri Chemical Demonstration Videotapes described below.

Test Bank The *Test Bank*, an extensive file of multiple-choice questions for each chapter, is available as a printed and bound book or on disk in Macintosh and IBM PC versions. Included with the computerized test banks is Exam-Record, which allows the instructor to record, curve, graph, and print out grades.

Overhead Transparencies A set of 94 overhead transparencies of four-color illustrations and photographs from the text is available for classroom or laboratory use.

Shakhashiri Chemical Demonstration Videotapes This set of fifty 3- to 5-minute classroom experiments bring the imagination and vitality of Bassam Shakhashiri to the classroom. Accompanying the videotapes is an **Instructor's Manual** containing a description of each demonstration as well as follow-up discussion questions.

Charles D. Winters

Acknowledgments

These reviewers carried out the very difficult work of providing sound and thoughtful criticism:

Oren P. Anderson, Colorado State University
Jerry A. Driscoll, University of Utah
Elisheva Goldstein, California State Polytechnic University, Pomona
Wyman Grindstaff, Southwest Missouri State University
Cynthia T. Hahn, Bristol Community College
Arthur H. Hayes, Rancho Santiago College
Gerard F. Judd, Phoenix College
Leslie N. Kinsland, University of Southwestern Louisiana
David F. Koster, Southern Illinois University
Anne Loeb, College of Lake County
E. Jerome Maas, Oakton Community College
Glenda B. Michaels, Western State College
Glenn H. Miller, University of California, Santa Barbara
Charles N. Millner, Jr., California State Polytechnic University, Pomona
D. K. Philbin, Allan Hancock College
Fred Redmore, Highland Community College
Stephen P. Ruis, American River College
C. Sankey Sherer, Shelton State Community College
Ruth Sherman, Los Angeles City College
Phil Silberman, Scottsdale Community College
Jan W. Simek, California Polytechnic State University, San Luis Obispo
Ronald Takata, Honolulu Community College
Hewitt G. Wight, California Polytechnic State University, San Luis Obispo

Many people at Saunders College Publishing worked hard to create this book. I want to express my appreciation to all of them and especially to Jennifer Bortel, Dena Digilio-Betz, Nancy Lubars, Shelly McCarthy, Kate Pachuta, John Vondeling, and Marjorie Waldron.

For their crucial support and encouragement, I thank Mark, Leona, Anne, and Carl.

Barry L. Runk from Grant Heilman Photography, Inc.

Contents Overview

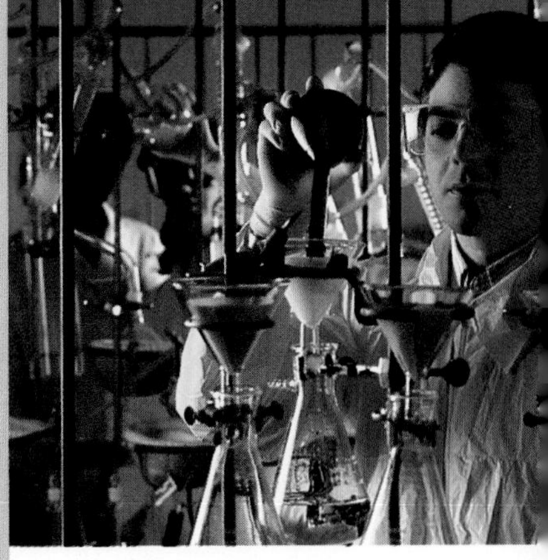

The Stock Market/Charles West

Table of Contents

Chip Clark

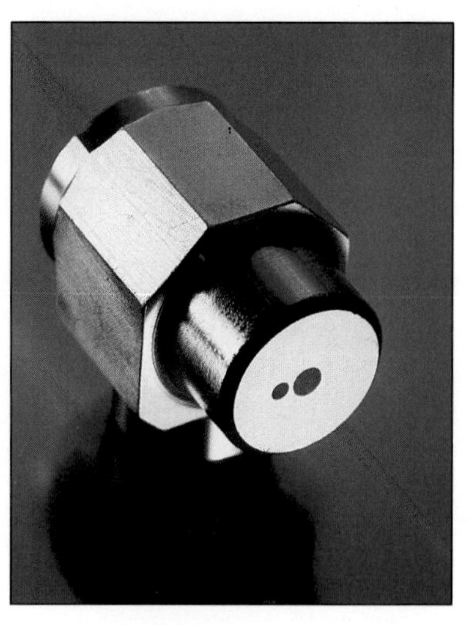

Telegraph Colour Library/FPG International

Charles D. Winters

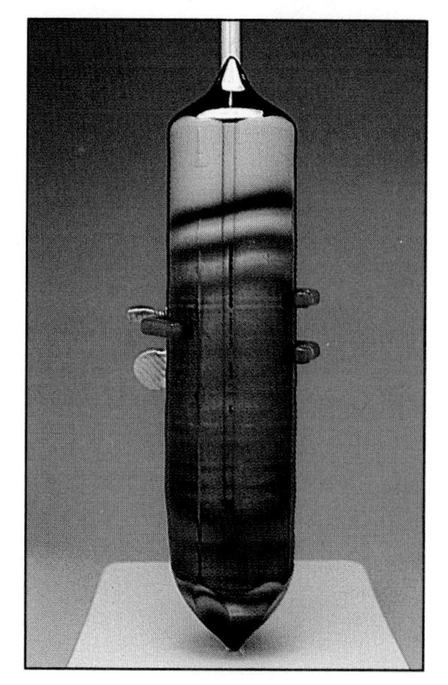

Charles D. Winters

Barry L. Runk from Grant Heilman Photography, Inc.

Phillip Wallick/The Stock Market

Charles D. Winters

Charles D. Winters

NASA

©*Kip Peticolas, Fundamental Photographs, New York*

(G. Glod/Superstock, Inc.)

1

Preparing to Learn Chemistry

The purpose of this book is to prepare you to take a college or university course in general chemistry confidently and enjoyably. To be successful in general chemistry, you need two kinds of preparation: knowledge of basic chemical principles and skill in the methods used to solve chemistry problems. This book is a summary of these principles and methods.

As you begin your work in this book, you'll find it helpful to have an overall sense of what chemistry is about and why it will be an important part of your college or university program. This chapter will describe some of the characteristics of chemistry that make it important, both as a source of useful materials and as an essential part of our understanding of the world.

As you begin to learn chemistry, it's important that you find study methods that work well for you. Your instructor will suggest some methods, and you'll learn others from other students and from this book. During the next several weeks, you'll be creating your own best approach to chemistry by finding out what works for you.

1-1 As a technology, chemistry provides most of the materials on which the quality of our lives depends

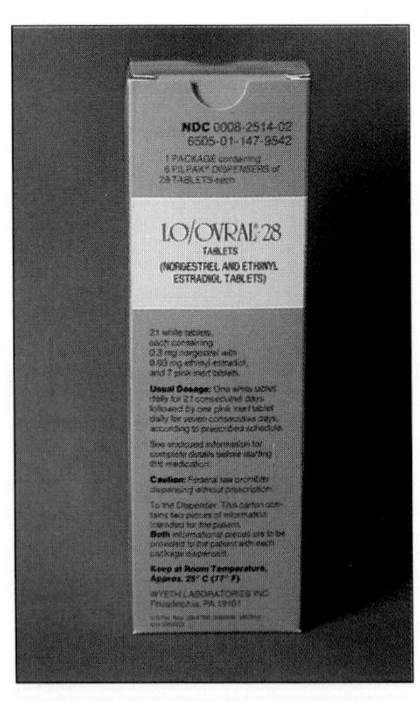

Oral contraceptives are among the thousands of modern medicines created by chemists. *(David R. Frazier Photolibrary)*

The ink on this page was prepared by chemists. So was the coating on the paper that made the ink print clearly. Chemists prepared the colored inks for the illustrations in this book, the photographic film for its pictures, the glue and thread for its binding, and the protective coating for its cover. Chemists designed and controlled the process that converted wood into paper for these pages.

As you go through your ordinary activities today, almost every material that you use or see will have been developed by chemists: soap; toothpaste; toothbrush; vitamins; the natural and synthetic fibers and the dyes in clothes; glass, metals, and plastics; all the parts in television sets, radios, stereos, and cars; gasoline; packaging materials and building materials; paints, lacquers, and varnishes. Chemistry is a fundamental part of the technology that determines the quality of our lives.

The most dramatic example of the importance of modern chemistry is its application to medicine. Surgery would be impossible without chemistry: Without the anesthetics created by chemists, surgical patients would die of shock from pain, and without the antiseptics created by chemists, the patients would die of infection. Every modern medicine, from aspirin to the latest anticancer agent, was created by chemists. Medicine today is based on chemistry, and so are agriculture, communication, and transportation. Chemistry is fundamental to the way we live.

Although chemistry is a fundamental part of our world, it's an almost invisible part. If you're now using a prescription medicine, you saw the doctor who prescribed the medicine and the pharmacist who sold it, but not the chemist who made it. When you buy gasoline, you see the person who sells it, but not the chemists and chemical engineers who formulated the blend for your engine and prepared it from crude oil. In the manufacture of most products, chemists make their contributions at the early stages, so, to the consumer, chemistry is an almost invisible technology. As a result, many people aren't familiar with chemistry and can become afraid that it will harm them. News reports increase these fears. We hear about one accident in one chemical plant but not about years of safe operations in hundreds of others, and we hear more about drugs that harm people than about those that help them.

The fact is that chemistry has contributed enormously to our well-being, and it has sometimes caused harm. In this respect, chemistry is like other fundamental technologies, such as agriculture or medicine: It's a powerful tool that can be used well or misused. One reason for including chemistry in college and university programs is so that educated people will understand chemistry well enough to be sure it's used for good purposes.

Ice melting is an example of a physical change. *(Charles D. Winters)*

1-2 As a science, chemistry explains and predicts changes in the form and composition of matter

Every year, more than 500 000 journal articles and patents report new work in chemistry; that's about one article or patent every minute. And the rate is increasing: Chemical knowledge doubles about every ten years.

Because chemistry is now an enormous and rapidly changing field of knowledge, it can be defined in many different ways. To start, we'll use this definition: Chemistry is an experimental science that explains and predicts changes in the form and composition of matter. As you learn more about chemistry, you'll see that many other definitions are possible; we'll consider some of them later in this book.

Matter is anything that takes up space and has mass.◆ The concept of matter is easy to understand because almost everything we see is made up of matter: rocks, water, trees, people, buildings, books, and so forth.

◆Weight is the force exerted on a mass by the earth's gravitational field. Scientists prefer to use the more general term, *mass*.

We know from everyday experience that matter undergoes changes: Water, at a low temperature, freezes to form ice; iron, exposed to water and air, turns to rust; food, when we digest it, changes into our body tissues. We see matter changing all the time. It's useful to distinguish two kinds of changes that matter can undergo. In **physical change,** the same substance changes from one form to another, and in **chemical change,** one substance is converted into another. Physical change is change of form. Chemical change is change of composition.

Sometimes it's easy to identify a change in matter as either a physical change or a chemical change and other times it's not. For example, we're used to thinking of water and ice as being two forms of the same substance, so it's easy to identify the freezing of water or the melting of ice as a physical change. Similarly, the conversion of food into our body tissues is such a dramatic change that we easily suppose it's a change in composition and, therefore, a chemical change. But in many cases it's not easy to be sure. Suppose I put a piece of butter in a pan and heat it. As the butter melts, I'll want to say that this is probably a physical change: The substance has changed from one form to another. But if I continue to heat the butter, it will undergo further changes, turning brown and then black and giving off a sharp-smelling smoke. It looks as if a new substance has been formed, so there's probably been a change in composition—a chemical change. Later in this book, you'll learn more exact

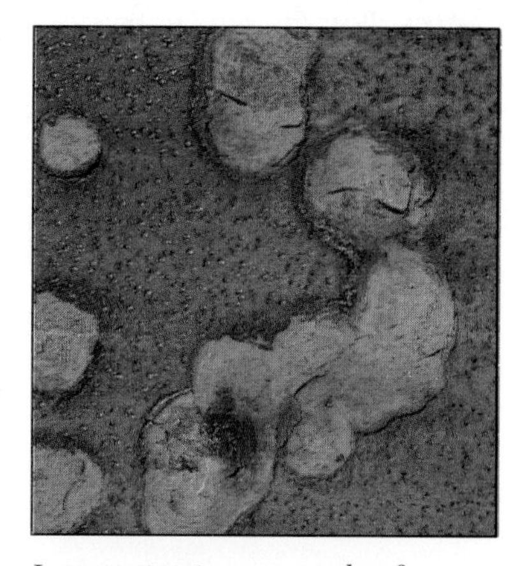

Iron rusting is an example of a chemical change. *(Charles D. Winters)*

ways other than simple observation to distinguish between physical and chemical changes.

It's obviously useful to understand the causes of physical and chemical changes. If we understand the process by which ice melts, for example, we can sometimes control it for our benefit; so we've learned to melt ice on streets and sidewalks by spreading salt on it. If we understand the rusting of iron, we can sometimes prevent it, and if we understand how food becomes body tissue, we may be able to improve nutrition. Diseases and aging are changes in living matter, and by understanding those changes, we may be able to prevent or reverse them. Understanding how matter changes is the key to solving many of our most important practical problems.

Another reason to understand physical and chemical changes is simple curiosity: As human beings, we have a need to find out why water sometimes freezes, why stars give off light, and why plants and animals grow, age, and die. This kind of curiosity has created science. By finding explanations to satisfy our curiosity, we've also found answers to practical problems, and in this way our science has become the basis for our technology.

(The Stock Market/Ted Horowitz)

1-3 All scientists work with observations, laws, hypotheses, and theories, but each scientist works in his or her own way

We live in a scientific culture and, in many direct and indirect ways, science affects almost every aspect of our lives. If your primary interest is in a technical field such as agriculture, architecture, or engineering, you may be interested in science mainly as a source of the materials and forms of energy used to solve technical problems. If you're interested in business, you may want to understand how science continually creates new industries, such as the computer industry and biotechnology. If your studies are mainly in the humanities or the social sciences, you may be interested in science as a source of philosophical knowledge about the nature of the world or as an example of human cooperation and competition. It's important for all educated persons to understand something about science because it's now a vital part of almost every human activity.

Modern science is so enormous, complex, and rapidly changing that no one can comprehend all of it in any detail, but it is possible to understand, in a general way, some of the unique characteristics that have made science so successful.

The unique aim of science is to find rational explanations for the behavior of the natural world. Scientists believe that there is an order in nature and that the order is discoverable by human reason. To describe that order, they've developed certain kinds of concepts called laws, hypotheses, and theories.

A **law** is a statement that summarizes observations.◆ For example, it's an everyday observation that when I lift an object and then let go of it, it will fall. I could summarize this kind of experience in a law: Objects above the surface of the earth that are free to move will spontaneously fall.

◆Legal laws and scientific laws are fundamentally different: A legal law requires or prohibits some form of human behavior, whereas a scientific law describes an observed regularity in nature.

A **hypothesis** is a statement of a possible explanation for a law. It's usually possible to imagine many explanations for the same events. For example, here are two possible explanations for our law of falling objects: (1) Objects fall because air pressure from the earth's atmosphere squeezes them downward, or (2) objects fall because all pieces of matter in the universe attract one another, and bigger pieces exert stronger attractions; the earth, as a very big piece of matter, pulls smaller pieces toward itself.

A **theory** is a hypothesis that is trusted because tests have confirmed its validity, but no theory is ever trusted completely. A hypothesis that is tested and fails has to be discarded. We must abandon our first hypothesis, for example, because experiments have shown that objects fall in a vacuum, where no air is present. The second hypothesis is now accepted as a theory, because it has been confirmed by experiments (two large masses hung close together on wires pull toward one another because of their mutual attraction) and other observations (the movements of the earth and moon have been shown to be influenced by the large mass of the sun).◆

(The Stock Market/Charles West)

◆An experiment is a procedure for making carefully controlled observations, and a laboratory is a place with the special equipment needed to conduct experiments.

An essential requirement of a scientific hypothesis or theory is that it must lead to predictions for testing it. Our first hypothesis, for example, predicts that objects should not fall in a vacuum, and this hypothesis has been found to be false by experiments that show this prediction is incorrect.

No scientific theory is ever completely secure, because, no matter how many times observations have confirmed a theory, there's always a possibility that it will be disproved by some later test.†

The function of laws, hypotheses, and theories is clear enough when we talk about science as a large field of knowledge, but when we look at the work of an individual scientist, many difficult questions appear. For example, in a scientist's thoughts, which comes first, observation, law, or hypothesis? It would be useless for a scientist to make purely random observations, but if a scientist set out to find only certain kinds of observations, wouldn't that mean that he or she already had a law or hypothesis in mind, and wouldn't that be unscientific prejudice? If a scientist started with a hypothesis, where did it come from? What kinds of observations should be allowed to prove or disprove a law, hypothesis, or theory?◆ These are difficult questions, and some of them do not yet have satisfactory answers.

◆Doesn't the observed fact that helium-filled balloons move spontaneously away from the surface of the earth invalidate our law of falling objects and the theory that explains it?

The difficulty is that the work of an individual scientist is a unique example of creativity, and no one fully understands the human creative process. All scientists work with observations, laws, hypotheses, and theories, but each works in his or her own way; there is no one scientific method. How each person succeeds in creating science or in learning it remains unknown, even to that person.

† Chemistry Insight

Because a theory that has been validated by many observations can lose its credibility if it's disproved only once, some philosophers of science have suggested that the essential function of observations in science is not to verify theories but to disprove them. These philosophers classify a theory as scientific only if it's capable of being disproved by observations. The philosopher best known for this point of view is Karl Popper (Austrian, born 1902).

1-4 What's the best way to learn chemistry?

During the next few weeks you'll be creating your own way of learning chemistry. To help you plan the way you want to learn, this Section will describe some important characteristics of this book and of chemistry as a subject, and will suggest some ways of learning chemistry that you may want to use.

This book is written at a fundamental level. It assumes that you haven't learned chemistry before and that you may not remember your high school science and mathematics well. Because it's designed for a one-term course, it presents only the most essential information on each topic; in your later courses in chemistry, you'll study these topics in greater depth.

You'll get oriented to chemistry more quickly if you understand that the subject matter has the following three characteristics:

(G. Glod/Superstock)

1. ***A Special Vocabulary.*** Much of your work with this book and in your later chemistry courses will be learning to understand and apply the special vocabulary of chemistry—words such as *atom, molecule, element, compound,* and so forth. Throughout your work in chemistry, concentrate on learning the vocabulary.
2. ***A High Standard of Exactness.*** Most of the special words used in chemistry have exact definitions, and it will be important for you to learn exactly what they are. This doesn't mean that you should memorize definitions without thinking about them, it means that you should understand clearly the definition of each technical word. Chemistry isn't a subject that you can sort of understand; you have to learn it carefully.
3. ***A Cumulative Structure.*** In this course and in your later courses in chemistry, the new concepts and skills in each topic will build on the concepts and skills you've already learned. How well you're able to understand each chapter will depend on how well you've learned the earlier ones. If you skip concepts you don't fully understand, you won't be able to learn later concepts that depend on them. Learn each topic thoroughly, and remember it.

Based on these characteristics of chemistry, here are three suggestions on how you can learn it well.

1. Build your notes from your book. Read the text assignment before you go to class so that you'll be able to understand the lecture fully. Take notes as you read; use one side of each page, then write your lecture notes on the blank sides, facing the appropriate part of your book notes.
2. Practice. Learning chemistry is like learning a new language: You need to keep what you've learned active in your mind so you can build new material on it. You'll forget quickly if you don't review often. In this book, the Summary at the end of each chapter gives you an easy way to review.
3. Test yourself. Don't wait for an exam to find out that you don't know the material well enough. Reviewing the Summary at the end of each chapter is one way to test yourself, and working the problems at the end of the chapter is another. If you find that you can't work a problem, reread the corresponding material in the chapter and try the problem again. If you're still stuck, ask your instructor for help right away—don't wait until you're reviewing for an exam. If you test yourself continuously, you'll learn the material well and you'll keep exams from becoming stressful.

The language of chemistry isn't difficult to understand, but becoming fluent in it takes effort. It's a satisfying subject to learn, because the effort you put in will be repaid directly in the knowledge and grades you receive.

Chapter Summary: Preparing to Learn Chemistry

The Summary at the end of each chapter collects most of the essential information from the chapter in a way that will make it easier for you to learn and review. Cover the Summary column with your hand or a piece of paper. Read each item in the Subject column and see if you can supply the corresponding information in the Summary column; when you can, put a check mark in the Check When Learned column. If an entry isn't clear, go back to the indicated Section and reread the appropriate material, then test yourself with the Summary again. If the concept you're working on still isn't clear, ask your instructor about it right away.

Section	Subject	Summary	Check When Learned
1-2	Chemistry	Experimental science that explains and predicts changes in the form and composition of matter.	☐
1-2	Matter	Anything that takes up space and has mass.	☐
1-2	Physical change	Change of form.	☐
1-2	Chemical change	Change of composition.	☐
1-3	Law	Statement that summarizes observations.	☐
1-3	Hypothesis	Statement of a possible explanation for a law.	☐
1-3	Theory	Hypothesis that is trusted because it has been confirmed by tests.	☐
1-4	Characteristics of chemistry	Special vocabulary; high standard of exactness; cumulative structure.	☐
1-4	Guidelines for learning chemistry	Build your notes from your book; practice; test yourself.	☐

2-1
It's more convenient to write very large or very small numbers as powers of ten.

2-2
Numbers expressed as powers of ten can be added, subtracted, multiplied, or divided.

2-3
The significant digits in a measured value show how precisely the measurement was made.

2-4
A defined value has infinite significant digits.

2-5
The answer to a calculation will be as precise as the least precise number used in the calculation.

2-6
To solve an algebraic equation, isolate the symbol for the unknown on one side of the equation.

Chapter Summary

Problems

(P. Rivera/Superstock)

2 Mathematical Background

The mathematical operations you'll need to know for your work in this book aren't complicated, and you've probably learned them before. But you may have forgotten some of them, so this chapter will review what you'll need to know before you start Chapter 3.◆

◆Your instructor may prefer to have you skip this chapter and refer back to parts of it as you need to for your later work in this book.

Calculations in chemistry often use very large or very small numbers, and these numbers are most conveniently expressed as powers of ten, in a form called **exponential notation.** In Sections 2-1 and 2-2 you'll learn how to write numbers in exponential notation and how to make calculations with numbers written in that form.

Chemical calculations usually deal with measured values, for example, with quantities that express masses or volumes. In working with measured values, it's necessary to show how precisely the measurements were made, and this is done by expressing the values with an appropriate number of **significant digits.** In Sections 2-3 through 2-5, you'll learn the procedures for using significant digits in writing numbers and in making calculations. These are important skills that you'll use in all of your later calculations in chemistry.

Some chemical calculations require algebra, and in Section 2-6 you'll review the steps for solving simple algebraic equations.

For your calculations in chemistry, you'll need a calculator with the usual functions of addition, subtraction, multiplication, and division, and with keys for exponents, logs and antilogs, and powers and roots; your instructor may recommend a particular model. Throughout this book, directions are included on how to use your calculator for the problems you're learning to solve.

After every few paragraphs in this book, you'll find an Exercise, a problem that deals with some aspect of the topic you've just been studying. It's very important that you work out each Exercise and compare your answer with the answer given in Appendix 1 at the end of the book. If you can't work an Exercise, it's a signal that you should go back over the material you've been studying before you try to go further.

2-1 It's more convenient to write very large or very small numbers as powers of ten

Scientists often work with very large or very small numbers. For example, an important number in chemistry, as we'll see later, is 602 000 000 000 000 000 000 000.

It's awkward to work with very large or very small numbers if they're written this way, with many zeroes and a decimal point, so scientists write them in a more convenient form. The number shown above, for example, can be written as 6.02×10^{23}. The 23, written as a superscript on the 10, is referred to as a **power of ten** or an **exponent.**◆ The expression 6.02×10^{23} is read "6.02 times 10 to the 23rd" and means that 6.02 is to be multiplied by 10 twenty-three times. Each multiplication by 10 moves the decimal point one place to the right.

◆To make them easy to review, definitions and other important pieces of information have been collected in the Summary at the end of each chapter.

Expressing a number as a power of ten is another way of showing where the decimal point in the number belongs; the advantage of using a power of ten is that it makes it easier to write or read a very large or small number. Since the power of ten and the decimal point are two different ways of expressing where the decimal point is, there are many different ways to write the same number. Each expression below represents the number 15 200.

$$1.52 \times 10^4 = 1.52 \times 10 \times 10 \times 10 \times 10 = 15\ 200$$

$$15.2 \times 10^3 = 15.2 \times 10 \times 10 \times 10 = 15\ 200$$

$$152 \times 10^2 = 152 \times 10 \times 10 = 15\ 200$$

$$1520 \times 10^1 = 1520 \times 10 = 15\ 200$$

$$15\ 200 \times 10^0 = 15\ 200 \times 1 = 15\ 200$$ (Any number raised to the zero power is equal to 1, so $10^0 = 1$.)

A number expressed as a power of ten is said to be written in **exponential notation.**

Scientists have agreed that they'll report exponential numbers in a particular form, called **scientific notation:** The number is written with one non-zero digit to the left of the decimal point. Thus 3.76×10^5 is in scientific notation and 37.6×10^4 is not.

Example 2.1

Write each of the following numbers in scientific notation: (a) 3175 (b) 29.6×10^2

Solution

(a) In scientific notation, the number is written with one non-zero digit to the left of the decimal point, so the decimal point in 3175 will be moved three places to the left; this has the effect of dividing the number by 10 three times, so the original value of the number will be restored by multiplying by 10 three times, that is, by multiplying by 10^3.

$$3175 = 3.175 \times 10 \times 10 \times 10 = 3.175 \times 10^3$$

(b) $29.6 \times 10^2 = 29.6 \times 10 \times 10 = 2.96 \times 10 \times 10 \times 10 = 2.96 \times 10^3$

◆Answers to Exercises are in Appendix 1.

Exercise 2.1◆

Write each of the following numbers in scientific notation: (a) 727 (b) 19.6×10 (c) 232×10^4

Example 2.2

Convert each of the following numbers from the exponential form to the decimal form: (a) 1.2×10^3 (b) 52.4×10^2 (c) 0.039×10^5

Solution

(a) To multiply by 10^3, move the decimal point three places to the right:

$$1.2 \times 10^3 = 1.2 \times 10 \times 10 \times 10 = 1200$$

(b) To multiply by 10^2, move the decimal point two places to the right:

$$52.4 \times 10^2 = 52.4 \times 10 \times 10 = 5240$$

(c) To multiply by 10^5, move the decimal point five places to the right:

$$0.039 \times 10^5 = 0.039 \times 10 \times 10 \times 10 \times 10 \times 10 = 3900$$

Exercise 2.2

Convert each of the following numbers from the exponential form to the decimal form: (a) 3.1×10^2 (b) 0.0523×10^3 (c) 101.2×10^3

A positive power of ten expresses a multiplication by ten, and a negative power of ten expresses a division by ten: $7.66 \times 10^{-3} = 7.66 \div 10 \div 10 \div 10 = 0.00766$. Each division by ten moves the decimal point one place to the left. Each of the following expressions represents the same number.

$$7.66 \times 10^{-3} = 7.66 \div 10 \div 10 \div 10 = 0.00766$$

$$0.766 \times 10^{-2} = 0.766 \div 10 \div 10 = 0.00766$$

$$0.0766 \times 10^{-1} = 0.0766 \div 10 = 0.00766$$

$$0.00766 \times 10^{0} = 0.00766 \times 1 = 0.00766$$

Example 2.3

Write each of the following numbers in scientific notation: (a) 0.351 (b) 0.00084 (c) 0.079×10^{-5} (d) 0.062×10^4

Solution

(a) In scientific notation the number is written with one non-zero digit to the left of the decimal point, so the decimal point in 0.351 will be moved one place to the right. Moving the decimal point one place to the right has the effect of multiplying the number by 10; the original value of the number is restored by dividing by 10, that is, by multiplying by 10^{-1}.

$$0.351 = 3.51 \div 10 = 3.51 \times 10^{-1}$$

(b) $0.00084 = 8.4 \div 10 \div 10 \div 10 \div 10 = 8.4 \times 10^{-4}$

(c) $$\begin{aligned} 0.079 \times 10^{-5} &= 0.079 \div 10 \div 10 \div 10 \div 10 \div 10 \\ &= 7.9 \div 10 \div 10 \div 10 \div 10 \div 10 \div 10 \div 10 = 7.9 \times 10^{-7} \end{aligned}$$

(d) A multiplication by 10 cancels a division by 10:

$$\begin{aligned} 0.062 \times 10^4 &= 0.062 \times 10 \times 10 \times 10 \times 10 \\ &= 6.2 \times 10 \times 10 \times \cancel{10} \times \cancel{10} \div \cancel{10} \div \cancel{10} \\ &= 6.2 \times 10 \times 10 \\ &= 6.2 \times 10^2 \end{aligned}$$

Exercise 2.3

Write each of the following numbers in scientific notation: (a) 0.0175 (b) 0.555×10^{-3} (c) 856×10^{-5} (d) 0.00913×10^2 (e) 102×10^{-2}

Example 2.4

Convert each of the following numbers from the exponential form to the decimal form: (a) 7.5×10^{-2} (b) 85.5×10^{-4} (c) 0.037×10^{-2}

Solution

(a) To multiply by 10^{-2}, move the decimal point two places to the left:

$$7.5 \times 10^{-2} = 7.5 \div 10 \div 10 = 0.075$$

(b) $85.5 \times 10^{-4} = 85.5 \div 10 \div 10 \div 10 \div 10 = 0.00855$

(c) $0.037 \times 10^{-2} = 0.037 \div 10 \div 10 = 0.00037$

Exercise 2.4

Convert each of the following numbers from the exponential form to the decimal form: (a) 6.6×10^{-3} (b) 0.005×10^{-2} (c) 22.6×10^{-3}

In writing numbers from 1 to 999, using scientific notation may not be an advantage: It's easier to write 567 than to write 5.67×10^2 or to write 11.6 rather than 1.16×10^1. Most scientists prefer not to use scientific notation for a number from 1 to 999. (In Section 2-3, we'll see that there is sometimes a good reason to write a number from 1 to 999 in scientific notation, even though writing it in the decimal form would be easier.)

The same is true for small numbers. It's more convenient to write 7.55×10^{-6} than 0.00000755, but it's easier to write 0.849 than 8.49×10^{-1}. Most scientists prefer to use scientific notation for a number that starts in the third place after the decimal point (write 9.77×10^{-3}, not 0.00977) but will prefer to use decimal notation for a number that starts in the first or second place after the decimal point (write 0.0856, not 8.56×10^{-2} and 0.352, not 3.52×10^{-1}).

Your instructor may have a preference as to whether you should write all numbers in scientific notation or use it only for very large or very small numbers.

2-2 Numbers expressed as powers of ten can be added, subtracted, multiplied, or divided

To add or subtract two numbers expressed in decimal form, we arrange them in a vertical column with their decimal points aligned:

$$\begin{array}{r} 15.33 \\ +2.21 \\ \hline 17.54 \end{array} \qquad \begin{array}{r} 328.79 \\ -25.31 \\ \hline 303.48 \end{array}$$

In exponential notation, the power of ten is another way of expressing where the decimal point in the number is located, so in adding or subtracting numbers in exponential notation, we first express them to the same power of ten; this has the same effect as aligning their decimal points.

To calculate $3.61 \times 10^3 + 2.222 \times 10^4$, we first express both numbers to the same power of ten. We could choose any power of ten, but it's convenient to choose one of the powers of ten that already appears in the problem, since then we only have to rewrite one of the numbers. We can rewrite 2.222×10^4 as 22.22×10^3, and add:

$$\begin{array}{r} 3.61 \times 10^3 \\ +2.222 \times 10^4 \\ \hline \end{array} \qquad \begin{array}{r} 3.61 \times 10^3 \\ +22.22 \times 10^3 \\ \hline 25.83 \times 10^3 = 2.583 \times 10^4 \end{array}$$

To calculate $13.288 \times 10^5 - 2.02 \times 10^4$, first express both numbers to the same power of ten:

$$\begin{array}{r} 13.288 \times 10^5 \\ -2.02 \;\; \times 10^4 \\ \hline \end{array} \qquad \begin{array}{r} 13.288 \times 10^5 \\ -0.202 \times 10^5 \\ \hline 13.086 \times 10^5 = 1.3086 \times 10^6 \end{array}$$

Exercise 2.5

Make the following calculations and express the answers in scientific notation:† (a) $5.23 \times 10^3 + 0.211 \times 10^4$ (b) $16.99 \times 10^{-5} - 2.1 \times 10^{-6}$

When numbers in exponential notation are multiplied, their exponents are added: $10^a \times 10^b = 10^{a+b}$.

$$(2 \times 10^2)(3 \times 10^3) = (2 \times 10 \times 10)(3 \times 10 \times 10 \times 10) = 6 \times 10^5$$

When numbers in exponential notation are divided, their exponents are subtracted: $10^a/10^b = 10^{a-b}$.

$$\frac{6 \times 10^5}{2 \times 10^2} = \frac{6 \times 10 \times 10 \times 10 \times \cancel{10} \times \cancel{10}}{2 \times \cancel{10} \times \cancel{10}} = 3 \times 10 \times 10 \times 10 = 3 \times 10^3$$

$$\frac{8 \times 10^{-4}}{2 \times 10^{-3}} = \frac{8 \div 10 \div \cancel{10} \div \cancel{10} \div \cancel{10}}{2 \div \cancel{10} \div \cancel{10} \div \cancel{10}} = 4 \times 10^{-1} = 0.4$$

Exercise 2.6

Use a calculator to solve the following problems, and express your answers in scientific notation. (a) $(1.5 \times 10^5)(2.4 \times 10^3)$ (b) $(6.2 \times 10^3)(1.5 \times 10^{-3})$ (c) $(1.8 \times 10^{-12})(3.0 \times 10^{-9})$ (d) $(4.8 \times 10^3)/(3.0 \times 10^3)$ (e) $(5.4 \times 10^{-6})/(7.5 \times 10^4)$ (f) $(3.6 \times 10^{-4})/(1.5 \times 10^{-3})$.

† Chemistry Insight

If you use your calculator to add or subtract numbers expressed in exponential form, it will automatically adjust the decimal point and the power of ten during the operation. On your calculator, the key used to enter powers of 10 is probably marked EXP or EE. To enter 5.23×10^3, press these keys: 5, ., 2, 3, EXP or EE, 3. You'll probably see a display in which the power of 10 appears as two digits on the far right: 5.23 03. Calculator operations with numbers expressed as powers of ten are carried out in the usual way. This is the sequence of keys to solve part (a) in Exercise 2.5: 5, ., 2, 3, EXP, 3, +, 0, ., 2, 1, 1, EXP, 4, =. To change a positive to a negative number, or vice versa, use the +/− key. Use this sequence to enter 16.99×10^{-5}: 1, 6, ., 9, 9, EXP, +/−, 5.

2-3 The significant digits in a measured value show how precisely the measurement was made

An environmental chemist measures the acidity of acid rain. *(Runk/Schoenberger from Grant Heilman Photography, Inc.)*

Much scientific work involves making and reporting measurements. An astronomer may measure the distance from the earth to a star and report it to other astronomers, an oceanographer may measure the temperature deep in the ocean and report it to other oceanographers, or an environmental chemist may measure the amounts of impurities in a sample of water and report the results to other environmental chemists.

Making and communicating a scientific measurement involves three steps:

1. The measurement is *made*, using an appropriate instrument.
2. The measurement is *recorded* by the person who made it.
3. The recorded measurement is *read* by someone else.

The person who records a measured value and the person who reads it must understand one another clearly. To make sure that they do understand one another, there are rules to govern how measured values will be written and how they'll be read.

One rule is that a number expressing a measurement must be written with a label to show the units used in making the measurement; for example, 3.57 *inches* shows that the measurement was made in inches, and 4.88 *pounds* shows that the measurement was made in pounds. (Scientists prefer metric units, described in Chapter 3.)

Another rule is that the record of a measurement should show how precisely it was made. The person who records a measurement has an obligation to record it in a way that shows its precision, and the person who reads the measurement has a right to rely on the degree of precision shown in the recorded value.

Significant digits are the digits in a measured value that show how precisely the measurement was made. A measurement reported as 7.12 inches (3 significant digits) shows that the measurement was made to hundredths of an inch, and a measurement reported as 1.4 pounds (2 significant digits) shows that the measurement was made to tenths of a pound.

Table 2.1 lists the rules for deciding how many significant digits there are in a measured value.

Exercise 2.7

Decide on the number of significant digits in each of the following measured values: (a) 0.0500 inches (b) 23 000 inches (c) 174 inches (d) 1020 inches (e) 10.0 inches

The last entry in Table 2.1 shows that zeroes ending a number before a decimal point present a problem of interpretation: We can't be sure, without other information, whether the person who recorded the zeroes intended them to be significant digits or not. Was the value 13 500 inches measured to five significant digits, or are the zeroes present only to show the position of the decimal point, so that there are only three significant digits in the value? Or was the value measured to four significant digits, so that one of the zeroes is significant and the other is not?

Table 2.1 Rules for Significant Digits

Number	Number of Significant Digits	Rule
7.59 inches	3	All non-zero digits are significant
90.05 inches	4	Zeroes between significant digits are significant.
15.300 inches	5	Zeroes that end a number after the decimal are significant.
0.0072 inches	2	Zeroes that begin a number are not significant.
13 500 inches	3	Zeroes that end a number before the decimal are assumed not to be significant, unless we're told otherwise.

In scientific writing, the meaning of a reported value must be clear to the person who reads it, so this kind of uncertainty can't be allowed. Scientific notation gives us a way to avoid the problem:

If 13 500 inches is measured to three significant digits, it should be reported as 1.35×10^4 inches.
If 13 500 inches is measured to four significant digits, it should be reported as 1.350×10^4 inches. (Zeroes that end a number after the decimal are significant.)
If 13 500 inches is measured to five significant digits, it should be reported as 1.3500×10^4 inches.

Exercise 2.8

Decide on the number of significant digits in each of the following measured values: (a) 0.060 inches (b) 2.50×10^{-5} inches (c) 2000 inches (d) 2.00×10^3 inches

Exercise 2.9

Write each of the following measured values in scientific notation, with three significant digits: (a) 0.300 inches (b) 3.050 inches (c) 125 000 inches

A measured value is a quantity, consisting of a number and a unit label, that reports a measurement. The number of significant digits in a measured value depends on the precision of the instrument used to make the measurement. For example, the value 4.14 pounds shows that the weighing device gave three significant digits. A less precise instrument might have given 4.1 pounds (two significant digits), and a more precise instrument might have given 4.142 pounds (four significant digits). See the Chemistry Insight box on page 16.†

When a measured quantity is given in a problem, we have to trust that it's written with the correct number of significant digits. To decide on the number of significant digits in a measured value, use the rules in Table 2.1. Using those rules, we can identify three significant digits in each of these measured values: 25.0 inches, 0.0371 inches, 201 inches.

† Chemistry Insight

There's a difference between the precision and the accuracy of a measured value. The **precision** of a measured value, shown by the number of significant digits it contains, expresses how reproducible the value is. If someone uses a ruler to measure a length and finds the length is 8.36 inches, we assume that anyone else using the same ruler would measure the same length as 8.36 ± 0.01 inches, that is, as 8.35, 8.36, or 8.37 inches. The **accuracy** of a measured value is an assessment of its correctness. Suppose that a one-foot ruler had been accidentally made so that the numbering of inches on it ran 1, 2, 3, 4, 5, 7, 8, 9, 10, 11, 12. All measurements taken with the ruler would be equally precise (reproducible), but they wouldn't all be equally accurate (correct): Measurements below the 7-inch line would be accurate, and measurements above the 7-inch line would be inaccurate by 1 inch.

Some numbers refer to objects that can't be divided, so the numbers must be whole numbers; that is, they must have only zeroes after the decimal point. For example, a report of 15 refrigerators in a dealer's inventory must mean 15.(infinite zeroes) refrigerators, and a report of 74 cattle in a rancher's herd must mean 74.(infinite zeroes) cattle. These numbers express quantities found by counting objects, so they're referred to as counted values. A **counted value** has an infinite number of zeroes after the decimal point.

In a counted value, the counted objects must be whole objects, meaning that the digits *after* the decimal point must be zeroes, but there are no restrictions on the digits *before* the decimal. If we're told that the population of the United States is 250 000 000 people, we know that this refers to whole people, but we can't know, without more information, whether these zeroes are significant digits. Reporting quantities in scientific notation avoids this kind of uncertainty: The quantity 2.50×10^8 people has three significant digits.

Exercise 2.10

Decide how many significant digits there are in each of the following values: (a) a table 62 inches long (b) a house containing 4 tables (c) a man weighing 172 pounds (d) a train with 74 railway cars (e) our solar system containing 9 planets

Some measured values are commonly expressed as fractions. For example, we express speeds as miles per hour, and *per* means *divided by*. "Fifty-five miles per hour" or "55 mph" is expressed mathematically as 55 miles/1 hour. If steak costs $2.50 per pound, the mathematical expression is $2.50/1 pound. As these examples show, in a measured value expressed as a fraction, the denominator is 1.

How many significant digits should we read in such a fraction? The number of significant digits in the numerator is read according to the usual rules: In 55 miles/1 hour, the value 55 miles has 2 significant digits, and in $2.50/1 pound, the value $2.50 has 3 significant digits. The 1 in the denominator is read as *exactly* 1, that is, as 1.(infinite number of zeroes), with an infinite number of significant digits.

2-4 A defined value has infinite significant digits

Relationships between currencies, for example, the relationship of the English pound to the American dollar, are measured relationships that constantly change. Relationships within currencies, for example, \$1 = 100 cents, are defined relationships that don't change. *(Eric Carle/Superstock)*

Every system of measurement contains defined values. A **defined value** is a statement of a defined relationship between two units of measurement; for example, the statement, "There are 12 inches in 1 foot," defines the relationship between feet and inches. In mathematical form, the statement is 12 inches = 1 foot. Other examples are 1 dollar = 100 cents, 1 mile = 5280 feet, 1 pound = 16 ounces, and 1 quart = 2 pints.

How many significant digits should we read in a defined value? The answer depends on understanding what it means to say that a value is defined: It means that the value has not been measured but agreed on by the people who will use the measurement system. If I want to know the length of my desk top, I'll have to get a ruler and measure it, but I don't measure the number of inches in one foot, I learn it as a definition. Similarly, the relationship of the American dollar to the English pound is a measured value that constantly changes and must be continuously remeasured, but the relationship 1 dollar = 100 cents is a definition that is agreed upon by the people who use the American monetary system.

A defined value is an exact numerical relationship that has an infinite number of significant digits. For example, the statement

1 foot = 12 inches means 1 foot (exactly) = 12 inches (exactly) or
1.(infinite number of zeroes) foot = 12.(infinite number of zeroes) inches.

The fact that defined values have infinite significant digits has important consequences in calculations, as we'll see in Chapter 3.

2-5 The answer to a calculation will be as precise as the least precise number used in the calculation

The answer to a calculation can be no more precise than the least precise number that went into the calculation; this is the mathematical version of the adage that a chain is only as strong as its weakest link. The following examples show how to determine how many significant digits should appear in the answer to a calculation.

In adding or subtracting quantities, report the answer to the same number of *decimal places* (not significant digits) as there are in the quantity with the least number of decimal places.

$$\begin{array}{rr} & 17.51 \\ & +2.9 \\ \hline & 20.41 \\ \text{report} & 20.4 \end{array} \qquad \begin{array}{rr} & 12.334 \\ & -3. \\ \hline & 9.334 \\ \text{report} & 9 \end{array}$$

Exercise 2.11

Write the answer to each problem with the appropriate number of significant digits: (a) 20.1 + 13.52 (b) 9.710 − 2.2

Exercise 2.12

Write the answer to each problem in scientific notation with the appropriate number of significant digits: (a) $8.84 \times 10^{-2} + 3.6 \times 10^{-3}$ (b) $9.56 \times 10^{4} - 2.7 \times 10^{3}$

In multiplying or dividing quantities, report your answer to the same number of significant digits as there are in the quantity with the fewest significant digits.

$$(2.47)(3.1) = 7.7 \quad \text{(3.1 has 2 significant digits)}$$

$$12.8/1.2 = 11 \quad \text{(1.2 has 2 significant digits)}$$

$$(3.74 \times 10^{5})(2.9 \times 10^{3}) = 1.1 \times 10^{9} \quad (2.9 \times 10^{3} \text{ has 2 significant digits})$$

$$(9.444 \times 10^{2})/(3.8 \times 10^{-2}) = 2.5 \times 10^{4} \quad (3.8 \times 10^{-2} \text{ has 2 significant digits})$$

Exercise 2.13

Decide how many significant digits should be used in the answer to each of the following calculations: (a) (1.5)(2) (b) $(6.66 \times 10^{-3})(2.8 \times 10^{3})$ (c) 3.7/2.333 (d) $(8.94 \times 10^{-4})/(3.5 \times 10^{-3})$

In deciding how to report the answer to a calculation to the appropriate number of significant digits, you also have to decide how to **round off** the number, that is, you have to decide what your last reported digit will be. Table 2.2 shows how numbers are rounded off.

Table 2.2 Rules for Rounding Off

Round	To Three Significant Digits	Rule
6.5421	6.54	If the number following the last significant digit is less than 5, leave that digit unchanged. (The last significant digit is 4 and the number following it is 2, so the 4 is left unchanged.)
6.5474	6.55	If the number following the last significant digit is greater than 5, increase that digit by 1. (The last significant digit is 4 and the number following it is 7, so the 4 is increased to 5.)
6.545002	6.55	If the number following the last significant digit is 5, followed by a number other than zero, increase that digit by 1. (The last significant digit is 4, and the sequence of numbers following it is 5002, so the 4 is increased to 5.)
6.54500 6.575	6.54 6.58	If the number following the last significant digit is 5, or 5 followed only by zeroes, leave the digit unchanged if it's even or increase it by 1 if it's odd. (In 6.54500, the last significant digit is 4 and the sequence of numbers following it is 500, so the 4 is left unchanged. In 6.575, the last significant digit is 7 and the number following it is 5, so the 7 is increased to 8.)

Exercise 2.14

Round off each of the following numbers to three significant digits: (a) 8.774×10^5 (b) 1.299×10^{-2} (c) 0.003345 (d) 3.55500.

2-6 To solve an algebraic equation, isolate the symbol for the unknown on one side of the equation

The solution to the equation

$$x - 3 = 12$$

is

$$x = 15$$

As this example shows, an algebraic equation is solved when the symbol for its unknown (x, in this case) has been isolated on one side of the equation and set equal to a number on the other side of the equation. By custom, the solution is written with the symbol for the unknown on the left side of the equation and the number it's equal to on the right.

One fundamental principle is used in solving algebraic equations: Any operation that is performed on one side of an equation must also be performed on the other side; *operation* means addition, subtraction, multiplication, or division. Performing the same operation on each side of an equation preserves the equality.

To solve an equation, perform the operations necessary to isolate the symbol for the unknown on the left side of the equation. In the equation

$$x - 3 = 12$$

the x is on the left side of the equation, but it's not alone. To eliminate the -3 from the left side of the equation we add 3 to it, since $-3 + 3 = 0$. To preserve the equality, we must also add 3 to the right side of the equation.

$$x - 3 + 3 = 12 + 3$$

Performing the arithmetic gives

$$x + 0 = 15$$

or

$$x = 15$$

The following examples show the use of subtraction, multiplication, and division to solve equations.

$x + 5 = 4$	$3x = 21$	$\frac{x}{2} = 25$
$x + 5 - 5 = 4 - 5$	$\frac{\cancel{3}x}{\cancel{3}} = \frac{21}{3}$	$\frac{\cancel{2}x}{\cancel{2}} = (2)(25)$
$x = -1$	$x = 7$	$x = 50$

The same methods are used if the equation contains only symbols, as shown below.

Solve $x - a = b$ for x.

$$x - a + a = b + a$$
$$x = b + a$$

Solve $ay = b$ for y.

$$\frac{\cancel{a}y}{\cancel{a}} = \frac{b}{a}$$

$$y = \frac{b}{a}$$

Solve $abz = cd$ for z.

$$\frac{\cancel{ab}z}{\cancel{ab}} = \frac{cd}{ab}$$

$$z = \frac{cd}{ab}$$

Exercise 2.15

Solve each of the following equations for x: (a) $x + 9 = 14$ (b) $\frac{x}{5} = 3$ (c) $3x = 24$

Exercise 2.16

Solve each of the following equations for y: (a) $y + m = n$ (b) $\frac{y}{p} = q$ (c) $ay = b$

Exercise 2.17

Solve $PV = nRT$ for (a) P and (b) T.

"Mr. Osborne, may I be excused? My brain is full."

Chapter Summary: Mathematical Background

Section	Subject	Summary	Check When Learned
2-1	Exponent	Power to which a number is raised; for example, 3 in 10^3.	☐
2-1	Exponential notation	Numbers written as powers of ten.	☐
2-1	Meaning of: positive power of ten	Multiply by ten that many times; move the decimal that many places to the right.	☐
	negative power of ten	Divide by ten that many times; move the decimal that many places to the left.	☐
2-1	Scientific notation	Numbers expressed as powers of ten, with one non-zero digit to the left of the decimal.	☐
2-2	Before adding or subtracting numbers in exponential notation	Express them to the same exponent.	☐
2-2	$10^a \times 10^b =$	10^{a+b}	☐
2-2	$10^a/10^b =$	10^{a-b}	☐
2-3	Reason for label on measured value	Shows the units of measurement.	☐
2-3	Significant digits	Digits in a measured value that show how precisely the measurement was made.	☐
2-3	Digits that are significant	Non-zero digits, zeroes that end number after the decimal, and zeroes between significant digits.	☐
2-3	Digits that are not significant	Zeroes that begin a number; zeroes that end a number before the decimal, unless otherwise specified.	☐
2-3	Significant digits in fractional expressions such as 55 miles/1 hour	Infinite in the denominator; apply the usual rules to the numerator.	☐
2-3	Counted value	Number that expresses a value found by counting indivisible objects, e.g., 15 eggs.	☐
2-3	Significant digits in counted value such as 15 eggs	Read as 15.(infinite zeroes), with infinite significant digits.	☐
2-4	Significant digits in defined values such as 1 foot = 12 inches	Infinite significant digits on each side of the equation.	☐
2-5	Significant digits in addition or subtraction	Round to the fewest decimal places in the calculation.	☐

Section	Subject	Summary	Check When Learned
2-5	Significant digits in multiplication or division	Round to the fewest significant digits in the calculation.	☐
2-5	Rounding off	Deciding what the last significant digit in a calculated answer will be.	☐
2-5	Rules for Rounding Off If number following last significant digit is:		
	greater than 5	Increase the digit by 1.	☐
	less than 5	Leave the digit unchanged.	☐
	exactly 5	Increase the digit by 1 if odd; leave it unchanged if even.	☐
2-6	Method for solving algebraic equations	Isolate the unknown on left side of equation by performing the same operations on each side.	☐

Problems

Answers to odd-numbered problems are in Appendix 1.

Exponential and Scientific Notation (Sections 2-1 and 2-2)

1. Write each of the following numbers in scientific notation: (a) 8375 (b) 0.000368 (c) 10.7×10^{-2} (d) 0.0050×10^{4}

2. Write each of the following numbers in scientific notation: (a) 9222 (b) 0.0000254 (c) 85.2×10^{5} (d) 0.000242×10^{-9}

3. Convert each of the following numbers from the exponential form to the decimal form: (a) 0.00713×10^{4} (b) 2.25×10^{-2} (c) 127.4×10^{-3}

4. Convert each of the following numbers from the exponential form to the decimal form: (a) 634×10^{-5} (b) 0.049×10^{4} (c) 5.4×10^{3}

5. Make the following calculations, and express the answers in scientific notation: (a) $16.4 \times 10^{3} + 3.59 \times 10^{4}$ (b) $73.6 \times 10^{-5} - 2.41 \times 10^{-4}$

6. Make the following calculations, and express the answers in scientific notation: (a) $5.05 \times 10^{4} + 21.4 \times 10^{3}$ (b) $84.4 \times 10^{-5} - 3.16 \times 10^{-4}$

7. Make the following calculations, and express the answers in scientific notation. (a) $(1.5 \times 10^{3})(3.0 \times 10^{-5})$ (b) $(9.9 \times 10^{-8})/(3.0 \times 10^{-4})$ (c) $\dfrac{(9.2 \times 10^{2})(3.3 \times 10^{-4})}{(6.6 \times 10^{-7})(2.0 \times 10^{-3})}$

8. Make the following calculations, and express the answers in scientific notation. (a) $(2.5 \times 10^{-2})(6.0 \times 10^{4})$ (b) $(4.6 \times 10^{-3})/(2.0 \times 10^{-3})$ (c) $\dfrac{(1.2 \times 10^{3})(5.0 \times 10^{2})}{(5.0 \times 10^{-2})(8.0 \times 10^{-4})}$

Significant Digits (Sections 2-3 through 2-5)

9. Decide on the number of significant digits in each of the following measured values: (a) 23.7 ft (b) 109 lb (c) 0.052 in. (d) 1.50 yd (e) 20 000 miles

10. Decide on the number of significant digits in each of the following measured values: (a) 703 yd (b) 0.0500 ft (c) 125 lb (d) 2001 miles (e) 4000 miles

11. Decide on the number of significant digits in each of the following measured values: (a) 1.63×10^4 ft (b) 3.050×10^{-3} lb (c) 2.0×10^3 in. (d) 2×10^3 in. (e) 0.07010×10^5 miles

12. Decide on the number of significant digits in each of the following measured values: (a) 3.50×10^3 ft (b) 0.033×10^4 in. (c) 3.0×10^{-3} ft (d) 7.010×10^2 lb (e) 3.00×10^{-3} ft

13. Decide on the number of significant digits in each of the following measured values: (a) 65 miles/1 hour (b) $4.75/1 lb

14. Decide on the number of significant digits in each of the following measured values: (a) $3.54/1 lb (b) $1.10/1 dozen

15. Decide on the number of significant digits in each of the following values: (a) 22 bicycles (b) 12 mountains (c) 15 counties

16. Decide on the number of significant digits in each of the following values: (a) 57 cities (b) 3 rivers (c) 17 typewriters

17. Write the measured value 370 000 miles in scientific notation, with three significant digits.

18. Write the measured value 1 230 000 lb in scientific notation, with four significant digits.

19. In the expression 1 mile = 5280 ft, how many significant digits are there in "1 mile" and in "5280 ft"? Explain your answer.

20. In the expression 1 dollar = 4 quarters, how many significant digits are there in "1 dollar" and in "4 quarters"? Explain your answer.

21. Round off each of the following numbers to three significant digits: (a) 7.537×10^4 (b) 0.05555 (c) 21.45 (d) 21.4502

22. Round off each of the following numbers to three significant digits: (a) 4.675 (b) 32.69 (c) 0.4915 (d) 7.6650

23. Report the answer to each calculation in scientific notation, to the appropriate number of significant digits.
(a) $23.59 + 2.3$
(b) $2.577 - 8.4$
(c) $9.2 \times 10^{-4} + 2.88 \times 10^{-3}$
(d) $(9.77 \times 10^{-3})(4.88 \times 10^4)$
(e) $\dfrac{(0.005001)(1.6 \times 10^4)}{(2.7884 \times 10^{-2})}$

24. Report the answer to each calculation in scientific notation, to the appropriate number of significant digits.
(a) $0.057 + 0.688$
(b) $0.044 - 1.59$
(c) $6.3 \times 10^5 + 4.4 \times 10^4$
(d) $(8.44 \times 10^2)(2.96 \times 10^{-3})$
(e) $\dfrac{(2.3 \times 10^{-2})(7.7 \times 10^{-3})}{(6.55 \times 10^{-4})}$

25. Express the answer to each of the following calculations in scientific notation, with the appropriate number of significant digits.
(a) $1.652 \times 10^{-3} + 2.93 \times 10^{-2}$
(b) $8.6 \times 10^3 - 20.0 \times 10^2$
(c) $(3.2 \times 10^{-4})(1.10 \times 10^4)$
(d) $(0.004)/(1.65 \times 10^{-2})$
(e) $\dfrac{(5.41 \times 10^5)(2.7 \times 10^2)}{(1.6 \times 10^{-2})(2 \times 10^3)}$

26. Express the answer to each of the following calculations in scientific notation, with the appropriate number of significant digits.
(a) $20.7 \times 10^4 + 6.2 \times 10^3$
(b) $7 \times 10^{-2} - 6.5 \times 10^{-3}$
(c) $(9.4 \times 10^2)(3.15 \times 10^1)$
(d) $(6.92 \times 10^{-5})/(3.1 \times 10^{-3})$
(e) $\dfrac{(1.165 \times 10^{-2})(3.977 \times 10^{-3})}{(2.72 \times 10^{-5})(2 \times 10^4)}$

Algebra (Section 2-6)

27. Solve each of the following equations for x. (a) $5 - x = 8$ (b) $15x = -75$ (c) $x/2 = 4$

28. Solve each of the following equations for y. (a) $-y - a = -b$ (b) $3y = a$ (c) $ay/b = c$

29. Solve $xy = z$ for (a) x and (b) z.

30. Solve $z/x = y$ for (a) x and (b) z.

3-1
Samples of matter can be classified as mixtures or pure substances; pure substances can be classified as compounds or elements.

3-2
Many substances can exist in the form of a solid, a liquid, or a gas.

3-3
Scientists use metric and SI units of measurement.

3-4
Length units can be used to express areas or volumes.

3-5
The density of a substance is its mass per unit volume; the common SI units are g/cm^3.

3-6
The Celsius and Kelvin temperature scales are used for scientific work.

3-7
Quantities of heat are measured in joules or calories.

3-8
Energy in the form of electromagnetic radiation is measured in terms of its wavelength, velocity, and frequency.

3-9
Dimensional analysis is a systematic way to convert one set of units to another.

3-10
What's the best way to use dimensional analysis?

Inside Chemistry:
What is heat?

Chapter Summary

Problems

(Fundamental Photographs)

3

Measuring Matter and Energy

The aim of scientists has been to understand the world, partly out of simple curiosity and partly to make it a better place for human beings to live. Through many centuries of successes and failures scientists have discovered and developed certain approaches that have led to useful concepts about how the world works; these approaches are called scientific methods.

Three of the most successful scientific methods are naming, classifying, and measuring. Naming things and classifying things make it much easier to think about them, and measuring things—describing their properties in numbers—makes it possible to understand them more exactly. In this chapter you'll learn the basic naming, classifying, and measuring systems used in chemistry.

The most fundamental classification in chemistry is the distinction between matter and energy: **Matter** is anything that takes up space and has mass, and **energy** is the capacity to move matter; whenever matter is moved, energy is used.†

Exercise 3.1◆

Can you think of anything that is matter but that can't be seen? Can you think of anything that you can see that isn't matter?

◆Answers to Exercises are in Appendix 1.

Matter comes in many forms, and it's useful to have a systematic way to distinguish among them. In Sections 3-1 and 3-2 you'll see how matter can be classified as mixtures or pure substances; as compounds or elements; and as solids, liquids, or gases. You're probably familiar with the metric system as a set of units for measuring the properties of matter. Sections 3-3 through 3-5 review the basic units of the metric system and introduce a form of the metric system called the SI, which has been adapted for scientific work.

Energy also has many forms, and this chapter considers two of them, heat and light. Sections 3-6 and 3-7 describe ways of measuring heat: the Celsius and Kelvin temperature scales and the units called calories and joules. Section 3-8 describes the measurement of light energy in terms of its velocity, frequency, and wavelength.

In Sections 3-9 and 3-10 you'll learn a method of calculation called **dimensional analysis,** which is used in solving many chemistry problems. Dimensional analysis has many advantages: It's not difficult to learn, it can be used to solve many different kinds of problems, it's fast and accurate, and it requires only simple mathematical operations.

Throughout this chapter you'll be learning many scientific words. It's important that you learn them well because they'll be the basic vocabulary for the rest of this book.

† Chemistry Insight

It's sometimes difficult to see that energy in a certain form could be used to move matter. We're familiar with energy in the form of sunlight, for example, but it takes a long line of reasoning to show that energy from sunlight, stored in petroleum deposits and released from burning gasoline, moves our cars.

3-1 Samples of matter can be classified as mixtures or pure substances; pure substances can be classified as compounds or elements

Most matter is in the form of complicated mixtures—soil and trees and people are familiar examples. **Heterogeneous mixtures** are those in which we can see that two or more substances are present; for example, it's easy to recognize that a mixture of water and sand is heterogeneous. **Homogeneous mixtures** are those in which we can't see two or more substances; salt dissolved in water, for example, looks like pure water.

Homogeneous mixtures that are liquids or gases are usually called **solutions.**◆ An example of a gaseous mixture or solution is air, which is about 80% nitrogen gas and 20% oxygen gas, with other gases present in small amounts.

◆It's possible to have solutions that are solids; homogeneous mixtures of gold and silver, for example, are used to make jewelry. Homogeneous mixtures, or solutions, of metals are called **alloys.**

Whether we call a mixture homogeneous or heterogeneous depends on what we can see. A mixture of salt and sugar looks homogeneous, but if you studied it through a microscope, you might be able to see that two substances were present and decide that it was heterogeneous.

When the pure substances powdered sulfur (top) and powdered iron (right) are combined, they form a heterogeneous mixture (left). *(Charles Steele)*

Exercise 3.2

If you shook equal amounts of oil and vinegar together, you might look at the mixture and describe it as homogeneous, but if you let the mixture stand and looked at it again, you might describe it as heterogeneous. Explain.

Mixtures are more complicated than pure substances, and harder to understand. For this reason much of the work in any chemical laboratory involves separating mixtures into pure substances. Some methods for separating mixtures are simple (for example, you could separate a mixture of sand and water by pouring it onto a filter that would trap the sand and let the water pass through) and some are complicated, requiring special and expensive equipment. However it's done, separating a mixture consists of physically separating the particles of its constituents from one another, like sorting laundry. For this reason a **mixture** can be defined as a sample of matter that can be separated into two or more substances by physical means.

A magnet can be used to separate powdered iron from its mixture with powdered sulfur. *(Charles Steele)*

The diagram in Figure 3.1 shows that pure substances can be classified as compounds or elements. A **compound** is a pure substance that can be decomposed into elements, and an element is a substance that can't be decomposed into simpler substances.

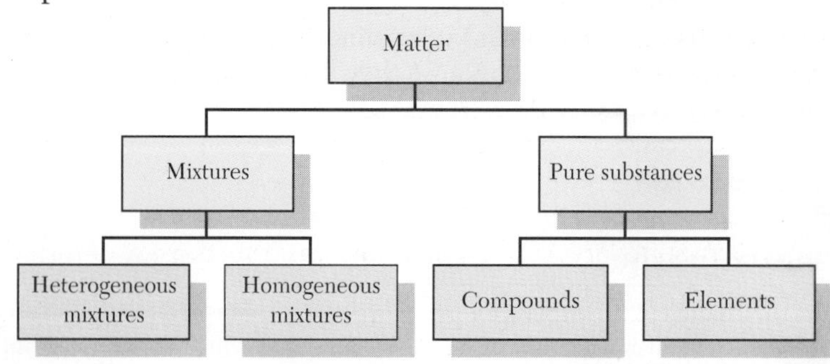

Figure 3.1
Forms of matter.

There are several ways to convert compounds into their elements. Some compounds decompose into their elements when they're heated. For example, a red compound of mercury and oxygen, when it's heated to high temperatures, forms oxygen gas and metallic mercury. Electricity breaks down many compounds into their elements. If an electric current is passed through a sample of water, for example, the water decomposes into two gases, hydrogen and oxygen, so water is classified as a compound. Later we'll see more examples of ways that compounds can be decomposed into their elements.

If a pure substance can be decomposed, it must be a compound since it's made up of more fundamental substances, its elements. If a pure substance can't be decomposed, we can't be sure, from that evidence alone, that it's an element, since we may not have tried a method of decomposition that will work.

Exercise 3.3

At a high temperature common table salt can be melted to form a liquid. If an electric current is passed through the liquid, two substances are formed: a soft, shiny metal, called sodium, and a pale green gas, called chlorine. From this evidence would you say that salt is an element or a compound? Explain.

Exercise 3.4

When sodium metal is melted and an electric current is passed through it, no change occurs. From this evidence would you say that sodium is a compound or an element? Explain.

Part of chemistry, called **analytical chemistry,** is directed toward understanding how compounds are put together, by breaking them down into simpler compounds or into their elements. **Synthetic chemistry** deals with the reverse process, making compounds by combining elements or by building them up from simpler compounds. One definition of chemistry is that it's the science that studies how the elements interact to form all of the many kinds of matter and also how these elements can interact to form new kinds of matter.

3-2 Many substances can exist in the form of a solid, a liquid, or a gas

When they're heated, many compounds and elements change from solid to liquid to gas, and they reverse these changes when they're cooled. The most familiar example is water: Solid ice melts to liquid water at 32°F, and liquid water boils to form water vapor (gas) at 212°F; water vapor cooled to 212°F condenses to form liquid water, and liquid water cooled to 32°F freezes to form ice.

Water in the solid, liquid, and gas states. *(Fundamental Photographs, New York)*

The physical conditions solid, liquid, and gas are referred to as **states of matter,** and conversions among them are referred to as **changes of state.** The following diagram summarizes the terms used to describe changes of state:

$$\text{solid} \underset{\text{freezing}}{\overset{\text{melting}}{\rightleftharpoons}} \text{liquid} \underset{\text{condensing}}{\overset{\text{boiling}}{\rightleftharpoons}} \text{gas (vapor)}$$

A **phase** is a region of matter separated by a visible boundary from the regions of matter next to it. An ice cube floating in carbonated water has three phases: ice, liquid water, and gas (carbon dioxide) bubbles. Different phases don't have to be different states. For example, oil floating on water consists of two different phases, both in the liquid state.

The phase concept can describe homogeneous and heterogeneous mixtures: A homogeneous mixture is a mixture in which only one phase is present, and a heterogeneous mixture is a mixture in which two or more phases are present.

Exercise 3.5

A glass contains sand and water. Including the glass and the air around it, how many phases are present?

3-3 Scientists use metric and SI units of measurement

(Jack Plekan/Fundamental Photographs, New York)

In the United States we use the traditional English system of measurements shown in Table 3.1; you probably memorized this system so long ago that you don't remember learning it. The English system is difficult to learn, because the names of the units and their numerical relationships to one another have almost no regularity.

Table 3.1 English Measurements

Mass	Length	Volume
1 pound (lb) = 16 ounces (oz)	1 foot (ft) = 12 inches (in.)	1 pint (pt) = 16 fluid ounces (fl oz)
1 ton = 2000 lb	1 yard (yd) = 3 ft	1 quart (qt) = 2 pt
	1 mile (mi) = 5280 ft	1 gallon (gal) = 4 qt

Chemists use a modified version of the metric system known as the SI (from the French, *le Système International d'Unités*). In your later work in chemistry or physics you may find it useful to distinguish between metric and SI units; this book uses a selection of units from the metric system and the SI that have been chosen because they're important and easy to learn.

In the metric system and the SI each category of measurement (mass, length, and volume) has a simplest unit (gram, meter, and liter). Other units within each category are related to the simplest unit by powers of ten, and each

power of ten is identified by a prefix attached to the name of the simplest unit. For example, the prefix *milli-* means 1×10^{-3}, so

$$1 \text{ milligram} = 1 \times 10^{-3} \text{ gram}$$
$$1 \text{ millimeter} = 1 \times 10^{-3} \text{ meter}$$
$$1 \text{ milliliter} = 1 \times 10^{-3} \text{ liter}$$

There are many metric and SI prefixes, but we'll need only those shown in Table 3.2. You must learn these prefixes, their abbreviations, and the powers of ten they stand for.◆

◆This Section gives the minimum information about metric and SI units that you'll need for your work in this book. A more complete description of the metric system and the SI appears in Appendix 2.

Table 3.2 Important SI Prefixes

Power of 10	Prefix	Abbreviation
1×10^{6}	mega	M
1×10^{3}	kilo	k
1×10^{-1}	deci	d
1×10^{-2}	centi	c
1×10^{-3}	milli	m
1×10^{-6}	micro	μ (Greek letter, pronounced *mew*)
1×10^{-9}	nano	n
1×10^{-12}	pico	p

Exercise 3.6

Without looking at Table 3.2, name the prefix identified by each of the following abbreviations: (a) μ (b) c (c) n (d) k (e) d

Exercise 3.7

Without looking at Table 3.2, write the abbreviation for the prefix that designates each of the following powers of ten: (a) 10^{-3} (b) 10^{-1} (c) 10^{3} (d) 10^{-2} (e) 10^{-9}

Table 3.3 lists the metric and SI units we'll use and their English equivalents. You must learn the information in this table.

Table 3.3 Important Metric and SI Units and Conversions◆

Mass	Length	Volume
1 lb = 454 g (grams)	1 in. = 2.54 cm (centimeters)	1.06 qt = 1 L (liter)
1 Mg = 1×10^{6} g	1 Mm = 1×10^{6} m	1 ML = 1×10^{6} L
1 kg = 1×10^{3} g	1 km = 1×10^{3} m	1 kL = 1×10^{3} L
1 dg = 1×10^{-1} g	1 dm = 1×10^{-1} m	1 dL = 1×10^{-1} L
1 cg = 1×10^{-2} g	1 cm = 1×10^{-2} m	1 cL = 1×10^{-2} L
1 mg = 1×10^{-3} g	1 mm = 1×10^{-3} m	1 mL = 1×10^{-3} L
1 μg = 1×10^{-6} g	1 μm = 1×10^{-6} m	1 μL = 1×10^{-6} L
1 ng = 1×10^{-9} g	1 nm = 1×10^{-9} m	1 nL = 1×10^{-9} L
1 pg = 1×10^{-12} g	1 pm = 1×10^{-12} m	1 pL = 1×10^{-12} L

◆Since everyday use of metric units is increasing in the United States, it's useful to remember these approximate relationships: A liter is about a quart, a pound is about 500 grams, a kilometer is about two thirds of a mile (1 km = 0.621 mi), and a meter is about one yard (1 yd = 0.914 m). For scientific work we need the more exact relationships in the table.

Exercise 3.8

Without looking at Table 3.3, remember (a) the number of quarts in a liter, (b) the number of grams in a pound, and (c) the number of centimeters in an inch.

Exercise 3.9

Without looking at Table 3.3, remember (a) the number of g in a cg, (b) the number of L in a mL, and (c) the number of m in a km.

3-4 Length units can be used to express areas or volumes

In the English system we say that a square with a side whose length is 1 foot has an area of 1 square foot, and a cube with a 1-foot edge has a volume of 1 cubic foot. In mathematical form:

area: $1 \text{ ft} \times 1 \text{ ft} = 1 \text{ ft}^2$ (1 ft^2 is read "1 square foot")

volume: $1 \text{ ft} \times 1 \text{ ft} \times 1 \text{ ft} = 1 \text{ ft}^3$ (1 ft^3 is read "1 cubic foot")

A rectangle with sides 2 feet and 3 feet has an area of $2 \text{ ft} \times 3 \text{ ft} = 6 \text{ ft}^2$, and a box with edges 1 foot, 2 feet, and 3 feet has a volume of $1 \text{ ft} \times 2 \text{ ft} \times 3 \text{ ft} = 6 \text{ ft}^3$.

Exercise 3.10

What's the area in square feet of a rectangle whose sides are 1 foot and 3 feet?

Exercise 3.11

What's the volume in cubic feet of a box whose edges are 3 feet, 3 feet, and 1 foot?

Length units in the metric system and the SI can also express areas or volumes. A square 2 kilometers on an edge has an area of $2 \text{ km} \times 2 \text{ km} = 4 \text{ km}^2$ (read "4 square kilometers"). A box with edges 2 centimeters, 1 centimeter, and 3 centimeters has a volume of $2 \text{ cm} \times 1 \text{ cm} \times 3 \text{ cm} = 6 \text{ cm}^3$ (read "6 cubic centimeters").

Figure 3.2 shows the use of English or SI length units to express areas or volumes.

The metric system and the SI have been designed to show a simple relationship between units for length and volume:

$$1 \text{ cm}^3 = 1 \text{ mL}$$

Exercise 3.12

What's the area in square decimeters of a rectangle whose sides are 1 decimeter and 4 decimeters?

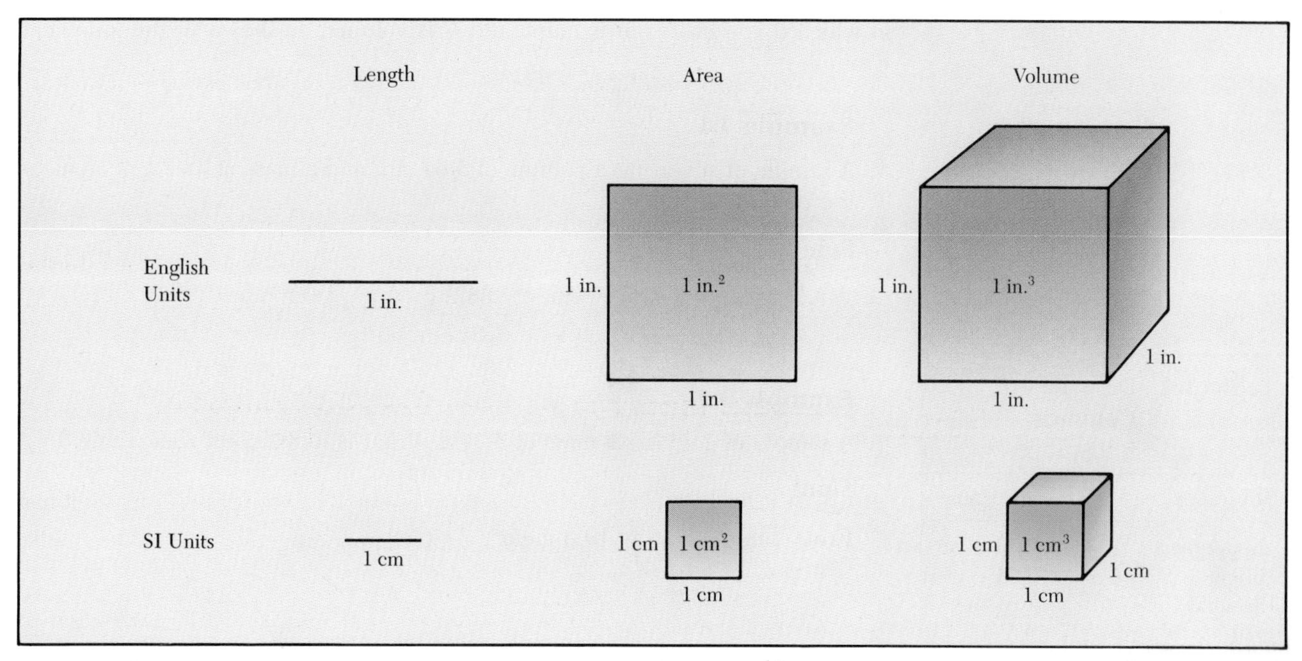

Figure 3.2
Using length units to express areas or volumes.

Exercise 3.13

The sides of a box are 1 centimeter, 2 centimeters, and 2 centimeters: (a) What is its volume in cubic centimeters? (b) What is its volume in milliliters?

3-5 The density of a substance is its mass per unit volume; the common SI units are g/cm³

From everyday experience we know that different objects of the same size, if they're made of different materials, can have very different masses: Picking up a block of wood is very different from picking up a block of concrete of the same size. As a more exact example, a cubic centimeter of aluminum has a mass of 2.70 g, and a cubic centimeter of gold has a mass of 19.3 g. More matter is packed in a cubic centimeter of gold than in a cubic centimeter of aluminum.

This difference between aluminum and gold can be expressed as a difference in density. The **density** of a substance is its mass per unit volume. In the SI the usual units for density are grams per cubic centimeter, g/cm^3, or grams per milliliter, g/mL: The density of aluminum is 2.70 g/cm^3, and the density of gold is 19.3 g/cm^3.

Density values for several common substances are given in Table 3.4. It's often convenient to represent density with the formula

$$d = \frac{m}{V}$$

in which d is density, m is mass, and V is volume, as shown in the following Examples.

Table 3.4 Densities of Common Substances*

Substance	Density, g/cm^3
aluminum	2.70
copper	8.94
diamond	3.51
gold	19.3
iron	7.86
lead	11.3
mercury	13.6
water	1.00**

*Most substances expand when they're heated and contract when they're cooled, so density depends on temperature; the densities in this table are at 25°C (77°F). (The temperature units °C, degrees Celsius, are described in the next Section of this chapter.) The number of significant digits in a density value is found by the method described in Section 2-3.

**When the metric system was designed, the cubic centimeter was defined as the volume that would be occupied by exactly 1 gram of water at its temperature of maximum density, 4.08°C (39.3°F). At 25°C (77°F) the density of water is 0.997 g/cm^3; in Table 3.4 this value is rounded off to 1.00 g/cm^3.

Example 3.1

A sample of silver has a volume of 3.64 cm^3 and a mass of 38.2 g. Calculate the density of silver.

Solution

$$d = \frac{m}{V} = \frac{38.2 \text{ g}}{3.64 \text{ cm}^3} = 10.5 \text{ g/cm}^3$$

Example 3.2

A sample of gold has a mass of 57.4 g. What is its volume? (See Table 3.4.)

Solution

From Table 3.4, d = 19.3 g/cm^3.

$$d = \frac{m}{V}$$

$$Vd = \frac{\cancel{V}m}{\cancel{V}}$$

$$Vd = m$$

$$\frac{V\cancel{d}}{\cancel{d}} = \frac{m}{d}$$

$$V = \frac{m}{d} = \frac{57.4 \cancel{\text{g}}}{19.3 \cancel{\text{g}}/\text{cm}^3} = 2.97 \text{ cm}^3$$

Exercise 3.14

A piece of copper in the shape of a rectangular solid has edges that measure 1.5 cm, 2.2 cm, and 2.5 cm. What is its mass? (See Table 3.4.)

Density values for liquids are sometimes given as *specific gravity* values. The **specific gravity** of a liquid is its density divided by the density of water. Since the density of water in SI units is 1.00 g/cm^3, dividing by this value has the effect of removing the units for density: The specific gravity for a liquid is the same as its density, without units. For mercury, for example,

$$\text{specific gravity} = \frac{\text{density of mercury}}{\text{density of water}} = \frac{13.6 \cancel{\text{g/cm}^3}}{1.00 \cancel{\text{g/cm}^3}} = 13.6$$

3-6 The Celsius and Kelvin temperature scales are used for scientific work

Energy, the capacity to do work, is a more difficult concept than matter. One reason is that we can't see energy, we can only see its effects. Another reason is that energy can take many forms; from everyday experience we're familiar with a few of them, such as muscular energy, gravitational energy, mechanical energy, and electrical energy.

One of the forms of energy we're most familiar with is heat. In this Section and the next one you'll learn how to describe two properties of heat, its intensity and its quantity.

Temperature is the measure of heat intensity. For everyday purposes in the United States, we usually use the Fahrenheit temperature scale, but for scientific work the Celsius and Kelvin scales are used.† The relationships among these scales are shown in Table 3.5.

Table 3.5 Temperature Scales◆

Reference Temperatures	Fahrenheit Scale	Celsius Scale◆	Kelvin Scale
Boiling point of water	212°F	100°C	373 K
Freezing point of water	32°F	0°C	273 K
Limit of lowest temperature	−459°F	−273°C	0 K

◆The boiling point and freezing point of water vary with pressure; for example, water boils at lower temperatures at higher altitudes, where the atmospheric pressure is lower. The temperatures in this table assume a standard pressure of one *atmosphere*. The *atmosphere* and other pressure units are described in Chapter 13.

◆We're beginning to use the Celsius scale for everyday temperature measurements, so it's useful to remember that a common room temperature, 70°F, is 21°C.

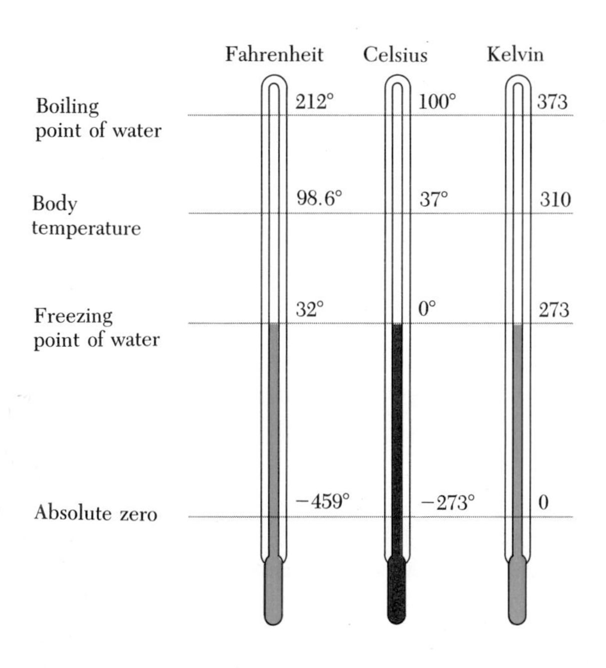

† Chemistry Insight

The Fahrenheit and Kelvin temperature scales are named for the men who created them: Gabriel Daniel Fahrenheit (1696–1736) was a German physicist, and William Thomson, 1st Baron Kelvin (1824–1907), was an English physicist and mathematician. The Celsius scale, formerly called the Centigrade scale, is named for Anders Celsius (1701–1744), a Swedish astronomer who first proposed a temperature scale divided into 100 degrees between points designating the freezing and boiling temperatures of water.

The Fahrenheit scale is read in units of **degrees Fahrenheit,** symbol °F; the Celsius scale is read in units of **degrees Celsius,** symbol °C; and the Kelvin scale is read in units of **kelvins,** symbol K.

The Celsius and Kelvin scales were designed to be useful for scientific work. The Celsius scale has the advantage that its reference temperatures for the freezing point and boiling point of water are easy to remember: 0°C and 100°C. The advantage of the Kelvin scale is that it has no negative temperatures, because the limit of lowest temperature, called *absolute zero,* is designated 0 K.◆

◆Why is there a limit of lowest temperature? One explanation is given in Section 13-3.

The following formulas are used to convert a temperature on one scale to the corresponding temperature on another scale:◆

◆The formula for converting between Fahrenheit and Celsius can be written in other forms, such as $(5/9)(F - 32) = C$ and $F - 32 = (9/5)C$, or $C = \frac{F - 32}{1.8}$ and $F = 1.8C + 32$. The form I've described seems easiest to remember, but use one of the other forms if you or your instructor prefer it.

$$\frac{F - 32}{180} = \frac{C}{100} \qquad K = C + 273$$

The following example shows the conversion of 125°F to K:

$$\frac{125 - 32}{180} = \frac{C}{100}$$

$$\frac{(100)(125 - 32)}{180} = \frac{C(\cancel{100})}{\cancel{100}}$$

$$\frac{(100)(93)}{180} = C$$

$$C = 52°C$$

$$K = C + 273 = 52 + 273 = 325 \text{ K}$$

In converting a temperature from one scale to another, express your answer to the same number of decimal places as there are in the value you began with: 125°F = 52°C = 325 K. In this book we'll use temperatures that are whole numbers, that is, temperatures with no digits after the decimal point.◆

◆For more precise calculations the relationship K = C + 273.15 is used.

Exercise 3.15

Convert 1 K to °F.

3-7 Quantities of heat are measured in joules or calories

There's a difference between intensity of heat, measured in temperature units, and quantity of heat. If a paper clip and a 10-pound iron bar are each at 100°C, the intensity of heat is the same in each object, but there's a much greater quantity of heat in the iron bar. In scientific work, the units used for measuring quantities of heat are the **calorie,** abbreviated cal, and the **joule,** symbol J.◆

◆Joule is pronounced "jool" and named for James P. Joule (1818–1889), an English physicist.

The calorie, which was the standard unit for expressing quantities of heat for many years, was originally defined as the amount of heat required to raise the temperature of one gram of water by one degree Celsius.◆ Recently, scientists have agreed to replace the calorie with the joule as the new standard. The calorie is now defined in terms of the joule:

◆The British Thermal Unit (BTU) is still occasionally used as a heat unit in engineering or manufacturing. The BTU is the amount of heat required to raise the temperature of one pound of water one degree Fahrenheit.

1 calorie = 4.184 joules or 1 cal = 4.184 J

Because this is a defined relationship, the number of significant digits on each side of the equation is infinite.

Although the joule is now the standard heat unit, calories are still commonly used because many people learned to use calories and want to continue with them. In the long run there's no advantage in using two heat units that are about the same size, so the joule will probably replace the calorie completely, but that may take many years.

Because the calorie and the joule are small units, it's often convenient to express quantities of heat in kilocalories (kcal) or kilojoules (kJ).◆

◆It's easy to get confused in talking about *calories*, as used in physics and chemistry, and *Calories*, as used in describing energies contained in various foods. The *Calorie* that measures food energy, abbreviated Cal, is actually the *kilocalorie*.

Exercise 3.16

Write the equation for the relationship between kilocalories and kilojoules.

3-8 Energy in the form of electromagnetic radiation is measured in terms of its wavelength, velocity, and frequency

Our most important information about the world comes to us in the form of energy we call light. Our senses of touch, taste, smell, and hearing give us very limited information and are useful only for short distances, but our sight—our sense that responds to light—gives us very detailed information about nearby objects and also about objects that are millions of miles away.

Because light is such an important part of our experience, thousands of scientists have investigated it since the beginning of modern science about three centuries ago; we'll see later that some of the most important concepts of modern chemistry have come from an understanding of light. This Section looks briefly at the theory of the nature of light now held by scientists and explains how various forms of light can be described mathematically.

Light consists of waves. Our eyes aren't sensitive enough to see individual waves of light, but we can imagine them as being similar to waves in water. Light waves have three basic characteristics: wavelength, velocity, and frequency.

Wavelength is the distance from a point on one wave to the corresponding point on an adjacent wave. By scientific convention, wavelength is represented by a letter of the Greek alphabet, **λ** (lambda), as shown below. Because wavelength is a distance, it's measured in length units, for example in meters per wave, usually expressed simply as meters. Light waves have very short wavelengths, which are often measured in nanometers per wave, usually expressed simply as nanometers.

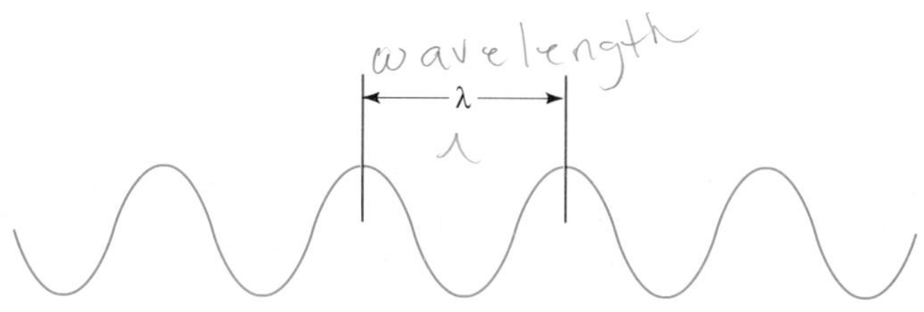

The **velocity** of waves is the speed with which they move. In the preceding drawing think of the waves as moving from left to right across the page. In a vacuum light waves travel at a constant and incredibly high rate, 3.00×10^8 meters per second (186 000 miles per second), and this velocity changes only slightly when light passes through media such as air or water. The constant velocity of light in a vacuum is represented by the symbol c.

Frequency is the number of waves that pass a fixed point in a certain amount of time, usually one second. In the drawing below imagine that the waves are moving from left to right, and imagine that you're counting the number of wave crests that pass the vertical line every second. That count is the frequency. The symbol for frequency is the Greek letter $\boldsymbol{\nu}$ (nu, pronounced *new*), and its units are waves per second, usually abbreviated to per second.

The easiest way to understand the meaning of the units used for wavelength, velocity, and frequency is to see that there is a simple relationship among these properties, as shown in Figure 3.3. Imagine that each of the sets of waves shown in the drawing is moving across the page from left to right at the same velocity. Because the waves in the upper set have a long wavelength, only a few wave crests pass the vertical line every second, and the frequency is low; because the waves in the lower set have a short wavelength, many wave crests pass the vertical line every second, and the frequency is high.

The mathematical relationship among velocity, wavelength, and frequency is $\mathbf{c} = \boldsymbol{\lambda\nu}$, and the corresponding units are

$$\mathbf{c} \quad = \quad \boldsymbol{\lambda} \qquad \boldsymbol{\nu}$$

$$\frac{\text{meters}}{\text{second}} = \left(\frac{\text{meters}}{\text{wave}}\right)\left(\frac{\text{waves}}{\text{second}}\right) \quad \text{or}$$

$$\frac{\text{meters}}{\text{second}} = (\text{meters})\left(\frac{1}{\text{second}}\right)$$

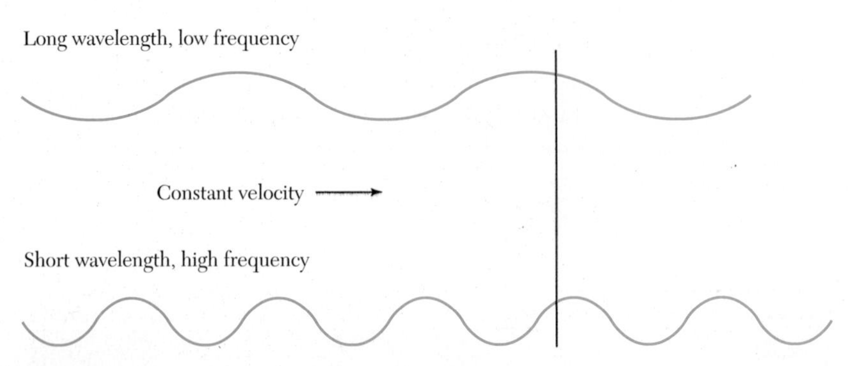

Figure 3.3
As wavelength increases, frequency decreases.

Remember that *per* means *divided by:* Meters/second is read "meters per second," and 1/second is read "per second." The units for frequency, 1/second, are abbreviated s^{-1} and read "per second."◆ These units have been given the name **hertz,** symbol **Hz.**◆ So each of these terms means the same thing: 1/second, 1/s, s^{-1}, per second, hertz, and Hz.

◆A negative exponent means division: $10^{-3} = 1/10^3$ and $s^{-1} = 1/s^1 = 1/s$.

◆The unit *hertz* is named for the German physicist Heinrich Rudolph Hertz (1857–1894).

Exercise 3.17

(a) Write *miles/hour* in words. (b) Write *per hour* in mathematical notation, two different ways (the abbreviation for hour is h). (c) *Miles per hour* can be written miles $hour^{-1}$ or mi h^{-1}. Write *meters per second* in these forms. (d) Write *3.00 × 10^8 meters per second* in the two forms used in part (c).

The wavelength and frequency of visible light correspond to its color, as shown in Figure 3.4. The shortest wavelength visible to the human eye is about 400 nm and corresponds to violet; the longest visible wavelength is about 800 nm and corresponds to red.

Example 3.3

Calculate the frequency of violet light, whose wavelength is 400 nm (3 significant digits), and express it in Hz.

Solution

Convert 400 nm to m, the length units used in the value for c.

$$400 \text{ nm} = 400 \times 10^{-9} \text{ m} = 4.00 \times 10^{-7} \text{ m}$$

Solve $c = \lambda\nu$ for ν.

$$c = \lambda\nu$$

$$\lambda\nu = c$$

$$\frac{\not\lambda\nu}{\not\lambda} = \frac{c}{\lambda}$$

$$\nu = \frac{c}{\lambda}$$

Substitute into this equation the values of c and λ, and solve for ν.

$$\nu = \frac{3.00 \times 10^8 \not\text{m}/\text{s}}{4.00 \times 10^{-7} \not\text{m}}$$

$$\nu = 7.50 \times 10^{14}/\text{s} = 7.50 \times 10^{14} \text{ s}^{-1}$$

$$\nu = 7.50 \times 10^{14} \text{ Hz}$$

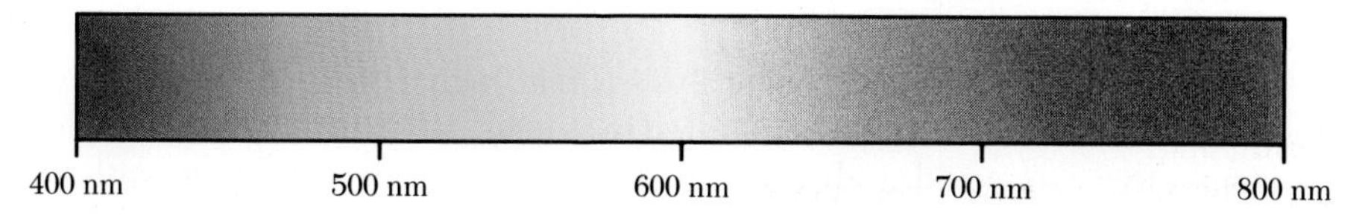

Figure 3.4
Wavelengths of visible light.

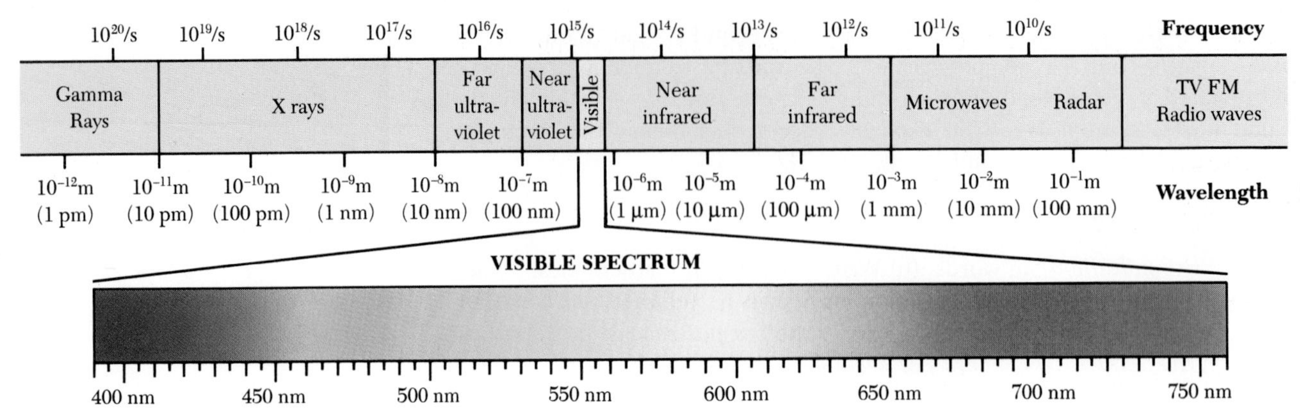

Figure 3.5
Wavelengths in the spectrum of electromagnetic radiation. *(After Ebbing, Darrell D.,* General Chemistry, *3rd ed. Copyright © 1990 by Houghton Mifflin Company.)*

Exercise 3.18

Below each vertical line in Figure 3.4 on page 37, write the corresponding frequency, in s^{-1}.

Through evolution our eyes have become sensitive to light in the wavelength range of about 400 nm to about 800 nm. We can't see radiation outside these wavelengths but we can detect it with instruments, and these instruments have shown that visible light is only a small portion of the spectrum of **electromagnetic radiation,** as shown in Figure 3.5.

Exercise 3.19

Using information from Figures 3.4 and 3.5, write a one-sentence definition of *visible light.*

Electromagnetic radiation outside the visible range has been put to important uses, as you'll recognize from the names on the spectrum in Figure 3.5. One useful generalization is that waves with shorter wavelengths penetrate objects more deeply. For this reason ultraviolet rays from the sun are harmful to our eyes, and X rays can pass through objects that are impenetrable to visible light.

3-9 Dimensional analysis is a systematic way to convert one set of units to another

In working with measured values, it's often useful to convert from one set of units to another, for example, to convert a distance in kilometers to miles or to convert an area in square feet to square yards. Most of the chemical calculations you'll do will include unit conversions, so it's important for your success in chemistry that you're able to make this kind of calculation quickly and accurately. The method used to make unit conversions is called **dimensional analysis.**◆

◆Also called the **factor-label method** or the **factor-unit method.**

The advantage of dimensional analysis is that it gives you a systematic way to solve either simple or very complicated problems. We'll begin with examples so simple that you could probably solve them with a little reasoning, without using a systematic method. Be sure you understand these examples well, so that you'll be able to use dimensional analysis to solve more complicated problems.

Example 3.4

How many inches are in 4.0 feet?

Solution

Step 1: State the problem in an abbreviated form.

Use the form shown below to identify the given quantity and the quantity to be found.

4.0 ft = ? in.   problem
given   to be found

Step 2: Write a statement (called a unit map) that shows the conversion to be made from the given units to the units to be found.

ft → in.   unit map

Step 3: Write a statement that shows the mathematical relationship between the units that are given and the units that are to be found.

1 foot =12 inches   equation relating units

Step 4: Convert the relationship between the given units and the units to be found into two factors.

Divide each side of the equation relating units by the other side.

1 foot = 12 inches

$$\frac{1\ \text{ft}}{12\ \text{in.}} = \frac{\cancel{12\ \text{in.}}}{\cancel{12\ \text{in.}}} \qquad \frac{1\ \text{ft}}{12\ \text{in.}} = 1 \qquad \text{factor}$$

$$\frac{\cancel{1\ \text{ft}}}{\cancel{1\ \text{ft}}} = \frac{12\ \text{in.}}{1\ \text{ft}} \qquad \frac{12\ \text{in.}}{1\ \text{ft}} = 1 \qquad \text{factor}$$

A **factor** is a fraction that shows the relationship between the given units and the units to be found.

Step 5: Choose one of the factors and multiply the given quantity by it to solve for the quantity to be found.

ft → in.   unit map

$$\frac{4.0\ \cancel{\text{ft}}}{}\left(\frac{12\ \text{in.}}{1\ \cancel{\text{ft}}}\right) = 48\ \text{in.} \qquad \text{calculation}$$

The answer should have two significant digits: 4.0 ft is a measured value with two significant digits; 1 ft and 12 in. are defined values, with infinite significant digits.

Multiplying by the incorrect factor gives incorrect or meaningless units:

$$\frac{4.0\ \text{ft}}{}\left(\frac{1\ \text{ft}}{12\ \text{in.}}\right) = 0.33\ \text{ft}^2/\text{in.} \qquad \text{incorrect calculation}$$

Step 6: Check your answer: Compare the units in your answer with the units to be found in your original statement of the problem and make certain you've found the answer you were after; decide whether your numerical answer is reasonable.

Our answer, 48 in., has the right units. Inches are smaller units than feet, so our numerical answer should be larger than the number we started with, and it is.

Example 3.5

How many nanometers are in 52.6 meters?

Solution

$52.6\ \text{m} = ?\ \text{nm}$	problem
$\text{m} \rightarrow \text{nm}$	unit map
$1\ \text{m} = 1 \times 10^9\ \text{nm}$	equation relating units
$\dfrac{1\ \text{m}}{1 \times 10^9\ \text{nm}} = 1 \qquad \dfrac{1 \times 10^9\ \text{nm}}{1\ \text{m}} = 1$	factors
$\text{m} \rightarrow \text{nm}$	unit map
$52.6\ \text{m}\left(\dfrac{1 \times 10^9\ \text{nm}}{1\ \text{m}}\right) = 5.26 \times 10^{10}\ \text{nm}$	calculation
Units are correct and answer is reasonable.	check

Exercise 3.20

(a) How many feet are in 5.00 miles? (b) How many kilometers are in 2.74×10^5 meters?

As shown in the following Example, some calculations require more than one factor.

Example 3.6

How many inches are in 3.29 miles?

Solution

The unit map and the calculation require more than one step.

$3.29\ \text{mi} = ?\ \text{in.}$	problem
$\text{mi} \rightarrow \text{ft} \rightarrow \text{in.}$	unit map
$1\ \text{mi} = 5280\ \text{ft} \quad \dfrac{1\ \text{mi}}{5280\ \text{ft}} = 1 \quad \dfrac{5280\ \text{ft}}{1\ \text{mi}} = 1$	equation relating units, factors
$1\ \text{ft} = 12\ \text{in.} \quad \dfrac{1\ \text{ft}}{12\ \text{in.}} = 1 \quad \dfrac{12\ \text{in.}}{1\ \text{ft}} = 1$	equation relating units, factors

Step	Question to Ask
1. State problem.	What quantities are given and to be found?
2. Write unit map.	What unit conversions should I make to go from the given units to the units to be found?
3. Write equations relating units.	What are the equations for the mathematical relationships between the units for each conversion on the unit map?
4. Write factors.	What two factors are formed from each equation relating units?
5. Multiply by correct factors and decide on significant digits.	Which factors will make the conversions shown on the unit map? How many significant digits should I use in the answer?
6. Check answer for reasonableness.	Are the units in the answer those which were to be found? Is the number in the answer reasonable?

Figure 3.6
Summary of dimensional analysis.

mi → ft → in. unit map

$$\frac{3.29\ \text{mi}}{}\left(\frac{5280\ \text{ft}}{1\ \text{mi}}\right)\left(\frac{12\ \text{in.}}{1\ \text{ft}}\right) = 2.08 \times 10^5\ \text{in.}$$ calculation

Units are correct and answer is reasonable. check

Figure 3.6 summarizes the steps and the reasoning used in dimensional analysis.

The following Examples show variations in the method of dimensional analysis.

Example 3.7

A cab company bought seven cabs for $64 780.66. What was the cost of each cab?

Solution

The relationship between the given units and the units to be found is contained in the problem.

1 cab = $? problem

cab → $ unit map

7 cabs = $64 780.66 equation relating units

$$\frac{7\ \text{cabs}}{\$64\ 780.66} = 1 \qquad \frac{\$64\ 780.66}{7\ \text{cabs}} = 1$$ factors

cabs → $ — unit map

$$\frac{1\ \text{cab}}{}\left(\frac{\$64\,780.66}{7\ \text{cabs}}\right) = \$9254.38$$ — calculation

Units are correct and answer is reasonable. — check

Example 3.8

How many cubic feet are in 25 cubic yards?

Solution

Length units can be cubed to give volume units.

$25\ \text{yd}^3 = ?\ \text{ft}^3$ — problem

$\text{yd}^3 \rightarrow \text{ft}^3$ — unit map

$1\ \text{yd} = 3\ \text{ft} \qquad (1\ \text{yd})^3 = (3\ \text{ft})^3$
$1\ \text{yd}^3 = 27\ \text{ft}^3$ — equation relating units

$$\frac{1\ \text{yd}^3}{27\ \text{ft}^3} = 1 \qquad \frac{27\ \text{ft}^3}{1\ \text{yd}^3} = 1$$ — factors

$\text{yd}^3 \rightarrow \text{ft}^3$ — unit map

$$\frac{25\ \text{yd}^3}{}\left(\frac{27\ \text{ft}^3}{1\ \text{yd}^3}\right) = 680\ \text{ft}^3 = 6.8 \times 10^2\ \text{ft}^3$$ — calculation

Units are correct and answer is reasonable. — check

Example 3.9

The density of lead is 11.3 g/cm^3. Calculate its density in lb/in^3.

Solution

$11.3\ \text{g/cm}^3 = ?\ \text{lb/in.}^3$ — problem

$\text{g/cm}^3 \rightarrow \text{lb/cm}^3 \rightarrow \text{lb/in.}^3$ — unit map

$454\ \text{g} = 1\ \text{lb}$ — equation relating units

$$\frac{454\ \text{g}}{1\ \text{lb}} = 1 \qquad \frac{1\ \text{lb}}{454\ \text{g}} = 1$$ — factors

$1\ \text{in.} = 2.54\ \text{cm} \qquad (1\ \text{in.})^3 = (2.54\ \text{cm})^3$
$1\ \text{in.}^3 = 16.4\ \text{cm}^3$ — equation relating units

$$\frac{1\ \text{in.}^3}{16.4\ \text{cm}^3} = 1 \qquad \frac{16.4\ \text{cm}^3}{1\ \text{in.}^3} = 1$$ — factors

$\text{g/cm}^3 \rightarrow \text{lb/cm}^3 \rightarrow \text{lb/in.}^3$ — unit map

$$\frac{11.3\ \text{g}}{1\ \text{cm}^3}\left(\frac{1\ \text{lb}}{454\ \text{g}}\right)\left(\frac{16.4\ \text{cm}^3}{1\ \text{in.}^3}\right) = 0.408\ \text{lb/in.}^3$$ — calculation

Units are correct and answer is reasonable. — check

Exercise 3.21

(a) The area of a field is 0.158 square miles. What is its area in square feet? (b) Calculate the density of water in pounds per pint.

3-10 What's the best way to use dimensional analysis?

Some of the steps in dimensional analysis are automatic and easy, and others require careful thinking. Here are some ways to help you concentrate on the thinking steps.

1. ***State Problem.*** In solving most problems, this will be the most crucial and most difficult step; if you can state the problem in the form used for dimensional analysis, you can probably solve it. It's not always easy to decide what quantity is given and what units are to be found. Consider this problem:

   What's the average mass of a fish if 14 of them have a mass of 33.6 pounds?

   If you're puzzled, begin by asking yourself what units the answer will have, and write those units down in a partial statement of the problem. In this problem, we're asked to find an average mass, so our answer will be in mass units, pounds. Write:

   $$= \text{? lb}$$

   Next ask yourself, What is the thing I'm supposed to find the mass of? The problem asks for the mass of a fish, that is, one fish. Write:

   $$1 \text{ fish} = \text{? lb}$$

   Concentrate your effort on the first step, because if you state the problem correctly, you can probably solve it.

2. ***Write Unit Map.*** For a one-step calculation (converting feet to inches, for example) this step is automatic. If more than one step is needed, you'll have to decide what route you'll take. In many problems, more than one route is possible. For example, the problem

   How many miles are in 8.05×10^5 centimeters?

   can be solved this way if you know the conversions for centimeters → inches → feet → miles

   $$\underline{8.05 \times 10^5 \cancel{\text{cm}}}\left(\frac{1 \cancel{\text{in.}}}{2.54 \cancel{\text{cm}}}\right)\left(\frac{1 \cancel{\text{ft}}}{12 \cancel{\text{in.}}}\right)\left(\frac{1 \text{ mile}}{5280 \cancel{\text{ft}}}\right) = 5.00 \text{ miles}$$

   or this way if you know the conversions for centimeters → meters → yards → miles

   $$\underline{8.05 \times 10^5 \cancel{\text{cm}}}\left(\frac{1 \cancel{\text{m}}}{100 \cancel{\text{cm}}}\right)\left(\frac{1 \cancel{\text{yd}}}{0.914 \cancel{\text{m}}}\right)\left(\frac{1 \text{ mile}}{1760 \cancel{\text{yd}}}\right) = 5.00 \text{ miles}$$

   The route you choose in making your unit map will depend on what you know. As you learn more chemistry, and as the problems you work get longer and more complicated, it will be more likely that your way of solving a problem will be different from someone else's. There may be many routes to the same correct answer.

If you've done the first two steps properly, the remaining steps, though they should be done carefully, will be automatic:

3. ***Write Equation(s) Relating Units.***
4. ***Write Factors.***
5. ***Choose Factor(s) and Multiply; Decide on Significant Digits.***
6. ***Check Answer.***

In introducing dimensional analysis, I've recommended that you write down all the parts of your solution to a problem: problem statement, unit map, and so forth. Your instructor may prefer that you include some parts and not others. As you get more practice, you'll probably find that you need to write

Inside Chemistry — What is heat?

Human beings have always been able to experience sensations from heat, but our understanding of the nature of heat is less than 200 years old. Our understanding of heat was made possible by the invention of scientific instruments to measure heat and by the acceptance of the belief that matter consists of extremely tiny particles.

Our direct sensations of heat vary so much from one person to another and from one moment to another that they can't be used as a basis for a scientific study of heat. The direct sensation of temperature, for example, is influenced by immediate past experience: Water at room temperature will feel cold to a person who is overheated and hot to a person who is chilled. Because direct sensations of temperature are subjective, scientifically useful information about temperature became available only after the invention of the thermometer, early in the seventeenth century.

(Guido A. Rossi/The Image Bank)

Heat, as we sense it directly, seems to flow from one place to another. For example, if one end of an iron bar is heated, the heat seems to flow through the bar from the heated end, much as water flows through a pipe. Because of this apparent fluid nature of heat, in the seventeenth and eighteenth centuries a theory was developed that heat is a fluid that can flow from one place to another and can be absorbed into substances in much the same way that water is absorbed into a sponge. In the late eighteenth century this fluid was named *caloric,* from the Latin word for heat, *calor.*

Although the theory that heat is a fluid successfully explained some properties of heat, it did not explain others. For example, if heat is a fluid, it should have mass, but repeated and careful measurements with delicate instruments showed that the masses of objects didn't increase when they were heated.

In the middle of the nineteenth century the theory that heat is a fluid was replaced by a theory describing heat as a property of a sample of matter that depends on the motions of its particles. According to this theory, any sample of matter—solid, liquid, or gas—consists of an enormous number of extremely small particles, and these particles are in rapid motion. When a sample of matter is heated, its particles move more rapidly and the temperature of the sample measures how fast they're moving. This theory, that heat is related to the motions of particles of matter, is the view currently held by scientists.

less and less, and eventually you may find that you can go directly from reading the problem to writing out the multiplication by factors. Don't be in a hurry to abbreviate what you write down, let it happen at its own pace. If you get stuck on a problem, try writing out each step carefully.

In solving a problem, your immediate aim is to get the answer, but that isn't the most important part of what you're doing. When you've solved a problem, ask yourself what you've learned from doing it that might help you work another problem. A problem that's been hard to solve has probably taught you more than one that's been easy.

It's a good idea to work problems with one or two other people. You'll save time and frustration, because you can help one another when you get stuck. But don't make a habit of having someone else solve problems for you. On an exam you'll have to solve them for yourself.

Chapter Summary: Measuring Matter and Energy

Section	Subject	Summary	Check When Learned
	Matter	Anything that takes up space and has mass.	☐
	Energy	Capacity to move matter.	☐
3-1	Mixture	Sample of matter that can be separated into two or more substances by physical means.	☐
3-1	Heterogeneous mixture	Mixture in which two or more substances can be seen.	☐
3-1	Homogeneous mixture	Mixture with uniform appearance, as if only one substance were present.	☐
3-1	Solution	Usual term for homogeneous mixture that's a liquid or a gas. (Solid solutions of metals are called alloys.)	☐
3-1	Compound	Pure substance that can be decomposed into simpler substances.	☐
3-1	Element	Pure substance that cannot be decomposed into simpler substances.	☐
3-1	Analytical chemistry	Analysis of compounds to determine how their elements are combined.	☐
3-1	Synthetic chemistry	Preparation of compounds from elements or other compounds.	☐
3-1	Chemistry	Study of how elements interact.	☐
3-2	States of matter	Solid, liquid, and gas.	☐
3-3	SI	Modified metric system used by chemists.	☐
3-3	Prefix symbols and meanings:		
	mega-	M, 10^6	☐
	kilo-	k, 10^3	☐

Section	Subject	Summary	Check When Learned
	deci-	d, 10^{-1}	☐
	centi-	c, 10^{-2}	☐
	milli-	m, 10^{-3}	☐
	micro-	μ, 10^{-6}	☐
	nano-	n, 10^{-9}	☐
	pico-	p, 10^{-12}	☐
3-3	Conversions between English and metric or SI units:		
	1 lb =	454 g	☐
	1 in. =	2.54 cm	☐
	1.06 qt =	1 L	☐
3-4	Area units, using cm	cm^2	☐
3-4	Volume units, using cm	cm^3	☐
3-4	1 cm^3 =	1 mL	☐
3-5	Density	Mass per unit volume; usual SI units are g/cm^3.	☐
3-5	Density formula	d = m/V, where m is mass, V is volume, and d is density.	☐
3-5	Specific gravity of a liquid	Density of the liquid divided by the density of water. In the SI, the specific gravity of a liquid has the same numerical value as its density, and no units.	☐
3-6	Temperature is a measurement of	Heat intensity.	☐
3-6	Units and symbols for these scales:		
	Fahrenheit	degrees Fahrenheit, °F	☐
	Celsius	degrees Celsius, °C	☐
	Kelvin	kelvins, K	☐
3-6	Lowest limit, freezing point of water, and boiling point of water on these scales:		
	Fahrenheit	−459°F, 32°F, 212°F	☐
	Celsius	−273°C, 0°C, 100°C	☐
	Kelvin	0 K, 273 K, 373 K	☐

Section	Subject	Summary	Check When Learned
3-6	Equation to convert between °F and °C	(F − 32)/180 = C/100	☐
3-6	Equation to convert between °C and K	K = C + 273	☐
3-7	Units used to measure quantity of heat	joule (J) or calorie (cal)	☐
3-7	1 cal =	4.184 J	☐
3-8	λ	Wavelength: distance from point on one wave to corresponding point on adjacent wave, in length units (e.g., meters or nanometers).	☐
3-8	c	Velocity of electromagnetic radiation (in a vacuum): 3.00×10^8 m/s.	☐
3-8	ν	Frequency: number of waves passing a point in one second; units are s^{-1} or Hz.	☐
3-8	Hz	Frequency units: Hz = hertz = s^{-1}.	☐
3-8	Relationship among λ, ν, and c	$c = \lambda\nu$	☐
3-9, 3-10	Outline of dimensional analysis	State problem. (What quantities are given and to be found?)	☐
		Write unit map. (What unit conversions should I make to go from the given units to the units to be found?)	☐
		Write equations relating units. (What are the equations for the mathematical relationships between the units for each conversion on the unit map?)	☐
		Write factors. (What two factors are formed from each equation relating units?)	☐
		Multiply by correct factors and decide on significant digits. (Which factors will make the conversions shown on the unit map? How many significant digits should I use in the answer?)	☐
		Check answer for reasonableness. (Are the units in the answer those that were to be found? Is the number in the answer reasonable?)	☐
3-9	Unit map	Diagram of unit conversions to be made in solving a problem by dimensional analysis (e.g., miles → ft → in.).	☐
3-9	Factor	Fraction that shows relationship between two units (e.g., 1 in./2.54 cm).	☐

Problems

Answers to odd-numbered problems are in Appendix 1.

Mixtures, Compounds, and Elements (Section 3-1)

1. In one sentence, state the difference between a mixture and a compound.
2. In one sentence, state the difference between a compound and an element.
3. Decide whether each of the following mixtures is homogeneous or heterogeneous: (a) tea (the liquid) (b) orange juice (c) vinegar
4. Decide whether each of the following materials is homogeneous or heterogeneous: (a) salt (b) a mixture of salt and pepper (c) concrete.
5. When limestone—a white solid—is heated to a high temperature, it changes into another white solid and a gas. From this evidence would you classify limestone as a compound or an element? Explain.
6. When iron is heated, it doesn't change into other substances, and when it's melted and an electric current is passed through it, no change occurs. From this evidence would you classify iron as a compound or an element? Explain.

States of Matter and Phases (Section 3-2)

7. If a piece of iron were heated to higher and higher temperatures, what changes of state would you expect it to undergo?
8. If a sample of nitrogen gas from the atmosphere were cooled to lower and lower temperatures, what changes of state would you expect it to undergo?
9. Imagine a piece of wood floating on water. Including the air above the wood and water, how many states of matter are present? How many phases?
10. Imagine that you stir a teaspoonful of sugar into a cup of water and watch it dissolve. Before the sugar dissolves, how many states of matter are present? How many phases? After the sugar dissolves, how many states of matter are present? How many phases?

Metric and SI Units (Sections 3-3 and 3-4)

11. Name the prefix identified by each of the following abbreviations: (a) p (b) M (c) d.
12. Name the prefix identified by each of the following abbreviations: (a) n (b) c (c) k.
13. Write the abbreviation for the prefix that designates each of the following powers of ten: (a) 1×10^{-2} (b) 1×10^{3} (c) 1×10^{-9}
14. Write the abbreviation for the prefix that designates each of the following powers of ten: (a) 1×10^{-1} (b) 1×10^{-12} (c) 1×10^{6}.
15. Write the equation that defines the relationship between the pound and the gram.
16. Write the equation that defines the relationship between the inch and the centimeter.
17. Write the equation that defines the relationship between (a) the liter and the deciliter and (b) the kilometer and the meter.
18. Write the equation that defines the relationship between (a) the gram and the centigram and (b) the nanometer and the meter.
19. What's the area in square inches of a rectangle with sides 2 in. and 4 in.?
20. What's the area in square centimeters of a rectangle with sides 3 cm and 2 cm?
21. What's the volume in cubic meters of a box whose edges are 1 m, 2 m, and 4 m?
22. What's the volume in cubic yards of a box whose edges are 2 yd, 2 yd, and 3 yd?
23. The dimensions of a rectangular block of wood are 2.75 cm, 4.66 cm, and 1.74 cm. What is its volume in cubic centimeters?
24. The dimensions of a block of marble are 2.04 m, 65.5 cm, and 1.49 m. Calculate its volume in cubic centimeters.

Density and Specific Gravity (Section 3-5)

25. The element osmium has the highest density of any substance on earth: One cubic centimeter of osmium has a mass of 22.5 g. What is the density of osmium?
26. A cubic kilometer of silver would have a mass of 1.03×10^{13} kg. Calculate the density of silver.
27. The density of platinum is 21.4 g/cm^3. What is the mass in grams of 6.25 cm^3 of platinum?
28. The density of chromium is 7.14 g/cm^3. What is the volume of 34.7 g of chromium?
29. The density of magnesium is 1.74 g/cm^3. What is the mass in kilograms of 1.00 cubic decimeter of magnesium?
30. The density of zinc is 7.14 g/cm^3. What is the volume in cubic meters of 1.00×10^2 kilograms of zinc?
31. The dimensions of a rectangular block of titanium metal are 1.32 cm, 2.72 cm, and 1.95 cm, and its mass is 31.5 g. Calculate the density of titanium.
32. A block of tin has a volume of 0.0840 cubic decimeters and a mass of 0.614 kg. Calculate the density of tin in grams per cubic centimeter.
33. Glycerol is a clear, sticky liquid used as a sweetening agent in candies, as an ingredient in cosmetics, and as a component of many other products. Its density is 1.26 g/cm^3. What is its specific gravity?
34. The element bromine is a red liquid with a suffocating odor. Its specific gravity is 3.10. What is its density?
35. Mercury in the only metal that's a liquid at ordinary room temperatures. Its specific gravity is 13.6. What is the volume in milliliters of 20.0 g of mercury?
36. From the information given in problem 35 calculate the mass in kilograms of 1.00 L of mercury.

Temperature Scales and Heat Units (Sections 3-6 and 3-7)

37. What is the limit of lowest temperature on the Celsius scale? On the Kelvin scale?
38. What is the boiling point of water on the Celsius scale? On the Kelvin scale?
39. Convert 75°F to °C.
40. Convert 15°C to °F.
41. Convert 225°C to K.
42. Convert 372 K to °C.
43. Convert 172 K to °F.
44. Convert −33°F to K.
45. Write the mathematical equation for the relationship between the calorie and the kilojoule.
46. Write the mathematical equation for the relationship between the joule and the kilocalorie.

Electromagnetic Radiation (Section 3-8)

47. Write the symbol and units for wavelength.
48. Write the symbol and units for frequency.
49. Red light has a longer wavelength than blue light. Which has the lower frequency, blue light or red light?
50. Green light has a higher frequency than yellow light. Which has the longer wavelength, yellow light or green light?
51. Write the equation that shows the relationship among velocity, frequency, and wavelength of electromagnetic radiation, and solve it for frequency.
52. Write the equation that shows the relationship among velocity, frequency, and wavelength of electromagnetic radiation, and solve it for wavelength.
53. Calculate the frequency of X rays whose wavelength is 1.00×10^{-10} m.
54. Calculate the wavelength of microwaves whose frequency is 1.00×10^{11} Hz.
55. Calculate the wavelength of infrared rays whose frequency is 1.00×10^{14} s^{-1}.
56. Calculate the frequency of ultraviolet rays whose wavelength is 100 nm (3 significant digits).

Dimensional Analysis (Sections 3-9 and 3-10)

57. How many inches are in 25.0 feet?
58. How many feet are in 137 inches?
59. How many picometers are in 25.0 meters?
60. How many meters are in 2.00×10^7 millimeters?
61. How many miles are in 3.75×10^4 inches?
62. How many kilometers are in 4.29×10^{12} centimeters?

63. How many kilograms are in 4.69×10^7 milligrams?

64. How many ounces are in 3.00 tons?

65. How many centimeters are in 2.50 feet?

66. How many tons are in 5.00×10^4 megagrams?

67. How many milliliters are in 1.00×10^5 centiliters?

68. How many milliliters are in 7.25×10^3 microliters?

69. How many square inches are in 9.65 square feet?

70. How many square centimeters are in 7.27 square meters?

71. How many square meters are in 3.00 square kilometers?

72. The hectare is a defined unit of land measurement equal to 10 000 square meters. How many square kilometers are in one hectare?

73. How many cubic centimeters are in 2.00 cubic kilometers?

74. How many cubic yards are in 1255 cubic inches?

75. How many joules are in 257 calories?

76. How many calories are in 85.7 joules?

77. How many kilocalories are in 3.66×10^4 joules?

78. How many kilojoules are in 9.49×10^5 calories?

79. Convert the value of c, 3.00×10^8 m/s, to miles per hour.

80. The light year, a distance unit used by astronomers, is the distance light travels in one year. Assuming that there are exactly 365 days in one year, calculate the length of one light year, in miles.

Image of iodine atoms. *(Bruce C. Schardt. Cover of* SCIENCE, *Volume 243, 24 Feb. © 1989 by the American Association for the Advancement of Science)*

4
Atoms and Chemical Symbols

Chemistry is based on the atomic theory, the belief that matter is composed of tiny particles, called atoms. ◆

◆ *Matter* here means matter as we ordinarily see it on our planet. At very high temperatures, as in the sun or in other stars, atoms are broken down into smaller particles.

For better and for worse the atomic theory has affected the life of almost everyone living today. Vitamins, plastics, explosives, paints, anesthetics, fuels, fertilizers, dyes, pesticides, synthetic fibers, and detergents are only a few of the many kinds of useful substances made by applications of the atomic theory. The atomic theory has become so much a part of our lives that, as you look around your room, almost everything you see will have been produced, at least in part, by processes of technology and manufacturing based on this theory.

In this chapter you'll look at the principles of the atomic theory and begin learning the symbolic language that chemists use to describe the behavior of atoms.

The chapter first discusses the origins of the atomic theory. In Section 4-1 you'll see how the remarkable insights of an English schoolteacher, John Dalton, created the atomic theory about 180 years ago. Dalton invented the modern language of chemical symbols; in Section 4-2 you'll learn some of the symbols in their modern form.

Sections 4-3 through 4-9 describe the more detailed understanding of atoms that scientists have developed since Dalton's day: How the masses of atoms can be expressed on a scale of atomic mass units; how atoms are composed of still smaller particles called protons, neutrons, and electrons; and how new atoms can be formed through the processes of nuclear fission and nuclear fusion.

In less than two centuries the atomic theory has grown from a set of simple ideas in the mind of one man into a belief held by all educated people. The atomic theory has also developed into forms of science and technology that determine the quality of our lives and that may determine whether we live or die. Modern biology, chemistry, physics, computer science, engineering, communications, agriculture, manufacturing, warfare, and medicine are based on the atomic theory. By learning the fundamentals of that theory, we can begin to understand one of the foundations of our world.

Figure 4.1
John Dalton. *(The Edgar Fahs Smith Collection for the History of Chemistry, Special Collections Department, Van Pelt Library, University of Pennsylvania)*

4-1 Modern atomic theory began with the work of John Dalton, an Englishman, at the start of the nineteenth century

If I take a piece of matter, say a piece of iron, and keep cutting it into smaller and smaller pieces, will I eventually get to pieces that can't be cut?

In the fifth century BC the Greek philosophers Leucippus and Democritus discussed this question. They argued that, if it were possible to keep cutting matter into smaller pieces indefinitely, then the pieces must finally disappear entirely. But this would mean that a solid object was made up of pieces that didn't exist. Since that's impossible, they concluded that, by cutting matter into smaller and smaller pieces, we must eventually arrive at particles that can't be further divided. **Atom** is derived from the Greek word *atomos,* which means *undivided.*

The later Greek philosopher Aristotle (384–322 BC) rejected the theory of atoms, and his views dominated science for almost 20 centuries. After about

AD 1600 a few scientists again began to consider that matter might consist of tiny, indestructible particles, but a scientific theory of atoms wasn't created until early in the nineteenth century. That theory was the work of an English schoolteacher, John Dalton (1766–1844).

In creating the modern atomic theory, Dalton took two crucial steps. First, he stated the fundamental principles of his theory clearly:◆

◆These are Dalton's principles, but not his exact words.

1. Matter is made of indestructible particles, called atoms.
2. All atoms of the same element are identical in size and mass. Atoms of different elements have different sizes and masses.
3. Compounds are formed by combinations of atoms.

Exercise 4.1◆

◆Answers to Exercises are in Appendix 1.

Which of Dalton's principles are the same as those held by Leucippus and Democritus, and which ones did he add?

Dalton's second crucial step was to change the way that scientists used chemical symbols. Before Dalton, scientists had used symbols for the elements simply to stand for their names. Dalton realized that it's difficult to think about atoms because they're too small to be seen, and to make it easier, he proposed that the symbol for an element should be understood to represent one atom of that element. By making this change, Dalton created the symbolic language of modern chemistry.

4-2 Each element has a unique symbol, which stands for one atom of the element

Modern chemistry is written in a symbolic language, and the alphabet for that language consists of the symbols for the elements. The names and symbols for all of the elements are shown in the table on the inside front cover of this book. Most of the symbols are abbreviations of the English names for the elements, and a few are abbreviations of their names in other languages. Lead, for example, has the symbol Pb, from its Latin name, *plumbum*.◆

◆Newly discovered elements are named by a method described in Section 4-5.

The names and symbols in Table 4.1 will be used often in this book and in your other courses in chemistry, so you should learn them. Your instructor may want to add to this list.

Table 4.1 Symbols for Common Elements◆

◆Notice that, in the symbol for an element, the first letter is capitalized and the second letter is not.

Element	Symbol	Element	Symbol	Element	Symbol
aluminum	Al	helium	He	oxygen	O
argon	Ar	hydrogen	H	phosphorus	P
barium	Ba	iodine	I	potassium	K
boron	B	iron	Fe	silicon	Si
bromine	Br	lead	Pb	sodium	Na
calcium	Ca	lithium	Li	sulfur	S
carbon	C	magnesium	Mg	tin	Sn
chlorine	Cl	mercury	Hg	zinc	Zn
copper	Cu	neon	Ne		
fluorine	F	nitrogen	N		

Elements: bromine, copper, aluminum, and silicon. *(Charles D. Winters)*

Exercise 4.2

Without looking at the table above, recall the names for the elements represented by these symbols: (a) K (b) Na (c) Cu (d) C (e) Ca

Following the principle established by John Dalton, the symbol for an element represents one atom of that element. The symbol sometimes also refers to the element in general, and Chapter 12 describes how it can have a special meaning in chemical calculations.

4-3 Atoms contain negatively charged particles, called electrons, and positively charged particles, called protons

Dalton created the principles of his atomic theory between 1800 and 1810. Over the following century scientists carried out a series of experiments, mostly in Germany, England, and the United States, leading to the view that atoms themselves must be made up of even smaller particles. This Section looks briefly at the research that led to the discoveries of two of those particles, the electron and the proton.

Because electrons are much too small to be seen, our understanding of them has had to be pieced together slowly, from indirect evidence. Between 1820 and 1920, the belief that electrons exist was built up by hundreds of scientists carrying out thousands of experiments and creating dozens of theories to explain their results. Some experiments failed or were performed badly and had to be redesigned and repeated, and many theories were proved wrong and had to be abandoned. But gradually, with many corrections, scientists developed, tested, and made secure the theory of the existence of electrons.

Many of the experiments that led to the belief in electrons were carried out with a device known as a **gas discharge tube.** During the last half of the nineteenth century scientists built many different kinds of discharge tubes to test many theories; three examples are shown in Figures 4.2 through 4.4.

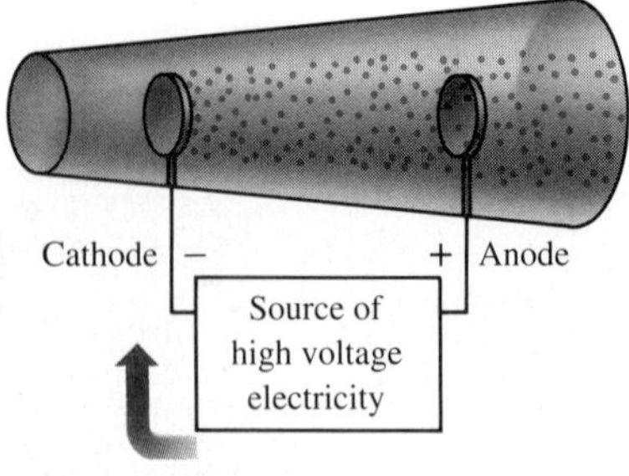

Figure 4.2
Gas discharge tube. The tube is a hollow glass cylinder from which the air has been pumped out. As electricity is forced onto the metal cathode, a glow flows from the cathode to the anode and on to the right end of the tube.

Figure 4.2 depicts a discharge tube in its simplest form. The tube is a sealed glass vessel from which most of the air has been removed, so that the space inside the tube is under high vacuum. At each end of the tube, a wire is sealed through the glass and is attached, inside the tube, to a metal disk called an electrode. Outside the tube these wires are connected to a source of high-voltage electricity in such a way that an electric current flows into the disk in the left end of the tube (called the cathode) and out of the disk at the right end (the anode). When the electric current is flowing, a glow appears between the cathode and the anode. The study of this glow first provided clues about the existence and nature of electrons.

The discharge tube shown in Figure 4.3 has special features used to study the glow: The inside surface of the right end of the tube is coated with a fluorescent material that emits light where rays from the cathode strike it, and a piece of black cardboard, in the shape of an X, is mounted between the cathode and the anode. When the electric current is flowing, the fluorescent screen emits light, except for an X-shaped portion in the center; this suggests that rays or streams of particles are being emitted from the cathode and are moving

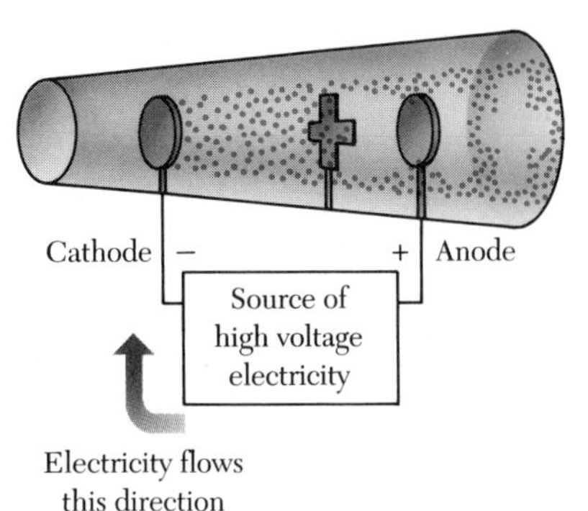

Figure 4.3
Gas discharge tube with obstacle between cathode and anode. Cathode rays flow from the cathode to the anode and on to the fluorescent screen on the right end of the tube. Where the screen is shielded by the X-shaped obstacle, no rays strike it.

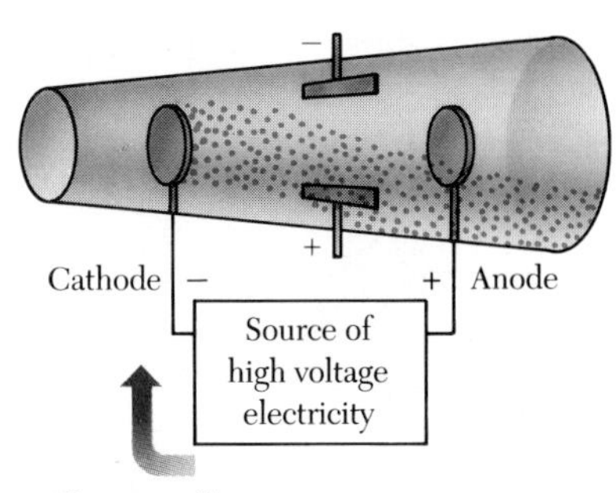

Figure 4.4
Gas discharge tube with added deflecting electrodes. A negative charge on the top electrode deflects the cathode rays downward, showing that the rays are negatively charged.

toward the anode. Where they strike the screen, it fluoresces, and where the screen is shielded by the X-shaped piece of cardboard, it stays dark. These rays are called **cathode rays.**

The discharge tube shown in Figure 4.4 has a second set of electrodes sealed into the top and bottom of the tube. When this set of electrodes receives an electrical charge, the cathode rays are bent. If, as shown in the drawing, the top electrode is negative and the bottom one is positive, the rays bend downward; if these charges are reversed, the rays bend upward. Since opposite electrical charges attract one another and like charges repel, these deflections of the particles streaming from the cathode suggest that they carry a negative charge.

These same effects are observed no matter what materials are used to make the discharge tube: Changing the metal used in the electrodes, the gas inside the tube, or the composition of the glass has no effect on the cathode rays. Emission of the same particles by different cathode materials suggests that the particles may be constituents of all elements, that is, of all atoms.

About the year 1900, these particles were named electrons. More complicated experiments gave more detailed information about them: An electron was found to have a mass of 9.11×10^{-28} g (about 1/1837 the mass of the hydrogen atom, which is the lightest atom) and an electrical charge of -1.602×10^{-19} coulomb.◆ We can summarize the results of a century of research on electrons with this statement: **Electrons** are particles with a very small negative electrical charge and a very small mass that are constituents of all atoms and that can, under some conditions, be separated from their atoms.

◆In the study of electricity, charge is measured in units called coulombs. A simpler way of designating electrical charge, used in chemistry, is described later in this Section.

Among the hundreds of scientists who did the work leading to this understanding, two deserve mention here. Joseph John Thomson (1856–1940, Nobel Prize 1906), at Cambridge University in England, showed that cathode rays are negatively charged and that they are the same no matter what material they come from; he also made measurements of the charge and mass of the electron. Robert A. Millikan (1868–1953, Nobel Prize 1923), at the University of Chicago, accurately measured the charge on the electron and made possible precise calculation of its mass.

Figure 4.5
Joseph John Thomson. *(Cavendish Laboratory, University of Cambridge)*

A more complicated line of research led to the conclusion that atoms also contain a particle carrying a positive charge. Experiments showed that this particle, called a **proton,** has a mass of 1.672×10^{-24} g and a charge of $+1.602 \times 10^{-19}$ coulomb.

Figure 4.6
Robert A. Millikan. *(Library of Congress)*

Exercise 4.3

Write a statement about the proton that is similar to the statement above, which begins in boldface, about the electron. Notice that the masses of the proton and the electron are very different.

In chemistry the important point about the charge on the proton and the charge on the electron is that they have the same magnitude but are opposite in sign. For this reason their charges are designated simply as 1+ and 1−.

Exercise 4.4

Every atom contains protons and electrons and is electrically neutral; that is, there is no net negative or positive charge on the whole atom. Write a statement about the relationship between the number of protons and the number of electrons in every atom.

4-4 Research on radioactivity has shown that atoms have a structure

In 1896 the French physicist Henri-Antoine Becquerel (1852–1908, Nobel Prize 1903) noticed that compounds of the element uranium darkened photographic plates, even when the plates were wrapped in paper. He investigated and found that samples of uranium give off spontaneous emissions and that these emissions contain a stream of electrons moving at very high speeds. Becquerel had discovered the phenomenon we now call radioactivity and started the atomic age.

The British physicist Ernest Rutherford (1871–1937, Nobel Prize 1908) also studied the natural emissions from samples of uranium and showed that they include a second kind of radiation, a stream of particles with a relatively large mass and a positive charge. He named this form of radiation **α rays** and called the individual particles **α particles.**◆ Later, an α particle was shown to have twice the positive charge, and four times the mass, of a proton. Rutherford named the form of radiation identified by Becquerel **β rays** and named the individual particles **β particles.**◆ Shortly after the turn of the century scientists discovered another form of emission, which they named **γ rays;**◆ these rays were later found to resemble X rays, with very short wavelengths.

◆α, alpha, is the first letter of the Greek alphabet.

◆β, beta, is the second letter of the Greek alphabet.

◆γ, gamma, is the third letter of the Greek alphabet.

We'll define **radioactivity** as the emission of radiation or particles from atoms. Some atoms emit spontaneously and are said to show **natural radioactivity,** others emit only when they're bombarded with streams of particles and are said to show **artificial** or **induced radioactivity.**

Exercise 4.5

Complete the following table.

Types of Radioactive Emissions

Name	Greek Symbol	Description
alpha	α	Stream of positive particles with twice the charge and four times the mass of a proton.
beta		
gamma		

Early research on radioactivity, performed by a few experimenters working with small pieces of laboratory equipment, has led to millions of experiments, carried out all over the world. Today some of these experiments require the efforts of thousands of scientists using equipment whose size is measured in miles.◆ Governments have spent billions of dollars on radioactivity research because it has led to the discovery of sources of gigantic quantities of energy for military and peaceful applications, and scientists have pursued this research because it has provided a detailed understanding of the structure of atoms. In this Section we'll look briefly at the most dramatic discoveries that have created the current picture of how an atom is put together.

◆The newest facility for this kind of research is the Superconducting Super Collider being built just south of Dallas, Texas. Its experimental chamber, shaped like a race track, will be 53 miles long.

In the first few years of the twentieth century physicists began to get evidence about how the protons and electrons in an atom are arranged. One clue came from the fact—shown by the experiments with gas discharge tubes described earlier—that the electrons seem to be loosely held in atoms. Another clue came from experiments in which cathode-ray beams were focused on thin pieces of metal foils: Amost all of the cathode rays passed right through the foil, suggesting that the atoms in the foil are themselves mostly empty space.

In 1908 Rutherford and his colleagues tried another series of experiments. They bombarded thin sheets of metal foils with α particles and found that some of the particles were sharply deflected by the foil. These results were completely unexpected. Because α particles are relatively heavy, it was assumed that they'd force their way through atoms with little or no deflection. When Rutherford learned the results of these experiments, he said that it was like being told that an artillery shell had ricocheted off a piece of tissue paper.

Because α particles are heavy and positively charged, Rutherford reasoned that it would take a heavy, positively charged object to deflect them. He proposed that, in an atom, the protons are packed together to form a small central core, with the electrons circling it at relatively great distances. This model fits the experimental evidence: Electrons, being far from the protons that attract them, are relatively easy to remove. Cathode rays and most α particles pass through the large empty spaces between the electrons and the protons, but a few α particles are deflected because they collide with the small cluster of protons at the atom's core. Rutherford called the core of the atom the **nucleus**, and his model was called the **nuclear atom.**

Figure 4.7
Ernest Rutherford. *(Special Collections Department, Van Pelt Library, University of Pennsylvania)*

The Fermi National Accelerator Laboratory at Batavia, Illinois. The largest circle is an experimental chamber that is 4 miles in circumference. *(Fermilab Visual Media Services)*

By the 1920's scientists suspected that atoms contain another fundamental particle, one that carries no electric charge. The existence of this third kind of particle was established in 1932: The **neutron** is electrically neutral, has about the same mass as the proton, and is in the nucleus. Because protons and neutrons make up the nuclei of atoms, they're collectively referred to as **nucleons.**

Table 4.2 summarizes information about the three subatomic particles.◆

◆For the purposes of chemistry we'll describe atoms in terms of protons, neutrons, and electrons. Physicists have found that these subatomic particles are themselves made up of still more fundamental particles, but this more complicated model of the atom isn't needed in chemistry.

Table 4.2 Properties of Subatomic Particles

Name	Location	Electric Charge	Mass
proton	nucleus	positive, 1+	1.672×10^{-24} g
neutron	nucleus	neutral, 0	1.675×10^{-24} g
electron	outside nucleus	negative, 1−	9.11×10^{-28} g

Exercise 4.6

An α particle is made up of protons and neutrons. How many of each does it contain? (See the description of the α particle at the beginning of this Section.)

4-5 The numbers of protons, neutrons, and electrons in an atom can be represented by its atomic number, Z, and its mass number, A

The **atomic number** for an atom, sometimes represented by the symbol Z, is the number of protons in its nucleus. Since an atom is electrically neutral, the atomic number is also the number of electrons in the atom. In the Periodic Table of the Elements on the inside front cover of this book, you'll see that the elements are numbered consecutively with whole numbers (integers) in increasing order from left to right and top to bottom: These are the elements' atomic numbers.†

A Periodic Table shows the atomic number for each element, so it tells you how many protons and electrons there are in one atom of each element. As

† Chemistry Insight

The elements with atomic numbers greater than 103 are newly discovered elements; these elements have been given temporary names and three-letter symbols until their final names are agreed on. The temporary name for a new element is constructed from its atomic number: Each digit is assigned a root name, and these roots, plus the suffix *-ium,* are combined to form the name for the element. The roots are *nil* for 0, *un* for 1, *bi* for 2, *tri* for 3, *quad* for 4, *pent* for 5, *hex* for 6, *sept* for 7, *oct* for 8, and *enn* for 9. Using this system, element 106 is named unnilhexium and assigned the symbol Unh.

you'll see later, the Periodic Table contains much more information—it's one of the most valuable resources you'll have in solving chemical problems. From this point on, unless you're directed otherwise, always assume you can use the Periodic Table at the front of the book in solving any Exercise or Problem.

Exercise 4.7

(a) How many protons are there in an oxygen atom? (b) How many electrons are there in a potassium atom?

The mass of one proton or one neutron is about 1837 times the mass of one electron, so almost the entire mass of an atom is in its protons and neutrons. The **mass number** for an atom, symbol A, is the sum of the numbers of protons and neutrons in its nucleus, that is, the number of its nucleons. The difference between A and Z is the number of neutrons in the atom:

A, mass number	− Z, atomic number	= neutrons
(protons + neutrons) −	(protons)	= neutrons

It's sometimes useful to include the atomic number and the mass number with the symbol for an element in this form $^{A}_{Z}E$, where E is the symbol for the element. For example, the symbol $^{39}_{19}K$ stands for a potassium atom with 19 protons, 19 electrons, and 20 neutrons.

Exercise 4.8

(a) How many protons, neutrons, and electrons are in $^{190}_{76}Os$? (b) Write the symbol, including A and Z, for a mercury atom with 80 electrons and 120 neutrons. (c) Name the element represented by $^{40}_{20}E$.

It's important to remember that protons and neutrons are in the nucleus and electrons are outside it. We can represent that arrangement with the kind of diagram shown below for the $^{40}_{18}Ar$ atom.

18 p^+	
22 n^0	18e^-
nucleus	electrons

Exercise 4.9

(a) Draw a diagram, as shown above, for an atom of tin with Z = 50 and A = 119.

(b) Name the element represented by the following diagram.

12 p^+	
12 n^0	12 e^-
nucleus	electrons

(c) Translate the following diagram into a symbol of the form $^{A}_{Z}E$.

29 p^+	
35 n^0	29 e^-
nucleus	electrons

4-6 The standard for atomic mass is the $^{12}_{6}C$ atom

In choosing a standard of mass, the most important consideration is convenience. For example, in the English system of measurement the pound became the standard many years ago, because it gave masses for common objects in numbers of convenient size.

Once a standard of mass has been agreed on, we express other masses relative to it. If I say that my mass is 170 pounds, I'm saying that my mass is 170 times the mass of the standard pound.

Atoms are so small that the most convenient standard to use for their masses is the mass of one of the atoms. The atom $^{12}_{6}C$ has been selected as the standard, with an assigned mass of exactly 12 **atomic mass units,** abbreviated amu.◆ The masses of other atoms are expressed in amu, relative to this standard, just as, in the English system, the masses of people are expressed in pounds, relative to the standard pound. For example, a magnesium atom whose mass is 24 amu is twice as heavy as the standard carbon atom.

◆Why is the standard carbon atom assigned a mass of *twelve* atomic mass units? The reason is that this choice results in a value of 1 amu for the mass of the lightest atom, $^{1}_{1}H$.

Table 4.3 shows the masses of the subatomic particles on the atomic-mass scale.

Table 4.3 Masses of Subatomic Particles

Particle	Mass
proton	1.007276 amu
neutron	1.008665 amu
electron	0.0005846 amu

Exercise 4.10

Table 4.3 shows that the mass of one proton is 1.007276 amu, and in Section 4-4 we saw that the mass of one proton is 1.672×10^{-24} g. Use this information to calculate the number of atomic mass units in one gram.

The masses of electrons are so small, compared with those of protons and neutrons, that they make no significant contribution to the masses of atoms. Because the mass of a proton or a neutron is very close to 1 amu (1.01 amu, to three significant digits), the mass of an atom in amu will be very close to its mass number. The atom $^{35}_{17}Cl$, for example, will have a mass of about 35 amu.

Exercise 4.11

An atom contains seven protons and seven neutrons. (a) Write its symbol, including A and Z. (b) What is its approximate atomic mass?

4-7 The atomic mass of an element is the average of the masses of its naturally occurring atoms

In creating his atomic theory, John Dalton supposed that all atoms of the same element would have the same mass, but it's been found that this isn't true. A sample of hydrogen from nature, for example, has two different atoms. Each of them has one proton and one electron—so the atomic number of each of them is 1—but they differ in their numbers of neutrons: One of the atoms contains no neutrons, and the other contains one neutron. These atoms are therefore $^{1}_{1}H$ and $^{2}_{1}H$.

Atoms with the same number of protons (and electrons) but different numbers of neutrons are called **isotopes.** Isotopes can also be defined as atoms of the same element with different masses or as atoms with the same atomic number but different mass numbers. One way of referring to an isotope is by its symbol, for example $^{1}_{1}H$ or $^{2}_{1}H$, and another way is by adding the mass number to the name of the element: hydrogen-1 or hydrogen-2.

A sample of an element taken from nature will be a mixture of its isotopes. For example, a sample of 100 000 oxygen atoms taken from the air has the composition shown in Table 4.4.◆

◆Scientists assume that when our planet and its atmosphere were formed, all the atoms were thoroughly mixed, so any sample of oxygen we take should have the same isotopic composition as any other sample.

Table 4.4 Isotopic Composition of Naturally Occurring Oxygen

Isotope	Number of Atoms in 100,000 Oxygen Atoms	Isotopic Percentage
$^{16}_{8}O$	99,759	99.759%
$^{17}_{8}O$	37	0.037%
$^{18}_{8}O$	204	0.204%
	100,000	100.000%

The **atomic mass** of an element, shown on the Periodic Table, is the average of the masses of its naturally occurring isotopes.◆ The units, which aren't shown on the Periodic Table, are amu. Because almost all of the atoms in naturally occurring oxygen are $^{16}_{8}O$, we'd expect that its average atomic mass would be very close to 16 amu, and it is: The atomic mass for oxygen to six significant digits is 15.9994 amu, rounded off on the Periodic Table to 16.0 amu.

◆There's a tradition, sometimes still followed, of referring to atomic *masses* as atomic *weights*. In this book, we'll always use *mass* instead of *weight*.

Exercise 4.12

You learned earlier that naturally occurring hydrogen consists of two isotopes, $^{1}_{1}H$ and $^{2}_{1}H$. From the atomic mass for hydrogen on the Periodic Table, which of these atoms do you think is more prevalent?

As you'll see later in this book, the atomic masses of the elements are used often in chemical calculations, so another important use of the Periodic Table is as a source of atomic masses for chemical calculations.

Exercise 4.13

From the atomic numbers and atomic masses on the Periodic Table, it looks as if atoms of the lighter elements have about equal numbers of protons and neutrons. How does this relationship change as the elements get heavier?

4-8 Changes in atomic nuclei can result in the conversion of one element into another

Becquerel began the atomic age by discovering the spontaneous radioactivity of samples of uranium, and Rutherford showed that one of the emissions from uranium was a stream of α particles. In this emission process, the nucleus of a uranium isotope with a mass of 238 amu loses an α particle, and a new element is formed.

We can understand this process better by expressing it in symbols. Uranium has an atomic number of 92, so its nucleus must contain 92 protons; the isotope with mass number 238 must contain 238 − 92 = 146 neutrons. We can represent this nucleus as $^{238}_{92}U$. An α particle consists of 2 protons and 2 neutrons, so its atomic number is 2 and its mass number is 4; the element whose atomic number is 2 is helium, so an α particle is the nucleus of the helium isotope whose mass is 4 amu: $^{4}_{2}He$.

When a $^{238}_{92}U$ nucleus loses an $^{4}_{2}He$ particle, it loses 2 protons and 2 neutrons, so the nucleus that's left must consist of 90 protons and 144 neutrons, and its mass number must be 90 + 144 = 234. The Periodic Table shows that the element with atomic number 90 has the symbol Th (its name is thorium) so we can represent the new nucleus as $^{234}_{90}Th$.

In this example of radioactivity the $^{238}_{92}U$ nucleus broke up to form $^{234}_{90}Th$ and $^{4}_{2}He$. We can express this process this way:

$$^{238}_{92}U \rightarrow {}^{234}_{90}Th + {}^{4}_{2}He$$

This expression is an example of an **equation,** a before-and-after representation of a process, using symbols. The process represented by an equation is referred to as a **reaction.** The substances that are present before the reaction are shown to the left of the arrow and are called **reactants,** and the substances that are present after the reaction are shown to the right of the arrow and are called **products.** The arrow represents change with time. Because the reactant and products in this process are nuclei, it's called a **nuclear reaction,** and its equation is a **nuclear equation.**◆

◆It's important to remember that a *reaction* is a process and an *equation* is a symbolic representation of the process. It's like the difference between you and your name.

In this reaction the $^{238}_{92}U$ nucleus breaks into two pieces, $^{234}_{90}Th$ and $^{4}_{2}He$. After the reaction is over, we have the same number of protons and neutrons we started with, but they're arranged differently; this fact can be expressed by saying that, in this process, the numbers of protons and neutrons are *conserved.* In terms of the symbolism used in the nuclear equation, the sum of the atomic numbers to the left of the arrow must equal the sum to the right (92 = 90 + 2), and the same must be true for the mass numbers (238 = 234 + 4).

The conversion of $^{238}_{92}U$ to $^{234}_{90}Th$ and $^{4}_{2}He$ is a natural process, that is, it occurs spontaneously in any sample of $^{238}_{92}U$. The conversion of one element into another can also be induced artificially, by bombarding a nucleus with another nucleus or with a subatomic particle. For example, the beryllium nucleus $^{9}_{4}Be$ doesn't undergo radioactive emission spontaneously, but when it's bombarded with an α particle ($^{4}_{2}He$), a reaction occurs in which a carbon nucleus ($^{12}_{6}C$) and a neutron ($^{1}_{0}n$) are formed:

$$^{9}_{4}Be + {}^{4}_{2}He \rightarrow {}^{12}_{6}C + {}^{1}_{0}n$$

It was this reaction that led to the discovery of the neutron.

A reaction in which one nucleus is artificially induced to change into another by bombardment is called a **transmutation reaction.** All of the elements with atomic numbers greater than 92 have been created by transmutation. Because these elements lie beyond uranium on the Periodic Table, they're referred to as **transuranium elements.**

Exercise 4.14

Bombardment of a plutonium-239 nucleus ($^{239}_{94}Pu$) with an α particle results in the formation of a neutron and a new nucleus. Write the nuclear equation for this process and label the products and reactants.

When a $^{235}_{92}U$ nucleus is hit by a neutron, one of several reactions may occur:

$$^{235}_{92}U + ^{1}_{0}n \rightarrow ^{142}_{56}Ba + ^{92}_{36}Kr + 2\ ^{1}_{0}n$$

$$^{235}_{92}U + ^{1}_{0}n \rightarrow ^{89}_{37}Rb + ^{144}_{55}Cs + 3\ ^{1}_{0}n$$

$$^{235}_{92}U + ^{1}_{0}n \rightarrow ^{90}_{38}Sr + ^{143}_{54}Xe + 3\ ^{1}_{0}n$$

In these equations, the 2 or 3 before the symbol for a neutron shows that two or three neutrons are released from each $^{235}_{92}U$ nucleus; in an equation a multiplier in front of a symbol is called a **coefficient.**◆

◆If a coefficient isn't shown, it's assumed to be 1.

A reaction in which one nucleus splits into two or more nuclei is called **nuclear fission.** In each of the fission processes shown above, one neutron causes a reaction that releases three neutrons. These reactions can be carried out so that each of the neutrons produced collides with another $^{235}_{92}U$ nucleus to give a series of reactions, known collectively as a **chain reaction,** as shown in Figure 4.8.

In **nuclear fusion** two lighter nuclei combine to form one heavier one, for example

$$^{2}_{1}H + ^{3}_{1}H \rightarrow ^{4}_{2}He + ^{1}_{0}n$$

Figure 4.8
Nuclear fission by a chain reaction.

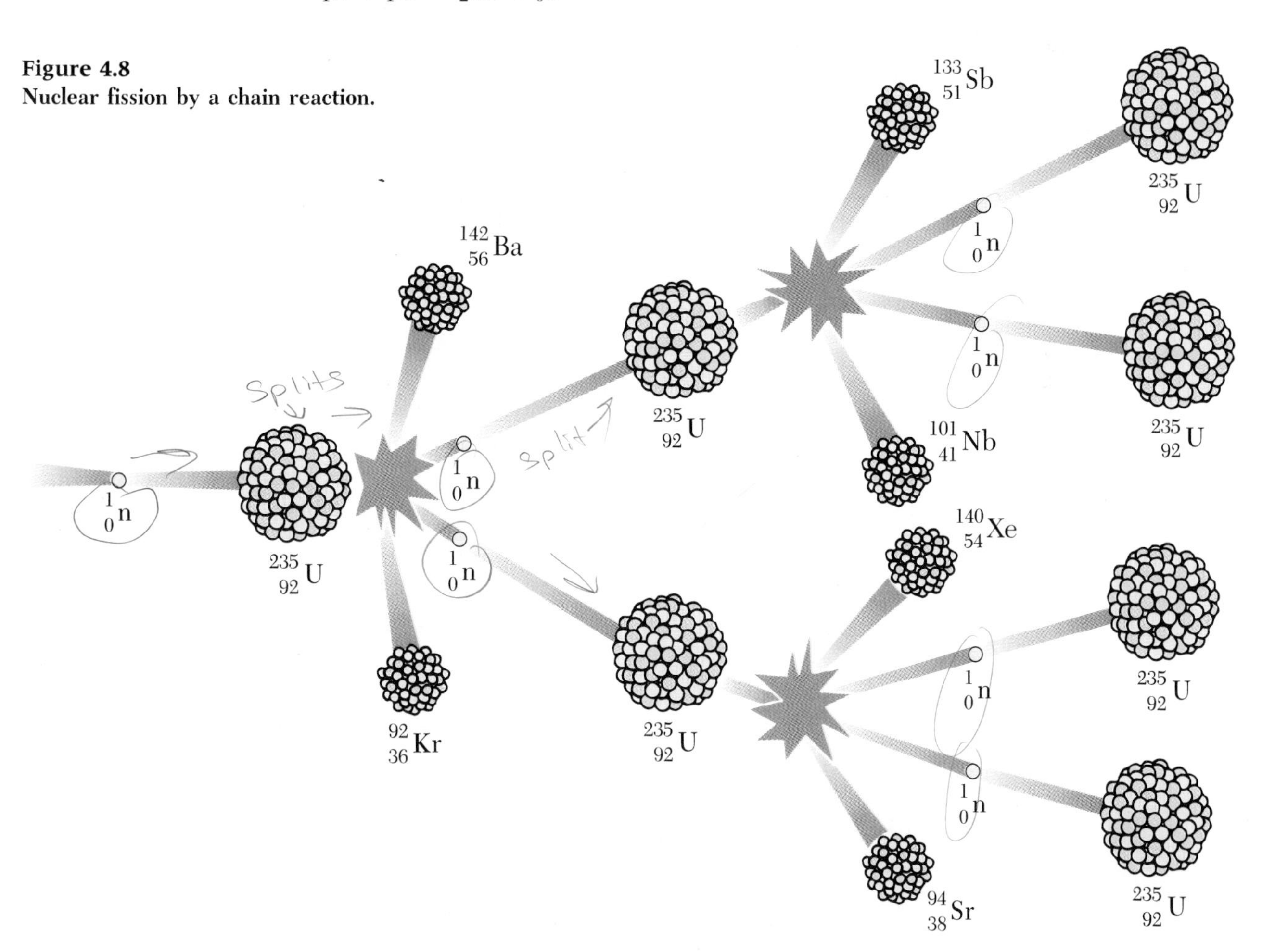

In fission and fusion reactions large amounts of energy are released. Fission reactions were used in the atomic bombs dropped on Hiroshima and Nagasaki in 1945 and are used in nuclear power plants today. Fusion reactions occur in the sun and other stars, and in hydrogen bombs. They're being investigated as a possible energy source for peaceful uses.

Inside Chemistry: How small is an atom?

Atoms are so much smaller than the smallest objects we can see that it's really impossible to imagine how tiny they are. The best we can do is to look at some amazing numbers.

The diameter of a sodium atom is 372 picometers. The period at the end of this sentence is about 1.00×10^8 pm in diameter, and we can use dimensional analysis to calculate how many sodium atoms would have to be lined up next to one another to equal the diameter of the period.

$1.00 \times 10^8 \text{ pm} = \text{? Na atoms}$

$\text{pm} \rightarrow \text{Na atoms}$

$1 \text{ Na atom} = 372 \text{ pm}$

$$\frac{1 \text{ Na atom}}{372 \text{ pm}} = 1 \qquad \frac{372 \text{ pm}}{1 \text{ Na atom}} = 1$$

$$\text{pm} \rightarrow \text{Na atom}$$

$$1.00 \times 10^8 \cancel{\text{pm}} \left(\frac{1 \text{ Na atom}}{372 \cancel{\text{pm}}}\right) = 2.68 \times 10^5 \text{ Na atoms}$$

It would take 268 000 sodium atoms, side by side, to span the width of one period.

We can also get a sense of how small atoms are by considering their masses. The mass of one sodium atom is 3.82×10^{-23} g; how many sodium atoms are there in 1.00 gram of sodium?

$1.00 \text{ g Na} = \text{? Na atoms}$

$\text{g Na} \rightarrow \text{Na atoms}$

$$\frac{1 \text{ Na atom}}{3.82 \times 10^{-23} \text{ g}} = 1 \qquad \frac{3.82 \times 10^{-23} \text{ g}}{1 \text{ Na atom}} = 1$$

$$\text{g Na} \rightarrow \text{Na atoms}$$

Figure 4.9
Image of iodine atoms on a platinum surface.
(Bruce C. Schardt. Cover of SCIENCE, *Volume 243, 24 Feb. © 1989 by the American Association for the Advancement of Science)*

$$1.00 \cancel{\text{g Na}} \left(\frac{1 \text{ Na atom}}{3.82 \times 10^{-23} \cancel{\text{g Na}}}\right) = 2.62 \times 10^{22} \text{ Na atoms}$$

These calculations show that, even in quantities of matter that seem very small to us, there are enormous numbers of atoms.

Although atoms are much too small to be seen, it's possible, by using complicated instruments, to generate images of them. One example is shown in Figure 4.9.

In a nuclear power plant, nuclear fission reactions are used to generate electrical energy. *(Richard Megna, Fundamental Photographs)*

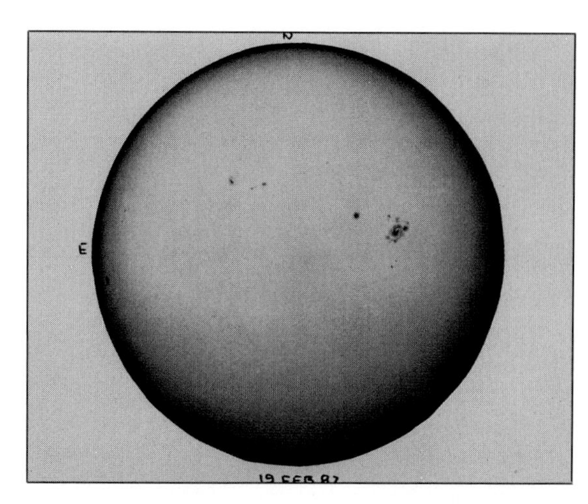

Fusion reactions are the source of the sun's energy. *(National Solar Observatory/National Optical Astronomy Observatories)*

Chapter Summary: Atoms and Chemical Symbols

Section	Subject	Summary	Check When Learned
4-1	John Dalton's atomic theory	Matter is made of indestructible particles called atoms. Atoms of the same element are identical and atoms of different elements are different. Compounds are formed by combinations of atoms.	☐
4-1	Meaning of chemical symbol	Symbol stands for one atom of element.	☐
4-2	Symbols and names for most important elements:		
	Al	aluminum	☐
	Ar	argon	☐
	Ba	barium	☐
	B	boron	☐
	Br	bromine	☐
	Ca	calcium	☐
	C	carbon	☐

Section	Subject	Summary	Check When Learned
	Cl	chlorine	☐
	Cu	copper	☐
	F	fluorine	☐
	He	helium	☐
	H	hydrogen	☐
	I	iodine	☐
	Fe	iron	☐
	Pb	lead	☐
	Li	lithium	☐
	Mg	magnesium	☐
	Hg	mercury	☐
	Ne	neon	☐
	N	nitrogen	☐
	O	oxygen	☐
	P	phosphorus	☐
	K	potassium	☐
	Si	silicon	☐
	Na	sodium	☐
	S	sulfur	☐
	Sn	tin	☐
	Zn	zinc	☐
4-3	Evidence that cathode rays are streams of particles	A fluorescent screen can be shielded from the rays by interposing a solid object, e.g., a piece of cardboard.	☐
4-3	Evidence that the particles carry a negative charge	The particles are repelled by a negatively charged electrode (cathode) and attracted by a positively charged one (anode).	☐
4-3	Evidence that the particles are in all matter	Cathode rays are produced no matter what materials are used to construct the discharge tube.	☐
4-3	Electron	Particle with a very small negative electrical charge and a very small mass; a constituent of all atoms.	☐
4-3	Joseph John Thomson	English scientist who showed that cathode rays were negatively charged particles and made measurements of their mass and charge.	☐

Section	Subject	Summary	Check When Learned
4-3	Robert A. Millikan	American scientist who measured accurately the charge on the electron.	☐
4-3	Proton	Particle with a very small positive electrical charge, equal in magnitude but opposite in sign to that on the electron, and with a much larger mass (about 2000 times as great) as the mass of the electron.	☐
4-3	Designations of charges on electron and proton	1− and 1+.	☐
4-3	Relationship between number of protons and number of electrons in any atom	The number of protons and the number of electrons in any atom are always equal; an atom is always electrically neutral.	☐
4-4	Radioactivity	Emission of α particles, β particles, or γ rays from an atom.	☐
4-4	Natural radioactivity	Spontaneous emission of radioactivity.	☐
4-4	Induced radioactivity	Radioactive emission caused by bombardment with subatomic fragments. The fragments may be subatomic particles, e.g., neutrons, or they may be nuclei, e.g., α particles.	☐
4-4	α particles	Form of radioactive emission consisting of helium nuclei, each containing 2 protons and 2 neutrons: ${}^{4}_{2}He$.	☐
4-4	β particles	Form of radioactive emission consisting of electrons moving at very high speeds.	☐
4-4	γ rays	Form of radioactive emission consisting of electromagnetic radiation with very short wavelengths.	☐
4-4	Evidence that atoms are mostly empty space	Streams of cathode rays or α particles pass through metal foils, except for a few α particles, which are scattered.	☐
4-4	Ernest Rutherford	Discoverer of α particles and of nuclear structure of atom.	☐
4-4	Evidence for nuclear atom	When a stream of α particles is directed toward a sheet of metal foil, most of them pass through and a few are scattered. The heavy, positively charged α particles must be scattered by a heavy, positively charged object, the nucleus.	☐
4-4	Neutron	Particle with no electrical charge and with about the same mass as the proton.	☐
4-4	Composition of nucleus	Contains protons and neutrons. The electrons surround the nucleus at relatively great distances.	☐
4-4	Nucleons	Protons and neutrons.	☐

Section	Subject	Summary	Check When Learned
4-5	Z	Atomic number, number of protons (or electrons) in atom.	☐
	A	Mass number, number of protons and neutrons in atom.	☐
	A − Z	Number of neutrons in atom.	☐
4-5	Meaning of $^{19}_{9}F$	Fluorine atom with 9 protons, 9 electrons, 10 neutrons.	☐
4-6	amu	Atomic mass units.	☐
4-6	Standard for atomic masses	$^{12}_{6}C$ atom, assigned a mass of exactly 12 amu.	☐
4-6	Approximate mass of $^{19}_{9}F$	19 amu.	☐
4-6	Isotopes	Atoms with the same number of protons (and electrons) but different numbers of neutrons. Atoms with same Z and different A.	☐
4-7	Atomic mass	Average mass of element's naturally occurring isotopes.	☐
4-6, 4-7	Two numbers for each element on Periodic Table	Atomic number and atomic mass.	☐
4-8	Equation	Before-and-after representation of a reaction, using symbols.	☐
4-8	Reactant	Substance that exists before reaction, shown on left in equation.	☐
4-8	Product	Substance that exists after reaction, shown on right in equation.	☐
4-8	Coefficient	Multiplier in equation showing how many of the product or reactant species are used or produced.	☐
4-8	Difference between reaction and equation	A reaction is a process and an equation is its representation in symbolic form.	☐
4-8	Nuclear reaction	Reaction in which reactants and products are nuclei.	☐
4-8	Nuclear equation	Equation that represents a nuclear reaction.	☐
4-8	Transmutation	Reaction in which one nucleus is converted to another by bombarding it with subatomic particles, e.g., neutrons, or fragments (α particles).	☐
4-8	Transuranium elements	Man-made elements beyond uranium on the Periodic Table.	☐
4-8	Nuclear fission	Reaction in which one nucleus splits into two or more nuclei and releases large amounts of energy.	☐
4-8	Chain reaction	Series in which each nuclear fission reaction causes several more.	☐
4-8	Nuclear fusion	Reaction in which two lighter nuclei combine to form one heavier one and release large amounts of energy.	☐

Problems

Assume you can use the Periodic Table at the front of the book unless you're directed otherwise. Answers to odd-numbered problems are in Appendix 1.

Early Atomic Theory (Section 4-1)

1. Democritus and Leucippus, in the fifth century BC, held the belief that matter ultimately consists of tiny, indivisible particles. Was that belief, as they held it, an observation, a hypothesis, or a theory? Explain.
2. Democritus and Leucippus, in the fifth century BC, held the belief that matter ultimately consists of tiny, indivisible particles. Are atoms indivisible? Explain.
3. One of John Dalton's principles was that matter is made of indestructible particles, called atoms. Would that principle now be regarded as true? Explain.
4. One of John Dalton's principles was that all atoms of the same element are identical in mass. Would that principle now be regarded as true? Explain.
5. One of John Dalton's principles was that atoms of different elements have different masses. Would that principle now be regarded as true? Explain.
6. In what crucial way did John Dalton change the use of chemical symbols?

Chemical Symbols (Section 4-2)

7. Write the symbol for each of these elements: (a) hydrogen (b) helium (c) tin (d) copper.
8. Write the symbol for each of these elements: (a) magnesium (b) mercury (c) lead (d) iron.
9. Write the name for each of these elements: (a) Ba (b) B (c) Br (d) Pb.
10. Write the name for each of these elements: (a) Ca (b) C (c) Cl (d) F.

Atomic Structure (Sections 4-3 and 4-4)

11. Describe evidence suggesting that cathode rays carry a negative charge.
12. Describe evidence suggesting that electrons are part of all kinds of matter.
13. Which particle has the larger mass, the proton or the electron?
14. The electron carries a negative charge and the proton carries a positive charge. Which charge is larger?
15. If an atom contains 23 protons, how many electrons does it contain?
16. If an atom contains 42 electrons, how many protons does it contain?
17. What is an α particle? What is an α ray?
18. What is a β particle? What is a β ray?
19. List these particles in order of increasing mass: α particle, β particle, proton.
20. What is the charge on each of these particles: α particle, β particle, neutron?
21. What evidence did Rutherford and his co-workers find for the nuclear structure of atoms?
22. Distinguish among these terms: (a) neutron (b) nucleus (c) nucleon.

Atomic Number, Mass Number, and Atomic Mass (Sections 4-5 through 4-7)

23. Describe the difference between the atomic number and the mass number for an atom.
24. Describe the difference between the mass number and the atomic mass for an atom.
25. What is the relationship between the atomic number for an atom and the number of protons it contains?
26. What is the relationship between the atomic number for an atom and the number of electrons the atom contains?
27. Which is true for an atom: (a) number of protons + number of neutrons = A or (b) number of protons + number of neutrons = Z?
28. An atom of the element tin contains 50 protons, 50 electrons, and 69 neutrons. Which of these particles makes no significant contribution to the mass of the atom? Why?

29. Why is the mass of an atom almost the same as its mass number?

30. Why are atomic numbers always integers (whole numbers)?

31. Why are mass numbers always integers (whole numbers)?

32. How many electrons are there in (a) a silicon atom (b) a boron atom?

33. Explain the significance of the symbol $^{39}_{19}K$.

34. An atom of bromine contains 35 protons, 35 electrons, and 45 neutrons. Write its symbol, including its atomic number and mass number.

35. Write the correct symbol for the element represented by E in $^{127}_{53}E$.

36. An atom of an element contains 16 protons. Write the symbol for the element.

37. Why is the mass of an atom used as the standard for atomic mass?

38. Which atom is the standard for atomic mass?

39. What units are used for the atomic masses on the Periodic Table? What is their abbreviation?

40. The mass of one proton is 1.672×10^{-24} g. If you had 1.00 lb of protons, how many protons would you have?

41. Which of the following statements is correct: (a) Atoms of the same element have the same A or (b) Atoms of the same element have the same Z?

42. Which of the following statements is correct: (a) Isotopes of the same element have the same mass number or (b) Isotopes of the same element have the same atomic number?

43. Suppose that the entire sample of boron in the world consisted of one atom of $^{9}_{5}B$, one atom of $^{10}_{5}B$, and one atom of $^{11}_{5}B$. What would the atomic mass of boron be?

44. Suppose that the entire sample of boron in the world consisted of one atom of $^{9}_{5}B$, two atoms of $^{10}_{5}B$, and three atoms of $^{11}_{5}B$. What would the atomic mass of boron be?

Nuclear Reactions (Section 4-8)

45. What is the difference between nuclear fission and nuclear fusion?

46. What are transuranium elements? What do all of the transuranium elements have in common?

47. What is a chain reaction?

48. What is the difference between a reaction and an equation?

49. In the equation shown below identify the reactants, products, and coefficient.

$$^{235}_{92}U + ^{1}_{0}n \rightarrow ^{139}_{56}Ba + ^{94}_{36}Kr + 3\ ^{1}_{0}n$$

50. A thorium-230 nucleus ($^{230}_{90}Th$) spontaneously breaks up into an α particle and a radium-226 nucleus ($^{226}_{88}Ra$). Write the nuclear equation for this reaction.

51. A uranium-235 nucleus ($^{235}_{92}U$) spontaneously breaks up into a thorium-231 nucleus ($^{231}_{90}Th$) and one other particle. Write the nuclear equation for this reaction.

52. Write the nuclear equation for the reaction of an aluminum-27 nucleus with a neutron to form an α particle and a sodium-24 nucleus.

Three elements from group 4A: lead (Pb), silicon (Si), and tin (Sn). *(Charles D. Winters)*

5
Introduction to the Periodic Table

5-1
On the Periodic Table elements are classified as metals, metalloids, or nonmetals.

5-2
On the Periodic Table elements are aligned into periods and groups and classified as representative or transition elements.

5-3
It's important to know the structure of the Periodic Table thoroughly.

Inside Chemistry:
How was the Periodic Table created?

Chapter Summary

Problems

In every chemistry classroom and laboratory, you'll see a Periodic Table of the Elements on the wall, because in learning chemistry and in solving chemistry problems, no other source of information is used as often. In its symbols and numbers, and most of all in its structure, the Periodic Table contains information about the chemical behavior of the elements and about the masses, compositions, and structures of their atoms.

This chapter will introduce you to the Periodic Table. You'll learn three of the ways that the Table classifies elements: as periods and groups; as representative and transition elements; and as metals, metalloids, and nonmetals. You'll also see how the basic structure of the Table was created about a century ago by a Russian chemist, Dmitri Ivanovich Mendeleev, and how it has evolved into its current form. In Chapter 6 you'll see how the structure of the Periodic Table is itself a model for the structures of atoms.

The Periodic Table will be the resource you'll turn to most often as you work problems and take exams in chemistry, so it's important that you feel confident about using it. Before you start reading the new material about the Table in this chapter, you may want to review the following summary of what you learned about it in Chapter 4.

The box for each element on the Periodic Table shows the symbol, atomic number, atomic mass, and the number of protons and electrons for one atom of the element. The entry for copper, for example, is

29 Cu 63.6

The atomic number, 29, shows that there are 29 protons (positively charged) in the nucleus of a copper atom and 29 electrons (negatively charged) around the nucleus. The atomic mass of copper, 63.6 amu (atomic mass units), is the average of the masses of its naturally occurring isotopes.◆

◆Analysis has shown that copper in nature is 68% $^{64}_{29}Cu$ and 32% $^{63}_{29}Cu$.

For review use the Periodic Table on the inside front cover of this book to answer each of the following Exercises.◆

◆Answers to Exercises are in Appendix 1.

Exercise 5.1

How many protons are there in one atom of each of the following elements: (a) mercury (b) sodium (c) tin?

Exercise 5.2

How many electrons are there in one atom of each of the following elements: (a) nitrogen (b) magnesium (c) potassium?

Exercise 5.3

Write the name for (a) the element whose atom contains 26 electrons and (b) the element whose atom contains 9 protons.

Exercise 5.4

Identify the number and location (in the nucleus or outside it) of each kind of subatomic particle in each of the following isotopes: (a) the isotope of boron with mass number 11 (b) the isotope of mercury with mass number 200

5-1 On the Periodic Table elements are classified as metals, metalloids, or nonmetals

The heavy, stair-step line on the right side of the Periodic Table separates the elements into metals, to the left of the line, and nonmetals, to the right. An element whose box has one of its sides lying on either side of the stair-step line is classified as a metalloid. There are three exceptions: Hydrogen is a nonmetal, and aluminum and polonium (Po) are metals.

We'll see later that the words *metal* and *nonmetal* can have special meanings in chemistry. For now, we'll define a **metal** as a solid, usually shiny substance that conducts heat and electricity well, and that can be formed into many shapes such as sheets, wires, and tubes. Not all of the metals have all of these properties (mercury, for example, is a liquid), but the description is accurate enough to be useful as a general definition.

We'll define a **nonmetal** as an element that's a poor conductor of heat and electricity. At room temperatures, eleven of the nonmetals (hydrogen, nitrogen, oxygen, fluorine, chlorine, helium, neon, argon, krypton, xenon, and

	1A	2A	B										3A	4A	5A	6A	7A	8A
1	H																	He
2	Li	Be											B	C	N	O	F	Ne
3	Na	Mg											Al	Si	P	S	Cl	Ar
4	K	Ca	Sc	Ti	V	Cr	Mn	Fe	Co	Ni	Cu	Zn	Ga	Ge	As	Se	Br	Kr
5	Rb	Sr	Y	Zr	Nb	Mo	Tc	Ru	Rh	Pd	Ag	Cd	In	Sn	Sb	Te	I	Xe
6	Cs	Ba	La*	Hf	Ta	W	Re	Os	Ir	Pt	Au	Hg	Tl	Pb	Bi	Po	At	Rn
7	Fr	Ra	Ac†	Rf	Ha	Unh	Uns	Uno	Une									

*	Ce	Pr	Nd	Pm	Sm	Eu	Gd	Tb	Dy	Ho	Er	Tm	Yb	Lu
†	Th	Pa	U	Np	Pu	Am	Cm	Bk	Cf	Es	Fm	Md	No	Lr

(a)

(b)

(c)

Figure 5.1
(a) **Metals.** Top: iron, cobalt, nickel, copper, zinc. Bottom: scandium, titanium, vanadium, chromium, manganese. (b) **Nonmetals.** Center: chlorine. Left to right: phosphorus, carbon, selenium, sulfur, iodine. (c) **Metalloids.** Top: boron. Bottom: silicon. *(Chip Clark)*

radon) are gases, five (carbon, phosphorus, sulfur, selenium, and iodine) are solids, and one (bromine) is a liquid. The solid nonmetals are more brittle than metals and can't be easily formed into sheets, wires, or tubes.

A **metalloid** has properties between those of metals and those of nonmetals. Silicon, for example, conducts heat and electricity better than sulfur, but not as well as copper, and silicon can be shaped more easily than sulfur, but not as easily as copper. All of the metalloids are solids at room temperatures.

Pictures of some metals, nonmetals, and metalloids are shown in Figure 5.1 a–c.

Exercise 5.5

Classify each of the following elements as a metal, nonmetal, or metalloid:
(a) potassium (b) phosphorus (c) arsenic (As)

5-2 On the Periodic Table elements are aligned into periods and groups and classified as representative or transition elements

The horizontal rows on the Periodic Table are called **periods** and the vertical rows are called **groups.** The first two tall groups on the left and the last six tall groups on the right are numbered 1A through 8A, and the ten short groups in the middle of the Table are designated B groups.◆ The periods on most Periodic Tables aren't numbered, but in this book they're numbered 1 through 7 to make it easier to refer to them.

◆This numbering system for groups is the one that's most commonly used in the United States. Another system, which numbers the groups 1 through 18 from left to right, is coming into use, and you may see Tables with that system.

The elements in groups 1A through 8A are called **representative elements.** Four of these groups have names: The elements in group 1A, except hydrogen, are called the **alkali metals;** group 2A is called the **alkaline-earth metals;** group 7A is called the **halogens;** and group 8A is called the **noble gases.**

The elements in the B groups are called **transition elements;** because all of them are metals, they're also called **transition metals.** The elements with atomic numbers 58 through 71 and 90 through 103 are usually shown as footnotes to the Table because if they're put in their appropriate places in the body of the Table, it becomes too long to fit well on a page. Elements 57 through 71 are called the **lanthanide series,** after the first member, lanthanum (La), and elements 89 through 103 are called the **actinide series,** after actinium (Ac). The elements in the lanthanide and actinide series are called **inner-transition elements.**

	1A	2A	B										3A	4A	5A	6A	7A	8A
1	H																	He
2	Li	Be											B	C	N	O	F	Ne
3	Na	Mg											Al	Si	P	S	Cl	Ar
4	K	Ca	Sc	Ti	V	Cr	Mn	Fe	Co	Ni	Cu	Zn	Ga	Ge	As	Se	Br	Kr
5	Rb	Sr	Y	Zr	Nb	Mo	Tc	Ru	Rh	Pd	Ag	Cd	In	Sn	Sb	Te	I	Xe
6	Cs	Ba	La*	Hf	Ta	W	Re	Os	Ir	Pt	Au	Hg	Tl	Pb	Bi	Po	At	Rn
7	Fr	Ra	Ac†	Rf	Ha	Unh	Uns	Uno	Une									

*	Ce	Pr	Nd	Pm	Sm	Eu	Gd	Tb	Dy	Ho	Er	Tm	Yb	Lu
†	Th	Pa	U	Np	Pu	Am	Cm	Bk	Cf	Es	Fm	Md	No	Lr

From left to right in the same period the properties of the elements change gradually from metallic to nonmetallic.

One of the most important messages contained in the arrangement of the Periodic Table is that elements in the same group have similar chemical properties; that is, they combine with other elements in similar ways.◆ In Chapters 7 through 10 we'll see that the similar behavior of elements in the same group is the basis for describing the formation of compounds.

◆ For this reason, groups are also called **families.**

Exercise 5.6

In each of the following sets, write the symbols for the elements that you'd expect to have similar chemical properties: (a) sulfur, strontium (Sr), selenium (Se) and (b) barium, boron, beryllium (Be).

Exercise 5.7

Name the elements in the third period and classify each of them as a metal, nonmetal, or metalloid.

Exercise 5.8

Name the first three members of groups 1A, 2A, 7A, and 8A.

5-3 It's important to know the structure of the Periodic Table thoroughly

You need to know the Periodic Table well because you'll be using it throughout the rest of this book and in any further chemistry courses you take. If you follow the steps below, you'll learn the Table thoroughly and save yourself a lot of work and time later on. Complete each step before you go on to the next one, and test yourself with the Exercises as you go.

1. Learn the names and symbols for the elements shown in Table 4.1, Section 4-2.

Exercise 5.9

(a) Write the symbol for the name: potassium, helium, sulfur. (b) Write the name for the symbol: Ba, Mg, Si.

2. Be sure you know the significance of the numbers shown for each element on the Periodic Table. The atomic number, above the symbol, is the number of protons, and also the number of electrons, in one atom of the element; this information will be important in Chapter 6 and later chapters. The atomic mass, below the symbol, is the average of the masses of the naturally occurring isotopes of the element, in amu (atomic mass units); this information will be important in Chapter 7 and later chapters.

Exercise 5.10

Identify the number of electrons in one atom of each of the following elements: neon, nitrogen, iron, chlorine.

3. Develop a mental image of the overall shape of the Periodic Table, including the locations of the representative and transition groups. The representative elements form two tall columns on the left side of the Table and six tall columns on the right side. The transition metals form a long, low block, ten elements wide and three or four elements deep, in the center of the Table. The inner-transition metals form a block of elements at the bottom of the Table, fourteen elements long and two deep.

Exercise 5.11

Without looking at the Periodic Table, draw an approximate outline of its shape. Make the Table large enough to fill most of an 8″ × 11″ sheet, turned sideways. Show the number of groups in each section. Shade in the sections for the transition metals.

4. Learn the number designations for the groups of representative elements. Across the top of the Table the groups of representative elements are numbered 1A through 8A. Learn the number designations for the periods: From top to bottom they're numbered 1 through 7.

Exercise 5.12

On the sketch of the Periodic Table you made for Exercise 5.11, number the groups of representative elements and the periods.

5. Learn which groups have these names associated with them: group 1A, alkali metals; group 2A, alkaline-earth metals; group 7A, halogens; group 8A, noble gases.

Exercise 5.13

On the sketch of the Periodic Table you made for Exercise 5.11, write in the names *noble gases*, *alkaline-earth metals*, *halogens*, and *alkali metals* over the appropriate groups.

6. Learn the locations of the lanthanide series and the actinide series. These are the footnoted elements. The lanthanide series begins with lanthanum (La), element number 57, and runs through element number 71; the actinide series begins with element number 89, actinium (Ac), and runs through element number 103.

Exercise 5.14

On the sketch of the Periodic Table you made for Exercise 5.11, write *actinide series* and *lanthanide series* across the appropriate blocks of elements.

Exercise 5.15

Using the Periodic Table at the front of the book, identify each of these elements as a member of the lanthanide or actinide series: Md (mendelevium), No (nobelium), Eu (europium).

7. Learn the location of the stair-step line that separates metals from nonmetals. The line begins on the left side of the box for boron and follows a stair-step path down and to the right.

Exercise 5.16

On the sketch of the Periodic Table you made for Exercise 5.11, write in the words *metal, metalloid,* and *nonmetal* to show the locations of these classes of elements.

◆The state of an element depends on the temperature and pressure; the states described here are at ordinary room conditions.

8. Learn which elements are solids, liquids, or gases.◆ This turns out to be easy: Bromine and mercury are liquids, the elements shown in the partial Periodic Table below are gases, and all the rest are solids.

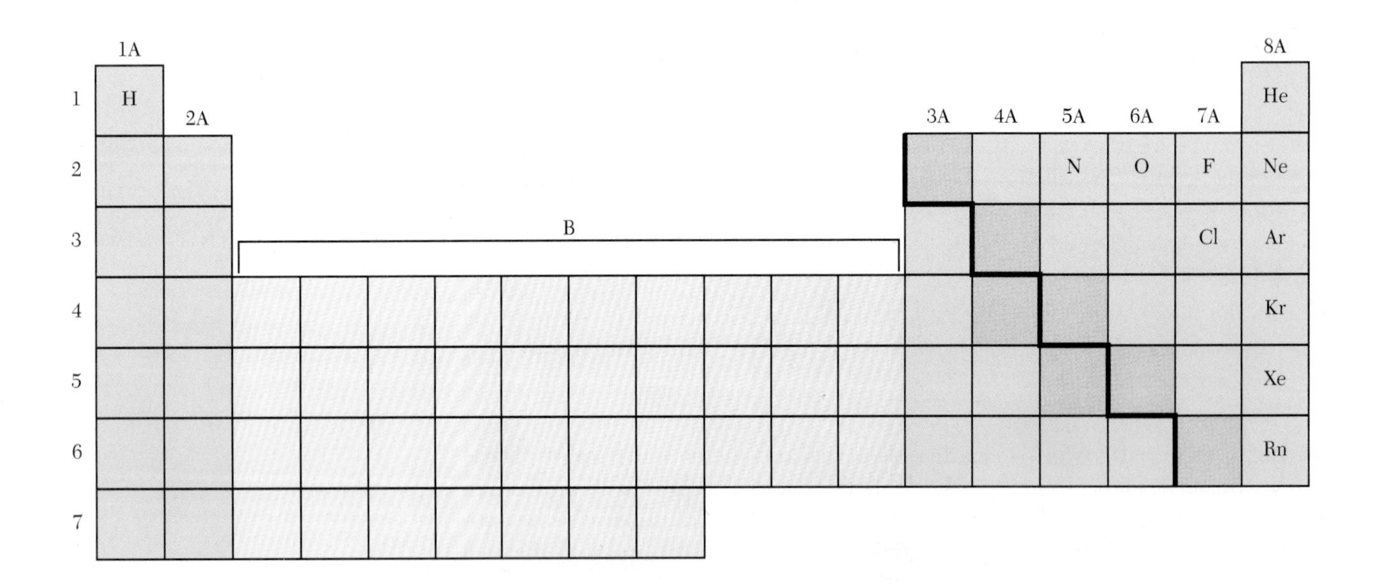

Exercise 5.17

Without looking at the partial Table above, identify each of the following elements as a solid, liquid, or gas: phosphorus, fluorine, silver, bromine, nitrogen, and neon.

Inside Chemistry

How was the Periodic Table created?

By the 1860's scientists had discovered 60 elements and were finding new elements at a rate of about one every other year. So much information about elements was accumulating that the need to find some way to organize it was becoming urgent, and several methods for classifying the elements had been proposed. But the chemists who created these classifications weren't successful in persuading other chemists to use them; most chemists thought that attempts to classify the elements were artificial and not very scientific.

Dmitri Ivanovich Mendeleev changed everyone's mind.† Mendeleev did his research by writing a description of the properties of each element on a separate card, then trying to arrange the cards to find patterns in the properties. Eventually he created a design in which elements were placed in vertical rows by similar properties and in horizontal rows by increasing atomic masses, essentially the form of the modern Periodic Table. His discovery can be summarized in a statement known as **the periodic law:** The properties of the elements are a regularly repeated function of their atomic masses.*

Why was Mendeleev's classification successful when the earlier systems were not?

*A German chemist, Julius Lothar Meyer (1830–1871), working independently of Mendeleev, created a similar table at about the same time. Because Mendeleev's work was more thorough and because he published his results first, he's given primary credit for creating the design of the modern Periodic Table.

Figure 5.2
Dmitri Ivanovich Mendeleev. *(Library of Congress)*

One reason was that Mendeleev understood the limitations of his table and admitted that he wasn't able to classify all of the elements in a simple scheme. Hydrogen wouldn't fit into any of his groups, so he put it in a group by itself. He was uncertain about how to classify some heavier elements, for example, silver, platinum, and gold. In fact, Mendeleev's table established only the groups of the modern Periodic Table that contain the representative elements, and many years passed before the table was successfully modified to accommodate the transition and inner-transition elements.

† Chemistry Insight

Mendeleev's ancestors were Russian and Mongolian. He was born in Tobolsk, Siberia, in 1834, the youngest of 14 children, and had a difficult childhood, which left him with permanently weak health. When he was still very young, his father, who had been principal of the local secondary school, went blind, and his mother supported the family by starting and operating a glass factory. When Mendeleev was 16, his father died of tuberculosis and the factory was destroyed by fire. His mother took him to Petrograd and enrolled him in school. Exhausted by her hardships, she died a few months later. Mendeleev finished his studies at Petrograd and did postgraduate work at Heidelberg and Paris. In 1869 he became professor of chemistry at the University of Petrograd, and in the same year he published his now famous paper "The relation of the properties to the atomic masses of the elements." He died in 1907.

Another reason for Mendeleev's success was that he was able to show that some elements didn't fit into his table well because their properties hadn't been determined correctly. His use of the table to correct earlier work established it as a very useful tool for chemists.

But the most important reason for Mendeleev's success was that he was able to predict the existence of elements that hadn't yet been discovered. When Mendeleev arranged all of the known elements in horizontal rows by increasing atomic masses, the elements with similar properties didn't always fall in the same vertical columns. Confident of his theory, he moved elements with similar properties into vertical alignment, leaving blank spaces in the table that he predicted would be filled by elements not yet found.

For example, Mendeleev's table left a blank space below aluminum. He predicted, in 1871, that the element that would be discovered to fit this space would be a metal with an atomic mass of about 68 amu and a density of about 5.93 g/mL, and that it would melt at an unusually low temperature. In 1875, the discovery of gallium was reported: It's a metal with atomic mass 69.7 amu, density 5.91 g/mL, and a melting temperature of 30.1°C—it will melt in the palm of your hand, as shown in Figure 5.3.

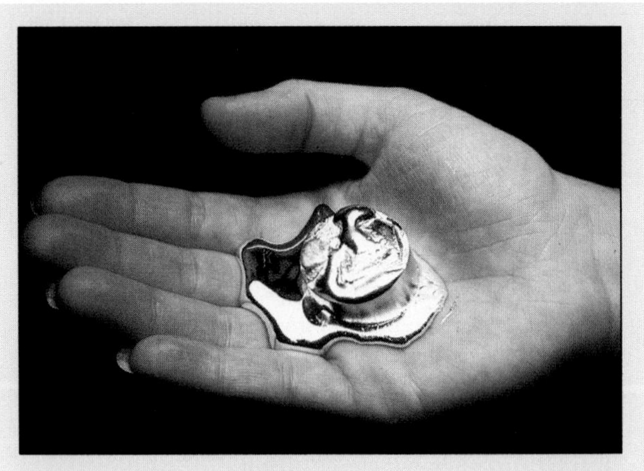

Figure 5.3
Gallium melting. *(Chip Clark)*

Chemists were enormously impressed with Mendeleev's ability to organize the large amounts of accumulated data about the known elements, to correct those data, and most dramatically, to point to the existence of elements not yet discovered. The Periodic Table, essentially in the form created by Mendeleev, became an immediate success and has been used by every chemist, and by every student of chemistry, since his day.†

† Chemistry Insight

On the Periodic Table we now use the elements are in order of increasing atomic *number*, giving a slightly different sequence from Mendeleev's arrangement by increasing atomic *mass*. The modern form of the periodic law says that the properties of the elements are a regularly repeated function of their atomic numbers.

Chapter Summary: Introduction to the Periodic Table

Section	Subject	Summary	Check When Learned
	Information in box for each element on Periodic Table	Symbol, atomic number (number of protons or electrons), atomic mass (average of masses of naturally occurring isotopes) in amu.	☐
5-1	Function of heavy stair-step line on right side of Table	Divides elements into metals to the left, and nonmetals to the right; elements on line, except Al and Po, are metalloids. Hydrogen is a nonmetal.	☐

Section	Subject	Summary	Check When Learned
5-1	Metal	Solid, shiny substance that conducts heat and electricity well and can be formed easily into complicated shapes.	
5-1	Nonmetal	Substance that's a poor conductor of heat and electricity and can't be formed easily into complicated shapes.	
5-1	Metalloid	Substance with properties between those of a metal and those of a nonmetal.	
5-2	Group	Vertical row on Periodic Table.	
5-2	Relationship of properties of elements in a group	Properties are similar.	
5-2	Period	Horizontal row on Periodic Table.	
5-2	Relationship of properties of elements in a period	From left to right, properties change from metallic to nonmetallic.	
5-2	Representative elements	Elements in groups 1A through 8A.	
5-2	Transition elements	Elements in B groups.	
5-2	Alkali metals	Group 1A elements except H: Li, Na, K, Rb, Cs.	
5-2	Alkaline-earth metals	Group 2A elements: Be, Mg, Ca, Sr, Ba, Ra.	
5-2	Halogens	Group 7A elements: F, Cl, Br, I, At.	
5-2	Noble gases	Group 8A elements: He, Ne, Ar, Kr, Xe, Rn.	
5-2	Lanthanide series	Elements 57 through 71, usually shown as a footnote to the Periodic Table.	
5-3	Elements that are gases	1A 2A 3A 4A 5A 6A 7A 8A 1H He 2 N O F Ne 3 Cl Ar 4 Kr 5 Xe 6 Rn	
5-3	Elements that are liquids	Br, Hg.	
5-3	Elements that are solids	All others.	

Problems

Assume you can use the Periodic Table at the front of the book unless you're directed otherwise. Answers to odd-numbered problems are in Appendix 1.

Atomic Number, Mass Number, and Atomic Mass (Introduction)

1. The atomic number for an element is 13. (a) How many electrons are in one atom of the element? (b) How many protons are in one atom of the element? (c) How many neutrons are in one atom of the isotope of the element with mass number 27? (d) Name the element.

2. For the element potassium: (a) How many electrons are in one atom? (b) How many protons are in one atom? (c) How many neutrons are in one atom of the isotope with mass number 39?

3. What is the atomic mass of each of these elements: (a) calcium (b) carbon (c) sodium?

4. Name the element with atomic mass (a) 28.1 amu (b) 24.3 amu (c) 207.2 amu.

5. Explain the significance of the symbol $^{32}_{16}S$.

6. An atom of fluorine contains 9 protons, 9 electrons, and 10 neutrons. Write its symbol, including its atomic number and mass number.

Metals, Metalloids, and Nonmetals (Section 5-1)

7. What are the properties of a typical metal? From your everyday experience, what evidence do you have that copper is a typical metal?

8. In its physical properties, how does hydrogen differ from the other elements in group 1A?

9. Name the only metalloid in the period that begins with lithium.

10. Classify each of these elements as a metal, metalloid, or nonmetal: (a) phosphorus (b) oxygen (c) helium (d) boron.

11. Which is the largest class of elements: metals, metalloids, or nonmetals?

12. Which is the smallest class of elements: metals, metalloids, or nonmetals?

Periods and Groups (Section 5-2)

13. How many periods are there?

14. Name the group of which calcium is a member.

15. Write the name and number of the group in which all of the elements are nonmetals. What physical characteristic do all of these elements have in common?

16. The first period contains 2 elements and the second contains 8; finish this list of the number of elements in each period: 2, 8, . . . Remember to include elements 57 through 71 and 89 through 103.

Representative and Transition Elements (Section 5-2)

17. Without looking at a Periodic Table, describe where the representative elements are on the Table.

18. Without looking at a Periodic Table, describe where the transition metals are on the Table.

19. Without looking at a Periodic Table, describe where the lanthanide and actinide series are.

20. Decide whether each of the following elements is a representative element, a transition element, or an inner-transition element: (a) xenon (b) erbium (c) radium.

21. Name the halogens, except At.

22. Which elements are liquids?

23. How many elements are in the transition-metal series that begins with scandium (Sc)? How many in the series that begins with yttrium (Y)?

24. How many elements are in the lanthanide series? In the actinide series?

In a laser, light is generated by electrons in atoms falling from higher to lower energy levels, as described in Section 6-1. *(Telegraph Colour Library/FPG International)*

6
Atomic Structure and the Periodic Table

During the first 30 years of this century, work that had been done separately by physicists and by chemists was brought together to create the theory of atomic structure on which chemical knowledge now depends. This chapter describes the creation of that theory.

By the end of the nineteenth century, chemists had adopted the form of the Periodic Table described in Chapter 5, with elements aligned vertically in groups with similar properties and arranged horizontally in periods in order of increasing atomic mass.[†] The structure of the Periodic Table was based almost entirely on the observed properties of the elements, and hardly at all on theory. At the end of the nineteenth century, chemists still used essentially the same form of the atomic theory that Dalton had stated at the beginning of the century: Atoms exist, have mass, and can combine with one another. In this elementary form the atomic theory says nothing about what structures atoms may have or why they enter into combinations.

By the end of the nineteenth century physicists had carried out research, working with gas-discharge tubes and radioactive substances, showing that atoms consist of even smaller particles, eventually identified as electrons, protons, and neutrons. In 1908 Rutherford proposed the nuclear model of the atom: The positively charged particles in an atom, and almost all of its mass, are in a tiny, central core or nucleus, and the negative charges are located outside the nucleus.

By about 1910, research by chemists on the properties of elements had led to the establishment of the structure of the Periodic Table, and research by physicists had led to the first significant proposal of a structure for atoms.

If Dalton's atomic theory is true, these structures must be related to one another. According to Dalton, the observed properties of the elements, especially the ways that they combine to form compounds, are determined by the behavior of their atoms; and the behavior of atoms is presumably determined by their structure, the arrangement of their subatomic particles.

The structure of the Periodic Table itself raises questions: Why do elements in the same group behave in similar ways? Why do the properties of elements change gradually across a period? Why do the representative elements behave differently from the transition elements? Why do these numbers of elements occur in the first six periods: 2, 8, 8, 18, 18, 32?

† Chemistry Insight

As mentioned on page 80, in the current form of the Periodic Table, the elements are arranged in sequence by increasing atomic *number*, and in a few places this sequence doesn't correspond with their increasing atomic *mass*. For example, the atomic mass of tellurium (Te, atomic number 52) is 127.6 amu, but the atomic mass of the next element, iodine, is 126.9 amu. The atomic numbers of elements are found by experiments in physics that are too complicated to describe in an introductory chemistry book. The first person to discover a method for determining atomic numbers was an English physicist, Henry Moseley, working at the University of Manchester from 1912 to 1914. When England entered World War I in 1914, Moseley enlisted in the army. He was sent to the front lines and killed by a sniper on August 10, 1915, three months before his twenty-eighth birthday.

The answer to these questions is this: The structure of the Periodic Table is related directly to the arrangement of electrons in atoms, because that arrangement determines the behavior of the atoms. In this chapter, we'll see how the arrangement of electrons in atoms was discovered and how it can be read from the structure of the Periodic Table.

6-1 An early theory described the atom as a miniature solar system, with electrons in circular orbits around the nucleus

In 1911 Ernest Rutherford proposed a model for the atom in which a heavy, positively charged nucleus is surrounded, at a relatively large distance, by electrons (Section 4-4). The electrons are held in the atom by the attraction between their negative charge and the positive charge of the nucleus.

Having accepted Rutherford's model, scientists began to ask more detailed questions about atomic structure: How are the particles in the nucleus held together? Are the electrons moving or stationary? If they're moving, do they move randomly or do they follow paths arranged in some regular pattern?

One hypothesis was that the atom had the structure of a miniature solar system, with the electrons in orbits around the nucleus. But there was a major objection to this model: Calculations based on traditional theories of physics showed that the electrons should steadily lose energy and be pulled in a spiral down into the nucleus. Figure 6.1 illustrates this prediction for the simplest atom, the hydrogen atom, with one proton and one electron.

In the meantime, however, a new theory of energy had been proposed that could be used to support the solar-system model of the atom.

In 1900 the German physicist Max Planck had suggested that the energy from a hot object, for example, a heated iron bar, isn't emitted continuously as it seems to be, but occurs in tiny, individual bursts, called **quanta.**◆ Because Planck's quantum theory contradicted traditional theories of energy, it met strong resistance, but in 1905 Albert Einstein showed that the quantum theory could be used to explain the properties of light. A beam of light behaves in some ways as if it were made up of waves and in other ways as if it were made up of a stream of particles.◆ Einstein showed that these two views could be reconciled if light were thought of as a stream of tiny bursts of energy, following Planck's theory. These quanta of light came to be called **photons.**

◆Each burst is a **quantum.** Planck (1858–1947) won the Nobel Prize in 1918.

◆This apparent ambiguity in the behavior of light is sometimes referred to as its **wave-particle duality.**

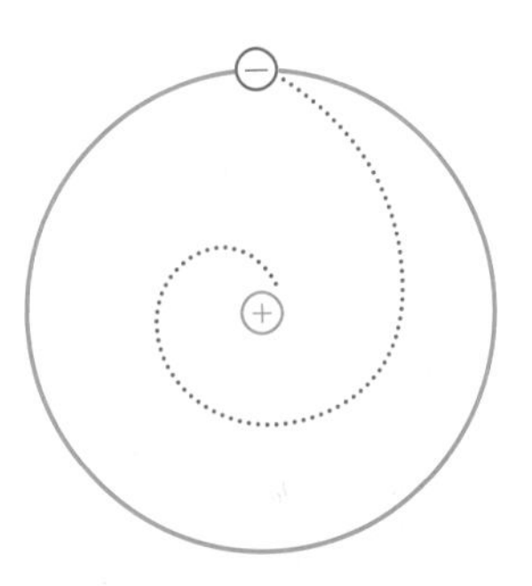

Figure 6.1
An early objection to the solar-system model of the atom. According to the theories of physics before 1900, the solar-system model for the hydrogen atom wouldn't work, because the electron should leave its orbit, lose energy continuously, and be pulled in a spiral into the nucleus.

Figure 6.2
Max Planck. *(Library of Congress)*

◆Answers to Exercises are in Appendix 1.

Exercise 6.1◆

Define quantum, quanta, quantum theory, and photon.

In 1913 the Danish physicist Niels Bohr showed that the quantum theory could be used to create a modified solar-system model for the hydrogen atom.◆ In the Bohr model, shown in Figure 6.4, the electron's orbit can only be at certain fixed distances from the nucleus. Each possible location for an orbit is called an **energy level,** and the energy levels are numbered 1, 2, 3, . . . outward from the nucleus, as shown in the figure. In theory, each atom has an infinite number of energy levels, but it's rarely necessary to consider more than the first six or seven.

◆Bohr (1885–1962) won the Nobel Prize in 1922.

Figure 6.3
Niels Bohr. *(American Institute of Physics, Niels Bohr Library, W.F. Meggers Collection)*

Exercise 6.2

Draw a Bohr model for a hydrogen atom, like the one in Figure 6.4, but with the electron in the third energy level.

To see how Bohr's model uses the quantum theory, we'll compare the energy changes that can occur in the Bohr atom with the energy changes that can occur in the more familiar system shown in Figure 6.5, a book resting on the edge of a table.

The book on the edge of the table, if it's given a slight push, will fall spontaneously to the floor, because of the attraction of the earth's gravity. Once the book begins to fall, it falls continuously, and while it's falling, it continuously releases energy. We could show that energy was being released by tying a string around the book, running the string over a pulley, and tying the other

Figure 6.5
Energy changes that occur when a book falls or is raised.

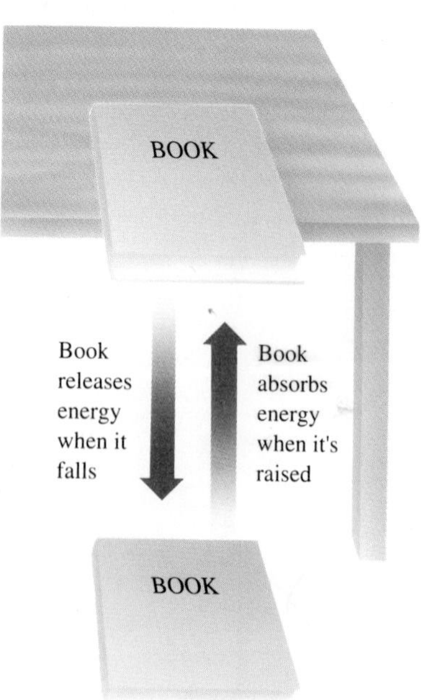
BOOK
Book releases energy when it falls
Book absorbs energy when it's raised
BOOK

Figure 6.4
The Bohr atom. The nucleus is surrounded by energy levels 1, 2, 3, ... The orbit of the electron can only be at one of the energy levels, not between them.

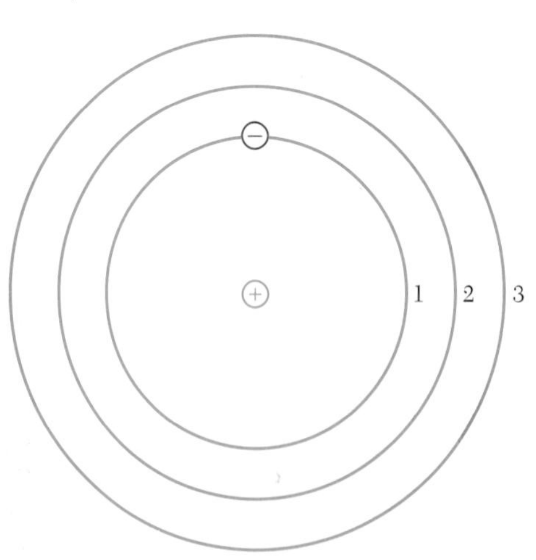
1
2
3

end of the string around an object with a mass smaller than the book's: As the book fell, the energy released would raise the lighter object, against the attraction of the earth's gravity.◆

To raise the book from the floor back onto the table would require using energy. With a string-and-pulley arrangement, we could use a heavier falling object to lift the book by mechanical energy, or we could raise the book by hand, using muscular energy. The law of conservation of energy requires that the amount of energy used to raise the book must equal the amount of energy released when the book falls, but the form of energy may change. For example, muscular or mechanical energy may be used to raise the book, and mechanical or heat energy may appear when it falls.

◆When the book falls without lifting another object, the energy released appears as heat. The amount of heat released is so small that we don't ordinarily notice it, but careful temperature measurements would show that the air is heated by friction as the book passes through it and that the floor is heated by the impact when the book smacks into it.

The book on the table has more energy than the book on the floor, and this energy of position is called **potential energy.** The higher the book is raised, the more potential energy it has.

Figure 6.6 shows the electron in a hydrogen atom moving from energy level 1 to energy level 3 and back again. In some ways, these changes are similar to the movement of the book between the floor and the table. The electron is attracted toward the nucleus, so it will move spontaneously from a higher energy level to a lower one; when the electron moves from a higher to a lower energy level, energy is released from the atom; and energy must be put into the atom to move the electron from a lower to a higher energy level. The amount of energy absorbed when the electron moves from energy level 1 to energy level 3 is the same as the amount of energy released when it falls back again (energy is conserved), but the energy absorbed and the energy released need not be in the same form.◆

◆When the electron is in the lowest energy level, the atom is said to be in its **ground state,** and when the electron is in a higher energy level, the atom is said to be in an **excited state.**

Atoms can absorb energy in the form of heat and emit it in the form of light, and the quantum theory, applied to the Bohr model of the atom, explains this process.◆ When a sample of hydrogen gas is heated, the electrons in some of the atoms absorb energy and are moved to higher energy levels. Suppose, for example, that an electron moves from energy level 1 to energy level 3. Because of the attraction of the nucleus, the electron will fall back to a lower energy level, and in falling it will emit light. The amount of energy in the light emitted

◆You've seen this conversion of energy when an iron bar is heated: At a certain temperature, it begins to glow.

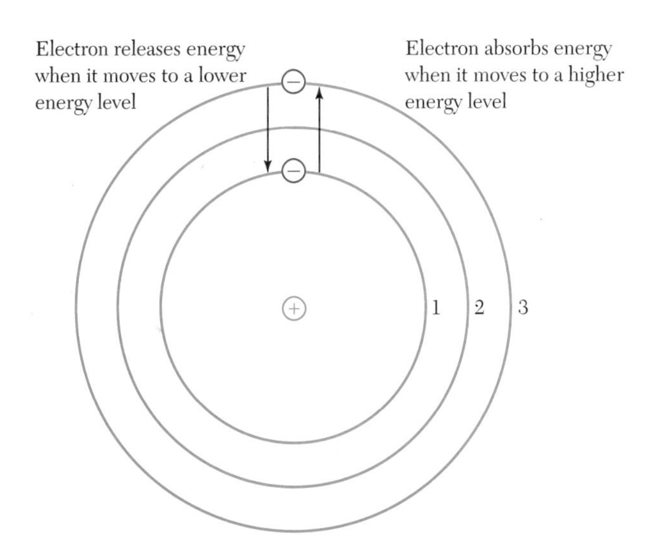
Electron releases energy when it moves to a lower energy level
Electron absorbs energy when it moves to a higher energy level
1
2
3

Figure 6.6
Energy changes that occur when an electron moves to a lower or higher energy level.

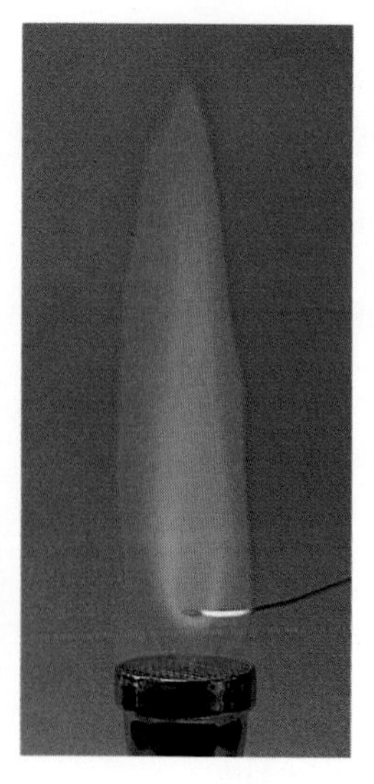

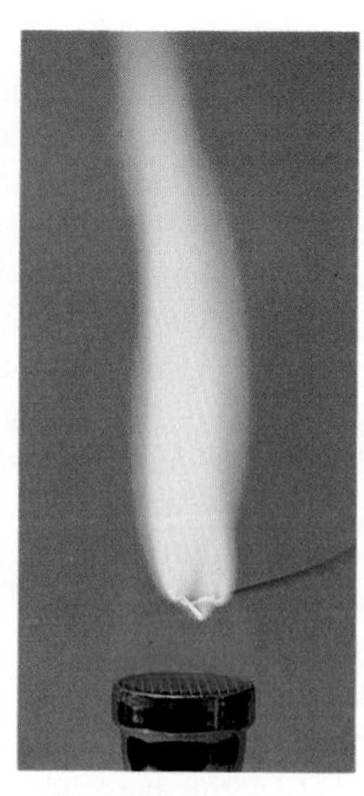
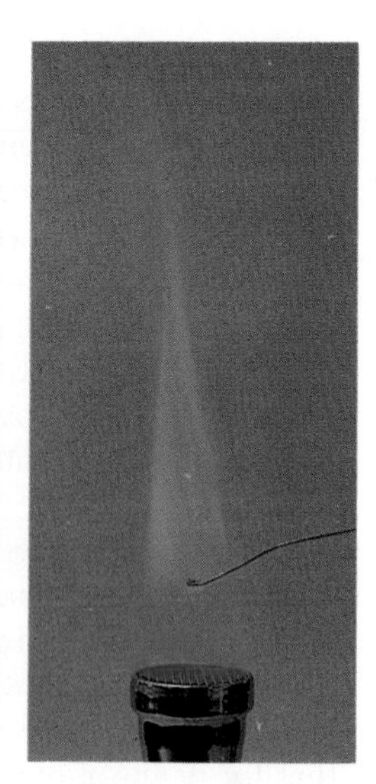

When atoms are heated their electrons absorb energy and move to higher energy levels. As the electrons fall back to lower energy levels, light is emitted. From left to right: barium, strontium, calcium, sodium, and potassium. *(Jim Morganthaler)*

In this laser electrons in neon atoms fall from higher to lower energy levels, causing the emission of light with a wavelength of 640 nm. *(Marna Clarke)*

will be determined by the difference in energy between the energy level the electron starts from and the energy level it falls to. Starting from energy level 3, the electron can fall to level 2 or 1, and it will emit light with different amounts of energy in making each of these transitions.

In the quantum theory light consists of photons. The wavelength of a beam of light is related to its frequency by $c = \lambda\nu$ (Section 3-8), and its frequency is related to the energy of its photons by

$$E = h\nu$$

where E is the energy of the photon in joules, ν is frequency in s^{-1}, and h is **Planck's constant, 6.63×10^{-34} J · s.**

Example 6.1

When the electron in a hydrogen atom moves from energy level 4 to energy level 1, the wavelength of emitted light is 4.86×10^{-7} m. Calculate the energy of the emitted photon.

Solution

$$\nu = \frac{c}{\lambda} = \frac{3.00 \times 10^{8}\ \cancel{m} \cdot s^{-1}}{4.86 \times 10^{-7}\ \cancel{m}} = 6.17 \times 10^{14}\ s^{-1}$$

$$E = h\nu = (6.63 \times 10^{-34}\ J \cdot \cancel{s})(6.17 \times 10^{14}\ \cancel{s}^{-1}) = 4.09 \times 10^{-19}\ J$$

Exercise 6.3

When the electron in a hydrogen atom moves from energy level 3 to energy level 1, the energy of the emitted photon is 1.94×10^{-18} J. Calculate its wavelength.

Bohr's model was an important step in understanding atomic structure, because it provided a simple picture of atoms that could be used to explain how they absorbed and emitted energy. In the next Section, we'll see that further applications of the quantum theory have created a more complex and surprising picture of the atom than the simple solar-system model.

6-2 In the modern view of the atom electrons are waves as well as particles, and their locations are uncertain

In 1905 Einstein used the quantum theory to show that the fundamental units of light have the properties of both waves and particles. In 1923 the French physicist Louis de Broglie suggested a more general and more surprising hypothesis: He proposed that all objects have wave properties.◆

◆De Broglie (1892–1987) won the Nobel Prize in 1929.

De Broglie's argument, in the form of sophisticated mathematics, showed that the wave properties of objects of ordinary size would be too small to be observable, but that these properties would become significant for objects as small as subatomic particles. In 1927 scientists in England and the United States carried out experiments that confirmed de Broglie's hypothesis by showing that electrons have significant wave properties.

In the late 1920's the German physicist Erwin Schrödinger created a mathematical model called the Schrödinger equation to describe the wave properties of electrons in atoms.◆ At the same time another German physicist, Werner Heisenberg, established a limit to the certainty with which the locations of electrons in atoms could be known. His calculations showed that the uncertainty in our knowledge of the position of an electron in an atom will always be about the same as the size of the atom.◆

◆Schrödinger (1887–1961) won the Nobel Prize in 1933.

◆Heisenberg (1901–1976) won the Nobel Prize in 1932. Heisenberg's theory doesn't say that the exact location of an electron in an atom is *unknown*, but that it's *unknowable*. His theory has special importance because it's one of a very few scientific principles that establish limits on our knowledge.

Figure 6.7
Louis de Broglie. *(American Institute of Physics, Niels Bohr Library)*

Figure 6.8
Erwin Schrödinger. *(Library of Congress)*

Figure 6.9
Werner Heisenberg. *(Library of Congress)*

The Schrödinger equation and Heisenberg's uncertainty principle are part of the branch of physics called **quantum mechanics,** which gives us the modern view of atomic structure. A calculation made with the Schrödinger equation will give a probability of finding an electron in a certain region of an atom. In drawing a picture of the atom, we can't truthfully represent the electron's precise location, because it will always be associated with a large measure of uncertainty. By agreement, scientists describe the approximate location of the electron as that region in which its probability of being located is 90% or more. Figure 6.10 is an interpretation of the quantum mechanical view of the ground state hydrogen atom, with the electron in the first energy level.

The quantum mechanical description of the atom is mathematical, but it can be translated, at least partially, into words and pictures. There are three basic terms in this description: energy level, sublevel, and orbital.

In Figure 6.10 the most darkly shaded region is an orbital. The general definition is that an **orbital** is a region, with a certain shape, in which there is a high (90% or greater) probability that an electron could be found. The orbital shown in Figure 6.10 has the shape of a spherical shell, with its center at the nucleus. The hydrogen atom depicted is in the ground state; that is, its electron is in the lowest energy level, the one nearest the nucleus.

As we saw in Section 6-2, the nucleus of every atom is surrounded by a series of energy levels numbered 1, 2, 3, . . . outward from the nucleus. Every energy level is made up of one or more sublevels, and every sublevel is made up of one or more orbitals. All of the orbitals in the same sublevel have the same shape.

The first energy level is the simplest: It consists of one sublevel, and that sublevel consists of one spherical orbital. In Figure 6.10, the darkly shaded region could be referred to as the first energy level, the sublevel of the first energy level, or the orbital of the sublevel of the first energy level. For the first energy level, each of these terms means the same thing.

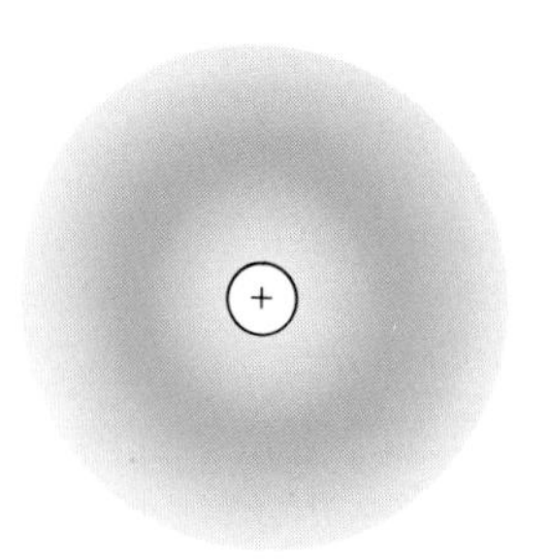

Figure 6.10
The hydrogen atom. In this drawing the hydrogen atom is shown in cross section, as if it had been sliced in two. In the quantum mechanical description of the ground state hydrogen atom the location of the electron is the region in which the probability of finding it is 90% or more. The shading in this drawing shows probability: The electron could be found at any distance from the nucleus, but the probability is 90% or more that it will be found in the most darkly shaded region.

Energy levels are numbered 1, 2, 3, . . . , and sublevels and their orbitals are represented by letters. An orbital with a spherical shape, such as the one in Figure 6.10, is designated by the letter s. In Figure 6.10, the darkly shaded region can be designated as energy level 1, as sublevel 1s, or as the s orbital in energy level 1. All three designations refer to the same thing.

Each energy level contains an s sublevel, made up of one s orbital: There is a 1s sublevel, a 2s sublevel, a 3s sublevel, and so forth. These s sublevels can be pictured as forming a series of spherical shells nesting one inside the other, as shown in Figure 6.11.

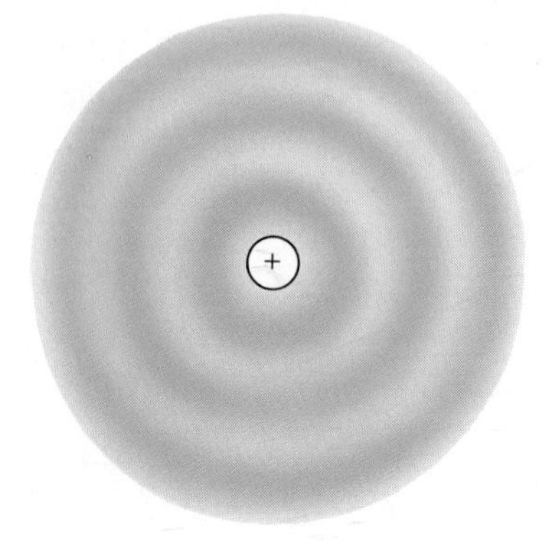

Figure 6.11
s orbitals. The atom is shown in cross section. Each energy level has an s orbital, in the shape of a spherical shell centered on the nucleus.

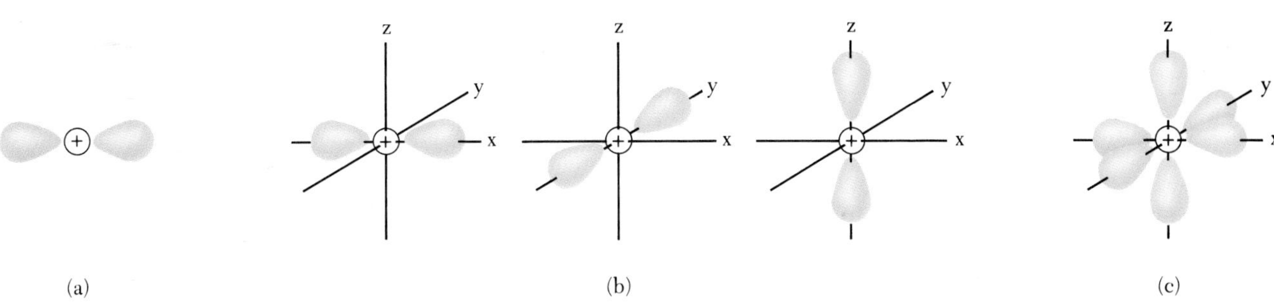

Figure 6.12
2p orbitals. The 2p sublevel consists of three p orbitals. Drawing (a) shows one p orbital; the orbital has two parts, one on each side of the nucleus. Drawing (b) shows all three 2p orbitals, each orbital lying along one axis in a coordinate system, with the nucleus at its center. Drawing (c) shows the three 2p orbitals on one coordinate system.

Energy level 2 consists of a 2s sublevel and a 2p sublevel. The 2p sublevel, shown in Figure 6.12, is made up of three p orbitals, at right angles to one another. Each p orbital has the shape of a dumbbell and consists of two parts, as shown in the figure.◆ This pattern is repeated in the higher energy levels: There is a 3p sublevel, a 4p sublevel, and so forth, each with the same shape but larger and farther from the nucleus.

Each higher energy level adds another sublevel. Energy level 3 consists of three sublevels: 3s, 3p, and 3d. A d sublevel has five d orbitals. Energy level 4 consists of four sublevels: 4s, 4p, 4d, and 4f. An f sublevel has seven f orbitals. Table 6.1 shows the sublevels and orbitals in the first four energy levels.

◆The letters used to designate orbitals were chosen by experimenters many years ago for reasons that are no longer important, but the designations are still used.

Table 6.1 Sublevels and Orbitals in the First Four Energy Levels

Energy Level	Sublevel	Number of Orbitals
1	s	1
2	s	1
	p	3
3	s	1
	p	3
	d	5
4	s	1
	p	3
	d	5
	f	7

Exercise 6.4

Without looking at Table 6.1, finish filling in the following table:

Sublevel	Number of Orbitals
s	
p	
d	
f	

Exercise 6.5

Without looking at Table 6.1, finish filling in the following table:

Energy Level	Available Sublevels
1	s
2	
3	
4	

The energy levels above 4 add more orbitals and sublevels, but they aren't significant for our purposes. In the energy levels above 4, we'll consider only the s, p, d, and f sublevels.

The d and f orbitals have complicated shapes that we won't consider. A significant limitation of the quantum mechanical model of the atom is that it can be translated only partially from the mathematical language in which it was created into useful descriptions in words or pictures.

Quantum mechanics is fundamentally important in chemistry for this reason: The distribution of the electrons among the orbitals of an atom determines its chemical behavior; that is, the distribution of the electrons in an atom determines how that atom will combine with other atoms. In the rest of this chapter, we'll see that a very simplified version of the quantum mechanical model can be used to show that the arrangements of electrons in atoms are related to the positions of the elements on the Periodic Table.

6-3 Each orbital can hold a maximum of two electrons

Quantum mechanics describes an atom as a nucleus, containing protons and neutrons, surrounded by electrons in energy levels. Each energy level consists of one or more sublevels, and each sublevel consists of one or more orbitals. An orbital is a region in which there is a high probability that an electron could be found.

Each orbital can hold a maximum of two electrons; that is, each orbital can hold 0, 1, or 2 electrons, but not more. The corresponding numbers of electrons that sublevels can hold are shown in the following chart. You should learn the maximum number of electrons for each sublevel; you'll need this information in Section 6-5.

Sublevel	Number of Orbitals × 2 =	Maximum Number of Electrons
s	1	2
p	3	6
d	5	10
f	7	14

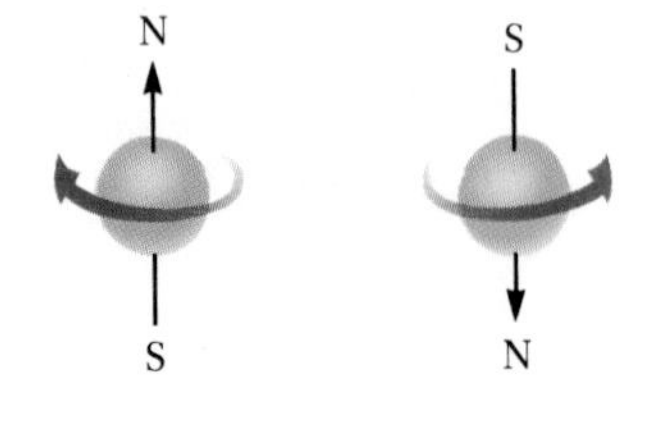

Figure 6.13
Spinning electrons. A spinning electron creates a magnetic field. The direction of the field depends on the direction in which the electron spins.

An orbital can hold a maximum of two electrons because of a property of electrons called **spin.** Each electron can be thought of as spinning on its axis, like the earth. The spinning electron creates a magnetic field, as shown in Figure 6.13, and the direction of the field depends on the direction of the spin. Quantum mechanical calculations show that the electrons in an orbital must have opposite spins; since only two spins are possible, an orbital can hold a maximum of two electrons.

6-4 The arrangement of the elements on the Periodic Table shows the order in which energy sublevels occur outward from the nucleus

We'll see in the next Section that the electrons in an atom occupy its energy sublevels in the order in which the sublevels occur outward from the nucleus. As shown in Figure 6.14, the order of the sublevels outward from the nucleus can be read from the structure of the Periodic Table.

Figure 6.14 shows that the sequence of sublevels on the Periodic Table has these regularities:

The representative (A-group) elements correspond to s and p sublevels. The s blocks are 2 elements long and the p blocks are 6 elements long.
The transition and inner-transition (B-group) elements correspond to d and f sublevels. The d (transition) blocks are 10 elements long and the f (inner-transition) blocks are 14 elements long.

Figure 6.14
Energy sublevels and the Periodic Table. To find the order in which energy sublevels occur outward from the nucleus, read left to right across each period: The order is 1s2s2p3s3p4s3d4p5s4d5p6s4f5d6p7s5f6d. As this sequence shows, some of the energy levels overlap one another. For example, the 4s sublevel is closer to the nucleus than the 3d sublevel, and the 6s sublevel is closer to the nucleus than either the 4f or the 5d sublevel.

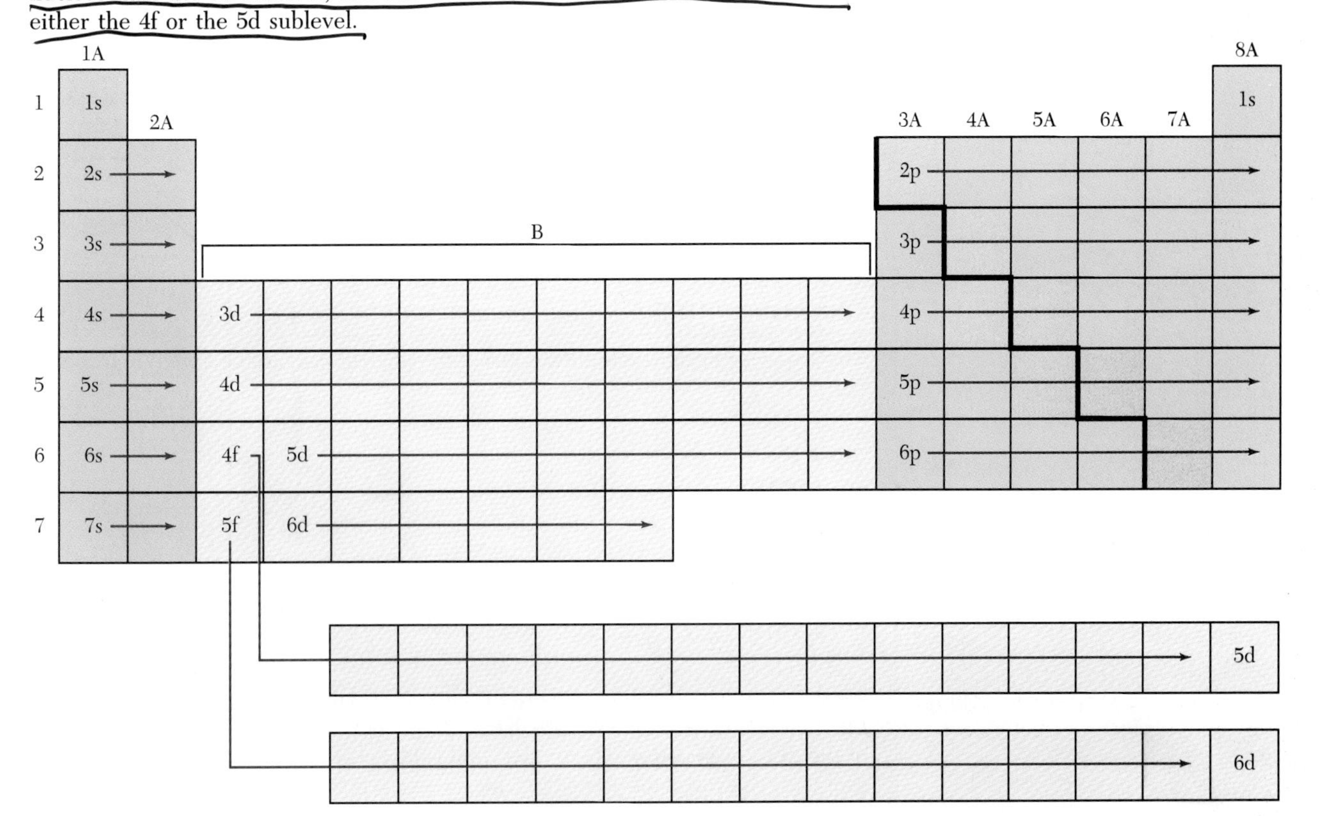

Before you leave this Section, learn the locations of the energy sublevels shown on the Periodic Table in Figure 6.14. This information will be fundamentally important for your later work in this book and in other chemistry courses.†

Exercise 6.6

Using only the Periodic Table at the front of the book, decide which energy sublevel corresponds to the position of each of the following elements: (a) hydrogen (b) sodium (c) rubidium (Rb) (d) neon (e) sulfur (f) iron (g) osmium (Os) (h) europium (Eu) (i) uranium (U).

6-5 Electrons occupy sublevels in the order in which the sublevels occur outward from the nucleus

The sequence of sublevels outward from the nucleus can be found from the structure of the Periodic Table, as described in Section 6-4. The sequence is 1s2s2p3s3p4s3d4p5s4d5p6s4f5d6p7s5f6d.

In any atom the electrons fill the sublevels in the order in which the sublevels occur outward from the nucleus.◆ For example, hydrogen, the first element on the Periodic Table, has atomic number 1, so a hydrogen atom has 1 electron, and that electron is in the sublevel closest to the nucleus, the 1s sublevel.

◆There are a few exceptions, as we'll see later, but they won't be important for our purposes.

The distribution of electrons in the sublevels of an atom is referred to as the atom's **electron configuration.** A system of notation, called spdf notation, is used to show electron configurations. The electron configuration for hydrogen is shown on the following page.

† Chemistry Insight

The sequence of sublevels outward from the nucleus can also be found from this diagram:

Start

1s

2s2p

3s3p3d

4s4p4d4f

5s5p5d5f

6s6p6d6f

7s7p7d7f

End

To use this diagram, begin at the right end of the top arrow and follow it downward and to the left; when you get to the left end of each arrow, go to the right end of the next lower arrow. The sequence of sublevels read in this way is the same sequence shown by the Periodic Table.

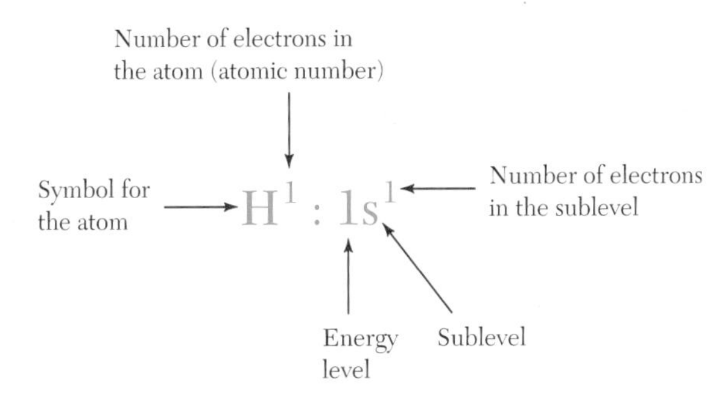

In spdf notation, the superscript on the symbol for the element is its atomic number, the number of electrons in one atom of the element. The number-and-letter designations for the sublevels are written in the order in which the sublevels occur outward from the nucleus, and the number of electrons present in a sublevel is shown as a superscript on the symbol for the sublevel. The notation above shows that the hydrogen atom contains one electron, and that electron is in the 1s sublevel.

Your goal in this Section will be to learn to use spdf notation to write the electron configuration for any element, using the Periodic Table at the front of the book.

Helium, the second element on the Periodic Table, has atomic number 2, so a helium atom has 2 electrons. Both electrons are in the 1s sublevel, and the electron configuration for helium is He^2: $1s^2$.

With 2 electrons the 1s sublevel is filled.◆ Lithium (the third element, atomic number 3) has 3 electrons; two of them fill the 1s sublevel, and the third occupies the next sublevel outward from the nucleus, the 2s sublevel. The electron configuration for lithium is Li^3: $1s^2 2s^1$.

◆The maximum number of electrons that can occupy an s, p, d, or f sublevel is 2, 6, 10, or 14, respectively; see Section 6-3.

This pattern continues through the entire Periodic Table: As each sublevel is filled, electrons occupy the next available sublevel outward from the nucleus. The electron configurations for all of the elements are shown in Table 6.2.

As shown on the right in Table 6.2, the electron configuration for an atom can be written in an abbreviated form. In this form, only the sublevels in the period in which the element occurs are shown, and all of the filled sublevels from earlier periods are represented by the symbol for the preceding noble gas. For example:

Complete electron configuration: P^{15}:$1s^2 2s^2 2p^6 3s^2 3p^3$

Abbreviated electron configuration: P^{15}:$[Ne]^{10} 3s^2 3p^3$

Filled sublevels through the second period

Sublevels across the third period

The electron configurations found from the positions of the elements on the Periodic Table are *predicted* electron configurations. There are laboratory methods for determining *actual* electron configurations, and, for most elements, the predicted and actual configurations are the same. The exceptions, shown in blue in Table 6.2, won't be important for our purposes.

Table 6.2 Electron Configurations for the Elements
The actual configuration, if it's different from the predicted configuration, is printed in blue

Complete Electron Configuration	Abbreviated Configuration
H^{1}: $1s^{1}$	
He^{2}: $1s^{2}$	
Li^{3}: $1s^{2}2s^{1}$	Li^{3}: $[He]^{2}2s^{1}$
Be^{4}: $1s^{2}2s^{2}$	Be^{4}: $[He]^{2}2s^{2}$
B^{5}: $1s^{2}2s^{2}2p^{1}$	B^{5}: $[He]^{2}2s^{2}2p^{1}$
C^{6}: $1s^{2}2s^{2}2p^{2}$	C^{6}: $[He]^{2}2s^{2}2p^{2}$
N^{7}: $1s^{2}2s^{2}2p^{3}$	N^{7}: $[He]^{2}2s^{2}2p^{3}$
O^{8}: $1s^{2}2s^{2}2p^{4}$	O^{8}: $[He]^{2}2s^{2}2p^{4}$
F^{9}: $1s^{2}2s^{2}2p^{5}$	F^{9}: $[He]^{2}2s^{2}2p^{5}$
Ne^{10}: $1s^{2}2s^{2}2p^{6}$	Ne^{10}: $[He]^{2}2s^{2}2p^{6}$
Na^{11}: $1s^{2}2s^{2}2p^{6}3s^{1}$	Na^{11}: $[Ne]^{10}3s^{1}$
Mg^{12}: $1s^{2}2s^{2}2p^{6}3s^{2}$	Mg^{12}: $[Ne]^{10}3s^{2}$
Al^{13}: $1s^{2}2s^{2}2p^{6}3s^{2}3p^{1}$	Al^{13}: $[Ne]^{10}3s^{2}3p^{1}$
Si^{14}: $1s^{2}2s^{2}2p^{6}3s^{2}3p^{2}$	Si^{14}: $[Ne]^{10}3s^{2}3p^{2}$
P^{15}: $1s^{2}2s^{2}2p^{6}3s^{2}3p^{3}$	P^{15}: $[Ne]^{10}3s^{2}3p^{3}$
S^{16}: $1s^{2}2s^{2}2p^{6}3s^{2}3p^{4}$	S^{16}: $[Ne]^{10}3s^{2}3p^{4}$
Cl^{17}: $1s^{2}2s^{2}2p^{6}3s^{2}3p^{5}$	Cl^{17}: $[Ne]^{10}3s^{2}3p^{5}$
Ar^{18}: $1s^{2}2s^{2}2p^{6}3s^{2}3p^{6}$	Ar^{18}: $[Ne]^{10}3s^{2}3p^{6}$
K^{19}: $1s^{2}2s^{2}2p^{6}3s^{2}3p^{6}4s^{1}$	K^{19}: $[Ar]^{18}4s^{1}$
Ca^{20}: $1s^{2}2s^{2}2p^{6}3s^{2}3p^{6}4s^{2}$	Ca^{20}: $[Ar]^{18}4s^{2}$
Sc^{21}: $1s^{2}2s^{2}2p^{6}3s^{2}3p^{6}4s^{2}3d^{1}$	Sc^{21}: $[Ar]^{18}4s^{2}3d^{1}$
Ti^{22}: $1s^{2}2s^{2}2p^{6}3s^{2}3p^{6}4s^{2}3d^{2}$	Ti^{22}: $[Ar]^{18}4s^{2}3d^{2}$
V^{23}: $1s^{2}2s^{2}2p^{6}3s^{2}3p^{6}4s^{2}3d^{3}$	V^{23}: $[Ar]^{18}4s^{2}3d^{3}$
Cr^{24}: $1s^{2}2s^{2}2p^{6}3s^{2}3p^{6}4s^{1}3d^{4}$	Cr^{24}: $[Ar]^{18}4s^{2}3d^{4}$
Cr^{24}: $1s^{2}2s^{2}2p^{6}3s^{2}3p^{6}4s^{2}3d^{5}$	Cr^{24}: $[Ar]^{18}4s^{1}3d^{5}$
Mn^{25}: $1s^{2}2s^{2}2p^{6}3s^{2}3p^{6}4s^{2}3d^{5}$	Mn^{25}: $[Ar]^{18}4s^{2}3d^{5}$
Fe^{26}: $1s^{2}2s^{2}2p^{6}3s^{2}3p^{6}4s^{2}3d^{6}$	Fe^{26}: $[Ar]^{18}4s^{2}3d^{6}$
Co^{27}: $1s^{2}2s^{2}2p^{6}3s^{2}3p^{6}4s^{2}3d^{7}$	Co^{27}: $[Ar]^{18}4s^{2}3d^{7}$
Ni^{28}: $1s^{2}2s^{2}2p^{6}3s^{2}3p^{6}4s^{2}3d^{8}$	Ni^{28}: $[Ar]^{18}4s^{2}3d^{8}$
Cu^{29}: $1s^{2}2s^{2}2p^{6}3s^{2}3p^{6}4s^{2}3d^{9}$	Cu^{29}: $[Ar]^{18}4s^{2}3d^{9}$
Cu^{29}: $1s^{2}2s^{2}2p^{6}3s^{2}3p^{6}4s^{1}3d^{10}$	Cu^{29}: $[Ar]^{18}4s^{1}3d^{10}$
Zn^{30}: $1s^{2}2s^{2}2p^{6}3s^{2}3p^{6}4s^{2}3d^{10}$	Zn^{30}: $[Ar]^{18}4s^{2}3d^{10}$
Ga^{31}: $1s^{2}2s^{2}2p^{6}3s^{2}3p^{6}4s^{2}3d^{10}4p^{1}$	Ga^{31}: $[Ar]^{18}4s^{2}3d^{10}4p^{1}$
Ge^{32}: $1s^{2}2s^{2}2p^{6}3s^{2}3p^{6}4s^{2}3d^{10}4p^{2}$	Ge^{32}: $[Ar]^{18}4s^{2}3d^{10}4p^{2}$
As^{33}: $1s^{2}2s^{2}2p^{6}3s^{2}3p^{6}4s^{2}3d^{10}4p^{3}$	As^{33}: $[Ar]^{18}4s^{2}3d^{10}4p^{3}$
Se^{34}: $1s^{2}2s^{2}2p^{6}3s^{2}3p^{6}4s^{2}3d^{10}4p^{4}$	Se^{34}: $[Ar]^{18}4s^{2}3d^{10}4p^{4}$
Br^{35}: $1s^{2}2s^{2}2p^{6}3s^{2}3p^{6}4s^{2}3d^{10}4p^{5}$	Br^{35}: $[Ar]^{18}4s^{2}3d^{10}4p^{5}$
Kr^{36}: $1s^{2}2s^{2}2p^{6}3s^{2}3p^{6}4s^{2}3d^{10}4p^{6}$	Kr^{36}: $[Ar]^{18}4s^{2}3d^{10}4p^{6}$
Rb^{37}: $1s^{2}2s^{2}2p^{6}3s^{2}3p^{6}4s^{2}3d^{10}4p^{6}5s^{1}$	Rb^{37}: $[Kr]^{36}5s^{1}$
Sr^{38}: $1s^{2}2s^{2}2p^{6}3s^{2}3p^{6}4s^{2}3d^{10}4p^{6}5s^{2}$	Sr^{38}: $[Kr]^{36}5s^{2}$

Table 6.2 *continued*

Complete Electron Configuration	Abbreviated Configuration
Y^{39}: $1s^22s^22p^63s^23p^64s^23d^{10}4p^65s^24d^1$	Y^{39}: $[Kr]^{36}5s^24d^1$
Zr^{40}: $1s^22s^22p^63s^23p^64s^23d^{10}4p^65s^24d^2$	Zr^{40}: $[Kr]^{36}5s^24d^2$
Nb^{41}: $1s^22s^22p^63s^23p^64s^23d^{10}4p^65s^24d^3$	Nb^{41}: $[Kr]^{36}5s^24d^3$
Nb^{41}: $1s^22s^22p^63s^23p^64s^23d^{10}4p^65s^14d^4$	Nb^{41}: $[Kr]^{36}5s^14d^4$
Mo^{42}: $1s^22s^22p^63s^23p^64s^23d^{10}4p^65s^24d^4$	Mo^{42}: $[Kr]^{36}5s^24d^4$
Mo^{42}: $1s^22s^22p^63s^23p^64s^23d^{10}4p^65s^14d^5$	Mo^{42}: $[Kr]^{36}5s^14d^5$
Tc^{43}: $1s^22s^22p^63s^23p^64s^23d^{10}4p^65s^24d^5$	Tc^{43}: $[Kr]^{36}5s^24d^5$
Ru^{44}: $1s^22s^22p^63s^23p^64s^23d^{10}4p^65s^24d^6$	Ru^{44}: $[Kr]^{36}5s^24d^6$
Ru^{44}: $1s^22s^22p^63s^23p^64s^23d^{10}4p^65s^14d^7$	Ru^{44}: $[Kr]^{36}5s^14d^7$
Rh^{45}: $1s^22s^22p^63s^23p^64s^23d^{10}4p^65s^24d^7$	Rh^{45}: $[Kr]^{36}5s^24d^7$
Rh^{45}: $1s^22s^22p^63s^23p^64s^23d^{10}4p^65s^14d^8$	Rh^{45}: $[Kr]^{36}5s^14d^8$
Pd^{46}: $1s^22s^22p^63s^23p^64s^23d^{10}4p^65s^24d^8$	Pd^{46}: $[Kr]^{36}5s^24d^8$
Pd^{46}: $1s^22s^22p^63s^23p^64s^23d^{10}4p^64d^{10}$	Pd^{46}: $[Kr]^{36}4d^{10}$
Ag^{47}: $1s^22s^22p^63s^23p^64s^23d^{10}4p^65s^24d^9$	Ag^{47}: $[Kr]^{36}5s^24d^9$
Ag^{47}: $1s^22s^22p^63s^23p^64s^23d^{10}4p^65s^14d^{10}$	Ag^{47}: $[Kr]^{36}5s^14d^{10}$
Cd^{48}: $1s^22s^22p^63s^23p^64s^23d^{10}4p^65s^24d^{10}$	Cd^{48}: $[Kr]^{36}5s^24d^{10}$
In^{49}: $1s^22s^22p^63s^23p^64s^23d^{10}4p^65s^24d^{10}5p^1$	In^{49}: $[Kr]^{36}5s^24d^{10}5p^1$
Sn^{50}: $1s^22s^22p^63s^23p^64s^23d^{10}4p^65s^24d^{10}5p^2$	Sn^{50}: $[Kr]^{36}5s^24d^{10}5p^2$
Sb^{51}: $1s^22s^22p^63s^23p^64s^23d^{10}4p^65s^24d^{10}5p^3$	Sb^{51}: $[Kr]^{36}5s^24d^{10}5p^3$
Te^{52}: $1s^22s^22p^63s^23p^64s^23d^{10}4p^65s^24d^{10}5p^4$	Te^{52}: $[Kr]^{36}5s^24d^{10}5p^4$
I^{53}: $1s^22s^22p^63s^23p^64s^23d^{10}4p^65s^24d^{10}5p^5$	I^{53}: $[Kr]^{36}5s^24d^{10}5p^5$
Xe^{54}: $1s^22s^22p^63s^23p^64s^23d^{10}4p^65s^24d^{10}5p^6$	Xe^{54}: $[Kr]^{36}5s^24d^{10}5p^6$
Cs^{55}: $1s^22s^22p^63s^23p^64s^23d^{10}4p^65s^24d^{10}5p^66s^1$	Cs^{55}: $[Xe]^{54}6s^1$
Ba^{56}: $1s^22s^22p^63s^23p^64s^23d^{10}4p^65s^24d^{10}5p^66s^2$	Ba^{56}: $[Xe]^{54}6s^2$
La^{57}: $1s^22s^22p^63s^23p^64s^23d^{10}4p^65s^24d^{10}5p^66s^24f^1$	La^{57}: $[Xe]^{54}6s^24f^1$
La^{57}: $1s^22s^22p^63s^23p^64s^23d^{10}4p^65s^24d^{10}5p^66s^25d^1$	La^{57}: $[Xe]^{54}6s^25d^1$
Ce^{58}: $1s^22s^22p^63s^23p^64s^23d^{10}4p^65s^24d^{10}5p^66s^24f^2$	Ce^{58}: $[Xe]^{54}6s^24f^2$
Ce^{58}: $1s^22s^22p^63s^23p^64s^23d^{10}4p^65s^24d^{10}5p^66s^24f^15d^1$	Ce^{58}: $[Xe]^{54}6s^24f^15d^1$
Pr^{59}: $1s^22s^22p^63s^23p^64s^23d^{10}4p^65s^24d^{10}5p^66s^24f^3$	Pr^{59}: $[Xe]^{54}6s^24f^3$
Nd^{60}: $1s^22s^22p^63s^23p^64s^23d^{10}4p^65s^24d^{10}5p^66s^24f^4$	Nd^{60}: $[Xe]^{54}6s^24f^4$
Pm^{61}: $1s^22s^22p^63s^23p^64s^23d^{10}4p^65s^24d^{10}5p^66s^24f^5$	Pm^{61}: $[Xe]^{54}6s^24f^5$
Sm^{62}: $1s^22s^22p^63s^23p^64s^23d^{10}4p^65s^24d^{10}5p^66s^24f^6$	Sm^{62}: $[Xe]^{54}6s^24f^6$
Eu^{63}: $1s^22s^22p^63s^23p^64s^23d^{10}4p^65s^24d^{10}5p^66s^24f^7$	Eu^{63}: $[Xe]^{54}6s^24f^7$
Gd^{64}: $1s^22s^22p^63s^23p^64s^23d^{10}4p^65s^24d^{10}5p^66s^24f^8$	Gd^{64}: $[Xe]^{54}6s^24f^8$
Gd^{64}: $1s^22s^22p^63s^23p^64s^23d^{10}4p^65s^24d^{10}5p^66s^24f^75d^1$	Gd^{64}: $[Xe]^{54}6s^24f^75d^1$
Tb^{65}: $1s^22s^22p^63s^23p^64s^23d^{10}4p^65s^24d^{10}5p^66s^24f^9$	Tb^{65}: $[Xe]^{54}6s^24f^9$
Dy^{66}: $1s^22s^22p^63s^23p^64s^23d^{10}4p^65s^24d^{10}5p^66s^24f^{10}$	Dy^{66}: $[Xe]^{54}6s^24f^{10}$
Ho^{67}: $1s^22s^22p^63s^23p^64s^23d^{10}4p^65s^24d^{10}5p^66s^24f^{11}$	Ho^{67}: $[Xe]^{54}6s^24f^{11}$
Er^{68}: $1s^22s^22p^63s^23p^64s^23d^{10}4p^65s^24d^{10}5p^66s^24f^{12}$	Er^{68}: $[Xe]^{54}6s^24f^{12}$

Table 6.2 *continued*

Complete Electron Configuration	Abbreviated Configuration
Tm^{69}: $1s^22s^22p^63s^23p^64s^23d^{10}4p^65s^24d^{10}5p^66s^24f^{13}$	Tm^{69}: $[Xe]^{54}6s^24f^{13}$
Yb^{70}: $1s^22s^22p^63s^23p^64s^23d^{10}4p^65s^24d^{10}5p^66s^24f^{14}$	Yb^{70}: $[Xe]^{54}6s^24f^{14}$
Lu^{71}: $1s^22s^22p^63s^23p^64s^23d^{10}4p^65s^24d^{10}5p^66s^24f^{14}5d^1$	Lu^{71}: $[Xe]^{54}6s^24f^{14}5d^1$
Hf^{72}: $1s^22s^22p^63s^23p^64s^23d^{10}4p^65s^24d^{10}5p^66s^24f^{14}5d^2$	Hf^{72}: $[Xe]^{54}6s^24f^{14}5d^2$
Ta^{73}: $1s^22s^22p^63s^23p^64s^23d^{10}4p^65s^24d^{10}5p^66s^24f^{14}5d^3$	Ta^{73}: $[Xe]^{54}6s^24f^{14}5d^3$
W^{74}: $1s^22s^22p^63s^23p^64s^23d^{10}4p^65s^24d^{10}5p^66s^24f^{14}5d^4$	W^{74}: $[Xe]^{54}6s^24f^{14}5d^4$
Re^{75}: $1s^22s^22p^63s^23p^64s^23d^{10}4p^65s^24d^{10}5p^66s^24f^{14}5d^5$	Re^{75}: $[Xe]^{54}6s^24f^{14}5d^5$
Os^{76}: $1s^22s^22p^63s^23p^64s^23d^{10}4p^65s^24d^{10}5p^66s^24f^{14}5d^6$	Os^{76}: $[Xe]^{54}6s^24f^{14}5d^6$
Ir^{77}: $1s^22s^22p^63s^23p^64s^23d^{10}4p^65s^24d^{10}5p^66s^24f^{14}5d^7$	Ir^{77}: $[Xe]^{54}6s^24f^{14}5d^7$
Pt^{78}: $1s^22s^22p^63s^23p^64s^23d^{10}4p^65s^24d^{10}5p^66s^24f^{14}5d^8$	Pt^{78}: $[Xe]^{54}6s^24f^{14}5d^8$
Pt^{78}: $1s^22s^22p^63s^23p^64s^23d^{10}4p^65s^24d^{10}5p^66s^14f^{14}5d^9$	Pt^{78}: $[Xe]^{54}6s^14f^{14}5d^9$
Au^{79}: $1s^22s^22p^63s^23p^64s^23d^{10}4p^65s^24d^{10}5p^66s^24f^{14}5d^9$	Au^{79}: $[Xe]^{54}6s^24f^{14}5d^9$
Au^{79}: $1s^22s^22p^63s^23p^64s^23d^{10}4p^65s^24d^{10}5p^66s^14f^{14}5d^{10}$	Au^{79}: $[Xe]^{54}6s^14f^{14}5d^{10}$
Hg^{80}: $1s^22s^22p^63s^23p^64s^23d^{10}4p^65s^24d^{10}5p^66s^24f^{14}5d^{10}$	Hg^{80}: $[Xe]^{54}6s^24f^{14}5d^{10}$
Tl^{81}: $1s^22s^22p^63s^23p^64s^23d^{10}4p^65s^24d^{10}5p^66s^24f^{14}5d^{10}6p^1$	Tl^{81}: $[Xe]^{54}6s^24f^{14}5d^{10}6p^1$
Pb^{82}: $1s^22s^22p^63s^23p^64s^23d^{10}4p^65s^24d^{10}5p^66s^24f^{14}5d^{10}6p^2$	Pb^{82}: $[Xe]^{54}6s^24f^{14}5d^{10}6p^2$
Bi^{83}: $1s^22s^22p^63s^23p^64s^23d^{10}4p^65s^24d^{10}5p^66s^24f^{14}5d^{10}6p^3$	Bi^{83}: $[Xe]^{54}6s^24f^{14}5d^{10}6p^3$
Po^{84}: $1s^22s^22p^63s^23p^64s^23d^{10}4p^65s^24d^{10}5p^66s^24f^{14}5d^{10}6p^4$	Po^{84}: $[Xe]^{54}6s^24f^{14}5d^{10}6p^4$
At^{85}: $1s^22s^22p^63s^23p^64s^23d^{10}4p^65s^24d^{10}5p^66s^24f^{14}5d^{10}6p^5$	At^{85}: $[Xe]^{54}6s^24f^{14}5d^{10}6p^5$
Rn^{86}: $1s^22s^22p^63s^23p^64s^23d^{10}4p^65s^24d^{10}5p^66s^24f^{14}5d^{10}6p^6$	Rn^{86}: $[Xe]^{54}6s^24f^{14}5d^{10}6p^6$
Fr^{87}: $1s^22s^22p^63s^23p^64s^23d^{10}4p^65s^24d^{10}5p^66s^24f^{14}5d^{10}6p^6 7s^1$	Fr^{87}: $[Rn]^{86}7s^1$
Ra^{88}: $1s^22s^22p^63s^23p^64s^23d^{10}4p^65s^24d^{10}5p^66s^24f^{14}5d^{10}6p^6 7s^2$	Ra^{88}: $[Rn]^{86}7s^2$
Ac^{89}: $1s^22s^22p^63s^23p^64s^23d^{10}4p^65s^24d^{10}5p^66s^24f^{14}5d^{10}6p^6 7s^25f^1$	Ac^{89}: $[Rn]^{86}7s^25f^1$
Ac^{89}: $1s^22s^22p^63s^23p^64s^23d^{10}4p^65s^24d^{10}5p^66s^24f^{14}5d^{10}6p^6 7s^26d^1$	Ac^{89}: $[Rn]^{86}7s^26d^1$
Th^{90}: $1s^22s^22p^63s^23p^64s^23d^{10}4p^65s^24d^{10}5p^66s^24f^{14}5d^{10}6p^6 7s^25f^2$	Th^{90}: $[Rn]^{86}7s^25f^2$
Th^{90}: $1s^22s^22p^63s^23p^64s^23d^{10}4p^65s^24d^{10}5p^66s^24f^{14}5d^{10}6p^6 7s^26d^2$	Th^{90}: $[Rn]^{86}7s^26d^2$
Pa^{91}: $1s^22s^22p^63s^23p^64s^23d^{10}4p^65s^24d^{10}5p^66s^24f^{14}5d^{10}6p^6 7s^25f^3$	Pa^{91}: $[Rn]^{86}7s^25f^3$
Pa^{91}: $1s^22s^22p^63s^23p^64s^23d^{10}4p^65s^24d^{10}5p^66s^24f^{14}5d^{10}6p^6 7s^25f^26d^1$	Pa^{91}: $[Rn]^{86}7s^25f^26d^1$
U^{92}: $1s^22s^22p^63s^23p^64s^23d^{10}4p^65s^24d^{10}5p^66s^24f^{14}5d^{10}6p^6 7s^25f^4$	U^{92}: $[Rn]^{86}7s^25f^4$

Table 6.2 *continued*

Complete Electron Configuration	Abbreviated Configuration
U^{92}: $1s^2 2s^2 2p^6 3s^2 3p^6 4s^2 3d^{10} 4p^6 5s^2 4d^{10} 5p^6 6s^2 4f^{14} 5d^{10} 6p^6 7s^2 5f^3 6d^1$	U^{92}: $[Rn]^{86} 7s^2 5f^3 6d^1$
Np^{93}: $1s^2 2s^2 2p^6 3s^2 3p^6 4s^2 3d^{10} 4p^6 5s^2 4d^{10} 5p^6 6s^2 4f^{14} 5d^{10} 6p^6 7s^2 5f^5$	Np^{93}: $[Rn]^{86} 7s^2 5f^5$
Np^{93}: $1s^2 2s^2 2p^6 3s^2 3p^6 4s^2 3d^{10} 4p^6 5s^2 4d^{10} 5p^6 6s^2 4f^{14} 5d^{10} 6p^6 7s^2 5f^4 6d^1$	Np^{93}: $[Rn]^{86} 7s^2 5f^4 6d^1$
Pu^{94}: $1s^2 2s^2 2p^6 3s^2 3p^6 4s^2 3d^{10} 4p^6 5s^2 4d^{10} 5p^6 6s^2 4f^{14} 5d^{10} 6p^6 7s^2 5f^6$	Pu^{94}: $[Rn]^{86} 7s^2 5f^6$
Am^{95}: $1s^2 2s^2 2p^6 3s^2 3p^6 4s^2 3d^{10} 4p^6 5s^2 4d^{10} 5p^6 6s^2 4f^{14} 5d^{10} 6p^6 7s^2 5f^7$	Am^{95}: $[Rn]^{86} 7s^2 5f^7$
Cm^{96}: $1s^2 2s^2 2p^6 3s^2 3p^6 4s^2 3d^{10} 4p^6 5s^2 4d^{10} 5p^6 6s^2 4f^{14} 5d^{10} 6p^6 7s^2 5f^8$	Cm^{96}: $[Rn]^{86} 7s^2 5f^8$
Cm^{96}: $1s^2 2s^2 2p^6 3s^2 3p^6 4s^2 3d^{10} 4p^6 5s^2 4d^{10} 5p^6 6s^2 4f^{14} 5d^{10} 6p^6 7s^2 5f^7 6d^1$	Cm^{96}: $[Rn]^{86} 7s^2 5f^7 6d^1$
Bk^{97}: $1s^2 2s^2 2p^6 3s^2 3p^6 4s^2 3d^{10} 4p^6 5s^2 4d^{10} 5p^6 6s^2 4f^{14} 5d^{10} 6p^6 7s^2 5f^9$	Bk^{97}: $[Rn]^{86} 7s^2 5f^9$
Cf^{98}: $1s^2 2s^2 2p^6 3s^2 3p^6 4s^2 3d^{10} 4p^6 5s^2 4d^{10} 5p^6 6s^2 4f^{14} 5d^{10} 6p^6 7s^2 5f^{10}$	Cf^{98}: $[Rn]^{86} 7s^2 5f^{10}$
Es^{99}: $1s^2 2s^2 2p^6 3s^2 3p^6 4s^2 3d^{10} 4p^6 5s^2 4d^{10} 5p^6 6s^2 4f^{14} 5d^{10} 6p^6 7s^2 5f^{11}$	Es^{99}: $[Rn]^{86} 7s^2 5f^{11}$
Fm^{100}: $1s^2 2s^2 2p^6 3s^2 3p^6 4s^2 3d^{10} 4p^6 5s^2 4d^{10} 5p^6 6s^2 4f^{14} 5d^{10} 6p^6 7s^2 5f^{12}$	Fm^{100}: $[Rn]^{86} 7s^2 5f^{12}$
Md^{101}: $1s^2 2s^2 2p^6 3s^2 3p^6 4s^2 3d^{10} 4p^6 5s^2 4d^{10} 5p^6 6s^2 4f^{14} 5d^{10} 6p^6 7s^2 5f^{13}$	Md^{101}: $[Rn]^{86} 7s^2 5f^{13}$
No^{102}: $1s^2 2s^2 2p^6 3s^2 3p^6 4s^2 3d^{10} 4p^6 5s^2 4d^{10} 5p^6 6s^2 4f^{14} 5d^{10} 6p^6 7s^2 5f^{14}$	No^{102}: $[Rn]^{86} 7s^2 5f^{14}$
Lr^{103}: $1s^2 2s^2 2p^6 3s^2 3p^6 4s^2 3d^{10} 4p^6 5s^2 4d^{10} 5p^6 6s^2 4f^{14} 5d^{10} 6p^6 7s^2 5f^{14} 6d^1$	Lr^{103}: $[Rn]^{86} 7s^2 5f^{14} 6d^1$
Rf^{104}: $1s^2 2s^2 2p^6 3s^2 3p^6 4s^2 3d^{10} 4p^6 5s^2 4d^{10} 5p^6 6s^2 4f^{14} 5d^{10} 6p^6 7s^2 5f^{14} 6d^2$	Rf^{104}: $[Rn]^{86} 7s^2 5f^{14} 6d^2$
Ha^{105}: $1s^2 2s^2 2p^6 3s^2 3p^6 4s^2 3d^{10} 4p^6 5s^2 4d^{10} 5p^6 6s^2 4f^{14} 5d^{10} 6p^6 7s^2 5f^{14} 6d^3$	Ha^{105}: $[Rn]^{86} 7s^2 5f^{14} 6d^3$
Unh^{106}: $1s^2 2s^2 2p^6 3s^2 3p^6 4s^2 3d^{10} 4p^6 5s^2 4d^{10} 5p^6 6s^2 4f^{14} 5d^{10} 6p^6 7s^2 5f^{14} 6d^4$	Unh^{106}: $[Rn]^{86} 7s^2 5f^{14} 6d^4$
Uns^{107}: $1s^2 2s^2 2p^6 3s^2 3p^6 4s^2 3d^{10} 4p^6 5s^2 4d^{10} 5p^6 6s^2 4f^{14} 5d^{10} 6p^6 7s^2 5f^{14} 6d^5$	Uns^{107}: $[Rn]^{86} 7s^2 5f^{14} 6d^5$
Uno^{108}: $1s^2 2s^2 2p^6 3s^2 3p^6 4s^2 3d^{10} 4p^6 5s^2 4d^{10} 5p^6 6s^2 4f^{14} 5d^{10} 6p^6 7s^2 5f^{14} 6d^6$	Uno^{108}: $[Rn]^{86} 7s^2 5f^{14} 6d^6$
Une^{109}: $1s^2 2s^2 2p^6 3s^2 3p^6 4s^2 3d^{10} 4p^6 5s^2 4d^{10} 5p^6 6s^2 4f^{14} 5d^{10} 6p^6 7s^2 5f^{14} 6d^7$	Une^{109}: $[Rn]^{86} 7s^2 5f^{14} 6d^7$

Table 6.3 shows the steps for finding the complete electron configuration for an atom of any element, using a Periodic Table.

Table 6.3 Steps for Finding the Electron Configuration for an Atom of any Element

Example: Write the complete electron configuration for bromine.

Step	Example
1. Find the element on the Periodic Table, and identify the number of electrons in one atom.	Bromine, Br, is the third element down in group 7A. Its atomic number is 35, so one atom has 35 electrons.
2. Identify the sublevel being filled.	Bromine is in the block of elements from Ga to Kr, in which electrons are occupying the 4p sublevel.
3. Find the number of electrons in that sublevel.	From Ga to Br there are five elements. Each element adds one electron to the 4p sublevel, so bromine is $4p^5$.
4. Write the complete electron configuration. Fill each sublevel with electrons until you get to the sublevel you identified in step 2.	Br^{35}: $1s^2 2s^2 2p^6 3s^2 3p^6 4s^2 3d^{10} 4p^5$
5. Check to be sure that the total number of electrons shown in sublevels is the same as the atomic number.	There are 35 electrons shown in sublevels, and the atomic number is 35.

Example 6.2

Write the complete electron configuration for (a) sulfur, (b) vanadium, V, and (c) erbium, Er.

Solution

(a) Sulfur, atomic number 16, has 16 electrons and is the fourth element in the block that is filling the 3p sublevel. S^{16}: $1s^2 2s^2 2p^6 3s^2 3p^4$.

(b) Vanadium, atomic number 23, has 23 electrons and is the third element in the block that is filling the 3d sublevel.
V^{23}: $1s^2 2s^2 2p^6 3s^2 3p^6 4s^2 3d^3$.

(c) Erbium, atomic number 68, has 68 electrons and is the twelfth element in the block that is filling the 4f sublevel.
Er^{68}: $1s^2 2s^2 2p^6 3s^2 3p^6 4s^2 3d^{10} 4p^6 5s^2 4d^{10} 5p^6 6s^2 4f^{12}$.

Exercise 6.7

Write the complete electron configuration for (a) oxygen, (b) zinc, and (c) americium, Am.

Example 6.3

Write the abbreviated electron configuration for iodine.

Solution

Iodine, atomic number 53, has 53 electrons and is the fifth element in the block that is filling the 5p sublevel. The preceding noble gas, Kr, atomic number 36, ends the fourth period, so we can use $[Kr]^{36}$ to stand for all of the filled sublevels through the fourth period.
I^{53}: $[Kr]^{36}5s^2 4d^{10} 5p^5$.

Exercise 6.8

Write the abbreviated electron configuration for barium, Ba.

The electron configurations that are found by using the Periodic Table show all of the electrons in an atom in the sublevels nearest to the nucleus. An atom with this configuration is said to be in the **ground state.** As we saw in Section 6-1, electrons in atoms can absorb energy and move from lower to higher energy levels. An atom in which the electrons are in some configuration other than the ground state is said to be in an **excited state.**

The electrons in a ground-state atom have the lowest possible potential energy because they're in the nearest available positions to the nucleus. When an atom changes from the ground state to an excited state, energy must be absorbed, and when it changes from an excited state to the ground state, energy must be emitted.

Example 6.4

Decide whether each of the following atoms is in the ground state or in an excited state: (a) N^7: $1s^2 2s^1 2p^4$ (b) Na^{11}: $1s^2 2s^2 2p^6 3s^1$

Solution

(a) The predicted (ground-state) electron configuration for nitrogen is N^7: $1s^2 2s^2 2p^3$. The configuration shown in the problem is different from the ground-state configuration (one electron has moved from the 2s sublevel to the 2p sublevel), so the atom is in an excited state.

(b) The predicted (ground-state) electron configuration for sodium is Na^{11}: $1s^2 2s^2 2p^6 3s^1$. This is the same configuration as that shown in the problem, so the atom is in the ground state.

Exercise 6.9

Decide whether each of the following atoms is in the ground state or in an excited state: (a) F^9: $1s^2 2s^2 2p^5$ (b) C^6: $1s^2 2s^1 2p^3$

6-6 An orbital diagram shows the distribution of electrons among the orbitals in a sublevel

An electron configuration in spdf notation shows the number of electrons in each sublevel, for example

$$C^6: 1s^2 2s^2 2p^2$$

It's sometimes useful to show how the electrons are distributed in the orbitals within sublevels, and this can be done by means of an **orbital diagram:** Each orbital is represented by a circle and each electron by a vertical arrow, as shown below.

electron configuration: C^6: $1s^2 2s^2 2p^2$
orbital diagram: C^6: $1s^2$ (↑↓) $2s^2$ (↑↓) $2p^2$ (↑)(↑)( )

The 1s and 2s sublevels each consist of one orbital, and the 2p sublevel consists of three orbitals. Each orbital can hold a maximum of two electrons, and they must have opposite spins; the two possible spin directions are symbolized by pointing the arrows up or down. Quantum mechanics specifies that electrons entering a sublevel will occupy all of the orbitals with one electron each before pairing electrons in an orbital. Following this requirement, the two electrons in the 2p sublevel occupy separate orbitals, as shown above. The orbital diagram shows that there are four paired and two unpaired electrons in the carbon atom.◆

◆The orbitals can be represented by lines instead of circles. Using this notation, the orbital diagram for carbon is
C^6: $1s^2$ ↑↓ $2s^2$ ↑↓ $2p^2$ ↑ ↑ __

Example 6.5

Write the electron configurations and orbital diagrams for nitrogen and oxygen.

Solution

N^7: $1s^2 2s^2 2p^3$

N^7: $1s^2$ (↑↓) $2s^2$ (↑↓) $2p^3$ (↑)(↑)(↑)

O^8: $1s^2 2s^2 2p^4$

O^8: $1s^2$ (↑↓) $2s^2$ (↑↓) $2p^4$ (↑↓)(↑)(↑)

Exercise 6.10

Write the electron configurations and orbital diagrams for fluorine and neon.

Example 6.6

(a) Write the electron configuration for nickel, Ni. (b) Write the orbital diagram for the 3d sublevel. (c) How many paired and unpaired electrons are in a nickel atom?

Solution

(a) Ni^{28}: $1s^2 2s^2 2p^6 3s^2 3p^6 4s^2 3d^8$

(b) The 3d sublevel consists of five orbitals, and they contain 8 electrons:

$3d^8$

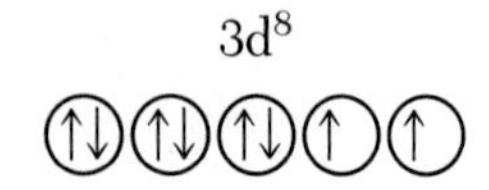

(c) All of the sublevels from the 1s through the 4s are filled with electrons, so all of those electrons are paired. The orbital diagram for the 3d sublevel shows that it contains 2 unpaired electrons, so the nickel atom contains 2 unpaired and 28 − 2 = 26 paired electrons.

Exercise 6.11

(a) Write the electron configuration for manganese, Mn. (b) Write the orbital diagram for the 3d sublevel. (c) How many paired and unpaired electrons are in a manganese atom?

6-7 In the electron configuration for a representative element, the electrons in the s and p sublevels of the outermost energy level have a special significance and are identified as valence electrons

Electron configurations are important, as we'll see in Chapters 7 through 10, because they provide the information needed to explain how atoms combine with one another.

In this book we'll work most often with the representative (A-group) elements. The electron configurations for the representative elements show complete regularity across a period and within a group. From left to right across a period, electrons in the atoms of the representative elements occupy first the s and then the p sublevels of the same energy level, and the number of the energy level is the same as the number of the period. For example, these are the abbreviated electron configurations for the representative elements across the fourth period:

K^{19}: $[Ar]^{18}4s^1$
Ca^{20}: $[Ar]^{18}4s^2$
(3d transition elements occur here)
Ga^{31}: $[Ar]^{18}4s^23d^{10}4p^1$
Ge^{32}: $[Ar]^{18}4s^23d^{10}4p^2$
As^{33}: $[Ar]^{18}4s^23d^{10}4p^3$
Se^{34}: $[Ar]^{18}4s^23d^{10}4p^4$
Br^{35}: $[Ar]^{18}4s^23d^{10}4p^5$
Kr^{36}: $[Ar]^{18}4s^23d^{10}4p^6$

Downward in a group each representative element has the same arrangement of electrons in s and p sublevels, at successively higher energy levels. For example, these are the abbreviated configurations for groups 1A and 6A:

Period	Group 1A	Group 6A
1	H^1: $1s^1$	
2	Li^3: $[He]^24s^1$	O^8: $[He]^22s^22p^4$
3	Na^{11}: $[Ne]^{10}3s^1$	S^{16}: $[Ne]^{10}3s^23p^4$
4	K^{19}: $[Ar]^{18}4s^1$	Se^{34}: $[Ar]^{18}4s^23d^{10}4p^4$
5	Rb^{37}: $[Kr]^{36}5s^1$	Te^{52}: $[Kr]^{36}5s^24d^{10}5p^4$
6	Cs^{55}: $[Xe]^{54}6s^1$	Po^{84}: $[Xe]^{54}6s^24f^{14}5d^{10}6p^4$
7	Fr^{87}: $[Rn]^{86}7s^1$	

Because of these regularities we can identify for each group of representative elements a **characteristic group electron configuration,** which shows the arrangement of electrons in the s and p sublevels of the outermost energy level:

Group	1A	2A	3A	4A	5A	6A	7A	8A
Characteristic group electron configuration	s^1	s^2	s^2p^1	s^2p^2	s^2p^3	s^2p^4	s^2p^5	s^2p^6

These group configurations give important information, so you should remember them.

The group configurations are important because they identify the electrons—called **valence electrons**—that determine how atoms of representative elements will combine with one another. In an atom of a representative element, the valence electrons are the electrons in the s and p sublevels of the outermost energy level.

To identify the valence electrons of a representative element quickly, find the element on the Periodic Table and combine its period number with its characteristic group electron configuration. For example, sulfur is in period 3 and in group 6A (characteristic configuration s^2p^4), so its valence electrons are $3s^23p^4$.

The characteristic group configurations and the valence electron configurations for all of the representative elements are shown in Figure 6.15. The valence electron configuration for helium ($1s^2$) doesn't fit the configuration for group 8A (s^2p^6), because the first energy level has no p sublevel.

Example 6.7

The abbreviated electron configuration for iodine is I^{53}: $[Kr]^{36}5s^24d^{10}5p^5$. Identify the valence electrons.

Figure 6.15
Valence electron configurations for the representative elements. Characteristic group configurations are shown in blue.

Period	1A s^1	2A s^2	3A s^2p^1	4A s^2p^2	5A s^2p^3	6A s^2p^4	7A s^2p^5	8A s^2p^6
1	H $1s^1$							He $1s^2$
2	Li $2s^1$	Be $2s^2$	B $2s^22p^1$	C $2s^22p^2$	N $2s^22p^3$	O $2s^22p^4$	F $2s^22p^5$	Ne $2s^22p^6$
3	Na $3s^1$	Mg $3s^2$	Al $3s^23p^1$	Si $3s^23p^2$	P $3s^23p^3$	S $3s^23p^4$	Cl $3s^23p^5$	Ar $3s^23p^6$
4	K $4s^1$	Ca $4s^2$	Ga $4s^24p^1$	Ge $4s^24p^2$	As $4s^24p^3$	Se $4s^24p^4$	Br $4s^24p^5$	Kr $4s^24p^6$
5	Rb $5s^1$	Sr $5s^2$	In $5s^25p^1$	Sn $5s^25p^2$	Sb $5s^25p^3$	Te $5s^25p^4$	I $5s^25p^5$	Xe $5s^25p^6$
6	Cs $6s^1$	Ba $6s^2$	Tl $6s^26p^1$	Pb $6s^26p^2$	Bi $6s^26p^3$	Po $6s^26p^4$	At $6s^26p^5$	Rn $6s^26p^6$
7	Fr $7s^1$	Ra $7s^2$						

Solution

This problem can be solved in two ways: (1) For a representative element the valence electrons are the electrons in the s and p sublevels of the outermost energy level. In the electron configuration given in the problem, these are the $5s^2 5p^5$ electrons. (2) Iodine is in period 5 and in group 7A (characteristic configuration $s^2 p^5$), so its valence electrons are $5s^2 5p^5$.

Exercise 6.12

The abbreviated electron configuration for arsenic is As^{33}: $[Ar]^{18} 4s^2 3d^{10} 4p^3$. Identify the valence electrons.

Example 6.8

Identify the valence electrons for magnesium.

Solution

This problem can be solved in two ways. (1) Write the abbreviated electron configuration for magnesium: Mg^{12}: $[Ne]^{10} 3s^2$. The valence electrons are the electrons in the s and p sublevels of the outermost energy level, the $3s^2$ electrons. (2) Magnesium is in period 3 and in group 2A (characteristic configuration s^2), so its valence electrons are $3s^2$.

Exercise 6.13

Identify the valence electrons for tin, Sn.

6-8 A Lewis symbol represents an atom of a representative element by its symbol, with a dot for each valence electron

The electron configuration for an atom of a representative element can often be shown conveniently in a very abbreviated form, which we'll use in this book, known as a Lewis symbol. ◆ A **Lewis symbol** represents an atom of a representative element by its chemical symbol, with a dot for each valence electron. Lewis symbols, also called **electron-dot symbols,** are used only for representative elements, not for transition elements.

◆ Named for Professor Gilbert N. Lewis (1875–1946) of the University of California at Berkeley, who made many important contributions to chemistry.

Table 6.4 shows three different ways to designate the electron configurations of the elements in the second period: in spdf notation, in orbital diagrams, and in Lewis symbols. The valence electrons for these elements are the electrons in the 2s and 2p sublevels. Valence electrons that are paired in orbitals are shown as paired dots in Lewis symbols. A nitrogen atom, for example, has two paired valence electrons ($2s^2$) and three unpaired valence electrons ($2p^3$), so its Lewis symbol is $\cdot\underset{\cdot}{\ddot{N}}\cdot$.

In writing a Lewis symbol, the grouping of the dots that represent electrons is significant, but their position and sequence around the symbol for the element are not. In the Lewis symbol for an oxygen atom, for example, the dots are grouped to show four paired and two unpaired electrons, but they can be arranged in any position or sequence: $:\underset{\cdot}{\dot{O}}:$, $\cdot\underset{\cdot\cdot}{\ddot{O}}\cdot$, $:\underset{\cdot}{\ddot{O}}\cdot$, $\cdot\underset{\cdot\cdot}{\dot{O}}:$, and so forth.

Table 6.4 Electron Configurations of Elements in the Second Period
In each spdf notation and orbital diagram, the valence electrons are shown in red.

spdf Notation	Orbital Diagram	Lewis Symbol
Li^3: $1s^2 2s^1$	Li^3: $1s^2$ (↑↓) $2s^1$ (↑)	Li·
Be^4: $1s^2 2s^2$	Be^4: $1s^2$ (↑↓) $2s^2$ (↑↓)	Be:
B^5: $1s^2 2s^2 2p^1$	B^5: $1s^2$ (↑↓) $2s^2$ (↑↓) $2p^1$ (↑)( )( )	·B:
C^6: $1s^2 2s^2 2p^2$	C^6: $1s^2$ (↑↓) $2s^2$ (↑↓) $2p^2$ (↑)(↑)( )	·Ċ:
N^7: $1s^2 2s^2 2p^3$	N^7: $1s^2$ (↑↓) $2s^2$ (↑↓) $2p^3$ (↑)(↑)(↑)	·N̈· (one dot below)
O^8: $1s^2 2s^2 2p^4$	O^8: $1s^2$ (↑↓) $2s^2$ (↑↓) $2p^4$ (↑↓)(↑)(↑)	·Ö· (two dots below)
F^9: $1s^2 2s^2 2p^5$	F^9: $1s^2$ (↑↓) $2s^2$ (↑↓) $2p^5$ (↑↓)(↑↓)(↑)	:F̈· (two dots below)
Ne^{10}: $1s^2 2s^2 2p^6$	Ne^{10}: $1s^2$ (↑↓) $2s^2$ (↑↓) $2p^6$ (↑↓)(↑↓)(↑↓)	:N̈e: (two dots below)

The arrangements shown in Table 6.4 above are the ones that most chemists customarily use.

Because each group of representative elements has a characteristic configuration of valence electrons, it also has a characteristic Lewis symbol, as shown below. You'll use these characteristic Lewis symbols often, so you should learn them.◆

◆The Lewis symbol for helium doesn't match the characteristic group symbol for group 8A. The first period has no p sublevel, helium is He^2: $1s^2$, and its Lewis symbol is He:.

Group	1A	2A	3A	4A	5A	6A	7A	8A
Characteristic group electron configuration	s^1	s^2	s^2p^1	s^2p^2	s^2p^3	s^2p^4	s^2p^5	s^2p^6
Characteristic group Lewis symbol	Z·	Z:	·Z:	·Ż:	·Z̈· (one dot below)	·Z̈· (two dots below)	:Z̈· (two dots below)	:Z̈: (two dots below)

Example 6.9

Use the Periodic Table at the front of the book to write Lewis symbols for (a) sodium (b) arsenic, As (c) chlorine.

Solution

(a) Sodium is in group 1A, characteristic Lewis symbol Z·, so its Lewis symbol is Na·.

(b) Arsenic is in group 5A, characteristic Lewis symbol $\cdot\overset{\cdot\cdot}{\underset{\cdot}{\mathrm{Z}}}\cdot$, so its Lewis symbol is $\cdot\overset{\cdot\cdot}{\underset{\cdot}{\mathrm{As}}}\cdot$.

(c) Chlorine is in group 7A, characteristic Lewis symbol $:\overset{\cdot\cdot}{\underset{\cdot\cdot}{\mathrm{Z}}}\cdot$, so its Lewis symbol is $:\overset{\cdot\cdot}{\underset{\cdot\cdot}{\mathrm{Cl}}}\cdot$.

Exercise 6.14

Use the Periodic Table at the front of the book to write Lewis symbols for (a) sulfur (b) krypton, Kr (c) potassium.

Inside Chemistry — What is the Nobel Prize?

Alfred Nobel (1833–1896) was a Swedish inventor and industrialist who earned a fortune, mainly from the development and production of explosives. When Nobel joined his father and brothers in the family business of manufacturing explosives, the most important commercial explosive was nitroglycerin, a transparent, oily liquid that detonates on rapid heating or concussion. Nitroglycerin detonates easily, and many workers, including one of Nobel's brothers, lost their lives in accidental explosions. Nobel found that he could lower the risk of accidental detonation by mixing nitroglycerin with an absorbent powder called diatomaceous earth; he patented the mixture and named it dynamite.

Nobel's profits from dynamite and other products made him wealthy. In his will he directed that most of his estate should be used to establish annual awards for achievement in five fields: chemistry, physics, physiology or medicine, literature, and peace; an award is now given also in economics. The first Nobel Prizes were awarded in 1901.

Candidates for Nobel Prizes are nominated to Swedish and Norwegian academic committees, who select the recipients. The awards, which consist of a gold medal, a diploma, and a gift of money, are presented in Stockholm, with the king of Sweden officiating, annually on December 10, the anniversary of Nobel's death.

Figure 6.16
The King of Sweden presents Sidney Altman of Yale University with the 1989 Nobel Prize for chemistry.
(Courtesy of T. Charles Erickson and Michael Marsland, Yale University, Office of Public Information. © Leonard Lessin/Waldo Feng/Mt. Sinai CORE)

Chapter Summary: Atomic Structure and The Periodic Table

Section	Subject	Summary	Check When Learned
6-1	Quantum theory	Energy from hot objects isn't emitted continuously, as it seems to be, but in many small bursts, called quanta (each burst is a quantum).	☐
6-1	Photon	The quantum of light energy.	☐
6-1	Energy level	A spherical region, at a certain distance from the nucleus and centered on it, which an electron could occupy. In theory, there is an infinite number of energy levels around a nucleus. According to the quantum theory, electrons can only be in these regions and not between them.	☐
6-1	Designation for energy levels	Numbered 1, 2, 3, . . . outward from the nucleus.	☐
6-1	Bohr model for atom	A solar-system model. Electrons can lie in circular orbits in energy levels.	☐
6-1	Importance of Bohr model	It applies quantum theory to atomic structure, explains why electrons don't spiral into the nucleus, and explains how elecrons absorb or emit energy when they move between energy levels.	☐
6-1	Energy changes when electron moves between energy levels	When an electron moves to a lower energy level, energy is emitted, and when it moves to a higher level, energy is absorbed. The energy may be in the form of electromagnetic radiation (light), and its wavelength will depend on which energy levels the electron moves between.	☐
6-1	$E = h\nu$	Equation used to calculate the energy of electromagnetic radiation. A photon with frequency ν (s^{-1}) will have energy E (J); h, Planck's constant, is 6.63×10^{-34} J · s.	☐
6-2	Schrödinger equation	Equation in quantum mechanics used to calculate the probability of finding an electron in a certain region around a nucleus.	☐
6-2	Heisenberg uncertainty principle	Mathematical principle that says the location of a moving electron in an atom can't be known with certainty.	☐
6-2	Quantum mechanics	Mathematical branch of physics that uses the Schrödinger equation and the uncertainty principle to describe approximate locations for electrons in atoms.	☐
6-2	Orbital	A region, with a certain shape, in which there is a high (90%) probability that an electron could be found.	☐
6-2	Sublevel	A group of orbitals of the same kind.	☐
6-2	Relationship among orbitals, sublevels, and energy levels	Orbitals make up sublevels, and sublevels make up energy levels.	☐

Section	Subject	Summary	Check When Learned
6-2	Relationship between the number for an energy level and the number of sublevels it contains	They're the same number: Energy level 1 contains one sublevel, energy level 2 contains two, and so on.	☐
6-2	Letters used to designate orbitals and sublevels	s, p, d, f	☐
6-2	Shapes of s orbitals	Spherical shells centered on the nucleus; see Figure 6.11.	☐
6-2	Shapes of p orbitals	Dumbbell-shaped regions lying along coordinate axes centered on the nucleus; see Figure 6.12.	☐
6-2	Numbers of orbitals in s, p, d, and f sublevels	The s sublevel has 1 orbital, the p has 3, the d has 5, and the f has 7.	☐
6-3	Maximum number of electrons in an orbital	2	☐
6-3	Reason for limitation on number of electrons in an orbital	Each electron can have one of two possible spins, and the electrons in an orbital must have opposite spins.	☐
6-3	Maximum number of electrons in s, p, d, and f sublevels	2, 6, 10, 14	☐
6-3	Order of sublevels across the Periodic Table	**Period** — **Sublevels** 1 — 1s 2 — 2s, 2p 3 — 3s, 3p 4 — 4s, 3d, 4p 5 — 5s, 4d, 5p 6 — 6s, 4f, 5d, 6p 7 — 7s, 5f, 6d	☐
6-4	Sublevels being filled by representative elements	s and p	☐

Section	Subject	Summary	Check When Learned
6-4	Sublevels being filled by transition and inner-transition elements	d (transition) and f (inner-transition).	☐
6-4	Sublevel being filled by series that begins with: Sc Y La Lu Ac Lr	 3d 4d 4f 5d 5f 6d	☐
6-4	Number of elements in an s series p series d series f series	 2 6 10 (The Periodic Table ends before the 6d series is complete.) 14	☐
6-5	Order in which sublevels fill	The order in which they occur outward from the nucleus, which is the same as the order in which they appear on the Periodic Table, as shown in Section 6-4: 1s2s2p3s3p4s3d4p5s4d5p6s4f5d6p7s5f6d.	☐
6-5	Electron configuration	The distribution of electrons in the sublevels of an atom.	☐
6-5	spdf notation	The system of notation used to describe electron configurations.	☐
6-5	Meaning of each letter and number in this notation: Cl^{17}: $1s^2 2s^2 2p^6 3s^2 3p^5$	number of electrons in the atom (atomic number) ↓; number of electrons in each sublevel ↓ ↓ ↓ ↓ ↓ symbol for the atom → Cl^{17}: $1s^2$ $2s^2$ $2p^6$ $3s^2$ $3p^5$ ↑ ↑ ↑ ↑ ↑ sublevels in order outward from nucleus (number shows energy level, letter shows sublevel)	☐
6-5	Abbreviated electron configuration	Configuration in which only the sublevels in the period in which the element occurs are shown, and the filled sublevels from earlier periods are represented by the symbol for the preceding noble gas, for example, Cl^{17}: $[Ne]^{10}3s^2 3p^5$.	☐

Section	Subject	Summary	Check When Learned
6-5	Meaning of each letter and number in this notation: Cl^{17}: $[Ne]^{10}3s^23p^5$	number of electrons in the atom (atomic number) ↓ ; number of electrons in each sublevel ↓ ↓ symbol for the atom → Cl^{17}: $[Ne]^{10}3s^2$ $3p^5$ ↗ ↑ ↑ abbreviation stands for 10 electrons in energy levels 1 and 2 ($1s^22s^22p^6$) ; sublevels in energy level 3 (number shows energy level, letter shows sublevel)	☐ ☐
6-5	Steps for finding electron configuration for an atom of any element	(1) Find element on Periodic Table and identify number of electrons in one atom (atomic number). (2) Identify sublevel being filled. (3) Find number of electrons in that sublevel. (4) Write complete electron configuration. Each sublevel before one being filled will be filled with electrons. (5) Make sure total number of electrons in sublevels is same as total number of electrons in atom.	☐
6-5	Ground state	Condition of an atom in which all of its electrons are in the lowest available sublevels.	☐
6-5	Excited state	Condition of an atom in which some of its electrons are not in the lowest available sublevels.	☐
6-6	Rule for distribution of electrons among orbitals in a sublevel	The electrons will occupy the orbitals with one electron each, before pairing electrons in an orbital.	☐
6-6	Orbital diagram	Description of arrangement of electrons in the orbitals of an atom, using arrows to show electron spins, e.g. N^7: $1s^2$ $2s^2$ $2p^3$ (↑↓) (↑↓) (↑)(↑)(↑)	☐
6-7	Valence electrons	Electrons in the s and p sublevels of the outermost energy level, in the atom of a representative element. For example, in F^9: $1s^22s^22p^5$, the valence electrons are the $2s^22p^5$ electrons.	☐
6-7	Characteristic electron configuration for the s and p sublevels of the outermost energy level, for group		
	1A	s^1	
	2A	s^2	

Section	Subject	Summary	Check When Learned
	3A 4A 5A 6A 7A 8A	s^2p^1 s^2p^2 s^2p^3 s^2p^4 s^2p^5 s^2p^6	☐
6-7	Method for identifying the valence electrons for a representative element	On the Periodic Table, identify the period and group the element is in. The period number is the number of the energy level for the valence electrons, and their configuration is given by the characteristic group configuration. For example, silicon, Si, is in period 3 and group 4A (characteristic configuration s^2p^2), so its valence electrons are $3s^23p^2$.	☐
6-8	Lewis symbol	Symbol for the atom of a representative element in which valence electrons are shown as dots.	☐
6-8	Characteristic Lewis symbol for group 1A 2A 3A 4A 5A 6A 7A 8A	 $Z\cdot$ $Z:$ $\cdot Z:$ $\cdot\overset{\cdot}{Z}:$ $\cdot\underset{\cdot}{\overset{\cdot\cdot}{Z}}\cdot$ $\cdot\underset{\cdot\cdot}{\overset{\cdot\cdot}{Z}}\cdot$ $:\underset{\cdot\cdot}{\overset{\cdot\cdot}{Z}}\cdot$ $:\underset{\cdot\cdot}{\overset{\cdot\cdot}{Z}}:$	☐
6-8	Method for finding the Lewis symbol for the atom of a representative element	On the Periodic Table identify the group the element is in. The characteristic Lewis symbol for the group shows the number and arrangement of electrons in the Lewis symbol for the element. For example, sulfur is in group 6A (characteristic Lewis symbol $\cdot\underset{\cdot\cdot}{\overset{\cdot\cdot}{Z}}\cdot$), so the Lewis symbol for sulfur is $\cdot\underset{\cdot\cdot}{\overset{\cdot\cdot}{S}}\cdot$.	☐
6-8	Only element whose Lewis symbol doesn't match the characteristic Lewis symbol for its group	Helium. The characteristic Lewis symbol for group 8A is $:\underset{\cdot\cdot}{\overset{\cdot\cdot}{Z}}:$, but the first period has no p sublevel, so helium is $He:$.	☐

Problems

Assume you can use the Periodic Table at the front of the book unless you're directed otherwise. Answers to odd-numbered problems are in Appendix 1.

The Quantum Theory and the Bohr Atom (Section 6-1)

1. An early theory proposed a solar-system model for the atom. Describe that model.
2. Based on the traditional theories of physics, what objection was raised to the solar-system model for the atom?
3. State the fundamental principle of quantum theory.
4. Define *quantum, quanta,* and *photon.*
5. Describe the Bohr model for the hydrogen atom.
6. Using the Bohr model for the hydrogen atom, explain why the electron will move spontaneously toward the nucleus but will not move spontaneously away from it.
7. Which energy level is closer to the nucleus, 3 or 5?
8. Which energy level is farther from the nucleus, 7 or 6?
9. Define *ground state* and *excited state* as they apply to the Bohr model for the hydrogen atom.
10. Define the terms in the equation $E = h\nu$, including their units.
11. An electron moving between two energy levels emitted light with a wavelength of 6.40×10^{-7} m. (a) Did the electron move from a lower energy level to a higher one or vice versa? (b) What was the energy of the emitted radiation?
12. An electron moving between two energy levels emitted light with a wavelength of 589 nm. (a) Did the electron move toward the nucleus or away from it? (b) What was the energy of the emitted radiation? (c) Wavelengths of emitted radiation are often expressed in nm rather than in m. What advantage is there in using nm?

Energy Levels, Sublevels, and Orbitals (Sections 6-2 and 6-3)

13. In your own words, define carefully the terms *energy level* and *sublevel,* and state the relationship between them.
14. What is an orbital? What is the relationship between orbitals and sublevels?
15. Which kind of sublevel is present in every energy level?
16. Which kind of sublevel is present in every energy level except the first energy level?
17. What sublevels are present in the fourth energy level?
18. Which energy level contains three sublevels?
19. Sketch the shape of the 2s orbital.
20. Sketch the shape of the 2p orbitals.
21. How many orbitals are in an s sublevel? In a p sublevel?
22. How many orbitals are in a d sublevel? In an f sublevel?
23. Fill in the numbers in this table:

Sublevel	Maximum Number of Electrons
s	
p	
d	
f	

24. Finish filling in this table:

Energy Level	Sublevels	Maximum Number of Electrons
1		
2	s, p	2 + 6 = 8
3		
4		

Sublevels and the Periodic Table (Section 6-4)

25. Which groups are filling s sublevels?
26. Which groups are filling p sublevels?
27. Which sublevel is being filled by the elements with atomic numbers 39 through 48?
28. Which sublevels are being filled by the inner-transition elements?
29. Outward from the nucleus, which sublevel follows each of these sublevels: (a) 2p (b) 4s (c) 4f?
30. What is the sequence of sublevels across the sixth period?

Electron Configurations (Section 6-5)

31. Define *electron configuration*.

32. Define *spdf notation*.

33. Write the complete electron configuration for (a) fluorine (b) iodine (c) potassium (d) zirconium (Zr) (e) praseodymium (Pr).

34. Write the complete electron configuration for (a) mercury (b) sulfur (c) neon (d) barium (e) plutonium (Pu).

35. Write the abbreviated electron configuration for (a) nitrogen (b) bismuth (Bi) (c) osmium (Os) (d) molybdenum (Mo) (e) holmium (Ho).

36. Write the abbreviated electron configuration for (a) magnesium (b) silicon (c) zinc (d) platinum (Pt) (e) californium (Cf).

37. Identify the element with this electron configuration: $1s^2 2s^2 2p^6 3s^2 3p^6 4s^2 3d^{10} 4p^6 5s^2 4d^{10} 5p^6 6s^1$.

38. Identify the element with this electron configuration: $[Ar]^{18} 4s^2 3d^3$.

39. For each of the following electron configurations, decide whether the corresponding atom is in the ground state or an excited state: (a) $1s^2 2s^2 2p^2$ (b) $[Xe]^{54} 6s^2 4f^3 6p^1$.

40. For each of the following electron configurations, decide whether the corresponding atom is in the ground state or an excited state: (a) $1s^2 2s^1 2p^0 3s^1$ (b) $[Ar]^{18} 4s^0 3d^2$.

Orbital Diagrams (Section 6-6)

41. Write the electron configuration and orbital diagram for (a) hydrogen (b) lithium.

42. Write the electron configuration and orbital diagram for (a) helium (b) boron.

43. How many unpaired electrons are in an atom of argon?

44. How many unpaired electrons are in an atom of holmium, Ho?

Valence Electrons and Lewis Symbols (Sections 6-7 and 6-8)

45. Identify the valence electrons for each of the following elements:
    (a) O^8: $1s^2 2s^2 2p^4$
    (b) Se^{34}: $1s^2 2s^2 2p^6 3s^2 3p^6 4s^2 3d^{10} 4p^4$
    (c) Bi^{83}: $[Xe]^{54} 6s^2 4f^{14} 5d^{10} 6p^3$

46. Identify the valence electrons for each of the following elements:
    (a) S^{16}: $[Ne]^{10} 3s^2 3p^4$
    (b) Rb^{37}: $[Kr]^{36} 5s^1$
    (c) Te^{52}: $[Kr]^{36} 5s^2 4d^{10} 5p^4$

47. Identify the characteristic valence-electron configuration for (a) group 3A (b) group 4A (c) group 6A

48. Identify the characteristic valence-electron configuration for (a) the alkali metals (b) the alkaline-earth metals (c) the halogens.

49. Identify the element with this valence-electron configuration: (a) $2s^1$ (b) $2s^2 2p^3$ (c) $7s^2$

50. An element has eight valence electrons; name its group.

51. Which of the following elements has six valence electrons: (a) silicon (b) antimony (Sb) (c) oxygen?

52. What is the relationship between the group number for an element and the number of its valence electrons?

53. Write the Lewis symbols for the elements whose electron configurations are given in problem 49.

54. Write the Lewis symbols for (a) oxygen (b) sulfur.

55. Write the Lewis symbol for each element across the third period.

56. Write the Lewis symbols for the first two elements in group 8A.

57. Choose the correct term or terms in parentheses: An element that has only s valence electrons must be a (metal, nonmetal, metalloid).

58. Choose the correct term or terms in parentheses: An element that has s and p valence electrons must be a (metal or nonmetal, metal or metalloid, nonmetal or metalloid).

59. Write the Lewis symbol for (a) calcium (b) nitrogen (c) bromine

60. (a) Write the complete electron configuration for selenium (Se). (b) Write its abbreviated electron configuration and circle the valence electrons. (c) Write the orbital diagram for its valence electrons. (d) Write its Lewis symbol.

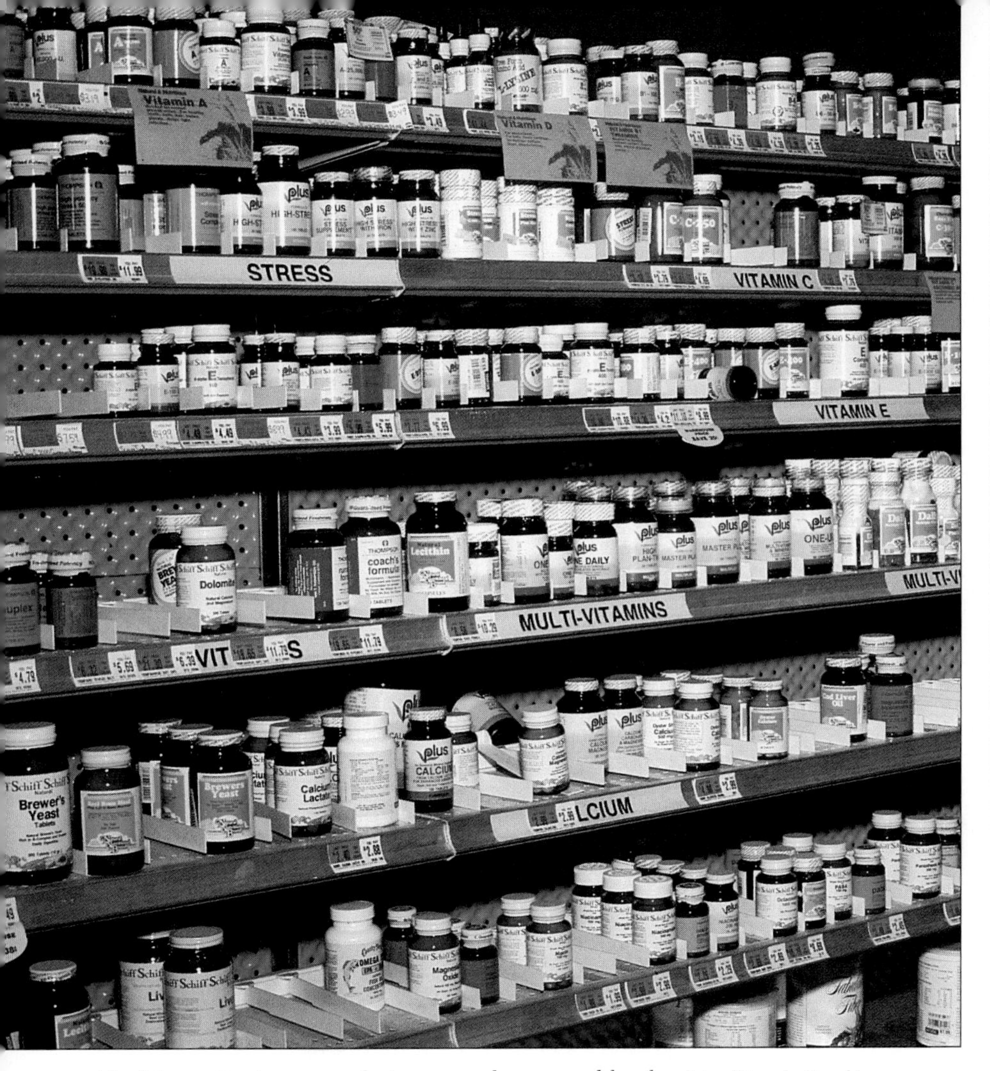

All of these vitamins are covalent compounds, prepared by chemists. *(Beverly March)*

7
Covalent Bonds

Elements combine to form compounds, and the process is sometimes dramatically visible. Iron exposed to the oxygen in the air, for example, is converted to powdery, brownish-red rust, a compound of iron and oxygen. A mixture of the gases hydrogen and oxygen, ignited by a spark, explodes and forms the liquid compound water. Chlorine, a pale green gas, reacts explosively with sodium, a shiny metal, to produce the white crystalline compound that chemists call sodium chloride, and that we know from everyday experience as table salt. These combinations of elements to form compounds are examples of **chemical reactions,** processes in which substances interact with one another to form new and different substances.

According to the atomic theory, when elements combine to form compounds, the atoms of the elements become attached to one another by forces called **chemical bonds.** Using this description of the behavior of atoms, we can also define a chemical reaction as a process in which chemical bonds are formed or broken. In this chapter we'll consider one type of chemical bond, the **covalent bond,** and in Chapter 8 we'll consider another type, the **ionic bond.**

A compound whose atoms are held together by covalent bonds is called a **covalent compound.** For more than two centuries, thousands of chemists have investigated millions of covalent compounds, slowly building up a detailed understanding of their bonding. In this chapter you'll learn how atoms combine to form covalent bonds and why covalent compounds are important enough to justify the very large investment of scientific work that's been made in them.

7-1 Two atoms form a covalent bond because they have less energy if they are bonded

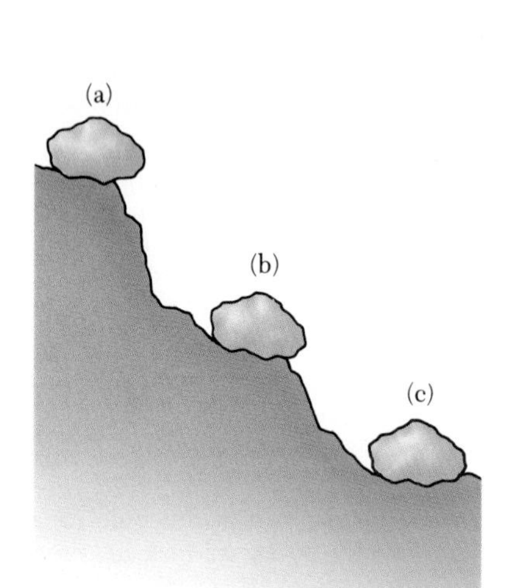

Figure 7.1
The stability of a rock depends on its position. The rock has the most potential energy and is least stable in position (a); it has less energy and is more stable in position (b); and it has least energy and is most stable in position (c). It can move spontaneously from a or b to c.

Figure 7.1 shows a rock in three possible positions: (a) at the top of a high hill, (b) at the top of a lower hill, and (c) in a valley below the hills. It will take energy to move the rock, against the force of gravity, from the valley to the top of the lower hill, and it will take more energy to move it from the valley to the top of the higher hill. Because energy is conserved, these same amounts of energy would be released if the rock rolled from the top of either hill to the bottom of the valley. The rock has less potential energy when it's in the valley than when it's at the top of either hill.

The rock can roll spontaneously from the top of either hill down into the valley, but it can't roll spontaneously from the valley to the top of a hill; in this sense the rock is most stable in the valley, less stable at the top of the lower hill, and least stable at the top of the higher hill. In general, we can say that a system becomes more stable as it releases energy.

Figure 7.2 shows two hydrogen atoms in three possible positions: (a) very close together, so that their 1s orbitals are almost completely overlapped and their nuclei are next to each other; (b) far apart, so that their orbitals aren't overlapped at all; and (c) at an intermediate distance, so that their orbitals overlap but their nuclei are not next to one another. The energy relationships in this diagram are similar to those in Figure 7.1. When the two hydrogen atoms are pushed very close together, their nuclei repel one another strongly, so the atoms will spontaneously move apart.◆ When the atoms are far enough apart that their orbitals don't overlap at all, the atoms will tend to move sponta-

◆ Particles with opposite electrical charges attract one another, and particles with like charges repel one another.

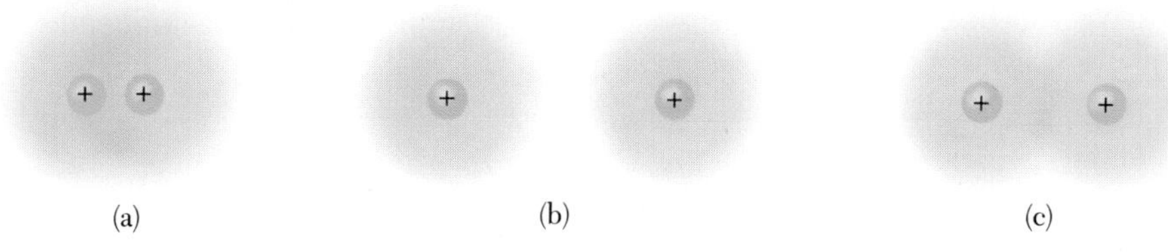

Figure 7.2
The stability of two hydrogen atoms depends on their relative positions. The two hydrogen atoms have the most energy and are least stable in position (a); they have less energy and are more stable in position (b). The two atoms have least energy and are most stable in position (c), which allows optimum overlap of orbitals and formation of a covalent bond.

neously together, because the nucleus of each atom attracts the electron of the other. As the orbitals of the two atoms overlap, each electron can move around both nuclei; a certain amount of overlap between the two atoms gives the optimum balance between the attractions of the atoms for one another and the repulsions between their nuclei. The sharing of two electrons allowed by this optimum overlap is a **covalent bond.**

A covalent bond forms when two orbitals from two different atoms overlap and two electrons share the overlapped orbitals. It forms because the bonded atoms are more stable—they have less total energy—than the unbonded atoms.

Figure 7.2 shows the formation of a covalent bond by the overlap of two s orbitals. Similar but more complicated drawings can show covalent bonding by overlap of an s orbital with a p orbital or of two p orbitals with one another.

7-2 Nonpolar covalent bonds are formed between atoms that share electrons equally, following the octet rule

Two hydrogen atoms will combine with one another to form the simplest kind of chemical bond, called a **single nonpolar covalent bond.** As shown in Figure 7.3, we can represent this process with a drawing that shows the formation of an overlap between the orbitals of the bonded atoms.

Figure 7.3
Two hydrogen atoms form a covalent bond. In each hydrogen atom the 1s electron orbits its nucleus. After the 1s orbitals overlap, both electrons orbit both nuclei.

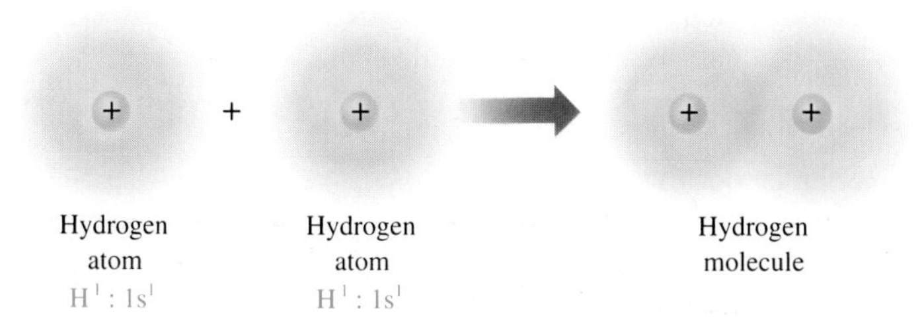

We can also represent this process more simply by writing a chemical equation, using Lewis symbols:

H·	+	H·	→	H:H
hydrogen atom		hydrogen atom		hydrogen molecule
	reactants			product

A **chemical reaction** is a process in which chemical bonds are formed or broken, and a **chemical equation** is a written description, using chemical symbols, of a chemical reaction. The substances that enter into a reaction are called **reactants,** and the substances that are produced from a reaction are called **products.** The arrow in a chemical equation can be translated into the word "form" or "forms." The general statement of a chemical reaction is "Reactants form products," and the specific statement of the preceding equation is "Two hydrogen atoms form one hydrogen molecule."◆

◆The meaning of chemical equations is discussed in more detail in Chapter 9.

The previous equation shows that the reactants, two hydrogen atoms, form the product, a hydrogen molecule. A **molecule** is a group of two or more atoms that are attached to one another by covalent bonds. An atom is represented by a **symbol,** and a molecule is represented by a group of symbols called a **formula.** In this equation the hydrogen atoms are represented by Lewis symbols, and the hydrogen molecule is represented by a Lewis formula, a combination of Lewis symbols.◆

◆Lewis symbols and formulas are also called **electron-dot** symbols and formulas.

Instead of showing the symbols for the two hydrogen atoms separately, we can write

2 H·	→	H:H
hydrogen atoms		hydrogen molecule

The number 2 in this equation is a coefficient. A **coefficient** is a multiplier, written before a symbol or formula in a chemical equation, to show the number of times the symbol or formula appears in the equation. If no coefficient is shown, the number is assumed to be 1; in the preceding equation the formula for the hydrogen molecule appears once, so no coefficient is written before it.

A **covalent bond** is a pair of electrons shared by two atoms. In the reaction shown above each hydrogen atom contributes one electron to form a covalent bond, and the bond is represented by the two dots in the formula for the hydrogen molecule.

The bond in the hydrogen molecule is called a **single bond** because it consists of one, and only one, pair of electrons. The two hydrogen atoms in the hydrogen molecule share their two electrons equally. A covalent bond (a pair of electrons) that is shared equally between two bonded atoms is called a **nonpolar covalent bond.**

The following diagram summarizes the meanings of the words used to describe the bond in the hydrogen molecule.

single	nonpolar	covalent	bond
There is one pair of electrons in the bond.	The electrons in the bond are shared equally between the two bonded atoms.	The bond consists of shared electrons.	

There are also **double** or **triple nonpolar covalent bonds:** two or three pairs of electrons shared equally between two atoms. The nitrogen molecule, for example, consists of two nitrogen atoms and contains a triple nonpolar covalent bond.◆ We can represent the formation of a nitrogen molecule from two nitrogen atoms by this equation:

◆Nitrogen molecules make up about 80% of the air.

$$2\ \cdot\underset{\cdot}{\ddot{N}}\cdot \quad \rightarrow \quad :N\vdots\vdots N:$$

nitrogen atoms — nitrogen molecule

A molecule of the compound ethylene, which consists of two carbon atoms and four hydrogen atoms, contains a double nonpolar covalent bond.◆ We can write this equation for the formation of an ethylene molecule from its atoms:

◆Ethylene is made from petroleum and is used to make the plastic polyethylene for products such as plastic trash bags.

$$4\,H\cdot \quad + \quad 2\ \cdot\dot{C}: \quad \rightarrow \quad \begin{array}{cc} H & H \\ \underset{\cdot\cdot}{\overset{\cdot\cdot}{C}}\!:\!: & \underset{\cdot\cdot}{\overset{\cdot\cdot}{C}} \\ H & H \end{array}$$

hydrogen atoms — carbon atoms — ethylene molecule

The bond between the two carbon atoms is a double nonpolar covalent bond.

How many covalent bonds will be formed between atoms, that is, how many electrons will they share? The fundamental rule of chemical bonding, called the **octet rule,** states that a hydrogen atom forming a covalent bond will share 2 electrons, and every other atom forming a covalent bond will share enough electrons to give it a total of 8 shared and unshared electrons in its valence shell. The application of the octet rule to the molecules of hydrogen, nitrogen, and ethylene is shown below. The circle around each atom encloses its valence electrons.

Each hydrogen atom shares 2 electrons.

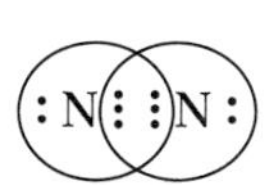

Each nitrogen atom has a total of 8 electrons: 2 unshared electrons and 6 shared electrons.

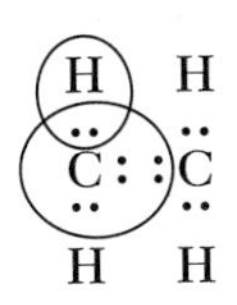

Each hydrogen atom shares 2 electrons and each carbon atom shares 8.

The octet rule can also be based on the electron configurations of the elements in group 8A, the noble gases, whose Lewis symbols are $He\!:$, $:\!\underset{\cdot\cdot}{\overset{\cdot\cdot}{Ne}}\!:$, $:\!\underset{\cdot\cdot}{\overset{\cdot\cdot}{Ar}}\!:$, $:\!\underset{\cdot\cdot}{\overset{\cdot\cdot}{Kr}}\!:$, $:\!\underset{\cdot\cdot}{\overset{\cdot\cdot}{Xe}}\!:$, $:\!\underset{\cdot\cdot}{\overset{\cdot\cdot}{Rn}}\!:$. A helium atom has 2 valence electrons, and an atom of each of the other noble gases has 8 valence electrons, so the octet rule can be stated this way: An atom that forms a covalent bond will share enough electrons to give it the same number of valence electrons as a noble-gas atom: 2 valence electrons for a bonded hydrogen atom and 8 valence electrons for any other bonded atom.

◆Answers to Exercises are in Appendix 1.

Exercise 7.1◆

In each of the following Lewis formulas, circle each atom and its electrons to show that it obeys the octet rule.

(a) $\mathrm{H:\underset{\cdot\cdot}{\overset{\cdot\cdot}{O}}:H}$ (b) $\mathrm{:\underset{\cdot\cdot}{\overset{\cdot\cdot}{F}}:\underset{\cdot\cdot}{\overset{\cdot\cdot}{F}}:}$ (c) $\mathrm{\underset{\cdot\cdot}{\overset{\cdot\cdot}{O}}::C::\underset{\cdot\cdot}{\overset{\cdot\cdot}{O}}}$ (d) $\mathrm{H:C\vdots\vdots C:H}$

7-3 Polar covalent bonds are formed between atoms that share electrons unequally, following the octet rule

◆The system used to name covalent compounds is described in Chapter 9.

A hydrogen atom will combine with a fluorine atom to form a molecule called hydrogen fluoride.◆

$$\underset{\text{hydrogen atom}}{\mathrm{H\cdot}} + \underset{\text{fluorine atom}}{\mathrm{\cdot\underset{\cdot\cdot}{\overset{\cdot\cdot}{F}}:}} \rightarrow \underset{\text{hydrogen fluoride molecule}}{\mathrm{H:\underset{\cdot\cdot}{\overset{\cdot\cdot}{F}}:}}$$

Each atom in the hydrogen fluoride molecule obeys the octet rule: The hydrogen atom shares 2 electrons, and the fluorine atom has 2 shared and 6 unshared electrons. The bond between the hydrogen and fluorine atoms is a single covalent bond (one pair of electrons shared by two atoms), but it isn't a *nonpolar* bond. It's described as a single *polar* covalent bond.

A **polar covalent bond** is a pair of electrons shared unequally between two atoms. A fluorine atom has a stronger attraction for shared electrons than a hydrogen atom does, so the electrons are pulled closer to the fluorine atom. Because electrons carry a negative charge, this displacement of electrons toward the fluorine atom gives the fluorine end of the bond a slight negative charge and leaves the hydrogen end with a slight positive charge: A polar bond has a negative end and a positive end. This polarity can be shown by using the symbols $\delta+$ and $\delta-$ for slight positive and negative charges, as shown in the Lewis formula below.◆

◆The symbol δ, the lowercase Greek letter delta, is used in mathematics to mean *partial*. The symbols $\delta+$ and $\delta-$ are read "partial positive charge" and "partial negative charge."

$$\overset{\delta+}{\mathrm{H}}:\overset{\delta-}{\underset{\cdot\cdot}{\overset{\cdot\cdot}{\mathrm{F}}}}:$$

As the example of the hydrogen fluoride molecule shows, different atoms have different powers of attraction for shared electrons. The ability of an atom to attract shared electrons is called its **electronegativity.** Electronegativity can be expressed in numbers, as shown on the partial Periodic Table in Figure 7.4.◆ On this scale, the fluorine atom, which has the highest electronegativity, is assigned a value of 4.0.

◆The concept of electronegativity, on which the modern theory of chemical bonding depends, was created by the American chemist Linus Pauling (born 1901), whose amazing achievements earned the Nobel Prize for Chemistry in 1954 and the Nobel Prize for Peace in 1963.

What's important in bonding is the electronegativity *difference* between two bonded atoms. If two bonded atoms have the same electronegativity (the same power of attraction for shared electrons), the bond will be nonpolar. For this reason a bond between two atoms of the same element will be nonpolar, as in $\mathrm{H:H}$, $\mathrm{:\underset{\cdot\cdot}{\overset{\cdot\cdot}{F}}:\underset{\cdot\cdot}{\overset{\cdot\cdot}{F}}:}$, and $\mathrm{:N\vdots\vdots N:}$. If two covalently bonded atoms have different electronegativities, the bond will be polar: The more electronegative atom will have a partial negative charge and the less electronegative atom will have a partial positive charge, as in hydrogen fluoride. (See Chemistry Insight Box on page 121.†)

Period	1A	2A	3A	4A	5A	6A	7A	8A
1	H 2.1							He
2	Li 1.0	Be 1.5	B 2.0	C 2.5	N 3.0	O 3.5	F 4.0	Ne
3	Na 0.9	Mg 1.2	Al 1.5	Si 1.8	P 2.1	S 2.5	Cl 3.0	Ar
4	K 0.8	Ca 1.0	Ga 1.6	Ge 1.8	As 2.0	Se 2.4	Br 2.8	Kr
5	Rb 0.8	Sr 1.0	In 1.7	Sn 1.8	Sb 1.9	Te 2.1	I 2.5	Xe
6	Cs 0.7	Ba 0.9	Tl 1.8	Pb 1.8	Bi 1.9	Po 2.0	At 2.2	Rn
7	Fr 0.7	Ra 0.9						

Figure 7.4
Electronegativity values for the representative elements. By convention, numerical values for electronegativity have no units. In general, electronegativity values for the representative elements increase upward in a group and to the right in a period. Electronegativity values aren't included for the elements in group 8A, the noble gases, because they very rarely form bonds.

In describing chemical bonds, it's sometimes useful to refer to numerical electronegativity values, but it's usually not necessary. Instead, we can use the following simple rules.

Atoms of the same element have the same electronegativity.
For the representative elements electronegativity generally increases upward in a group and to the right in a period.
Electronegativity increases in this order: $H < C < Cl = N < O < F$. These are elements whose bonding we'll consider often enough to make it worthwhile to remember the series.

The Lewis formulas below represent molecules with single, double, and triple polar covalent bonds. Each atom in each of these molecules follows the octet rule. The more electronegative atom carries a partial negative charge, and the less electronegative atom carries a partial positive charge.

$\overset{\delta+}{H}:\overset{\delta-}{\underset{\cdot\cdot}{\ddot{Cl}}}:$ A molecule of hydrogen chloride, a toxic, corrosive gas with a sharp, unpleasant odor. A solution of hydrogen chloride in water is sold commercially as muriatic acid and is used to clean outdoor tile and masonry.

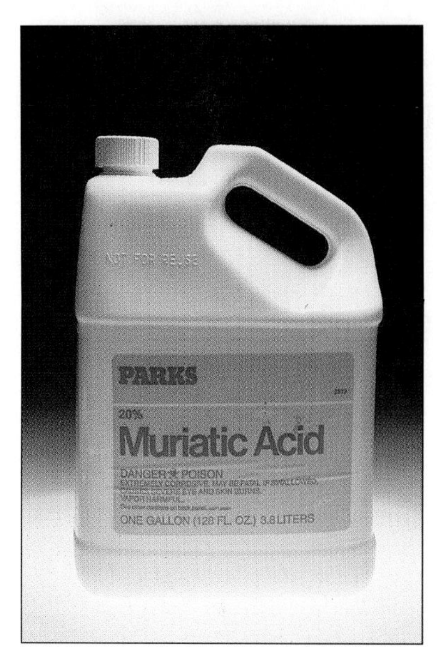

(Charles D. Winters)

† Chemistry Insight

In a tug-of-war between two people, what matters is not how strong each of them is, but what the difference in strength is between them. A weak person will lose to a strong person, but in a contest between two very strong persons or two very weak persons, neither person will win. Two hydrogen atoms, each with electronegativity 2.1, will share electrons equally, and so will two fluorine atoms, each with electronegativity 4.0. But a hydrogen atom and a fluorine atom will share bonding electrons unequally, with the fluorine atom getting the larger share.

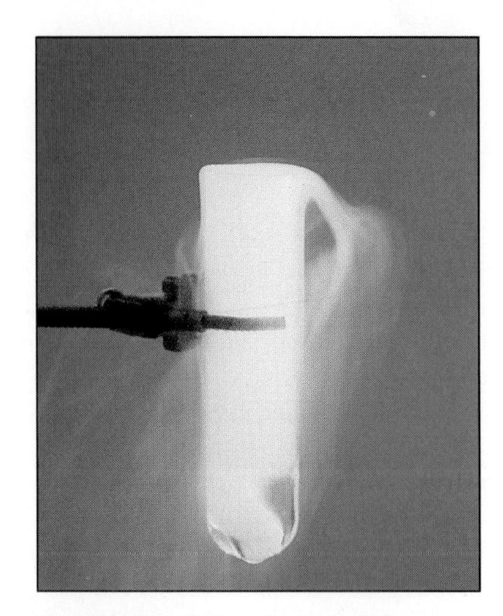

Dry ice is solid carbon dioxide, CO_2. *(Paul Silverman, Fundamental Photographs, New York)*

Two products of combustion are carbon monoxide, CO, and carbon dioxide, CO_2. *(Barry L. Runk from Grant Heilman Photography, Inc.)*

$\overset{\delta-}{\ddot{\text{O}}}::\overset{\delta+}{\text{C}}::\overset{\delta-}{\ddot{\text{O}}}$ (with two further dots below each O) A molecule of carbon dioxide, containing two double polar covalent bonds. Most of the food you eat consists of molecules that contain carbon atoms. Chemical reactions in your cells convert these carbon atoms to carbon dioxide molecules, which are carried by your bloodstream to your lungs and exhaled.

$:\overset{\delta+}{\text{C}}\vdots\vdots\overset{\delta-}{\text{O}}:$ A molecule of carbon monoxide, containing a triple polar covalent bond. Flammable substances such as wood, coal, and gasoline consist of molecules that contain carbon atoms. If these substances are burned with a plentiful supply of oxygen, the carbon atoms are converted to carbon dioxide molecules. If they're burned with a limited supply of oxygen, some of the carbon atoms are converted to carbon monoxide molecules.

Exercise 7.2

In each of the following Lewis formulas, identify each bond as single, double, or triple and as nonpolar or polar:

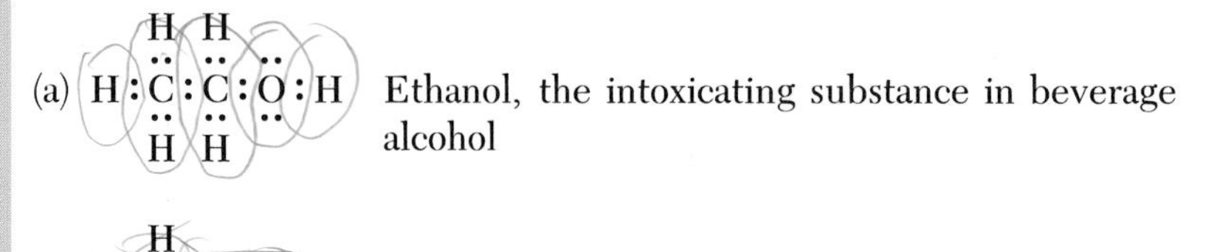

(a) H:C:C:O:H (with H above and below each C, and two electron pairs on O) Ethanol, the intoxicating substance in beverage alcohol

(b) H:C::O (with H above C, and two electron pairs on O) Formaldehyde, used by biologists to preserve specimens

(c) $\text{H}:\text{C}\vdots\vdots\text{N}:$ Hydrogen cyanide, a poisonous gas

Exercise 7.3

In each of the Lewis formulas shown in Exercise 7.2, circle each atom and its electrons to show that it follows the octet rule.

Exercise 7.4

To the Lewis formula in part (a) of Exercise 7.2, add $\delta+$ and $\delta-$ to show the direction of polarity of the oxygen–hydrogen bond. Do the same for the carbon–oxygen bond in part (b) and the carbon–nitrogen bond in part (c).

7-4 A molecule can be represented by a Lewis formula, a line formula, or a molecular formula

We'll use three kinds of formulas to represent molecules: Lewis formulas, line formulas, and molecular formulas.

A **Lewis formula,** as we've seen, shows valence electrons as dots and may use $\delta+$ and $\delta-$ to show the direction of bond polarity. For example, the Lewis formula for carbon dioxide can be written $\ddot{\underset{..}{\text{O}}}::\text{C}::\ddot{\underset{..}{\text{O}}}$ or $\overset{\delta-}{\ddot{\underset{..}{\text{O}}}}::\overset{\delta+}{\text{C}}::\overset{\delta-}{\ddot{\underset{..}{\text{O}}}}$.

A **line formula** is a modified version of a Lewis formula that uses a line instead of a pair of dots to represent the two electrons in a covalent bond. Line formulas are often used instead of Lewis formulas because they're easier to read, and we'll usually use line formulas in this book.◆ In a line formula $\delta+$ and $\delta-$ may be added to show bond polarity. The line formula for carbon dioxide can be written $\ddot{\underset{\cdot\cdot}{\text{O}}}=\text{C}=\ddot{\underset{\cdot\cdot}{\text{O}}}$ or $\overset{\delta-}{\ddot{\underset{\cdot\cdot}{\text{O}}}}=\overset{\delta+}{\text{C}}=\overset{\delta-}{\ddot{\underset{\cdot\cdot}{\text{O}}}}$.

◆ Line formulas are also called **structural** formulas.

A **molecular formula** shows only the number of atoms of each element in the molecule. A subscript on each symbol shows the number of atoms of that element present in the molecule; if no subscript appears, one atom is present. The molecular formula for carbon dioxide is CO_2.

Other examples of Lewis, line, and molecular formulas are shown in Table 7.1.

Table 7.1 Lewis Formulas, Line Formulas, and Molecular Formulas

Lewis Formula	Line Formula	Molecular Formula
H:H	H—H	H_2
$:\ddot{\underset{\cdot\cdot}{\text{F}}}:\ddot{\underset{\cdot\cdot}{\text{F}}}:$	$:\ddot{\underset{\cdot\cdot}{\text{F}}}—\ddot{\underset{\cdot\cdot}{\text{F}}}:$	F_2
$\text{H}:\ddot{\underset{\cdot\cdot}{\text{F}}}:$	$\text{H}—\ddot{\underset{\cdot\cdot}{\text{F}}}:$	HF
$\text{H}:\ddot{\underset{\cdot\cdot}{\text{O}}}:\text{H}$	$\text{H}—\ddot{\underset{\cdot\cdot}{\text{O}}}—\text{H}$	H_2O
:N⋮⋮N:	:N≡N:	N_2

Exercise 7.5

Write the line formula (without $\delta+$ and $\delta-$) and the molecular formula for each of these Lewis formulas from Exercise 7.2:

(a) $\text{H}:\underset{\text{H}}{\overset{\text{H}}{\text{C}}}:\underset{\text{H}}{\overset{\text{H}}{\text{C}}}:\ddot{\underset{\cdot\cdot}{\text{O}}}:\text{H}$ (b) $\text{H}:\overset{\text{H}}{\text{C}}::\ddot{\underset{\cdot\cdot}{\text{O}}}$ (c) H:C⋮⋮N:

Molecular formulas are used often because they have the advantage of being very concise, but they have a significant disadvantage: They require you to know or to figure out what the bonding in the molecule is. For this reason it's important that you know how to convert a molecular formula into the corresponding line formula, as described in the following Section.

7-5 To write a line formula from a molecular formula, combine Lewis symbols according to the octet rule

The most common examples of covalent bonding occur between atoms of only a few elements, those shown on the Periodic Table in Figure 7.5. Your aim in this Section will be to learn the procedure that converts a molecular formula for a simple compound of these elements into the corresponding line formula. To illustrate the procedure, we'll construct the line formula for carbon dioxide, CO_2.

	1A	2A	3A	4A	5A	6A	7A	8A
1	H							
2			B	C	N	O	F	
3				Si	P	S	Cl	
4				Ge	As	Se	Br	
5					Sb	Te	I	
6								
7								

Figure 7.5
Representative elements that commonly show covalent bonding.

1. **Write the Lewis symbol for each atom and count the number of valence electrons shown in all of these Lewis symbols.**
   Carbon (group 4A) has 4 electrons in its Lewis symbol, and oxygen (group 6A) has 6, so the atoms we have to work with are $\cdot\dot{\mathrm{C}}:$ and 2 $\cdot\ddot{\underset{\cdot\cdot}{\mathrm{O}}}\cdot$, and the total number of valence electrons is 16.
2. **Write the symbols for the atoms in this pattern: At the center, write the symbol (not the Lewis symbol) for the element that has only one atom in the molecule. Around it write the symbols (not Lewis symbols) for the other atoms in the molecule. Connect each outside atom to the central atom by a single bond.**
   The CO_2 molecule consists of 1 C atom and 2 O atoms. Write

   O C O

   Connect each O atom to the C atom by a single bond.

   O—C—O

   Each single bond uses 2 of the 16 electrons we began with.
3. **Put the remaining electrons around the outside atoms as unshared electrons, to give each outside atom a total of 8 shared and unshared electrons. If there are electrons left over, put them as unshared electrons on the central atom.**
   We began with 16 electrons and we've used 4 for the two C–O bonds, so we have 12 left. Put 6 electrons (3 pairs) around each oxygen atom; this gives each oxygen atom 8 electrons (2 shared and 6 unshared).

   $$:\ddot{\underset{\cdot\cdot}{\mathrm{O}}}\text{—C—}\ddot{\underset{\cdot\cdot}{\mathrm{O}}}:$$

   We've used all of the 16 electrons we started with, so we have none to put on the central carbon atom.
4. **If the central atom doesn't have 8 electrons, change an unshared electron pair on an outer atom to a shared pair (covalent bond) with the central atom, and continue this process until the central atom has 8 electrons. Make the bonding as symmetrical as possible.**
   If we change one unshared electron pair to a bond, we get

   $$\ddot{\underset{\cdot\cdot}{\mathrm{O}}}\text{=C—}\ddot{\underset{\cdot\cdot}{\mathrm{O}}}:$$

This gives the carbon atom a share in 6 electrons (3 bonds). To give the carbon atom a share in 8 electrons, we have our choice of writing either

$$:\mathrm{O}\equiv\mathrm{C}-\ddot{\underset{\cdot\cdot}{\mathrm{O}}}: \qquad \text{or} \qquad \ddot{\underset{\cdot\cdot}{\mathrm{O}}}=\mathrm{C}=\ddot{\underset{\cdot\cdot}{\mathrm{O}}}$$

The formula on the right is preferred because it's more symmetrical. If we add symbols to show bond polarity, we have

$$\overset{\delta-}{\ddot{\underset{\cdot\cdot}{\mathrm{O}}}}=\overset{\delta+}{\mathrm{C}}=\overset{\delta-}{\ddot{\underset{\cdot\cdot}{\mathrm{O}}}}$$

In the Examples below, line formulas are worked out according to the preceding steps.

Example 7.1

Write the line formula for carbon monoxide, CO, and show the direction of polarity for its bond.

Solution

The atoms are $\cdot\dot{\mathrm{C}}:$ and $\cdot\ddot{\underset{\cdot\cdot}{\mathrm{O}}}\cdot$, with a total of 10 valence electrons.

C—O	Uses 2 electrons.
$:\mathrm{C}-\ddot{\underset{\cdot\cdot}{\mathrm{O}}}:$	Oxygen has 8 electrons but carbon has only 4. All 10 electrons have been used. (You could also write $:\underset{\cdot\cdot}{\ddot{\mathrm{C}}}-\mathrm{O}:$ or $:\ddot{\mathrm{C}}-\ddot{\mathrm{O}}:$.)
$:\mathrm{C}=\ddot{\mathrm{O}}:$	Oxygen has 8 electrons and carbon has 6.
$:\mathrm{C}\equiv\mathrm{O}:$	Oxygen has 8 electrons and carbon has 8.
$:\overset{\delta+}{\mathrm{C}}\equiv\overset{\delta-}{\mathrm{O}}:$	Oxygen is to the right of carbon in the same period, so oxygen is more electronegative than carbon.

Example 7.2

Write the line formula for CF_4, and show the direction of polarity for each bond.

Solution

The atoms are 1 $\cdot\dot{\mathrm{C}}:$ and 4 $:\ddot{\underset{\cdot\cdot}{\mathrm{F}}}\cdot$, with a total of 32 valence electrons.

$\mathrm{F}-\underset{\mathrm{F}}{\overset{\mathrm{F}}{\mathrm{C}}}-\mathrm{F}$	Uses 8 electrons.
$:\ddot{\underset{\cdot\cdot}{\mathrm{F}}}-\underset{:\underset{\cdot\cdot}{\mathrm{F}}:}{\overset{:\ddot{\mathrm{F}}:}{\mathrm{C}}}-\ddot{\underset{\cdot\cdot}{\mathrm{F}}}:$	Uses 32 electrons. Each atom follows the octet rule.

$$\begin{array}{c} \delta- \\ :\ddot{F}: \\ \delta- \quad | \delta+ \quad \delta- \\ :\ddot{F}-C-\ddot{F}: \\ | \\ :\ddot{F}: \\ \delta- \end{array}$$

Fluorine is to the right of carbon in the same period, so fluorine is more electronegative than carbon.

Example 7.3

Write the line formula for water, H_2O.

Solution

The atoms are 2 $H\cdot$ and $\cdot\ddot{\underset{\cdot\cdot}{O}}\cdot$, with a total of 8 valence electrons.

H—O—H Uses 4 electrons.

$H-\ddot{\underset{\cdot\cdot}{O}}-H$ Each of the hydrogen atoms has 2 electrons, satisfying the octet rule, so it's not necessary to add unshared electrons to them. The remaining 4 electrons are added to the oxygen atom, giving it a total of 8 electrons.

Example 7.4

Write the line formula for H_2SO_4.

Solution

The atoms are 2 $H\cdot$, $\cdot\ddot{\underset{\cdot\cdot}{S}}\cdot$, and 4 $\cdot\ddot{\underset{\cdot\cdot}{O}}\cdot$, with a total of 32 valence electrons.

$$\begin{array}{c} O \\ | \\ H-O-S-O-H \\ | \\ O \end{array}$$

Uses 12 electrons. As this example shows, if a molecule contains hydrogen and oxygen atoms, they'll often be bonded to one another. You could show the two hydrogen atoms bonded to any two oxygen atoms; for example,

$$\begin{array}{c} O \\ | \\ O-S-O-H \\ | \\ O \\ | \\ H \end{array}$$

$$\begin{array}{c} :\ddot{O}: \\ | \\ H-\ddot{\underset{\cdot\cdot}{O}}-S-\ddot{\underset{\cdot\cdot}{O}}-H \\ | \\ :\underset{\cdot\cdot}{O}: \end{array}$$

Uses 32 electrons. Each atom follows the octet rule.

Exercise 7.6

Write the line formula for each of the following molecular formulas, and show the direction of polarity for each bond: (a) HBr (b) OF_2 (c) $HClO_4$

A few simple molecules don't follow the octet rule, so their line formulas can't be found by the preceding procedure. One example is N_2O, in which there are 17 valence electrons. Using our procedure, the closest we can come to a reasonable line formula is ·O—N=O. A more surprising exception is the oxygen molecule, O_2. Using our procedure, we'd predict the line formula O=O, but the actual bonding in oxygen, based on its properties, is ·O—O·.

7-6 The line formula for a molecule usually does not show its actual shape

A line formula shows the arrangement of bonds in a molecule, but it usually does not show the actual shape of the molecule.

If a molecule has only two atoms, its shape will be linear, because two points determine a straight line: H—H, :F—F:, and :C≡O: are linear molecules. Since only one shape is possible for a molecule that consists of two bonded atoms, the line formula correctly predicts the shape of the molecule.

If a molecule has three atoms, it may be linear or bent. In the procedure used to construct the line formula, the three atoms are written in a linear arrangement, for example, H—O—H and O=C=O, but the actual shape of the molecule may not be linear. The carbon dioxide molecule, in fact, is linear, and the water molecule is bent, with an angle of 104°—a little more than a right angle, 90°—between the two H—O bonds.

O
H H
104°

Molecules with more atoms can have more complicated shapes. For example, the molecule CF_4 has a tetrahedral shape, as shown in Figure 7.6.

The shape of a molecule can be important because it can determine how the molecule will interact with other molecules. In this book the only molecule whose shape will be important for us is the water molecule, whose properties are described in Chapters 14 through 16.

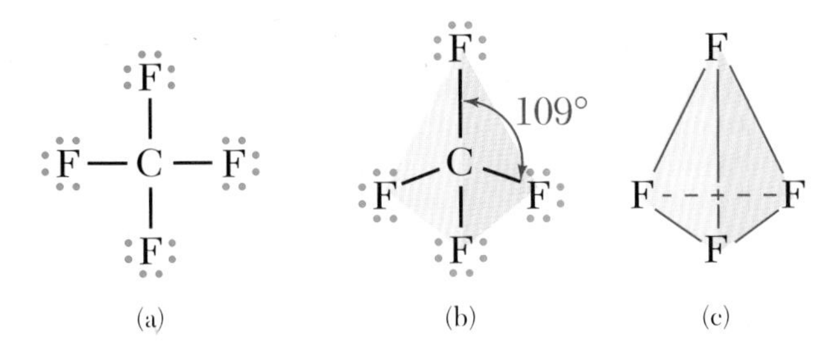

Figure 7.6
The molecule CF_4 has a tetrahedral shape. (a) The line formula for CF_4 is written with the four fluorine atoms at the corners of a square—see Example 7.2. (b) The actual shape of the molecule is tetrahedral, with the carbon atom at the center and an angle of 109° between each pair of C—F bonds. (c) A tetrahedron is a triangular pyramid, and each of its four faces is an equilateral triangle. To picture the shape of the CF_4 molcecule, think of each fluorine atom as being at a corner of the pyramid.

7-7 The mass of a molecule can be found by adding up the masses of its atoms

A molecule is a combination of atoms, and the mass of a molecule is the sum of the masses of its atoms. The mass of a molecule is called its **molecular mass.** In this book, we'll calculate molecular masses to three significant digits, unless there's a reason to be more precise.◆

◆Molecular masses will be important in Chapters 11, 12, 15, and 16.

Example 7.5

Calculate the molecular mass of water, H_2O.

Solution

From the Periodic Table at the front of the book, the atomic masses are 1.01 amu for H and 16.0 amu for O.

$$\begin{aligned} 2 \text{ H atoms} &= 2 \times 1.01 \text{ amu} = 2.02 \text{ amu} \\ 1 \text{ O atom} &= 1 \times 16.0 \text{ amu} = \underline{16.0 \text{ amu}} \\ \text{molecular mass of } H_2O &= 18.02 \text{ amu} = 18.0 \text{ amu, to three significant digits} \end{aligned}$$

Exercise 7.7

Calculate the molecular mass of carbon dioxide, CO_2.

Two hydrogen atoms will bond with one another to form an H_2 molecule, so hydrogen has both an *atomic mass,* 1.01 amu, and a *molecular mass,* 2.02 amu. The same is true for nitrogen: The *atomic mass* of N is 14.0 amu and the *molecular mass* of N_2 is 28.0 amu.

Exercise 7.8

The formula for a molecule of iodine is I_2. What is the atomic mass of iodine? What is its molecular mass?

7-8 There are millions of covalent compounds, with an enormous range of properties

Almost everything you see around you is made of covalent compounds. They make up wood, paper, ink, dyes, paints, plastics, and the natural or synthetic fibers in cloth. Plants and animals consist almost entirely of covalent compounds.

During the past two centuries, chemists have discovered the formulas for several million of these compounds, and no one knows how many millions more there may be. Most of these compounds contain only five elements: carbon, hydrogen, oxygen, nitrogen, and sulfur.

How are so many compounds possible from so few elements? The main reason is that carbon atoms have the unique ability to bond with themselves and with other atoms to form chains of atoms that can be several hundred thousand atoms long. In a molecule 100 000 atoms long, changing the sequence of two atoms can make a different molecule, so the number of possible molecules is enormous.

The millions of known covalent compounds have an almost endless variety of properties: Some are gases, some are liquids, and some are solids; some are transparent, some are white, and some are colored; some are soluble in water and some are insoluble; some are flammable and some smother flames; some are foods or medicines and some are poisons. The variety of the properties and uses of covalent compounds is nearly as great as the variety of their molecules.

Table 7.2, on the following pages, gives examples of a few important covalent compounds. As the line formulas in Table 7.2 show, covalently bonded atoms can form not only chains but rings, so molecules that consist of more than a few atoms can have complicated structures. The line formulas for these molecules are worked out by more complicated methods than the method described in Section 7-5. In these line formulas some bonds appear much longer than others. The bonds in these molecules in fact have similar lengths, but in drawing a two-dimensional line formula to represent a three-dimensional molecule, it's sometimes necessary to distort bond lengths. Even in these complicated molecules, however, each bond can be classified as single, double, or triple and as nonpolar or polar. The nearly limitless variety of molecules arises from the formation of covalent bonds between atoms of only a few elements, following the octet rule. ◆

◆ Because there are so many carbon compounds and because they're so important, the study of these compounds is a special field, called **organic chemistry.** The study of the carbon compounds that make up living systems is another large and special field, **biochemistry.**

Early chemists describe the first dirt molecule

(THE FAR SIDE, copyright 1986 Universal Press Syndicate. Reprinted with permission.)

Table 7.2 Examples of Important Covalent Compounds

Line Formula, Name, and Molecular Formula	Properties and Uses
H–C(H)(H)–H Methane, CH_4	Colorless, odorless, flammable gas. The natural gas used as a fuel in cooking and heating is about 85% methane. Methane is a major constituent of the atmospheres of Jupiter, Saturn, Uranus, and Neptune.
H–C(H)(H)–C(=O)–C(H)(H)–H Acetone, C_3H_6O	Colorless liquid with a characteristic pungent odor—you may know it as the solvent used to remove fingernail polish. Used commercially in large quantities as a solvent for fats, oils, waxes, resins, rubber, plastics, lacquers, varnishes, and rubber cements. Also used to manufacture other compounds.

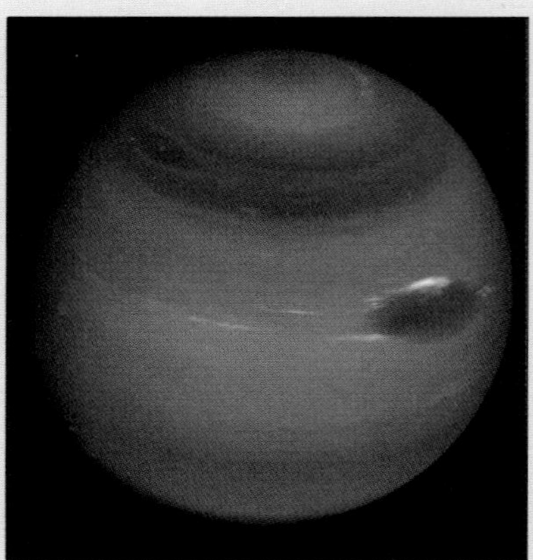

Much of the atmosphere of Neptune is methane gas. *(NASA/JPL)*

Nail polish remover is acetone. *(Charles D. Winters)*

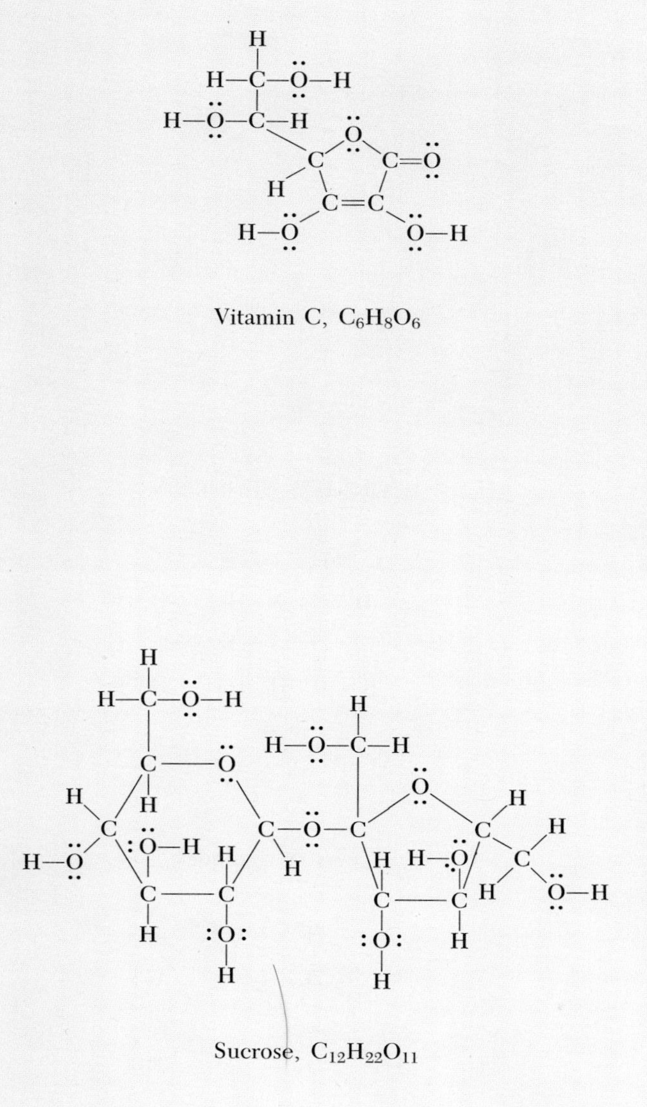

Vitamin C, $C_6H_8O_6$

White, crystalline solid with a sharp taste. Vitamin C occurs in citrus fruits. A diet lacking Vitamin C leads to scurvy, a disease characterized by bleeding gums, spots on the skin, and increasing weakness.

Crystals of vitamin C. *(David R. Frazier, Photolibrary)*

Sucrose, $C_{12}H_{22}O_{11}$

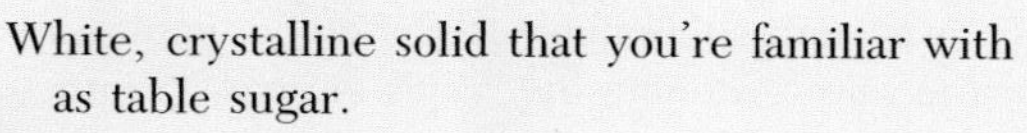

White, crystalline solid that you're familiar with as table sugar.

Crystals of table sugar, sucrose. *(Paul Silverman, Fundamental Photographs, New York)*

(continued)

Table 7.2 Examples of Important Covalent Compounds *(continued)*

Line Formula, Name, and Molecular Formula	Properties and Uses
Naphthalene, $C_{10}H_8$	White, crystalline solid with a characteristic odor. You may have seen and smelled it as moth balls.  Mothballs are naphthalene. *(Paul Silverman, Fundamental Photographs, New York)*
Methionine, $C_5H_{11}NO_2S$	White, crystalline solid. Methionine is one of the essential amino acids, compounds that form proteins and must be present in the diet of human beings for good health. 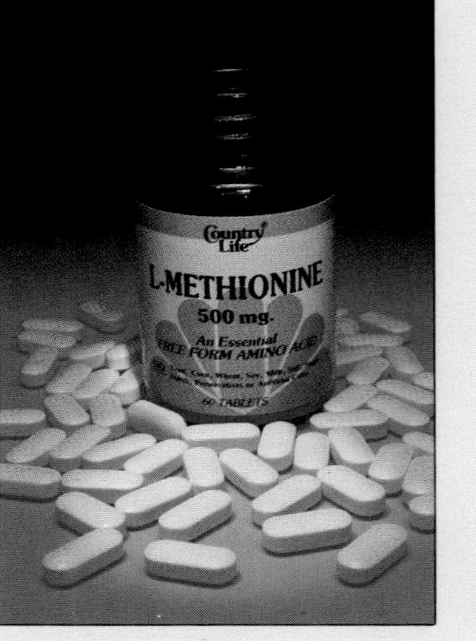 Methionine, a building block for proteins, is essential to human nutrition. *(Charles D. Winters)*

H　:O:
H—C—O—N—O:
　:O:
H—C—O—N—O:
　:O:
H—C—O—N—O:
H

Nitroglycerin, $C_3H_5N_3O_9$

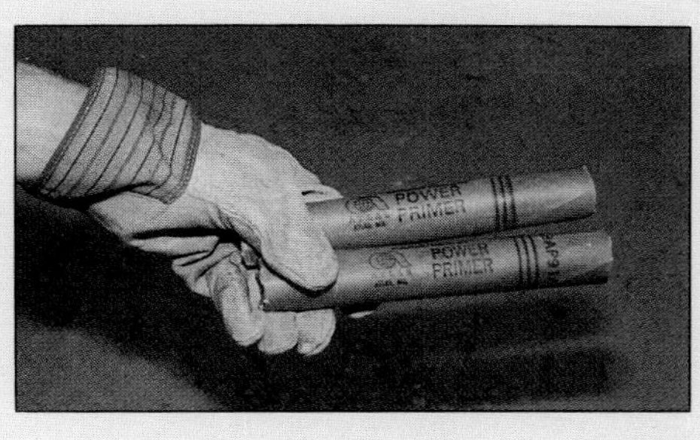

Transparent, oily liquid with a sweet taste; explodes with heat or concussion. Used as an explosive and in the manufacture of dynamite, which is 75% nitroglycerin, mixed with a solid so it can be handled more safely. Taken orally or absorbed through the skin, it causes dilation of blood vessels and is used as a medicine for that effect.

Dynamite is made by absorbing liquid nitroglycerin in a powder and forming the product into sticks wrapped in paper; see the Inside Chemistry essay at the end of Chapter 6. *(David R. Frazier, Photolibrary)*

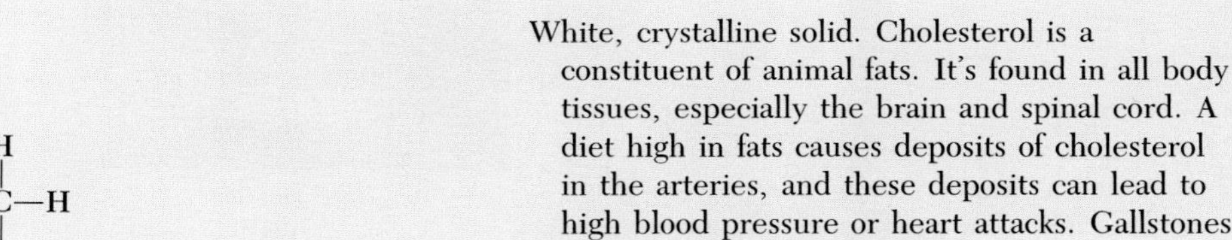

Cholesterol, $C_{27}H_{46}O$

White, crystalline solid. Cholesterol is a constituent of animal fats. It's found in all body tissues, especially the brain and spinal cord. A diet high in fats causes deposits of cholesterol in the arteries, and these deposits can lead to high blood pressure or heart attacks. Gallstones are almost pure cholesterol.

Cholesterol crystals, artificially colored. *(Courtesy of P. Dieppe from* Atlas of Clinical Rheumatology, *Gower Medical Publishing, London, UK, 1986.)*

(continued)

Table 7.2 Examples of Important Covalent Compounds *(continued)*

Line Formula, Name, and Molecular Formula	Properties and Uses
Cocaine, $C_{17}H_{21}NO_4$	White, crystalline solid. A dangerously addictive narcotic. 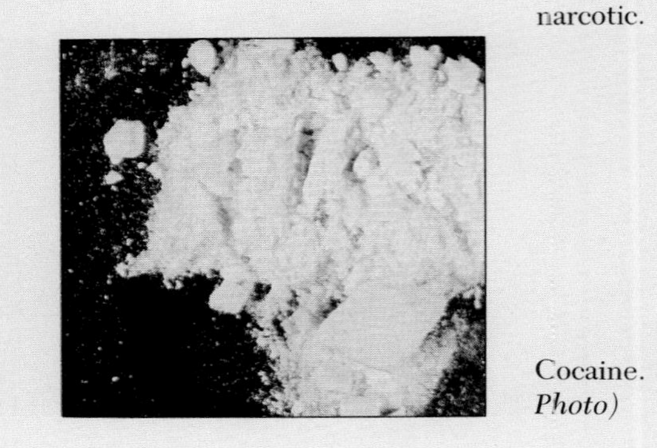 Cocaine. *(Martin/Custom Medical Stock Photo)*
Aspirin, $C_9H_8O_4$	White, crystalline solid. Used as a pain reliever and fever reducer. Many doctors recommend that adults take an aspirin a day to reduce the risk of heart attack. 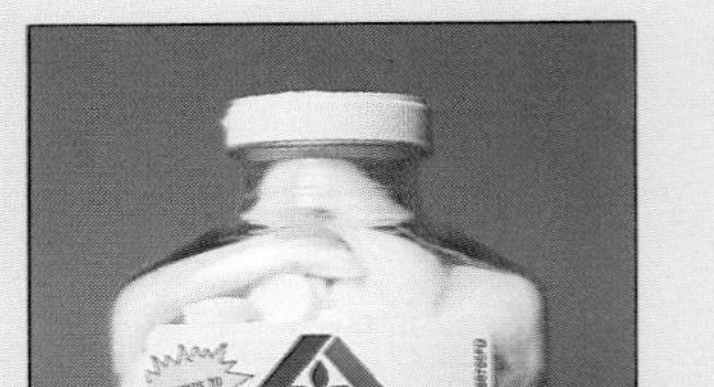 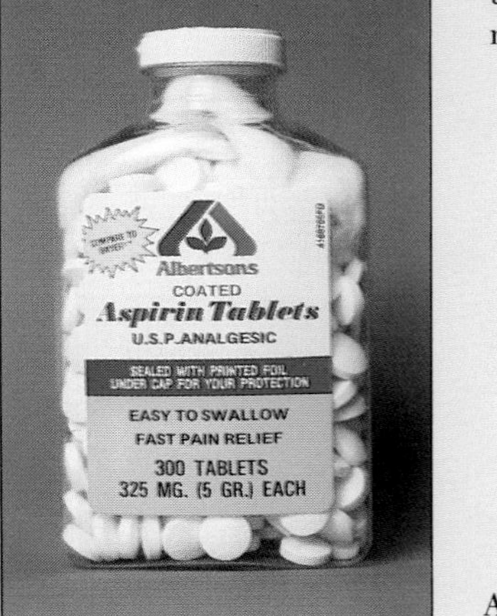 Aspirin. *(David R. Frazier, Photolibrary)*

Inside Chemistry

What are steroids?

As shown in Table 7.2, the line formula for cholesterol is

Cholesterol is one member of the family of compounds called **steroids;** in all of the compounds in this family the molecules have the system of four rings shown in the formula for cholesterol. This ring system can be represented by an abbreviated outline:

Ring system common to all steroids

Cholesterol, $C_{27}H_{46}O$

Thousands of steroids are known, and many of them have important biological activities. These are four examples:

Progesterone, $C_{21}H_{30}O_2$, a female sex hormone, prepares the lining of the uterus for implantation of an ovum.

Testosterone, $C_{19}H_{28}O_2$, a male sex hormone, causes the appearance of male secondary sex characteristics.

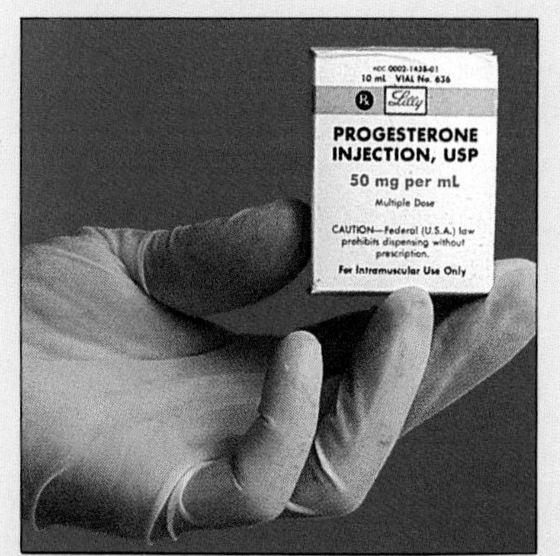

The steroid progesterone is a female sex hormone. *(Custom Medical Stock Photo, Inc.)*

Cortisone, $C_{21}H_{28}O_5$, is an antiinflammatory drug.

Norethindrone, $C_{20}H_{26}O_2$, is an essential component of oral contraceptives.

Steroids exert their powerful biological effects in very small amounts. The typical amount of norethindrone in a contraceptive pill, for example, is 1.0 mg.

Chapter Summary: Covalent Bonds

Section	Subject	Summary	Check When Learned
	Chemical bond	Force that holds atoms together in combination.	☐
	Chemical reaction	Process in which substances interact with one another to form new and different substances. Process in which chemical bonds are broken or formed.	☐
7-1, 7-2	Covalent bond	Two electrons shared between two atoms.	☐
7-1	Cause of formation of covalent bond	A covalent bond consists of two electrons shared by orbital overlap between two atoms. A certain amount of orbital overlap lowers the total energy of the atoms by balancing their attractive and repulsive interactions and makes them more stable.	☐
7-2	Single covalent bond	Bond in which two, and only two, electrons (one pair) are shared between two atoms.	☐
7-2	Double covalent bond	Bond in which four electrons (two pairs) are shared between two atoms.	☐
7-2	Triple covalent bond	Bond in which six electrons (three pairs) are shared between two atoms.	☐
7-2	Octet rule	Two versions: (1) A hydrogen atom that forms a covalent bond will share 2 electrons, and every other atom that forms a covalent bond will share enough electrons to give it a total of 8 shared and unshared electrons in its valence shell. (2) An atom that forms a covalent bond will share enough electrons to give it the same number of valence electrons as a noble-gas atom: 2 valence electrons for a bonded hydrogen atom and 8 valence electrons for any other bonded atom.	☐

Section	Subject	Summary	Check When Learned
7-2	Molecule	Group of two or more atoms bonded to one another.	☐
7-2	Formula	Collection of symbols representing a molecule.	☐
7-2	Reactants	Substances that enter into a chemical reaction.	☐
7-2	Products	Substances that are produced from a chemical reaction.	☐
7-2	Chemical equation	A description of a chemical reaction, using symbols and formulas. The general statement of an equation is, "Reactants form products."	☐
7-2	Coefficient	A multiplier in a chemical equation that shows the number of times a symbol or formula appears in the equation, for example, the 2 in $\underset{\text{hydrogen atoms}}{2\,H\cdot} \rightarrow \underset{\text{hydrogen molecule}}{H:H}$ If no coefficient is shown, the multiplier is assumed to be 1.	☐
7-2, 7-3	Nonpolar covalent bond	Bond in which electrons are shared equally between two atoms with the same electronegativity.	☐
7-3	Polar covalent bond	Bond in which electrons are shared unequally between two atoms with different electronegativities.	☐
7-3	$\delta+$ or $\delta-$	Symbols, using lowercase Greek delta, which show partial positive or partial negative charge on each end of a polar covalent bond.	☐
7-3	Electronegativity	Ability of an atom to attract shared electrons toward itself.	☐
7-3	Variation of electronegativity among representative elements on the Periodic Table	In general, electronegativity increases upward in a group and to the right in a period.	☐
7-3	Order of increasing electronegativity among Cl, F, O, H, N, C	$H < C < Cl = N < O < F$	☐
7-4	Lewis formula	Combination of Lewis symbols that shows the bonding in a molecule and represents each valence electron by a dot; for example, $\underset{\cdot\cdot}{\overset{\cdot\cdot}{O}}::C::\underset{\cdot\cdot}{\overset{\cdot\cdot}{O}}$.	☐
7-4	Line formula	Modified version of Lewis formula in which each pair of electrons in a bond is represented by a line, for example, $\underset{\cdot\cdot}{\overset{\cdot\cdot}{O}}=C=\underset{\cdot\cdot}{\overset{\cdot\cdot}{O}}$.	☐
7-4	Molecular formula	Formula that shows only the number of atoms of each element in a molecule. The subscript on the symbol shows the number of atoms (no subscript means 1 atom)—for example, CO_2.	☐

Section	Subject	Summary	Check When Learned
7-5	Steps in writing line formula from molecular formula	1. Write the Lewis symbol for each atom and count the number of valence electrons shown in all of these Lewis symbols. 2. Write the symbols for the atoms in this pattern: At the center, write the symbol (not the Lewis symbol) for the atom that has only one atom in the molecule. Around it write the symbols (not Lewis symbols) for the other atoms in the molecule. Connect each outside atom to the central atom by a single bond. 3. Put the remaining electrons around the outside atoms as unshared electrons, to give each outside atom a total of 8 shared and unshared electrons. If there are electrons left over, put them as unshared electrons on the central atom. 4. If the central atom doesn't have 8 electrons, change an unshared electron pair on an outer atom to a shared pair (covalent bond) with the central atom, and continue this process until the central atom has 8 electrons. Make the bonding as symmetrical as possible.	☐
7-6	Relationship between line formula and shape of molecule	A molecule with only two atoms necessarily has a linear line formula and a linear shape. There is no necessary relationship between the line formula and shape of a molecule with more than two atoms.	☐
7-6	Shape of water molecule	Bent, with an angle slightly greater than a right angle between the two H—O bonds: $H\diagup\ddot{\underset{..}{O}}\diagdown H$.	☐
7-7	Molecular mass	Mass of a molecule, the sum of the masses of its atoms, in amu.	☐

Problems

Assume you can use the Periodic Table at the front of the book unless you're directed otherwise. Answers to odd-numbered problems are in Appendix 1.

Symbols, Formulas, and Equations (Sections 7-1 through 7-4)

1. In your own words, write a one-sentence statement that explains the differences among the terms *symbol, formula,* and *equation*.

2. In your own words, write a one-sentence statement that explains the difference between a *reactant* and a *product*.

3. In your own words, explain the difference between a *coefficient* and a *subscript*.

4. In the following equation, identify the coefficient for each reactant:

$$\cdot\dot{C}: + 2\,\cdot\ddot{\underset{..}{O}}\cdot \rightarrow \ddot{\underset{..}{O}}{=}C{=}\ddot{\underset{..}{O}}$$

5. In your own words, write a one-sentence statement that explains the difference between a *reaction* and an *equation*.

6. An equation can be thought of as a before-and-after picture of a reaction. Do the products correspond to the before or the after part of the picture?

7. Translate this equation into a sentence:

$$\cdot\ddot{\underset{\cdot\cdot}{F}}: + \cdot\ddot{\underset{\cdot\cdot}{F}}: \rightarrow :\ddot{\underset{\cdot\cdot}{F}}—\ddot{\underset{\cdot\cdot}{F}}:$$

8. Using Lewis symbols and line formulas, translate this sentence into an equation: Two hydrogen atoms and one oxygen atom form one water molecule.

9. What is the difference between (a) an atom and a symbol (b) a formula and a molecule?

10. What is a chemical bond?

11. In your own words, write a one-sentence statement that explains why a covalent bond forms.

12. In your own words, define *covalent bond.*

Octet Rule (Sections 7-2 and 7-3)

13. In the following formula, circle each atom and its electrons to show that it obeys the octet rule. Identify each bond as single, double, or triple.

```
            H
            |
H—C≡C—C=C—H
               |
               H
```

14. In the following formula, circle each atom and its electrons to show that it obeys the octet rule. Identify each bond as single, double, or triple.

```
    H  :O:
    |   ‖   ..
H—C—C—O—H
    |       ..
    H
```

15. In the line formula for Vitamin C in Table 7.2, circle one of the carbon atoms and its electrons and one of the oxygen atoms and its electrons to show that they obey the octet rule.

16. In the line formula for methionine in Table 7.2, circle the sulfur atom and its electrons and the nitrogen atom and its electrons to show that they obey the octet rule.

Electronegativity (Section 7-3)

17. Using the Periodic Table at the front of the book, predict which atom in each pair will have the higher electronegativity: (a) C or O (b) S or O (c) Rb or I.

18. Using the Periodic Table at the front of the book, predict which atom in each pair will have the higher electronegativity: (a) F or Cl (b) P or Cl (c) F or P.

19. In your own words, write a one-sentence statement that describes the difference between a nonpolar covalent bond and a polar covalent bond.

20. In each of the following formulas, add $\delta+$ and $\delta-$ to show the direction of bond polarity.

(a) $:\ddot{\underset{\cdot\cdot}{Br}}—\ddot{\underset{\cdot\cdot}{Cl}}:$ (b) $:\ddot{\underset{\cdot\cdot}{Br}}—\ddot{\underset{\cdot\cdot}{Br}}:$ (c) $H—\ddot{\underset{\cdot\cdot}{Br}}:$

21. In the following formula, add $\delta+$ and $\delta-$ to show the direction of polarity for each bond.

```
       :O:
   ..   ‖   ..
H—O—N—O:
   ..       ..
```

22. Some bonds are more polar than others. For each of the following pairs, decide which bond you would expect to be more polar and explain your decision: (a) a C–N bond or a C–O bond (b) an Si–F bond or an Si–Cl bond.

Line Formulas and Molecular Formulas (Section 7-5)

23. Write line formulas for these molecular formulas and add $\delta+$ and $\delta-$ to show the direction of polarity for each bond: (a) Br_2 (b) BrF (c) $SiCl_4$.

24. Write line formulas for these molecular formulas and add $\delta+$ and $\delta-$ to show the direction of polarity for each bond: (a) AsH_3 (b) HIO_3 (c) H_2Te.

25. In the following table, add the line formula for each molecular formula.

	Nonpolar Bond	**Polar Bond**
Single bond	H_2	HF
Double bond	C_2H_4	CO_2
Triple bond	N_2	CO

26. Write line formulas for (a) CH_4 (b) C_2H_4 (c) C_2H_2.

27. Write line formulas for (a) N_2 (b) N_2H_2 (c) N_2H_4.

28. See the partial Periodic Table in Figure 7.5. For each element in the table, write the line formula for the molecule that you would expect to be formed between one atom of that element and the appropriate number of hydrogen atoms.

29. Using Z to stand for any element in group 4A, the general formula for the compounds of hydrogen with the elements in group 4A is ZH_4. Write a similar general formula for the compounds of hydrogen with the elements in group 5A.

30. Using Z to stand for any element in group 4A, the general formula for the compounds of hydrogen with the elements in group 4A is ZH_4. Write a similar general formula for the compounds of hydrogen with the elements in group 7A.

Molecular Mass (Section 7-7)

31. Calculate the molecular mass of (a) CH_4 (b) H_2SO_4.

32. Calculate the molecular mass of these compounds from Table 7.2: (a) nitroglycerin, $C_3H_5N_3O_9$ (b) cholesterol, $C_{27}H_{46}O$.

Ionic compounds. Clockwise from upper left: KCl, NaCl, CaF_2, Fe_2O_3, $CuBr_2$. *(Charles D. Winters)*

8 Ionic Bonds

Table salt, sodium chloride, NaCl. *(Barry L. Runk from Grant Heilman Photography, Inc.)*

The forces by which atoms attach to one another in compounds are called chemical bonds. In Chapter 7 we saw that atoms of nonmetals can attach to one another by sharing pairs of electrons, following the fundamental rule of chemical bonding, the octet rule. A shared pair of electrons is a covalent bond, and a group of atoms held together by covalent bonds is a molecule.

In this chapter we'll see that atoms of metals can react with atoms of nonmetals by transferring electrons from the metal atoms to the nonmetal atoms, to form positively and negatively charged particles called **ions.** The formation of ions from atoms of the representative elements also follows the octet rule. The attraction of oppositely charged ions for one another is described as an **ionic bond,** and compounds that consist of ions are called **ionic compounds.**

Ionic compounds have very different properties from covalent compounds. Many ionic compounds are in everyday use, and some of them (sodium chloride is the most familiar example) are essential in human nutrition.

8-1 Three important properties of atoms—atomic radius, ionization energy, and electron affinity—change in regular ways across the Periodic Table

Our description of ionic bonding will be based on three important properties of atoms: atomic radius, ionization energy, and electron affinity.

Atomic Radius

The boundary of an atom is the region of space occupied by the electrons in its outermost energy level; according to quantum mechanics, locations of electrons can't be known precisely, so the sizes of atoms can't be measured exactly. One way to designate an approximate size for an atom is shown in Figure 8.1: The radius of a hydrogen atom is taken to be half the distance between the nuclei in a hydrogen molecule. An atomic radius measured in this way is called a covalent radius.

The atomic radii of the representative elements vary in a regular way across the Periodic Table, as shown in Figure 8.2. As we'd expect, atomic radii in-

Figure 8.1
Covalent radius of the hydrogen atom. The internuclear distance in the hydrogen molecule, H_2, is 75 pm, so the covalent radius of one hydrogen atom can be designated as half of this distance, 38 pm.

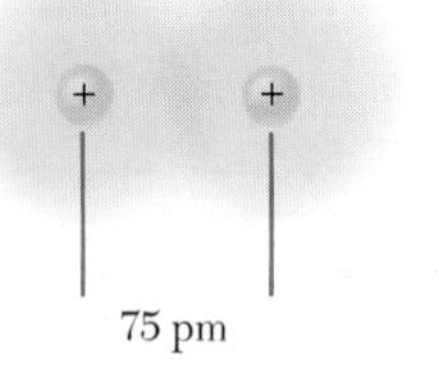

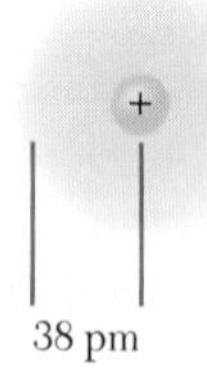

	1A							8A
1	H 38	2A	3A	4A	5A	6A	7A	He 5
2	Li 152	Be 111	B 88	C 77	N 70	O 66	F 64	Ne 70
3	Na 186	Mg 160	Al 143	Si 117	P 110	S 104	Cl 99	Ar 94
4	K 231	Ca 197	Ga 122	Ge 122	As 121	Se 117	Br 114	Kr 109
5	Rb 244	Sr 215	In 162	Sn 140	Sb 141	Te 137	I 133	Xe 130
6	Cs 262	Ba 217	Tl 171	Pb 175	Bi 146	Po 165	At	Rn 140

Figure 8.2
Atomic radii for some representative elements. The radii are given in picometers. In general, atomic radii decrease upward in a group and to the right in a period.

crease downward in a group because, with each successive period, electrons are added to a new energy level farther from the nucleus. It's surprising at first to find that atomic radii in general *decrease* from left to right in a period. The cause of this decrease in radius is the increase in the number of protons in the nucleus. Left to right across a period, more electrons are added to the same energy level, and more protons are added to the nucleus. As more protons are added to the nucleus, the positive nuclear charge increases and the nucleus pulls more strongly on all of the negative electrons, drawing them closer.

Ionization Energy

The **ionization energy** is the amount of energy required to remove the outermost electron from an atom. Imagine that you can reach into an atom, hold its nucleus with one hand, and with your other hand pull the outermost electron off the atom. The amount of energy you'd use to remove the outermost electron would be the atom's ionization energy.

The electron that's farthest from the nucleus is the easiest to remove, for two reasons. One reason is that the force of attraction between the nucleus and an electron becomes weaker as the distance between them becomes greater.◆ The second reason is that the inner electrons weaken the attraction of the outer electrons toward the nucleus. All electrons have negative charges, so they repel one another. Between the outer electrons and the nucleus are inner electrons that repel the outer electrons, weakening their attachment to the atom. This repulsion of outer electrons by inner electrons is called the **shielding effect** because it operates to shield outer electrons from the pull of the nuclear charge.

◆The attractive force between two particles with opposite electrical charges decreases very rapidly as the distance between them increases.

Figure 8.3 is a diagram of the process in which the ionization energy is absorbed to remove the outermost electron from a lithium atom.

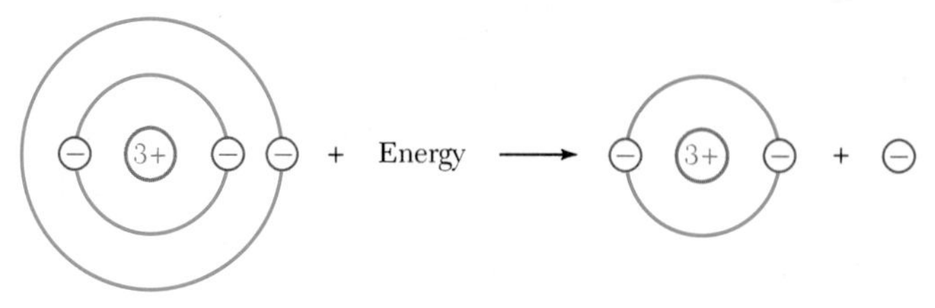

Figure 8.3
Ionization energy and shielding effect for lithium. The electron configuration for lithium is Li^3: $1s^2 2s^1$. It takes energy to pull the outermost (2s) electron off of the lithium atom, because the charge on the nucleus and the charge on the electron attract one another. The 1s electrons repel the 2s electron, weakening its attraction toward the nucleus, and this repulsion is called the **shielding effect** of the 1s electrons.

With a few exceptions that are unimportant for our purposes, ionization energy increases upward in a group and to the right in a period. These trends result from changes in nuclear charge, atomic radius, and the shielding effect. Upward in a group the atomic number decreases, so the number of protons in the nucleus decreases and the nuclear charge decreases. But the outermost electron is closer to the nucleus and is shielded by fewer interior electrons, so the attraction of the nucleus for the outermost electron increases and ionization energy increases. From left to right in a period the atomic number increases, so the number of protons in the nucleus increases and the nuclear charge increases. The outermost electron is closer to a higher nuclear charge and is shielded by the same number of interior electrons, so the attraction of the nucleus for the outermost electron increases and ionization energy increases.

Electron Affinity

The positive charge on the nucleus of an atom is strong enough to attract and hold an extra electron. The **electron affinity** is the amount of energy released when an electron is added to the outermost energy level of an atom. Figure 8.4 shows the process in which a lithium atom adds an electron and releases energy. Across the Periodic Table electron affinity generally increases from left to right and from bottom to top, and these trends can be explained by changes in nuclear charge, atomic radius, and the shielding effect. From left to right across a period nuclear charge increases, atomic radius decreases, and the shielding

Figure 8.4
Electron affinity for lithium. The nucleus of the lithium atom can attract and hold a fourth electron. The energy released when the fourth electron is added is called the **electron affinity.**

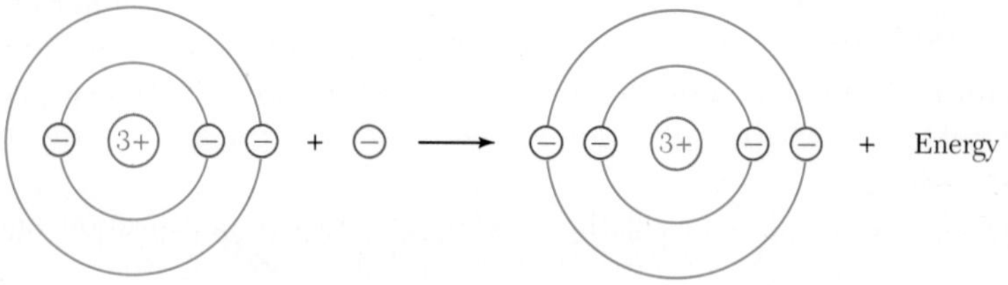

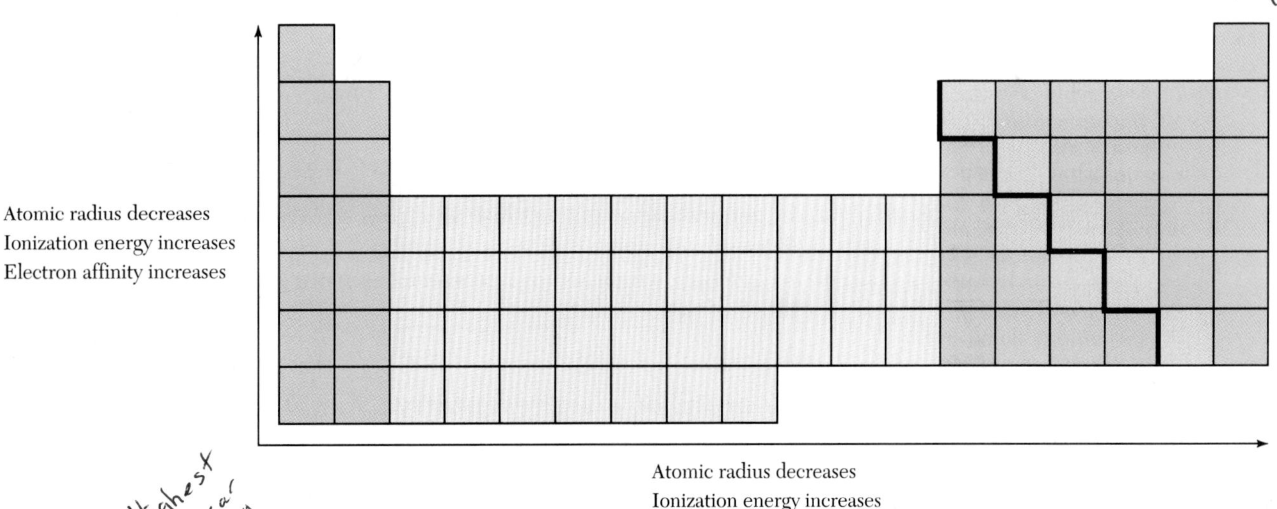

Figure 8.5
General trends across the Periodic Table in the atomic radius, ionization energy, and electron affinity.

effect remains constant, so the attractive power of the nucleus for another electron increases. Upward in a group nuclear charge decreases, but atomic radius and the shielding effect both decrease, and the net result is an increasing attraction for an added electron.

Figure 8.5 summarizes the trends in atomic radius, ionization energy, and electron affinity across the Periodic Table.◆ We can also summarize the periodic trends in atomic radius, ionization energy, and electron affinity this way: Upward in a group and to the right in a period, atoms are smaller, it's harder to remove an electron, and it's easier to add an electron.

Use the Periodic Table at the front of the book to answer the following Exercises.◆

◆Ionization energy and electron affinity can be measured quantitatively and expressed in energy units, but for our purposes in this book, we can use the simpler approach of working with their qualitative variations across the Periodic Table.

◆Answers to Exercises are in Appendix 1.

Exercise 8.1

Arrange each set of three elements in order of increasing atomic radius: (a) calcium, krypton (Kr), arsenic (As) (b) carbon, silicon, nitrogen

Exercise 8.2

Arrange these elements in order of increasing nuclear charge: potassium, phosphorus, fluorine

Exercise 8.3

Arrange these elements in order of increasing shielding effect experienced by their outermost electrons: carbon, nitrogen, phosphorus.

Exercise 8.4

Arrange these elements in order of decreasing ionization energy: potassium, chlorine, argon (Ar).

Exercise 8.5

Arrange these elements in order of decreasing electron affinity: selenium (Se), bromine, chlorine.

Exercise 8.6

(a) In general, which atoms lose electrons more easily, those of metals or those of nonmetals?
(b) In general, which atoms add electrons more easily, those of metals or those of nonmetals?

8-2 An atom of a representative metal will easily lose electrons until it has the same configuration of valence electrons as an atom of the preceding noble gas on the Periodic Table

The ionization energy—the amount of energy required to remove the outermost electron from an atom—increases from left to right on the Periodic Table. Metals are on the left on the Periodic Table, so they have low ionization energies. Metals lose electrons easily.

When a metal atom loses an electron, it forms a positively charged species called a **cation.** The equation in Lewis symbols below shows the loss of one electron from a sodium atom to form a cation with a 1+ charge.

$Na\cdot$	$\rightarrow$	Na^+	+	e^-
sodium atom		sodium cation		electron
$1s^22s^22p^63s^1$		$1s^22s^22p^6$		
11 protons = 11+		11 protons = 11+		
11 electrons = 11−		10 electrons = 10−		
total charge = 0		total charge = 1+		
radius = 186 pm		radius = 95 pm		

It's important to understand clearly why the loss of an electron from an atom produces a species with a positive charge. The sodium atom has 11 protons (11 positive charges) in its nucleus and 11 electrons (11 negative charges) around the nucleus. Because the number of protons and electrons (+ and − charges) is the same, the sodium atom, like all atoms, is electrically neutral: Its overall charge is 0. When the sodium atom loses an electron, the cation that is produced has 11 protons and 10 electrons, so its overall charge is 11+ + 10− = 1+. For every electron lost from an atom, the corresponding cation will have one positive charge.

When the outermost electron is lost from a sodium atom, the remaining 10 electrons in the sodium cation are much closer to the nucleus and much more strongly held. You can see why by looking at the electron configurations of the Na atom and the Na^+ cation. In the Na atom ($1s^22s^22p^63s^1$) the outermost electron is in the *third* energy level, and in the Na^+ cation ($1s^22s^22p^6$) the outermost electrons are in the *second* energy level, much closer to the nucleus. The change in radius is dramatic: The radius of the Na atom is 186 pm and the radius of the Na^+ cation is 95 pm.

The outermost electron in a sodium atom is loosely held and easily removed, but the remaining 10 electrons in the sodium cation are much closer to the nucleus and held so strongly that they are, except under extraordinary conditions, unremovable. For this reason a sodium atom easily loses one electron to form a 1+ cation but resists losing more than one electron.

There are important relationships, shown below, among the electron configurations of a sodium atom, a sodium cation, and a neon atom. When a sodium atom loses one electron, the 1+ cation that is formed has the same number of electrons and the same electron configuration as an atom of neon, which is the next element before sodium on the Periodic Table.

Na·	Na^+	:Ne: (with dots above and below)
sodium atom	sodium cation	neon atom
$1s^22s^22p^63s^1$	$1s^22s^22p^6$	$1s^22s^22p^6$

The Lewis symbol for the sodium atom shows its $3s^1$ electron, and that electron disappears in the symbol for the sodium cation. The Lewis symbol for the neon atom shows its $2s^22p^6$ electrons, and these eight electrons are referred to as an *octet*.

The relationship between the electron configurations of a sodium atom and a neon atom is shown on the partial Periodic Table in Figure 8.6. As the arrows in this figure show, the loss of one electron by an atom of any alkali metal will produce a cation with the same configuration of valence electrons as an atom of the preceding noble gas on the Periodic Table. In each alkali-metal cation formed, the remaining electrons will be strongly held. For this reason, any alkali-metal (group 1A) atom will easily lose one electron to form a 1+ cation—Li^+, Na^+, K^+, Rb^+, Cs^+, and Fr^+—but will resist losing more than one electron. The general equation is

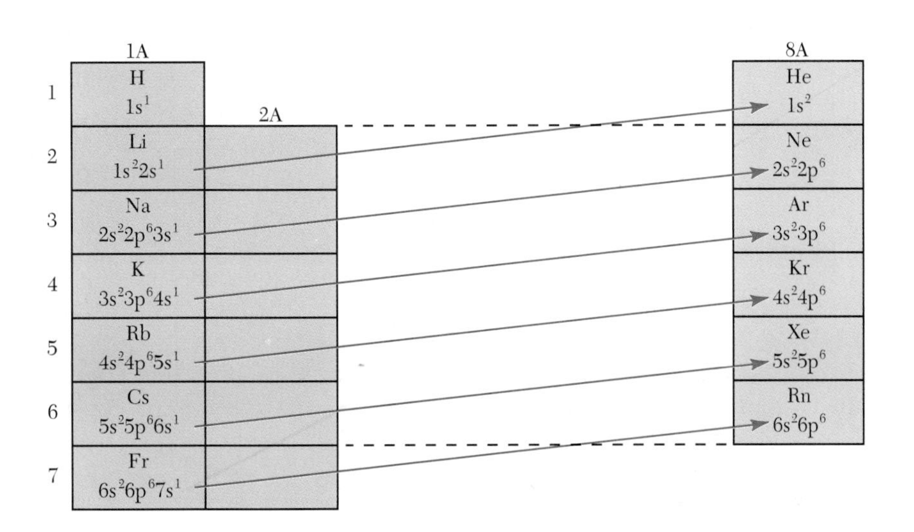

Figure 8.6
Partial electron configurations of the alkali metals and the noble gases. When an atom of an alkali metal (group 1A) loses one electron, the cation that's formed has the same configuration of valence electrons as an atom of the preceding noble gas (group 8A) on the Periodic Table.

Exercise 8.7

Write the equation in Lewis symbols for the formation of a 1+ cation from a potassium atom.

We can use similar reasoning to show that an atom of any alkaline-earth metal (group 2A) will easily lose two electrons to form a 2+ cation. Figure 8.7 shows the relationships between the electron configurations of the alkaline-earth metals and the noble gases.

When an atom of an alkaline-earth metal loses two electrons, the 2+ cation formed has the same configuration of valence electrons as an atom of the preceding noble gas on the Periodic Table. The electrons in the cation will be tightly held. For this reason, an atom of any alkaline-earth (group 2A) metal will easily lose two electrons to form a 2+ cation—Be^{2+}, Mg^{2+}, Ca^{2+}, Sr^{2+}, Ba^{2+}, and Ra^{2+}—but will resist losing more than two electrons. The general equation is

M:	→	M^{2+}	+	$2\ e^-$
alkaline-earth metal atom		alkaline-earth metal cation		2 electrons

Exercise 8.8

Write the equation in Lewis symbols for the formation of a 2+ cation from a calcium atom.

Exercise 8.9

How many protons and how many electrons are in the cation Sr^{2+}?

The examples of the metals in groups 1A and 2A point to a general rule for the formation of cations from atoms of the representative metals: An atom of a representative metal will easily lose electrons to form a cation that has the same configuration of valence electrons as an atom of the preceding noble gas on the

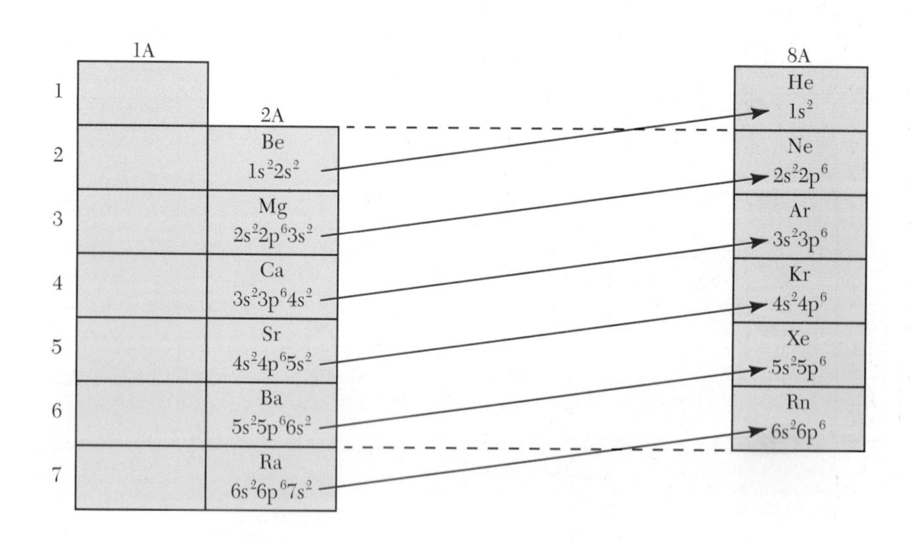

Figure 8.7
Partial electron configurations of the alkaline-earth metals and the noble gases. When an atom of an alkaline-earth metal (group 2A) loses two electrons, the cation that's formed has the same configuration of valence electrons as an atom of the preceding noble gas (group 8A) on the Periodic Table.

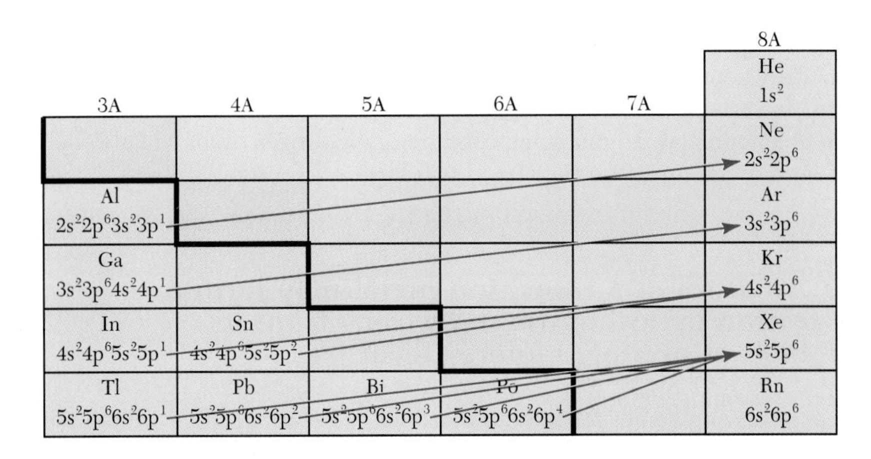

Figure 8.8
Partial electron configurations for the metals in groups 3A through 6A and the noble gases. When an atom of one of the metals loses all of the electrons in its outermost energy level, the cation that's formed has the same configuration of valence electrons as an atom of the preceding noble gas (group 8A) on the Periodic Table.

Periodic Table. This rule is very useful, as you'll see in much of your later work in this book, because with it you can use a Periodic Table to predict the charge on the cation of a representative metal: The charge will be the same as the group number. Because the atoms of the noble gases, except helium, have eight valence electrons, this rule can be thought of as a form of the octet rule.

The application of the preceding rule to the metals in groups 3A through 6A is shown in the partial Periodic Table in Figure 8.8. The atoms of some of these metals also form cations in which only the p electrons from the outermost energy level have been lost. An atom of tin, for example, can form a 2+ cation by losing its $5p^2$ electrons or a 4+ cation by losing its $5s^25p^2$ electrons.◆

◆The behavior of each of the representative metals in forming cations is discussed in more detail in Chapter 10.

Exercise 8.10

Write the equation in Lewis symbols for the formation of a 3+ cation from an aluminum atom.

You should learn the symbols and names for the cations of the representative metals shown in Table 8.1, because we'll use them often later in this book. The names follow a simple pattern. If a metal atom forms only one cation, the name for the cation is the same as the name for the atom: Na is a sodium atom and Na^+ is a sodium ion. If a metal atom forms more than one cation, the name for each cation contains a Roman numeral that shows the charge on the cation: Sn is a tin atom, Sn^{2+} is a tin(II) ion, and Sn^{4+} is a tin(IV) ion.◆ As these examples show, in naming a cation, the shorter term *ion* is usually used.◆

◆In writing a name that contains a Roman numeral, don't leave a space before the first parenthesis: The name of Sn^{2+} is tin(II) ion, not tin (II) ion.

◆An older system for naming cations used Latinized names for the elements, with the suffix *-ous* for the cation with the lower charge and the suffix *-ic* for the cation with the higher charge. In this system Sn^{2+} is stannous ion and Sn^{4+} is stannic ion. You'll still see these names occasionally.

Table 8.1 Symbols and Names for the Important Cations of the Representative Elements

Group 1A	Group 2A	Group 3A	Group 4A
H^+ hydrogen ion	Mg^{2+} magnesium ion	Al^{3+} aluminum ion	Sn^{2+} tin(II) ion
Li^+ lithium ion	Ca^{2+} calcium ion		Sn^{4+} tin(IV) ion
Na^+ sodium ion	Sr^{2+} strontium ion		Pb^{2+} lead(II) ion
K^+ potassium ion	Ba^{2+} barium ion		Pb^{4+} lead(IV) ion

Hydrogen is a special case. We saw in Chapter 7 that a hydrogen atom will form a covalent bond with another hydrogen atom or with an atom of a nonmetal. In forming a covalent bond a hydrogen atom behaves as if it were an atom of a nonmetal. Under some conditions a hydrogen atom will also behave as if it were an atom of a metal and will form a 1+ cation:

$$H\cdot \rightarrow H^+ + e^-$$

8-3 An atom of a transition metal may form a cation by losing the s electrons from its highest occupied energy level

The symbols and names for the ten transition-metal cations that will be important for your work in this book are shown in Table 8.2. You should learn them.

Table 8.2 Symbols and Names for the Important Cations of the Transition Metals

1+ Cations	2+ Cations	1+ or 2+ Cations	2+ or 3+ Cations
Ag^+ silver ion	Cd^{2+} cadmium ion	Cu^+ copper(I) ion	Fe^{2+} iron(II) ion
	Ni^{2+} nickel ion	Cu^{2+} copper(II) ion	Fe^{3+} iron(III) ion
	Zn^{2+} zinc ion	Hg_2^{2+} mercury(I) ion	
		Hg^{2+} mercury(II) ion	

For seven of the cations shown in Table 8.2—Ag^+, Cd^{2+}, Cu^+, Fe^{2+}, Hg^{2+}, Ni^{2+}, and Zn^{2+}—cation formation occurs by the loss of the s electrons from the highest energy level of the corresponding atom. For example, the silver atom, whose abbreviated electron configuration is $[Kr]^{36}5s^14d^{10}$, loses its $5s^1$ electron to form the Ag^+ cation, and the zinc atom, whose abbreviated electron configuration is $[Ar]^{18}4s^23d^{10}$, loses its $4s^2$ electrons to form the Zn^{2+} cation.

Exercise 8.11

Write the abbreviated electron configuration for an iron atom and circle the electrons that are lost in forming the iron(II) ion.

A description of the formation of the other three ions shown in the table—Cu^{2+}, Fe^{3+}, and Hg_2^{2+}—requires a more advanced discussion of bonding than we'll use in this book. The mercury(I) ion, Hg_2^{2+}, is unique in consisting of two metal cations bonded to one another.

8-4 An atom of a nonmetal, except carbon, will easily gain electrons until it has the same configuration of valence electrons as an atom of the next noble gas on the Periodic Table

The electron affinity—the energy released when an electron is added to an atom—increases to the right on the Periodic Table. Nonmetals are on the right on the Periodic Table, so nonmetals have high electron affinities. Nonmetals, except carbon and the noble gases, gain electrons easily.

When a nonmetal atom gains an electron, it forms a negatively charged species called an **anion.** The equation in Lewis symbols below shows the addition of one electron to a fluorine atom to form an anion with a 1− charge.

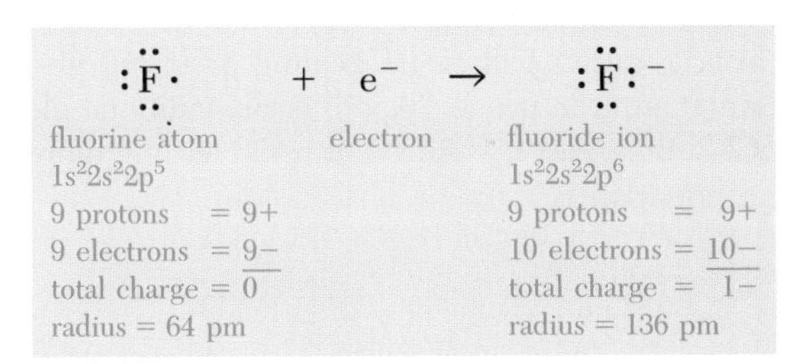

The anion formed from the fluorine atom is called the *fluoride ion.* As you'll see in further examples in this Section, an anion that is formed by adding one or more electrons to an atom of a nonmetal is named by changing the suffix of the name of the nonmetal to *-ide.*

The fluorine atom contains 9 protons and 9 electrons, so it's electrically neutral. The fluoride ion has 9 protons and 10 electrons, so it carries a 1− charge. For every electron gained by an atom, the corresponding anion will have one negative charge.

Adding an electron to an atom causes it to expand, so the anionic radius is always greater than the atomic radius, and the change can be very large: The radius of the fluorine atom is 64 pm, and the radius of the fluoride ion is 136 pm.

There are important relationships, shown below, among the electron configurations of a fluorine atom, a fluoride ion, and a neon atom. When a fluorine atom gains one electron, the fluoride ion that is formed has the same number of electrons, and the same configuration of valence electrons, as an atom of neon, the next noble gas on the Periodic Table.

:F·	:F:$^-$	:Ne:
fluorine atom	fluoride ion	neon atom
$1s^22s^22p^5$	$1s^22s^22p^6$	$1s^22s^22p^6$

The relationship between the electron configurations of a fluorine atom and a neon atom is shown on the partial Periodic Table in Figure 8.9. As shown in

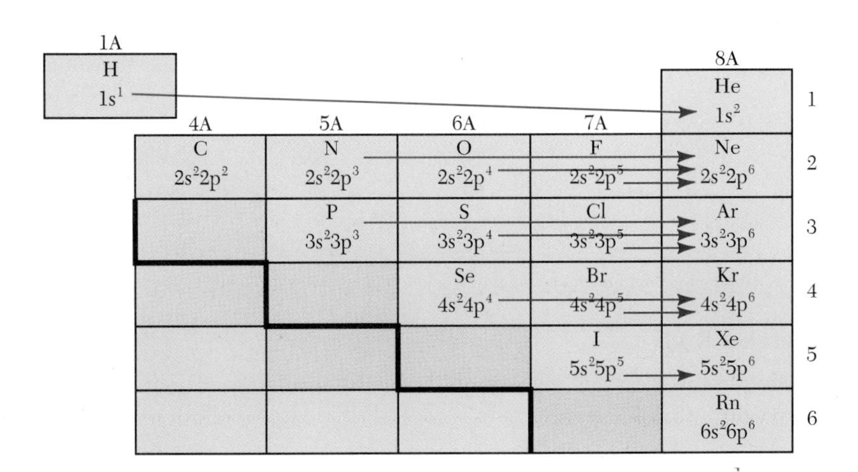

Figure 8.9
Partial electron configurations of the nonmetals and the noble gases. When a halogen atom (group 7A) or a hydrogen atom (group 1A) gains one electron, the anion that's formed has the same configuration of valence electrons as the atom of the next noble gas (group 8A) on the Periodic Table. Similarly, atoms of the nonmetals in group 6A add two electrons and atoms of the nonmetals in group 5A add three electrons to achieve noble-gas electron configurations.

this figure, the gain of one electron by an atom of any nonmetal in group 7A (the halogens) will produce an anion with the same configuration of valence electrons as an atom of the next noble gas on the Periodic Table.

A noble-gas electron configuration is very stable, so any atom or ion that has a noble-gas configuration will resist gaining or losing electrons. For this reason, any nonmetal atom in group 7A will easily gain one electron to form a 1− anion—F^-, Cl^-, Br^-, and I^-—but will resist adding more than one electron. The general equation is

$$:\underset{\cdot\cdot}{\overset{\cdot\cdot}{X}}\cdot \quad + \quad e^- \quad \rightarrow \quad :\underset{\cdot\cdot}{\overset{\cdot\cdot}{X}}:^-$$

halogen atom | electron | halide ion

Exercise 8.12

Write the equation in Lewis symbols for the formation of an iodide ion from an iodine atom.

As shown in Figure 8.9, a hydrogen atom will gain one electron to form an anion with the same electron configuration as a helium atom. The equation is

$$H\cdot \quad + \quad e^- \quad \rightarrow \quad H:^-$$

hydrogen atom | electron | hydride ion

We saw earlier that a hydrogen atom can share its electron in a covalent bond or lose its electron to form the H^+ cation. The reactions by which a hydrogen atom can lose, gain, or share its electron are of great importance, as you'll see in Chapters 9, 10, 11, 12, 14, 15, and 16.

We can use similar reasoning to describe the formation of anions from atoms of the nonmetals in groups 6A and 5A. The general equations are

$$\cdot\underset{\cdot\cdot}{\overset{\cdot\cdot}{Y}}\cdot \quad + \quad 2\,e^- \quad \rightarrow \quad :\underset{\cdot\cdot}{\overset{\cdot\cdot}{Y}}:^{2-}$$

atom of group 6A nonmetal | 2 electrons | anion of group 6A nonmetal

$$\cdot\underset{\cdot}{\overset{\cdot\cdot}{Z}}\cdot \quad + \quad 3\,e^- \quad \rightarrow \quad :\underset{\cdot\cdot}{\overset{\cdot\cdot}{Z}}:^{3-}$$

atom of group 5A nonmetal | 3 electrons | anion of group 5A nonmetal

Exercise 8.13

Write the equation in Lewis symbols for the formation of an oxide ion from an oxygen atom.

Exercise 8.14

Write the equation in Lewis symbols for the formation of a nitride ion from a nitrogen atom. How many protons and how many electrons are in a nitride ion?

This is the general rule for the formation of an anion from an atom of a nonmetal: An atom of a nonmetal, except carbon, will easily gain electrons until it has the same configuration of valence electrons as an atom of the next noble gas on the Periodic Table. As shown in the preceding discussion, you can use this rule to predict the charge on the anion of a nonmetal.

An anion can be represented by a Lewis symbol, for example, $:\ddot{\underset{\cdot\cdot}{O}}:^{2-}$, or by a symbol without electrons, for example, O^{2-}. Table 8.3 shows the symbols and names for the anions of the nonmetals; you should learn them because you'll be using them often later in this book.

Table 8.3 Symbols and Names for the Anions of the Nonmetals

Group 1A	Group 5A	Group 6A	Group 7A
H^- hydride ion	N^{3-} nitride ion	O^{2-} oxide ion	F^- fluoride ion
	P^{3-} phosphide ion	S^{2-} sulfide ion	Cl^- chloride ion
		Se^{2-} selenide ion	Br^- bromide ion
			I^- iodide ion

8-5 Atoms of metals react with atoms of nonmetals to form cations and anions, following the octet rule

Atoms of metals easily lose electrons and atoms of nonmetals easily gain electrons, so it's not surprising to learn that a common type of chemical reaction is the transfer of electrons from metal atoms to nonmetal atoms to form cations and anions. For example, a sodium atom will react with a fluorine atom:

$$\underset{\text{sodium atom}}{Na\cdot} \quad + \quad \underset{\text{fluorine atom}}{:\ddot{\underset{\cdot\cdot}{F}}\cdot} \quad \rightarrow \quad \underset{\text{sodium ion}}{Na^+} \quad + \quad \underset{\text{fluoride ion}}{:\ddot{\underset{\cdot\cdot}{F}}:^-}$$

The sodium ion and the fluoride ion produced by this reaction have opposite electrical charges, so they attract one another. The attraction between a cation and an anion is an **ionic bond.** A compound that consists of cations and anions attracted to one another by ionic bonds is called an **ionic compound.**

A magnesium atom will react with two fluorine atoms. Each magnesium atom loses two electrons and each fluorine atom gains one:

$$\underset{\text{magnesium atom}}{Mg:} \quad + \quad \underset{\text{fluorine atoms}}{2\ :\ddot{\underset{\cdot\cdot}{F}}\cdot} \quad \rightarrow \quad \underset{\text{magnesium ion}}{Mg^{2+}} \quad + \quad \underset{\text{fluoride ions}}{2\ :\ddot{\underset{\cdot\cdot}{F}}:^-}$$

Two sodium atoms will react with one oxygen atom. Each sodium atom loses one electron and each oxygen atom gains two:

$$\underset{\text{sodium atoms}}{2\,\mathrm{Na}\cdot} \quad + \quad \underset{\text{oxygen atom}}{\cdot\underset{\cdot\cdot}{\overset{\cdot\cdot}{\mathrm{O}}}\cdot} \quad \rightarrow \quad \underset{\text{sodium ions}}{2\,\mathrm{Na}^{+}} \quad + \underset{\text{oxide ion}}{:\underset{\cdot\cdot}{\overset{\cdot\cdot}{\mathrm{O}}}:^{2-}}$$

As you can see from these examples, the numbers of metal atoms and nonmetal atoms that will react with one another to form cations and anions will be determined by the numbers of valence electrons in the atoms. In general, the atom of a representative metal will lose the electrons in its outermost energy level, and the resulting cation will have the same configuration of valence electrons as an atom of the preceding noble gas on the Periodic Table. The atom of a nonmetal, in forming an anion, will gain enough electrons in its outermost energy level to attain the same configuration of valence electrons as an atom of the next noble gas on the Periodic Table. The number of electrons lost by the metal atoms must equal the number of electrons gained by the nonmetal atoms.

Example 8.1

Using Lewis symbols, write an equation for the reaction between aluminum atoms and fluorine atoms to form the corresponding ions. Use coefficients to show the appropriate numbers of atoms and ions.

Solution

Write the Lewis symbols for the reactant atoms. Aluminum is in group 3A and has 3 valence electrons, and fluorine is in group 7A and has 7 valence electrons.

$$\cdot\mathrm{Al}: + :\underset{\cdot\cdot}{\overset{\cdot\cdot}{\mathrm{F}}}\cdot \rightarrow$$

Write the Lewis symbols for the product ions. The aluminum atom will lose its 3 valence electrons, and the fluorine atom will gain 1 electron.

$$\cdot\mathrm{Al}: + :\underset{\cdot\cdot}{\overset{\cdot\cdot}{\mathrm{F}}}\cdot \rightarrow \mathrm{Al}^{3+} + :\underset{\cdot\cdot}{\overset{\cdot\cdot}{\mathrm{F}}}:^{-}$$

Decide how many atoms of each reactant are needed. Each aluminum atom loses 3 electrons and each fluorine atom gains 1, so 3 fluorine atoms will react with 1 aluminum atom. Insert the necessary coefficients in the equation.

$$\cdot\mathrm{Al}: + 3\ :\underset{\cdot\cdot}{\overset{\cdot\cdot}{\mathrm{F}}}\cdot \rightarrow \mathrm{Al}^{3+} + 3\ :\underset{\cdot\cdot}{\overset{\cdot\cdot}{\mathrm{F}}}:^{-}$$

Example 8.2

Use Lewis symbols to write equations for the reactions between (a) magnesium atoms and oxygen atoms (b) sodium atoms and nitrogen atoms (c) aluminum atoms and oxygen atoms

Solution

(a) $\mathrm{Mg}: + \cdot\underset{\cdot\cdot}{\overset{\cdot\cdot}{\mathrm{O}}}\cdot \rightarrow$

$$\mathrm{Mg}: + \cdot\underset{\cdot\cdot}{\overset{\cdot\cdot}{\mathrm{O}}}\cdot \rightarrow \mathrm{Mg}^{2+} + :\underset{\cdot\cdot}{\overset{\cdot\cdot}{\mathrm{O}}}:^{2-}$$

(b) $Na\cdot + \cdot\ddot{N}\cdot \rightarrow$

$Na\cdot + \cdot\ddot{N}\cdot \rightarrow Na^{+} + :\ddot{N}:^{3-}$

$3\ Na\cdot + \cdot\ddot{N}\cdot \rightarrow 3\ Na^{+} + :\ddot{N}:^{3-}$

(c) $\cdot Al: + \cdot\ddot{O}\cdot \rightarrow$

$\cdot Al: + \cdot\ddot{O}\cdot \rightarrow Al^{3+} + :\ddot{O}:^{2-}$

$2\ \cdot Al: + 3\ \cdot\ddot{O}\cdot \rightarrow 2\ Al^{3+} + 3\ :\ddot{O}:^{2-}$

Exercise 8.15

Use Lewis symbols to write equations for the reactions between (a) sodium atoms and chlorine atoms (b) aluminum atoms and nitrogen atoms (c) calcium atoms and bromine atoms

8-6 The formula for an ionic compound can be determined from the charges on its ions

The equation in Lewis symbols for the formation of an ionic compound shows the relative numbers of cations and anions in the compound. For example, the equation

$$Na\cdot + :\ddot{Cl}\cdot \rightarrow Na^{+} + :\ddot{Cl}:^{-}$$

shows that the compound produced in this reaction contains one sodium ion for each chloride ion. The *Lewis formula* for the compound is

$$Na^{+}\left[:\ddot{Cl}:\right]^{-}$$

The **Lewis formula** for an ionic compound shows the electron distribution in its ions, the charges on the ions, and the number of ions. The purpose of the brackets is to make it easy to see that the dots representing electrons belong to the chloride ion.

The **condensed formula** for an ionic compound shows only the relative numbers of ions present in the compound. The numbers of ions are shown as subscripts on the symbols for the elements. Examples of condensed formulas are NaCl, MgF_2, and Al_2O_3.◆ Condensed formulas are usually used instead of Lewis formulas for ionic compounds, and we'll use them often in this book.

◆Do not include the charges on the ions in the condensed formula for an ionic compound. Write NaCl, not $Na^{+}Cl^{-}$.

For your later work in this book, you'll need to be able to predict the condensed formula for the ionic compound that will be formed between two elements. One way to determine the formula, as we've just seen, is by writing the equation in Lewis symbols for the formation of ions from the atoms of the elements. In the equation the coefficients for the products are the numbers that will be the subscripts in the condensed formula for the ionic compound.◆

◆Lewis symbols aren't used for transition elements, so this method can't be used to predict formulas for their compounds.

Example 8.3

Write the condensed formula for the ionic compound that will be formed between potassium and sulfur.

Solution

$$K\cdot + \cdot\ddot{\underset{\cdot\cdot}{S}}\cdot \rightarrow$$

$$K\cdot + \cdot\ddot{\underset{\cdot\cdot}{S}}\cdot \rightarrow K^+ + :\ddot{\underset{\cdot\cdot}{S}}:^{2-}$$

$$2\,K\cdot + \cdot\ddot{\underset{\cdot\cdot}{S}}\cdot \rightarrow 2\,K^+ + :\ddot{\underset{\cdot\cdot}{S}}:^{2-}$$

The products in this equation show that, in the compound, there are 2 potassium ions for each sulfide ion, so its condensed formula is K_2S.

Exercise 8.16

Write the condensed formula for the ionic compound that will be formed between sodium and phosphorus.

A quicker way to determine a condensed formula is to choose the smallest subscripts that will make the total charges on all of the cations equal the total charges on all of the anions.◆ For example, in the compound formed between sodium and chlorine, the ions will be Na^+ and Cl^-, so the charge on 1 cation will equal the charge on 1 anion and the condensed formula will be NaCl.◆ Similarly, in the compound formed between magnesium and fluorine, the ions will be Mg^{2+} and F^-, so the charge on 1 cation will equal the charges on 2 anions and the condensed formula will be MgF_2. Other examples are shown in Figure 8.10.◆

◆Using this method depends on knowing the charges on the cations and anions shown in Tables 8.1, 8.2, and 8.3.

◆Choose the *smallest* subscripts that make the charges equal. The formula is NaCl, not Na_2Cl_2 or some other formula with higher subscripts.

◆Because this method doesn't depend on Lewis symbols, it can be used for transition elements as well as representative elements.

Exercise 8.17

Write the condensed formula for the ionic compound that will be formed between (a) Ba^{2+} and Cl^- (b) sodium and oxygen (c) tin(IV) ion and sulfide ion.

It's also important for you to be able to carry out the reverse of the process just described, that is, to be able to identify the ions present in an ionic compound by looking at its condensed formula. Make a habit of letting each condensed formula you see remind you of the symbols for the corresponding ions shown in Tables 8.1, 8.2, and 8.3.

Figure 8.10
Determining condensed formulas for ionic compounds. Each condensed formula is found by choosing the smallest subscripts that will make the total charges on all of the cations equal the total charges on all of the anions.

	F^-	O^{2-}	N^{3-}
K^+	KF	K_2O	K_3N
Ca^{2+}	CaF_2	CaO	Ca_3N_2
Fe^{3+}	FeF_3	Fe_2O_3	FeN

Example 8.4

What ions are present in CaS?

Solution

As shown in Table 8.1, the ion that corresponds to Ca is Ca^{2+}, and as shown in Table 8.2, the ion that corresponds to S is S^{2-}.

Exercise 8.18

What ions are present in SrI_2?

8-7 A chemical bond can be covalent, ionic, or a mixture of the two

We've seen that chemical bonds can be classified as covalent or ionic and that the class of covalent bonds can be subdivided into nonpolar covalent and polar covalent. Whether the bond between two atoms will be nonpolar covalent, polar covalent, or ionic will depend on the electronegativity *difference* between the bonded atoms.

If the electronegativity difference between two bonded atoms is zero, they will share the bonding electrons equally, and their bond will be nonpolar covalent. Two atoms of the same element will have the same electronegativity, so a bond between them will be nonpolar covalent; one example is the bond in the chlorine molecule, :C̤̈l—C̤̈l:. If atoms of two different elements happen to have the same electronegativity, a bond between them will also be nonpolar covalent. For example, the Periodic Table in Figure 7.4 shows that nitrogen and chlorine have the same electronegativity, so each of the bonds in the molecule

```
        ..
       :Cl:
        |
        N
   ..  / ..\  ..
  :Cl        Cl:
   ..         ..
```

will be nonpolar covalent.

If there is a small electronegativity difference between two bonded atoms, they will share the bonding electrons unequally, with the more electronegative atom getting a greater share, and their bond will be polar covalent. An example is $\overset{\delta+}{\text{H}}$—$\overset{\delta-}{\text{C̤̈l}}$:.

If there is a large electronegativity difference between two bonded atoms, the bonding electrons will be transferred completely from the less electronegative atom to the more electronegative atom, and their bond will be ionic. A large electronegativity difference will occur between two atoms if one of them has a large ionization energy and a low electron affinity and the other has a low ionization energy and a high electron affinity, that is, if one of them is an atom of a metal and the other is an atom of a nonmetal. An example is Na^+ [:C̤̈l:]$^-$.

As this description suggests, there are not, in fact, three distinct types of bonds. Instead, there is a continuous range of bonding from nonpolar covalent at one extreme to pure ionic at the other. In between are bonds that are slightly polar, halfway between nonpolar and ionic, or highly polar. This book uses the name *polar covalent* for the entire range between the two extremes.

Numerical values for electronegativity can define classes of bonds quantitatively, but our discussions won't require quantitative definitions.◆ We'll continue to describe bonds as nonpolar covalent, polar covalent, or ionic, and, if necessary, to modify the description with such phrases as "slightly polar," "highly polar," and "ionic with substantial covalent character."

◆For example, a bond can be classified as ionic if the electronegativity difference between the bonded atoms is greater than 1.7.

With the following simple rules, you'll be able to make reliable predictions about the nature of chemical bonds.

In general, electronegativity increases upward in a group and to the right in a period.

A bond between atoms of nonmetals will be covalent. If the bond is between two atoms of the same element, it will be nonpolar covalent; if it's between atoms of two different elements, it will be polar covalent.◆ The direction of bond polarity may be predictable from the electronegativity trends previously described.

◆The exceptions, such as the N–Cl bond described earlier in this Section, are rare enough to make the general rule useful.

A bond between atoms of a metal and a nonmetal will be ionic. Unless the nonmetal is N, O, F, or Cl, the bond will have substantial covalent character.

Exercise 8.19

Use the preceding rules and the Periodic Table at the front of the book to decide whether the bond between rubidium (Rb) and bromine will be covalent or ionic and explain your answer.

Exercise 8.20

Use the preceding rules and the Periodic Table at the front of the book to decide whether the bond between phosphorus and chlorine will be covalent or ionic and explain your answer.

8-8 Some ions contain more than one nonmetal atom

The compound $NaNO_3$ contains both ionic and covalent bonds. It has an ionic bond between the ions Na^+ and NO_3^-. The anion NO_3^- is made up of more than one atom and is described as a **polyatomic anion,** the prefix *poly-* meaning *many.*◆ In NO_3^-, the nonmetal atoms are bonded to one another, as we'd expect, by polar covalent bonds.

◆An ion that is derived from only one atom can be described as a **monatomic ion** (*mono-* means *one*). Na^+ is a monatomic cation and Cl^- is a monatomic anion.

The line formula for a polyatomic ion is found by the same procedure used to find the line formula for a covalent molecule (see the description in Section 7-5), with the additions underlined. We'll apply this procedure to the NO_3^- ion.

1. **Write the Lewis symbol for each atom and count the number of valence electrons in all of these symbols. Add one electron for each charge on a polyatomic anion or subtract one electron for each charge on a polyatomic cation.**

Nitrogen (group 5A) has 5 electrons in its Lewis symbol and oxygen (group 6A) has 6, so the atoms we have to work with in NO_3^- are $\cdot\ddot{\underset{\cdot}{N}}\cdot$ and 3 $\cdot\ddot{\underset{\cdot\cdot}{O}}\cdot$, with a total of 23 electrons. We also have one additional electron to make the 1− charge on the anion, so the final total of electrons is 24.

2. **Write the symbols for the atoms in this pattern: For the element with only one atom in the ion, write that atom's symbol at the center, and write the symbols for the other atoms in the ion around it. Connect each outside atom to the central atom by a single bond.**

```
    O
    |
    N
   / \
  O   O
```

3. **Put the remaining electrons around the outside atoms as unshared electrons, to give each outside atom a total of 8 shared and unshared electrons. If there are electrons left over, put them as unshared electrons on the central atom.**

```
    ..
   :O:
    |
    N
 .. / \ ..
:O     O:
 ..     ..
```

4. **If the central atom doesn't have 8 electrons, change an unshared electron pair on an outer atom to a shared pair (covalent bond) with the central atom, and continue this process until the central atom has 8 electrons. Make the bonding as symmetrical as possible. Add the charge on the ion.**

```
   :O: ⁻
    ||
    N
 .. / \ ..
:O     O:
 ..     ..
```

The line formula for $NaNO_3$ will be written

```
       [    :O:    ] ⁻
       [    ||     ]
Na⁺    [    N      ]
       [ .. / \ .. ]
       [:O     O:  ]
       [ ..     .. ]
```

The formulas and names for the common polyatomic ions are shown in Table 8.4 on page 160.◆ You should learn them, because you'll be using them often later in this book.

◆Your instructor may want to modify this table.

Using the preceding procedure, you should be able to write the line formula for any polyatomic ion in this table except acetate ion, which has the more complicated formula shown here.

```
    H  :O:
    |   ||    ..
H — C — C — O: ⁻
    |         ..
    H
```

acetate ion

Table 8.4 Formulas and Names for Common Polyatomic Ions

1+ Cations		1− Anions		2− Anions		3− Anions	
NH_4^+	ammonium ion	HCO_3^-	hydrogen carbonate ion	CO_3^{2-}	carbonate ion	PO_4^{3-}	phosphate ion
		HSO_3^-	hydrogen sulfite ion	SO_3^{2-}	sulfite ion		
		HSO_4^-	hydrogen sulfate ion	SO_4^{2-}	sulfate ion		
		OH^-	hydroxide ion	O_2^{2-}	peroxide ion		
		NO_2^-	nitrite ion				
		NO_3^-	nitrate ion				
		ClO^-	hypochlorite ion				
		ClO_2^-	chlorite ion				
		ClO_3^-	chlorate ion				
		ClO_4^-	perchlorate ion				
		CN^-	cyanide ion				
		$C_2H_3O_2^-$	acetate ion				

Example 8.5

Write the line formula for hydrogen sulfate ion.

Solution

As shown in Table 8.4, the condensed formula is HSO_4^-. The atoms are $\text{H}\cdot$, $\cdot\ddot{\underset{\cdot\cdot}{\text{S}}}\cdot$, and 4 $\cdot\ddot{\underset{\cdot\cdot}{\text{O}}}\cdot$. Adding one electron for the negative charge gives a total of 32 electrons.

```
          O
          |
    H—O—S—O
          |
          O
```

As this example shows, if a polyatomic ion contains a hydrogen atom and an oxygen atom, we can assume that they'll be bonded to one another. In this formula, it doesn't matter which oxygen atom we choose.

```
              ..
             :O: ⁻
       ..     |    ..
    H—O—S—O:
       ..     |    ..
             :O:
              ..
```

We've used all 32 electrons. The hydrogen atom has 2 valence electrons and each sulfur or oxygen atom has 8 electrons, so the formula is complete.

Exercise 8.21

Write the line formula for hydrogen carbonate ion.

The condensed formulas for compounds containing polyatomic ions can be determined by choosing subscripts that make the total charges on all of the cations equal the total charges on all of the anions, as shown in Figure 8.11.

	CN^-	SO_4^{2-}	PO_4^{3-}
NH_4^+	NH_4CN	$(NH_4)_2SO_4$	$(NH_4)_3PO_4$
Sn^{2+}	$Sn(CN)_2$	$SnSO_4$	$Sn_3(PO_4)_2$
Al^{3+}	$Al(CN)_3$	$Al_2(SO_4)_3$	$AlPO_4$

Figure 8.11
Condensed formulas for ionic compounds that contain polyatomic ions. Each condensed formula is found by choosing the smallest subscripts that will make the total charges on all of the cations equal the total charges on all of the anions. If the subscript for a polyatomic ion is greater than 1, the part of the formula that represents the ion is enclosed in parentheses.

Example 8.6

Write the condensed formula for the compound that will be formed between calcium ion and hydroxide ion.

Solution

From Tables 8.1 and 8.4 the ions are Ca^{2+} and OH^-. The charges on 2 anions will equal the charge on 1 cation, so the condensed formula is $Ca(OH)_2$.

Exercise 8.22

Write the condensed formula for the compound that will be formed between (a) sodium ion and chlorate ion (b) potassium ion and sulfite ion.

You should also be able to identify the monatomic or polyatomic ions that are present in a condensed formula. Make a habit of letting each condensed formula you see remind you of the corresponding ions.

Example 8.7

What ions are present in Li_3PO_4?

Solution

The ions are Li^+ and PO_4^{3-}, as shown in Tables 8.1 and 8.4.

Exercise 8.23

What ions are present in (a) Na_2O (b) Na_2O_2?

8-9 The formula for a covalent compound represents its molecule, and the formula for an ionic compound represents its formula unit

A covalent compound consists of individual fundamental units, its molecules. A sample of water, for example, is a collection of individual water molecules, as shown in Figure 8.12. Each water molecule consists of two hydrogen atoms

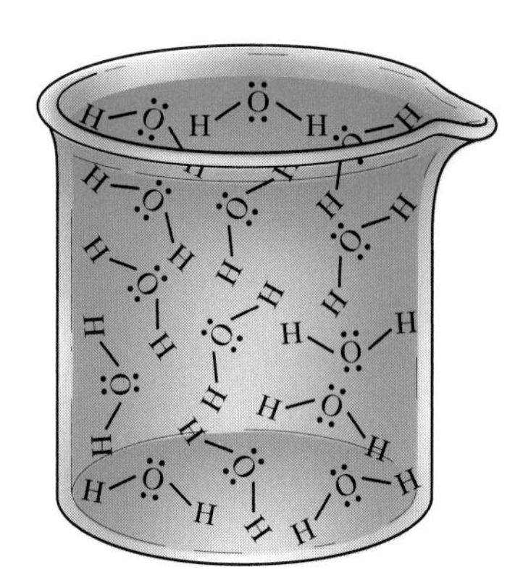

Figure 8.12
A sample of water is a collection of water molecules. Each molecule is an individual unit made up of 2 hydrogen atoms and 1 oxygen atom. The formula H_2O represents one molecule.

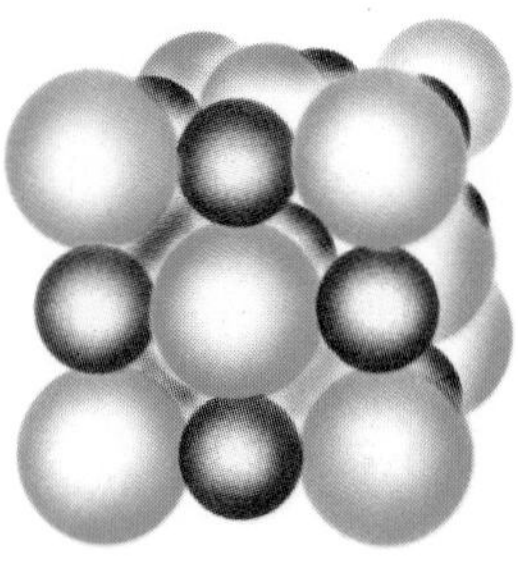

Figure 8.13
A crystal of sodium chloride consists of alternating sodium and chloride ions. Each sodium ion (red) is attracted to its surrounding chloride ions (green), and each chloride ion is attracted to its surrounding sodium ions. There are no individual NaCl molecules. The formula NaCl represents one formula unit.

attached to one oxygen atom by single polar covalent bonds; the formula for water, H_2O, represents one molecule.

In an ionic compound there is no molecule. For example, a sample of the ionic compound sodium chloride (table salt) consists of white crystals, and each of these crystals has many millions of sodium ions and an equal number of chloride ions. As shown in Figure 8.13, the ions are arranged so that a sodium ion always alternates with a chloride ion, in three dimensions. Each positive sodium ion in the crystal is surrounded by several negative chloride ions and is equally attracted to each of them; each negative chloride ion is surrounded by several positive sodium ions and is equally attracted to each of them. In a crystal of sodium chloride one sodium ion and one chloride ion are not bonded only to each other, so there is no molecule of sodium chloride.

Sodium chloride has equal numbers of sodium ions and chloride ions, and the formula NaCl expresses this one-to-one relationship—it does not represent a molecule of sodium chloride. The formula NaCl is said to represent one *formula unit* of sodium chloride. The formula for an ionic compound represents its **formula unit,** a theoretical unit whose ions are in the ratio that they have in the compound.

Example 8.8

Decide whether each of the following formulas represents a molecule or a formula unit: (a) CO (b) MgO

Solution

(a) Carbon and oxygen are nonmetals, so they form a covalent bond. The formula CO represents a molecule.

(b) Magnesium is a metal and oxygen is a nonmetal, so they form an ionic bond. The formula MgO represents a formula unit.

Exercise 8.24

Decide whether each of the following formulas represents a molecule or a formula unit: (a) NO_2 (b) $Fe(NO_3)_3$

8-10 The mass of a formula unit can be found by adding up the masses of its atoms

The molecular mass for a covalent compound is the mass of one molecule, found by adding up the masses of the atoms in the molecule (Section 7-7). We can't calculate a molecular mass for an ionic compound because an ionic compound doesn't consist of molecules, but we can calculate its **formula mass,** the sum of the masses of the ions in its formula. We'll calculate formula masses to three significant digits, unless there's a reason to be more precise.◆

◆Formula masses will be important in Chapters 11, 12, and 16.

Example 8.9

Calculate the formula mass for Al_2O_3.

Solution

From the Periodic Table at the front of the book, the atomic masses are 27.0 amu for Al and 16.0 amu for O.

$$2\ Al^{3+} \text{ ions} = 2 \times 27.0 \text{ amu} = 54.0 \text{ amu}$$
$$3\ O^{2-} \text{ ions} = 3 \times 16.0 \text{ amu} = 48.0 \text{ amu}$$
$$\text{formula mass for } Al_2O_3 = 102.0 \text{ amu, round to } 102 \text{ amu}$$

◆As we saw in Section 4-6, the mass of one electron is only 5.846×10^{-4} amu. In forming a cation or anion, the loss or gain of electrons causes so small a change in mass that we can always use the mass of the atom as the mass of the ion.

Exercise 8.25

Calculate the formula mass for $BaBr_2$.

A polyatomic ion is a group of atoms covalently bonded to one another, but since it carries a positive or negative charge, it isn't a molecule. We can assign a mass to it—the sum of the masses of its atoms—and that mass will be the formula mass for the ion.◆

◆In a polyatomic ion, as in a monatomic ion, the electrons that have been lost or gained to cause the charge on the ion have a negligible mass, which is ignored in calculating the mass of the ion.

Exercise 8.26

Calculate the formula mass for phosphate ion.

8-11 Many familiar substances contain ionic compounds

The most familiar pure ionic compound is table salt, sodium chloride. An average adult human being contains about 160 g of sodium chloride and loses about 200 mg every day. To maintain good health, this loss must be replaced, so human beings need a source of salt in their diet; meats contain enough salt to provide the needed amount, but a pure vegetarian diet has to be supplemented with salt.

In the United States, about 43 million tons of salt is manufactured every year, mostly from the evaporation of sea water. About three quarters of this salt is used in chemical manufacturing processes.

We use a few other simple ionic compounds in everyday life: $NaHCO_3$ (commonly called sodium bicarbonate) and $MgCO_3$ neutralize stomach acid and relieve indigestion; SnF_2 is a source of fluoride ion, an ingredient in toothpastes that strengthens teeth; and Na_3PO_4 is used to make strong household cleaning solutions. The coloring agents in glasses and glazes are also ionic compounds, as shown in Table 8.5.

Minerals, the substances that make up rocks, are another familiar group of ionic compounds. Table 8.6 on page 165 shows the classification scheme used

Sodium hydrogen carbonate, $NaHCO_3$, also called sodium bicarbonate or baking soda. *(R. Mathena, Fundamental Photographs, New York)*

Table 8.5 Ionic Compounds in Colored Glasses and Glazes

Compound	Color	Compound	Color
CoO	blue	FeO	blue or blue-green
Co_2O_3	violet	Fe_2O_3	yellow-brown
Cr_2O_3	green	MnO_2	violet
CrO_3	yellow to orange	Mn_2O_3	purple
CuO	greenish blue	V_2O_3	yellowish green
Cu_2O	greenish blue	V_2O_5	yellow

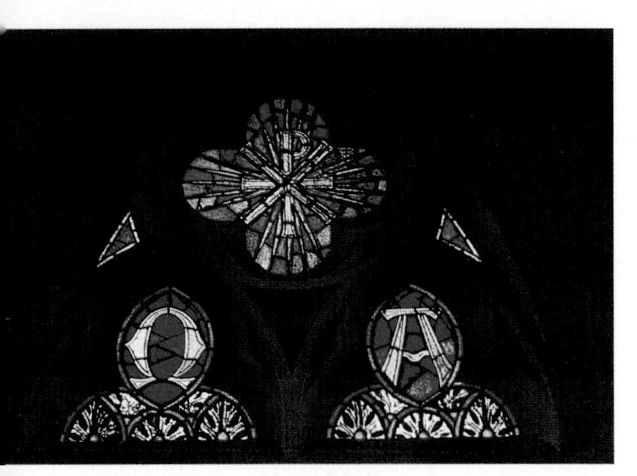

Stained glass window. *(Runk/Schoenberger from Grant Heilman Photography, Inc.)*

(R. Glander/Superstock)

Glazes on pottery.† *(P. Amranand/Superstock)*

for minerals, based on their chemical composition. About 2000 minerals have so far been identified.

In Table 8.6 the formulas for kernite and carnotite show water molecules as part of each compound. Compounds that contain water molecules as part of their composition are called **hydrates.** A description of the way in which water molecules are attached to the other atoms or ions in a hydrate requires a more advanced theory of bonding and won't be dealt with in this book.

† Chemistry Insight

A glaze is a thin layer of glass added to the surface of a clay object for protection and decoration. Glasses and glazes are complicated mixtures of covalent and ionic compounds, melted together under carefully controlled conditions. The exact color that an ionic compound will give to a glass or glaze depends on the other components of the mixture and on the heating conditions used to melt them together.

Table 8.6 Chemical Classification of Minerals

Class	Example	Occurrence
Element	S_8, sulfur	Louisiana and Texas
Arsenate	$Cu_2(AsO_4)(OH)$, olivenite	Arizona and Utah
Borate	$Na_2B_4O_7 \cdot 4H_2O$, kernite	California; the main ore for borax
Carbonate	$CaCO_3$, calcite	Every state, as limestone
Halide	HgCl, calomel	Arkansas and Texas
Hydroxide	$Mg(OH)_2$, brucite	New York, Pennsylvania, and Nevada
Molybdate	$CaMoO_4$, powellite	Michigan, Nevada, and Utah
Nitrate	KNO_3, niter	Kentucky and Tennessee
Oxide	Fe_2O_3, hematite	Many locations, large deposits in Alabama and Minnesota; the main ore for iron
Phosphate	$LiFePO_4$, triphylite	California, New Hampshire, and South Dakota
Silicate	$Mg_3Si_2O_5(OH)_4$, serpentine	California, New Jersey, New York, and Vermont
Sulfate	Na_2SO_4, thenardite	Arizona, California, and Nevada
Sulfide	PbS, galena	Many locations, including Illinois, Kansas, Missouri, and Oklahoma
Tungstate	$CaWO_4$, scheelite	Arizona, California, Connecticut, and Nevada
Uranate Vanadate	$K_2(UO_2)_2(VO_4)_2 \cdot 3H_2O$, carnotite	Arizona, California, New Mexico, and Utah; the ore for uranium

Sulfur, S_8. *(Runk/Schoenberger from Grant Heilman Photography, Inc.)*

Serpentine, $Mg_3Si_2O_5(OH)_4$. *(Paul Silverman, Fundamental Photographs, New York)*

Limestone, $CaCO_3$, along the Verde River in Arizona *(J. Cowlin)*

Galena, PbS. *(Charles D. Winters)*

Inside Chemistry How does fluoride ion help create healthier teeth?

In the United States, the average number of dental cavities in school-age children has decreased by about 50% in the last fifty years. The three most important reasons for this improvement are better nutrition; better oral hygiene; and the introduction of fluoride ion into drinking water, toothpastes, and mouthwashes.

Dental cavities occur when bacteria in the mouth produce compounds, called acids, that destroy tooth enamel. (Acids are discussed in Chapters 15 and 16.) Fluoride ion helps prevent tooth decay in two ways: by slowing the rate at which oral bacteria produce acids and by strengthening tooth enamel.

Tooth enamel is a complicated ionic compound, called hydroxyapatite, made up of calcium ions, phosphate ions, and hydroxide ions; its formula is $Ca_{10}(PO_4)_6(OH)_2$. If fluoride ions are present in the bloodstream when enamel is being formed, hydroxide ions are replaced by fluoride ions, and hydroxyapatite is converted to fluoroapatite:

$$\underset{\text{hydroxyapatite}}{Ca_{10}(PO_4)_6(OH)_2} + 2\ F^- \rightarrow \underset{\text{fluoroapatite}}{Ca_{10}(PO_4)_6F_2} + 2\ OH^-$$

Fluoroapatite resists destruction by acids, so enamel that contains fluoride ion resists the formation of cavities.

Because fluoride ion helps prevent tooth decay, many communities in the United States add fluoride ion to their drinking water in the form of sodium fluoride, NaF, or other compounds. Fluoridation of drinking water began in Grand Rapids, Michigan, in 1945; today, about 120 million people in the United States, and about 130 million people in other countries, drink fluoridated water. When fluoridated water is swallowed, the fluoride ions pass out of the digestive tract into the bloodstream and are then incorporated into newly forming enamel.

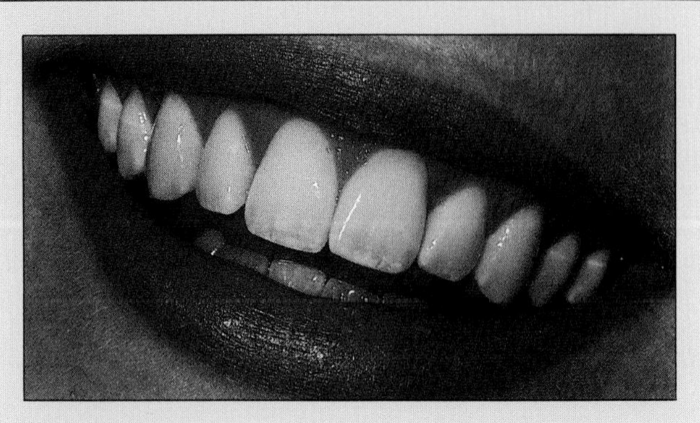

Healthy, attractive teeth are now common in part because of the addition of fluoride ion to toothpastes and drinking water. *(Sylvia Miller/Custom Medical Stock Photo, Inc.)*

Fluoridated toothpastes and mouthwashes help prevent cavities because fluoride ions in the mouth can replace hydroxide ions in hydroxyapatite on the surface of tooth enamel. Tin(II) fluoride, SnF_2, and other compounds are used to provide fluoride ion in toothpastes and mouthwashes.

Scientists disagree on whether fluoride ion to prevent tooth decay should be supplied only in toothpastes and mouthwashes or should be provided in fluoridated drinking water as well. More studies on this issue are being carried out, and you may hear about them in the news.

Chapter Summary: Ionic Bonds

Section	Subject	Summary	Check When Learned
	Ion	Charged particle formed when an atom or group of atoms gains or loses electrons.	☐
	Ionic bond	Attractive force between oppositely charged ions.	☐
	Ionic compound	Compound made up of ions; a covalent compound is made up of molecules.	☐

Section	Subject	Summary	Check When Learned
8-1	(Covalent) atomic radius	Half the distance between the bonded nuclei of two atoms of the same element, usually measured in pm.	☐
8-1	Periodic trends in atomic radius	Decreases upward in a group and to the right in a period.	☐
8-1	Ionization energy	Energy required to remove outermost electron from atom.	☐
8-1	Periodic trends in ionization energy	Increases upward in a group and to the right in a period.	☐
8-1	Shielding effect	Repulsion of electrons in outer energy levels by electrons in inner energy levels, in effect shielding the outer electrons from the nuclear charge.	☐
8-1	Electron affinity	Energy released when an electron is added to the outermost energy level of an atom.	☐
8-1	Periodic trends in electron affinity	Increases upward in a group and to the right in a period.	☐
8-2	Class of elements that lose electrons easily	Metals.	☐
8-2, 8-8	Cation	Positively charged ion formed when an atom or group of atoms loses one or more electrons.	☐
8-2	Change in radius when atom forms cation	Radius decreases.	☐
8-2	Charge on cations of group 1A metals group 2A metals	1+ 2+	☐
8-2	General equation for formation of cation of alkali metal	$M\cdot \rightarrow M^{+} + e^{-}$	☐
8-2	General equation for formation of cation of alkaline-earth metal	$M: \rightarrow M^{2+} + 2\ e^{-}$	☐
8-2	Rule for forming cations of representative metals	An atom of a representative metal will easily lose electrons to form a cation that has the same configuration of valence electrons as an atom of the preceding noble gas on the Periodic Table.	☐
8-2	Explanation of formation of more than one ion by some representative metals	The atoms of some representative metals in groups 3A through 5A may lose only the p or the s and p electrons in their outermost energy levels. For example, tin, outermost energy level $5s^2 5p^2$, may form Sn^{2+} or Sn^{4+}.	☐

Section	Subject	Summary	Check When Learned
8-2	Rule for naming monatomic cations	If a metal atom forms only one cation, the name for the cation is the name for the atom; if a metal atom forms more than one cation, the name for the cation is the name for the atom, followed by a Roman numeral, in parentheses, showing the charge on the cation.	☐
8-2	Name:		
	H^+	hydrogen ion	
	Li^+	lithium ion	
	Na^+	sodium ion	
	K^+	potassium ion	
	Mg^{2+}	magnesium ion	
	Ca^{2+}	calcium ion	
	Sr^{2+}	strontium ion	
	Ba^{2+}	barium ion	
	Al^{3+}	aluminum ion	
	Sn^{2+}	tin(II) ion	
	Sn^{4+}	tin(IV) ion	
	Pb^{2+}	lead(II) ion	
	Pb^{4+}	lead(IV) ion	☐
8-3	Explanation of formation of some cations of transition metals	The transition-metal atom loses the electrons in its highest occupied energy level. For example, a zinc atom, $[Ar]^{18}4s^2 3d^{10}$, loses its $4s^2$ electrons to form Zn^{2+}.	☐
8-3	Name:		
	Ag^+	silver ion	
	Cd^{2+}	cadmium ion	
	Ni^{2+}	nickel ion	
	Zn^{2+}	zinc ion	
	Cu^+	copper(I) ion	
	Cu^{2+}	copper(II) ion	
	Hg_2^{2+}	mercury(I) ion	
	Hg^{2+}	mercury(II) ion	
	Fe^{2+}	iron(II) ion	
	Fe^{3+}	iron(III) ion	☐
8-4	Class of elements that gain electrons easily	Nonmetals.	☐
8-4, 8-8	Anion	Negatively charged ion formed when an atom or group of atoms gains one or more electrons.	☐
8-4, 8-8	Monatomic anion	Anion derived from one atom, for example, F^-.	☐
8-4	Rule for naming a monatomic anion	Change the suffix of the element's name to -ide; for example, F^- is fluoride ion.	☐

Section	Subject	Summary	Check When Learned
8-4	Change in radius when atom forms anion	Radius increases.	☐
8-4	Rule for forming anions of representative elements	An atom of a nonmetal, except carbon, will easily gain electrons until it has the same configuration of valence electrons as an atom of the next noble gas on the Periodic Table.	☐
8-4	General equation for formation of anion of group 5A group 6A group 7A	 $\cdot\ddot{\underset{\cdot}{Z}}\cdot + 3\,e^- \rightarrow :\ddot{\underset{\cdot\cdot}{Z}}:^{3-}$ $\cdot\ddot{\underset{\cdot\cdot}{Y}}\cdot + 2\,e^- \rightarrow :\ddot{\underset{\cdot\cdot}{Y}}:^{2-}$ $:\ddot{\underset{\cdot\cdot}{X}}\cdot + e^- \rightarrow :\ddot{\underset{\cdot\cdot}{X}}:^{-}$	☐
8-4	Charge on anion of nonmetal of group 5A group 6A group 7A	 3− 2− 1−	☐
8-4	Equation for formation of anion from hydrogen atom	$H\cdot + e^- \rightarrow H:^-$	☐
8-4	Name: H^- F^- Cl^- Br^- I^- O^{2-} S^{2-} Se^{2-} N^{3-} P^{3-}	 hydride ion fluoride ion chloride ion bromide ion iodide ion oxide ion sulfide ion selenide ion nitride ion phosphide ion	☐
8-5	Rule for reaction of metal atom with nonmetal atom to form ions	In general, the atom of a representative metal will lose the electrons in its outermost energy level. The atom of a nonmetal will gain enough electrons in its outermost energy level to attain the same electron configuration as an atom of the next noble gas on the Periodic Table. The number of electrons lost by the metal atoms must equal the number of electrons gained by the nonmetal atoms.	☐

Section	Subject	Summary	Check When Learned
8-6	For an ionic compound, the distinction between Lewis formula and condensed formula	The Lewis formula shows the numbers of ions, the charges on the ions, and the valence electrons, for example $Na^+ [:\ddot{\underset{..}{Cl}}:]^-$. The condensed formula shows only the numbers of ions, for example NaCl.	☐
8-6	Method for deciding condensed formula for compound that will be formed between a metal and a nonmetal	Identify the corresponding cation and anion. In the formula use the lowest subscripts that will make the total charges on the cations equal the total charges on the anions.	☐
8-7	General rule describing bonding between two nonmetals or between a metal and a nonmetal	The bonding between two nonmetals will be nonpolar covalent or polar covalent, and the bonding between a metal and a nonmetal will be ionic. Both kinds of bonding follow the octet rule.	☐
8-7	Rules for predicting type of bond that will occur between two atoms	(1) In general, electronegativity increases upward in a group and to the right in a period. (2) A bond between atoms of nonmetals will be covalent; if it's between atoms of two different elements, it will be polar covalent. The direction of bond polarity may be predictable from the electronegativity trends previously described. (3) A bond between atoms of a metal and a nonmetal will be ionic. Unless the nonmetal is N, O, F, or Cl, the bond will have substantial covalent character.	☐
8-8	Polyatomic ion	Ion that contains more than one nonmetal atom, for example NH_4^+ or NO_3^-.	☐
8-8	Steps for writing formula for a polyatomic ion	(1) Write the Lewis symbol for each atom and count the number of electrons in all of these symbols. Add one electron for each charge on a polyatomic anion or subtract one electron for each charge on a polyatomic cation. (2) Write the symbols for the atoms in this pattern: For the element with only one atom in the ion, write that atom's symbol at the center, and write the symbols for the other atoms in the ion around it. Connect each outside atom to the central atom by a single bond. (3) Put the remaining electrons around the outside atoms as unshared electrons, to give each outside atom a total of 8 shared and unshared electrons. If there are electrons left over, put them as unshared electrons on the central atom. (4) If the central atom doesn't have 8 electrons, change an unshared electron pair on an outer atom to a shared pair (covalent bond) with the central atom, and continue this process until the central atom has 8 electrons. Make the bonding as symmetrical as possible. Add the charge on the ion.	☐

Section	Subject	Summary	Check When Learned
8-8	Name:		
	NH_4^+	ammonium ion	
	OH^-	hydroxide ion	
	NO_2^-	nitrite ion	
	NO_3^-	nitrate ion	
	$C_2H_3O_2^-$	acetate ion	
	ClO^-	hypochlorite ion	
	ClO_2^-	chlorite ion	
	ClO_3^-	chlorate ion	
	ClO_4^-	perchlorate ion	
	HCO_3^-	hydrogen carbonate ion	
	HSO_3^-	hydrogen sulfite ion	
	HSO_4^-	hydrogen sulfate ion	
	CN^-	cyanide ion	
	O_2^{2-}	peroxide ion	
	SO_3^{2-}	sulfite ion	
	SO_4^{2-}	sulfate ion	
	CO_3^{2-}	carbonate ion	
	PO_4^{3-}	phosphate ion	☐
8-8	Method for writing the condensed formula for a compound that contains a polyatomic ion	Identify the ions. In the formula use the lowest subscripts that will make the total charges on the cations equal the total charges on the anions.	☐
8-9	Molecule	Electrically neutral group of atoms bonded to one another by covalent bonds; the fundamental unit of a covalent compound.	☐
8-9	Formula unit	For an ionic compound, the formula that shows the numerical relationship of cations to anions in the compound, for example, NaCl, K_2O, or $MgBr_2$. For a polyatomic ion, the condensed formula for the ion, for example NH_4^+ or NO_2^-.	☐
8-10	Formula mass	The sum of the masses of the atoms or ions in a formula that represents a formula unit.	☐
8-11	Hydrate	Compound that contains one or more water molecules as part of its composition, for example, $CaCl_2 \cdot 2H_2O$.	☐

Problems

Assume you can use the Periodic Table at the front of the book unless you're directed otherwise. Answers to odd-numbered problems are in Appendix 1.

Ionic and Covalent Bonding

1. In your own words, describe the difference between an ionic bond and a covalent bond.

2. In your own words, define the terms *nonpolar covalent bond, polar covalent bond,* and *ionic bond.*

Atomic Radius, Ionization Energy, and Electron Affinity (Section 8-1)

3. In your own words, define the term *atomic radius.*

4. The distance between the bonded nuclei in a chlorine molecule is 1.98×10^{-10} m. Calculate the covalent atomic radius of a chlorine atom, in picometers.

5. Using the Periodic Table at the front of the book, predict which element in each pair will have the larger atomic radius: (a) Ne or Kr (b) Ca or Ga (c) As or Cl.

6. In your own words, explain why atomic radius decreases upward in a group and to the right in a period.

7. In your own words, explain the concept of ionization energy.

8. In your own words, explain the concept of the shielding effect.

9. Using the Periodic Table at the front of the book, predict which element in each pair will have the smaller ionization energy: (a) Ne or Kr (b) Ca or Ga (c) As or Cl.

10. Why is the ionization energy of potassium less than that of sodium?

11. Why is the ionization energy of potassium less than that of bromine?

12. In the same period, which class of elements will have the lower ionization energy, metals or nonmetals?

13. Groups 4A, 5A, and 6A contain metals and nonmetals. Which class will have the higher ionization energy, the metals or the nonmetals?

14. In the same period, which class of elements will have the higher ionization energy, metals or metalloids?

15. In your own words, explain the concept of electron affinity.

16. Why is the electron affinity of fluorine greater than that of iodine?

17. Why is the electron affinity of fluorine greater than that of lithium?

18. Is this statement correct: "The ionization energy is the amount of energy required to take the outermost electron off of an atom, and the electron affinity is the amount of energy released when that electron is put back on the atom"? Explain.

Cations and Anions (Sections 8-2 through 8-6)

19. In your own words, define the terms *ion, cation,* and *anion.*

20. In your own words, explain the difference between a monatomic ion and a polyatomic ion.

21. How many protons and electrons are in a potassium atom? In a potassium ion?

22. How many protons and electrons are in a sulfur atom? In a sulfide ion?

23. Write the abbreviated electron configuration for the magnesium atom and circle the electrons that are lost in forming the magnesium ion.

24. Write the abbreviated electron configuration for cadmium and circle the electrons that are lost in forming the cadmium ion.

25. Write an equation in Lewis symbols showing the loss of three electrons from an aluminum atom to form a cation. Use the octet rule to explain why this is the stable cation for aluminum.

26. Write an equation in Lewis symbols showing the formation of the stable cation you'd expect from an atom of radium (Ra).

27. Why is it easy to remove one electron from a potassium atom but difficult to remove two?

28. In what way is a sodium ion similar to a neon atom?

29. Complete this sentence by choosing the correct word in parentheses: A monatomic cation is always (larger, smaller) than its atom, and a monatomic anion is always (larger, smaller) than its atom.

30. Write an equation in Lewis symbols showing the addition of two electrons to a selenium (Se) atom to form a selenide ion. Use the octet rule to explain why this is the stable anion for selenium.

31. Write an equation in Lewis symbols showing the formation of the stable anion you'd expect from an atom of iodine.

32. What do Cl^-, Ar, and K^+ have in common?

33. Write the symbol for the noble-gas atom that has the same electron configuration as the calcium ion.

34. Write the symbol for the noble-gas atom that has the same electron configuration as the sulfide ion.

35. Write an equation in Lewis symbols showing the formation of the product you'd expect from the reaction of magnesium atoms with nitrogen atoms.

36. Write an equation in Lewis symbols showing the formation of the product you'd expect from the reaction of iodine atoms with calcium atoms.

37. Name (a) Pb^{4+} (b) Hg^{2+} (c) Ba^{2+}.

38. Name (a) H^- (b) H^+ (c) Br^-.

39. Write the symbols for (a) sodium ion (b) mercury(I) ion (c) nickel ion.

40. Write the symbols for (a) iodide ion (b) sulfide ion (c) aluminum ion.

41. Fill in the formulas in the following table:

	H^-	S^{2-}	N^{3-}
Na^+			
Sn^{2+}			
Sn^{4+}			

42. Fill in the formulas in the following table:

	I^-	O^{2-}	P^{3-}
Cu^+			
Ba^{2+}			
Al^{3+}			

43. Predict the formulas for the compounds that will be formed between (a) peroxide ion and cadmium ion (b) hypochlorite ion and silver ion (c) sodium ion and acetate ion.

44. Predict the formulas for the compounds that will be formed between (a) ammonium ion and sulfite ion (b) lead(II) ion and hydroxide ion (c) magnesium ion and carbonate ion.

45. Write the symbols for the ions present in (a) KCl (b) CdS (c) CaH_2.

46. Write the symbols for the ions present in (a) $ZnBr_2$ (b) FeO (c) SnS_2.

Covalent and Ionic Bonds (Section 8-7)

47. Why is a covalent bond between two atoms of the same element always nonpolar?

48. Which bond would you expect to be more polar, a bond between a fluorine atom and an oxygen atom or a bond between a fluorine atom and a sulfur atom? Why?

49. Which bond would you expect to be more polar, a bond between a fluorine atom and a chlorine atom or a bond between a fluorine atom and a bromine atom? Why?

50. In your own words, explain why OF_2 is a covalent compound but MgF_2 is an ionic compound.

Polyatomic Ions (Section 8-8)

51. Write line formulas for (a) ammonium ion and (b) nitrite ion.

52. Write line formulas for (a) iodate ion and (b) phosphate ion.

53. Predict the formula for the compound that will be formed between (a) zinc ion and chlorate ion (b) lead(II) ion and acetate ion (c) sodium ion and hydroxide ion.

54. Predict the formula for the compound that will be formed between (a) copper(I) ion and chloride ion (b) hydrogen ion and perchlorate ion (c) cadmium ion and hydrogen carbonate ion.

55. Predict the formula for the compound that will be formed between sodium ion and (a) sulfide ion (b) sulfite ion (c) sulfate ion.

56. Predict the formula for the compound that will be formed between calcium ion and (a) carbonate ion (b) hydrogen carbonate ion (c) cyanide ion.

57. Write the symbols or formulas for the ions present in (a) $CaSO_4$ (b) $Al(ClO_2)_3$ (c) $(NH_4)_2CO_3$.

58. Write the symbols or formulas for the ions present in (a) NH_4CN (b) $NH_4C_2H_3O_2$ (c) $Cu_3(PO_4)_2$.

Molecules and Formula Units (Section 8-9)

59. Decide whether each of the following formulas represents a molecule or a formula unit: (a) H_2O (b) Na_2O (c) CO_2 (d) PbO_2 (e) SO_2.

60. Calculate the formula or molecular mass for (a) Ag_2CO_3 (b) SO_3 (c) SO_3^{2-}.

9-1
The name for a binary covalent compound is formed by combining the names for its elements.

9-2
To name an ionic compound, name its cation, then its anion.

9-3
To be complete, a chemical equation must be balanced in atoms and charges.

9-4
Equations can describe real or imaginary reactions.

9-5
It's often convenient to write an equation with condensed or molecular formulas.

9-6
In an equation the label (s), (ℓ), or (g) shows that a reactant or product is a solid, liquid, or gas.

9-7
An equation can be modified to show that a reaction absorbs or releases energy.

9-8
Reactions can be understood more easily by classifying them into a few basic types.

9-9
To learn to name compounds and write equations quickly and accurately, practice in a variety of ways.

Inside Chemistry:
What are generic drugs?

Chapter Summary

Problems

Iron, Fe, in the form of steel wool, burns brightly in pure oxygen, O_2. The equation for the reaction is $4\ Fe + 3\ O_2 \rightarrow 2\ Fe_2O_3$. *(Charles Steele)*

9

Naming Compounds and Writing Equations

You've already learned to read and write the language of chemistry: You first learned the symbols for atoms and ions, then learned to combine them into formulas, and then learned to combine the formulas into equations. In this chapter, you'll add to those skills. In Sections 9-1 and 9-2 you'll learn to name covalent and ionic compounds, and in Sections 9-3 through 9-9 you'll learn to write equations for several basic types of chemical reactions.

It's important that you be able to read and write the language of chemistry quickly and accurately because it will be the language you'll use for all of your later work in this book. The best way to learn the skills in this chapter is to practice them thoroughly by working many Exercises and problems.

9-1 The name for a binary covalent compound is formed by combining the names for its elements

Because it requires some memorization, learning how to name chemical compounds isn't easy, but it's very important. To be successful in learning chemistry, you must know how to translate the formula for a compound into its name or its name into its formula. The system used for naming compounds is referred to as **chemical nomenclature.**

Different kinds of compounds are named by different methods. In this Section, you'll learn how to name binary covalent compounds. A **binary compound** contains only two elements—for example, HCl, SF_6, or P_2O_5.

Name the elements in a binary covalent compound in the order in which they appear in the formula.

Name a binary covalent compound of hydrogen by writing "hydrogen," then the name of the second element, with its suffix changed to *-ide*.

Hydrogen with Group 6A Elements		Hydrogen with Group 7A Elements	
H_2O	hydrogen oxide	HF	hydrogen fluoride
H_2S	hydrogen sulfide	HCl	hydrogen chloride
H_2Se	hydrogen selenide	HBr	hydrogen bromide
H_2Te	hydrogen telluride	HI	hydrogen iodide

To name any other binary covalent compound, name each element, change the suffix of the second element to -ide, and add a prefix to the name of each element to show how many atoms of that element are in one molecule of the compound.

Prefix	Meaning
mono-	one
di-	two
tri-	three
tetra-	four
penta-	five
hexa-	six
hepta-	seven
octa-	eight
nona-◆	nine
deca-	ten

◆The prefix *ennea-* has been recommended to replace *nona-*, but nona- is more commonly used.

Examples are

CO	carbon monoxide	P_2O_5	diphosphorus pentoxide
CO_2	carbon dioxide	Cl_2O_7	dichlorine heptoxide
CCl_4	carbon tetrachloride	P_4O_{10}	tetraphosphorus decoxide

As you can see in these examples, the prefix mono- is used for the second element but not for the first: CO is not monocarbon monoxide but carbon monoxide. You'll also see that when a prefix is added, a vowel may be dropped from the end of the prefix to make spelling and pronunciation easier: not carbon monooxide but carbon monoxide, not diphosphorus pentaoxide but diphosphorus pentoxide.

◆Answers to Exercises are in Appendix 1.

Exercise 9.1◆

Write the name from the formula or the formula from the name: (a) carbon disulfide (b) IBr (c) sulfur hexafluoride (d) P_2S_5

A few compounds are always referred to by their common or traditional names: The formal name for H_2O is hydrogen oxide, but the compound is always called water. The common names for NH_3 and CH_4 are **ammonia** and **methane.**◆

◆Household ammonia, used for cleaning, is a solution of the gas NH_3 in water. Methane, also called natural gas, is used in many homes for heating and cooking.

9-2 To name an ionic compound, name its cation, then its anion◆

◆Before you start this Section, be sure you know the names and symbols or formulas for the ions in Tables 8.1 through 8.4.

Once you know the names for the cations and anions, writing the name for an ionic compound is easy: Write the name for the cation, then the name for the anion, as shown in the following examples:

$NaCl$	sodium chloride
$CaSO_4$	calcium sulfate
$CuSO_3$	copper(II) sulfite
$Al(HCO_3)_3$	aluminum hydrogen carbonate
$Hg(CN)_2$	mercury(II) cyanide
$Ba(OH)_2$	barium hydroxide

To write the formula for an ionic compound from its name, first write the symbol or formula for the cation, then write the symbol or formula for the anion. Combine appropriate numbers of cations and anions into the formula for the compound so that the total charges on the cations equal the total charges on the anions. For example, to write the formula for tin(IV) nitrate, first write

$$Sn^{4+} \qquad NO_3^-$$

We'll need four nitrate ions for each tin(IV) ion, so the formula will be $Sn(NO_3)_4$.

(David R. Frazier Photolibrary)

Exercise 9.2

Write the name for the formula or the formula for the name: (a) KI (b) barium nitrate (c) $FeBr_2$ (d) copper(I) phosphate (e) $(NH_4)_2CO_3$

Clockwise from upper left: potassium chloride, KCl; sodium chloride, NaCl; calcium fluoride, CaF_2; iron(III) oxide, Fe_2O_3; and copper(II) bromide, $CuBr_2$. *(Charles D. Winters)*

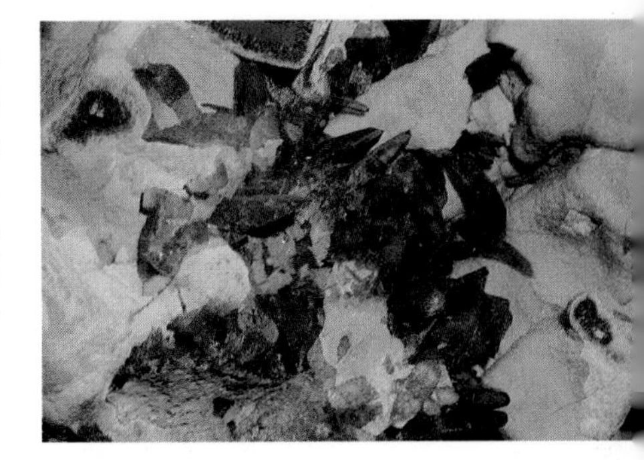

Barium sulfate, $BaSO_4$, in the form of the mineral barite. *(Brian Parker/Tom Stack & Associates)*

If you know the basic rules of nomenclature and the names for the common ions, you can sometimes use them to figure out the names or formulas for ions or compounds you haven't seen before. For example, BrO_3^- isn't among the ions whose names and formulas you've learned, but ClO_3^-, chlorate ion, is. Because of the similarity of BrO_3^- to ClO_3^- (both bromine and chlorine are in group 7A, the halogens), it's reasonable to suppose that BrO_3^- would be called bromate ion, and it is. Similarly, NaIO is comparable to NaClO, sodium hypochlorite, and is called sodium hypoiodite.

Exercise 9.3

(a) Name $Ca(BrO_4)_2$. (b) Write the formula for zinc iodite.

Some unfamiliar ions that contain hydrogen can also be identified by comparing them with these pairs of ions you already know:

SO_3^{2-} sulfite ion	HSO_3^- hydrogen sulfite ion
SO_4^{2-} sulfate ion	HSO_4^- hydrogen sulfate ion
CO_3^{2-} carbonate ion	HCO_3^- hydrogen carbonate ion

In each pair, the second member can be thought of as being derived from the first by adding a hydrogen ion:

$$H^+ + SO_3^{2-} \rightarrow HSO_3^-$$

Adding the hydrogen ion reduces the charge on the original anion by one negative charge. From this pattern, you can figure out the name for HS^-: Since S^{2-} is sulfide ion, HS^- is hydrogen sulfide ion.

Calcium carbonate, $CaCO_3$, in the form of the mineral calcite. *(Allen B. Smith/Tom Stack & Associates)*

Exercise 9.4

(a) Name $Ba(HSO_4)_2$. (b) Write the formula for barium hydrogen selenide.

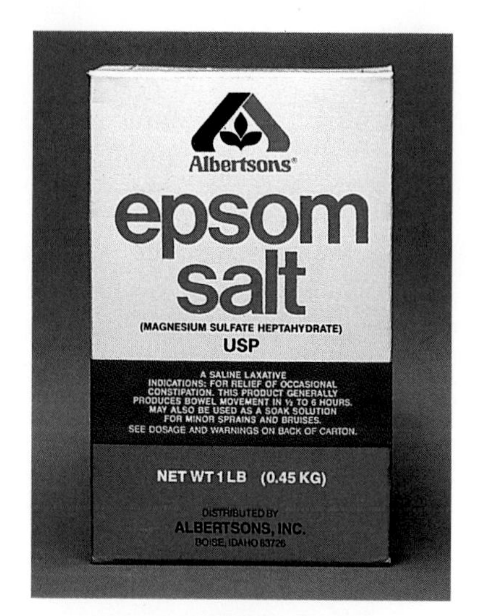

Epsom salt is magnesium sulfate heptahydrate, $MgSO_4 \cdot 7H_2O$. *(David R. Frazier Photolibrary)*

Copper(II) sulfate pentahydrate, $CuSO_4 \cdot 5H_2O$. *(Charles D. Winters)*

◆ Use the *lowest* coefficients. The equation $2\cdot\dot{\mathrm{C}}: + 4\cdot\ddot{\underset{..}{\mathrm{O}}}\cdot \rightarrow 2\ \ddot{\underset{..}{\mathrm{O}}}{=}\mathrm{C}{=}\ddot{\underset{..}{\mathrm{O}}}$ is balanced, but it's not in the preferred form.

In naming a hydrate, the number of water molecules is designated by one of the prefixes used in naming covalent compounds. For example, $CaCl_2 \cdot 2H_2O$ is calcium chloride dihydrate, and $MgSO_4 \cdot 7H_2O$ is magnesium sulfate heptahydrate.

Exercise 9.5

Name $CuSO_4 \cdot 5H_2O$.

9-3 To be complete, a chemical equation must be balanced in atoms and charges

In a chemical reaction, no atoms are created or destroyed, they're only rearranged into different groupings. For this reason, to be an accurate description of a reaction, a chemical equation must show the same number of atoms of each element in the reactants and in the products. For example, in the equation

$$2\,\mathrm{H}\cdot + \cdot\ddot{\underset{..}{\mathrm{O}}}\cdot \rightarrow \mathrm{H}{-}\ddot{\underset{..}{\mathrm{O}}}{-}\mathrm{H}$$

the water molecule produced contains two hydrogen atoms and one oxygen atom, so these same numbers of atoms must be present in the reactants. An equation that shows the same number of atoms of each element in the reactants and in the products is said to be **balanced.**

Exercise 9.6

Decide whether each of the following equations is balanced:

(a) $\cdot\dot{\mathrm{C}}: + \cdot\ddot{\underset{..}{\mathrm{O}}}\cdot \rightarrow \ddot{\underset{..}{\mathrm{O}}}{=}\mathrm{C}{=}\ddot{\underset{..}{\mathrm{O}}}$

(b) $\cdot\dot{\mathrm{C}}: + \cdot\ddot{\underset{..}{\mathrm{O}}}\cdot \rightarrow :\mathrm{C}{\equiv}\mathrm{O}:$

To balance an equation, insert into it the lowest coefficients that will show the same number of atoms of each element on each side. For example, the equation shown in part (a) of Exercise 9.6 is unbalanced because it shows one oxygen atom on the left but two oxygen atoms, in a carbon dioxide molecule, on the right. We can balance it by inserting 2 as a coefficient on the left:◆

$$\cdot\dot{\mathrm{C}}: + 2\cdot\ddot{\underset{..}{\mathrm{O}}}\cdot \rightarrow \ddot{\underset{..}{\mathrm{O}}}{=}\mathrm{C}{=}\ddot{\underset{..}{\mathrm{O}}}$$

Exercise 9.7

Balance this equation: $\mathrm{H}\cdot + \cdot\dot{\mathrm{C}}: \rightarrow \mathrm{H}{-}\underset{}{\overset{\mathrm{H}\,|}{\mathrm{C}}}{=}\overset{\mathrm{H}\,|}{\mathrm{C}}{-}\mathrm{H}$

In an equation that shows ions the charges must also be balanced: The sum of the charges on each side of the equation must be the same. For example, the equation

$$\text{Mg:} + \text{:}\ddot{\underset{\cdot\cdot}{\text{F}}}\cdot \rightarrow \text{Mg}^{2+} + \text{:}\ddot{\underset{\cdot\cdot}{\text{F}}}\text{:}^{-}$$

is unbalanced, even though it shows the same number of atoms of each element on each side. In this equation, the sum of the charges on the left is zero, and the sum of the charges on the right is 1+:

$$\begin{array}{lcl} \text{Mg:} + \text{:}\ddot{\underset{\cdot\cdot}{\text{F}}}\cdot & \rightarrow & \text{Mg}^{2+} + \text{:}\ddot{\underset{\cdot\cdot}{\text{F}}}\text{:}^{-} \\ 0 \;+\; 0 = 0 & & 2+ \;+\; 1- = 1+ \end{array}$$

We can balance the charges by inserting a coefficient of 2 on the right:◆

◆By balancing the charges, you're balancing the number of electrons shown in the equation.

$$\begin{array}{lcl} \text{Mg:} + \text{:}\ddot{\underset{\cdot\cdot}{\text{F}}}\cdot & \rightarrow & \text{Mg}^{2+} + 2\text{:}\ddot{\underset{\cdot\cdot}{\text{F}}}\text{:}^{-} \\ 0 \;+\; 0 = 0 & & 2+ \;+\; 2(1-) = 0 \end{array}$$

To balance the atoms, insert a 2 on the left:

$$\text{Mg:} + 2\text{:}\ddot{\underset{\cdot\cdot}{\text{F}}}\cdot \rightarrow \text{Mg}^{2+} + 2\text{:}\ddot{\underset{\cdot\cdot}{\text{F}}}\text{:}^{-}$$

Exercise 9.8

Balance this equation: $\text{Na}\cdot + \cdot\ddot{\underset{\cdot\cdot}{\text{S}}}\cdot \rightarrow \text{Na}^{+} + \text{:}\ddot{\underset{\cdot\cdot}{\text{S}}}\text{:}^{2-}$

Because an equation isn't complete unless it's balanced, the term *equation* in this book will mean *balanced equation.*

9-4 Equations can describe real or imaginary reactions

Just as English sentences can describe real or imaginary events, so chemical equations can describe real or imaginary reactions. For example, the equation

$$2\,\text{H}\cdot + \cdot\ddot{\underset{\cdot\cdot}{\text{O}}}\cdot \rightarrow \text{H—}\ddot{\underset{\cdot\cdot}{\text{O}}}\text{—H}$$

describes the imaginary reaction of two hydrogen atoms with one oxygen atom to form a water molecule. The reaction described by this equation is imaginary because hydrogen and oxygen don't exist in nature in the forms in which they're shown in the equation. In nature, hydrogen and oxygen exist as diatomic (two-atom) molecules, H—H and $\cdot\ddot{\underset{\cdot\cdot}{\text{O}}}\text{—}\ddot{\underset{\cdot\cdot}{\text{O}}}\cdot$. The equation for the real reaction is

$$2\,\text{H—H} + \cdot\ddot{\underset{\cdot\cdot}{\text{O}}}\text{—}\ddot{\underset{\cdot\cdot}{\text{O}}}\cdot \rightarrow 2\,\text{H—}\ddot{\underset{\cdot\cdot}{\text{O}}}\text{—H}$$

An equation describes a real reaction if it shows the reactants and products in the forms in which they actually occur, otherwise, it shows an imaginary reaction. Metals occur in nature as atoms, so they should be shown in that form in equations for real reactions. Many nonmetals and metalloids, however, occur as molecules, and should be shown as molecules in equations for real reactions; for our purposes, the most important examples are the seven nonmetals that occur as diatomic molecules:

$$\mathrm{H{-}H},\ :\mathrm{N{\equiv}N}:,\ \cdot\ddot{\underset{..}{\mathrm{O}}}{-}\ddot{\underset{..}{\mathrm{O}}}\cdot,\ :\ddot{\underset{..}{\mathrm{F}}}{-}\ddot{\underset{..}{\mathrm{F}}}:,\ :\ddot{\underset{..}{\mathrm{Cl}}}{-}\ddot{\underset{..}{\mathrm{Cl}}}:,\ :\ddot{\underset{..}{\mathrm{Br}}}{-}\ddot{\underset{..}{\mathrm{Br}}}:,\ \text{and}\ :\ddot{\underset{..}{\mathrm{I}}}{-}\ddot{\underset{..}{\mathrm{I}}}:$$

When you're first learning chemistry, it's often convenient to use equations for imaginary reactions, as we did in Chapters 7 and 8, because they're simpler than equations for real reactions. In your further work in this book, you'll more often be using equations for real reactions.

Example 9.1

Does the equation

$\mathrm{Mg}: + 2:\ddot{\underset{..}{\mathrm{F}}}\cdot \rightarrow \mathrm{Mg}^{2+} + 2:\ddot{\underset{..}{\mathrm{F}}}:^{-}$ represent an imaginary reaction or a real one?

Solution

Fluorine occurs in nature as diatomic molecules, but it's shown in this equation as atoms, so the equation represents an imaginary reaction.

Example 9.2

Write the equation for the real reaction that corresponds to the imaginary reaction in Example 9.1.

Solution

$$\mathrm{Mg}: + :\ddot{\underset{..}{\mathrm{F}}}{-}\ddot{\underset{..}{\mathrm{F}}}: \rightarrow \mathrm{Mg}^{2+} + 2:\ddot{\underset{..}{\mathrm{F}}}:^{-}$$

Exercise 9.9

Decide whether each of the following equations represents a real reaction or an imaginary one. If it represents an imaginary reaction, write an equation for the corresponding real one.

(a) $2\mathrm{K}\cdot + \mathrm{H{-}H} \rightarrow 2\mathrm{K}^{+} + 2\mathrm{H}:^{-}$

(b) $2\mathrm{K}\cdot + \cdot\ddot{\underset{..}{\mathrm{O}}}\cdot \rightarrow 2\mathrm{K}^{+} + :\ddot{\underset{..}{\mathrm{O}}}:^{2-}$

9-5 It's often convenient to write an equation with condensed or molecular formulas

An equation can be written in Lewis symbols and line formulas or in ordinary symbols and condensed or molecular formulas. For example, the equation

$$\mathrm{Mg}: + :\ddot{\underset{..}{\mathrm{F}}}{-}\ddot{\underset{..}{\mathrm{F}}}: \rightarrow \mathrm{Mg}^{2+} + 2:\ddot{\underset{..}{\mathrm{F}}}:^{-}$$

can also be written

$$Mg + F_2 \rightarrow Mg^{2+} + 2F^- \quad \text{or}$$
$$Mg + F_2 \rightarrow MgF_2$$

Equations written in ordinary symbols and condensed or molecular formulas are simpler and can be written or read more quickly, so they're often preferred.

Exercise 9.10

Rewrite the following equation, using molecular formulas:

$$\text{H—H} + :\ddot{\underset{..}{\text{F}}}\text{—}\ddot{\underset{..}{\text{F}}}: \rightarrow 2\ \text{H—}\ddot{\underset{..}{\text{F}}}:$$

Equations written in ordinary symbols and condensed or molecular formulas are balanced in the usual way, by inserting appropriate coefficients. For example, the unbalanced equation

$$H_2 + O_2 \rightarrow H_2O$$

is balanced by inserting the coefficient 2 on each side:◆

$$2\ H_2 + O_2 \rightarrow 2\ H_2O$$

◆Do *not* balance an equation by changing subscripts. It might seem reasonable at first glance to balance the equation shown to the left by writing $H_2 + O \rightarrow H_2O$, but the change from O_2 to O is not legitimate, because it changes the description of one of the reactants.

Exercise 9.11

Balance this equation:

$$N_2 + O_2 \rightarrow NO$$

If the equation shows ions, the charges are also balanced in the usual way. For example, the equation

$$Li + O_2 \rightarrow Li^+ + O^{2-}$$

is unbalanced. We can balance the atoms by inserting the coefficient 2 on the right:

$$Li + O_2 \rightarrow Li^+ + 2\ O^{2-}$$

The equation is still unbalanced, because, as shown below, the sum of the charges on the left is 0 and the sum of the charges on the right is 3−:

$$\begin{array}{lcl} Li + O_2 & \rightarrow & Li^+ + 2\ O^{2-} \\ 0 + 0 = 0 & & 1+ \ + 2(2-) = 3- \end{array}$$

We can balance the charges by inserting the coefficient 4 on each side:

$4\ Li + O_2$	$\rightarrow$	$4\ Li^+ + 2\ O^{2-}$
$4(0) + 0 = 0$		$4(1+) + 2(2-) = 0$

Exercise 9.12

Balance this equation:

$$H^+ + SO_4^{2-} \rightarrow H_2SO_4$$

Condensed and molecular formulas are convenient, but they have the disadvantage of not showing whether the bonding in a compound is ionic or covalent. To overcome this disadvantage, you need to be able to translate an equation written in ordinary symbols and condensed or molecular formulas into the corresponding equation written in Lewis symbols and line formulas.

Example 9.3

Translate this equation into line formulas:

$$3\ H_2 + N_2 \rightarrow 2\ NH_3$$

Solution

$$3\ H{-}H + :N{\equiv}N: \rightarrow 2\ H{-}\underset{\cdot\cdot}{\overset{\overset{\displaystyle H}{|}}{N}}{-}H$$

Exercise 9.13

Translate this equation into Lewis symbols and line formulas:

$$4\ Na + O_2 \rightarrow 2\ Na_2O$$

Because equations in ordinary symbols and condensed or molecular formulas are easier to work with, we'll usually use them in the rest of this book. In your further work in this book, assume you should write equations with ordinary symbols and condensed or molecular formulas unless you're directed otherwise.

9-6 In an equation the label (s), (ℓ), or (g) shows that a reactant or product is a solid, liquid, or gas

In scientific work, the forms of matter that we call solid, liquid, and gas are referred to as **physical states** or **states of matter.**◆ It's sometimes useful to know whether a reactant or product is in the solid, liquid or gas state, and this information may be shown in a chemical equation by the designations (s), (ℓ), and (g). Here is one example:

◆The gas state is discussed in Chapter 13, and the liquid and solid states are discussed in Chapter 14.

$$Mg(s) + F_2(g) \rightarrow MgF_2(s)$$

In writing an equation it's sometimes possible to predict what state a reactant or product will be in. If the reaction occurs under ordinary room conditions, the following elements will be in the states described:

Gases: H_2, N_2, O_2, F_2, Cl_2.◆

Liquids: Hg and Br_2.

Solids: All others.

◆The noble gases are also gases under ordinary room conditions, but they undergo almost no reactions.

Under ordinary room conditions, all of the ionic compounds we'll consider are solids, and water, of course, is a liquid. No useful generalization can be made about other covalent compounds.

Mercury, Hg. *(Charles D. Winters)*

Example 9.4

Assume that the reaction

$$2\ K + Br_2 \rightarrow 2\ KBr$$

occurs under ordinary room conditions. Rewrite the equation, adding (s), (ℓ), or (g) as appropriate for each reactant and product.

Solution

From the rules given, under ordinary room conditions, the element K is a solid, the element Br_2 is a liquid, and the ionic compound KBr is a solid.

$$2\ K(s) + Br_2(\ell) \rightarrow 2\ KBr(s)$$

Exercise 9.14

Assume that the reaction

$$4\ Fe + 3\ O_2 \rightarrow 2\ Fe_2O_3$$

occurs under ordinary room conditions. Rewrite the equation, adding (s), (ℓ), or (g) as appropriate for each reactant and product.

9-7 An equation can be modified to show that a reaction absorbs or releases energy

Bromine, Br_2. *(Charles D. Winters)*

Every chemical reaction either absorbs or releases energy. In photosynthesis, for example, plants absorb sunlight as part of a series of reactions by which they convert carbon dioxide and water into the complicated covalent compounds called carbohydrates. In the discharge of dry cells and batteries chemical reactions generate electrical energy, and in combustion reactions—for example, in the burning of coal, oil, or natural gas—chemical reactions produce heat and light.

Energy can be thought of as a reactant if it's absorbed in a reaction or as a product if it's released, and the word "energy" can be added to the appropriate side of the equation. In the combustion of natural gas, for example, the main reaction is the reaction of methane with oxygen, and energy is released as heat and light:

$$CH_4 + 2\ O_2 \rightarrow CO_2 + 2\ H_2O + \text{energy}$$

If it's useful to be more specific, the form of energy can be shown:

$$2\ NaCl + \text{electricity} \rightarrow 2\ Na + Cl_2$$

$$H_2 + F_2 \rightarrow 2\ HF + \text{heat}$$

A reaction that absorbs heat is said to be **endothermic,** and a reaction that releases heat is said to be **exothermic.**

Exercise 9.15

The reaction

$$2\ H_2 + O_2 \rightarrow 2\ H_2O$$

is exothermic. Rewrite the equation, adding the word "heat" in the appropriate place.

Explosion of the dirigible *Hindenberg* at Lakehurst, New Jersey, on May 6, 1937. The dirigible was filled with hydrogen, and the explosion reaction was $2\ H_2(g) + O_2(g) \rightarrow 2\ H_2O(g) + \text{heat}$. The reaction was so intensely exothermic that the steel girders that gave the dirigible its structure bent and melted from the heat. *(The Bettman Archive)*

9-8 Reactions can be understood more easily by classifying them into a few basic types

In the 200 years since modern chemistry began, scientists have understood tens of millions of reactions well enough to write equations for them. Throughout the rest of your work with this book, you'll read and write hundreds of equations. To make it easier to understand reactions and their equations, we'll classify them for now into the four types shown in Table 9.1, and we'll add other types later.

Table 9.1 Types of Chemical Reactions

Class	Characteristics	Examples
Combination	Two or more reactants form one product.	$H_2 + Cl_2 \rightarrow 2\ HCl$
Decomposition	One reactant forms two or more products.	$2\ KClO_3 \rightarrow 2\ KCl + 3\ O_2$
Combustion	Nonmetals or their compounds react with oxygen to produce their oxides and release heat.	$C + O_2 \rightarrow CO_2 + \text{heat}$ $CH_4 + 2\ O_2 \rightarrow CO_2 + 2H_2O + \text{heat}$
Replacement	One element replaces another in a compound.	$Fe_2O_3 + 2\ Al \rightarrow Al_2O_3 + 2\ Fe$

Exercise 9.16

Identify the appropriate classes for the reactions shown by the following equations. A reaction may be in more than one class.

(a) $2\ H_2 + O_2 \rightarrow 2\ H_2O$

(b) $CaCO_3 \rightarrow CaO + CO_2$

9-9 To learn to name compounds and write equations quickly and accurately, practice in a variety of ways

The only way to become proficient in the language of chemistry is by practice. When you see the symbol for an atom or ion, think of its name, and vice versa. When you see the formula for a compound, think of its name, and vice versa.

Practice by working many simple and complicated problems, and check your answers. Some examples follow, and many more problems are available at the end of this chapter.

Example 9.5

The unbalanced equation

$$Sn + F_2 \rightarrow SnF_4$$

is a partial description of an exothermic reaction. (a) Classify the reaction. (b) Balance the equation. (c) Name each reactant and product. (d) Modify the equation to show that the reaction is exothermic and to show the state of each reactant and product, assuming that the reaction occurs under ordinary room conditions.

Solution

(a) Combination reaction.

(b) through (d): $Sn(s) + 2\ F_2(g) \rightarrow SnF_4(s) + \text{heat}$

tin fluorine tin(IV) fluoride

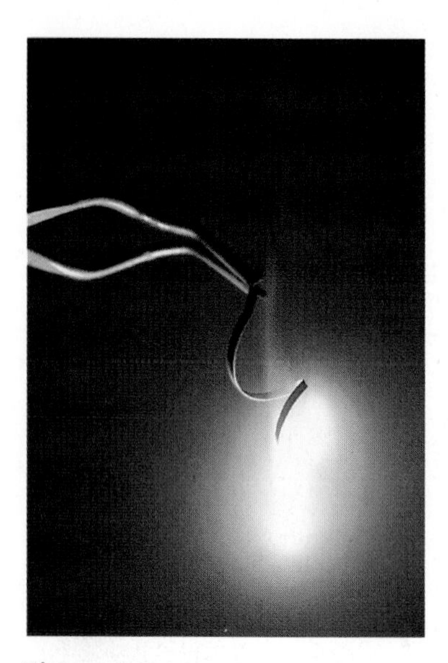

Figure 9.1
Magnesium metal burning in air.
(Charles D. Winters)

Exercise 9.17

Figure 9.1 shows magnesium metal burning in air. Air is about 80% nitrogen and about 20% oxygen, with much smaller amounts of other substances.

(a) Write a balanced equation for the reaction of magnesium with oxygen. Name each reactant and product and designate its physical state under ordinary room conditions. Classify the reaction.
(b) Do the same for the reaction of magnesium with nitrogen.

Exercise 9.18

Figure 9.2 shows the reaction of sodium with chlorine. Write a balanced equation for this reaction. Name each reactant and product and designate its physical state under ordinary room conditions. Classify the reaction.

Figure 9.2
Sodium reacting with chlorine.
(Chip Clark)

Inside Chemistry

What are generic drugs?

One of the miracles we take for granted in modern developed countries is the availability of thousands of drugs to prevent or cure disease or to relieve pain. In your family there is probably at least one person whose life has been prolonged or improved by modern medicines.

All modern drugs are prepared by chemists. A few drugs are relatively simple ionic compounds—lithium carbonate (Li_2CO_3), for example, is used to control manic-depressive cycles—but most of them are complex covalent compounds of carbon, hydrogen, oxygen, and nitrogen. Table 7.2 shows five examples of complex covalent compounds that are drugs: aspirin, cocaine, methionine, nitroglycerin, and Vitamin C.

Many drugs are sold under a variety of names. For example, the anti-inflammatory compound ibuprofen, whose line formula is

```
      H
      |
 H—C—H  H    H        H          :O:
      |     |     \  C—C  /           ||   ..
 H—C———C—C        C———C—C—O—H
      |     |     /  C=C  \      |        ..
 H—C—H  H   H        H   H—C—H
      |                            |
      H                            H
```

is sold by various drug companies under these trademark names: Adran, Advil, Anco, Amibufen, Anflagen, Apsifen, Artril 300, Bluton, Brufanic, Brufen, Brufort, Buburone, Butylenin, Dolgin, Dolgirid, Dolgit, Dolocyl, Dolo-Dolgit, Ebufac, Emodin, Eopbron, Femadon, Fenbid, Haltran, Ibu-Attritin, Ibumetin, Ibuprocin, Ibu-slo, Ibutid, Ibutop, Inabrin, Inoven, Lamidon, Lebrufen, Liptan, Lobufen, Medipren, Mono-Attritin, Motrin, Mynosedin, Napacetin, Nobfelon, Nobfen, Nobgen, Novogent N, Nuprin, Nurofen, Opturem, Paxofen, Prontalgin, Rebugen, Recidol, Roidenin, Seclodin, Suspren, Tabalon, Trendar, Urem. Because ibuprofen is the compound common to all of these commercial drug preparations, it's referred to as their **generic drug.** Prices for the same generic drug, sold under different brand names, can vary, so you can save money by choosing the least expensive brand of a generic drug.

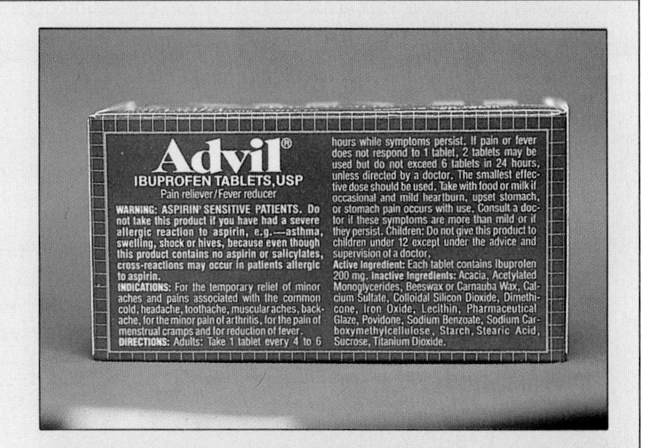

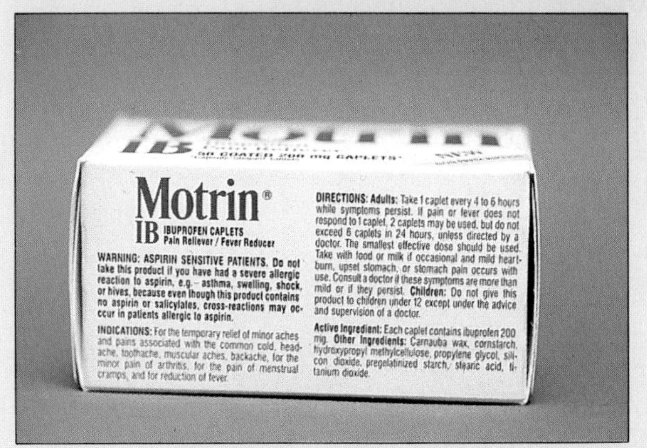

The generic drug ibuprofen is sold under many trade names, including ADVIL® and MOTRIN®. *(David Frazier Photolibrary)*

Chapter Summary: Naming Compounds and Writing Equations

Section	Subject	Summary	Check When Learned
9-1	Binary compound	Compound that contains only two elements, for example, HCl, H_2O, or Al_2O_3.	☐
9-1	Rule for naming a binary covalent compound that contains hydrogen	Write "hydrogen," then the name of the second element, with its suffix changed to -ide. Examples: HCl, hydrogen chloride; H_2S, hydrogen sulfide.	☐
9-1	Rule for naming other binary covalent compounds	Name the elements in the order in which they appear in the formula. Change the suffix of the second element to -ide and add a prefix to the name of each element to show how many atoms of that element are in one molecule.	☐
9-1	Prefixes to show these numbers of atoms:		
	one	mono-	☐
	two	di-	☐
	three	tri-	☐
	four	tetra-	☐
	five	penta-	☐
	six	hexa-	☐
	seven	hepta-	☐
	eight	octa-	☐
	nine	nona- or ennea-	☐
	ten	deca-	☐
9-1	Common names for:		
	H_2O	Water.	☐
	NH_3	Ammonia.	☐
	CH_4	Methane.	☐
9-2	Rule for naming an ionic compound	Name the cation, then the anion. Examples are NaCl, sodium chloride, and $CuSO_4$, copper(II) sulfate.	☐
9-2	Rule for designating the number of water molecules in a hydrate.	Use the same prefixes that designate numbers of atoms in covalent compounds. For example, $CuSO_4 \cdot 5H_2O$ is copper(II) sulfate pentahydrate.	☐
9-3	Meaning of *balanced* as it applies to a chemical equation	The equation contains the same number of atoms of each element in the reactants and in the products. If the equation contains ions, the sum of the charges in the reactants is the same as the sum of the charges in the products.	☐

Section	Subject	Summary	Check When Learned
9-3	Procedure for balancing an equation	If the equation contains ions, insert the lowest coefficients that will make the sum of the charges the same on each side. Insert the lowest coefficients that will make the number of atoms of each element the same on each side.	☐
9-4	Distinction between the equation for an imaginary reaction and the equation for a real one	An equation describes a real reaction if it shows the substances in the forms in which they actually occur; otherwise it describes an imaginary reaction.	☐
9-4	Forms in which these substances ordinarily occur:		
	metals	atoms	☐
	hydrogen	H_2	☐
	nitrogen	N_2	☐
	oxygen	O_2	☐
	fluorine	F_2	☐
	chlorine	Cl_2	☐
	bromine	Br_2	☐
	iodine	I_2	☐
9-5	Advantage of writing an equation in ordinary symbols and condensed or molecular formulas, rather than in Lewis symbols and line formulas	Equations written in ordinary symbols and condensed or molecular formulas are simpler and can be written or read more quickly.	☐
9-6	Physical states or states of matter	Solid, liquid, and gas.	☐
9-6	Notation used in an equation to show that a substance is a: solid liquid gas	(s) For example: (ℓ) $2\ KClO_3(s) \rightarrow 2\ KCl(s) + 3\ O_2(g)$ (g)	☐

Section	Subject	Summary	Check When Learned
9-6	Elements in these states under ordinary room conditions:		
	gases	H_2, N_2, O_2, F_2, Cl_2 and the noble gases.	☐
	liquids	Hg, Br_2.	☐
	solids	All others.	☐
9-6	State of ionic compounds under room conditions	Solid.	☐
9-6	State of covalent compounds under room conditions	No simple generalization can be made. Water is a liquid.	☐
9-7	Notation to show that energy is absorbed or released during a reaction	If energy is absorbed, write "energy" on the reactant side. If energy is released, write "energy" on the product side.	☐
9-7	Meaning of:		
	endothermic	An endothermic reaction absorbs heat.	☐
	exothermic	An exothermic reaction releases heat.	☐
9-7	Notation to show that heat is absorbed or released during a reaction.	If heat is absorbed, write "heat" on the reactant side. If heat is released, write "heat" on the product side.	☐
9-8	Characteristics of types of chemical reactions:		
	combination	Two or more reactants form a product.	☐
	decomposition	One reactant forms two or more products.	☐
	combustion	Nonmetals or their compounds react with oxygen to produce their oxides and release heat.	☐
	replacement	One element replaces another in a compound.	☐

Problems

Assume you can use the Periodic Table at the front of the book unless you're directed otherwise. Answers to odd-numbered problems are in Appendix 1.

Naming Compounds (Sections 9-1 and 9-2)

1. Decide whether each of these compounds is ionic or covalent: (a) H_2O (b) Na_2O (c) FeO.
2. Decide whether each of these compounds is ionic or covalent: (a) PF_3 (b) FeF_3 (c) NH_3.
3. Name these compounds: (a) SnH_4 (b) SnO_2 (c) CH_4.
4. Name these compounds: (a) KCl (b) $KClO_2$ (c) $KClO_4$.
5. Name these compounds: (a) Al_2S_3 (b) P_2O_5 (c) ZnI_2.
6. Name these compounds: (a) NaH (b) NH_4Cl (c) ICl.
7. Name these compounds: (a) $Cu(NO_3)_2$ (b) $Mg_3(PO_4)_2$ (c) AlN.
8. Name these compounds: (a) NO_2 (b) $LiNO_2$ (c) N_2O_4.
9. Name these compounds: (a) $NaC_2H_3O_2$ (b) $CaCO_3$ (c) $Ca(OH)_2$.
10. Name these compounds: (a) $Sr(HCO_3)_2$ (b) $Sr(HS)_2$ (c) $Sr(HSO_3)_2$.
11. Write formulas for these compounds: (a) carbon disulfide (b) cadmium sulfate (c) copper(I) iodide.
12. Write formulas for these compounds: (a) carbon tetrachloride (b) calcium acetate (c) zinc hypochlorite.
13. Write formulas for these compounds: (a) iron(II) cyanide (b) hydrogen sulfide (c) mercury(II) hydrogen carbonate.
14. Write formulas for these compounds: (a) sodium phosphate (b) hydrogen bromide (c) copper(II) oxide.
15. Write formulas for these compounds: (a) barium chlorate (b) barium bromate (c) barium iodate.
16. Write formulas for these compounds: (a) sodium nitride (b) nitrogen trichloride (c) magnesium oxide monohydrate.
17. Write formulas for these compounds: (a) lithium hydroxide (b) sodium perchlorate (c) ammonium sulfide.
18. Write formulas for these compounds: (a) iron(III) acetate (b) iron(II) sulfide (c) iron(III) sulfide.
19. Write formulas for these compounds: (a) nickel sulfite (b) nickel sulfide (c) nickel sulfate.
20. Write formulas for these compounds: (a) silver nitrite (b) silver nitride (c) silver nitrate.

Balancing Equations (Sections 9-3 through 9-5)

21. Balance these equations:
   (a) $H\cdot + \cdot\ddot{\underset{\cdot}{N}}\cdot \rightarrow H{-}\underset{\underset{H}{|}}{\ddot{N}}{-}H$
   (b) $H\cdot + \cdot\ddot{\underset{\cdot}{N}}\cdot \rightarrow H{-}\ddot{N}{=}\ddot{N}{-}H$
   (c) $H\cdot + \cdot\ddot{\underset{\cdot}{N}}\cdot \rightarrow H{-}\underset{\underset{H}{|}}{\ddot{N}}{-}\underset{\underset{H}{|}}{\ddot{N}}{-}H$
22. Balance these equations:
   (a) $Na\cdot + \cdot\ddot{\underset{\cdot\cdot}{O}}\cdot \rightarrow Na^+ + :\ddot{\underset{\cdot\cdot}{O}}:^{2-}$
   (b) $Mg: + \cdot\ddot{\underset{\cdot\cdot}{O}}\cdot \rightarrow Mg^{2+} + :\ddot{\underset{\cdot\cdot}{O}}:^{2-}$
   (c) $\dot{Al}: + \cdot\ddot{\underset{\cdot\cdot}{O}}\cdot \rightarrow Al^{3+} + :\ddot{\underset{\cdot\cdot}{O}}:^{2-}$
23. Balance these equations:
   (a) $H_2 + I_2 \rightarrow HI$
   (b) $H_2 + N_2 \rightarrow NH_3$
   (c) $Li + H_2 \rightarrow LiH$
24. Balance these equations:
   (a) $IBr \rightarrow I_2 + Br_2$
   (b) $SO_2 + O_2 \rightarrow SO_3$
   (c) $H_2O + SO_3 \rightarrow H_2SO_4$

25. Balance these equations:
    (a) $Fe + O_2 \rightarrow Fe_2O_3$
    (b) $Zn + O_2 \rightarrow ZnO$
    (c) $K + O_2 \rightarrow K_2O$

26. Balance these equations:
    (a) $H_2O + CO_2 \rightarrow H_2CO_3$
    (b) $H_2O + SO_2 \rightarrow H_2SO_3$
    (c) $H_2O + P_2O_5 \rightarrow H_3PO_4$

27. Balance these equations:
    (a) $C_2H_2 + O_2 \rightarrow CO_2 + H_2O$
    (b) $C_2H_4 + O_2 \rightarrow CO_2 + H_2O$
    (c) $C_2H_6 + O_2 \rightarrow CO_2 + H_2O$

28. Balance these equations:
    (a) $Na_2O + H_2O \rightarrow NaOH$
    (b) $CaO + H_2O \rightarrow Ca(OH)_2$
    (c) $Ni(OH)_2 \rightarrow NiO + H_2O$

29. Balance the equations for these reactions:
    (a) Chlorine reacts with magnesium to form magnesium chloride.
    (b) Calcium reacts with nitrogen to form calcium nitride.
    (c) Methane reacts with fluorine to form carbon tetrafluoride and hydrogen fluoride.

30. Balance the equations for these reactions:
    (a) Oxygen reacts with fluorine to form oxygen difluoride.
    (b) Strontium reacts with fluorine to form strontium fluoride.
    (c) Ammonia reacts with fluorine to form nitrogen trifluoride and hydrogen fluoride.

31. Balance the equations for these reactions:
    (a) Hydrogen ion combines with carbonate ion to form hydrogen carbonate ion.
    (b) Hydrogen ion combines with sulfite ion to form hydrogen sulfite ion.
    (c) Hydrogen ion combines with sulfide ion to form hydrogen sulfide ion.

32. Balance the equations for these reactions:
    (a) Calcium carbonate forms calcium oxide and carbon dioxide.
    (b) Barium sulfite forms barium oxide and sulfur dioxide.
    (c) Phosphorus trichloride reacts with chlorine to form phosphorus pentachloride.

33. Complete and balance these equations:
    (a) $Na\cdot + :\underset{\cdot\cdot}{\overset{\cdot\cdot}{Br}}-\underset{\cdot\cdot}{\overset{\cdot\cdot}{Br}}: \rightarrow$
    (b) $Ca: + :\underset{\cdot\cdot}{\overset{\cdot\cdot}{Cl}}-\underset{\cdot\cdot}{\overset{\cdot\cdot}{Cl}}: \rightarrow$
    (c) $\cdot Al: + :\underset{\cdot\cdot}{\overset{\cdot\cdot}{F}}-\underset{\cdot\cdot}{\overset{\cdot\cdot}{F}}: \rightarrow$

34. Complete and balance these equations:
    (a) $H-H + :\underset{\cdot\cdot}{\overset{\cdot\cdot}{Cl}}-\underset{\cdot\cdot}{\overset{\cdot\cdot}{Cl}}: \rightarrow$
    (b) $H-H + \cdot\underset{\cdot\cdot}{\overset{\cdot\cdot}{O}}-\underset{\cdot\cdot}{\overset{\cdot\cdot}{O}}\cdot \rightarrow$
    (c) $H-H + Li\cdot \rightarrow$

35. Complete and balance these equations to show the formation of the corresponding ionic compounds:
    (a) $Sn^{2+} + ClO_3^- \rightarrow$
    (b) $Cu^{2+} + CO_3^{2-} \rightarrow$
    (c) $Ni^{2+} + P^{3-} \rightarrow$

36. Complete and balance these equations to show the formation of the corresponding ionic compounds:
    (a) $Fe^{3+} + PO_4^{3-} \rightarrow$
    (b) $Al^{3+} + S^{2-} \rightarrow$
    (c) $Sn^{4+} + Cl^- \rightarrow$

37. Write equations for these reactions:
    (a) The reaction of potassium with bromine.
    (b) The reaction of barium with bromine.
    (c) The reaction of aluminum with bromine.

38. Write equations for these reactions:
    (a) The reaction of nickel with oxygen.
    (b) The reaction of lithium with oxygen.
    (c) The reaction of aluminum with oxygen.

39. Write equations for these reactions:
    (a) The reaction of sulfur with oxygen. Assume that the product is sulfur dioxide and that sulfur exists as sulfur atoms.
    (b) The reaction of sulfur with oxygen. Assume that the product is sulfur trioxide and that sulfur exists as sulfur atoms.
    (c) The reaction of sulfur with fluorine. Assume that the product is sulfur hexafluoride and that sulfur exists as sulfur atoms.

40. Write equations for these reactions:
    (a) The reaction of sulfur with oxygen. Assume that the product is sulfur dioxide and that sulfur exists as S_8 molecules.
    (b) The reaction of sulfur with oxygen. Assume that the product is sulfur trioxide and that sulfur exists as S_8 molecules.
    (c) The reaction of sulfur with fluorine. Assume that the product is sulfur hexafluoride and that sulfur exists as S_8 molecules.

Identifying Real and Imaginary Reactions (Section 9-4)

41. Decide whether each of the following equations represents a real reaction or an imaginary one. If it represents an imaginary reaction, write the equation for the corresponding real one.
    (a) $\cdot\dot{\text{C}}: + 4\,:\ddot{\text{F}}\cdot \rightarrow$ CF_4 (Lewis structure: central C bonded to four $:\ddot{\text{F}}:$)
    (b) $O_2 + 2\ F_2 \rightarrow 2\ OF_2$
    (c) $2\ Na\cdot + \cdot\ddot{\text{O}}\cdot \rightarrow 2\ Na^+ + :\ddot{\text{O}}:^{2-}$

42. Decide whether each of the following equations represents a real reaction or an imaginary one. If it represents an imaginary reaction, write the equation for the corresponding real one.
    (a) $3\ H{-}H + :N{\equiv}N: \rightarrow 2\ H{-}\ddot{N}{-}H$ (with H on N)
    (b) $H\cdot + :\ddot{\text{Br}}\cdot \rightarrow H{-}\ddot{\text{Br}}:$
    (c) $C + O_2 \rightarrow CO_2$

Writing Equations in Line Formulas or Condensed Formulas (Section 9-5)

43. Rewrite these equations, using molecular formulas:
    (a) $3\ H{-}H + :N{\equiv}N: \rightarrow 2\ H{-}\ddot{N}{-}H$ (with H on N)
    (b) $H{-}H + :\ddot{\text{Br}}{-}\ddot{\text{Br}}: \rightarrow 2\ H{-}\ddot{\text{Br}}:$
    (c) $2\,:C{\equiv}O: + \cdot\ddot{\text{O}}{-}\ddot{\text{O}}\cdot \rightarrow 2\ \ddot{\text{O}}{=}C{=}\ddot{\text{O}}$

44. Rewrite these equations, using line formulas:
    (a) $CO + 2\ H_2 \rightarrow CH_3OH$
    (b) $2\ C_2H_2 + 5\ O_2 \rightarrow 4\ CO_2 + 2\ H_2O$
    (c) $CH_4 + Cl_2 \rightarrow CH_3Cl + HCl$

Modifying Equations to Show States of Matter and Energy Changes (Sections 9-6 and 9-7)

45. Rewrite each of the following equations to show the physical state of each reactant and product under ordinary room conditions. If appropriate information is available, modify the equation to show an energy change.
    (a) $Hg + F_2 \rightarrow HgF_2$ (energy released)
    (b) $I_2 + Cl_2 \rightarrow 2\ ICl(s)$
    (c) $2\ KClO_3 \rightarrow 2\ KCl + O_2$ (endothermic)

46. Rewrite each of the following equations to show the physical state of each reactant and product under ordinary room conditions. If appropriate information is available, modify the equation to show an energy change.
    (a) $Mg + Br_2 \rightarrow MgBr_2$
    (b) $2\ CO + O_2 \rightarrow 2\ CO_2$ (exothermic)
    (c) $C + O_2 \rightarrow CO_2$ (combustion reaction)

Classifying Reactions (Section 9-8)

47. Classify each reaction in problem 27.
48. Classify each reaction in problem 28.
49. Classify each reaction in problem 43.
50. Classify each reaction in problem 45.

Naming Compounds and Writing Equations (Sections 9-1 through 9-8)

51. Under ordinary room conditions sodium undergoes a highly exothermic reaction with fluorine. Write an equation for this reaction. Include designations of the states of the reactants and products and show that the reaction is exothermic. Under the equation, name each reactant and product.

52. Methanol, CH_3OH, also called wood alcohol, is sometimes used as a fuel in internal combustion engines. Write an equation for its combustion. Assume that methanol is a liquid and that the other reactants and products are in their usual states under ordinary room conditions. Show the heat change that occurs in the reaction. Under the equation, name each reactant and product.

53. Under ordinary room conditions zinc undergoes an exothermic reaction with chlorine. Write an equation for this reaction. Include designations of the states of the reactants and products and show that the reaction is exothermic. Under the equation, name each reactant and product.

54. Ethanol, C_2H_5OH, the intoxicating substance in beverage alcohol, can be burned. Write an equation for its combustion. Assume that ethanol is a liquid and that the other reactants and products are in their usual states under ordinary room conditions. Show the heat change that occurs in the reaction. Under the equation, name each reactant and product.

55. Write an equation for the combustion of hydrogen sulfide; one of the products is sulfur dioxide. Show the state of each reactant and product under ordinary room conditions, assuming that hydrogen sulfide and sulfur dioxide are gases.

56. Iron reacts with fluorine to form iron(III) fluoride. Write an equation for this reaction. Show the state of each reactant and product under ordinary room conditions.

57. Using M as the symbol for any alkali metal:
   (a) Write an equation for the reaction of any alkali metal with chlorine. Show the state of each reactant and product under ordinary room conditions.
   (b) Write the corresponding equation in Lewis symbols and line formulas.

58. Using M as the symbol for any alkaline-earth metal:
   (a) Write an equation for the reaction of any alkaline-earth metal with chlorine. Show the state of each reactant and product under ordinary room conditions.
   (b) Write the corresponding equation in Lewis symbols and line formulas.

59. Using M as the symbol for any alkali metal:
   (a) Write an equation for the reaction of any alkali metal with oxygen. Show the state of each reactant and product under ordinary room conditions.
   (b) Write the corresponding equation in Lewis symbols and line formulas.

60. Using M as the symbol for any alkaline-earth metal:
   (a) Write an equation for the reaction of any alkaline-earth metal with oxygen. Show the state of each reactant and product under ordinary room conditions.
   (b) Write the corresponding equation in Lewis symbols and line formulas.

When red mercury(II) oxide is heated to 500°C, it decomposes into mercury and oxygen: $2\,HgO(s) \rightarrow 2\,Hg(\ell) + O_2(g)$. *(Chip Clark)*

10
The Periodic Table as a Guide to Formulas and Reactions

One important test of your knowledge of basic chemistry will be your ability to predict formulas for simple compounds of the representative elements. Another important test will be your ability to predict equations for the reactions by which these simple compounds are formed. In Chapters 7, 8, and 9 you learned the fundamental principles used in making these predictions.

In this chapter you'll develop your skills further by learning to use the Periodic Table to predict formulas and reactions quickly and easily. We'll look at each group of the representative elements to identify their typical compounds and the reactions that form those compounds. We'll also consider some of the physical properties of the representative elements and a few of their important uses.

The behavior of the transition elements is more complicated, and we'll consider it only briefly.

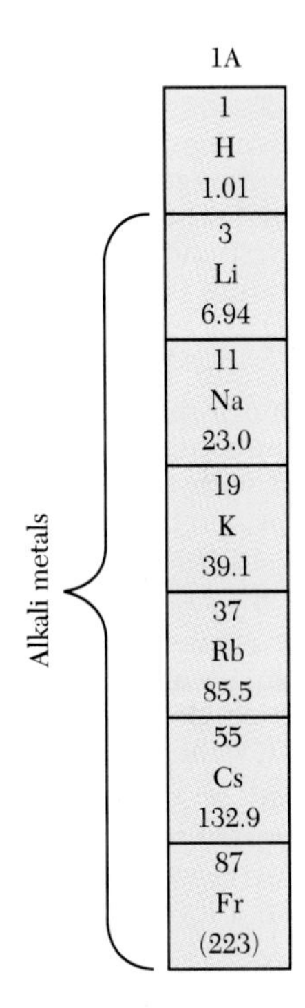

10-1 The elements in group 1A, except hydrogen, always form 1+ cations

Group 1A consists of hydrogen and the alkali metals—lithium, sodium, potassium, rubidium, cesium, and francium. At ordinary temperatures and pressures hydrogen is a colorless, odorless gas, and lithium, sodium, potassium, rubidium, and cesium are silvery-white metals that are relatively soft, about the consistency of hard butter. Figure 10.1 shows a sample of sodium. The bulk appearance of francium and many of its other properties aren't known because all of the isotopes of francium are radioactive and have short lifetimes, so only very small samples of the element have been prepared.

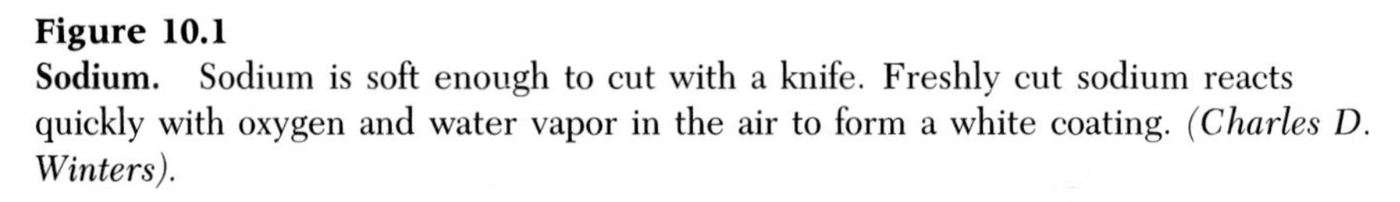

Figure 10.1
Sodium. Sodium is soft enough to cut with a knife. Freshly cut sodium reacts quickly with oxygen and water vapor in the air to form a white coating. *(Charles D. Winters).*

Table 10.1 Group 1A Elements

Element	Symbol	Discovery	Melting Point, °C	Boiling Point, °C
Hydrogen	H	England, 1766	−259	−253
Lithium	Li	Sweden, 1817	180	1336
Sodium	Na	England, 1807	98	883
Potassium	K	England, 1807	54	774
Rubidium	Rb	Germany, 1861	39	696
Cesium	Cs	Germany, 1860	29	705
Francium	Fr	France, 1939	?	?

Table 10.1 summarizes information about the discovery, melting point, and boiling point of each of the elements in group 1A.◆ Among the melting points and boiling points of metals, those of the alkali metals are low. For comparison, iron melts at 1525° and boils at 3000°C.

◆Melting and boiling temperatures depend on pressure. The temperatures here are for an ordinary pressure of one atmosphere.

Hydrogen differs from the other elements in group 1A in being a gas under ordinary room conditions, and it also shows differences in its chemical properties. We'll consider its unique properties at the end of this Section.

The reactions of the elements in group 1A can be understood by considering their electron configurations and electronegativity, shown in Table 10.2. The alkali metals have the lowest electronegativity, and the lowest ionization energy, among all of the elements on the Periodic Table. As a result, atoms of alkali metals easily lose electrons and become positive ions.◆ The atoms of alkali metals always become 1+ cations, because their characteristic valence electron configuration is s^1. The s^1 electron in the outermost energy level is easily lost, but the remaining electrons, in filled energy levels, can't be removed by ordinary chemical processes.

◆Electronegativity (Section 7-3) is a measure of the ability of an atom to draw bonding electrons to itself, and ionization energy (Section 8-1) is the amount of energy required to remove the outermost electron from an atom or ion.

From the information that the alkali metals always exist in their compounds as 1+ cations, we can predict formulas for their compounds. Knowing the formula for a compound often makes it possible to predict the reaction by which

Table 10.2 Electron Configurations and Electronegativity Values for the Group 1A Elements
Characteristic valence electron configuration for group 1A: s^1.

Element	Abbreviated Electron Configuration	Electronegativity
Hydrogen	$1s^1$	2.1
Lithium	$[He]^2 2s^1$	1.0
Sodium	$[Ne]^{10} 3s^1$	0.9
Potassium	$[Ar]^{18} 4s^1$	0.8
Rubidium	$[Kr]^{36} 5s^1$	0.8
Cesium	$[Xe]^{54} 6s^1$	0.7
Francium	$[Rn]^{86} 7s^1$	0.7

Figure 10.2
Predicting formulas and reactions.

PREDICTING FORMULAS AND REACTIONS	
Example: Predict the formula for the compound which will be formed between sodium and chlorine, and predict the equation for the reaction by which the compound will be formed from its elements.	
Steps	**Example**
1. Decide whether the compound will be covalent or ionic.	Sodium (Na) is a metal and chlorine (Cl) is a nonmetal, so the compound will be ionic.
2. If the compound is covalent, predict its formula from the principles in Chapter 7. If the compound is ionic, predict its formula from the principles in Chapter 8.	The ions are Na^+ and Cl^-, so the formula will be NaCl.
3. Write the symbols and formulas for the reactants and products for the formation of the compound from its elements.	The element sodium exists as Na atoms, and the element chlorine exists as Cl_2 molecules: $Na + Cl_2 \longrightarrow NaCl$
4. Balance the equation.	$2\,Na + Cl_2 \longrightarrow 2\,NaCl$

the compound will be formed from its elements. Figure 10.2 shows the reasoning used to make these predictions.

Example 10.1

(a) Predict the formula for the compound that will be formed between potassium and iodine. (b) Predict the equation for the reaction by which the compound will be formed from its elements.

Sodium, a silvery metal, reacts with chlorine, a green gas, to form sodium chloride, a white, crystalline solid. *(Chip Clark)*

Solution

(a) Potassium (K) is a metal and iodine (I) is a nonmetal, so the compound will be ionic. The ions are K^+ and I^-, so the formula will be KI.

(b) Potassium exists as K and iodine exists as I_2, so the reactants and products will be

$$K + I_2 \rightarrow KI$$

The balanced equation will be

$$2\ K + I_2 \rightarrow 2\ KI$$

Exercise 10.1◆

(a) Predict the formula for the compound that will be formed between lithium and bromine. (b) Predict the equation for the reaction by which the compound will be formed from its elements.

◆Answers to Exercises are in Appendix 1.

The preceding Example and Exercise illustrate two important principles: (1) Elements in the same group form compounds with similar formulas. (2) Similar compounds are formed by similar reactions. These principles have limitations, but they can often be used to predict formulas and reactions quickly and easily.

For example, if we know that the formula for the compound formed by sodium and chlorine is NaCl, we can use the principle that elements in the same group form compounds with similar formulas to predict the formulas for 34 other compounds formed between elements in group 1A, which contains sodium, and elements in group 7A, which contains chlorine:

		Group 7A				
		F	Cl	Br	I	At
	H	HF	HCl	HBr	HI	HAt
	Li	LiF	LiCl	LiBr	LiI	LiAt
	Na	NaF	NaCl	NaBr	NaI	NaAt
Group 1A	K	KF	KCl	KBr	KI	KAt
	Rb	RbF	RbCl	RbBr	RbI	RbAt
	Cs	CsF	CsCl	CsBr	CsI	CsAt
	Fr	FrF	FrCl	FrBr	FrI	FrAt

From the example of the equation for the formation of sodium chloride and from the principle that similar compounds are formed by similar reactions, we can write an equation for the formation of each compound shown in the chart above, keeping in mind that hydrogen exists as H_2. Examples are

$$H_2 + F_2 \rightarrow 2\ HF$$
$$2\ Li + Cl_2 \rightarrow 2\ LiCl$$
$$2\ Na + Br_2 \rightarrow 2\ NaBr$$
$$2\ K + I_2 \rightarrow 2\ KI$$

Figure 10.3
Reaction of potassium with water.
(Charles Steele)

Example 10.2

The formula for sodium oxide is Na_2O. Predict the formula for potassium oxide.

Solution

Sodium and potassium are both in group 1A and should form compounds with similar formulas. The formula for potassium oxide should be K_2O.

Exercise 10.2

The equation for the formation of sodium oxide from its elements is

$$4\ Na + O_2 \rightarrow 2\ Na_2O$$

Predict the equation for the formation of potassium oxide from its elements.

The alkali metals are highly reactive elements: They react rapidly with most of the nonmetals and with many compounds, often with the release of large quantities of heat. For example, the alkali metals burn brightly when exposed to pure oxygen. As shown in Exercise 10.2, the equation for the reaction of sodium with oxygen is

$$4\ Na + O_2 \rightarrow 2\ Na_2O$$

and the corresponding equations for the other alkali metals are similar.

The alkali metals react violently with water, releasing large quantities of heat. The reaction that occurs when potassium is dropped into water, shown in Figure 10.3, can be represented by this equation:

$$2\ K(s) + 2\ H_2O(\ell) \rightarrow 2\ KOH(aq) + H_2(g)$$

The symbol **aq** stands for the word **aqueous,** an adjective which means *in water*. The formula KOH(aq) is read "aqueous potassium hydroxide," and it shows that the potassium hydroxide is dissolved in water to form a solution.◆ The reaction of an alkali metal with water is dangerous because the heat released by the reaction can ignite the hydrogen produced in the reaction as it mixes with air, causing an explosion.

◆Aqueous solutions are described in detail in Chapters 15 and 16.

Exercise 10.3

Write the equation you predict for the reaction of sodium with water and show the state of each reactant and product.

Exercise 10.4

Complete the following equation:

$$Li(s) + H_2O(\ell) \rightarrow$$

Because the alkali metals react with oxygen and water, they have to be protected from air, so they're usually stored under mineral oil or kerosene. When an alkali metal is exposed to air, it quickly loses its silvery luster and turns a dull white. In Figure 10.1 on page 196 you can see that the piece of sodium being cut has a white coating.

Reactions that form other, more complicated compounds of the alkali metals, for example Na_3PO_4 or K_2CO_3, are described in Chapter 16.

Hydrogen has a higher electronegativity than the other elements in group 1A and a higher ionization energy, so it loses its electron less easily. In most of its compounds hydrogen doesn't form a 1+ cation, but instead shares its electron in a covalent bond. As we saw in Chapter 8, a hydrogen atom can also gain an electron to form a hydride ion, H^-.

Exercise 10.5

The formula for sodium hydride is NaH. (a) Write the equation for its formation from the elements. (b) Write the formula for potassium hydride. (c) Complete the following equation:

$$Li(s) + H_2(g) \rightarrow$$

Exercise 10.6

The formula for ammonia is NH_3. Predict the formula for the compound of hydrogen with arsenic (As).

Some of the elements in group 1A have important functions in nature or important uses in science or technology. Hydrogen is the most abundant element in the universe. On earth, it's part of almost every compound in living things. The pure element is used as a fuel in high-temperature welding, as a reactant in the hydrogenation of fats and oils for cooking, and for a variety of other purposes.† The proper balance of sodium and potassium ions is necessary for proper nerve function. Lithium carbonate is used as a medication to regulate manic-depressive cycles. Other compounds of the elements in group 1A have many special uses in scientific research and in technological applications.

† Chemistry Insight

Physically, the difference between a fat and an oil is that a fat is a solid at room temperatures and an oil is a liquid. Fats and oils are made up of compounds that contain long chains of carbon atoms. In general, the carbon chains of oils contain carbon–carbon double bonds and the carbon chains of fats do not. Traditionally, a molecule that contained a double bond was described as **unsaturated,** and a molecule that contained many double bonds was described as **polyunsaturated.** Unsaturated compounds can react with hydrogen to convert the double bonds to single bonds by the process called **hydrogenation,** for example,

$$-\overset{|}{C}=\overset{|}{C}- + H_2 \rightarrow -\underset{H}{\overset{|}{\underset{|}{C}}}-\underset{H}{\overset{|}{\underset{|}{C}}}-$$

and this reaction can be used to convert cooking oils to cooking fats. Until recent years cooks in the United States preferred fats to oils, so huge quantities of oils were converted to fats by hydrogenation. Because saturated fats cause an increase in blood cholesterol and polyunsaturated oils do not, many cooks now prefer oils, and hydrogenation has become a less important commercial process.

(*Jim Morganthaler*)

Figure 10.4
Calcium. Exposed to the air, calcium reacts quickly with oxygen, nitrogen, and water vapor to form a gray coating. *(Chip Clark)*

Alkaline earth metals
↓
2A

2A
4 Be 9.01
12 Mg 24.3
20 Ca 40.1
38 Sr 87.6
56 Ba 137.3
88 Ra 226

10-2 The elements in group 2A form 2+ cations

The elements in group 2A, the alkaline-earth metals, have the silvery-white appearance we associate with most metals, but they tarnish quickly on exposure to air because of the formation of oxides and nitrides. Figure 10.4 shows the change in calcium on exposure to air. The alkaline-earth metals are harder than the alkali metals and have higher melting points and boiling points, as shown in Table 10.3.

The reactions of the alkaline-earth metals are very similar to those of the alkali metals. Electron configurations and electronegativity values for the alkaline-earth metals are shown in Table 10.4. The alkaline-earth metals form cations in their compounds because, as a group, their electronegativity and ionization energy are low. The atom of an alkaline-earth metal always forms a 2+ cation because the s^2 electrons in the outermost energy level of the atom are easily lost, and the electrons in the filled interior energy levels are strongly held. Compared with the other metals in group 2A, beryllium and magnesium have relatively high electronegativity values. The electronegativity difference between beryllium and some nonmetals is small enough that their bonding has significant covalent character.

Table 10.3 Group 2A Elements

Element	Symbol	Discovery	Melting point, °C	Boiling Point, °C
Beryllium	Be	France and Germany, 1828	1300	2970
Magnesium	Mg	England, 1808	651	1100
Calcium	Ca	England, 1808	850	1440
Strontium	Sr	England, 1808	757	1366
Barium	Ba	England, 1808	850	1140
Radium	Ra	France, 1898	700	1737

Table 10.4 Electron Configurations and Electronegativity Values for the Group 2A Elements
Characteristic valence electron configuration for group 2A: s^2

Element	Abbreviated Electron Configuration	Electronegativity
Beryllium	$[He]^2 2s^2$	1.5
Magnesium	$[Ne]^{10} 3s^2$	1.2
Calcium	$[Ar]^{18} 4s^2$	1.0
Strontium	$[Kr]^{36} 5s^2$	1.0
Barium	$[Xe]^{54} 6s^2$	0.9
Radium	$[Rn]^{86} 7s^2$	0.9

Example 10.3

(a) Predict the formula for the compound that will be formed between magnesium and chlorine. (b) Predict the equation for the reaction by which the compound will be formed from its elements.

Solution

(a) Magnesium (Mg) is a metal and chlorine (Cl) is a nonmetal, so the compound will be ionic. The ions are Mg^{2+} and Cl^-, so the formula will be $MgCl_2$.

(b) Magnesium exists as Mg and chlorine exists as Cl_2, so the reactants and products will be

$$Mg + Cl_2 \rightarrow MgCl_2.$$

The equation is balanced as written here.

Exercise 10.7

(a) Predict the formula for the compound that will be formed between calcium and fluorine. (b) Predict the equation for the reaction by which the compound will be formed from its elements.

Exercise 10.8

(a) Predict the formula for the compound that will be formed between magnesium and oxygen. (b) Predict the equation for the reaction by which the compound will be formed from its elements. (c) Predict the formula for the compound that will be formed between barium and oxygen.

The alkaline-earth metals, except beryllium, react with water. The alkaline-earth metals are, as a group, less reactive than the alkali metals—they react more slowly and in some cases not at all. For example, at room temperatures calcium reacts only slowly with water, and beryllium does not react. The reaction that occurs when calcium is dropped into water (see Figure 10.5) can be represented by this equation:

$$Ca(s) + 2\ H_2O(\ell) \rightarrow Ca(OH)_2(aq) + H_2(g)$$

Figure 10.5
Reaction of calcium with water.
(Chip Clark)

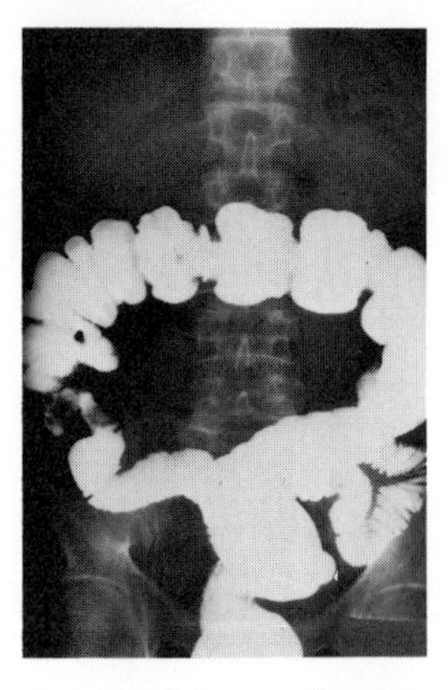

Gastrointestinal X-ray photograph. *(Marna Clark)*

◆The more electrons an atom or ion has, the more opaque it is to X rays. Beryllium is so transparent to X rays that it's used to make the windows in X-ray tubes through which the rays are emitted.

Exercise 10.9

Write an equation for the reaction you predict between strontium and water. Show the state of each reactant and product.

Reactions by which other, more complicated compounds of the alkaline-earth metals are formed, for example $Mg(NO_3)_2$ or $Ca_3(PO_4)_2$, will be described in Chapter 16.

Magnesium and calcium are abundant in the earth's rocks. Limestone is calcium carbonate, $CaCO_3$. The other elements in group 2A are much less common.

The alkaline-earth metals and their compounds have a variety of special uses, as shown by the following few examples. Beryllium and magnesium are used in alloys for special purposes. Magnesium hydroxide, $Mg(OH)_2$, also called milk of magnesia, is an antacid and laxative. Calcium sulfate is a constituent of plaster and cement. Strontium compounds are used to produce red light in fireworks. In preparation for gastrointestinal X-ray photography, patients drink suspensions of barium salts because barium is opaque to X rays and the suspension forms a clear image of the gastrointestinal tract.◆

10-3 In group 3A boron forms covalent compounds with nonmetals, and aluminum forms ionic compounds that contain the Al^{3+} ion

3A
5 B 10.8
13 Al 27.0
31 Ga 69.7
49 In 114.8
81 Tl 204.4

In group 3A (see Table 10.5) the first element, boron, is a metalloid, but the other four elements—aluminum, gallium, indium, and thallium—are metals. Boron exists as yellow crystals or a brown powder, whereas the other four elements in group 3A have the typical grayish-white appearance of metals. Among the elements in group 3A, aluminum is the most important for our purposes.

As shown in Table 10.6, the characteristic valence electron configuration for group 3A is s^2p^1. Boron, with significantly higher electronegativity than the other members of group 3A, forms covalent bonds with nonmetals. A boron atom shares its three s^2p^1 electrons to form covalent bonds. The other members of group 3A form ionic bonds with nonmetals. An aluminum atom loses its three s^2p^1 electrons to form the Al^{3+} cation. In some compounds of gallium, indium,

Indium and thallium. *(Chip Clark)*

Table 10.5 Group 3A Elements

Element	Symbol	Discovery	Melting point, °C	Boiling Point, °C
Boron	B	England and France, 1808	2150	No clear value
Aluminum	Al	Germany, 1827	660	1800
Gallium	Ga	France, 1875	30	1983
Indium	In	Germany, 1863	155	2000
Thallium	Tl	England, 1861	303	1457

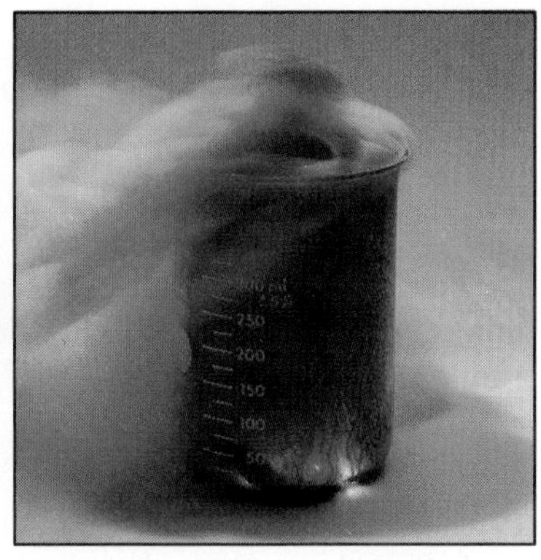

Reaction of aluminum with bromine to form aluminum bromide. *(Charles D. Winters)*

and thallium, all three s^2p^1 valence electrons are lost to form a 3+ cation, and in some compounds only the p^1 electron is lost, to form a 1+ cation.◆

◆We'll find similar behavior in groups 4A and 5A.

Example 10.4

(a) Predict the formula for the compound that will be formed between aluminum and bromine. (b) Predict the equation for the formation of the compound from its elements.

Solution

(a) Aluminum (Al) is a metal and bromine (Br) is a nonmetal, so the compound will be ionic. The ions are Al^{3+} and Br^-, so the formula will be $AlBr_3$.

(b) Aluminum exists as Al and bromine exists as Br_2, so the reactants and products will be

$$Al + Br_2 \rightarrow AlBr_3$$

The balanced equation will be

$$2\ Al + 3\ Br_2 \rightarrow 2\ AlBr_3$$

Table 10.6 Electron Configurations and Electronegativity Values for the Group 3A Elements
Characteristic valence electron configuration for group 3A: s^2p^1.

Element	Abbreviated Electron Configuration	Electronegativity
Boron	$[He]^2 2s^2 2p^1$	2.0
Aluminum	$[Ne]^{10} 3s^2 3p^1$	1.5
Gallium	$[Ar]^{18} 4s^2 4p^1$	1.6
Indium	$[Kr]^{36} 5s^2 5p^1$	1.7
Thallium	$[Xe]^{54} 6s^2 6p^1$	1.8

Exercise 10.10

Write an equation for the reaction you predict between aluminum and fluorine.

Exercise 10.11

(a) Predict the formula for the compound that will be formed between aluminum and oxygen. (b) Predict the equation for its formation from the elements.

Exercise 10.12

Predict formulas for the two oxides of gallium.

Exercise 10.13

Name these compounds: BCl_3, $AlCl_3$, $TlCl_3$, $TlCl$.

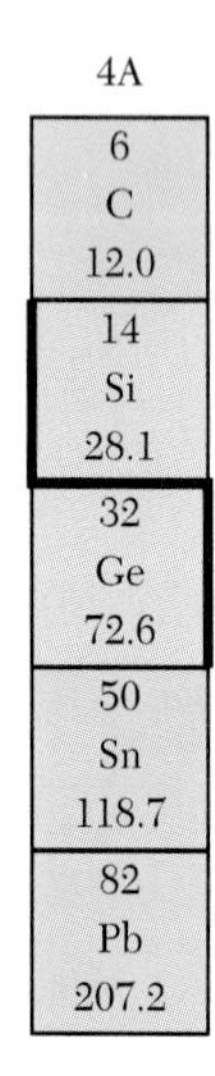

4A
6
C
12.0
14
Si
28.1
32
Ge
72.6
50
Sn
118.7
82
Pb
207.2

Aluminum is the third most abundant element in the earth's crust. It occurs, as aluminosilicate minerals, with the two most abundant elements, silicon and oxygen. The other elements in group 3A are relatively scarce.

The elements in group 3A and their compounds have a variety of uses. Boron is used to make alloys for special purposes. Aluminum alloys are light, strong, and resistant to corrosion, so they have many building and manufacturing applications. Gallium and indium are used in semiconductors. Thallium compounds are extremely poisonous and have been used to poison animal pests.

10-4 Downward in group 4A there is a change from nonmetal to metalloid to metal, with corresponding changes in bonding behavior

Group 4A contains a nonmetal, two metalloids, and two metals (see Table 10.7). Carbon, the nonmetal, occurs in three forms◆: as a black powder, called carbon black; as graphite, a slippery gray powder; and as diamond.◆ The metalloids silicon and germanium have an appearance between that of metals and that of nonmetals. Silicon occurs as black or gray lustrous crystals, and germanium occurs as grayish-white lustrous crystals. At ordinary temperatures tin occurs as a soft, silvery-white lustrous metal, and below −40°C it crumbles into a gray powder. Lead is a soft, bluish-white or silvery metal that is lustrous but quickly becomes dull on exposure to air.

◆Different forms of the same element are called **allotropes.** Elements that have more than one form are said to be **allotropic** and to show **allotropy.**

◆The structure of diamond is described in Section 14-6.

Table 10.7 Group 4A Elements

Element	Symbol	Discovery	Melting Point, °C	Boiling Point, °C
Carbon	C	Ancient	3570	3470◆
Silicon	Si	Sweden, 1824	1414	2355
Germanium	Ge	Germany, 1886	935	2700
Tin	Sn	Ancient	232	2260
Lead	Pb	Ancient	327	1740

◆This is the **sublimation point,** not the boiling point, for carbon. When solid carbon is heated to 3470°, it **sublimes,** that is, it passes directly from the solid state to the gas state, without forming a liquid.

Group 4A. Clockwise from the bottom: carbon, C; silicon, Si; lead, Pb, and tin, Sn. *(Charles D. Winters)*

As shown in Table 10.8, carbon has substantially higher electronegativity than the other elements in group 4A. Carbon is a nonmetal, and in forming compounds with other nonmetals, it forms covalent bonds. As mentioned in Chapter 7, there are more compounds of carbon than of any other element because of the ability of carbon atoms to bond with one another to form chains and rings.

Example 10.5

(a) Predict the formula for the compound that will be formed between one carbon atom and chlorine atoms. (b) Predict the equation for its formation from the elements; assume that carbon exists as carbon atoms.

Solution

(a) Carbon (C) is a nonmetal and chlorine (Cl) is a nonmetal, so the compound will be covalent. The atoms are $\cdot\dot{\mathrm{C}}:$ and $:\underset{..}{\ddot{\mathrm{Cl}}}\cdot$. Combining these

Table 10.8 Electron Configurations and Electronegativity Values for the Group 4A Elements
Characteristic valence electron configuration for group 4A: s^2p^2.

Element	Abbreviated Electron Configuration	Electronegativity
Carbon	$[He]^2 2s^2 2p^2$	2.5
Silicon	$[Ne]^{10} 3s^2 3p^2$	1.8
Germanium	$[Ar]^{18} 4s^2 4p^2$	1.8
Tin	$[Kr]^{36} 5s^2 5p^2$	1.8
Lead	$[Xe]^{54} 6s^2 6p^2$	1.8

atoms according to the octet rule, as described in Chapter 7, gives the formula CCl_4.

(b) Carbon exists as C and chlorine exists as Cl_2, so the reactants and product will be

$$C + Cl_2 \rightarrow CCl_4$$

The balanced equation will be

$$C + 2\ Cl_2 \rightarrow CCl_4$$

Silicon forms many simple compounds whose formulas can be predicted from the formulas for the corresponding compounds of carbon, as shown in the following exercises.

Exercise 10.14

(a) Predict the formula for the compound that will be formed between one silicon atom and chlorine atoms. (b) Predict the equation for its formation from the elements; assume that silicon exists as silicon atoms.

Exercise 10.15

The formula for methane is CH_4. The compound formed between one silicon atom and hydrogen atoms is called silane. Predict its formula.

With nonmetals germanium forms covalent compounds that have the expected formulas.

Exercise 10.16

Predict the formula for the compound formed between one germanium atom and chlorine atoms.

Tin and lead form 2+ and 4+ cations. The characteristic electron configuration for group 4A is s^2p^2. Loss of all four electrons from the outermost energy level produces a 4+ cation, and loss of only the p^2 electrons produces a 2+ cation.

Exercise 10.17

Write the equation you predict for the formation of lead(IV) fluoride from its elements.

Exercise 10.18

Write the equation you predict for the formation of tin(II) oxide from its elements.

Carbon and silicon are widespread in nature. Carbon occurs in most of the compounds that make up living things. In the earth's crust it occurs as limestone (calcium carbonate, $CaCO_3$), and in the atmosphere it occurs as carbon dioxide. Silicon is the second most abundant element in the earth's crust. With oxygen, the most abundant element, it forms a large number of compounds,

Calcite, $CaCO_3$.
(Gary Milburn/Tom Stack & Associates)

called silicates, that make up most of the earth's minerals. Silicates consist of chains or rings of alternating silicon and oxygen atoms, joined by covalent bonds, with metal cations bonded to some of the oxygen atoms. These chains or rings can contain hundreds or thousands of atoms. The two formulas below show examples of such structures.

```
                Na+      Na+
                ..       ..
               :O:-     :O:-
          ..    |   ..   |    ..
Na+  -:O — Si — O — Si — O:-  Na+
          ..    |   ..   |    ..
               :O:-     :O:-
                ..       ..
                Na+      Na+

                  ..  Ca2+ ..
               -:O          O:-
                 ..  \    / ..
                       Si
            ..  :O:  /    \  :O:  ..
         -:O     |            |    O:-
Ca2+      ..  \  |            | /  ..     Ca2+
          ..    Si     ..    Si    ..
         -:O  /    \   O   /    \  O:-
          ..           ..          ..
```

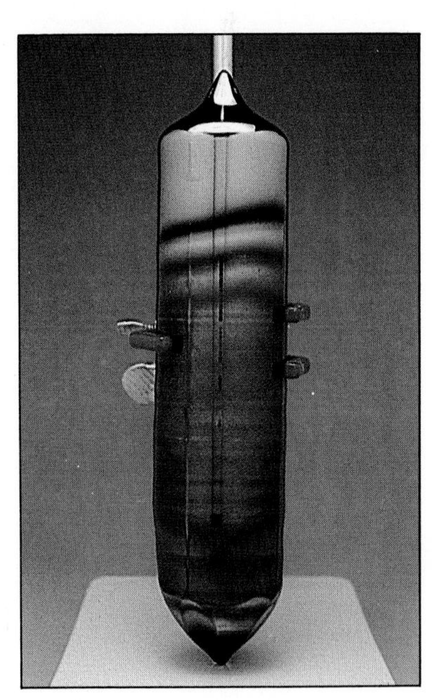

A solid cylinder of nearly pure silicon. *(Charles D. Winters)*

Exercise 10.19

In each formula above: (a) Circle one silicon atom and its bonds to show that it shares eight electrons. (b) Circle one oxygen atom and its shared and unshared electrons to show that it has a total of eight shared and unshared electrons.

The other elements in group 4A are scarce in nature.

The elements in group 4A and their compounds have an enormous variety of uses, as shown by the following few examples. Almost all of the pharmaceutical compounds now in use are compounds of carbon, designed and prepared by chemists. Very pure silicon is used to make the chips for integrated circuits, and germanium is used to make semiconductor devices. Tin is used as a protective coating for iron. Lead is used to make the electrodes in automobile storage batteries. ◆

◆ Section 17-11 describes the chemical reactions by which batteries generate an electric current.

Red phosphorus (left) and white phosphorus. *(Charles Steele)*

Clockwise from the left: arsenic, As; antimony, Sb; and bismuth, Bi. *(Charles D. Winters)*

5A
7 N 14.0
15 P 31.0
33 As 74.9
51 Sb 121.8
83 Bi 209.0

10-5 Downward in group 5A there is a change from nonmetal to metalloid to metal, with corresponding changes in bonding behavior

The behavior of the elements in group 5A follows closely the pattern of behavior in group 4A: Nitrogen and phosphorus are nonmetals, arsenic and antimony are metalloids, and bismuth is a metal.

Under ordinary room conditions nitrogen is a colorless, odorless gas. The most common form of phosphorus, called white phosphorus, consists of colorless crystals with the melting and boiling points shown in Table 10.9. There are two other allotropic forms, called black phosphorus and red phosphorus. Arsenic consists of gray, shiny, metallic-looking crystals. Antimony is silvery-white, lustrous, and hard, and it tarnishes slowly in air. Bismuth is a highly lustrous, grayish-white metal with a reddish tinge.

Downward in group 5A changes in bonding behavior follow the expected pattern: Electronegativity decreases, as shown in Table 10.10, and ionization energy decreases, so bonding with nonmetals changes from covalent to ionic.

The behavior of nitrogen is typical for a nonmetal. With metals nitrogen forms ionic compounds containing the nitride ion, N^{3-}, and with nonmetals it forms covalent compounds containing single, double, or triple covalent bonds.

Table 10.9 Group 5A Elements

Element	Symbol	Discovery	Melting Point, °C	Boiling Point, °C
Nitrogen	N	England, 1772	−210	−196
Phosphorus	P	Germany, 1669	44	280
Arsenic	As	Germany, 1250	No clear melting or boiling point; sublimes at 615°C	
Antimony	Sb	Ancient	631	1380
Bismuth	Bi	France, 1753	271	1438

Table 10.10 Electron Configurations and Electronegativity Values for the Group 5A Elements
Characteristic valence electron configuration for group 5A: s^2p^3.

Element	Abbreviated Electron Configuration	Electronegativity
Nitrogen	$[He]^2 2s^2 2p^3$	3.0
Phosphorus	$[Ne]^{10} 3s^2 3p^3$	2.1
Arsenic	$[Ar]^{18} 4s^2 4p^3$	2.0
Antimony	$[Kr]^{36} 5s^2 5p^3$	1.9
Bismuth	$[Xe]^{54} 6s^2 6p^3$	1.9

Example 10.6

(a) Predict the formula for the compound that will be formed between sodium and nitrogen. (b) Predict the equation for its formation from the elements.

Solution

(a) Sodium (Na) is a metal and nitrogen (N) is a nonmetal, so the compound will be ionic. The ions are Na^+ and N^{3-}, so the formula will be Na_3N.

(b) Sodium exists as Na and nitrogen exists as N_2, so the reactants and product will be

$$Na + N_2 \rightarrow Na_3N$$

The balanced equation will be

$$6\ Na + N_2 \rightarrow 2\ Na_3N$$

Example 10.7

Predict the formula for the compound that will be formed between nitrogen and chlorine.

Solution

Nitrogen (N) is a nonmetal and chlorine (Cl) is a nonmetal, so the compound will be covalent. The atoms are $\cdot\ddot{N}\cdot$ (with one dot below) and $:\ddot{Cl}\cdot$ (with two dots below). Combining these atoms according to the octet rule, as described in Chapter 7, gives the formula NCl_3.

Many of the compounds of phosphorus are similar to those of nitrogen. Use that similarity to answer the following Exercises.

Exercise 10.20

Write the formula you predict for the compound formed between phosphorus and (a) hydrogen (b) potassium (c) iodine.

White phosphorus reacts with chlorine; the product is PCl_5. *(Charles Steele)*

Phosphorus forms some compounds that don't follow the octet rule. One example is PCl_5, in which the phosphorus atom has ten bonding electrons. Its line formula is

```
        ..
   ..  :Cl:  ..
  :Cl\  |  /Cl:
   ..  \| /  ..
   ..   P    ..
  :Cl/    \Cl:
   ..       ..
```

Exercise 10.21

Name PCl_3 and PCl_5.

Arsenic and antimony are metalloids, and they have properties between those of metals and those of nonmetals. With very electronegative nonmetals, for example with fluorine, they form ionic compounds that have the expected formulas: AsF_3 and SbF_3. With less electronegative nonmetals, for example with hydrogen or iodine, they form covalent compounds that have the expected formulas: AsH_3 and SbH_3, AsI_3 and SbI_3. Like phosphorus, they form compounds that don't follow the octet rule: AsF_5 and SbF_5. As shown in Figure 10.6, the reaction between antimony and a nonmetal can be highly exothermic. The same is true for arsenic.

Bismuth, the only metal in group 5A, has the abbreviated electron configuration $[Xe]^{54}6s^2 6p^3$. It forms compounds that correspond to the loss of the $6p^3$ electrons, for example $Bi(NO_3)_3$ and Bi_2O_3 (shown in Figure 10.7), and it forms a few compounds that correspond to the loss of all of the $6s^2 6p^3$ electrons, for example BiF_5. With less electronegative nonmetals, for example with iodine, bismuth forms compounds that are more covalent than ionic, for example BiI_3.

Exercise 10.22

(a) Name the covalent compounds $AsCl_3$ and SbI_3. (b) Name the ionic compound $Bi(NO_3)_3$.

Figure 10.6
Reaction of antimony with bromine. *(Jim Morganthaler)*

Figure 10.7
Bismuth and bismuth(III) oxide. *(Chip Clark)*

Two elements in group 5A are abundant in nature. Nitrogen makes up about 80% of the atmosphere, and phosphorus, the tenth most abundant element, occurs as phosphate minerals, mostly calcium phosphate, $Ca_3(PO_4)_2$. Phosphorus forms a large number of compounds that contain oxygen. The behavior is like that of silicon: Phosphorus atoms and oxygen atoms can form large and complicated chains or rings. Two examples are shown below.

$Na^+ \; ^-:\ddot{O}-P(=O)(-O^-)-\ddot{O}-P(=O)(-O^-)-\ddot{O}:^- \; Na^+$ with Na^+ Na^+

K^+ ring structure of three P and three O atoms, each P bearing $=O$ and $-O^- K^+$

Exercise 10.23

Write the condensed formula for the compound shown on the left above.

The elements in group 5A and their compounds have many important uses. Nitrogen from the air is converted commercially to ammonia, which is used in large quantities as a fertilizer and as a raw material in the manufacture of explosives such as nitroglycerin.◆ The conversion of atmospheric nitrogen to the soluble nitrogen compounds required by growing plants also occurs in the soil, through the action of bacteria. Large quantities of phosphorus, made from calcium phosphate, are converted to phosphoric acid, H_3PO_4, which is used as an industrial chemical in a wide variety of processes. The main uses of arsenic, antimony, and bismuth are as alloys, for special applications.

◆The formula for nitroglycerin is shown in Table 7.2.

10-6 In group 6A oxygen and sulfur form covalent compounds with nonmetals, and with metals they form ionic compounds containing the anions O^{2-} and S^{2-}

6A
8 O 16.0
16 S 32.1
34 Se 79.0
52 Te 127.6
84 Po (209)

The elements in group 6A are shown in Table 10.11. Elemental oxygen occurs as a colorless, odorless gas with the formula O_2 and as ozone, O_3, a colorless gas that in small concentrations has a pleasant, characteristic odor.◆ Sulfur, which has several allotropic forms, is most often seen as a yellow powder. Selenium also has several allotropes, including a red crystalline form and a gray metallic

◆Environmental problems caused by loss of ozone from the earth's atmosphere are described in the Inside Chemistry essay at the end of Chapter 13.

Table 10.11 Group 6A Elements

Element	Symbol	Discovery	Melting Point, °C	Boiling Point, °C
Oxygen	O	England and Sweden, 1774	−218	−183
Sulfur	S	Ancient	215	445
Selenium	Se	Sweden, 1817	217	685
Tellurium	Te	Austria, 1782	450	990
Polonium	Po	France, 1898	254	962

Table 10.12 Electron Configurations and Electronegativity Values for the Group 6A Elements
Characteristic valence electron configuration for group 6A: s^2p^4.

Element	Abbreviated Electron Configuration	Electronegativity
Oxygen	$[He]^{2}2s^{2}2p^{4}$	3.5
Sulfur	$[Ne]^{10}3s^{2}3p^{4}$	2.5
Selenium	$[Ar]^{18}4s^{2}4p^{4}$	2.4
Tellurium	$[Kr]^{36}5s^{2}5p^{4}$	2.1
Polonium	$[Xe]^{54}6s^{2}6p^{4}$	2.0

form. Tellurium occurs as gray crystals or powder. Twenty-two isotopes are known for polonium, all of them radioactive, with short lifetimes, and less is known about its properties than about those of the other elements in group 6A. Table 10.12 shows the electron configurations and electronegativity values for the group 6A elements.

Earlier in this chapter we discussed the formulas for the typical compounds of oxygen with the elements of groups 1A through 5A. Among all of the elements, oxygen has the second-highest electronegativity (only fluorine, at 4.0, is higher), and each oxygen atom has six electrons in its outermost energy level. An oxygen atom easily gains two electrons, to complete its valence octet. With metals oxygen forms ionic compounds that contain oxide ion, O^{2-}, and with nonmetals oxygen forms covalent compounds in which oxygen atoms have single, double, or triple covalent bonds.

Exercise 10.24

(a) Predict the formula for the compound that will be formed between zinc and oxygen. (b) Predict the equation for its formation from the elements.

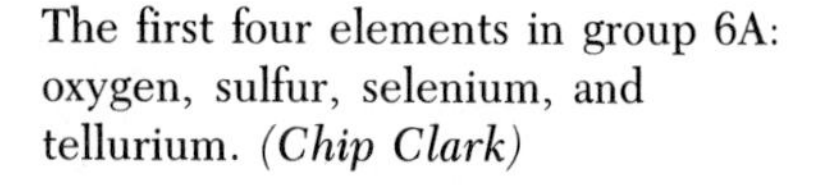
The first four elements in group 6A: oxygen, sulfur, selenium, and tellurium. *(Chip Clark)*

Sulfur atoms form bonds as if they were less electronegative versions of oxygen atoms. With metals sulfur forms ionic compounds that contain sulfide ion, S^{2-}, and with nonmetals sulfur forms covalent compounds in which sulfur atoms have single or double, but not triple, covalent bonds.

Tarnish on silver is silver sulfide, Ag_2S. *(Chip Clark)*

Exercise 10.25

Write the formula you predict for the compound of sulfur with (a) sodium (b) magnesium (c) aluminum.

Exercise 10.26

The compound of oxygen with fluorine is oxygen difluoride. Write the formula you predict for the compound that will be formed between sulfur and chlorine.

Exercise 10.27

Figure 10.8 shows sulfur burning. Write an equation for this reaction, assuming that sulfur exists as sulfur atoms and that the product is sulfur dioxide.

Compounds of selenium, tellurium, and polonium are less common and less important, for our purposes, than those of oxygen and sulfur. Most of their formulas can be predicted from the corresponding formulas for compounds of oxygen or sulfur.

Exercise 10.28

Write the formula you predict for the compound of selenium with hydrogen.

Oxygen is the most abundant element in our environment—if we include air, water, and the earth's crust, it makes up about 48% of our surroundings. Sulfur is abundant in minerals, for example as calcium sulfate dihydrate ($CaSO_4 \cdot 2H_2O$), in deposits as elemental sulfur (see Figure 10.9), and in other forms. The other elements in group 6A aren't abundant in nature.

Figure 10.8
Reaction of sulfur with oxygen.
(Leon Lewandowski)

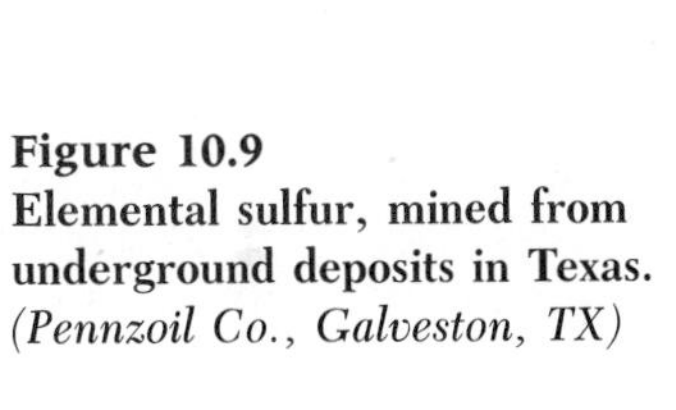

Figure 10.9
Elemental sulfur, mined from underground deposits in Texas.
(Pennzoil Co., Galveston, TX)

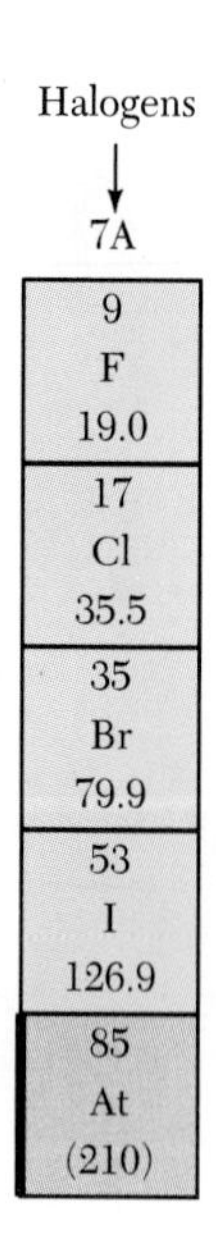

About two thirds of the oxygen used commercially serves to remove carbon impurities from iron ore in the process of making steel. The carbon is removed by reactions with oxygen that form carbon monoxide and carbon dioxide. Sulfur is used to make large quantities of sulfuric acid, H_2SO_4, a compound needed for many industrial processes. Selenium is used to make red glasses and enamels. Tellurium provides special properties in alloys.

10-7 The elements in group 7A form 1− anions with metals and single covalent bonds with nonmetals

Downward in groups 5A, 6A, and 7A there's a consistent change in physical properties: At ordinary temperatures and pressures the elements at the top of these groups are gases and those at the bottom are solids.◆ In group 7A the first two elements, fluorine and chlorine, are gases. Fluorine is pale yellow and chlorine is pale green. Bromine is a dark red liquid, and iodine is a bluish-black crystalline solid. The bulk properties of astatine aren't known because all of its isotopes are radioactive and only very small quantities of the element have been prepared. (See Table 10.13.)

As shown in Table 10.14, almost all of the chemical properties of the elements in group 7A, the halogens, can be predicted from two of their characteristics: Their atoms have high electronegativity, and the outermost energy levels of their atoms are one electron short of being filled. As a result, halogen atoms react with metals to form 1− anions and with nonmetals to form single covalent bonds.◆

◆The reasons for the differences in the physical states of the elements downward in groups 5A through 7A are discussed in Chapter 14.

◆*Halogen* means acid-former. The elements in group 7A are called halogens because their compounds with hydrogen dissolve in water to form acids, an important class of compounds described in Chapters 15 and 16.

Exercise 10.29

Write the formula you predict for the compound of bromine with (a) potassium (b) strontium (c) aluminum.

Exercise 10.30

Write the equation you predict for the reaction of chlorine with nickel.

The halogens form compounds with one another. Examples of these interhalogen compounds are ClF, IBr, and IF_5. As this last example shows, some interhalogen compounds don't obey the octet rule.

Table 10.13 Group 7A Elements

Element	Symbol	Discovery	Melting Point, °C	Boiling Point, °C
Fluorine	F	France, 1886	−223	−187
Chlorine	Cl	Sweden, 1774	−187	−35
Bromine	Br	France, 1826	−7	59
Iodine	I	France, 1811	114	183
Astatine	At	United States, 1940	unknown	unknown

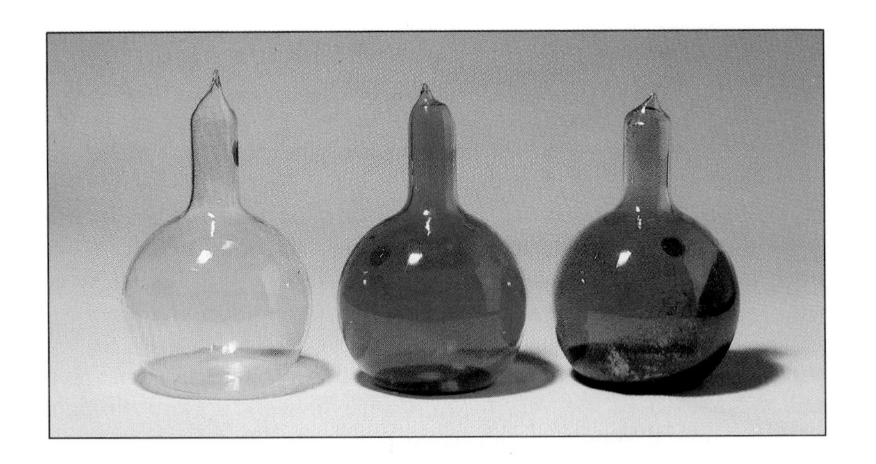

Figure 10.10
Chlorine, bromine, and iodine. The middle flask contains bromine vapor above liquid bromine, and the flask on the right contains iodine vapor above solid iodine. *(Leon Lewandowski)*

Exercise 10.31

Name ClF, IBr, and IF_5.

Exercise 10.32

Write the equation for the reaction you predict for the formation of ClF from its elements.

More complicated compounds that contain halogens, for example NaClO and $Al(BrO_3)_3$, aren't formed by direct reactions of their elements. In Chapter 16 we'll see how some of these compounds are formed.

Fluorine is abundant in minerals as calcium fluoride, CaF_2, and chlorine is abundant in minerals and in the oceans as sodium chloride. The other halogens are scarce in nature.

The halogens and their compounds have many uses, but only a few will be mentioned here. Fluorine is used commercially to prepare uranium hexafluoride, UF_6, an important step in the process of obtaining $^{235}_{92}U$ for nuclear reactors. Tin(II) fluoride, SnF_2, is a fluoridating agent in toothpastes.◆ Chlorine is used in huge quantities to prepare plastics and other commercial compounds and to purify water supplies. Large quantities of silver bromide, AgBr, and silver iodide, AgI, go into the production of photographic film.

◆The function of fluorides in toothpastes is described in the Inside Chemistry essay at the end of Chapter 8.

Table 10.14 Electron Configurations and Electronegativity Values for the Group 7A Elements
Characteristic valence electron configuration for group 7A: s^2p^5.

Element	Abbreviated Electron Configuration	Electronegativity
Fluorine	$[He]^2 2s^2 2p^5$	4.0
Chlorine	$[Ne]^{10} 3s^2 3p^5$	3.0
Bromine	$[Ar]^{18} 4s^2 4p^5$	2.8
Iodine	$[Kr]^{36} 5s^2 5p^5$	2.5
Astatine	$[Xe]^{54} 6s^2 6p^5$	2.2

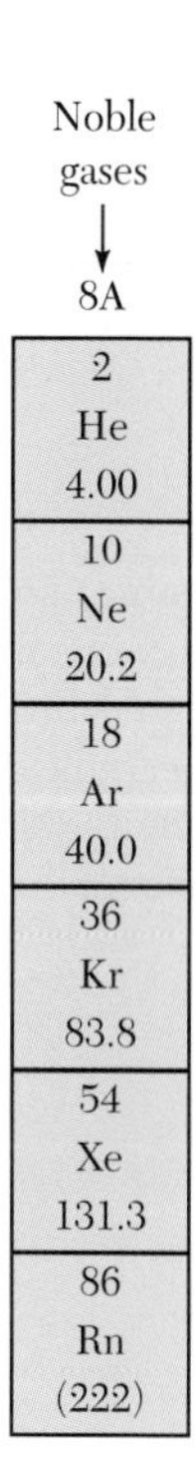

10-8 Contrary to the octet rule, the elements in group 8A form a few compounds

All of the members of group 8A, the noble gases, are colorless, odorless gases. Helium, which is obtained from deposits in natural-gas wells, has the lowest melting point and boiling point of any known substance. Liquid helium is used in research to cool other substances to very low temperatures. Neon, argon, krypton, and xenon occur in small quantities in the air, and they have a few special uses. A mixture of nitrogen and argon fills incandescent light bulbs, and neon is used in advertising signs.

Radon is produced in soil by the radioactive decay of uranium, and all of its isotopes are radioactive. Because radon is a gas, it seeps upward through the soil and can enter buildings through cracks in masonry. Radon may accumulate in some houses and other buildings to levels that are significantly hazardous. The Environmental Protection Agency estimates that as many as 20 000 deaths may occur every year in the United States from lung cancer caused by inhaling radioactive radon.

The atoms of the noble gases have electron configurations in which their outermost energy levels are filled with electrons, two electrons for a helium atom and eight electrons for an atom of each of the other elements in the group. According to the octet rule, these filled outer energy levels should mean that the atoms of these elements wouldn't form ionic or covalent bonds with atoms of other elements. That is, from the octet rule we'd predict that the elements of group 8A wouldn't form compounds.

But several compounds containing krypton, xenon, and radon have been prepared. Examples for xenon are XeF_2, XeF_4 (shown in Figure 10.11), XeF_6, XeO_3, XeO_4, $XeOF_4$, XeO_2F_2, $NaHXeO_4$, and Ba_2XeO_6. Most of these compounds contradict the octet rule because they require too many electrons around the xenon atom. For example, the line formula for XeF_2 is

$$:\ddot{\underset{\cdot\cdot}{\text{F}}}-\overset{\cdot\cdot\,\cdot\cdot}{\underset{\cdot\cdot}{\text{Xe}}}-\ddot{\underset{\cdot\cdot}{\text{F}}}:$$

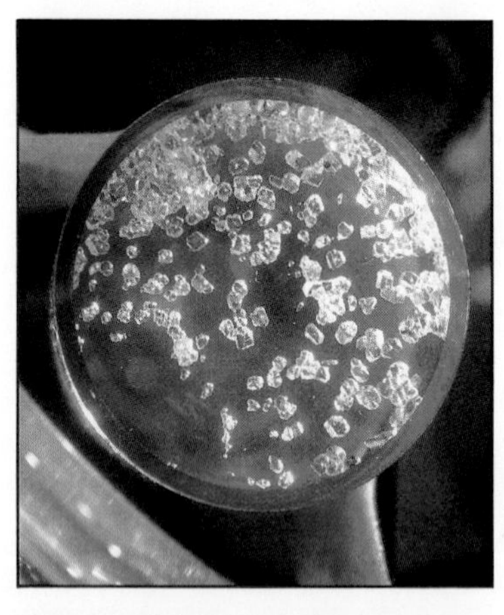

Figure 10.11
Crystals of xenon tetrafluoride, XeF_4. *(Argonne National Laboratory)*

Exercise 10.33

The binary compounds of xenon with fluorine or oxygen are named in the same way as other binary compounds of nonmetals. Name XeF_2.

Exercise 10.34

The compound Ba_2XeO_6 is called barium perxenate. What is the charge on the perxenate ion?

The electron configurations of the noble gases suggest that, because their outermost energy levels are filled, their atoms would not form bonds by attracting electrons to themselves. For this reason we can assume that when a bond between a noble-gas atom and another atom is formed, electrons from the noble-gas atom are attracted toward the other atom. On the basis of this assumption, answer the following Exercises.

Exercise 10.35

In the compounds of xenon whose formulas are shown above, the only elements whose atoms are bonded directly to xenon atoms are fluorine and

Table 10.15 Group 8A Elements
Characteristic valence electron configuration for group 8A: s^2p^6.

Element	Symbol	Discovery	Abbreviated Electron Configuration	Melting Point, °C	Boiling Point, °C
Helium	He	France, 1868	$1s^2$	−272◆	−269
Neon	Ne	England, 1898	$[He]^2 2s^2 2p^6$	−249	−246
Argon	Ar	England, 1894	$[Ne]^{10} 3s^2 3p^6$	−189	−186
Krypton	Kr	England, 1898	$[Ar]^{18} 4s^2 4p^6$	−169	−152
Xenon	Xe	England, 1898	$[Kr]^{36} 5s^2 5p^6$	−140	−109
Radon	Rn	Germany, 1900	$[Xe]^{54} 6s^2 6p^6$	−71	−62

◆Helium melts at −272°C under high pressure.

oxygen. Why would atoms of these two elements, rather than others, form bonds with xenon atoms?

Exercise 10.36

Why are compounds known for krypton, xenon, and radon, but not for helium, neon, and argon?

The theories of bonding that explain the existence of noble-gas compounds belong in more advanced chemistry books and won't be discussed here. We'll treat these compounds simply as exceptions to the octet rule.

We can use these exceptions to see something important about scientific methods. All of the noble gases had been discovered by 1900. Early attempts to prepare compounds of the noble gases were unsuccessful, and chemists began to doubt that such compounds could exist. In the early years of the twentieth century, the octet rule was shown to be successful in explaining bonding in most compounds, and the prediction from the octet rule was that atoms of the noble gases would not form bonds. Chemists came to believe, on the basis of both experiments and theory, that no compounds of the noble gases were possible, and they stopped looking for them. It wasn't until 1962 that a chemist questioned the established view and showed that noble-gas compounds could be prepared.†

One function of a scientific theory is to predict what we ought to find. The octet rule, as we've seen, is a simple theory that successfully predicts formulas for millions of ionic and covalent compounds. But the more simple and successful a theory is, the less likely we are to question it. By trusting a theory, we're guided toward some discoveries, but we're also guided away from others. For this reason, it's important that some scientists should work to affirm theories and that other scientists should work to question them.

† Chemistry Insight

The chemist who first prepared compounds of the noble gases was Neil Bartlett of the University of British Columbia. Before Bartlett's work the elements in group 8A were called the **inert gases,** to show that they formed no compounds. They're now called the **noble gases** because *noble* has been used in chemistry historically to identify elements that undergo relatively few reactions. Gold and silver, for example, are comparatively unreactive and are called **noble metals.**

Transition metals (B groups)

21 Sc 45.0	22 Ti 47.9	23 V 50.9	24 Cr 52.0	25 Mn 54.9	26 Fe 55.8	27 Co 58.9	28 Ni 58.7	29 Cu 63.6	30 Zn 65.4
39 Y 88.9	40 Zr 91.2	41 Nb 92.9	42 Mo 95.9	43 Tc (98)	44 Ru 101.1	45 Rh 102.9	46 Pd 106.4	47 Ag 107.9	48 Cd 112.4
57 La* 138.9	72 Hf 178.5	73 Ta 180.9	74 W 183.9	75 Re 186.2	76 Os 190.2	77 Ir 192.2	78 Pt 195.1	79 Au 197.0	80 Hg 200.6
89 Ac† (227)	104 Unq	105 Unp	106 Unh	107 Uns	108 Uno	109 Une			

*Lanthanides	58 Ce 140.1	59 Pr 140.9	60 Nd 144.2	61 Pm (145)	62 Sm 150.4	63 Eu 152.0	64 Gd 157.3	65 Tb 158.9	66 Dy 162.5	67 Ho 164.9	68 Er 167.3	69 Tm 168.9	70 Yb 173.0	71 Lu 175.0
†Actinides	90 Th 232.0	91 Pa (231)	92 U 238.0	93 Np (237)	94 Pu (244)	95 Am (243)	96 Cm (247)	97 Bk (247)	98 Cf (251)	99 Es (252)	100 Fm (257)	101 Md (258)	102 No (259)	103 Lr (260)

10-9 The chemical behavior of the transition metals is different from that of the representative elements

More than half of the elements on the Periodic Table are transition metals. They're given much less consideration than the representative elements in introductory chemistry textbooks because collectively they make up a much

The 3d transition metals in their order on the Periodic Table. From left to right and top to bottom: scandium, titanium, vanadium, chromium, manganese, iron, cobalt, nickel, copper, and zinc. *(Chip Clark)*

smaller part of our environment and because a full description of their compounds requires advanced theories of bonding. In this Section we'll look briefly at a few of their important characteristics.

All of the transition elements are metals. Mercury is a liquid at room temperatures, but the other transition metals are lustrous solids. Most of them are silvery-white, but a few are colored. Examples are copper, which is light red, and zinc, which is light blue.◆

◆The bulk properties of some of the transition metals aren't known, because only very small quantities of the elements have been prepared. In general, this is true for the elements whose atomic masses on the Periodic Table are in parentheses; all of the isotopes of these elements are radioactive.

As we saw in Chapter 6, the atoms of the representative elements are filling s or p sublevels, and the atoms of the transition elements are filling d or f sublevels. The sublevel being filled in each transition series is shown in Table 10.16.

Exercise 10.37

Use a Periodic Table to answer these questions about the element with atomic number 47: (a) Identify it as a transition or inner-transition metal. (b) Identify the electron sublevel it's filling. (c) Write its abbreviated electron configuration. (d) Name it.

Bonding by the transition metals doesn't follow the octet rule. The atoms of the transition metals form cations, as we'd expect, but the charges on the cations often aren't predictable from the rules used for the atoms of the representative metals. As shown in Chapter 8, many transition metals form more than one cation.

Exercise 10.38

Write the name for the formula or the formula for the name: (a) $ZnCO_3$ (b) $FeCO_3$ (c) copper(II) sulfite

The transition metals form many compounds, called **coordination compounds,** in which the atom or ion of a transition metal forms covalent bonds with atoms of other elements, with anions, or with molecules. The chemical species to which the transition-metal atom or ion is bonded are called **ligands.** The following are examples of coordination compounds:

1. A coordination compound exists in which six cyanide ions are covalently bonded to one iron(II) ion, to form an anion with a 4− charge. In the com-

Table 10.16 Electron Configurations of the Transition Metals

Classification	Atomic Numbers	Sublevel Filled
Transition	21 through 30	3d sublevel
Transition	39 through 48	4d sublevel
Transition	71 through 80	5d sublevel
Transition	103 through 106	Begin 6d sublevel
Inner-transition	57 through 70	4f sublevel
Inner-transition	89 through 102	5f sublevel

pound this charge is balanced by four potassium ions. We can imagine the compound being formed in steps from its parts:

$$Fe^{2+} + 6\ CN^- \rightarrow Fe(CN)_6^{4-}$$

$$Fe(CN)_6^{4-} + 4\ K^+ \rightarrow K_4[Fe(CN)_6]$$

In the formula for the compound the ligands are shown in parentheses, and the combination of the ligands with the metal ion is shown in brackets.

2. A coordination compound exists in which six ammonia molecules are covalently bonded to one cobalt(III) ion, to form a cation with a 3+ charge. In the compound this charge is balanced by three chloride ions. The imaginary reactions are

$$Co^{3+} + 6\ NH_3 \rightarrow Co(NH_3)_6^{3+}$$

$$Co(NH_3)_6^{3+} + 3\ Cl^- \rightarrow [Co(NH_3)_6]Cl_3$$

The formulas and names for coordination compounds can be complicated, and a description of their bonding requires more advanced theories than we'll use in this book.

Exercise 10.39

Write equations for the imaginary formation of a coordination compound by these steps: combination of one iron(II) ion with six water molecules to form a product and combination of this product with an appropriate number of nitrate ions to form the compound.

Most of the ionic compounds of the representative elements, and most of their simple covalent compounds, are white or colorless—sodium chloride and water are familiar examples. By contrast, many compounds of the transition metals are colored, both in the solid state and in aqueous solution. Examples are shown in Figure 10.12.

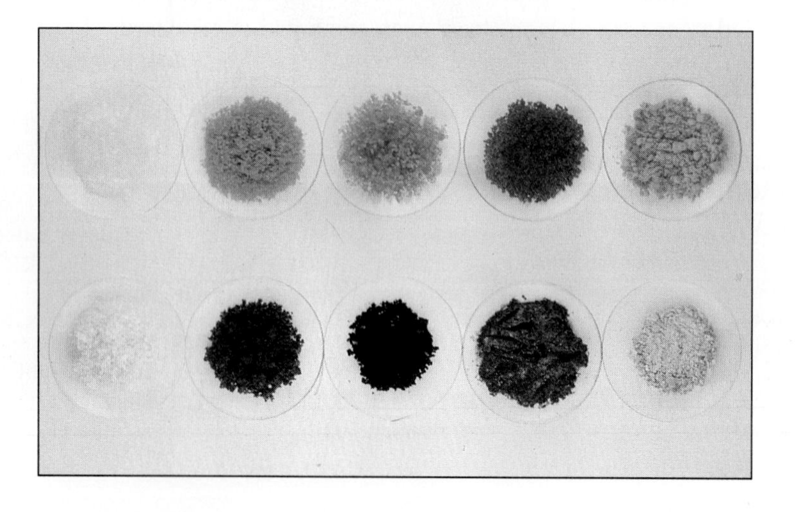

Figure 10.12
Crystals of the chlorides of the 3d transition metals, scandium through zinc. *(Chip Clark)*

From everyday experience you're familiar with important uses of some transition metals: iron for structures; chromium, nickel, and zinc for protective and decorative coatings; mercury in thermometers; silver and copper as electrical conductors; and silver, gold, and platinum for jewelry.

Some compounds of the transition metals have vital biological functions, and some have important commercial applications. Hemoglobin, the compound in blood that transports oxygen from the lungs to all body tissues, is a coordination compound of iron. Vitamin B_{12} is a coordination compound of cobalt. The images on a color television screen are formed by flashes of colored light from transition-metal compounds, coated on the back of the screen. The light flashes are emitted by electrons falling from higher to lower energy levels, as described in Chapter 6.

Inside Chemistry

Why was aluminum once a precious metal?

More than 3 000 000 tons of aluminum products are manufactured in the United States every year, and the metal is now so cheap that we use it in such everyday applications as cans for beverages and siding for houses. But a little more than a century ago aluminum was one of the precious metals, with a price comparable to that of silver or gold. Why was aluminum once so expensive, and why is it now so cheap?

Aluminum was once expensive, not because it's scarce in nature, but because the cost of obtaining the pure metal from its minerals was very high. Aluminum is, in fact, the most abundant metal in the earth's crust, and the third most abundant element, after oxygen and silicon. In its minerals, aluminum exists as Al^{3+} ions; to make metallic aluminum, electrons must be added to the ions to form atoms:

$$Al^{3+} + 3\ e^- \rightarrow Al$$

Historically, the price of aluminum metal has depended on the cost of carrying out this reaction.

The first preparations of pure aluminum metal were carried out in the 1820's. In these preparations aluminum from its minerals was converted to aluminum chloride, and the aluminum chloride was subjected to a replacement reaction with metallic potassium:

$$Al^{3+} \text{ in minerals} \rightarrow AlCl_3$$

$$AlCl_3 + 3\ K \rightarrow Al + 3\ KCl$$

Because of the cost of potassium these reactions were expensive, and the price of aluminum in 1855 was $254 per kilogram ($115 per pound).

(Gabe Palmer/The Stock Market)

In the 1850's a French chemist, Sainte-Claire Deville, devised a method for preparing aluminum in which sodium replaced the more expensive potassium:

$$AlCl_3 + 3\ Na \rightarrow Al + 3\ NaCl$$

By 1859 the cost of aluminum had dropped to $37 per kilogram ($17 per pound).

In the late 1880's two scientists—Charles Martin Hall in the United States and Paul L. T. Héroult in France—independently discovered a much cheaper process for producing aluminum metal from its ores. In the Hall-Héroult process, aluminum is prepared by melting a mixture containing aluminum oxide, Al_2O_3, and passing an electric current through the molten mixture. The electric current is a stream of electrons, and these electrons convert aluminum ions to aluminum atoms:

$$Al^{3+} + \underset{\text{electrons from electric current}}{3\ e^-} \rightarrow Al$$

This process produces aluminum at a much lower cost: By 1909 the price of aluminum had dropped to 66 cents per kilogram (30 cents per pound), 0.26% of its cost in 1855.

Chapter Summary: The Periodic Table as a Guide to Formulas and Reactions

Use a Periodic Table as you work through this Summary.

Section	Subject	Summary	Check When Learned
10-1	The only group 1A element that is a gas at ordinary room conditions	Hydrogen.	☐
10-1	Collective name for group 1A elements, except hydrogen	Alkali metals.	☐
10-1	In general, the group 1A elements have (high, medium, or low) electronegativity	Low.	☐
10-1	In general, the group 1A elements have (high, medium, or low) ionization energy	Low.	☐
10-1	Characteristic electron configuration for group 1A elements	s^1	☐

Section	Subject	Summary	Check When Learned
10-1	Lewis symbols for atoms of group 1A elements	H·, Li·, Na·, K·, Rb·, Cs·, Fr·	☐
10-1	Form in which alkali metals always exist in their compounds	1+ cations.	☐
10-1	Steps in predicting the formula for a compound and in predicting the equation for its formation from the elements	(1) Decide whether the compound will be covalent or ionic. (2) If the compound is covalent, predict its formula from the principles in Chapter 7. If the compound is ionic, predict its formula from the principles in Chapter 8. (3) Write the symbols or formulas for the reactants and products for the formation of the compound from its elements. (4) Balance the equation.	☐
10-1	Relationship among formulas for compounds of elements in the same group	Elements in the same group form compounds with similar formulas. Examples: NaCl and KCl, Na_2O and K_2O.	☐
10-1	Relationship among reactions by which similar compounds are formed	Similar compounds are formed by similar reactions. Examples: $2\ Na + Cl_2 \rightarrow 2\ NaCl$ $2\ K + Cl_2 \rightarrow 2\ KCl$	☐
10-1	Equation for reaction of an alkali metal with a halogen; use M for the alkali metal and X for the halogen	$2\ M + X_2 \rightarrow 2\ MX$	☐
10-1	Equation for reaction of an alkali metal with oxygen; use M for the alkali metal	$4\ M + O_2 \rightarrow 2\ M_2O$	☐
10-1	Equation for reaction of an alkali metal with water; use M for the alkali metal	$2\ M(s) + H_2O(\ell) \rightarrow 2\ MOH(aq) + H_2(g)$	☐
10-1	Meaning of (aq)	Aqueous, *dissolved in water.*	☐

Section	Subject	Summary	Check When Learned
10-1	Difference between H and other group 1A elements in bonding with nonmetals	H usually forms single, polar, covalent bonds; alkali metals always form 1+ cations.	☐
10-1	Equation for reaction of an alkali metal with hydrogen	$2\ M + H_2 \rightarrow 2\ MH$	☐
10-2	Name for group 2A elements	Alkaline-earth metals.	☐
10-2	Classification of group 2A elements as metal, metalloid, or nonmetal	All are metals.	☐
10-2	Characteristic electron configuration for group 2A elements	s^2	☐
10-2	Characteristic Lewis symbol for group 2A elements; use M as symbol	M:	☐
10-2	Form in which group 2A elements always exist in their compounds	2+ cations.	☐
10-2	Equation for reaction of an alkaline-earth metal with a halogen; use M for the alkaline-earth metal and X for the halogen	$M + X_2 \rightarrow MX_2$	☐
10-2	Equation for reaction of an alkaline-earth metal with oxygen; use M for the alkaline-earth metal	$2\ M + O_2 \rightarrow 2\ MO$	☐

Section	Subject	Summary	Check When Learned
10-2	Equation for reaction of an alkaline-earth metal with water; use M for the alkaline-earth metal	$M(s) + 2\ H_2O(\ell) \rightarrow M(OH)_2(aq) + H_2(g)$	☐
10-3	Classification of group 3A elements as metal, metalloid, or nonmetal	B, metalloid; others, metal.	☐
10-3	Characteristic electron configuration for group 3A elements	s^2p^1	☐
10-3	Characteristic Lewis symbol for group 3A elements; use Z as symbol	·Z:	☐
10-3	Form of bonding by boron and aluminum with nonmetals	B, covalent; Al^{3+}	☐
10-3	Name: BF_3, AlF_3	Boron trifluoride, aluminum fluoride.	☐
10-3	Equation for the reaction of aluminum with a halogen; use X for the halogen	$2\ Al + 3\ X_2 \rightarrow 2\ AlX_3$	☐
10-3	Equation for reaction of aluminum with oxygen	$4\ Al + 3\ O_2 \rightarrow 2\ Al_2O_3$	☐
10-4	Classification of group 4A elements as metal, metalloid, or nonmetal	C, nonmetal; Si and Ge, metalloids; Sn and Pb, metals.	☐
10-4	Characteristic electron configuration for group 4A elements	s^2p^2	☐

Section	Subject	Summary	Check When Learned
10-4	Characteristic Lewis symbol for group 4A elements; use Z as symbol	·Ż:	☐
10-4	Form of bonding by each group 4A element with nonmetals	C, Si, and Ge, covalent; Sn^{2+}, Sn^{4+}, Pb^{2+}, Pb^{4+}.	☐
10-4	Name: CF_4, PbF_4	Carbon tetrafluoride, lead(IV) fluoride.	☐
10-5	Classification of group 5A elements as metal, metalloid, or nonmetal	N and P, nonmetals; As and Sb, metalloids; Bi, metal.	☐
10-5	Characteristic electron configuration for group 5A elements	s^2p^3	☐
10-5	Characteristic Lewis symbol for group 5A elements; use Z as symbol	·Ẓ̈·	☐
10-5	Form of bonding by nitrogen and phosphorus with metals	N^{3-}, nitride ion; P^{3-}, phosphide ion.	☐
10-5	Form of bonding by nitrogen and phosphorus with nonmetals	Covalent bonds.	☐
10-6	Classification of group 6A elements as metal, metalloid, or nonmetal	O, S, and Se are nonmetals; Te is a metalloid; Po is a metal.	☐
10-6	Characteristic electron configuration for group 6A elements	s^2p^4	☐
10-6	Characteristic Lewis symbol for group 6A elements; use Z as symbol	·Z̤̈·	☐

Section	Subject	Summary	Check When Learned
10-6	Forms of bonding by O and S with metals	O^{2-}, oxide ion, and O_2^{2-}, peroxide ion; S^{2-}, sulfide ion.	☐
10-6	Forms of bonding by O and S with nonmetals	O, single, double, or triple covalent bonds; S, single or double covalent bonds.	☐
10-7	Classification of group 7A elements as metal, metalloid, or nonmetal	F, Cl, Br, and I are nonmetals; At is a metalloid.	☐
10-7	Name for group 7A elements	Halogens.	☐
10-7	Characteristic electron configuration for halogens	s^2p^5	☐
10-7	Characteristic Lewis symbol for halogens; use Z as symbol	:Z̤̈·	☐
10-7	In general, the group 7A elements have (high, medium, or low) electronegativity	High.	☐
10-7	Element with highest electronegativity	Fluorine.	☐
10-7	Form of bonding of group 7A elements with metals	1− anions.	☐
10-7	Form of bonding of group 7A elements with nonmetals	Single covalent bonds.	☐
10-7	Interhalogen	Compound formed between two halogens, e.g., ICl.	☐
10-7	Name: NaCl, ICl	Sodium chloride; iodine monochloride.	☐
10-8	Name for group 8A elements	Noble gases.	☐

Section	Subject	Summary	Check When Learned
10-8	Classification of group 8A elements as metals, metalloids, or nonmetals	All are nonmetals.	☐
10-8	Characteristic electron configuration for group 8A elements	s^2p^6, except for He, which is s^2.	☐
10-8	Lewis symbol for helium	He:	☐
10-8	Characteristic Lewis symbol for Ne, Ar, Kr, Xe, and Rn; use Z for symbol	:Z: (with dots above and below)	☐
10-8	Unusual character of compounds of group 8A elements	In the atoms of the elements in group 8A, the outermost energy levels are filled with electrons, so according to the octet rule, they should form no bonds. But a few compounds of some of the group 8A elements do exist.	☐
10-9	Classification of transition elements as metal, metalloid, or nonmetal	All are metals.	☐
10-9	Characteristic difference in electron configuration between representative and transition elements	Representative elements are filling s and p sublevels, and transition elements are filling d and f sublevels.	☐
10-9	Distinction between *transition metal* and *inner-transition* metal	Used as specific terms, *transition metal* refers to elements that are filling d sublevels and *inner-transition metal* refers to elements that are filling f sublevels. Used as a general term, *transition metal* refers to both of these specific classes.	☐

Section	Subject	Summary	Check When Learned
10-9	Characteristic bonding behavior of transition metals that distinguishes them collectively from the representative elements	In general, bonding among the representative elements follows the octet rule, and bonding by the transition metals does not.	☐
10-9	Coordination compound	Compound in which an atom or cation of a transition metal forms covalent bonds with atoms or anions of nonmetals or with molecules.	☐
10-9	Ligand	Species that bonds with atom or cation of transition metal in a coordination compound.	☐
10-9	Characteristic visual property of compounds of transition metals	Many are colored.	☐

Problems

Assume you can use the Periodic Table at the front of the book unless you're directed otherwise. Answers to odd-numbered problems are in Appendix 1.

Group 1A Elements (Section 10-1)

1. In its physical properties, how does hydrogen differ from the other elements in group 1A?
2. Less is known about francium than about the other alkali metals. Why?
3. What is the characteristic valence electron configuration for the elements in group 1A?
4. Write the Lewis symbol for (a) hydrogen (b) sodium (c) potassium.
5. Which element has the lowest electronegativity?
6. Write the symbol for (a) lithium ion (b) sodium ion (c) potassium ion.
7. Name (a) LiI (b) K_2O (c) Na_2SO_4.
8. Write the formula for (a) potassium carbonate (b) sodium phosphate (c) lithium sulfide.
9. From the formula for sodium chloride, predict the formula for potassium fluoride. Explain your reasoning.
10. Predict the equation for the reaction by which potassium fluoride will be formed from its elements.
11. The sample of sodium shown in Figure 10.1 has a white coating, formed by the reaction of sodium with oxygen, water vapor, and nitrogen in the air. Write the equation for the reaction you predict between sodium and (a) oxygen (b) water (c) nitrogen.
12. When sodium is exposed to the air, it forms a white coating, as shown in Figure 10.1. One of the compounds in the coating is sodium carbonate, formed in two steps: (a) a reaction of sodium with oxygen to form sodium oxide and (b) a reaction of sodium oxide with carbon dioxide to form sodium carbonate. Write equations for these reactions.

13. Write an equation for each of the following reactions and name the product: (a) the reaction of lithium with nitrogen (b) the reaction of sodium with bromine (c) the reaction of potassium with oxygen.

14. Write an equation for the reaction of lithium with water and show the state of each reactant and product under ordinary room conditions. Why is this reaction dangerous?

15. The formula for a compound is Na_2Z. (a) In what group is element Z? (b) If the formula mass of Na_2Z is 78.1 amu, what is Z?

16. The formula for a compound is ZH_3. (a) In what group is element Z? (b) If the molecular mass of ZH_3 is 77.9 amu, what is Z?

17. Write an equation for the reaction you predict between potassium and hydrogen and name the product.

18. Write the Lewis symbol for (a) a hydrogen atom (b) a hydrogen ion (c) a hydride ion.

Group 2A Elements (Section 10-2)

19. What is the characteristic valence electron configuration for the elements in group 2A?

20. Write the Lewis symbol for (a) magnesium (b) calcium (c) strontium.

21. Write the symbol for (a) magnesium ion (b) calcium ion (c) strontium ion.

22. Which element in group 2A has the highest electronegativity?

23. Name (a) CaI_2 (b) $Mg(CN)_2$ (c) $Ba(ClO_2)_2$.

24. Write the formula for (a) strontium sulfate (b) magnesium hydride (c) calcium phosphate.

25. Write the formula you predict for the compound that will be formed between calcium and (a) bromine (b) oxygen (c) nitrogen.

26. Write the equation for the reaction you predict for the formation from the elements of (a) magnesium chloride (b) magnesium oxide (c) magnesium nitride. Show the state of each reactant and product under ordinary room conditions.

27. Write an equation for the reaction you predict between barium and water. Show the state of each reactant and product.

28. In which compound would you expect the bonds to be more covalent, beryllium iodide or barium fluoride? Explain your answer.

29. The formula for a compound is MSO_4, where M is an alkaline-earth metal, and its formula mass is 184 amu. Name the compound.

30. The formula for a compound is Mg_3Z_2. In what group is element Z? Explain your answer.

Group 3A Elements (Section 10-3)

31. What is the characteristic valence electron configuration for the elements in group 3A?

32. Write the Lewis symbols for boron and aluminum.

33. (a) Write the electron configuration for aluminum. (b) Use the electron configuration to explain why an aluminum atom forms an Al^{3+} cation.

34. (a) Write the abbreviated electron configuration for gallium. (b) Use the electron configuration to explain why a gallium atom may form either a Ga^+ cation or a Ga^{3+} cation.

35. Explain why BF_3 is covalent but AlF_3 is ionic.

36. Many compounds of boron, including BF_3, are exceptions to the octet rule. Write the line formula for BF_3.

37. Name (a) BI_3 (b) $AlCl_3$ (c) $Al_2(CO_3)_3$.

38. Write the formula for (a) aluminum sulfide (b) aluminum hydride (c) aluminum hydroxide.

39. Write the formula you predict for the compound that will be formed between aluminum and (a) bromine (b) sulfur (c) nitrogen.

40. Write the equation for the reaction you predict for the formation from the elements of (a) aluminum oxide (b) aluminum fluoride.

41. In which compound would you expect the bonds to be more ionic, aluminum oxide or aluminum sulfide? Explain your answer.

42. In which compound would you expect the bonds to be more covalent, sodium iodide or aluminum iodide? Explain your answer.

Group 4A Elements (Section 10-4)

43. What is the characteristic valence electron configuration for the elements in group 4A?

44. Write the Lewis symbol for (a) carbon (b) silicon (c) lead.

45. (a) Write the abbreviated electron configuration for lead. (b) Use the electron configuration to explain why a lead atom may form either a Pb^{2+} cation or a Pb^{4+} cation.

46. Explain why CF_4 is covalent but PbF_4 is ionic.

47. Write the line formula for (a) CH_4 (b) CCl_4.

48. Write the line formula for (a) CO_2 (b) CO.

49. Name (a) CH_4 (b) CCl_4 (c) SnI_4 (d) PbO.

50. Name (a) $SiBr_4$ (b) $PbSO_3$ (c) $Sn(NO_2)_4$.

51. Write the formula for (a) carbon disulfide (b) tin(II) oxide (c) lead(IV) sulfide.

52. Write the formula for (a) tin(IV) acetate (b) lead(II) hydrogen carbonate (c) tin(IV) hydride.

53. Write the formulas you predict for the two compounds that will be formed between tin and sulfur. Name the compounds.

54. Write the equation for the reaction you predict for the formation from the elements of (a) lead(IV) fluoride (b) tin(IV) oxide.

55. In which compound would you expect the bonds to be more ionic, tin(IV) fluoride or tin(IV) iodide? Explain your answer.

56. In which compound would you expect the bonds to be more covalent, carbon tetraiodide or lead(IV) iodide? Explain your answer.

Group 5A Elements (Section 10-5)

57. What is the characteristic valence electron configuration for the elements in group 5A?

58. Write the Lewis symbol for (a) nitrogen (b) phosphorus (c) arsenic.

59. Write the symbol for (a) nitride ion (b) phosphide ion.

60. (a) Write the abbreviated electron configuration for bismuth. (b) Use the electron configuration to explain why a bismuth atom may form either a Bi^{3+} cation or a Bi^{5+} cation.

61. Name (a) NH_3 (b) NH_4Br (c) $AsCl_3$.

62. Write the formula for (a) dinitrogen tetroxide (b) phosphorus tribromide (c) calcium nitride.

63. The compounds of phosphorus, arsenic, and antimony with hydrogen are called phosphine, arsine, and stibine. Write the formulas you predict for them.

64. Write line formulas for (a) ammonium ion (b) nitrogen trichloride (c) nitrogen.

65. In which compound would you expect the bonds to be more ionic, $BiCl_3$ or NCl_3? Explain your answer.

66. If a polar covalent bond is formed between a phosphorus atom and a nitrogen atom, which atom will carry the partial positive charge? Explain your answer.

Group 6A Elements (Section 10-6)

67. What is the characteristic valence electron configuration for the elements in group 6A?

68. Write the Lewis symbol for (a) oxygen (b) sulfur (c) oxide ion (d) sulfide ion.

69. Write the electron configuration for (a) an oxygen atom (b) an oxide ion.

70. Arrange these elements in order of increasing electronegativity: fluorine, sulfur, oxygen.

71. Under ordinary room conditions, what is the physical state of oxygen? What is the physical state of sulfur?

72. Write the line formula for (a) oxygen (b) ozone.

73. If a polar covalent bond is formed between a sulfur atom and an oxygen atom, which atom will carry the partial negative charge? Explain your answer.

74. Write the formulas you know or predict for the compounds of hydrogen with each of the elements in group 6A.

75. Name (a) CaO (b) K_2S (c) H_2S.

76. Write the formula for (a) aluminum sulfide (b) magnesium selenide (c) iron(III) oxide.

77. Write line formulas for (a) hydrogen sulfide (b) sulfite ion.

78. Write the line formula for carbon disulfide.

Group 7A Elements (Section 10-7)

79. What is the characteristic valence electron configuration for the elements in group 7A?

80. (a) Write the Lewis symbol for a chlorine atom. (b) Write the Lewis symbol for a chloride ion. (c) Write the line formula for a chlorine molecule.

81. (a) Write the electron configuration for fluorine. (b) Write the electron configuration for fluoride ion.

82. (a) Write line formulas for ClF and BrF, and in each formula show the direction of bond polarity. (b) Will the bond polarity be greater in ClF or in BrF? Explain your answer.

83. Name (a) HF (b) NaF (c) ClF.

84. Write the formula for (a) iodine (b) copper(II) iodide (c) phosphorus triiodide.

85. (a) Write the formula you predict for the compound that will be formed between barium and bromine. (b) Write the name for the compound.

86. Write the equation for the reaction you predict between zinc and fluorine.

87. Complete the following equation and show the expected physical state of the product under ordinary room conditions.

$$K(s) + F_2(g) \rightarrow$$

88. Write the equation for the formation of hydrogen iodide from its elements.

Group 8A Elements (Section 10-8)

89. What is the characteristic valence electron configuration for the elements in group 8A?

90. Write the Lewis symbol for (a) argon (b) helium.

91. Under ordinary room conditions, what is the physical state of all of the elements in group 8A?

92. (a) Write the formula for krypton difluoride. (b) Name XeF_6.

93. In krypton difluoride, how many shared and unshared electrons are around the central atom?

94. In xenon tetrafluoride, how many shared and unshared electrons are around the central atom?

95. The compound XeO_3 is not an exception to the octet rule. Write its line formula and name it.

96. The compound XeO_4 is not an exception to the octet rule. Write its line formula and name it.

Elements in Groups 1A Through 8A (Sections 10-1 through 10-8)

97. Write the formulas for the compounds that you predict would be formed between hydrogen and each of the elements in the third period.

98. Write the formulas for the compounds that you predict would be formed between fluorine and each of the elements in the third period.

99. Write line formulas for the covalent compounds in problem 97.

100. Write line formulas for the covalent compounds in problem 98.

Transition Metals (Section 10-9)

101. Use the Periodic Table to answer these questions about the element with atomic number 28. (a) Identify it as a metal, nonmetal, or metalloid. (b) Identify it as a representative or transition element. (c) Identify it as a transition or inner-transition metal. (d) Identify the electron sublevel it's filling. (e) Write its abbreviated electron configuration. (f) Name it.

102. Without using the Periodic Table, answer the following questions about the element whose abbreviated electron configuration is $[Xe]^{54}6s^2 4f^7$. (a) What is its atomic number? (b) Is it a representative or transition element? (c) Is it a metal, nonmetal, or metalloid? (d) Is it a transition or inner-transition metal?

103. Write formulas for (a) copper(II) phosphate (b) iron(II) hydrogen sulfite.

104. Write names for (a) $Hg(ClO_3)_2$ (b) $Fe(CN)_3$.

105. Write equations for the imaginary stepwise formation of $[Cr(NH_3)_6]Br_3$.

106. Write equations for the imaginary stepwise formation of $[Cr(NH_3)_6][Co(CN)_6]$, starting with Co^{3+}.

One-mole samples of several common elements: (left to right) aluminum (foil), mercury, sulfur, chromium (in front of sulfur), lead, magnesium, and copper. *(Charles D. Winters)*

11 The Mole

11-1
One mole of a substance is its atomic mass, formula mass, or molecular mass in grams.

11-2
One mole of a substance contains 6.02×10^{23} atoms, ions, formula units, or molecules.

11-3
The percentage composition for a compound can be calculated from its formula.

11-4
The formula for a compound can be determined from its percentage composition and its formula mass or molecular mass.

Inside Chemistry:
Why is a mole called a mole?

Chapter Summary

Problems

We've seen that a variety of units can be used to describe the quantity of an element or a compound: The mass of an atom or a molecule is usually expressed in atomic mass units, and a larger quantity of an element or a compound can be described in mass units such as grams or milligrams. Quantities of elements or compounds that are liquids or gases can also be described in volume units such as liters or milliliters.

In this chapter you'll learn to use a new unit to describe the quantity of an element or a compound—the unit that chemists call **the mole.** It's important that you understand the mole well because it will be used in problems throughout the rest of this book and in all of your later chemistry work.◆

◆As you continue with this chapter, you'll need to have the details of dimensional analysis clear in your mind. Before you start on the next Section, be sure you know the material in Sections 3-9 and 3-10.

11-1 One mole of a substance is its atomic mass, formula mass, or molecular mass in grams

In learning to use the mole, it's convenient to begin by defining the mole in two ways. We'll consider one definition in this Section and the other in Section 11-2.

Letting X represent the symbol or formula for any substance,

1 mole of X = the atomic mass, formula mass, or molecular mass of X in grams

The abbreviation for mole is *mol.* The preceding equation gives a general definition of the mole that can be applied to any particular substance. To find the mass of one mole of a substance:

1. Identify its symbol or formula.
2. Determine its atomic mass, formula mass, or molecular mass in amu.
3. Change the units from amu to grams.

Example 11.1

Write the specific form of the mole definition for each of the following substances, in the form 1 mol X = xxx g X: (a) iron (b) chlorine (c) sodium chloride (d) water

Solution

(a) The element iron consists of atoms, its symbol is Fe, and it has an atomic mass of 55.8 amu, so 1 mol Fe = 55.8 g Fe.

(b) The element chlorine consists of molecules, its formula is Cl_2, and it has a molecular mass of 71.0 amu, so 1 mol Cl_2 = 71.0 g Cl_2.◆

◆The elements that exist as diatomic molecules are H_2, N_2, O_2, F_2, Cl_2, Br_2, and I_2.

(c) The compound sodium chloride consists of formula units, its formula is NaCl, and it has a formula mass of 58.5 amu, so 1 mol NaCl = 58.5 g NaCl.

(d) The compound water consists of molecules, its formula is H_2O, and it has a molecular mass of 18.0 amu, so 1 mol H_2O = 18.0 g H_2O.

From left to right, molar quantities of mercury(II) oxide (HgO), sucrose (table sugar, $C_{12}H_{22}O_{11}$), sulfur (S), copper(II) sulfate ($CuSO_4$), sodium chloride (table salt, NaCl), and copper (Cu). *(Chip Clark)*

We can apply the mole definition to imaginary substances. For example, the element chlorine exists as Cl_2 molecules, but we can imagine Cl atoms. The atomic mass of Cl is 35.5 amu, and the molecular mass of Cl_2 is 71.0 amu, so

$$1 \text{ mol Cl} = 35.5 \text{ g Cl} \qquad \text{and} \qquad 1 \text{ mol } Cl_2 = 71.0 \text{ g } Cl_2$$

The expression "one mole of chlorine" might mean either one mole of chlorine atoms or one mole of chlorine molecules. In this and similar ambiguous cases, if no other information is available to help in making the decision, it's better to assume that the expression refers to the real substance and not the imaginary one.

One mole of sulfate ions is an imaginary quantity, since sulfate ions can't exist by themselves—they have to have cations with them. But we can apply the mole definition: The formula for sulfate is SO_4^{2-} and its formula mass is 96.1 amu, so

$$1 \text{ mol } SO_4^{2-} = 96.1 \text{ g } SO_4^{2-}$$

Exercise 11.1◆

Write the specific form of the mole definition for each of the following substances, in the form 1 mol X = xxx g X: (a) NH_3 (b) $Ca(NO_3)_2$ (c) NO_2^- (d) oxygen (e) oxygen atoms (f) zinc

◆Answers to Exercises are in Appendix 1.

The general definition of the mole can be converted to a specific definition for a particular substance. The specific definition, in the form of an equation, can be used to create factors for dimensional analysis, as shown in the following Examples.

Example 11.2

How many moles of copper are in 5.00 grams of copper?

Solution

5.00 g Cu = ? mol Cu

g Cu → mol Cu

1 mol Cu = 63.6 g Cu (The atomic mass of Cu is 63.6 amu.)

$$\frac{1 \text{ mol Cu}}{63.6 \text{ g Cu}} = 1 \qquad \frac{63.6 \text{ g Cu}}{1 \text{ mol Cu}} = 1$$

$$\text{g Cu} \rightarrow \text{mol Cu}$$
$$\left(\frac{5.00 \cancel{\text{g Cu}}}{}\right)\left(\frac{1 \text{ mol Cu}}{63.6 \cancel{\text{g Cu}}}\right) = 0.0786 \text{ mol Cu}$$

Example 11.3

How many grams of $(NH_4)_3PO_4$ are in 0.250 mol?

Solution

0.250 mol $(NH_4)_3PO_4$ = ? g $(NH_4)_3PO_4$

mol $(NH_4)_3PO_4$ → g $(NH_4)_3PO_4$

1 mol $(NH_4)_3PO_4$ = 149 g $(NH_4)_3PO_4$ (The formula mass of $(NH_4)_3PO_4$ is 149 amu.)

$$\frac{1 \text{ mol } (NH_4)_3PO_4}{149 \text{ g } (NH_4)_3PO_4} = 1 \qquad \frac{149 \text{ g } (NH_4)_3PO_4}{1 \text{ mol } (NH_4)_3PO_4} = 1$$

$$\text{mol } (NH_4)_3PO_4 \rightarrow \text{g } (NH_4)_3PO_4$$
$$\left(\frac{0.250 \cancel{\text{mol } (NH_4)_3PO_4}}{}\right)\left(\frac{149 \text{ g } (NH_4)_3PO_4}{1 \cancel{\text{mol } (NH_4)_3PO_4}}\right) = 37.2 \text{ g } (NH_4)_3PO_4$$

Example 11.4

How many moles of iodine are in 3.00 kilograms?

Solution

3.00 kg I_2 = ? mol I_2

$$\text{kg } I_2 \rightarrow \text{g } I_2 \rightarrow \text{mol } I_2$$
$$\left(\frac{3.00 \cancel{\text{kg } I_2}}{}\right)\left(\frac{1 \times 10^3 \cancel{\text{g } I_2}}{1 \cancel{\text{kg } I_2}}\right)\left(\frac{1 \text{ mol } I_2}{254 \cancel{\text{g } I_2}}\right) = 11.8 \text{ mol } I_2$$

Use dimensional analysis to solve the following Exercises.

Exercise 11.2

The formula for sucrose (table sugar) is $C_{12}H_{22}O_{11}$. (a) Calculate the mass in grams of 0.500 mol of sucrose. (b) Calculate the number of moles of sucrose in 3.65 g of sucrose.

Exercise 11.3

Calculate the mass in grams of (a) 0.200 mol of $NaHCO_3$ (b) 1.50 mol of SF_6 (c) 1.55 mol of chloride ion.

11-2 One mole of a substance contains 6.02×10^{23} atoms, ions, formula units, or molecules

One mole of any substance must contain the same number of its fundamental particles (atoms, ions, formula units, or molecules) as one mole of any other substance. To see why, we'll consider the examples of water and sodium chloride.

Water consists of H_2O molecules, mass 18.0 amu, and sodium chloride consists of NaCl formula units, mass 58.5 amu; so the relationship of the mass of one molecule of water to the mass of one formula unit of sodium chloride is 18.0 to 58.5. The following table compares mass relationships for larger numbers of molecules and formula units.

Molecules of H_2O	Formula Units of NaCl	Mass of H_2O	Mass of NaCl	Relationship of Mass of H_2O to Mass of NaCl
1	1	(1)(18.0 amu)	(1)(58.5 amu)	18.0 to 58.5
2	2	(2)(18.0 amu)	(2)(58.5 amu)	18.0 to 58.5
10	10	(10)(18.0 amu)	(10)(58.5 amu)	18.0 to 58.5
1000	1000	(1000)(18.0 amu)	(1000)(58.5 amu)	18.0 to 58.5

The table shows that, as long as the number of H_2O molecules is the same as the number of NaCl formula units, the relationship of their masses must be 18.0 to 58.5. The reverse must also be true: If a mass of H_2O and a mass of NaCl have the relationship 18.0 to 58.5, then the number of H_2O molecules and NaCl formula units must be the same. Since 1 mol H_2O = 18.0 g H_2O and 1 mol NaCl = 58.5 g NaCl, the relationship of the masses of one mole of each compound is 18.0 to 58.5, so there must be the same number of H_2O molecules in one mole of H_2O as there are NaCl formula units in one mole of NaCl.

We can apply this reasoning to all elements and compounds: One mole of He must contain the same number of atoms as one mole of Cu, one mole of Cl_2 must contain the same number of molecules as one mole of NH_3, one mole of Fe must contain the same number of atoms as there are molecules in one mole of H_2O, and so forth. The number of fundamental particles in one mole of any substance has been found to be 6.02×10^{23}, referred to as **Avogadro's number.**◆

◆Avogadro's number is named for Amadeo Avogadro (1776–1856), an Italian physicist and chemist.

The second definition of the mole is that, for any substance,

$$1 \text{ mol of X} = 6.02 \times 10^{23} \text{ atoms, ions, formula units, or molecules of X}$$

This equation gives a general definition of the mole that can be applied to any particular substance.◆

◆The mole defines a certain number of things, so it's similar to the dozen. Just as 1 dozen X = 12 X, so 1 mol X = 6.02×10^{23} X.

The following Example shows the conversion of the general definition of the mole to the specific definition for a particular substance.

Example 11.5

Write the mole definition, in terms of Avogadro's number, for (a) iron (b) chlorine (c) sodium chloride (d) water.

Solution

(a) The element iron consists of atoms and has the symbol Fe, so 1 mol Fe = 6.02×10^{23} Fe atoms.

(b) The element chlorine consists of molecules and has the formula Cl_2, so 1 mol Cl_2 = 6.02×10^{23} Cl_2 molecules.

(c) The compound sodium chloride consists of formula units and has the formula NaCl, so 1 mol NaCl = 6.02×10^{23} NaCl formula units.

(d) The compound water consists of molecules and has the formula H_2O, so 1 mol H_2O = 6.02×10^{23} H_2O molecules.

We can apply this second mole definition to imaginary forms of matter. For example, chlorine exists as Cl_2 molecules, but we can imagine Cl atoms, and we can apply the mole definition to either molecules or atoms:

$$1 \text{ mol } Cl_2 = 6.02 \times 10^{23} \text{ } Cl_2 \text{ molecules}$$

$$1 \text{ mol Cl} = 6.02 \times 10^{23} \text{ Cl atoms}$$

Exercise 11.4

Write the specific definition of the mole, as Avogadro's number of particles, for each of the following substances. (a) oxygen (b) oxygen atoms (c) CH_4 (d) $FeCl_3$ (e) oxide ion

This second definition of the mole is used in dimensional analysis in the usual way, as shown in the following Examples.

Example 11.6

How many magnesium atoms are in 0.392 mole of magnesium?

Solution

0.392 mol Mg = ? Mg atoms

mol Mg → Mg atoms

1 mol Mg = 6.02×10^{23} Mg atoms

$$\frac{1 \text{ mol Mg}}{6.02 \times 10^{23} \text{ Mg atoms}} = 1 \qquad \frac{6.02 \times 10^{23} \text{ Mg atoms}}{1 \text{ mol Mg}} = 1$$

mol Mg → Mg atoms

$$\left(\frac{0.392 \cancel{\text{mol Mg}}}{}\right)\left(\frac{6.02 \times 10^{23} \text{ Mg atoms}}{1 \cancel{\text{mol Mg}}}\right) = 2.36 \times 10^{23} \text{ Mg atoms}$$

Example 11.7

A sample of $Ca(NO_3)_2$ contains 3.33×10^{24} formula units. What is its mass in kilograms?

Solution

3.33×10^{24} $Ca(NO_3)_2$ formula units = ? kg $Ca(NO_3)_2$

$Ca(NO_3)_2$ formula units → mol $Ca(NO_3)_2$ → g $Ca(NO_3)_2$ → kg $Ca(NO_3)_2$

$$\left(\frac{3.33 \times 10^{24}\ \cancel{Ca(NO_3)_2\ formula\ units}}{}\right)\left(\frac{1\ mol\ Ca(NO_3)_2}{6.02 \times 10^{23}\ \cancel{Ca(NO_3)_2\ formula\ units}}\right)$$

formula mass

$$\left(\frac{164\ g\ \cancel{Ca(NO_3)_2}}{1\ mol\ \cancel{Ca(NO_3)_2}}\right)\left(\frac{1\ kg\ \cancel{Ca(NO_3)_2}}{1 \times 10^3\ g\ \cancel{Ca(NO_3)_2}}\right) = 0.907\ kg\ Ca(NO_3)_2$$

Example 11.8

How many chlorine atoms are in 2.50 tons of CCl_4?

Solution

2.50 tons CCl_4 = ? Cl atoms

tons CCl_4 → lb CCl_4 → g CCl_4 → mol CCl_4 → CCl_4 molecules → Cl atoms

$$\left(\frac{2.50\ \cancel{tons\ CCl_4}}{}\right)\left(\frac{2000\ \cancel{lb\ CCl_4}}{1\ \cancel{ton\ CCl_4}}\right)\left(\frac{454\ \cancel{g\ CCl_4}}{1\ \cancel{lb\ CCl_4}}\right)\left(\frac{1\ \cancel{mol\ CCl_4}}{154\ \cancel{g\ CCl_4}}\right)$$

$$\left(\frac{6.02 \times 10^{23}\ \cancel{CCl_4\ molecules}}{1\ \cancel{mol\ CCl_4}}\right)\left(\frac{4\ Cl\ atoms}{1\ \cancel{CCl_4\ molecule}}\right) = 3.55 \times 10^{28}\ Cl\ atoms$$

Solve the following Exercise by dimensional analysis.

Exercise 11.5

Calculate (a) the mass in grams of one million water molecules (b) the mass in grams of one water molecule (c) the number of potassium ions in 25.0 grams of K_2CO_3.

11-3 The percentage composition for a compound can be calculated from its formula

Throughout the rest of this book you'll see that the mole has many important applications. One of these applications, which will be described in Section 11-4, is in determining the formula for a compound from its percentage composition.

The **percentage composition** for a compound is a statement of the percentage, by mass, of each element in the compound. For example, for sodium carbonate, Na_2CO_3,

the percentage composition is	and this means that, in 100 g of Na_2CO_3 there will be
43.4% Na	43.4 g of Na
11.3% C	11.3 g of C
45.3% O	45.3 g of O
100.0%	100.0 g

If the formula for a compound is known, its percentage composition can be calculated from this relationship:

For each element in a compound, % element =

$$\frac{\text{mass of element in formula}}{\text{mass of formula}} \times 100$$

The following calculation for sodium carbonate illustrates this procedure. The atomic and formula masses are

Na = 23.0 amu C = 12.0 amu O = 16.0 amu Na_2CO_3 = 106 amu

$$\%\ \text{Na} = \frac{2 \times \text{Na}}{\text{Na}_2\text{CO}_3} \times 100 = \frac{2 \times 23.0\ \cancel{\text{amu}}}{106\ \cancel{\text{amu}}} \times 100 = 43.4\%\ \text{Na}$$

$$\%\ \text{C} = \frac{1 \times \text{C}}{\text{Na}_2\text{CO}_3} \times 100 = \frac{1 \times 12.0\ \cancel{\text{amu}}}{106\ \cancel{\text{amu}}} \times 100 = 11.3\%\ \text{C}$$

$$\%\ \text{O} = \frac{3 \times \text{O}}{\text{Na}_2\text{CO}_3} \times 100 = \frac{3 \times 16.0\ \cancel{\text{amu}}}{106\ \cancel{\text{amu}}} \times 100 = 45.3\%\ \text{O}$$

100.0% total

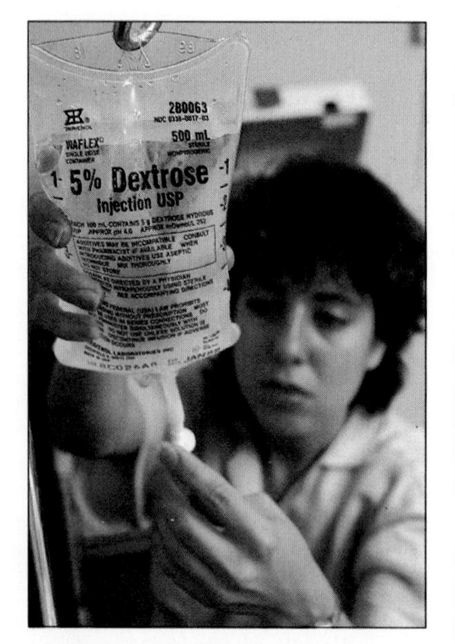

Glucose ($C_6H_{12}O_6$), also called dextrose, is administered intravenously as a source of nutrition for hospital patients who are unable to eat. *(Ed Wheeler/The Stock Market)*

Example 11.9

Calculate the percentage composition for calcium phosphate, $Ca_3(PO_4)_2$.

Solution

$$\%\ \text{Ca} = \frac{3 \times \text{Ca}}{\text{Ca}_3(\text{PO}_4)_2} \times 100 = \frac{3 \times 40.1\ \cancel{\text{amu}}}{310\ \cancel{\text{amu}}} \times 100 = 38.8\%\ \text{Ca}$$

$$\%\ \text{P} = \frac{2 \times \text{P}}{\text{Ca}_3(\text{PO}_4)_2} \times 100 = \frac{2 \times 31.0\ \cancel{\text{amu}}}{310\ \cancel{\text{amu}}} \times 100 = 20.0\%\ \text{P}$$

$$\%\ \text{O} = \frac{8 \times \text{O}}{\text{Ca}_3(\text{PO}_4)_2} \times 100 = \frac{8 \times 16.0\ \cancel{\text{amu}}}{310\ \cancel{\text{amu}}} \times 100 = 41.2\%\ \text{O}$$

100.0%

Exercise 11.6

The formula for blood sugar (chemical name glucose) is $C_6H_{12}O_6$. Calculate its percentage composition.

11-4 The formula for a compound can be determined from its percentage composition and its formula mass or molecular mass

The most important piece of information about a compound is its formula. In this Section we'll look briefly at part of the process involved in determining the formula for an unknown compound.

More than one kind of formula can be written for a compound. We've seen, for example, that the molecular formula for water is H_2O, its Lewis formula is $H:\underset{\cdot\cdot}{\ddot{O}}:H$, and its line formula is $H{-}\underset{\cdot\cdot}{\ddot{O}}{-}H$. The Lewis and line formulas give more information than the molecular formula, but the molecular formula has the advantage of being very brief.

The simplest formula, also called the empirical formula, for a compound is even briefer than its molecular formula and gives less information. The **simplest formula** for a compound is a formula that shows only the smallest whole-number relationships of atoms in the compound. If the molecular formula for a compound is known, its simplest formula can be found by dividing all of the subscripts in the molecular formula by the largest whole number (integer) that will divide evenly into each of them, as shown in the following examples.

Molecular Formula		Simplest Formula
C_2H_2	Divide each subscript by 2	CH
N_2H_4	Divide each subscript by 2	NH_2
C_5H_{10}	Divide each subscript by 5	CH_2
$C_6H_{12}O_6$	Divide each subscript by 6	CH_2O
H_2O	Divide each subscript by 1	H_2O

As the last entry in this table shows, the simplest formula for a compound will be the same as its molecular formula if the only integer by which all of the subscripts in the molecular formula are evenly divisible is 1.

Exercise 11.7

Write the simplest formula for the following compounds. (a) potassium oxalate, $K_2C_2O_4$ (b) benzene, C_6H_6 (c) caffeine, $C_8H_{10}N_4O_2$ (d) sodium chloride

The simplest formula for a compound is its molecular formula divided by a whole number, and the simplest-formula mass for the compound will be its molecular mass divided by that same whole number. For example, for ethylene, C_2H_4, these relationships are:

Molecular Formula	Simplest Formula
C_2H_4	$\frac{C_2H_4}{2} = CH_2$
Molecular Mass	**Simplest-formula mass**
28.0 amu	$\frac{28.0 \text{ amu}}{2} = 14.0 \text{ amu}$

Exercise 11.8

Naphthalene, a compound used in mothballs, has the molecular formula $C_{10}H_8$ and the molecular mass 128 amu (its line formula is shown in Table 7.2). What is its simplest-formula mass?

Moth balls are naphthalene, $C_{10}H_8$. *(Runk/Schoenberger from Grant Heilman)*

There are laboratory methods for finding the percentage composition and the formula mass or molecular mass of an unknown compound. The simplest formula for the compound can then be determined from its percentage composition, and the molecular formula for the compound can be determined from its simplest formula and its formula mass or molecular mass.

For example, butane, used as a fuel in some cigarette lighters, has been found to have the molecular mass 58.0 amu and the percentage composition 82.7% C and 17.3% H. From this information we can determine its molecular formula by the following steps.

1. **Calculate the number of moles of each element in 100 g of the compound.** We know from its percentage composition that 100 g of butane must contain 82.7 g of C and 17.3 g of H.

$$\left(\frac{82.7\ \cancel{\text{g C}}}{}\right)\left(\frac{1\ \text{mol C}}{12.0\ \cancel{\text{g C}}}\right) = 6.90\ \text{mol C}$$

$$\left(\frac{17.3\ \cancel{\text{g H}}}{}\right)\left(\frac{1\ \text{mol H}}{1.01\ \cancel{\text{g H}}}\right) = 17.0\ \text{mol H}$$

2. **Convert these numbers of moles to whole numbers.** First, divide each of the numbers by the smaller one. In this example, the smaller number is 6.90.

$$\frac{6.90\ \text{mol C}}{6.90} = 1.00\ \text{mol C}$$

$$\frac{17.0\ \text{mol H}}{6.90} = 2.46\ \text{mol H}$$

Sometimes this division will produce whole numbers. If, as in this case, it doesn't, then multiply these numbers of moles by the smallest integer that will convert them to whole numbers. If we multiply each number of moles by 2:

$$1.00\ \text{mol C} \times 2 = 2.00\ \text{mol C}$$

$$2.46\ \text{mol H} \times 2 = 4.92\ \text{mol H, which is very nearly } 5.00$$

3. **From the numbers of moles of the elements, determine the simplest formula and the simplest-formula mass.** This compound has 2 moles of C for every 5 moles of H. Since 1 mole of C and 1 mole of H contain the same number of atoms, it must be true that this compound has 2 atoms of C for every 5 atoms of H and that the simplest formula for the compound must be C_2H_5. We can then calculate the simplest-formula mass as 29.0 amu.

4. **Divide the molecular mass by the simplest-formula mass and multiply the simplest formula by this number to determine the molecular formula.**

$$\frac{\text{molecular mass}}{\text{simplest-formula mass}} = \frac{58.0\ \cancel{\text{amu}}}{29.0\ \cancel{\text{amu}}} = 2 = \frac{\text{molecular formula}}{\text{simplest formula}}$$

$$\text{molecular formula} = 2 \times \text{simplest formula} = 2 \times C_2H_5 = C_4H_{10}$$

Determining a molecular formula from a percentage composition and a formula mass or molecular mass uses the following sequence of steps:

percentages of elements
↓
moles of elements
↓
atoms of elements
↓
simplest formula
↓
molecular formula

Lactic acid, $C_3H_6O_3$, is produced in muscle tissue during exercise. *(The Stock Market/George J. Contorakes Inc.)*

Example 11.10

Lactic acid, a compound produced in muscle tissue during exercise, has been found to have the percentage composition 40.00% C, 6.71% H, and 53.29% O. Its molecular mass is 90.1 amu. What is its molecular formula?

Solution

$$\left(\frac{40.00\ \cancel{g\ C}}{}\right)\left(\frac{1\ \text{mol C}}{12.0\ \cancel{g\ C}}\right) = 3.33\ \text{mol C}$$

$$\left(\frac{6.71\ \cancel{g\ H}}{}\right)\left(\frac{1\ \text{mol H}}{1.01\ \cancel{g\ H}}\right) = 6.64\ \text{mol H}$$

$$\left(\frac{53.29\ \cancel{g\ O}}{}\right)\left(\frac{1\ \text{mol O}}{16.0\ \cancel{g\ O}}\right) = 3.33\ \text{mol O}$$

$$\frac{3.33\ \text{mol C}}{3.33} = 1.00\ \text{mol C}$$

$$\frac{6.64\ \text{mol H}}{3.33} = 1.99\ \text{mol H}$$

$$\frac{3.33\ \text{mol O}}{3.33} = 1.00\ \text{mol O}$$

These calculations show that in this compound there is 1.00 mol of C for 1.99 mol of H for 1.00 mol of O, so there must be 1 atom of C for 2 atoms of H for 1 atom of O, and the simplest formula must be CH_2O. The simplest-formula mass is 30.0 amu.

$$\frac{\text{molecular mass}}{\text{simplest-formula mass}} = \frac{90.1\ \cancel{\text{amu}}}{30.0\ \cancel{\text{amu}}} = 3 = \frac{\text{molecular formula}}{\text{simplest formula}}$$

$$\text{molecular formula} = 3 \times \text{simplest formula} = 3 \times CH_2O = C_3H_6O_3$$

Exercise 11.9

The percentage composition for Vitamin C is 40.9% C, 4.6% H, and 54.5% O, and its molecular mass is 176 amu. What is its molecular formula?

Exercise 11.10

Caffeine has the percentage composition 49.48% C, 5.19% H, 28.85% N, and 16.48% O, and its molecular mass is 194.2 amu. What is its molecular formula?

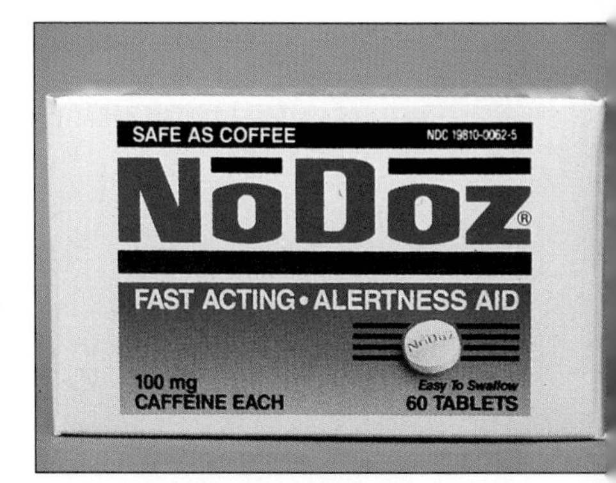

Caffeine is the active ingredient in NODOZ®. *(David R. Frazier)*

Inside Chemistry

Why is a mole called a mole?

In the middle of the seventeenth century European philosophers and scientists began to propose explanations for natural phenomena based on the belief that matter consists of tiny, invisible particles; these explanations eventually led to our modern atomic theory. In proposing these explanations, French philosophers and scientists created the word *molécule,* to designate one of the small particles of matter. The word is from the Latin *molecula,* meaning small mass, from the Latin *moles,* meaning mass. When the concept of the mole was created, at the start of our century, the name *mole* was chosen because of its relationships to *molecule* and to the Latin word for mass.

Some chemical elements also have names from Latin sources. The following list gives the origins of the names for a few of the elements.

Aluminum from the Latin *alumen,* meaning alum. The compound called alum (potassium aluminum sulfate, $KAl(SO_4)_2$) occurs naturally as white crystals. Because alum absorbs water strongly, medical practitioners in ancient times used it as an astringent to absorb moisture from tissues and promote healing, for example, by applying powdered alum to small cuts.

Calcium from the Latin *calx,* meaning chalk. Chalk is calcium carbonate, $CaCO_3$.

Carbon from the Latin *carbon,* meaning charcoal. Charcoal contains a large amount of carbon, both as the element and in many complicated covalent compounds.

Gallium from the Latin *Gallia,* meaning France. The element was discovered in France.

Hydrogen from the Greek *hydros,* meaning water, and *-gen,* meaning producer. When hydrogen is burned, it produces water: $2\ H_2(g) + O_2(g) \rightarrow 2\ H_2O(\ell)$.

Iodine from the Greek *ioeides,* meaning violet colored. Iodine vapors have a violet color.

Lithium from the Greek *lithos,* meaning stone. Lithium occurs in minerals that are found naturally in lumps or stones.

Nobelium for Alfred Nobel, the Swedish industrialist who established the Nobel Prize, and for the Nobel Institute in Stockholm, Sweden, where the element was claimed to have been produced for the first time.

Oxygen from the Greek *oksys,* meaning acidic, and *-gen,* meaning producer. An early theory to explain the properties of the class of compounds called *acids* proposed that all acids contain oxygen. The theory is no longer accepted. (Acids are discussed in Chapters 15 and 16.)

Radium from the Latin *radius,* meaning ray. Radium is radioactive and emits rays.

Silicon from the Latin *silex,* meaning flint or hard stone. Silicon occurs in many minerals, including those that make up flint.

Uranium from the name for the planet Uranus. The planet was discovered in 1781 and the element was discovered in 1789. In Greek mythology Uranos was the god of heaven.

One-mole samples of several common elements. Left to right: aluminum (as aluminum foil), mercury, sulfur, chromium (in front of sulfur), lead, magnesium, and copper. *(Charles D. Winters)*

Chapter Summary: The Mole

Section	Subject	Summary	Check When Learned
11-1	Abbreviation for mole	mol	☐
11-1, 11-2	General mole definitions	1 mol X = atomic, formula, or molecular mass of X in g 1 mol X = 6.02×10^{23} atoms, ions, formula units, or molecules of X	☐
11-1, 11-2	Mole definitions for substance that exists as atoms	1 mol X = atomic mass of X in g 1 mol X = 6.02×10^{23} X atoms	☐
11-1, 11-2	Mole definitions for substance that exists as formula units	1 mol X = formula mass of X in g 1 mol X = 6.02×10^{23} X formula units	☐
11-1, 11-2	Mole definitions for substance that exists as molecules	1 mol X = molecular mass of X in g 1 mol X = 6.02×10^{23} X molecules	☐
11-1	Elements that exist as diatomic molecules	H_2, N_2, O_2, F_2, Cl_2, Br_2, I_2	☐
11-1	Rule for deciding meaning of ambiguous expressions such as "one mole of oxygen"	"One mole of oxygen" might refer either to oxygen molecules or oxygen atoms. When in doubt, assume that the expression refers to the form in which the substance occurs in nature, in this case, oxygen molecules.	☐
11-2	Name for 6.02×10^{23}	Avogadro's number.	☐
11-3	Percentage composition for a compound	Statement of the percentage, by mass, of each element in the compound.	☐
11-3	Formula for calculating percentage of element in a compound	$\% \text{ element} = \frac{\text{mass of element in formula}}{\text{mass of formula}} \times 100$	☐
11-4	Simplest formula	Formula that shows only the smallest whole-number relationships of atoms in the compound. For the molecular formula $C_6H_{12}O_6$, the simplest formula is CH_2O.	☐
11-4	Method for finding simplest formula from molecular formula	Divide subscripts in molecular formula by largest integer that will divide into each of them to give a whole number.	☐

Section	Subject	Summary	Check When Learned
11-4	Steps for finding simplest formula from percentage composition and molecular mass	(1) Calculate the number of moles of each element in 100 g of the compound. (2) Convert these numbers of moles to whole numbers. (3) From the numbers of moles of the elements, determine the simplest formula. (4) Divide the molecular mass by the simplest-formula mass and multiply the simplest formula by this number to determine the molecular formula.	☐
11-4	Outline of steps for finding simplest formula from percentage composition and molecular mass	Percentages of elements → moles of elements → atoms of elements → simplest formula → molecular formula.	☐

Problems

Assume you can use the Periodic Table at the front of the book unless you're directed otherwise. As you use atomic masses and molecular masses in these problems, round them off to three significant digits. Answers to odd-numbered problems are in Appendix 1.

The Mole (Sections 11-1 and 11-2)

1. Calculate the number of grams in each of the following quantities: (a) 1.00 mol of Cr (b) 0.275 mole of HNO_3 (c) 1.22 mol of O_2.
2. Calculate the number of grams in each of the following quantities: (a) 1.00 mol of He (b) 0.364 mol of K_2O (c) 3.52 mol of Br_2.
3. Calculate the number of grams in each of the following quantities: (a) 0.500 mol of nitrogen (b) 1.00 mol of hydrogen sulfide (c) 1.30 mol of sodium phosphate.
4. Calculate the number of grams in each of the following quantities: (a) 0.151 mol of aluminum (b) 0.642 mol of iodine (c) 1.75 mol of carbon disulfide.
5. Calculate the number of moles in each of the following quantities: (a) 1.67 g of Mg (b) 0.443 g of PbO (c) 15.2 g of O_2.
6. Calculate the number of moles in each of the following quantities: (a) 2.85 g of S (b) 3.33 g of $HgBr_2$ (c) 6.79 g of Br_2.
7. Calculate the number of moles in each of the following quantities: (a) 3.52 g of Li (b) 0.711 g of SO_3 (c) 1.27 g of $CaSO_4$.
8. Calculate the number of moles in each of the following quantities: (a) 0.849 g of F_2 (b) 2.16 g of P_2O_5 (c) 1.50 g of NH_4Cl.
9. Calculate the number of moles in each of the following quantities: (a) 1.00×10^{24} Li atoms (b) 3.44×10^{20} N atoms (c) 2.78×10^{22} N_2 molecules (d) 1.75×10^{23} formula units of $SrCl_2$.
10. Calculate the number of moles in each of the following quantities: (a) 2.65×10^{25} Br atoms (b) 1.77×10^{25} Br_2 molecules (c) 3.60×10^{19} CO_2 molecules (d) 2.55×10^{24} $MgCO_3$ formula units.
11. Calculate the number of moles in each of the following quantities: (a) 2.67×10^{23} chromium atoms (b) 3.55×10^{24} water molecules (c) 1.25×10^{22} formula units of sodium chloride.
12. Calculate the number of moles in each of the following quantities: (a) 7.42×10^{22} barium atoms (b) 4.50×10^{20} ammonia molecules (c) 5.00×10^{23} formula units of barium nitrate.

13. Calculate the number of moles in each of the following quantities: (a) 29.3 g of nitrogen (b) 16.7 g of carbon (c) 3.42 lb of carbon tetrachloride.

14. Calculate the number of moles in each of the following quantities: (a) 13.4 g of iodine (b) 3.00 g of aluminum (c) 29.0 lb of phosphorus pentachloride.

15. Calculate the number of moles in each of the following quantities: (a) 5.19 g of iron (b) 3.76 g of methane (c) 0.0145 g of phosphorus trichloride.

16. Calculate the number of moles in each of the following quantities: (a) 0.217 g of chlorine (b) 0.0495 g of sodium cyanide (c) 1.27×10^3 g of ammonium sulfate.

17. Calculate the number of moles in each of the following quantities: (a) 2.00 kg of calcium nitrate (b) 3.00×10^{23} boron atoms (c) 1.65×10^{22} molecules of dinitrogen pentoxide (d) 7.39×10^{24} formula units of zinc nitrate.

18. Calculate the number of moles in each of the following quantities: (a) 255 mg of copper(II) sulfate pentahydrate (b) 4.67×10^{24} potassium atoms (c) 8.26×10^{25} molecules of sulfur hexafluoride (d) 9.47×10^{24} formula units of iron(III) chlorate.

19. Calculate the number of moles in each of the following quantities: (a) 2.91 cg of nickel (b) 5.02×10^{24} carbon atoms (c) 4.51×10^{-3} kg of potassium bromide (d) 1.58×10^{21} molecules of carbon monoxide.

20. Calculate the number of moles in each of the following quantities: (a) 7.75 μg of beryllium (b) 1.47×10^{24} molecules of hydrogen (c) 9.15 mg of oxygen difluoride (d) 3.76 pg of zinc hydride.

21. Make each of the following conversions: (a) 6.82 mol of sulfur trioxide to grams (b) 0.240 mol of sodium dihydrogen phosphate to milligrams (c) 2.80 tons of chlorine to moles (d) 0.232 kilograms of manganese(II) sulfite to formula units.

22. Make each of the following conversions: (a) 3.92 mol of nitrogen trichloride to grams (b) 1.00 mol of calcium hydrogen carbonate to milligrams (c) 0.500 tons of sodium sulfate to moles (d) 14.2 grams of hydrogen to molecules.

23. Make each of the following conversions: (a) 9.59×10^{24} formula units of copper(II) nitrate to grams (b) 6.66 μg of silver to atoms (c) 1.00 mg of carbon tetrachloride to molecules (d) 3.56×10^{-5} mol of potassium chlorate to micrograms.

24. Make each of the following conversions: (a) 8.78 kg of sulfur dioxide to molecules (b) 4.91×10^{24} atoms of nickel to kilograms (c) 1.00 lb of water to molecules (d) 8.56×10^{23} formula units of aluminum oxide to grams.

25. Calculate (a) the number of oxygen molecules in 1.00 g of oxygen (b) the mass in grams of one iron atom (c) the number of water molecules in 1 drop of water; assume that there are 20 drops in one milliliter.

26. Calculate (a) the number of copper atoms in 1.00 g of copper (b) the mass in grams of one molecule of carbon dioxide (c) the number of gold atoms in 1.00 cm^3 of gold; the density of gold is 19.3 g/cm^3.

Percentage Composition (Section 11-3)

27. Calculate the percentage composition for vanillin, the compound responsible for the flavor of vanilla. Its formula is $C_8H_8O_3$.

28. The compound isoamyl acetate, $C_7H_{14}O_2$, has the odor of bananas. Calculate its percentage composition.

29. Calculate the percentage composition for calcium hypochlorite.

30. Calculate the percentage composition for iron(III) cyanide.

31. Calculate the percentage composition for diphosphorus pentasulfide.

32. Calculate the percentage composition for carbon tetraiodide.

33. Calculate the percentage composition for zinc nitrite.

34. Calculate the percentage composition for copper(II) sulfide.

35. Calculate the percentage composition for silver carbonate.

36. Calculate the percentage composition for magnesium nitride.

Simplest Formula (Section 11-4)

37. Write the simplest formula for (a) sodium peroxide, Na_2O_2 (b) sodium oxalate, $Na_2C_2O_4$ (c) sodium sulfide.

38. Calculate the simplest-formula mass for (a) fructose (fruit sugar), $C_6H_{12}O_6$ (b) dinitrogen tetroxide (c) oxalic acid, $H_2C_2O_4$.

39. The line formula for the compound disilane is

$$\begin{array}{ccccc} & H & & H & \\ & | & & | & \\ H- & Si & - & Si & -H \\ & | & & | & \\ & H & & H & \end{array}$$

(a) What is its molecular mass? (b) What is its simplest formula? (c) What is its simplest-formula mass?

40. The compound tetrafluorethylene, used to make polytetrafluoroethylene (Teflon®), has the line formula

$$\begin{array}{ccc} :\ddot{F}: & & :\ddot{F}: \\ | & & | \\ C & = & C \\ | & & | \\ :\underset{..}{F}: & & :\underset{..}{F}: \end{array}$$

(a) What is its molecular mass? (b) What is its simplest formula? (c) What is its simplest-formula mass?

41. Lysine is an essential amino acid, a compound required for good human nutrition. Its molecular formula is $C_6H_{14}N_2O_2$. What is its simplest-formula mass?

42. Arginine is an amino acid, a compound that in living organisms is incorporated into proteins. Its molecular formula is $C_6H_{14}N_4O_2$. What is its simplest-formula mass?

43. Galactose, a sugar, has the simplest formula CH_2O, and its molecular mass is 180 amu. What is its molecular formula?

44. Ethylenediamine, used in making polymers, has the simplest formula CH_4N, and its molecular mass is 60.1 amu. What is its molecular formula?

Percentage Composition and Formula (Section 11-4)

45. A compound has a formula mass of 102 amu, and its percentage composition is 13.6% Li, 23.5% C and 62.7% O. Write its formula and name.

46. A compound has a formula mass of 92.0 amu, and its percentage composition is 30.4% N and 69.6% O. Write its formula and name.

47. A compound has a formula mass of 239 amu, and its percentage composition is 86.62% Pb and 13.38% O. Name it.

48. A compound has a formula mass of 170 amu, and its percentage composition is 63.50% Ag, 8.25% N, and 28.25% O. Name it.

49. The compound Ibuprofen, the active ingredient in Advil® and similar drugs, has a molecular mass of 206 amu, and its percentage composition is 75.69% C, 8.80% H, and 15.51% O. What is its molecular formula?

50. The compound maleic acid, which is used to retard rancidity in fats and oils, has the molecular mass 116 amu and the percentage composition 41.39% C, 3.47% H, and 55.14% O. Write its molecular formula.

51. Hydrazine, a compound used as a rocket fuel, has a molecular mass of 32.0 amu, and its percentage composition is 12.58% H and 87.41% N. What is its molecular formula?

52. Ethylene glycol, used as an antifreeze in car radiators, has a molecular mass of 62.07 amu, and its percentage composition is 38.70% C, 9.74% H, and 51.56% O. What is its molecular formula?

53. Benzene, a compound used to make many useful chemicals, has a molecular mass of 78.11 amu, and its percentage composition is 92.25% C and 7.75% H. What is its molecular formula?

54. Terephthalic acid, a compound used to make plastic films and sheets, has a molecular mass of 166 amu, and its percentage composition is 57.83% C, 3.64% H, and 38.52% O. What is its molecular formula?

A piece of magnesium metal burns in air, forming white magnesium oxide. *(Charles D. Winters)*

12

Calculations Based on Chemical Equations: Stoichiometry

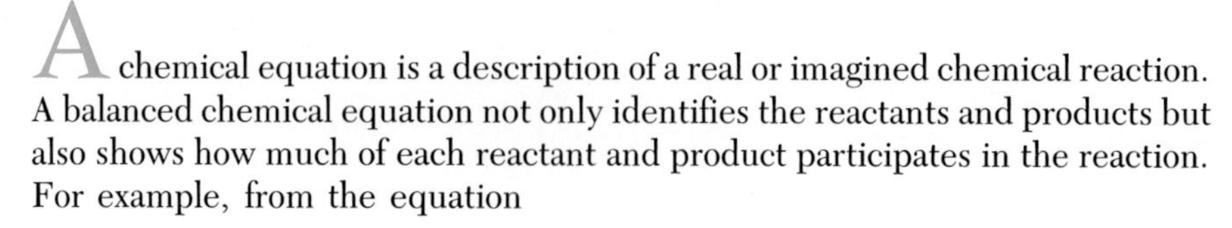

A chemical equation is a description of a real or imagined chemical reaction. A balanced chemical equation not only identifies the reactants and products but also shows how much of each reactant and product participates in the reaction. For example, from the equation

$$2\ Na + Cl_2 \rightarrow 2\ NaCl$$

we can see that, in the reaction of sodium with chlorine, it will take two sodium atoms and one chlorine molecule to produce two formula units of sodium chloride.

The units of a balanced chemical equation—atoms, ions, formula units, or molecules—are useful in the theoretical description of a reaction, but they aren't practical for carrying out a reaction in a laboratory or in an industrial plant because these units are much too small to measure or handle directly. For laboratory work quantities of reactants and products are usually measured in mass units, often in grams; in industrial work quantities of reactants and products may be measured in much larger units, such as kilograms or tons.

The masses of reactants and products that will be involved in carrying out a chemical reaction on a laboratory scale or on an industrial scale can be calculated by applying dimensional analysis to the equation for the reaction. The calculation of masses of reactants and products by means of dimensional analysis is called **stoichiometry.**

In a hydrogen flame, hydrogen combines with oxygen from the air to produce water; in this picture the water is condensing on the inside of the inverted beaker. The reaction is

$$2\ H_2(g) + O_2(g) \rightarrow 2\ H_2O(\ell)$$

(Charles D. Winters)

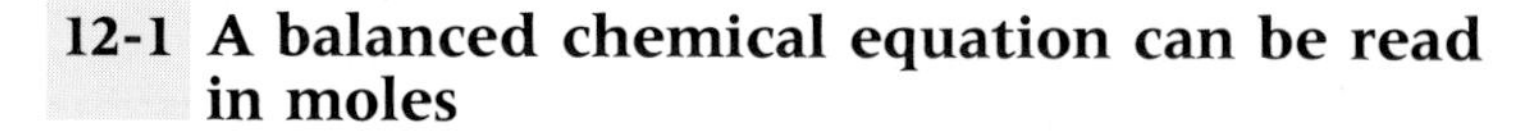

12-1 A balanced chemical equation can be read in moles

We've seen that one way of reading a chemical equation is in terms of atoms, ions, formula units, or molecules. For example, the equation

$$2\ H_2 + O_2 \rightarrow 2\ H_2O$$

shows that, in the reaction of hydrogen with oxygen, it will take two hydrogen molecules and one oxygen molecule to produce two water molecules.

It must also be true that it will take *four* hydrogen molecules and *two* oxygen molecules to form *four* water molecules, and the same will be true for any multiple of the numbers of molecules shown in the balanced chemical equation, as in the following example:

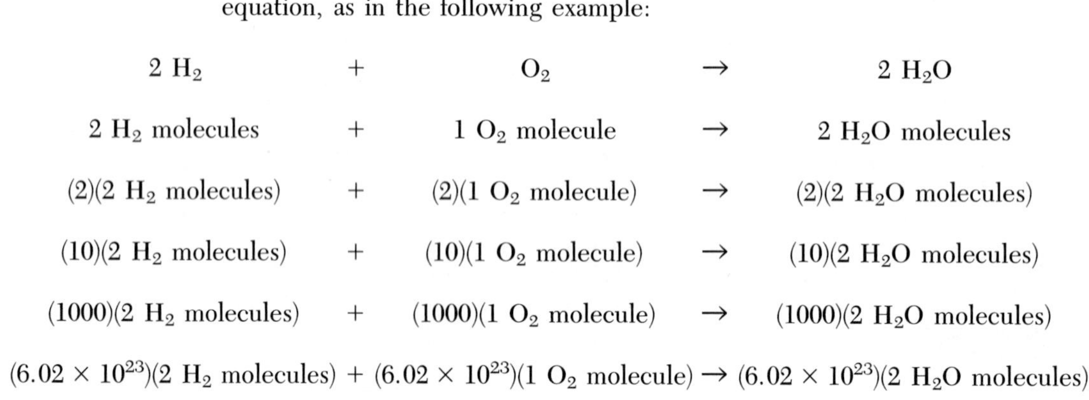

$2\ H_2$	+	O_2	→	$2\ H_2O$
$2\ H_2$ molecules	+	$1\ O_2$ molecule	→	$2\ H_2O$ molecules
$(2)(2\ H_2$ molecules)	+	$(2)(1\ O_2$ molecule)	→	$(2)(2\ H_2O$ molecules)
$(10)(2\ H_2$ molecules)	+	$(10)(1\ O_2$ molecule)	→	$(10)(2\ H_2O$ molecules)
$(1000)(2\ H_2$ molecules)	+	$(1000)(1\ O_2$ molecule)	→	$(1000)(2\ H_2O$ molecules)
$(6.02 \times 10^{23})(2\ H_2$ molecules)	+	$(6.02 \times 10^{23})(1\ O_2$ molecule)	→	$(6.02 \times 10^{23})(2\ H_2O$ molecules)

The last entry in the preceding table is important because it establishes a relationship between the chemical equation read in molecules and the same equation read in moles:

$$2\ H_2 \quad + \quad O_2 \quad \rightarrow \quad 2\ H_2O$$

$$(6.02 \times 10^{23})(2\ H_2\ \text{molecules}) + (6.02 \times 10^{23})(1\ O_2\ \text{molecule}) \rightarrow (6.02 \times 10^{23})(2\ H_2O\ \text{molecules})$$

$$\text{means } 2\ \text{mol}\ H_2 \quad + \quad 1\ \text{mol}\ O_2 \quad \rightarrow \quad 2\ \text{mol}\ H_2O$$

We could apply this same reasoning to any balanced chemical equation to show that it could be read either in terms of particles (atoms, ions, formula units, or molecules) or in moles, as shown in the following examples.

$$2\ Na \quad + \quad Cl_2 \quad \rightarrow \quad 2\ NaCl$$

$$2\ \text{Na atoms} + 1\ Cl_2\ \text{molecule} \rightarrow 2\ \text{NaCl formula units}$$

$$2\ \text{mol Na} \quad + \quad 1\ \text{mol}\ Cl_2 \quad \rightarrow \quad 2\ \text{mol NaCl}$$

$$2\ Mg \quad + \quad O_2 \quad \rightarrow \quad 2\ Mg^{2+} \quad + \quad 2\ O^{2-}$$

$$2\ \text{Mg atoms} + 1\ O_2\ \text{molecule} \rightarrow 2\ Mg^{2+}\ \text{ions} + 2\ O^{2-}\ \text{ions}$$

$$2\ \text{mol Mg} \quad + \quad 1\ \text{mol}\ O_2 \quad \rightarrow 2\ \text{mol}\ Mg^{2+} + 2\ \text{mol}\ O^{2-}$$

Sodium, a silvery metal, reacts with chlorine, a green gas, to form sodium chloride, a white, crystalline solid. *(Chip Clark)*

Exercise 12.1◆

Below the following equation, write the moles of reactants and products it describes:

$$CH_4 + 4\ Br_2 \rightarrow CBr_4 + 4\ HBr$$

◆ Answers to Exercises are in Appendix 1.

12-2 The masses of reactants and products in a chemical reaction are calculated from the mole relationships in its balanced chemical equation

The mole relationships among the reactants and products in a balanced chemical equation can be expressed in simple mathematical equations:

$$\underset{2\ \text{mol}\ H_2}{2\ H_2} \ + \ \underset{1\ \text{mol}\ O_2}{O_2} \ \rightarrow \ \underset{2\ \text{mol}\ H_2O}{2\ H_2O}$$

$$2\ \text{mol}\ H_2 = 1\ \text{mol}\ O_2 \qquad 2\ \text{mol}\ H_2 = 2\ \text{mol}\ H_2O \qquad 1\ \text{mol}\ O_2 = 2\ \text{mol}\ H_2O$$

These mathematical equations can be used as sources of factors for dimensional analysis, as in the following Example.

Example 12.1

How many moles of hydrogen will react with five moles of oxygen, according to the following equation?

$$2\ H_2 + O_2 \rightarrow 2\ H_2O$$

Solution

$5 \text{ mol } O_2 = ? \text{ mol } H_2$

$\text{mol } O_2 \rightarrow \text{mol } H_2$

$2 \text{ mol } H_2 = 1 \text{ mol } O_2$ (From the balanced chemical equation, read in moles.)

$$\frac{2 \text{ mol } H_2}{1 \text{ mol } O_2} = 1 \qquad \frac{1 \text{ mol } O_2}{2 \text{ mol } H_2} = 1$$

$$\text{mol } O_2 \rightarrow \text{mol } H_2$$

$$\left(\frac{5 \cancel{\text{mol } O_2}}{}\right)\left(\frac{2 \text{ mol } H_2}{1 \cancel{\text{mol } O_2}}\right) = 10 \text{ mol } H_2$$

In practical laboratory work, quantities of reactants and products are usually measured in grams or milligrams. The following example shows a typical problem in laboratory work.

Example 12.2

How many grams of hydrogen will react with 5.00 grams of oxygen, according to the following equation?

$$2 H_2 + O_2 \rightarrow 2 H_2O$$

Solution

$5.00 \text{ g } O_2 = ? \text{ g } H_2$

$\text{g } O_2 \rightarrow \text{mol } O_2 \rightarrow \text{mol } H_2 \rightarrow \text{g } H_2$

$$\frac{1 \text{ mol } O_2}{32.0 \text{ g } O_2} = 1 \qquad \frac{32.0 \text{ g } O_2}{1 \text{ mol } O_2} = 1$$ (From the mole definition)

$$\frac{2 \text{ mol } H_2}{1 \text{ mol } O_2} = 1 \qquad \frac{1 \text{ mol } O_2}{2 \text{ mol } H_2} = 1$$ (From the balanced chemical equation, read in moles)

$$\frac{1 \text{ mol } H_2}{2.02 \text{ g } H_2} = 1 \qquad \frac{2.02 \text{ g } H_2}{1 \text{ mol } H_2} = 1$$ (From the mole definition)

$$\text{g } O_2 \rightarrow \text{mol } O_2 \rightarrow \text{mol } H_2 \rightarrow \text{g } H_2$$

$$\left(\frac{5.00 \cancel{\text{g } O_2}}{}\right)\left(\frac{1 \cancel{\text{mol } O_2}}{32.0 \cancel{\text{g } O_2}}\right)\left(\frac{2 \cancel{\text{mol } H_2}}{1 \cancel{\text{mol } O_2}}\right)\left(\frac{2.02 \text{ g } H_2}{1 \cancel{\text{mol } H_2}}\right) = 0.631 \text{ g } H_2$$

The solution to this problem is an example of **stoichiometry,** the calculation of mass relationships in a chemical reaction. Any mass units can be used to measure the quantities of reactants or products, but the balanced chemical equation always expresses relationships in moles. For this reason, the center of every stoichiometric calculation will be the mole relationship found in the chemical equation. Every stoichiometric calculation follows this pattern:

$$\text{g } O_2 \rightarrow \text{mol } O_2 \rightarrow \text{mol } H_2 \rightarrow \text{g } H_2$$

$$\left(\frac{5.00 \cancel{\text{g } O_2}}{}\right)\left(\frac{1 \cancel{\text{mol } O_2}}{32.0 \cancel{\text{g } O_2}}\right)\left(\frac{2 \cancel{\text{mol } H_2}}{1 \cancel{\text{mol } O_2}}\right)\left(\frac{2.02 \text{ g } H_2}{1 \cancel{\text{mol } H_2}}\right) = 0.631 \text{ g } H_2$$

Translate the given quantity of reactant or product into moles, using one of the mole definitions.

Convert the moles of the given reactant or product to moles of the reactant or product to be found, using the mole relationships in the balanced chemical equation.

Translate the moles of the reactant or product to be found into the units required for the answer, using one of the mole definitions.

In abbreviated form, the pattern for every stoichiometric calculation is:

Translate given units into moles. → Convert moles of given reactant or product to moles of reactant or product to be found. → Translate moles into units required for answer.

Example 12.3

According to the following equation, how many grams of ammonia can be made from 1.65 kilograms of nitrogen?

$$3 \, H_2 + N_2 \rightarrow 2 \, NH_3$$

Solution

$1.65 \text{ kg } N_2 = ? \text{ g } NH_3$

$$\text{kg } N_2 \rightarrow \text{g } N_2 \rightarrow \text{mol } N_2 \rightarrow \text{mol } NH_3 \rightarrow \text{g } NH_3$$

$$\left(\frac{1.65 \cancel{\text{kg } N_2}}{}\right)\left(\frac{1 \times 10^3 \cancel{\text{g } N_2}}{1 \cancel{\text{kg } N_2}}\right)\left(\frac{1 \cancel{\text{mol } N_2}}{28.0 \cancel{\text{g } N_2}}\right)\left(\frac{2 \cancel{\text{mol } NH_3}}{1 \cancel{\text{mol } N_2}}\right)\left(\frac{17.0 \text{ g } NH_3}{1 \cancel{\text{mol } NH_3}}\right)$$

Translate into moles. Convert moles. Translate into units needed for answer.

$$= 2.00 \times 10^3 \text{ g } NH_3$$

Example 12.4

How many oxygen molecules will be needed to react with 505 grams of sodium to form sodium oxide?

Solution

The equation for this reaction is

$$4 \, Na + O_2 \rightarrow 2 \, Na_2O$$

$505 \text{ g Na} = ? \; O_2 \text{ molecules}$

$$\text{g Na} \rightarrow \text{mol Na} \rightarrow \text{mol } O_2 \rightarrow O_2 \text{ molecules}$$

$$\left(\frac{505 \cancel{\text{g Na}}}{}\right)\left(\frac{1 \cancel{\text{mol Na}}}{23.0 \cancel{\text{g Na}}}\right)\left(\frac{1 \cancel{\text{mol } O_2}}{4 \cancel{\text{mol Na}}}\right)\left(\frac{6.02 \times 10^{23} \; O_2 \text{ molecules}}{1 \cancel{\text{mol } O_2}}\right)$$

$$= 3.30 \times 10^{24} \; O_2 \text{ molecules}$$

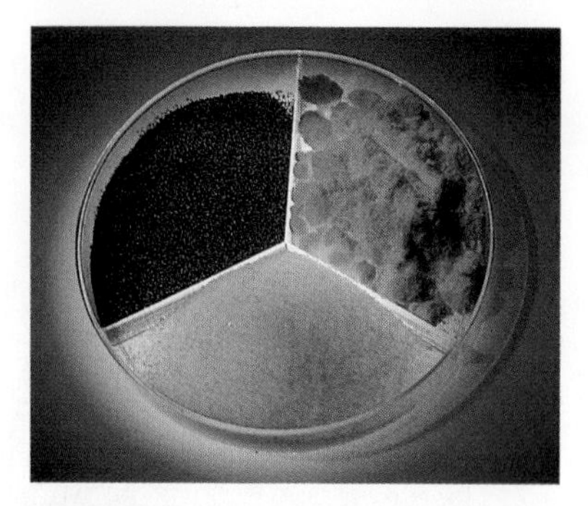

Magnesium, a metal (left), reacts with sulfur, a nonmetal (right), to produce the ionic compound magnesium sulfide (bottom): $Mg(s) + S(s) \rightarrow MgS(s)$ *(Charles D. Winters)*

Solve the following Exercises by dimensional analysis.

Exercise 12.2

In the combustion of natural gas, methane reacts with oxygen to form carbon dioxide and water. How many grams of oxygen will react with 25.5 grams of methane? The equation is

$$CH_4(g) + 2\ O_2(g) \rightarrow CO_2(g) + 2\ H_2O(g)$$

Exercise 12.3

According to the following equation, how many milligrams of magnesium sulfide can be formed from 2.85 milligrams of sulfur?

$$Mg(s) + S(s) \rightarrow MgS(s)$$

12-3 Stoichiometry is the basis for planning chemical reactions

Stoichiometry has many applications, but its most common use is in planning the amounts of reactants to be used in chemical reactions. As the following Examples and Exercises show, stoichiometry provides the method for planning the amounts of reactants for reactions on a laboratory scale or on a much larger industrial scale.

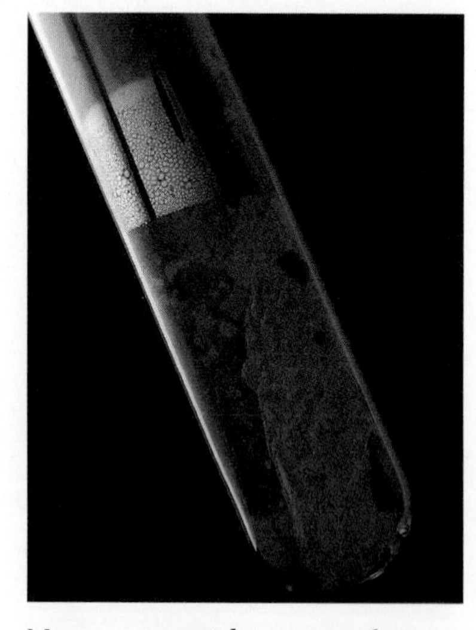

Mercury, a metal, reacts with oxygen, a nonmetal, to produce the red ionic compound mercury(II) oxide: $2\ Hg(\ell) + O_2(g) \rightarrow 2\ HgO(s)$ *(Charles D. Winters)*

Example 12.5

How many grams of Hg will be needed to prepare 5.00 grams of HgO? The equation is

$$2\ Hg + O_2 \rightarrow 2\ HgO$$

Solution

5.00 g HgO = ? g Hg

g HgO → mol HgO → mol Hg → g Hg

$$\left(\frac{5.00\ \cancel{g\ HgO}}{}\right)\left(\frac{1\ \cancel{mol\ HgO}}{217\ \cancel{g\ HgO}}\right)\left(\frac{2\ \cancel{mol\ Hg}}{2\ \cancel{mol\ HgO}}\right)\left(\frac{201\ g\ Hg}{1\ \cancel{mol\ Hg}}\right) = 4.63\ g\ Hg$$

Example 12.6

When potassium chlorate is heated, it decomposes to form potassium chloride and oxygen:

$$2\ KClO_3 \rightarrow 2\ KCl + 3\ O_2$$

A chemist wants to prepare 70.0 milligrams of oxygen. How many milligrams of potassium chlorate should she use?

Solution

$70.0 \text{ mg } O_2 = ? \text{ mg } KClO_3$

$$\text{mg } O_2 \rightarrow \text{g } O_2 \rightarrow \text{mol } O_2 \rightarrow \text{mol } KClO_3$$

$$\left(\frac{70.0 \cancel{\text{mg } O_2}}{}\right)\left(\frac{1 \cancel{\text{g } O_2}}{1 \times 10^3 \cancel{\text{mg } O_2}}\right)\left(\frac{1 \cancel{\text{mol } O_2}}{32.0 \cancel{\text{g } O_2}}\right)\left(\frac{2 \cancel{\text{mol } KClO_3}}{3 \cancel{\text{mol } O_2}}\right)$$

$$\rightarrow \text{g } KClO_3 \rightarrow \text{mg } KClO_3$$

$$\left(\frac{123 \cancel{\text{g } KClO_3}}{1 \cancel{\text{mol } KClO_3}}\right)\left(\frac{1 \times 10^3 \text{ mg } KClO_3}{1 \cancel{\text{g } KClO_3}}\right) = 179 \text{ mg } KClO_3$$

Exercise 12.4

According to the equation

$$2 \text{ Al} + 3 \text{ Br}_2 \rightarrow 2 \text{ AlBr}_3$$

how many grams of bromine will be needed to produce 4.00 grams of aluminum bromide?

Exercise 12.5

Almost every modern agricultural crop is grown in soil that has been enriched with ammonia or its compounds. To meet the demands for fertilizers and other applications, about 30 billion pounds of ammonia is manufactured every year in the United States by using this reaction:

$$3 \text{ H}_2(g) + \text{N}_2(g) \rightarrow 2 \text{ NH}_3(g)$$

A plant that manufactures ammonia has an order for 5.75 tons. How many tons of hydrogen and nitrogen should be bought as raw materials to fill the order?

Liquid ammonia being injected into soil. *(Grant Heilman/Grant Heilman Photography, Inc.)*

12-4 When a chemical reaction is carried out, the reactant that is used up first determines how much product is formed

A balanced chemical equation shows ideal quantities of reactants and products. For example, the equation

$$2 \text{ H}_2 + \text{O}_2 \rightarrow 2 \text{ H}_2\text{O}$$

shows that exactly 2 moles of hydrogen will react with exactly 1 mole of oxygen to form exactly 2 moles of water.

In an actual reaction the quantities of reactants used may not have the ideal relationship shown in the balanced equation. Suppose, for example, that the preceding reaction were carried out with 3.59 mole of hydrogen and 1.15 mole of oxygen. How much water would be formed?

The reasoning that's used to solve this kind of problem can be seen by comparing a chemical reaction with a manufacturing process, as shown in the following Example.

Example 12.7

Suppose you operate a factory that produces bicycles by attaching wheels to frames; the process can be described by this equation:

$$2 \text{ wheels} + 1 \text{ frame} \rightarrow 1 \text{ bicycle}$$

Your inventory contains 175 frames and 280 wheels. How many bicycles can you make, and what will be left in your inventory when you've made as many bicycles as you can?

Solution

Outline the solution to this problem by organizing its parts in a table:

	Have	Need	Left Over	Bicycles Produced
Wheels	280			
Frames	175			

Calculate the number of wheels you need for the number of frames you have, and vice versa:

frames → wheels

$$\left(\frac{175 \cancel{\text{frames}}}{}\right)\left(\frac{2 \text{ wheels}}{1 \cancel{\text{frame}}}\right) = 350 \text{ wheels}$$

You would need 350 wheels to match the 175 frames you have.

wheels → frames

$$\left(\frac{280 \cancel{\text{wheels}}}{}\right)\left(\frac{1 \text{ frame}}{2 \cancel{\text{wheels}}}\right) = 140 \text{ frames}$$

You would need 140 frames to match the 280 wheels you have.

	Have	Need	Left Over	Bicycles Produced
Wheels	280	350		
Frames	175	140		

Since you have 175 frames and need only 140, you'll have 175 − 140 = 35 frames left over. You have 280 wheels and need 350, so you'll use up all of the wheels and have none left over.

	Have	Need	Left Over	Bicycles Produced
Wheels	280	350	0	
Frames	175	140	35	

As you make bicycles, you'll run out of wheels before you run out of frames: When your inventory of wheels has run down to 0, you'll still have 35 frames. The number of bicycles you can make will be limited by the number of wheels in your inventory. Calculate the number of bicycles you can make from 280 wheels:

wheels → bicycles

$$\left(\frac{280 \cancel{\text{wheels}}}{}\right)\left(\frac{1 \text{ bicycle}}{2 \cancel{\text{wheels}}}\right) = 140 \text{ bicycles}$$

	Have	Need	Left Over	Bicycles Produced
Wheels	280	350	0	140
Frames	175	140	35	

The completed table shows this information: 280 wheels and 175 frames can be used to make 140 bicycles; there will be 0 wheels and 35 frames left over.

(Tim Bieber/The Image Bank)

The same reasoning can be used to solve chemical problems, as shown in the following Examples.

Example 12.8

The reaction described by

$$2\ H_2 + O_2 \rightarrow 2\ H_2O$$

was carried out, using 3.59 mol of H_2 and 1.15 mol of O_2. How many moles of water were produced, and how many moles of hydrogen and oxygen were left over?

Solution

	Have	Need	Left Over	H_2O Produced
H_2	3.59 mol			
O_2	1.15 mol			

Calculate the moles of O_2 you need for the moles of H_2 you have, and vice versa:

mol H_2 $\rightarrow$ mol O_2

$$\left(\frac{3.59\ \cancel{\text{mol H}_2}}{}\right)\left(\frac{1\ \text{mol O}_2}{2\ \cancel{\text{mol H}_2}}\right) = 1.80\ \text{mol O}_2$$

mol O_2 $\rightarrow$ mol H_2

$$\left(\frac{1.15\ \cancel{\text{mol O}_2}}{}\right)\left(\frac{2\ \text{mol H}_2}{1\ \cancel{\text{mol O}_2}}\right) = 2.30\ \text{mol H}_2$$

	Have	Need	Left Over	H_2O Produced
H_2	3.59 mol	2.30 mol		
O_2	1.15 mol	1.80 mol		

We have 3.59 mol of H_2 and need 2.30 mol, so 3.59 − 2.30 = 1.29 mol will be left over; we have 1.15 mol of O_2 and need 1.80 mol, so there will be none left over.

	Have	Need	Left Over	H_2O Produced
H_2	3.59 mol	2.30 mol	1.29 mol	
O_2	1.15 mol	1.80 mol	0 mol	

We'll run out of O_2 before we run out of H_2, so the amount of water we can produce will be determined by the amount of O_2 we have. Calculate the amount of water that can be produced from 1.15 mol of O_2:

$$\text{mol } O_2 \rightarrow \text{mol } H_2O$$

$$\left(\frac{1.15 \cancel{\text{mol } O_2}}{}\right)\left(\frac{2 \text{ mol } H_2O}{1 \cancel{\text{mol } O_2}}\right) = 2.30 \text{ mol } H_2O$$

	Have	Need	Left Over	H_2O Produced
H_2	3.59 mol	2.30 mol	1.29 mol	
O_2	1.15 mol	1.80 mol	0 mol	2.30 mol

In the reaction described in Example 12.8, oxygen is called the **limiting reactant** because it limits how much of the product will be formed. The reaction stops when the oxygen is used up, even though there is hydrogen left over. Another limiting-reactant problem is shown in the following Example.

Example 12.9

How many grams of ammonia can be made from 25.2 g of N_2 and 20.6 g of H_2, according to the following equation?

$$N_2 + 3\ H_2 \rightarrow 2\ NH_3$$

Solution

	Have	Need	Left Over	NH_3 Produced
N_2	25.2 g			
H_2	20.6 g			

$$\text{g } N_2 \rightarrow \text{mol } N_2 \rightarrow \text{mol } H_2 \rightarrow \text{g } H_2$$

$$\left(\frac{25.2 \cancel{\text{g } N_2}}{}\right)\left(\frac{1 \cancel{\text{mol } N_2}}{28.0 \cancel{\text{g } N_2}}\right)\left(\frac{3 \cancel{\text{mol } H_2}}{1 \cancel{\text{mol } N_2}}\right)\left(\frac{2.02 \text{ g } H_2}{1 \cancel{\text{mol } H_2}}\right) = 5.45 \text{ g } H_2$$

$$\text{g } H_2 \rightarrow \text{mol } H_2 \rightarrow \text{mol } N_2 \rightarrow \text{g } N_2$$

$$\left(\frac{20.6 \cancel{\text{g } H_2}}{}\right)\left(\frac{1 \cancel{\text{mol } H_2}}{2.02 \cancel{\text{g } H_2}}\right)\left(\frac{1 \cancel{\text{mol } N_2}}{3 \cancel{\text{mol } H_2}}\right)\left(\frac{28.0 \text{ g } N_2}{1 \cancel{\text{mol } N_2}}\right) = 95.2 \text{ g } N_2$$

	Have	Need	Left Over	NH_3 Produced
N_2	25.2 g	95.2 g	0 g	
H_2	20.6 g	5.45 g	15.2 g	

$$\text{g } N_2 \rightarrow \text{mol } N_2 \rightarrow \text{mol } NH_3 \rightarrow \text{g } NH_3$$

$$\left(\frac{25.2 \cancel{\text{g } N_2}}{}\right)\left(\frac{1 \cancel{\text{mol } N_2}}{28.0 \cancel{\text{g } N_2}}\right)\left(\frac{2 \cancel{\text{mol } NH_3}}{1 \cancel{\text{mol } N_2}}\right)\left(\frac{17.0 \text{ g } NH_3}{1 \cancel{\text{mol } NH_3}}\right) = 30.6 \text{ g } NH_3$$

	Have	Need	Left Over	NH_3 Produced
N_2	25.2 g	95.2 g	0 g	30.6 g
H_2	20.6 g	5.45 g	15.2 g	

Acetylene gas, C_2H_2, is used in the oxyacetylene torch for welding. The combustion reaction is

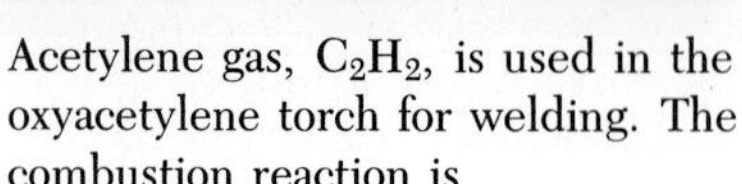

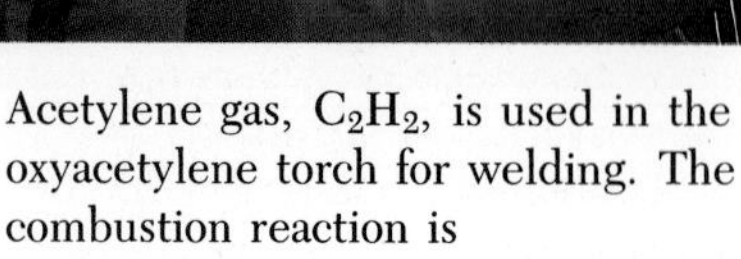

$$2\ C_2H_2(g) + 5\ O_2(g) \rightarrow 4\ CO_2(g) + 2\ H_2O(g).$$

(Randy Duchaine/The Stock Market)

Exercise 12.6

The combustion of acetylene gas is shown by the equation

$$2\ C_2H_2(g) + 5\ O_2(g) \rightarrow 4\ CO_2(g) + 2\ H_2O(g)$$

If this reaction were carried out using 10.0 g of acetylene and 5.00 g of oxygen, which would be the limiting reactant? How much of the nonlimiting reactant would be left over?

Exercise 12.7

A reaction was carried out using 2.00 mg of silver and 3.00 mg of fluorine to produce silver fluoride, according to this equation:

$$2\ Ag(s) + F_2(g) \rightarrow 2\ AgF(s)$$

How much of the product was formed?

Inside Chemistry — What was phlogiston?

In the first half of the eighteenth century, chemists came to believe that the process of combustion could be explained by assuming that burning objects released into the air a substance called phlogiston. The phlogiston theory successfully explained the following important regularities in the behavior of burning objects:

Observation	Explanation, According to the Phlogiston Theory
Some objects burn better than others.	An object burns well in proportion to the amount of phlogiston it contains. Wood and coal, for example, contain large amounts of phlogiston and burn well, while iron and rock contain no phlogiston and do not burn.
When an object burns, it leaves a residue—a piece of coal, for example, leaves ashes.	A piece of coal consists of ashes plus phlogiston. When coal burns, the phlogiston leaves and the ashes remain.
When an object burns, its residue has a smaller mass than the original object; for example, the mass of a piece of coal is larger than the mass of its ashes.	The decrease in mass occurs because the piece of coal loses its phlogiston.

What happens when coal burns? The phlogiston theory gave one explanation and the modern theory of combustion gives another. *(Barry L. Runk from Grant Heilman Photography)*

Toward the end of the eighteenth century, the phlogiston theory was increasingly questioned and then replaced by the oxygen theory of combustion which we now believe. The replacement came about because the oxygen theory gave a better explanation of the mass relationships in combustion reactions.

The essential differences between the phlogiston theory and the oxygen theory can be seen by comparing their explanation of the combustion of a piece of coal. When a piece of coal burns, gas is given off, ashes are left, and heat is produced:

$$\text{coal} \rightarrow \text{ashes} + \text{gas} + \text{heat}$$

According to the phlogiston theory, a piece of coal consists of noncombustible minerals, which form ashes, and phlogiston. When the coal burns, phlogiston is released:

coal → ashes + phlogiston(g) + heat

(coal = minerals + phlogiston; ashes = minerals)

According to the oxygen theory, a piece of coal consists of noncombustible minerals, which form ashes, and compounds of carbon and hydrogen. When the coal burns, oxygen from the air combines with the compounds of carbon and hydrogen to form carbon dioxide gas and water vapor:

coal + O_2 → ashes + $CO_2(g)$ + $H_2O(g)$ + heat

(coal = minerals + compounds of C and H; ashes = minerals)

Scientists can make experimental tests to decide between the phlogiston and oxygen theories because these theories predict different mass changes during combustion: According to the phlogiston theory, the mass of the coal should equal the mass of the ashes plus the mass of the gas (phlogiston) released; according to the oxygen theory, the mass of the coal should be less than the mass of the ashes plus the mass of the gases (CO_2 and H_2O) released. Experiments confirm the prediction of the oxygen theory.

Chapter Summary: Calculations Based on Chemical Equations: Stoichiometry

Section	Subject	Summary	Check When Learned
	Stoichiometry	Calculations of masses of reactants and products in a chemical reaction by dimensional analysis.	☐
12-1	Equations for mole relationships in $3\,H_2 + N_2 \rightarrow 2\,NH_3$	3 mol H_2 = 1 mol N_2 3 mol H_2 = 2 mol NH_3 1 mol N_2 = 2 mol NH_3	☐
12-2	Steps in stoichiometry	(1) Translate given units into moles. (2) Convert moles of given reactant or product to moles of reactant or product to be found. (3) Translate moles into units required for answer.	☐
12-4	Limiting reactant	Reactant from which least amount of product can be formed; reactant that is used up first.	☐
12-4	Method for solving a limiting-reactant problem for $A + B \rightarrow C + D$	Fill in the appropriate quantities in the columns in this table, from left to right: **Have** **Need** **Left Over** **C or D Produced** A B	☐

Problems

Assume you can use the Periodic Table at the front of the book unless you're directed otherwise. As you use atomic masses and molecular masses in these problems, round them off to three significant digits. Answers to odd-numbered problems are in Appendix 1.

Reading Balanced Chemical Equations in Moles (Section 12-1)

1. Below the following equation, write the moles of reactants and products it describes.
$$H_2(g) + Cl_2(g) \rightarrow 2\ HCl(g)$$

2. Below the following equation, write the moles of reactants and products it describes.
$$S(s) + O_2(g) \rightarrow SO_2(g)$$

3. Below the following equation, write the moles of reactants and products it describes.
$$2\ SO_2(g) + O_2(g) \rightarrow 2\ SO_3(g)$$

4. Below the following equation, write the moles of reactants and products it describes.
$$2\ Na(s) + H_2(g) \rightarrow 2\ NaH(s)$$

5. Calcium reacts with bromine to form calcium bromide. Write the balanced equation for this reaction, using condensed and molecular formulas; assume that the reaction occurs under ordinary room conditions, and show the state of each reactant and product. Below the equation, write the moles of reactants and products it describes.

6. Potassium reacts with chlorine to form potassium chloride. Write the balanced equation for this reaction, using condensed and molecular formulas; assume that the reaction occurs under ordinary room conditions, and show the state of each reactant and product. Below the equation, write the moles of reactants and products it describes.

7. Aluminum reacts with oxygen to form aluminum oxide. Write the balanced equation for this reaction, using condensed and molecular formulas; assume that the reaction occurs under ordinary room conditions, and show the state of each reactant and product. Below the equation, write the moles of reactants and products it describes.

8. Iron reacts with fluorine to form iron(III) fluoride. Write the balanced equation for this reaction, using condensed and molecular formulas; assume that the reaction occurs under ordinary room conditions, and show the state of each reactant and product. Below the equation, write the moles of reactants and products it describes.

9. Write the mathematical equation for the relationship between moles of HgO and moles of O_2 shown in this equation:
$$2\ HgO(s) \rightarrow 2\ Hg(\ell) + O_2(g)$$

10. Write the mathematical equation for the relationship between moles of O_2 and moles of CO_2 shown in this equation:
$$2\ C_2H_2(g) + 5\ O_2(g) \rightarrow 4\ CO_2(g) + 2\ H_2O(g)$$

11. Write the mathematical equation for the relationship between moles of Br_2 and moles of $C_2H_4Br_2$ shown in this equation:
$$C_2H_4 + Br_2 \rightarrow C_2H_4Br_2$$

12. Write the mathematical equation for the relationship between moles of Al and moles of Fe_2O_3 shown in this equation:
$$Fe_2O_3 + 2\ Al \rightarrow 2\ Fe + Al_2O_3$$

13. Write the mathematical equation for the relationship between moles of O_2 and moles of BaO shown in this equation:
$$2\ Ba + O_2 \rightarrow 2\ BaO$$

14. Write the mathematical equation for the relationship between moles of F_2 and moles of AlF_3 shown in this equation:
$$2\ Al + 3\ F_2 \rightarrow 2\ AlF_3$$

Calculating Moles of Reactants and Products (Section 12-2)

15. According to the equation
$$2\ SO_2(g) + O_2(g) \rightarrow 2\ SO_3(g)$$
how many moles of oxygen will react with 4.00 mol of sulfur dioxide?

16. According to the equation
$$4\ K(s) + O_2(g) \rightarrow 2\ K_2O(s)$$
how many moles of potassium will react with 3.00 mol of oxygen?

17. According to the equation
$$Zn(s) + Cl_2(g) \rightarrow ZnCl_2(s)$$
how many moles of zinc chloride will be formed from 0.500 mol of zinc?

18. According to the equation
$$CaCO_3(s) \rightarrow CaO(s) + CO_2(g)$$
how many moles of calcium carbonate will be needed to produce 0.250 mol of carbon dioxide?

19. According to the equation
$$NH_3(g) + 3\ Cl_2(g) \rightarrow NCl_3(g) + 3\ HCl(g)$$
how many moles of ammonia will react with 1.25 mol of chlorine?

20. According to the equation
$$2\ NO_2(g) \rightarrow N_2O_4(g)$$
how many moles of dinitrogen tetroxide will be formed from 0.562 mol of nitrogen dioxide?

21. According to the equation
$$2\ Hg(\ell) + O_2(g) \rightarrow 2\ HgO(s)$$
how many moles of mercury(II) oxide can be formed from 4.17 mol of mercury?

22. According to the equation
$$4\ Al(s) + 3\ O_2(g) \rightarrow 2\ Al_2O_3(s)$$
how many moles of aluminum oxide can be formed from 1.65 mol of oxygen?

Stoichiometry (Sections 12-1 through 12-3)

23. Calculate the mass of SO_3 that will be produced from 3.00 g of S, according to the following equation:
$$2\ S(s) + 3\ O_2(g) \rightarrow 2\ SO_3(g)$$

24. Calculate the mass of Na_2O that will be produced from 4.00 g of Na, according to the following equation:
$$4\ Na(s) + 2\ O_2(g) \rightarrow 2\ Na_2O(s)$$

25. Calculate the mass of BaO that will be produced from 3.00 g of Ba, according to the following equation:
$$2\ Ba(s) + O_2(g) \rightarrow 2\ BaO(s)$$

26. Calculate the mass of NaBr that will be produced from 1.00 g of Na, according to the following equation:
$$2\ Na(s) + Br_2(\ell) \rightarrow 2\ NaBr(s)$$

27. Calculate the mass of Ca that will react with 2.50 g of F_2, according to the following equation:
$$Ca(s) + F_2(g) \rightarrow CaF_2(s)$$

28. Calculate the mass of Al_2S_3 that will be produced from 9.65 g of Al, according to the following equation:
$$2\ Al(s) + 3\ S(s) \rightarrow Al_2S_3(s)$$

29. Calculate the mass of $KClO_3$ that will be required to produce 8.75 g of O_2, according to the following equation:
$$2\ KClO_3(s) \rightarrow 2\ KCl(s) + 3\ O_2(g)$$

30. Calculate the mass of HI that will be produced in a reaction that produces 4.93 g of NI_3, according to the following equation:
$$NH_3(g) + 3\ I_2(s) \rightarrow NI_3(s) + 3\ HI(g)$$

31. Calculate the mass of CaO that will be formed in a reaction that produces 12.5 g of CO_2, according to the following equation:
$$CaCO_3(s) \rightarrow CaO(s) + CO_2(g)$$

32. Calculate the mass of LiH that will be formed from 1.00 g of H_2, according to the following equation:
$$2\ Li(s) + H_2(g) \rightarrow 2\ LiH(s)$$

33. How many grams of chlorine will be needed to react with 4.04 g of potassium to produce potassium chloride?

34. How many grams of oxygen will be needed to react with 3.00 g of aluminum to produce aluminum oxide?

35. How many grams of calcium will be needed to react with 2.00 g of fluorine?

36. How many grams of zinc will be needed to react with 6.00 g of oxygen?

37. How many grams of Na_2SO_4 will be produced from 2.35 kg of NaOH, according to the following equation?
$$2\ NaOH + H_2SO_4 \rightarrow Na_2SO_4 + 2\ H_2O$$

38. How many milligrams of $Pb(NO_3)_2$ will react with 1.74 cg of K_2SO_4, according to the following equation?
$$K_2SO_4 + Pb(NO_3)_2 \rightarrow PbSO_4 + 2\ KNO_3$$

39. How many milligrams of ZnO will be formed from 1.55 mg of Zn, according to the following equation?
$$2\ Zn(s) + O_2(g) \rightarrow 2\ ZnO(s)$$

40. How many milligrams of LiF will be formed from 0.512 mg of F_2, according to the following equation?
$$2\ Li(s) + F_2(g) \rightarrow 2\ LiF(s)$$

41. How many milligrams of PbO will be required to react with 0.472 g of CO_2, according to the following equation?
$$PbO(s) + CO_2(g) \rightarrow PbCO_3(s)$$

42. How many grams of F_2 will be required to react with 1.29 kg of CH_4, according to the following equation?
$$CH_4(g) + 4\ F_2(g) \rightarrow CF_4(g) + 4\ HF(g)$$

43. How many grams of Al_2O_3 will be produced from 1.10 kg of Fe_2O_3, according to the following equation?
$$Fe_2O_3(s) + 2\ Al(s) \rightarrow 2\ Fe(s) + Al_2O_3(s)$$

44. How many milligrams of Cl_2 will be needed to react with 0.748 g of Al, according to the following equation?
$$2\ Al(s) + 3\ Cl_2(g) \rightarrow 2\ AlCl_3(s)$$

45. How many grams of HBr will be produced from 2.00 lb of Br_2, according to the following equation?
$$H_2(g) + Br_2(\ell) \rightarrow 2\ HBr(g)$$

46. How many pounds of MgF_2 will be produced from 232 g of Mg, according to the following equation?
$$Mg(s) + F_2(g) \rightarrow MgF_2(s)$$

47. Carbon monoxide reacts with oxygen to form carbon dioxide. How many grams of oxygen will react with 6.50 g of carbon monoxide?

48. Chromium reacts with oxygen to form chromium(III) oxide. How many grams of chromium will react with 2.75 g of oxygen?

49. Ethylene, C_2H_4, reacts with oxygen to form carbon dioxide and water. How many molecules of carbon dioxide will be formed from 0.225 ton of ethylene?

50. Butane, C_4H_{10}, reacts with oxygen to form carbon dioxide and water. How many molecules of oxygen will react with 22.0 grams of butane?

51. Chlorine reacts with methane to produce carbon tetrachloride and hydrogen chloride. How many grams of methane will be used in a reaction that produces 1.00×10^{23} molecules of hydrogen chloride?

52. Fluorine reacts with ethane, C_2H_6, to produce hexafluoroethane, C_2F_6, and hydrogen fluoride. How many grams of fluorine will be needed to react with 1.00×10^{24} molecules of ethane?

Limiting Reactant (Section 12-4)

53. The following reaction was carried out, using 2.00 mol of Be and 2.00 mol of O_2. How many moles of BeO were produced?

$$2\ Be(s) + O_2(g) \rightarrow 2\ BeO(s)$$

54. The following reaction was carried out, using 2.50 mol of Cu and 3.50 mol of S. How many moles of CuS were produced?

$$Cu(s) + S(s) \rightarrow CuS(s)$$

55. The following reaction was carried out, using 1.00 mol of Al and 1.00 mol of Cl_2. How many moles of $AlCl_3$ were produced?

$$2\ Al(s) + 3\ Cl_2(g) \rightarrow 2\ AlCl_3(s)$$

56. The following reaction was carried out, using 3.75 mol of Fe and 1.50 mol of O_2. How many moles of Fe_2O_3 were produced?

$$4\ Fe(s) + 3\ O_2(g) \rightarrow 2\ Fe_2O_3(s)$$

57. The following reaction was carried out, using 5.00 g of N_2 and 5.00 g of H_2. How many grams of NH_3 were produced?

$$N_2 + 3\ H_2 \rightarrow 2\ NH_3$$

58. The following reaction was carried out, using 3.00 g of H_2 and 3.00 g of O_2. How many grams of H_2O were produced?

$$2\ H_2 + O_2 \rightarrow 2\ H_2O$$

59. The following reaction was carried out, using 5.00 g of Se and 5.00 g of O_2. How many grams of SeO_2 were produced?

$$Se(s) + O_2(g) \rightarrow SeO_2(s)$$

60. The following reaction was carried out, using 2.00 g of CO and 3.00 g of O_2. How many grams of CO_2 were produced?

$$2\ CO(g) + O_2(g) \rightarrow 2\ CO_2(g)$$

61. The following reaction was carried out, using 9.57 g of P_4 and 3.43 g of O_2. How many grams of P_2O_5 were produced?

$$P_4(s) + 5\ O_2(g) \rightarrow 2\ P_2O_5(s)$$

62. The following reaction was carried out, using 1.75 g of O_2 and 2.85 g of F_2. How many grams of OF_2 were produced?

$$O_2(g) + 2\ F_2(g) \rightarrow 2\ OF_2(g)$$

63. Magnesium reacts with nitrogen to form magnesium nitride. How many grams of magnesium nitride will be formed from 0.0443 kg of magnesium and 7.05×10^4 mg of nitrogen?

64. Arsenic reacts with chlorine to form arsenic trichloride. How many milligrams of arsenic trichloride will be formed from 1.00 g of arsenic and 1.00 g of chlorine?

65. In a reaction of 2.00 g of calcium with 3.00 g of oxygen, what mass of product will be formed?

66. In a reaction of 1.00 g of lithium with 2.00 g of bromine, what mass of product will be formed?

13-1
The quantity of gas in a sample can be described by stating its mass or by stating its volume, temperature, and pressure.

13-2
A gas sample can be described as a collection of tiny, rapidly moving particles with large regions of empty space between them.

13-3
Three mathematical laws—Avogadro's Law, Boyle's Law, and Charles's Law—describe the relationships between the volume of a gas and its mass, pressure, and temperature.

13-4
Gas-law calculations can be based on memorized equations or on reasoning.

13-5
Avogadro's Law, Boyle's Law, and Charles's Law can be combined into one law.

13-6
The most general and most useful gas law is the ideal-gas equation, PV = nRT.

13-7
Gas volumes can be used in stoichiometry.

Inside Chemistry:
Why is it important to prevent the loss of ozone from the atmosphere?

Chapter Summary

Problems

(Phillip Wallick/The Stock Market)

13 Gases

We live at the bottom of an ocean of gases, called the atmosphere, and we depend on those gases for our survival. But, because most gases are invisible, we're usually much less aware of them in our surroundings than we are of liquids and solids.

Gases are difficult for a chemist to work with: They leak out of any container that isn't tightly sealed, most of them can't be seen, and many of them are poisonous or explosive. Special equipment is needed to measure quantities of gases and to move them from one container to another. Because gases are difficult to work with, early experimenters in chemistry preferred to work with liquids and solids.

When chemists carried out the first systematic studies of gases, in the seventeenth and eighteenth centuries, they found that the behavior of gases is simpler than that of liquids and solids, and easier to understand. In a liquid or a solid, the atoms, molecules, or ions are packed close together, and this crowding allows the particles to interact with one another in complicated ways.◆ In a gas, the particles are far apart, and they interact with one another only very slightly; as a result, gases behave more simply than liquids or solids. It was the study of the simple properties of gases that led John Dalton to the creation of the atomic theory.

◆These interactions are described in Chapter 14.

Gases are important in our lives, and they provide simple models of basic chemical theories. For these reasons, the study of gases is an important part of any introduction to the fundamentals of chemistry.

13-1 The quantity of gas in a sample can be described by stating its mass or by stating its volume, temperature, and pressure

Imagine that we have a sample of oxygen gas contained in a cylinder by a piston, as shown in Figure 13.1.

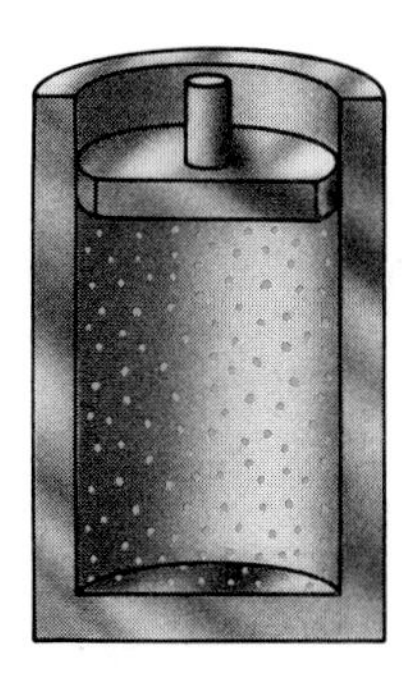

Figure 13.1
Sample of oxygen contained in a cylinder by a piston.

One way to describe the quantity of gas in the sample is to state its mass. For the quantities of gases used in laboratory work, the units usually used are grams or milligrams. Let's assume that the mass of oxygen in our sample is 50.0 mg.

Another way to describe the quantity of gas in the sample is to state its volume. But, as you know from everyday experience, the volume of a sample of gas depends on its temperature and pressure. A gas expands if it's heated and contracts if it's cooled. If the pressure on a gas is increased (for example, by pushing down on the piston in the system shown in Figure 13.1) the volume of the gas decreases, and if the pressure is decreased (by pulling up on the piston) the volume of the gas increases. To express the quantity of gas in a sample by specifying its volume, we also have to specify the temperature and pressure.

The volume units used for laboratory work are usually milliliters or liters, and the temperature units are usually °C or K; you're familiar with these units from Chapter 3. We'll assume that our oxygen sample has a volume of 42.4 mL and a temperature of 25°C.

The units most commonly used to express gas pressures in chemical work are the atmosphere, the torr, and the kilopascal.

The use of the **atmosphere** as a unit of pressure is based on the fact that the gases in the earth's atmosphere exert pressure. Because the density of the atmosphere varies with elevation, atmospheric pressure also varies with elevation. For example, the atmosphere is less dense at the top of a mountain than at its base, so the atmospheric pressure is less. A pressure of one atmosphere is the pressure exerted by the earth's atmosphere at sea level.

The pressure unit **torr** is based on the use of a mercury barometer to measure atmospheric pressure. Figure 13.2 is a picture of a mercury barometer. A glass tube, about 30 inches long and closed at one end, is filled with mercury, and the tube is then inverted so that its open end is under the surface of mercury in an open container.◆ The mercury inside the tube will drop a short distance. The pressure exerted downward by the atmosphere on the surface of the mercury in the open container will be balanced by the pressure exerted downward by the column of mercury in the tube. At sea level, the height (h) of the mercury in the tube will be about 760 millimeters. If the atmospheric pressure becomes less, the level of the mercury in the tube will fall below 760 mm, and if the atmospheric pressure becomes greater, the level will rise above 760 mm.

◆Mercury is poisonous. This abbreviated description of the steps in making a barometer doesn't include the precautions necessary to handle mercury safely.

The pressure unit torr is defined as one millimeter of mercury,◆ and

◆The torr is named for Evangelista Torricelli (1608–1647), Italian physicist. The torr is sometimes expressed as **mm of Hg**.

1 atmosphere (atm) = 760 torr

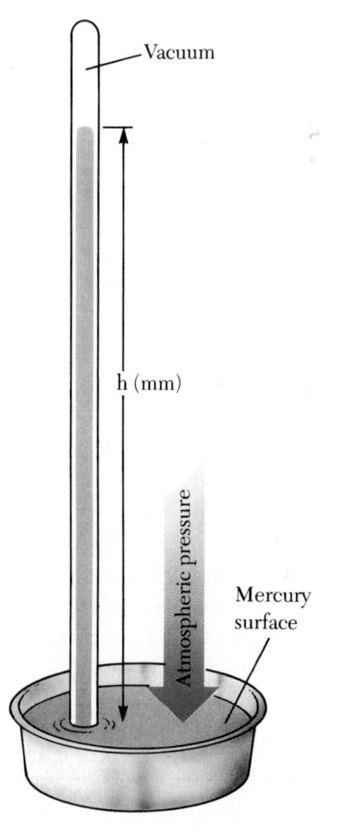

Figure 13.2
Mercury barometer. The earth's atmosphere exerts pressure downward on the surface of the mercury in the open container, and this pressure is balanced by the pressure exerted downward by the mass of the column of mercury in the glass tube. The atmospheric pressure is read as the height (h) of the mercury column above the surface of the mercury in the open container; at sea level, the height will be about 760 mm.

The pressure units atmosphere and torr have been used for many years and are still common. More recently the pascal (Pa) has been defined as the SI unit for pressure.◆ The relationship between the atmosphere and the pascal is

1 atmosphere (atm) = 101 kilopascals (kPa)

◆The pascal is named for Blaise Pascal (1623–1662), French philospher and mathematician.

We'll assume that the quantity of oxygen in our sample can be described either as 50.0 mg or as a volume (V) of 42.4 mL at a temperature (T) of 25°C and a pressure (P) of 684 torr, as shown in Figure 13.3.

Exercise 13.1◆

Convert the mass, volume, temperature, and pressure in the preceding oxygen sample to moles, liters, kelvins, and atmospheres.◆

◆Answers to Exercises are in Appendix 1.

◆Remember that oxygen is one of the diatomic gases (H_2, N_2, O_2, F_2, Cl_2) so 1 mol O_2 = 32.0 g O_2.

13-2 A gas sample can be described as a collection of tiny, rapidly moving particles with large regions of empty space between them

Imagine that two or three drops of perfume are put in a saucer in a room, with the doors and windows closed. After a few minutes, the liquid perfume will have evaporated to form a gas, and the scent of the perfume will be detectable in every part of the room.

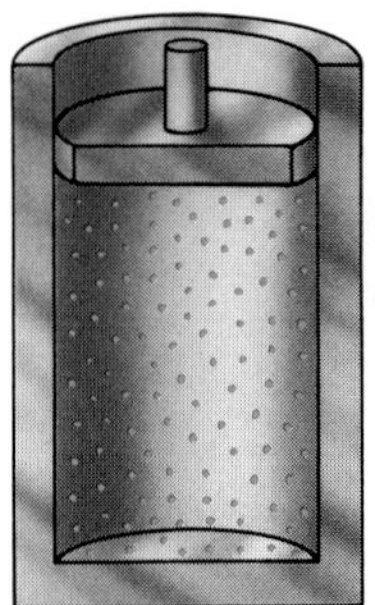

Figure 13.3
Sample of oxygen contained in a cylinder by a piston.

This simple experiment shows that the particles of the perfume in the gas state must be very far apart and in rapid, random motion. The particles of perfume gas must be moving rapidly and in all directions, because it takes only a few minutes for them to reach every corner of the room. The particles in the gas state must be relatively far apart, because the same particles that were originally confined to two or three drops of liquid have spread as a gas to occupy the entire volume of the room. Because the particles of perfume gas are in rapid, random motion, we can also conclude that they must constantly be colliding with one another, with all of the objects in the room, and with the walls, ceiling, and floor of the room itself.

These conclusions about perfume in the gas state can be extended to give a general description of all gases in terms of the behavior of their particles. This description, summarized in the following three statements, is called **the kinetic theory of gases.**◆

◆*Kinetic* is from a Greek word that means *moving*.

1. A gas sample consists of a large number of particles (atoms or molecules). The total volume of the particles is negligible compared with the total volume of the sample.◆
2. The particles of a gas are in rapid, random motion, and they constantly collide with one another and with the walls of their container. The collisions are assumed to be perfectly elastic.◆
3. The number of particles in a gas sample and their motion are responsible for the volume, temperature, and pressure of the sample.

◆In our example, the volume of the particles of perfume in the liquid state, two or three drops, is negligibly small compared with the volume occupied by the perfume in the gas state, the whole volume of the room.

◆For simplicity it's assumed that, in these collisions, the particles rebound without losing any of their momentum, like ideal billiard balls on a frictionless table. More advanced theories recognize that these collisions aren't completely elastic.

To illustrate the kinetic theory of gases, we can use the sample of oxygen gas shown in Figure 13.3. We can calculate the number of particles—in this case, oxygen molecules—in the sample. The sample contains 50.0 mg of O_2:

$50.0 \text{ mg } O_2 = ? \; O_2 \text{ molecules}$

$$\text{mg } O_2 \rightarrow \text{g } O_2 \rightarrow \text{mol } O_2 \rightarrow O_2 \text{ molecules}$$

$$\left(\frac{50.0 \, \cancel{\text{mg } O_2}}{}\right)\left(\frac{1 \, \cancel{\text{g } O_2}}{1 \times 10^3 \, \cancel{\text{mg } O_2}}\right)\left(\frac{1 \, \cancel{\text{mol } O_2}}{32.0 \, \cancel{\text{g } O_2}}\right)\left(\frac{6.02 \times 10^{23} \, O_2 \text{ molecules}}{1 \, \cancel{\text{mol } O_2}}\right)$$

$$= 9.41 \times 10^{20} \, O_2 \text{ molecules}$$

As stated in the kinetic theory, our sample consists of a large number of particles.

As shown in Figure 13.3, the volume of our sample of oxygen gas is 42.4 mL. How much of this volume is taken up by the oxygen molecules themselves, and how much of it is the empty space between the oxygen molecules? To answer this question, we can assume that when oxygen is in the *liquid* state, the oxygen molecules are packed next to one another, with no space between them.◆ The density of *liquid* oxygen is 1.14 g/mL, so the volume of 50.0 mg of *liquid* oxygen would be

◆This assumption isn't strictly correct, but it's accurate enough to give useful numbers for our purposes.

$$\text{mg } O_2 \rightarrow \text{g } O_2 \rightarrow \text{mL } O_2$$

$$\left(\frac{50.0 \, \cancel{\text{mg } O_2}}{}\right)\left(\frac{1 \, \cancel{\text{g } O_2}}{1 \times 10^3 \, \cancel{\text{mg } O_2}}\right)\left(\frac{1 \text{ mL } O_2}{1.14 \, \cancel{\text{g } O_2}}\right) = 0.0439 \text{ mL } O_2$$

The percentage of the volume of our gas sample that is taken up by the O_2 molecules themselves is

$$\frac{\text{volume of } O_2 \text{ molecules}}{\text{volume of gas sample}} \times 100 = \frac{\text{volume of liquid oxygen}}{\text{volume of gas sample}} \times 100$$

$$= \frac{0.0439 \, \cancel{\text{mL}}}{42.4 \, \cancel{\text{mL}}} \times 100 = 0.104\%$$

In our sample of oxygen gas only 0.104% of the volume is occupied by the O_2 molecules, and the rest is empty space between the molecules. The percentage of empty space is

$$100\% - 0.104\% = 99.9\%$$

In the words of the kinetic theory, the total volume of the gas particles—oxygen molecules—is negligible compared with the total volume of the gas sample.

According to the kinetic theory, the particles in a sample of gas are in rapid motion, so they're constantly colliding with the walls of their container. The bombardment of a surface by gas particles is measured as the pressure exerted by the gas. In our sample of oxygen gas, confined in a cylinder by a piston, many millions of oxygen molecules bombard the inside surface of the cylinder and the face of the piston each second. The total effect of these collisions is the pressure exerted by the oxygen gas, 684 torr.

If more gas molecules are crowded into a container, the pressure will increase, because the number of collisions of gas particles with the walls of the container will increase. This is what happens when you blow up a balloon or inflate a tire.

For example, imagine that you have a partially inflated balloon, as shown in the first drawing in Figure 13.4. The volume of the balloon is determined by a

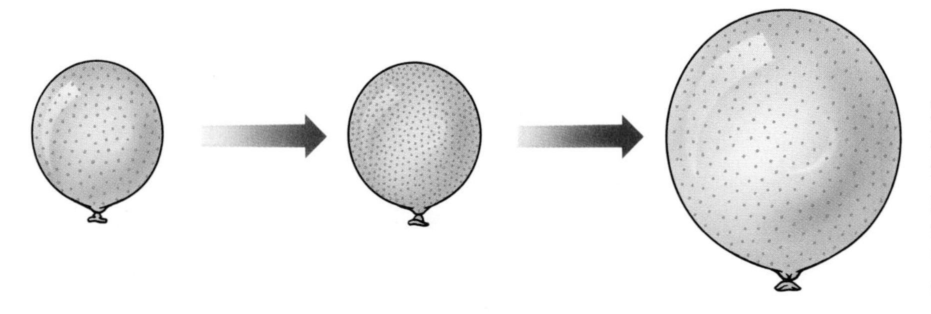

Figure 13.4
Inflating a balloon. As more air is forced into the balloon, the pressure exerted by the air molecules on the inside surface of the rubber membrane increases, and the balloon expands.

balance among three pressures: the pressure exerted by air molecules colliding with the *inside* surface of the rubber membrane, which pushes the membrane outward; the pressure exerted by air molecules bombarding the *outside* surface of the membrane, which pushes it inward; and the pressure exerted by the elasticity of the stretched rubber, which pulls the membrane inward.

If you force more air into the balloon, the increased number of air molecules inside the balloon will cause a larger number of collisions of air molecules with the inside surface of the membrane. The pressure on the *inside* surface will increase, and the balloon will expand. At some new and larger volume, the three pressures—the outward pressure of the air inside the balloon, the inward pressure of the air outside the balloon, and the inward pressure from the stretched rubber—will again be in balance.

The principles of the kinetic theory can also explain why a balloon expands if you warm it. The temperature of a gas is a measure of the average speed of the motions of its particles. As the air in a balloon absorbs heat, the air molecules move more rapidly. As the molecules move more rapidly, they collide more frequently with the inside surface of the rubber membrane, and each collision occurs with a stronger impact. As a result, the inside pressure on the membrane increases, and the balloon expands.

Exercise 13.2

Explain each of the following events in terms of the kinetic theory: (a) A tire is punctured and goes flat. (b) A tire expands as the car is driven to higher altitudes. (At higher altitudes the air is less dense, so fewer air molecules bombard the outside surface of the tire each second.)

13-3 Three mathematical laws—Avogadro's Law, Boyle's Law, and Charles's Law—describe the relationships between the volume of a gas and its mass, pressure, and temperature

Avogadro's Law

Imagine that we have a sample of hydrogen gas contained in a cylinder by a piston, as shown in the first drawing in Figure 13.5. The number of moles (n) of hydrogen is 1.00, its volume (V) is 250 mL, and its temperature (T) is 300 K.

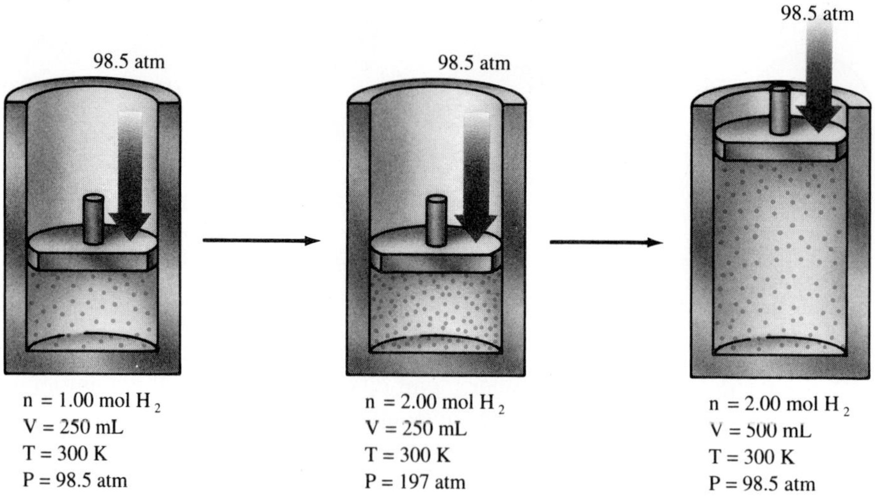

98.5 atm 98.5 atm 98.5 atm

$n = 1.00$ mol H_2
$V = 250$ mL
$T = 300$ K
$P = 98.5$ atm

$n = 2.00$ mol H_2
$V = 250$ mL
$T = 300$ K
$P = 197$ atm

$n = 2.00$ mol H_2
$V = 500$ mL
$T = 300$ K
$P = 98.5$ atm

Figure 13.5
Avogadro's Law. Doubling the number of moles (n) of gas in a sample, at constant temperature (T) and pressure (P), doubles the volume (V) of the sample.

The pressure exerted *downward* on the piston is 98.5 atm, and this pressure is exactly balanced by the pressure (P) exerted *upward* on the piston by the sample of hydrogen gas.

Now imagine that we suddenly double the amount of gas in the sample, from 1.00 mol to 2.00 mol, by injecting another mole of hydrogen into the cylinder. The immediate effect, as shown in the middle drawing, must be to double the pressure on the underside of the piston, since we now have twice as many gas molecules bombarding the face of the piston every second. Because the *downward* pressure on the piston has stayed the same while the *upward* pressure has doubled, the piston will rise. As the piston rises, the volume of the sample will increase, the hydrogen molecules will have more room to move, and there will be fewer collisions per second with the piston face. As a result, the pressure of the gas will decrease as the piston rises. Eventually, as shown in the drawing on the right, the pressure exerted upward by the gas will again be the same as the pressure exerted downward on the piston, and the piston will come to rest at a new position. The new volume will be twice the original volume.

If we double the number of moles of gas in a sample, at constant temperature and pressure, the volume of the sample will double. Similarly, if we reduce the number of moles of gas in a sample by half, the volume of the sample will decrease by half (imagine going from the system shown on the right in Figure 13.5 to the system shown on the left). **Avogadro's Law** states that, at constant temperature and pressure, the volume of a gas sample is directly proportional to the number of moles of gas in the sample.◆ The law can also be stated mathematically:

◆The law is named for Amadeo Avogadro (1776–1856), the Italian physicist and chemist for whom Avogadro's number is also named.

$$\frac{V_1}{V_2} = \frac{n_1}{n_2}$$

V_1 is the volume of a gas sample that contains n_1 moles of gas, and

V_2 is the volume of a sample that contains n_2 moles, at constant T and P

In Figure 13.5,

$$V_1 = 250 \text{ mL} \qquad n_1 = 1.00 \text{ mol}$$

$$V_2 = 500 \text{ mL} \qquad n_2 = 2.00 \text{ mol}$$

We can show that these values are consistent with Avogadro's Law by solving the law for V_2 and substituting the values for V_1, n_1, and n_2 into the equation:

$$\frac{V_1}{V_2} = \frac{n_1}{n_2}$$

$$V_2 = \frac{V_1 n_2}{n_1} = \frac{(250 \text{ mL})(2.00 \cancel{\text{mol}})}{(1.00 \cancel{\text{mol}})} = 500 \text{ mL}$$

The equation for Avogadro's Law contains symbols for four variables. If we have numerical values for any three of them, we can solve for the fourth.

Exercise 13.3

A sample of 0.350 mol of helium gas had a volume of 2.25 L. The temperature and pressure were held constant, more helium was added, and the new volume was 3.75 L. How many moles of helium were added?

Boyle's Law

In the middle of the seventeenth century, the English chemist Robert Boyle (1627–1691) carried out a series of experiments that revealed the relationship between the volume of a gas sample and its pressure. To understand Boyle's discovery, imagine that we begin again, as shown in the drawing in Figure 13.6, with 1.00 mol of hydrogen gas contained in a cylinder by a piston. The volume of the sample is 250 mL, its temperature is 300 K, and its pressure is 98.5 atm. Now imagine that the downward pressure on the piston is doubled. The piston will move downward. As it does so, the molecules of hydrogen will be crowded into a smaller space, and the frequency of collisions of hydrogen molecules with the face of the piston will increase; that is, the pressure of the

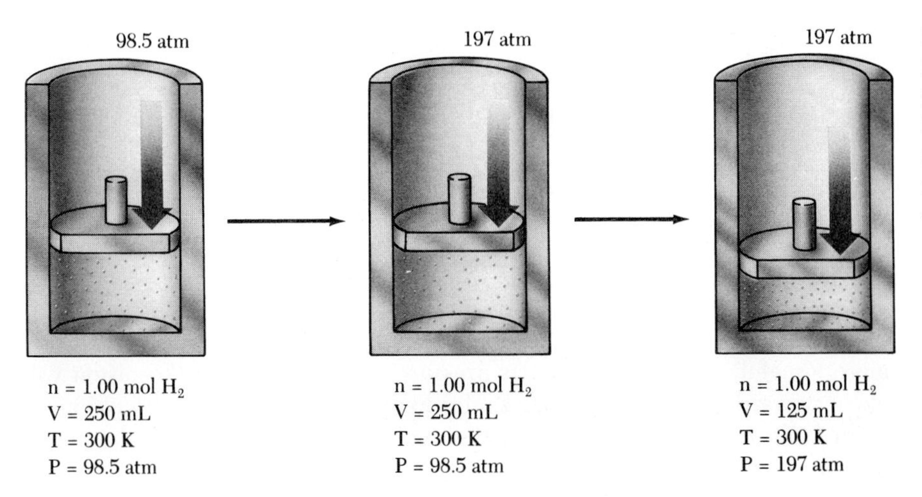

Figure 13.6
Boyle's Law. Doubling the pressure on a gas sample, while holding the number of moles of gas (n) and the temperature (T) constant, reduces the volume (V) of the sample by half.

◆The average speed at which the hydrogen molecules move depends on the temperature. As the piston moves downward, at constant temperature, the same number of hydrogen molecules, moving at the same average speed, is crowded into a smaller space, so the molecules must collide more frequently with the walls of the cylinder and the face of the piston.

hydrogen gas will increase.◆ At some new and lower position of the piston, the increased downward pressure will again be balanced by the upward pressure of the gas, as shown in the right-hand drawing in Figure 13.6.

Boyle's Law states that, at constant temperature, the volume of a gas sample is inversely proportional to its pressure. In mathematical form,

$$\frac{V_1}{V_2} = \frac{P_2}{P_1}$$

V_1 is the volume of a gas sample at pressure P_1, and V_2 is its volume at pressure P_2, at constant T and n

In our example,

$$V_1 = 250 \text{ mL} \qquad P_1 = 98.5 \text{ atm}$$

$$V_2 = 125 \text{ mL} \qquad P_2 = 197 \text{ atm}$$

$$\frac{V_1}{V_2} = \frac{P_2}{P_1}$$

$$V_2 = \frac{V_1 P_1}{P_2} = \frac{(250 \text{ mL})(98.5 \not{\text{atm}})}{(197 \not{\text{atm}})} = 125 \text{ mL}$$

Exercise 13.4

Imagine that you have the sample of hydrogen gas shown on the left in Figure 13.6. Now imagine that the pressure exerted downward by the piston is changed to 49.2 atm. Calculate the new volume of the sample.◆

◆In solving problems about gases, it often helps to draw diagrams or pictures to guide your thinking. In solving the Exercises in this Section, you may want to draw piston-and-cylinder pictures like the ones in Figures 13.5 and 13.6.

Exercise 13.5

A sample of 2.50 mol of nitrogen gas, confined in a cylinder by a piston at 40°C and a pressure of 375 torr, had a volume of 130 L. The pressure was changed, and the sample had a new volume of 52.5 L, at the same temperature. What was the new pressure, in atmospheres?

Charles's Law

Imagine again that we have 1.00 mol of hydrogen gas contained in a cylinder by a piston, as shown in Figure 13.7; the volume of the sample is 250 mL, its temperature is 300 K, and its pressure is 98.5 atm. Now imagine that we double the temperature of our sample of hydrogen, from 300 K to 600 K, while holding the number of moles of gas and the pressure constant. As the temperature increases, the average speed at which the hydrogen molecules are moving will increase. This increased speed will cause an increased frequency of collisions of hydrogen molecules with the face of the piston and make each collision more forceful. As a result, the gas pressure will increase and the piston will move upward. As the piston rises, the hydrogen molecules will have more room to move, so the frequency of their collisions with the face of the piston (the pressure of the gas) will decrease. The piston will again come to rest when

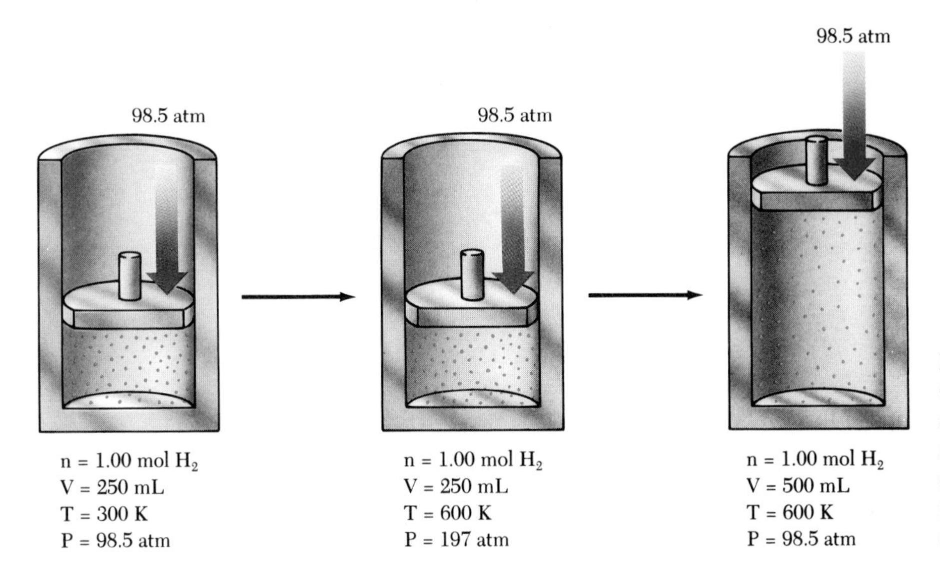

Figure 13.7
Charles's Law. Doubling the absolute temperature (T) of a gas sample, while holding the number of moles of gas (n) and the pressure (P) constant, doubles the volume (V) of the sample.

the upward pressure of hydrogen on the piston equals the downward pressure on it. The new volume will be twice the original volume.

Charles's Law establishes the relationship between the volume of a gas sample and its temperature: At constant pressure, the volume of a gas sample is directly proportional to its *absolute* temperature.◆ In mathematical form,

◆The law is named for its discoverer, J. A. C. Charles (1746–1823), French physicist.

$$\frac{V_1}{V_2} = \frac{T_1}{T_2}$$

V_1 is the volume of a gas sample at temperature T_1, and V_2 is its volume at temperature T_2, at constant P and n

The mathematical form of Charles's Law is true only for temperatures expressed on the absolute scale, and not on the Celsius or Fahrenheit scales.

An illustration of Charles's Law: As air-filled balloons are immersed in liquid nitrogen at 77 K, they shrink to much smaller volumes; when the balloons are removed from the liquid nitrogen they warm up and expand to their original volumes. *(Charles D. Winters)*

In the example shown in Figure 13.7:

$$V_1 = 250 \text{ mL} \qquad T_1 = 300 \text{ K}$$

$$V_2 = 500 \text{ mL} \qquad T_2 = 600 \text{ K}$$

$$\frac{V_1}{V_2} = \frac{T_1}{T_1}$$

$$V_2 = \frac{V_1T_2}{T_1} = \frac{(250 \text{ mL})(600 \not{K})}{(300 \not{K})} = 500 \text{ mL}$$

Exercise 13.6

A sample of fluorine gas had a volume of 175 mL at a temperature of 325 K and a pressure of 700 torr. To what temperature would the sample have to be cooled to change its volume to 150 mL, at the same pressure?

Graphs based on Charles's Law show that absolute zero, the lowest limit of temperature, occurs at about −273°C. For example, the graph in Figure 13.8 shows plots of volume versus temperature for samples of oxygen gas and carbon dioxide gas, at constant pressure. If the line for each gas is extended to a theoretical volume of zero, the corresponding temperature is about −273°C. ◆

◆ In this introduction to the study of gases, I've presented three simple gas laws—Avogadro's Law, Boyle's Law, and Charles's Law. There are other simple gas laws, and you may learn about them in your later chemistry work.

Figure 13.8
Graph of Charles's Law for O_2 and CO_2. Each blue line is a plot of actual measurements of temperature and volume for a sample of gas, at constant pressure. If these plots are extended, as shown by the red lines, they converge on a theoretical volume of 0 mL at about −273°C.

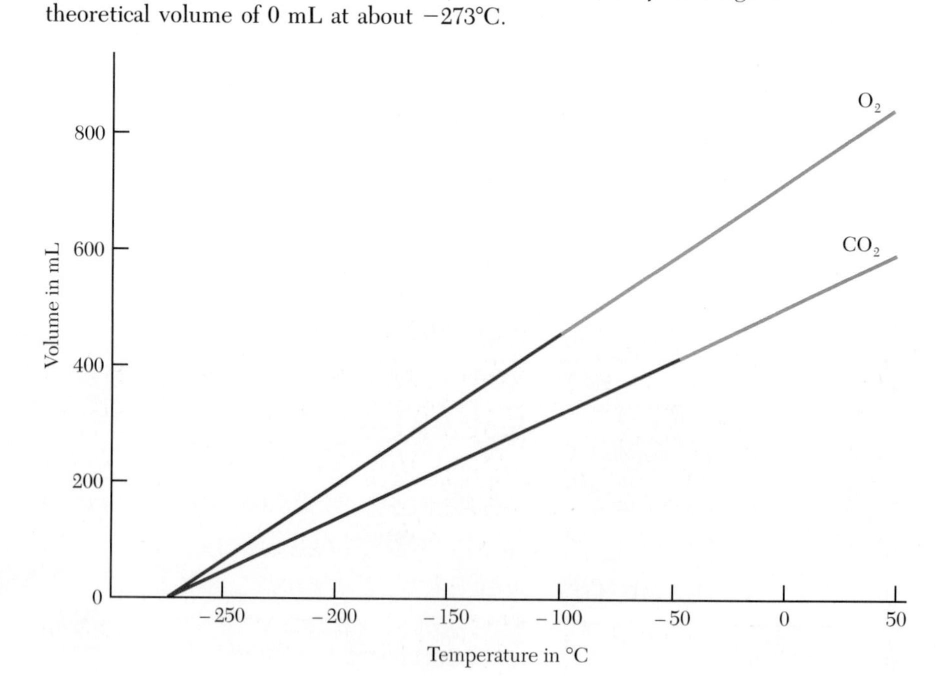

13-4 Gas-law calculations can be based on memorized equations or on reasoning

Avogadro's Law, Boyle's Law, and Charles's Law describe relationships among the number of moles (n) of gas in a sample, its volume (V), its pressure (P), and its temperature (T).

Avogadro's Law: $\frac{V_1}{V_2} = \frac{n_1}{n_2}$ at constant T and P

Boyle's Law: $\frac{V_1}{V_2} = \frac{P_2}{P_1}$ at constant T and n

Charles's Law: $\frac{V_1}{V_2} = \frac{T_1}{T_2}$ T in K, at constant P and n

You can use these laws to make calculations in one of two ways: by substituting values into the preceding equations or by reasoning from what you know about the behavior of gases. The following Example illustrates the difference between these two approaches.

Example 13.1

A sample of helium gas has a volume of 75.0 mL at 750 torr and 23.6°C. What will its volume be at 750 torr and 41.4°C?

Solution

The first step in solving this problem is to recognize that it describes a gas sample with a constant number of moles (n) of gas and a constant pressure (P). We know that n is constant because there is no mention in the problem that any helium is added to or taken from the sample during its temperature change. We know that P is constant because the pressure is 750 torr before and after the temperature change.

At the next step we have a choice of how to proceed. One approach treats the Example as a problem in mathematics. We identify it as a Charles's-Law problem, a change in V and T at constant n and P. Then we write the equation for Charles's Law, rearrange it to solve for V_2, and substitute values from the problem into it, remembering to change the temperatures from Celsius to Kelvin.

$$\frac{V_1}{V_2} = \frac{T_1}{T_2} \qquad V_2 = \frac{V_1T_2}{T_1} = \frac{(75.0\ \text{mL})(314\ \cancel{\text{K}})}{(297\ \cancel{\text{K}})} = 79.3\ \text{mL}$$

In the other approach we think about the problem, not in terms of mathematics, but in terms of what we know about how gases behave. According to the problem, the temperature of the gas sample is increasing. Gases expand when they're heated, so the final volume of the sample will be larger than its initial volume. To calculate the final volume, we'll multiply the initial volume by a fraction, made by putting one temperature over the other. There are two possibilities:

$$\frac{(75.0\ \text{mL})(297\ \cancel{\text{K}})}{(314\ \cancel{\text{K}})} = 70.9\ \text{mL} \qquad \text{or} \qquad \frac{(75.0\ \text{mL})(314\ \cancel{\text{K}})}{(297\ \cancel{\text{K}})} = 79.3\ \text{mL}$$

We know that the second calculation must be the correct one, because it shows a volume increase.

Each of these approaches to a gas-law calculation has its advantages. Working from a memorized equation avoids the effort of reasoning. Working by reasoning avoids memorization of equations, avoids the steps of rearranging the equation to solve for a specific unknown, and minimizes the possibility that a mistake in mathematics will lead to a wrong answer.

The two approaches can also be combined, so that a calculation is set up with a memorized equation and checked by reasoning. This method, illustrated in Example 13.2, is probably the one most commonly used to solve gas-law problems.◆

◆Your instructor may have a preferred approach to gas-law calculations.

Example 13.2

A sample of methane gas (natural gas, CH_4) had a volume of 1.44 L at 22.0°C and 2.62 atm. The pressure on the sample was increased to 5.11 atm, at 22.0°C. Calculate the new volume of the sample.

Solution

The volume (V) and pressure (P) of the sample are changing while its number of moles (n) and its temperature (T) are constant, so this is a Boyle's-Law problem.

$$\frac{V_1}{V_2} = \frac{P_2}{P_1} \qquad V_2 = \frac{V_1 P_1}{P_2} = \frac{(1.44\ \text{L})(2.62\ \cancel{\text{atm}})}{(5.11\ \cancel{\text{atm}})} = 0.738\ \text{L}$$

We can decide whether the answer is reasonable, based on what we know about the behavior of gases. The pressure on the sample was increased, so its volume should have decreased. Our answer shows a final volume smaller than the initial volume, so the answer is a reasonable one.

It's easy to make a mistake in rearranging a gas-law equation or in substituting numbers into it. In solving our Example, we could have made a mistake and written

$$V_2 = \frac{(1.44\ \text{L})(5.11\ \cancel{\text{atm}})}{(2.62\ \cancel{\text{atm}})} = 2.99\ \text{L}$$

This calculation is mathematically correct (the multiplication, division, and cancellation of units are done correctly), but it isn't scientifically correct because the result contradicts what we know about the behavior of gases: When the pressure on a gas sample is increased, its volume should decrease, but our calculation shows a volume increase. Reasoning points out the mistake and lets us correct it.

Chlorine gas, Cl_2. *(Charles Steele)*

As you work the following Exercises, use your understanding of the behavior of gases to verify that your calculations are reasonable.

Exercise 13.7

A sample of 0.549 mol of chlorine gas had a volume of 1.40 L at 30.0°C and 1.25 atm. Calculate the volume of 0.862 mol of chlorine at the same temperature and pressure.

Exercise 13.8

A sample of 2.25 mol of acetylene gas (C_2H_2, used in welding) had a volume of 78.4 L at 310 K and 675 torr. Calculate its volume at the same pressure and 410 K.

Exercise 13.9

A sample of 12.5 g of carbon monoxide gas (CO) had a volume of 12.0 L at 350 K and 540 torr. Calculate its volume at 350 K and 730 torr.

13-5 Avogadro's Law, Boyle's Law, and Charles's Law can be combined into one law

Avogadro's Law, Boyle's Law, and Charles's Law describe the changes in the volume (V) of a gas sample that occur when we change the number of moles (n) of gas, the pressure (P), or the temperature (T).

Avogadro's Law	Boyle's Law	Charles's Law
$\frac{V_1}{V_2} = \frac{n_1}{n_2}$	$\frac{V_1}{V_2} = \frac{P_2}{P_1}$	$\frac{V_1}{V_2} = \frac{T_1}{T_2}$

These equations can be combined into one equation, called the **combined gas law:**

$$\frac{V_1}{V_2} = \frac{n_1P_2T_1}{n_2P_1T_2}$$ or, in a form that is easier to remember, $$\frac{P_1V_1}{n_1T_1} = \frac{P_2V_2}{n_2T_2}$$

The combined gas law can be used in place of any of the three laws it contains.◆ The following Example uses it instead of Boyle's Law.

◆The combined gas law is usually given as a combination of Boyle's Law and Charles's Law: $P_1V_1/T_1 = P_2V_2/T_2$. I've included Avogadro's Law, to make the combined law more useful.

Example 13.3

A sample of nitrogen gas has a volume of 3.75 L at 130°C and 1.10 atm. Calculate its volume at 130°C and 0.900 atm.

Solution

In this problem, the temperature remains constant, so $T_1 = T_2$, and the number of moles of gas remains constant, so $n_1 = n_2$. Because of these equalities, T_1 and T_2 cancel one another and disappear from the equation, and the same is true for n_1 and n_2.

$$\frac{P_1V_1}{n_1T_1} = \frac{P_2V_2}{n_2T_2}$$

Solve for V_2:

$$V_2 = \frac{V_1P_1\cancel{n_2}\cancel{T_2}}{P_2\cancel{n_1}\cancel{T_1}} = \frac{V_1P_1}{P_2} = \frac{(3.75\ \text{L})(1.10\ \cancel{\text{atm}})}{(0.900\ \cancel{\text{atm}})} = 4.58\ \text{L}$$

Use the combined gas law to solve the following problem.

Exercise 13.10

A sample of oxygen gas had a volume of 85.0 mL at 290 K and 1.00 atm. The gas was heated at constant pressure, and its volume changed to 125 mL. What was its new temperature?

The combined gas law contains eight variables, and it can be used to solve for any one of them, if values for the other seven are known.

Example 13.4

A sample of 0.650 mol of hydrogen gas, contained in a cylinder by a piston, had a volume of 13.8 L. The temperature of the gas was 25.0°C and the pressure on it was 1.15 atm. The temperature was changed to 20.0°C, the pressure was changed to 1.50 atm, and 0.150 mol of hydrogen was added to the sample. Calculate its new volume.

Solution

$$\frac{P_1V_1}{n_1T_1} = \frac{P_2V_2}{n_2T_2}$$

$$n_2 = n_1 + 0.150 \text{ mol} = 0.650 \text{ mol} + 0.150 \text{ mol} = 0.800 \text{ mol}$$

$$V_2 = \frac{V_1P_1n_2T_2}{P_2n_1T_1} = \frac{(13.8 \text{ L})(1.15 \cancel{\text{atm}})(0.800 \cancel{\text{mol}})(293 \cancel{\text{K}})}{(1.50 \cancel{\text{atm}})(0.650 \cancel{\text{mol}})(298 \cancel{\text{K}})} = 12.8 \text{ L}$$

Reasoning can show whether we've set this calculation up correctly. The problem describes a pressure increase, so the effect of the pressure change should be a volume decrease. In our calculation, we multiply the initial volume by 1.15 atm/1.50 atm, which will lower the volume, so we've set the calculation up correctly. Similarly, the number of moles of gas is being increased, causing a volume increase, so we multiply by 0.800 mol/0.650 mol. The temperature is being decreased, causing a volume decrease, so we multiply by 293 K/298 K.

Use the combined gas law to solve the following problem, and use reasoning to check your calculation.

Exercise 13.11

A sample of 0.500 mol of helium gas had a volume of 12.5 L at 280 K and 700 torr. The pressure and temperature were changed. The new temperature was 310 K and the new volume was 12.0 L. What was the new pressure?

13-6 The most general and most useful gas law is the ideal-gas equation, PV = nRT

The left and right sides of the combined gas law refer to a sample of gas under two different sets of conditions of P, V, n, and T:

$$\frac{P_1V_1}{n_1T_1} = \frac{P_2V_2}{n_2T_2}$$

The two sides of this equation are equal, meaning that, if we have any values of P, V, n, and T for a gas sample, the value of PV/nT must always be the same:

$$\frac{P_1V_1}{n_1T_1} = \frac{P_2V_2}{n_2T_2} = \frac{P_3V_3}{n_3T_3} = \frac{P_4V_4}{n_4T_4} = \cdots$$

Each of these fractions has the same value, and the symbol R represents their common value:

$$\frac{P_1V_1}{n_1T_1} = R \qquad \frac{P_2V_2}{n_2T_2} = R \qquad \frac{P_3V_3}{n_3T_3} = R \qquad \frac{P_4V_4}{n_4T_4} = R$$

One equation summarizes all of these relationships:

$$\frac{PV}{nT} = R$$

This equation, called the **ideal-gas equation,** is usually written in this form:

$$PV = nRT$$

An **ideal gas** is a gas that behaves exactly as the kinetic theory and the gas laws predict. In the problems in this book, we'll assume that all gases behave as if they were ideal gases.◆

◆Real gases don't always show ideal-gas behavior. More complicated forms of the gas laws are used to describe the nonideal behavior of real gases.

We can find the numerical value of R and its units from the information in Figure 13.9, which shows that 1.00 mol of any ideal gas confined in a cylinder by a piston at a pressure of 1.00 atm and a temperature of 0°C (273 K) will have a volume of 22.4 L. By substituting these values of P, V, n, and T in the ideal-gas equation, we can calculate the value of R:

$$PV = nRT$$

$$R = \frac{PV}{nT} = \frac{(1.00 \text{ atm})(22.4 \text{ L})}{(1.00 \text{ mol})(273 \text{ K})} = \frac{(0.0821 \text{ atm})(1 \text{ L})}{(1 \text{ mol})(1 \text{ K})}$$

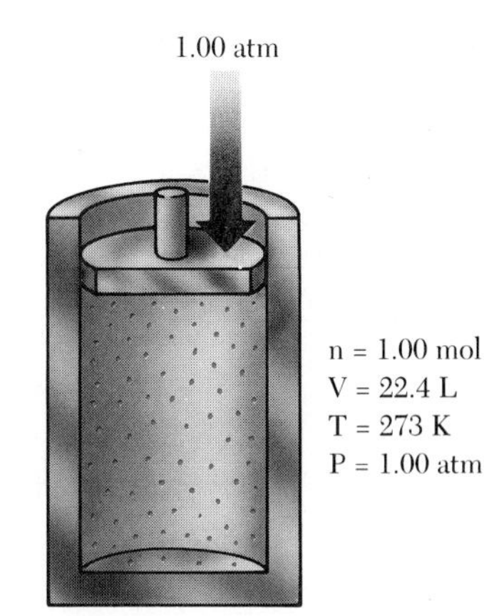

Figure 13.9
Volume of 1.00 mol of an ideal gas at 273 K and 1.00 atm. At 273 K (0°C) and 1.00 atm, 1.00 mol of an ideal gas will have a volume of 22.4 L.

The value of R is abbreviated 0.0821 atm–L/mol–K. The units, atm–L/mol–K, are necessary for R to function in the ideal-gas equation when the equation is solved for one of its variables, as shown in the following Examples.

Example 13.5

A sample of 0.500 mol of hydrogen gas had a volume of 8.00 L at a pressure of 2.63 atm. What was its temperature?

Solution

$$PV = nRT$$

$$T = \frac{PV}{nR} = \frac{(2.63 \cancel{\text{atm}})(8.00 \cancel{\text{L}})}{(0.500 \cancel{\text{mol}})(0.0821 \cancel{\text{atm}}\text{–}\cancel{\text{L}}/\cancel{\text{mol}}\text{–K})} = 512 \text{ K}$$

Example 13.6

A sample of neon gas had a volume of 8.37×10^3 mL at a pressure of 587 torr and a temperature of 42.0°C. What was the mass of the sample in grams?

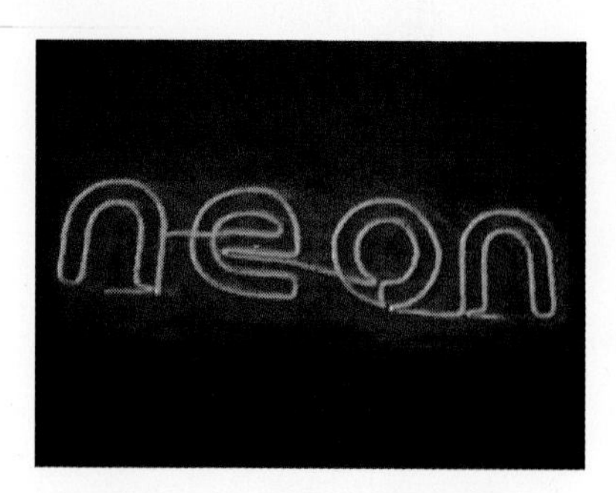

Neon gas is used in advertising signs. *(The World of Chemistry)*

Solution

P = 587 torr = 0.772 atm
V = 8.37×10^3 mL = 8.37 L
T = 42.0°C = 315 K

$$PV = nRT$$

$$n = \frac{PV}{RT} = \frac{(0.772\ \cancel{atm})(8.37\ \cancel{L})}{(0.0821\ \cancel{atm}\text{-}\cancel{L}/mol\text{-}\cancel{K})(315\ \cancel{K})} = 0.250\ mol$$

mol Ne → g Ne

$$\left(\frac{0.250\ \cancel{mol\ Ne}}{}\right)\left(\frac{20.2\ g\ Ne}{1\ \cancel{mol\ Ne}}\right) = 5.05\ g\ Ne$$

Use the ideal-gas equation to solve the following Exercises.

Exercise 13.12

A sample of 2.50 mol of oxygen gas had a volume of 40.0 L at 290 K. Calculate its pressure.

Exercise 13.13

Calculate the volume of 5.00 g of chlorine gas at 125°C and 600 torr.

In discussing the behavior of samples of gases it's often useful to compare them under the same conditions of temperature and pressure. The temperature 0°C and the pressure 1.00 atm are used as standard conditions for gases and are referred to as **standard temperature and pressure,** abbreviated STP. As shown in Figure 13.9 on page 281, 1.00 mol of an ideal gas occupies 22.4 L at STP; for this reason, 22.4 L is called the **molar volume** of an ideal gas at STP.

13-7 Gas volumes can be used in stoichiometry

Avogadro's law shows that, at constant temperature and pressure, the volume of a gas sample will vary directly with the number of moles of gas in the sample:

$$\frac{V_1}{V_2} = \frac{n_1}{n_2} \qquad \text{at constant T and P}$$

In other words, at constant temperature and pressure, the volume of a gas sample is a direct measure of the number of moles of gas in the sample.

As we saw in Chapter 12, a balanced chemical equation can be read in moles, and the corresponding molar relationships can be used to provide factors for stoichiometry. For example, the equation

$$3\ H_2(g) + N_2(g) \rightarrow 2\ NH_3(g)$$

can be read 3 mol H_2 1 mol N_2 2 mol NH_3

Because the volumes of gas samples, at constant temperature and pressure, are measures of the numbers of moles of gases in the samples, quantities of reactants and products that are gases can also be expressed in volume units. For example, the equation

$$3\ H_2(g) + N_2(g) \rightarrow 2\ NH_3(g)$$

3 L H_2	1 L N_2	2 L NH_3
3 mL H_2	1 mL N_2	2 mL NH_3
3 cm^3 H_2	1 cm^3 N_2	2 cm^3 NH_3

or in any other volume units. The relationships of the volume units shown in the balanced chemical equation can be used in stoichiometry, just as molar relationships are used.◆

◆Volume units can be used to describe stoichiometric relationships only for reactants or products that are gases, and not for those that are liquids or solids.

Example 13.7

Assuming that the temperature and pressure are constant, how many liters of ammonia can be produced from 4.25 L of hydrogen, according to the following equation?

$$3\ H_2(g) + N_2(g) \rightarrow 2\ NH_3(g)$$

Solution

This equation can be read in moles, and, because at constant T and P the volume of a gas sample is a measure of the number of moles of gas in the sample, the equation can also be read in volume units:

$$3\ H_2(g) + N_2(g) \rightarrow 2\ NH_3(g)$$

3 L H_2	1 L N_2	2 L NH_3

These volume relationships are used in stoichiometry just as molar relationships are used:

4.25 L H_2 = ? L NH_3

$$\text{L } H_2 \rightarrow \text{L } NH_3$$

$$\left(\frac{4.25\ \cancel{\text{L } H_2}}{}\right)\left(\frac{2\ \text{L } NH_3}{3\ \cancel{\text{L } H_2}}\right) = 2.83\ \text{L } NH_3$$

Example 13.8

Assuming that the temperature and pressure are constant, how many milliliters of chlorine will be needed to produce 50.0 mL of hydrogen chloride, according to this equation:

$$H_2(g) + Cl_2(g) \rightarrow 2\ HCl(g)$$

Solution

At constant T and P, the quantities of gases shown in a balanced chemical equation can be read in any volume units:

$$H_2(g) + Cl_2(g) \rightarrow 2\ HCl(g)$$

1 mL H_2	1 mL Cl_2	2 mL HCl

$50.0 \text{ mL HCl} = ? \text{ mL Cl}_2$

$$\text{mL HCl} \rightarrow \text{mL Cl}_2$$

$$\left(\frac{50.0 \cancel{\text{mL HCl}}}{}\right)\left(\frac{1 \text{ mL Cl}_2}{2 \cancel{\text{mL HCl}}}\right) = 25.0 \text{ mL Cl}_2$$

Solve the following Exercise by a stoichiometric calculation using gas volumes.

Exercise 13.14

The reaction

$$2\ H_2(g) + O_2(g) \rightarrow 2\ H_2O(g)$$

was carried out at constant temperature and pressure, using 53.6 mL of hydrogen. How many milliliters of oxygen were required for the reaction?

The gas laws can be combined with stoichiometry to make calculations in which quantities of reactants or products that are gases are expressed in volume units, in moles, or in mass units, as shown in the following Examples and Exercises.

Example 13.9

At STP, how many grams of oxygen will react with 135 mL of sulfur dioxide, according to the following equation?

$$2\ SO_2(g) + O_2(g) \rightarrow 2\ SO_3(g)$$

Solution

At STP, 1.00 mol of an ideal gas has a volume of 22.4 L (Section 13-6), so, at STP,

$$1.00 \text{ mol of } O_2 = 22.4 \text{ L } O_2$$

$135 \text{ mL SO}_2 = ? \text{ g O}_2$

$$\text{mL SO}_2 \rightarrow \text{L SO}_2 \rightarrow \text{L O}_2 \rightarrow \text{mol O}_2 \rightarrow \text{g O}_2$$

$$\left(\frac{135 \cancel{\text{mL SO}_2}}{}\right)\left(\frac{1 \cancel{\text{L SO}_2}}{1 \times 10^3 \cancel{\text{mL SO}_2}}\right)\left(\frac{1 \cancel{\text{L O}_2}}{2 \cancel{\text{L SO}_2}}\right)\left(\frac{1 \cancel{\text{mol O}_2}}{22.4 \cancel{\text{L O}_2}}\right)\left(\frac{32.0 \text{ g O}_2}{1 \cancel{\text{mol O}_2}}\right)$$

$$= 0.0964 \text{ g O}_2$$

The problem can also be solved this way:

$$\text{mL SO}_2 \rightarrow \text{L SO}_2 \rightarrow \text{mol SO}_2 \rightarrow \text{mol O}_2 \rightarrow \text{g O}_2$$

$$\left(\frac{135 \cancel{\text{mL SO}_2}}{}\right)\left(\frac{1 \cancel{\text{L SO}_2}}{1 \times 10^3 \cancel{\text{mL SO}_2}}\right)\left(\frac{1 \cancel{\text{mol SO}_2}}{22.4 \cancel{\text{L SO}_2}}\right)\left(\frac{1 \cancel{\text{mol O}_2}}{2 \cancel{\text{mol SO}_2}}\right)\left(\frac{32.0 \text{ g O}_2}{1 \cancel{\text{mol O}_2}}\right)$$

$$= 0.0964 \text{ g O}_2$$

Exercise 13.15

How many grams of fluorine will be needed to produce 575 mL of carbon tetrafluoride at STP, according to the following equation?

$$CH_4(g) + 4\ F_2(g) \rightarrow CF_4(g) + 4\ HF(g)$$

Example 13.10

When potassium chlorate is heated it decomposes into potassium chloride and oxygen:

$$2\ KClO_3(s) \rightarrow 2\ KCl(s) + 3\ O_2(g)$$

At 28°C and 735 torr, how many liters of oxygen will be produced by decomposition of 5.24 g of potassium chlorate?

Solution

Use stoichiometry to find the number of moles of oxygen produced.

$5.24\ g\ KClO_3 = ?\ mol\ O_2$

$$g\ KClO_3 \rightarrow mol\ KClO_3 \rightarrow mol\ O_2$$

$$\left(\frac{5.24\ \cancel{g\ KClO_3}}{}\right)\left(\frac{1\ \cancel{mol\ KClO_3}}{123\ \cancel{g\ KClO_3}}\right)\left(\frac{3\ mol\ O_2}{2\ \cancel{mol\ KClO_3}}\right) = 0.0639\ mol\ O_2$$

Use PV = nRT to convert 0.0639 mol of O_2 to liters.

$$K = C + 273 = 28 + 273 = 301\ K$$

$$\left(\frac{735\ \cancel{torr}}{}\right)\left(\frac{1\ atm}{760\ \cancel{torr}}\right) = 0.967\ atm$$

$$V = \frac{nRT}{P} = \frac{(0.0639\ \cancel{mol})(0.821\ \cancel{atm}\text{–}L/\cancel{mol}\text{–}\cancel{K})(301\ \cancel{K})}{(0.967\ \cancel{atm})} = 1.63\ L$$

Exercise 13.16

According to the following equation, how many milligrams of carbon dioxide will be produced by the reaction of 275 mL of oxygen with carbon monoxide at 125°C and 645 torr?

$$2\ CO(g) + O_2(g) \rightarrow 2\ CO_2(g)$$

Inside Chemistry

Why is it important to prevent the loss of ozone from the atmosphere?

Light from the sun includes ultraviolet radiation (Section 3-8), which is potentially harmful to animals because it can penetrate and damage delicate animal tissues. In human beings ultraviolet rays can cause blindness and skin cancer.

Two molecular forms of oxygen in the atmosphere absorb much of the harmful ultraviolet radiation from the sun before it can reach the surface of the earth. Diatomic oxygen, O_2, which makes up about 20% of the atmosphere, absorbs radiation with shorter wavelengths in the

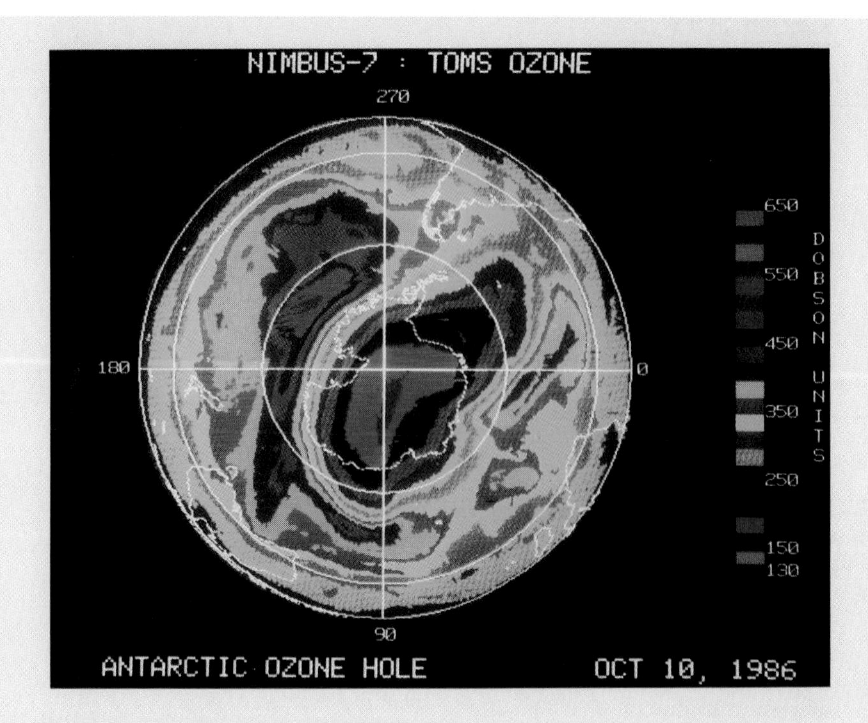

A computer map of ozone concentrations, measured by satellite, over the southern hemisphere. The purple area identifies a low concentration of ozone over Antarctica. *(NASA/ Goddard Space Flight Center)*

ultraviolet range. Triatomic oxygen, O_3, which occurs in small amounts in the upper atmosphere, absorbs longer-wavelength ultraviolet radiation. The triatomic form of oxygen is called *ozone.*

Evidence gathered during the past ten years shows that the amount of ozone in the upper atmosphere is decreasing. Because a decrease in the ozone layer increases our risk of harm from ultraviolet radiation, there has been an urgent research effort to find the cause of ozone depletion.

This research has shown that the loss of ozone in the upper atmosphere is probably being caused by gases called chlorofluorocarbons, often referred to as CFCs. Two examples are CCl_2F_2 and $CClF_3$:

CCl_2F_2 (Cl—C—Cl with F above and below C) $\qquad$ $CClF_3$ (F—C—Cl with F above and below C)

Throughout the past three decades, great quantities of CFCs have been used for many purposes. There are probably CFCs in your home now, as the circulating fluid in the refrigerator and as the propellant gas in spray cans of such products as deodorant, shaving cream, and spray paint. When CFCs escape from their containers, they move upward through the atmosphere and eventually reach the ozone layer.

In the upper atmosphere CFCs may destroy ozone in several ways. In one example, the process occurs in two steps. The first step is the reaction of a CFC molecule with light to form a chlorine atom:

$$CClF_3 + \text{light} \longrightarrow CF_3 + Cl$$

$CClF_3$ — CFC molecule; CF_3; Cl — chlorine atom

The second step is the reaction of the chlorine atom with an ozone molecule:

$$Cl + O_3 \longrightarrow ClO + O_2$$

Cl — chlorine atom; O_3 — ozone molecule; ClO — chlorine monoxide molecule; O_2 — oxygen molecule

The two-step process shown in the equations above is especially effective in destroying ozone because the molecule of chlorine monoxide that is produced in the second step can be converted back into a chlorine atom. In the upper atmosphere, there are oxygen atoms. A molecule of chlorine monoxide will react with an oxygen atom:

$$:\ddot{\underset{..}{Cl}}-\ddot{\underset{..}{O}}\cdot \; + \; \cdot\ddot{\underset{..}{O}}\cdot \longrightarrow \; :\ddot{\underset{..}{Cl}}\cdot \; + \; \cdot\ddot{\underset{..}{O}}-\ddot{\underset{..}{O}}\cdot$$

ClO	O	Cl	O_2
chlorine monoxide molecule	oxygen atom	chlorine atom	oxygen molecule

The chlorine atom that's released in this reaction can react with another ozone molecule. Because a chlorine atom can recycle in this way, one chlorine atom from one CFC molecule can destroy many ozone molecules.

The evidence presented by environmental chemists to show that CFCs are probably depleting the ozone layer has led to important steps toward eliminating this potential hazard. In 1990, one hundred nations agreed, through the United Nations Environment Program, that a ban on the use of CFCs will begin in the year 2000.

Chapter Summary: Gases

Section	Subject	Summary	Check When Learned
13-1	Ways to describe the quantity of gas in a sample	State the mass of the sample or state its volume, temperature, and pressure.	☐
13-1	Pressure units:		
	atmosphere (atm)	Pressure exerted by the earth's atmosphere at sea level.	☐
	torr	Defined by 1 atm = 760 torr.	☐
	mm of Hg	Same as torr.	☐
	pascal (Pa)	Defined by 1 atm = 101 kPa.	☐
13-2	Description of gases according to kinetic theory	(1) A gas consists of particles (atoms or molecules). The total volume of the particles is negligible compared with the total volume of the gas. (2) The particles are in rapid, random motion, and they constantly collide with one another and with the walls of their container. The collisions are assumed to be perfectly elastic. (3) The number of particles and their motion are responsible for the volume, temperature, and pressure of the sample.	☐
13-3	Avogadro's Law	$V_1/V_2 = n_1/n_2$ at constant T and P. V_1 is the volume of a gas sample that contains n_1 mol of gas, and V_2 is the volume of a sample that contains n_2 mol.	☐
13-3	Boyle's Law	$V_1/V_2 = P_2/P_1$ at constant n and T. V_1 is the volume of a gas sample at pressure P_1, and V_2 is the volume of the sample at pressure P_2.	☐
13-3	Charles's Law	$V_1/V_2 = T_1/T_2$ at constant n and P. V_1 is the volume of a gas sample at temperature T_1, and V_2 is the volume of the sample at temperature T_2. Temperatures must be in K.	☐
13-5	Combined gas law	$P_1V_1/n_1T_1 = P_2V_2/n_2T_2$. Temperatures must be in K.	☐

Section	Subject	Summary	Check When Learned
13-6	Ideal gas	Gas that behaves exactly as the kinetic theory and the gas laws predict.	☐
13-6	Ideal gas equation, including meaning and units for each term	PV = nRT. P is pressure in atm; V is volume in L; n is quantity of gas in mol; R is gas constant, 0.0821 atm–L/mol–K; T is temperature in K.	☐
13-6	STP	Standard temperature and pressure: 273 K and 1 atm.	☐
13-6	Molar volume of an ideal gas	The volume, 22.4 L, occupied by 1.00 mol of an ideal gas at STP.	☐
13-7	Basis for using gas volumes in stoichiometry	At constant T and P, the volume of a gas sample is directly proportional to the number of moles of gas in the sample, so the quantities of gases in a balanced chemical equation can be read either in moles or in any volume units.	☐

Problems

Assume you can use the Periodic Table at the front of the book unless you're directed otherwise. Answers to odd-numbered problems are in Appendix 1.

Units for Mass, Volume, Temperature, and Pressure (Section 13-1)

1. Make these conversions: (a) 345 mg of O_2 to g of O_2 (b) 6.88 g of O_2 to mol of O_2 (c) 763 mg of O_2 to mol of O_2.
2. Make these conversions: (a) 394 mL to L (b) 1.44 L to mL (c) 265 cm^3 to L.
3. Make these conversions: (a) 122°F to °C (b) 34°C to K (c) −137°C to K.
4. Make these conversions: (a) 0.494 atm to torr (b) 855 torr to atm (c) 337 kPa to atm.
5. A sample of chlorine gas had a mass of 2.00 g and a volume of 561 mL at a pressure of 950 torr and a temperature of 33°C. Convert these units to mol, L, atm, and K.
6. A sample of hydrogen gas had a mass of 0.777 kg and a volume of 13.7 dL at a pressure of 1213 mm of Hg and a temperature of −15°C. Convert these units to mol, L, atm, and K.
7. A sample of ammonia gas had a mass of 933 mg and a volume of 255 cL at a pressure of 825 torr and a temperature of 135°C. Convert these units to mol, L, atm, and K.
8. A sample of methane gas had a mass of 4.29 lb and a volume of 921 $in.^3$ at a temperature of 242°F and a pressure of 953 mm of Hg. Convert these units to mol, L, atm, and K.
9. Imagine that a mercury barometer is carried from the top of a mountain to its base. Would you expect the height of the column of mercury in the glass tube to increase or decrease? Explain.
10. Imagine that a mercury barometer is placed in a sealed chamber and that the air in the chamber is then removed by a pump. As the air is pumped out, would you expect the column of mercury in the glass tube to rise or fall? Explain.

Kinetic Theory of Gases (Section 13-2)

11. Imagine that a sample of gas is contained in a cylinder by a piston. If you push down on the piston, you'll feel increasing resistance the farther down you push it. Use the kinetic theory to explain this fact.

12. Using the kinetic theory, explain why a tire expands as it's inflated.

13. Using the kinetic theory, explain why a balloon collapses when it's punctured.

14. Using the kinetic theory, explain why cooling the air in a balloon will cause the balloon to shrink.

15. Using the kinetic theory, explain why a balloon expands as it rises through the earth's atmosphere.

16. In a common experiment in basic physics, the air is pumped out of a one-gallon metal can, and the can collapses. Use the kinetic theory to explain why the can collapses.

17. In theory, all of the oxygen in a room could move to one corner, so that anyone in the room would suffocate. Use the kinetic theory to explain why this danger is not significant.

18. If you inflate two identical balloons to the same size, one with helium and the other with air, both balloons will eventually deflate as gas leaks out of them through the rubber membrane, but the helium-filled balloon will deflate more rapidly than the air-filled balloon. Can you suggest a reason?

19. Methanol (wood alcohol) is a colorless liquid that has a density of 0.792 g/mL and boils at 65°C. At 215°C and 1.25 atm, 355 mg of methanol gas has a volume of 356 mL. Assuming that, in the liquid state, the molecules are packed close to one another, so the volume of the liquid is the volume of the molecules, calculate the percentage of the methanol gas that is empty space.

20. Ethanol (beverage alcohol) is a colorless liquid that has a specific gravity of 0.798 and boils at 78.5°C. At 300°C and 0.800 atm, 500 mg of ethanol gas has a volume of 635 mL. Assuming that, in the liquid state, the molecules are packed close to one another, so the volume of the liquid is the volume of the molecules, calculate the percentage of the ethanol gas that is empty space.

Avogadro's Law, Boyle's Law, and Charles's Law (Sections 13-3 and 13-4)

21. Use the kinetic theory to explain Avogadro's Law.

22. Use the kinetic theory to explain Boyle's Law.

23. Use the kinetic theory to explain Charles's Law.

24. Write a mathematical statement of this law: At constant volume, the pressure of a sample of gas is directly proportional to its absolute temperature.

25. Use the kinetic theory to explain the law stated in problem 24.

26. If the pressure on a sample of gas is doubled and its absolute temperature is doubled, what will be the change in its volume?

27. A sample of 0.232 mol of fluorine gas had a volume of 8.49 L at 525 torr and 35°C. Calculate its volume at 1.25 atm and 35°C.

28. A sample of 0.488 mol of neon gas had a volume of 17.4 L at 0.921 atm and 128°C. The temperature of the sample was changed while its pressure was held constant, and its new volume was 844 mL. What was its new temperature, in °C?

29. A sample of 0.232 mol of helium gas had a volume of 2.80 L at 355 K and 2.41 atm. Calculate its volume at 255 K and 2.41 atm.

30. A sample of 0.741 mol of oxygen gas had a volume of 12.7 L at 315 K and 1.51 atm. The temperature of the sample was increased to 415 K, at constant pressure, and the sample expanded to a new volume. To what value would the pressure have to be increased, at constant temperature, to change the volume of the sample from its new value back to its original value?

31. A sample of 0.0500 mol of ammonia gas had a volume of 235 mL. The temperature and pressure were held constant, and some ammonia was removed from the sample; its new volume was 175 mL. How many grams of ammonia were removed?

32. A sample of oxygen gas had a volume of 0.755 ft^3 at 25°C and 85.0 kPa. Calculate its volume in mL at 25°C and 794 torr.

33. A sample of nitrogen gas had a volume of 466 mL at 23°F. Calculate its volume at 46°F, at the same pressure.

34. Use the mathematical law you wrote for problem 24 to solve the following problem. A sample of helium gas had a temperature of 352 K at a pressure of 135 kPa. Calculate its temperature at a pressure of 165 kPa, at constant volume.

Combined Gas Law (Section 13-5)

35. Solve the combined gas law for (a) V_2 (b) n_2 (c) P_2 (d) T_2.

36. Show that, at constant P and T, the combined gas law is identical with Avogadro's Law.

37. Show that, at constant n and T, the combined gas law is identical with Boyle's Law.

38. Show that, at constant n and P, the combined gas law is identical with Charles's Law.

39. Show that the combined gas law contains the mathematical law you wrote for problem 24.

40. A sample of hydrogen gas had a volume of 6.72 L at 0.454 atm and 299 K. Calculate its volume at 7.49 atm and 275 K.

41. A sample of chlorine gas had a volume of 355 mL at 85°C and 688 torr. The temperature of the sample was changed and the pressure was changed to 852 torr; the new volume was 365 mL. What was the new temperature?

42. A sample of 1.25 mol of helium gas had a volume of 27.0 L at 303 K and 1.15 atm. The pressure was changed to 1.45 atm, the temperature was changed to 276 K, and 0.50 mol of helium was added to the sample. Calculate its new volume.

43. A sample of fluorine gas had a volume of 224 mL at 25°C and 450 torr. Calculate its volume at 75°C and 550 torr.

44. A sample of 3.65 g of carbon tetrachloride gas had a volume of 235 mL at 1.33 atm and a certain temperature. The pressure was increased to 1.57 atm and 2.47 g of carbon tetrachloride was added to the sample, at the same temperature. Calculate the new volume of the sample.

Ideal-Gas Equation (Section 13-6)

45. What is an ideal gas?

46. Solve the ideal-gas equation for (a) P (b) V (c) n (d) T.

47. If the units used in the ideal-gas equation were mol, torr, K, and mL, what would be the numerical value and the units for R?

48. Use the kinetic theory to explain why 1.00 mol of any ideal gas, with a volume of 22.4 L at 273 K, will exert the same pressure, 1.00 atm.

49. A sample of 0.405 mol of helium gas had a volume of 1.30 L at 315 K. What was its pressure?

50. A sample of 2.15 mol of hydrogen gas had a volume of 0.889 L at 0.955 atm. What was its temperature?

51. A sample of carbon dioxide gas had a volume of 2.65 L at 325 K and 1.52 atm. How many moles of carbon dioxide were in the sample?

52. A sample of nitrogen gas had a volume of 0.775 L at 265 K and 1.77 atm. How many grams of nitrogen were in the sample?

53. Calculate the temperature in °C of 2.55 g of neon gas with a volume of 227 mL at a pressure of 235 kPa.

54. A sample of chlorine gas had a volume of 2.49 L at 455 torr and 385 K. How many molecules were in the sample?

55. Calculate the density, in g/mL, of 0.353 mol of fluorine gas at 285 K and 1.45 atm. (Calculate the mass and volume of the sample, then calculate its density.)

56. Calculate the density, in g/mL, of 0.213 mol of phosphorus trichloride gas at 355°C and 769 torr. (Calculate the mass and volume of the sample, then calculate its density.)

57. A gas sample had a mass of 325 mg and a volume of 102 mL at 0.899 atm and 412 K. Calculate the molecular mass of the gas. (Calculate the number of moles of gas in the sample. From the fact that this number of moles has a mass of 325 mg, you can calculate the mass in grams of one mole.)

58. A sample of one of the noble gases had a mass of 4.75 g and occupied a volume of 1.10 L at 25°C and 955 torr. Which noble gas was it? (See the hint in parentheses in problem 57.)

Gas Stoichiometry (Section 13-7)

59. Assuming that the temperature and pressure are constant, how many mL of PCl_5 can be produced from 225 mL of PCl_3 and 355 mL of Cl_2 according to the following equation?

$$PCl_3(g) + Cl_2(g) \rightarrow PCl_5(g)$$

60. Assuming that the temperature and pressure are constant, how many liters of sulfur dioxide will be required to react with 3.50 L of oxygen, according to the following equation?

$$2\ SO_2(g) + O_2(g) \rightarrow 2\ SO_3(g)$$

61. In the reaction of hydrogen gas with oxygen gas to produce liquid water, how many milliliters of oxygen will be required to react with 165 mL of hydrogen, at constant temperature and pressure?

62. In the reaction of hydrogen gas with nitrogen gas to produce gaseous ammonia, how many liters of hydrogen will be needed to produce 435 mL of ammonia, at constant temperature and pressure?

63. At STP, how many liters of hydrogen chloride will be produced by the reaction of 1.15 mol of hydrogen, according to the following equation?

$$H_2(g) + Cl_2(g) \rightarrow 2\ HCl(g)$$

64. At STP, how many milliliters of hydrogen fluoride will be produced in a reaction that produces 12.7 g of silicon tetrafluoride, according to the following equation?

$$SiH_4(g) + 4\ F_2(g) \rightarrow SiF_4(g) + 4\ HF(g)$$

65. Ammonia reacts with fluorine to produce nitrogen trifluoride and hydrogen fluoride; at STP, all of the reactants and products are gases. How many moles of nitrogen trifluoride will be produced by the reaction of 15.2 L of fluorine, at STP?

66. At STP, solid calcium reacts with oxygen gas to produce solid calcium oxide. How many milligrams of calcium oxide will be produced by the reaction of 295 mL of oxygen?

67. How many moles of chlorine will be needed to react with 3.75 L of carbon monoxide at 35°C and 1.58 atm, according to the following equation?

$$CO(g) + Cl_2(g) \rightarrow COCl_2(g)$$

68. How many milligrams of carbon dioxide will be produced by the reaction of 374 mL of oxygen at 146°C and 725 torr, according to the following equation?

$$C(s) + O_2(g) \rightarrow CO_2(g)$$

69. At 943°C and 3.51 atm, methane reacts with oxygen to produce carbon dioxide and water; all of the reactants and products are gases. How many milliliters of oxygen will be needed to react with 2.00 g of methane?

70. At 413°C and 815 torr, chlorine reacts with fluorine to produce chlorine monofluoride; the reactants and the product are gases. How many milligrams of chlorine monofluoride will be produced by the reaction of 3.15 L of fluorine?

14-1
The behavior of liquids and solids can be explained by the existence of mutually attractive forces between their particles.

14-2
In all liquids and solids the synchronized movement of electrons in adjacent particles causes attractions between the particles.

14-3
Molecules with positive and negative ends attract one another.

14-4
In the liquid and solid states of some compounds hydrogen atoms act as links between molecules.

14-5
In a piece of metal the outermost electrons from each atom move among all of the atoms, holding them together.

14-6
Solids in which the interparticle forces are covalent or ionic bonds have very high melting points.

14-7
By predicting the interparticle forces in solids or liquids, you can sometimes predict their relative melting points or boiling points.

Inside Chemistry:
Why doesn't all of the water in a lake freeze?

Chapter Summary

Problems

Water is the substance we're most familiar with in the three states of matter—solid, liquid, and gas. *(Tony Stone Worldwide/Tom Dietrick)*

14 Liquids, Solids, and Changes of State

In Chapter 13 we saw that many of the properties of gases can be explained by the kinetic theory, which describes a gas as a collection of a large number of tiny particles in rapid, random motion. In this chapter we'll see that a similar description can explain many of the properties of liquids and solids.

The kinetic theory of gases assumes that there are no attractive or repulsive forces between gas particles. This assumption cannot be made for a liquid or a solid: There are attractive forces between the particles in a liquid or a solid, and these forces are responsible for the characteristic properties of the liquid and solid states. In this chapter, we'll consider the nature and effects of these forces.

14-1 The behavior of liquids and solids can be explained by the existence of mutually attractive forces between their particles

We know from everyday experience that gases, liquids, and solids show differences in their behavior. One difference is that a piece of a solid has a fixed shape, while a sample of a liquid or a gas adopts the shape of its container. Another difference is that a sample of a liquid or a solid has a fixed volume, while a gas sample adopts the volume of its container.

These differences among gases, liquids, and solids can be explained by differences in the behavior of their particles, as illustrated in Figure 14.1.

Figure 14.1(a) shows that the particles in a solid are in fixed, ordered positions.◆ All of the particles in a solid are held in their positions by forces of attraction between each particle and the particles around it. Because all of the particles are held in their positions, a piece of a solid has a fixed shape and a fixed volume. The ordered arrangement of the particles in a solid is called a **lattice.**

◆This statement is true for solids that are pure substances; it's not true for solids that are mixtures.

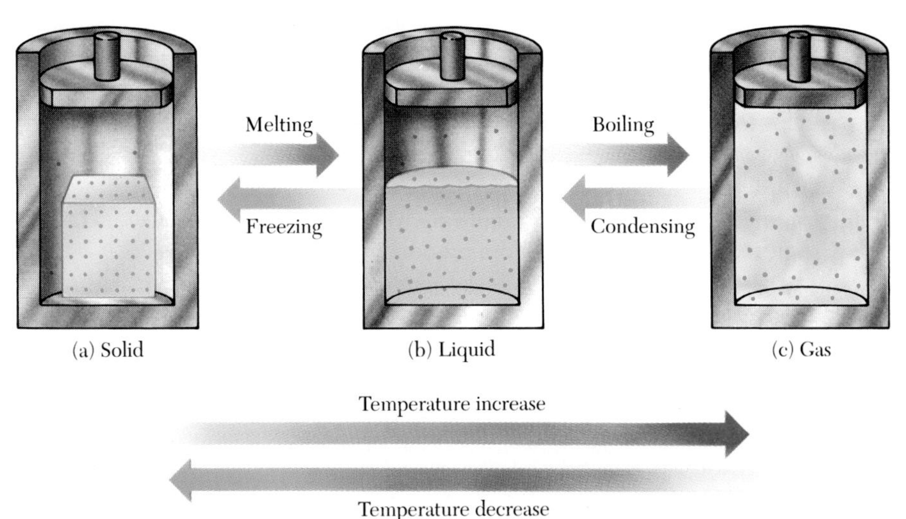

Figure 14.1
Particles in the solid, liquid, and gas states. (a) In the solid state attractive forces hold the particles together in fixed positions. A few particles have sublimed to form a gas around the solid. (b) In the liquid state attractive forces hold the particles together, but they move freely among one another. A few particles have evaporated to form a gas above the liquid. (c) In the gas state there are no attractive forces, and the particles move independently.

Although each particle in a lattice is in a fixed position, it isn't motionless. Each particle vibrates rapidly, and the rate of vibration is proportional to the temperature—as the temperature increases, the particles vibrate more rapidly.

Some of the particles in a solid vibrate more rapidly than others. At any temperature, some of the particles at the surface of a piece of a solid will vibrate rapidly enough to overcome the forces that attract them to the other particles in the solid. These particles will break free from the solid and form a gas; this process is called **sublimation.**◆

◆Figure 10.10 shows a sample of solid iodine with iodine vapor (gas) above it. The vapor has been formed by sublimation.

If the temperature becomes high enough, all of the particles in a piece of a solid will vibrate so rapidly that they break loose from their fixed positions. When this happens, the lattice collapses and a liquid is formed. The temperature at which the solid changes to liquid is called the **melting point** of the solid. The melting point of a solid measures the strength of the attractive forces holding its particles together in the lattice—the stronger the attractive forces, the higher the melting point.

A liquid flows to fit the shape of its container. *(Garry Gay/The Image Bank)*

In a liquid the attractive forces between the particles are strong enough to hold the particles together, but not strong enough to hold them in fixed positions. As a result all of the particles in a liquid constantly move among one another. As shown in Figure 14.1(b), at any moment some regions of a liquid contain particles with an ordered arrangement similar to a lattice, while other regions contain particles in disorder. In the next moment all the particles will have changed positions, and new regions of order and disorder will have appeared.◆ Because the particles in a liquid are attracted to one another, a sample of a liquid has a fixed volume; because the particles are free to move among one another, a sample of a liquid can flow to fit the shape of its container.

◆The pattern of particles in the liquid state has been described as "islands of order in a sea of disorder."

In a liquid, as in a solid, the rate at which the particles move is proportional to the temperature. Some of the particles move faster than others. At any temperature, some of the particles at the surface of the liquid will move fast enough to overcome the forces that attract them to the other particles in the liquid. These rapidly moving particles will break free from the liquid and form a gas; this is the process of **evaporation.**

If the temperature of a liquid becomes high enough, particles in the body of the liquid will move so fast that they overcome their mutually attractive forces and form bubbles of gas. The temperature at which the liquid changes to gas in the body of the liquid is called the **boiling point** of the liquid.◆ The boiling point of a liquid measures the strength of the attractive forces that hold its particles together—the stronger the attractive forces, the higher the boiling point.

◆Boiling points and melting points vary with pressure. In discussing boiling points and melting points in this book, we'll always assume a pressure of one atmosphere.

As we saw in Chapter 13, the particles in a gas move completely independently of one another. In a gas sample there are no attractive forces between the particles. As a result, the random motion of the particles causes the sample to adopt the shape and volume of any container it's in, as shown in Figure 14.1(c).◆

◆As mentioned in Chapter 13, the assumption that there are no attractive forces between gas particles is true only for ideal gases. In this book we'll assume that all gases behave as if they were ideal gases.

Figure 14.1 shows that the changes of state caused by a temperature increase can be reversed by a temperature decrease. The reverse of boiling is **condensing,** and the reverse of melting is **freezing.**

From the preceding general description of solids and liquids you can see that their behavior depends on the attractive forces between their particles. The remaining Sections of this chapter describe the several kinds of interparticle forces that occur in liquids and solids.

14-2 In all liquids and solids the synchronized movement of electrons in adjacent particles causes attractions between the particles

The outer part of all atoms and molecules is a cloud of moving electrons. When the atoms or molecules are packed close together, as they are in the liquid and solid states, the movements of electrons in adjacent particles influence one another, creating attractive forces between the particles.

To see how electrons in adjacent particles can influence one another, we'll use the example of the element neon. The nucleus of a neon atom is surrounded by ten moving electrons. At any given instant it's unlikely that these ten electrons will be distributed exactly evenly around the nucleus. Imagine that, at one instant, the electrons are distributed unevenly, as in the following drawing, which uses darker shading to show a greater concentration of electrons on the right side of the atom.

δ+ Ne δ−

Because the electrons are distributed unevenly, the right side of the neon atom, where the electron density is higher, will carry a slight negative charge, and the left side, where the positive charge of the nucleus is shielded by fewer electrons, will carry a slight positive charge. At each successive instant, the electrons will be distributed differently, and in each distribution there are likely to be more electrons on one side of the nucleus than on another. The following drawing shows that the overall effect of the continuous movement of the electrons will be the rapid circulation of slight negative and positive charges over the surface of the atom.

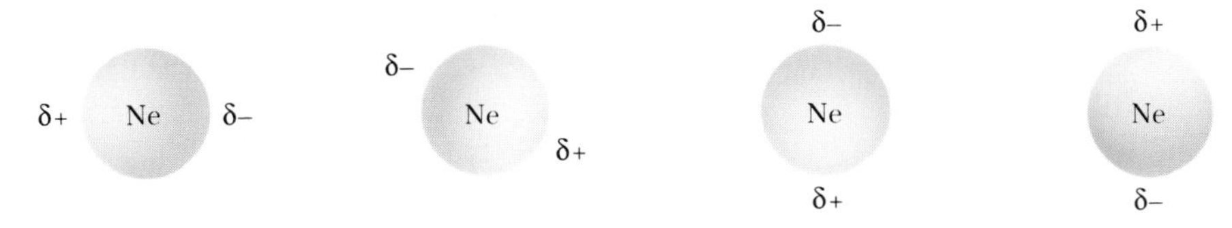

An object that carries a negative charge at one end and a positive charge at the other end is called a **dipole.** We can describe the formation and movement of the slight negative and positive charges in a neon atom by saying that the neon atom forms a series of **instantaneous dipoles.**

If two neon atoms are next to one another, their instantaneous dipoles will influence one another. Because like charges repel, the slight negative charge on the surface of one atom will repel the slight negative charge on the surface of the adjacent atom. As a result, the movements of the electrons around the two atoms will become synchronized, as shown in the following drawing. As shown

δ+ Ne δ− δ+ Ne δ−

δ− Ne δ+ δ+ Ne δ−

in this drawing, when the movements of the dipoles of the two atoms are synchronized, the slight negative charge on the surface of one atom will always be next to the slight positive charge on the surface of the adjacent atom. Because opposite charges attract, there will be an attractive force between the atoms. The attractive forces between instantaneous dipoles in liquids and solids are called **London forces,** after the German physicist, Fritz London (1900–1954), who discovered them.

London forces are present in all liquids and solids because all liquids and solids consist of closely packed particles that contain moving electrons. As you'll see in the later Sections of this chapter, other kinds of attractive forces can occur between the particles in some liquids and solids. London forces are the only attractive forces that are present in *all* liquids and solids.

Particles with larger numbers of electrons have stronger London forces. With a greater number of electrons in each particle, there is a greater possibility that the electrons will be distributed more unevenly, creating larger positive and negative charges on each particle and larger London forces between particles. Stronger London forces in liquids and solids cause higher boiling points and melting points. For this reason, the boiling points and melting points of the noble gases increase downward on the Periodic Table, as shown in Table 14.1.

◆Answers to Exercises are in Appendix 1.

Exercise 14.1◆

In liquid nitrogen and in liquid oxygen the particles are N_2 and O_2 molecules, and the only interparticle (intermolecular) forces that are present are London forces. Which liquid would you expect to have the higher boiling point? Why?

Exercise 14.2

Under ordinary room conditions chlorine is a gas and bromine is a liquid. Why?

Table 14.1 Boiling Points and Melting Points of the Noble Gases

Noble Gas	Number of Electrons	Melting Point, K	Boiling Point, K
Helium	2	1*	4
Neon	10	24	27
Argon	18	84	87
Krypton	36	104	121
Xenon	54	133	164
Radon	86	202	211

*Under high pressure.

14-3 Molecules with positive and negative ends attract one another

In a molecule of hydrogen chloride, the covalent bond between the hydrogen atom and the chlorine atom is polar because chlorine is more electronegative than hydrogen. The electrons in the bond are held more closely by the chlorine atom, so the chlorine end of the bond carries a partial negative charge and the hydrogen end carries a partial positive charge:

$$\overset{\delta+}{\mathrm{H}}-\overset{\delta-}{\underset{\cdot\cdot}{\overset{\cdot\cdot}{\mathrm{Cl}}}}:$$

Because the hydrogen chloride molecule is linear, the displacement of the bonding electrons toward the chlorine atom produces a slight negative charge on the chlorine end of the molecule and a slight positive charge on the hydrogen end of the molecule: The molecule is a dipole. If two molecules of hydrogen chloride are close together, their ends carrying the same charge will repel one another and their ends carrying opposite charges will attract one another. As a result, the molecules will align themselves head-to-tail:

$$\begin{array}{c} \overset{\delta+}{\mathrm{H}}-\overset{\delta-}{\underset{\cdot\cdot}{\overset{\cdot\cdot}{\mathrm{Cl}}}}: \\ :\underset{\delta-}{\underset{\cdot\cdot}{\overset{\cdot\cdot}{\mathrm{Cl}}}}-\underset{\delta+}{\mathrm{H}} \end{array}$$

The mutually attractive forces that arise between molecules that are dipoles are called **dipole–dipole interactions.**

In solid hydrogen chloride, the dipole–dipole interactions between the molecules hold them head-to-tail in a lattice, as shown in Figure 14.2(a). In

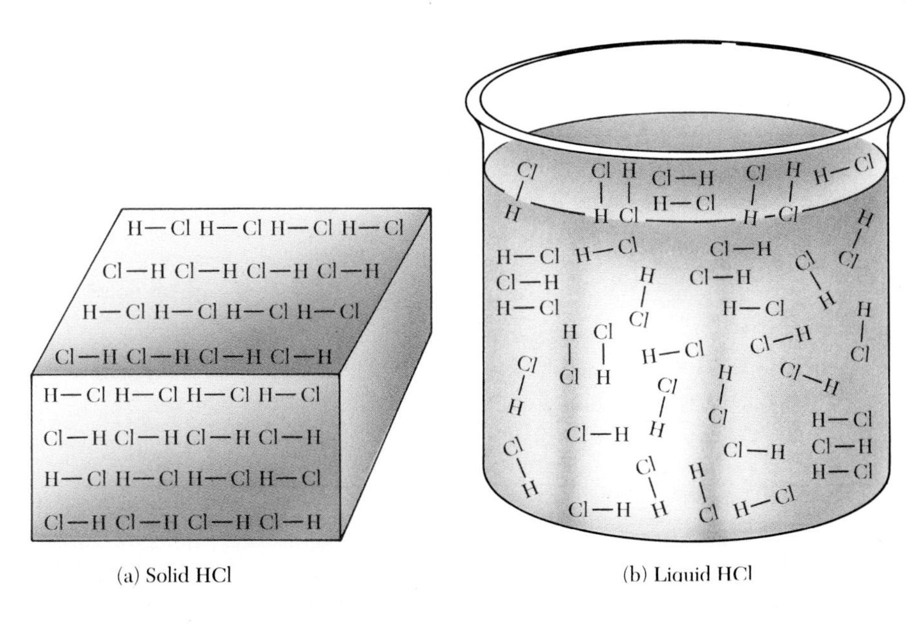

Figure 14.2
Solid and liquid hydrogen chloride. (a) In the solid state the molecules are aligned head-to-tail in a lattice because of their dipole–dipole interactions. (b) In the liquid state at each instant some of the molecules are aligned head-to-tail and some are not.

liquid hydrogen chloride, shown in Figure 14.2(b), the molecules are free to move among one another; at any instant, some of the molecules are in a highly ordered arrangement, similar to a lattice, and others are not.

Some diatomic molecules (molecules that contain only two atoms) are dipoles, and some are not. It's easy to decide whether a diatomic molecule is a dipole: If the bond is polar, the molecule is a dipole. For example, the molecules HF, CO, and IBr are dipoles, and the molecules H_2, N_2, and I_2 are not.

Exercise 14.3

Decide whether each of these molecules is a dipole: (a) O_2 (b) NO (c) Br_2.

Exercise 14.4

(a) What intermolecular forces are present in liquid hydrogen?
(b) What intermolecular forces are present in liquid hydrogen chloride?
(c) Which substance would you expect to have the higher boiling point, liquid hydrogen or liquid hydrogen chloride? Why?

◆The decision as to whether a molecule with more than two atoms will be a dipole depends on knowing the shape of the molecule. As mentioned in Section 7-6, a discussion of the shapes of molecules belongs in more advanced courses in chemistry.

If a molecule consists of more than two atoms, it's more difficult to predict whether it will be a dipole.◆ We'll limit our discussions of dipoles to diatomic molecules, with one very important exception—the water molecule.

As we saw in Section 7-6, the water molecule is bent, with an angle of 104° between the two H—O bonds. Oxygen is more electronegative than hydrogen, so each H—O bond is polar, with a partial positive charge on the hydrogen atom and a partial negative charge on the oxygen atom:

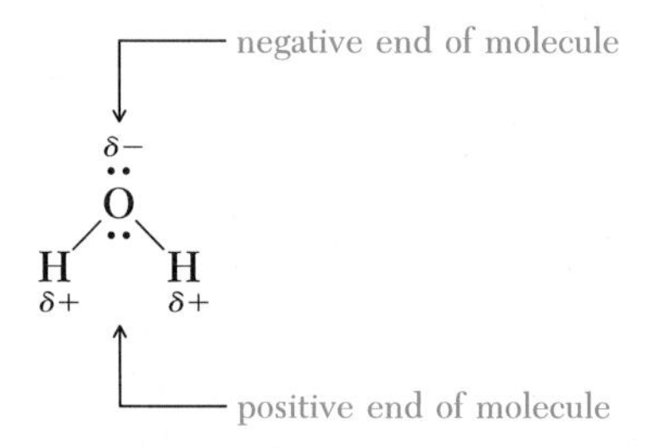

Because the bonds are polar and the molecule is bent, the water molecule has a negative end (the oxygen end) and a positive end (the hydrogen end): The water molecule is a dipole.◆

◆The fact that the water molecule is a dipole will be important in the discussion of aqueous solutions in Chapters 15 and 16.

Exercise 14.5

Write line formulas for two water molecules next to one another, showing the bent shape of each molecule and arranging them as you would expect them to be aligned because of the charges on their dipoles.

Exercise 14.6

Use line formulas to show how you would expect a water molecule and a molecule of hydrogen chloride to align themselves next to one another.

14-4 In the liquid and solid states of some compounds hydrogen atoms act as links between molecules

In a molecule of hydrogen fluoride the hydrogen atom is attached to the fluorine atom by a single polar covalent bond:

```
δ+  δ−
     ••
H—F:
     ••
```

In liquid or solid hydrogen fluoride the hydrogen atom in one HF molecule can also form a second, weaker bond with the fluorine atom in an adjacent molecule. The weaker bond is called a **hydrogen bond.** In the following examples each hydrogen bond is represented by a row of dots.

```
                                            ••          ••
                          ••             H—F:···H—F:
     ••         ••     H···:F:              ••          ••
  H—F:···H—F:          |   |                :
     ••         ••    :F:···H               H
                       ••                   |
                                           :F:
                                            ••
```

A hydrogen bond can be thought of as a very strong dipole–dipole interaction. Because the electronegativity of fluorine is much greater than that of hydrogen, the bond between the hydrogen atom and the fluorine atom in the hydrogen fluoride molecule is highly polar. As a result, each end of the molecule carries a relatively large partial charge, and the dipole–dipole interaction between two adjacent molecules is very strong. The unusually strong dipole–dipole interaction is a hydrogen bond.

A hydrogen bond can also be thought of as a very weak covalent bond. In forming a hydrogen bond, the hydrogen atom in one molecule attaches itself to an unshared pair of electrons on the fluorine atom in an adjacent molecule; considered in this way, a hydrogen bond is a very weak covalent bond.

Water molecules form hydrogen bonds. Each hydrogen atom in a water molecule is attached to the oxygen atom by a single polar covalent bond. The electronegativity of oxygen is much greater than that of hydrogen so the bonds are highly polar, and each end of the water dipole carries a relatively large partial charge. Because water molecules are highly polar, when water molecules are close together, hydrogen bonds form between them, as shown in the following examples:

```
              ••               ••
  H           O             O           H
   \         / ••\         / ••\       /
    :O:···H       H       H     H···:O:···H     H
   /                                   \    ••  /
  H                                     H    O
                                             ••
```

> **Exercise 14.7**
>
> Write the line formula for a water molecule, showing its bent shape. Use line formulas to show how each of the hydrogen atoms in this molecule can be linked by a hydrogen bond to another water molecule.

Exercise 14.8

Use line formulas to show hydrogen bonding (a) by the hydrogen atom of a hydrogen fluoride molecule to a water molecule and (b) by a hydrogen atom of a water molecule to a hydrogen fluoride molecule.

To form a hydrogen bond between two molecules, three conditions must be met:

One of the molecules must contain a hydrogen atom that's attached by a covalent bond to an atom of a highly electronegative element.
The other molecule must have an unshared electron pair on the atom of a highly electronegative element.
The two molecules must be close together.

These conditions are met, and hydrogen bonds occur, in the liquid or solid state of compounds whose molecules contain hydrogen atoms covalently bonded to atoms of nitrogen, oxygen, or fluorine.

Example 14.1

The line for ammonia is

```
    ..
H—N—H
    |
    H
```

(a) Would you predict that hydrogen bonds occur in liquid ammonia? Explain. (b) If your answer to (a) is yes, use line formulas to show an example of hydrogen bonding between two ammonia molecules.

Solution

(a) Yes. Each ammonia molecule contains hydrogen atoms covalently bonded to a nitrogen atom.

```
        H           H
        |           |
(b) H—N:···H—N:
        |           |
        H           H
```

Exercise 14.9

The line formula for methane is

```
    H
    |
H—C—H
    |
    H
```

(a) Would you predict that hydrogen bonds occur in liquid methane? Explain. (b) If your answer to (a) is yes, use line formulas to show an example of hydrogen bonding between two methane molecules.

Exercise 14.10

The line formula for methanol (wood alcohol) is

```
    H
    |  ..
H—C—O—H
    |  ..
    H
```

(a) Would you predict that hydrogen bonds occur in liquid methanol? Explain. (b) If your answer to (a) is yes, use line formulas to show an example of hydrogen bonding between two methanol molecules.

The concept of hydrogen bonding can sometimes explain the difference in boiling points between two liquids or the difference in melting points between two solids, as shown in the following Example and Exercise.

Example 14.2

Water boils at 100°C and hydrogen sulfide (H_2S) boils at −60°C. Suggest an explanation.

Solution

The molecules in liquid water are held together by hydrogen bonds, and the molecules in liquid hydrogen sulfide are not. (A molecule of hydrogen sulfide doesn't contain a hydrogen atom covalently bonded to nitrogen, oxygen, or fluorine, so it can't form hydrogen bonds.) Because of hydrogen bonding, the intermolecular forces in water are stronger than those in liquid hydrogen sulfide, so water has a higher boiling point.

Exercise 14.11

Ammonia (NH_3) melts at 195 K and methane (CH_4) melts at 90 K. Suggest an explanation.

In Chapters 15 and 16 we'll see that the concept of hydrogen bonding can also explain important properties of aqueous solutions.

14-5 In a piece of metal the outermost electrons from each atom move among all of the atoms, holding them together

The common properties of metals are familiar from everyday experience: Metals are solids; they can be worked into a variety of shapes, such as sheets, wires, and tubes; and they're good conductors of electricity. In this Section we'll see that these properties of metals can be explained as consequences of the interparticle forces that hold metals together. These forces are called **metallic bonding.**

From a chemical point of view, as we saw in Chapter 8, metals are elements whose atoms lose their outermost electrons easily, to form cations. This view of metals is the basis for the theory of metallic bonding. In a piece of metal the metal atoms are packed close together, and according to the theory of metallic bonding, the outermost energy levels of the atoms overlap, allowing the electrons in the outermost energy level of each atom to move among all of the overlapped energy levels, throughout the piece of metal. The sharing of their electrons holds the metal atoms together.

Metals can be worked into many shapes because metallic bonding creates strong, nondirectional forces between metal atoms. *(Charles D. Winters)*

Metallic bonding in a piece of aluminum is shown in Figure 14.3 on page 302.

Metallic bonding and covalent bonding are similar. In both forms of bonding, the orbitals of atoms overlap and the atoms share electrons through the

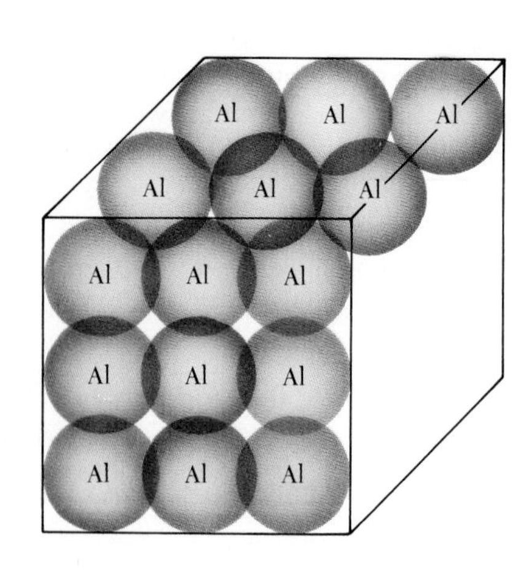

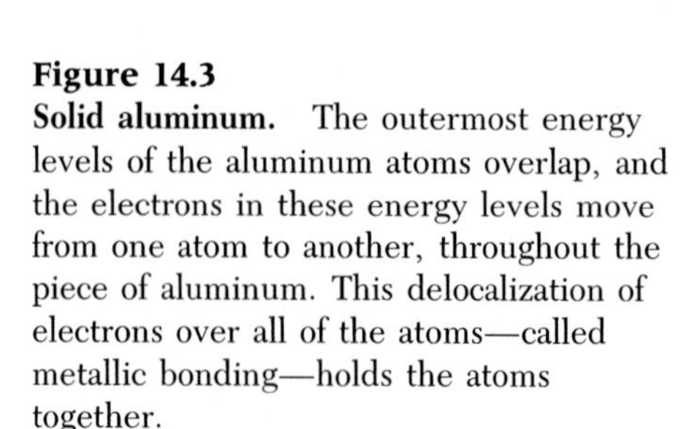

Figure 14.3
Solid aluminum. The outermost energy levels of the aluminum atoms overlap, and the electrons in these energy levels move from one atom to another, throughout the piece of aluminum. This delocalization of electrons over all of the atoms—called metallic bonding—holds the atoms together.

overlapped orbitals. In ordinary covalent bonding, electron sharing occurs between only two atoms, but in metallic bonding it occurs among all of the atoms throughout an entire piece of metal. Because the electrons in a piece of metal are free to move among many atoms, the electrons are said to be **delocalized.**◆

◆A piece of metal can be described, at the atomic level, as an orderly array of cations immersed in a sea of moving electrons.

Exercise 14.12

The electron configuration for aluminum is $1s^22s^22p^63s^23p^1$. Circle the part of the notation that represents the electrons shared in metallic bonding.

Copper, a typical metal, can be formed into many shapes. *(Charles D. Winters)*

The concept of metallic bonding explains why metals are pliable. The atoms in a piece of metal can be moved with respect to one another, without breaking the bonding, because the outermost energy level of each atom can overlap with the outermost energy levels of adjacent atoms in any direction. As a result, pieces of metal can be formed into many shapes.

Metallic bonding also explains why metals conduct electricity. If a stream of electrons—an electric current—flows into one end of a metal wire, the electrons are free to move through the wire and out its other end. The movement of electrons through the wire is a flow of electricity.

Exercise 14.13

Use the concept of metallic bonding to explain why it's easy to bend a wire but hard to pull it apart.

Among the 83 elements that are metals, the strength of metallic bonding varies widely. Mercury, for example, melts at −39°C and boils at 357°C, while platinum melts at 1174°C and boils at 4530°C.

14-6 Solids in which the interparticle forces are covalent or ionic bonds have very high melting points

The strongest attractive forces that can occur between the particles in a solid are covalent or ionic bonds. Because these interparticle forces are very strong, the solids in which they occur—called **covalent solids** and **ionic solids**—have very high melting points.◆

◆Covalent and ionic liquids, which exist only at very high temperatures, are unusual forms of matter that aren't important for our purposes.

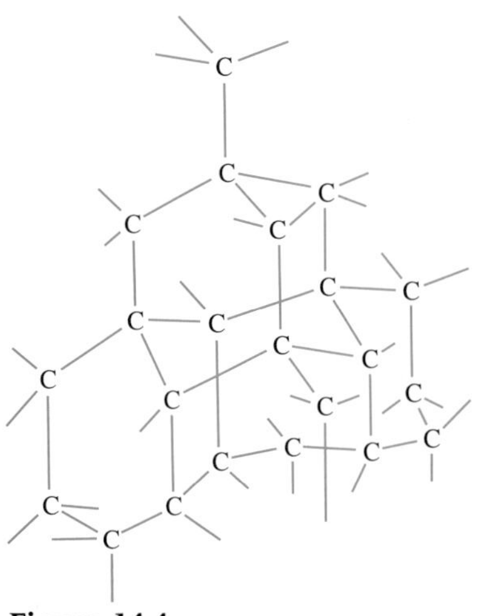

(Gemological Institute of America)

Figure 14.4
Diamond. Each carbon atom in a diamond is attached to four other carbon atoms by single nonpolar covalent bonds. Because these bonds are very strong interparticle forces, diamond has a very high melting point, 3550°C.

Diamond is a simple example of a covalent solid.◆ In a diamond, as shown in Figure 14.4, each carbon atom is attached to four other carbon atoms by single nonpolar covalent bonds. The resulting structure is very strong: Diamond, the hardest substance known, melts at 3550°C.

◆As mentioned in Section 10-4, diamond is one of the allotropic forms of carbon.

Quartz is an example of a covalent solid that contains two kinds of atoms. In the quartz lattice, shown in Figure 14.5, alternating silicon and oxygen atoms are attached to one another by single polar covalent bonds. Quartz melts at 1610°C.

As you can see in Figures 14.4 and 14.5, the bonds in covalent solids form a network. For this reason covalent solids are also called **network solids.**

Figure 14.5
Quartz. In quartz each silicon atom is attached to four oxygen atoms and each oxygen atom is attached to two silicon atoms, by single polar covalent bonds. Because the bonds are strong interparticle forces, quartz has a high melting point, 1610°C.

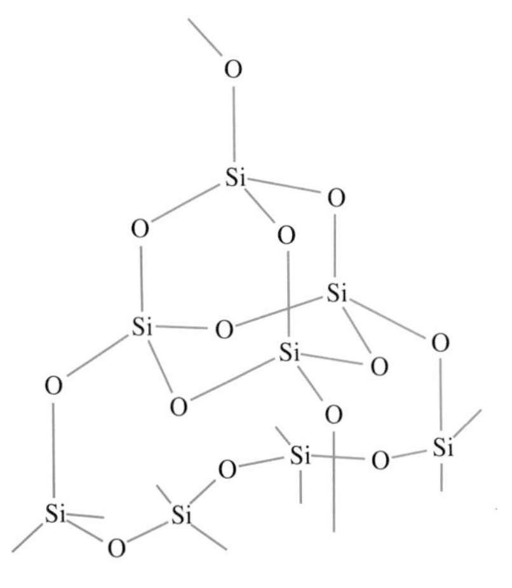

(Gemological Institute of America)

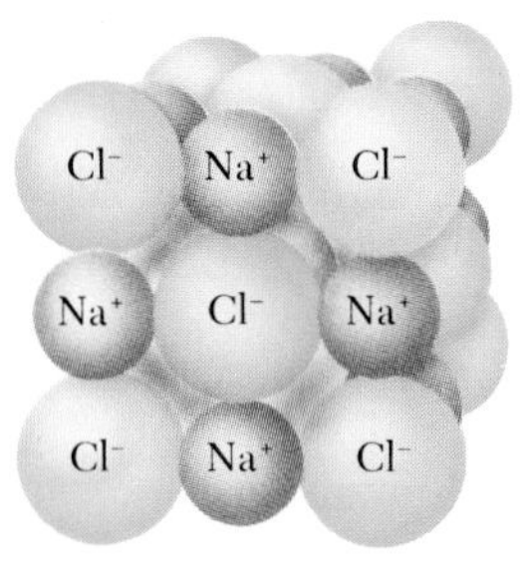

Figure 14.6
Sodium chloride. Each sodium ion is surrounded by chloride ions, and each chloride ion is surrounded by sodium ions. Because ionic bonds are strong interparticle forces, the melting point of sodium chloride is high, 804°C.

Crystals of sodium chloride, NaCl, magnified 40×. *(Runk/Schoenberger from Grant Heilman)*

The most familiar example of an ionic solid is sodium chloride, whose structure is shown in Figure 14.6. In a crystal of sodium chloride, as in all ionic solids, each cation is surrounded by anions and each anion is surrounded by cations. The attractions between the ions throughout the crystal create a very strong structure: Sodium chloride melts at 804°C.

Other ionic solids have similarly high melting points. Some examples are calcium carbonate ($CaCO_3$) 825°C, potassium sulfate (K_2SO_4) 1067°C, and magnesium fluoride (MgF_2) 1248°C.

Exercise 14.14

Silicon carbide, SiC, is a very hard substance that melts above 2200°C. Of the three substances whose structures are shown in Figures 14.4, 14.5, and 14.6, which would you expect silicon carbide to resemble most closely? Explain your answer.

A large crystal of silicon carbide, SiC. *(Charles D. Winters)*

14-7 By predicting the interparticle forces in solids or liquids, you can sometimes predict their relative melting points or boiling points

In the preceding Sections we've seen that several kinds of attractive forces can occur between the particles—atoms, ions, or molecules—in solids and liquids. Table 14.2 summarizes these forces.

As shown in Table 14.2, different kinds of interparticle forces have different strengths. The stronger the interparticle forces in a solid, the higher its melting point, and the stronger the interparticle forces in a liquid, the higher its boiling point.

Because melting points and boiling points depend on the strengths of interparticle forces, you can sometimes predict whether a solid or a liquid will have a high or low melting point or boiling point by predicting the kinds of forces

Table 14.2 Summary of Attractive Forces That Can Occur Between Particles in Liquids and Solids

Interparticle Force	Occurs Between	Description	Relative Strength
London forces	All atoms, ions, or molecules	Attractions between opposite charges on instantaneous dipoles, caused by the synchronized movement of electrons in adjacent particles	Variable from weak to strong; proportional to the number of electrons per particle
Dipole–dipole interactions	Molecules that are dipoles	Attractions between opposite charges on adjacent dipoles	Weak
Hydrogen bonding	Molecules that contain H bonded to N, O, or F	H atom in one molecule is bonded weakly to N, O, or F in an adjacent molecule	Moderate
Metallic bonding	Metal atoms	Electron sharing (delocalization) throughout all of the atoms in a piece of metal	Strong
Covalent bonds	Atoms in a network solid	Electron sharing between adjacent atoms	Strong
Ionic bonds	Cations and anions	Attractions between opposite charges on adjacent ions	Strong

that hold its particles together. Table 14.3 on page 306 shows questions you can ask to predict the kinds of interparticle forces that will be present in a solid or a liquid.

The following Examples show predictions of melting points and boiling points, based on the predictions of interparticle forces shown in Table 14.3.

Example 14.3

For each pair, predict which substance will have the higher melting point, and explain your answer: (a) Ne or Na (b) NaF or F_2 (c) NH_3 or CH_4

Solution

(a) Ne is atomic and Na is metallic, so Na should have a much higher melting point. (The actual values are −249°C for Ne and 98°C for Na.)
(b) NaF is ionic and F_2 is nondipole molecular, so NaF should have the higher melting point. (The actual values are 993°C for NaF and −223°C for F_2.)
(c) NH_3 and CH_4 each contain 10 electrons, so the strength of their London forces should be about the same. NH_3 can form hydrogen bonds and CH_4 cannot, so NH_3 should have the higher melting point. (The actual values are −78°C for NH_3 and −183°C for CH_4.)

Example 14.4

The compound butane consists of nondipole molecules with the formula C_4H_{10}, and the compound octane consists of nondipole molecules with the formula C_8H_{18}. At room temperatures one of these compounds is a liquid and the other is a gas. Predict which compound is a liquid, and explain your prediction.

Table 14.3 Questions To Ask in Predicting the Kinds of Interparticle Forces That Will Be Present in a Solid or a Liquid

Question	If answer is yes, these attractive forces are present, in addition to London forces.*	Prediction for melting point and boiling point	Example (followed by melting and boiling points, in °C)
Metallic? Is the substance a metal?	Metallic bonding	High	Iron, Fe (1555, 3000)
Ionic? Is the substance ionic?	Ionic bonding	High	Sodium chloride, NaCl (804, no definite boiling point)
Covalent? Does the substance consist of atoms held together by a network of covalent bonds?	Covalent bonding	High	Diamond, C (3550, no definite boiling point)
Dipole molecular? Does the substance consist of molecules that are dipoles?	Dipole–dipole interactions	Low	Hydrogen chloride, HCl (−114, −85)
H-bonded? In the molecule, is H attached by a covalent bond to N, O, or F?	Hydrogen bonding	Medium	Water, H_2O (0, 100)
Nondipole molecular? Does the substance consist of molecules that are not dipoles?	Only London forces	Low if number of electrons is small, high if number of electrons is large	Hydrogen, H_2 (−257, −253) Iodine, I_2 (114, 183)
Atomic? Does the substance consist of atoms with no covalent bonds between them?	Only London forces	Low	Neon, Ne (−249, −246)

*As we saw in Section 14-2, London forces are present in *all* liquids and solids.

Solution

Because the particles in each of these compounds are nondipole molecules, the only interparticle (intermolecular) forces will be London forces. Octane (C_8H_{18}) has more carbon atoms and more hydrogen atoms than butane (C_4H_{10}), so octane has more electrons and will have the higher boiling point. Because octane has the higher boiling point, it will be the liquid at room temperatures. (The actual boiling points are −0.5°C for butane and 126°C for octane.)

Exercise 14.15

For each pair, predict which substance will have the higher melting point, and explain your answer: (a) Cl_2 or NaCl (b) F_2 or ClF (c) H_2 or O_2

Exercise 14.16

The following compounds have increasing boiling points in the order shown: PH_3 (phosphine), AsH_3 (arsine), and NH_3 (ammonia). Suggest an explanation.

Octane, C_8H_{18}, is used as the basis for the octane rating of gasolines. *(Larry Lefever/Grant Heilman)*

Inside Chemistry Why doesn't all of the water in a lake freeze?

When most liquids freeze, the crystals of solid that form sink in the liquid, but when water freezes, the ice crystals that form rise in the liquid. This unusual behavior of ice, shown in Figure 14.7, occurs because ice is less dense than water, whereas for most substances the solid is more dense than the liquid.

A lake starts to freeze if the air temperature drops below 0°C, and it freezes first on the surface, where the water is exposed to the cold air. As ice forms on the surface of the water, it floats, and because ice is a poor conductor of heat, the ice insulates the water below it and prevents the lake from freezing solid. If ice were denser than water, the ice would sink as it formed, so that all of the water would eventually be exposed to the cold air, and the lake would freeze solid.

The unusual behavior of ice that causes it to form an insulating layer on the surface of a lake has important biological consequences. Because lakes don't usually freeze solid, the plants and animals that live in them are able to survive even when air temperatures stay below freezing for many weeks.

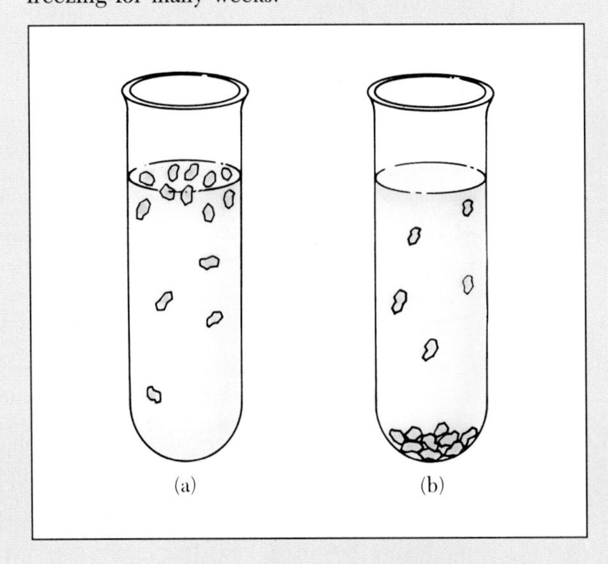

Figure 14.7
Freezing samples of water and carbon tetrachloride. (a) If a test tube containing water is cooled below 0°C, the ice crystals that form will rise to the top of the water because ice is less dense than water. (b) If a test tube containing carbon tetrachloride, CCl_4—a clear, colorless liquid that freezes at −23°C—is cooled below its freezing point, the crystals of carbon tetrachloride that form will sink to the bottom of the liquid because the solid is denser than the liquid. The behavior of carbon tetrachloride is typical of that for most liquids.

Chapter Summary: Liquids, Solids, and Changes of State

Section	Subject	Summary	Check When Learned
	Most important difference between a gas and a liquid or a solid, from the viewpoint of kinetic theory	In a liquid or a solid there are attractive forces between the particles; in a gas it's assumed that there are no attractive forces.	☐
14-1	Reason for fixed volume and shape of a piece of a solid	The particles in a solid are attracted to one another by forces that hold them in fixed, ordered positions.	☐
14-1	Lattice	Ordered arrangement of particles in a solid.	☐
14-1	Relationship of particle motion to temperature, in a solid	Each particle vibrates rapidly in its lattice position. The rate of vibration is proportional to the temperature.	☐
14-1	Sublimation	Movement of particles from the solid state to the gas state.	☐
14-1	Melting	Breakdown of lattice and conversion of solid to liquid.	☐
14-1	Freezing	Reverse of melting: formation of a solid from a liquid.	☐
14-1	Melting point	Temperature at which melting occurs.	☐
14-1	Relationship between melting point and strength of interparticle forces	The stronger the attractive interparticle forces in a solid, the higher its melting point.	☐
14-1	Reason for fixed volume and variable shape of a sample of a liquid	The particles in a liquid are attracted to one another by forces that hold them together, but the particles are free to move among one another.	☐
14-1	Relationship of particle motion to temperature, in a liquid	The rate of particle motion is proportional to the temperature.	☐
14-1	Evaporation	Movement of particles from the liquid state to the gas state, at the surface of a liquid.	☐
14-1	Boiling	Formation of gas bubbles in the body of a liquid.	☐
14-1	Condensation	Reverse of evaporation or of boiling: formation of a liquid from a gas.	☐
14-1	Boiling point	Temperature at which boiling occurs.	☐

Section	Subject	Summary	Check When Learned
14-1	Relationship between boiling point and strength of interparticle forces	The stronger the attractive interparticle forces in a liquid, the higher its boiling point.	☐
14-1	Reasons for variable volume and variable shape of a sample of gas	Because there are no attractive forces between the particles of a gas (an ideal gas), the random movement of the particles in a gas sample cause it to fill any container it's in.	☐
14-2	Dipole	Object that carries a negative charge at one end and a positive charge at the other end.	☐
14-2	London forces	Attractions between opposite charges in instantaneous dipoles, caused by the synchronized movement of electrons in adjacent particles.	☐
14-2	Kinds of liquids and solids in which London forces occur	All liquids and solids.	☐
14-2	Relationship between numbers of electrons in particles and strength of London forces	Particles with larger numbers of electrons have stronger London forces; stronger London forces cause higher melting points and boiling points.	☐
14-3	Dipole–dipole interactions	Attractions between opposite charges on adjacent dipoles.	☐
14-3	Kinds of liquids and solids in which dipole–dipole interactions occur	Liquids and solids consisting of molecules that are dipoles.	☐
14-3	Rule for deciding whether a diatomic molecule is a dipole	If the bond is polar, the molecule is a dipole. For example, HCl is a dipole and Cl_2 is not.	☐
14-3	Distribution of charges in water molecule.	Because oxygen is more electronegative than hydrogen, and because the water molecule is bent, the molecule is a dipole. The oxygen end carries a partial negative charge, and the hydrogen end carries a partial positive charge: negative end of molecule → O (δ−); H (δ+), H (δ+) ← positive end of molecule	☐

Section	Subject	Summary	Check When Learned
14-4	Hydrogen bond	In a liquid or a solid, a weak bond that forms between a hydrogen atom in one molecule and a nitrogen, oxygen, or fluorine atom in an adjacent molecule. This example shows a hydrogen bond between two HF molecules: $H-\underset{\cdot\cdot}{\ddot{F}}:\cdots H-\underset{\cdot\cdot}{\ddot{F}}:$	☐
14-4	Conditions for forming a hydrogen bond between two molecules	(1) One molecule must contain an H atom attached by a covalent bond to an atom of a highly electronegative element. (2) The other molecule must have an unshared electron pair on the atom of a highly electronegative element. (3) The two molecules must be close together. Summary: Hydrogen bonds occur in liquids or solids whose molecules contain H atoms covalently bonded to N, O, or F.	☐
14-5	Metallic bonding	Electron sharing (delocalization) throughout all of the atoms in a piece of metal.	☐
14-6	Covalent or network solid	Solid in which all of the atoms are attached to one another by a network of covalent bonds. An example is diamond, C.	☐
14-6	Ionic solid	Solid that consists of ions attached to one another by ionic bonds. An example is sodium chloride, NaCl.	☐
14-6	Characteristic properties of covalent and ionic solids	They typically have very high melting points.	☐
14-7	Relative strengths of interparticle forces:		
	London forces	Variable from weak to strong; proportional to the number of electrons per particle.	
	dipole–dipole interactions	Weak.	
	hydrogen bonding	Moderate.	
	metallic bonding	Strong.	
	covalent bonds	Strong.	
	ionic bonds	Strong.	☐
14-7	Questions to ask to predict the kinds of interparticle forces that will be present in a solid or a liquid	Metallic? (Is the substance a metal?) Ionic? (Is the substance ionic?) Covalent? (Does the substance consist of atoms held together by a network of covalent bonds?) Dipole molecular? (Does the substance consist of molecules that are dipoles?) H-bonded? (In the molecule, is H attached by a covalent bond to N, O, or F?) Nondipole molecular? (Does the substance consist of molecules that are not dipoles?) Atomic? (Does the substance consist of atoms with no covalent bonds between them?)	☐

Problems

Assume you can use the Periodic Table at the front of the book unless you're directed otherwise. Answers to odd-numbered Problems are in Appendix 1.

States of Matter and Changes of State (Section 14-1)

1. The compound naphthalene, $C_{10}H_8$, whose line formula is shown in Table 7.2, is a white solid at ordinary room temperatures. Naphthalene is used to make mothballs: When small balls of naphthalene are mixed with stored clothes or blankets, the balls slowly disappear, releasing vapors of naphthalene, which repel moths. What process is occurring as the balls disappear?
2. On sunny days piles of snow may become smaller, even though the temperature stays below 0°C. Suggest an explanation.
3. If a beaker containing water is placed on a table in a room, the water level in the beaker will slowly fall until all of the water disappears. Explain this process.
4. If a beaker containing water is placed on a table under a bell jar—a heavy glass dome—as shown below, the water level in the beaker will slowly fall until it reaches a certain level, and then stop. Explain.

5. In your own words, define *melting point* and *boiling point* in terms of (a) the observed behavior of solids and liquids and (b) the behavior of the particles in solids and liquids.
6. Distinguish between the terms *melting point* and *freezing point*.
7. Which of the three states of matter is the most highly ordered? Which is the most disordered?
8. As the temperature of a solid drops, how do the motions of its particles change?
9. In your own words, explain why a liquid adopts the shape of its container but a solid does not.
10. In your own words, explain why a gas adopts the volume of its container but a liquid does not.
11. In your own words, distinguish between evaporation and boiling.
12. Describe the processes of condensation and freezing in terms of the behavior of the particles in gases, liquids, and solids.

London Forces (Section 14-2)

13. Why does argon have stronger London forces than neon?
14. Oxygen, O_2, boils at 90 K and ozone, O_3, boils at 161 K. Suggest an explanation.
15. Chlorine freezes at −187°C and bromine freezes at −7°C. Suggest an explanation.
16. Nitrogen, oxygen, fluorine, and neon are the last four elements in the second period on the Periodic Table; their respective boiling points are 77 K, 90 K, 86 K, and 27 K. Suggest an explanation for the fact that the boiling point for neon is significantly lower than that for nitrogen, oxygen, or fluorine.

17. Although there are exceptions, in general melting points and boiling points increase downward on the Periodic Table. Suggest an explanation for this general trend.

18. No covalent compound with a molecular mass greater than about 150 amu is a gas at room temperatures. Suggest an explanation.

Dipole–Dipole Interactions (Section 14-3)

19. Explain why a molecule of hydrogen chloride is designated as a dipole, while a neon atom is designated as an *instantaneous* dipole.

20. Why don't two molecules of hydrogen chloride align themselves as shown below?

```
     ..
  H—Cl:
     ..
     ..
  H—Cl:
     ..
```

21. Bromine boils at 59°C, and hydrogen bromide boils at −67°C. Suggest an explanation.

22. Write line formulas for two molecules of carbon monoxide next to one another, arranging them as you would expect them to be aligned because of the charges on their dipoles.

23. Using line formulas, show how you would expect six water molecules to arrange themselves around a sodium ion. Draw the sodium ion about the size of one of the water molecules.

24. Using line formulas, show how you would expect six water molecules to arrange themselves around a chloride ion. Draw the chloride ion about the size of one of the water molecules.

25. Using line formulas, show how you would expect six water molecules to arrange themselves around a molecule of hydrogen chloride. Draw the HCl molecule about twice the size of one of the water molecules, and show three water molecules around each end of the HCl molecule.

26. Write line formulas for two molecules of iodine monobromide next to one another, arranging them as you would expect them to be aligned because of the charges on their dipoles.

Hydrogen Bonding (Section 14-4)

27. Write line formulas for four water molecules, and arrange them to show three different ways that they can form hydrogen bonds with one another.

28. Name the interparticle (intermolecular) forces in (a) liquid hydrogen (b) liquid fluorine (c) liquid hydrogen fluoride.

29. Using line formulas, show how one water molecule could form hydrogen bonds with four other water molecules.

30. Using line formulas, show how one ammonia molecule could form hydrogen bonds with four water molecules.

31. Methane, CH_4, has a lower boiling point than silane, SiH_4; phosphine, PH_3, has a lower boiling point than ammonia, NH_3. Suggest an explanation.

32. The two compounds shown below contain the same number of electrons, but compound (a) boils at −24°C, and compound (b) boils at 79°C. Suggest an explanation.

```
        H       H                 H  H
        |  ..   |                 |  |  ..
(a) H—C—O—C—H        (b) H—C—C—O—H
        |  ..   |                 |  |  ..
        H       H                 H  H
```

33. Explain why boiling points increase in this order: hydrogen, oxygen, water, hydrogen peroxide.

34. Explain why boiling points increase in this order: hydrogen, nitrogen, ammonia, hydrazine (N_2H_4).

Metallic Bonding (Section 14-5)

35. Each alkali metal has a much higher melting point than the noble gas in the same period. Why?

36. Draw a picture of a lithium atom: Use the symbol Li to represent its nucleus, and use two concentric circles to represent its occupied energy levels. Use six of these pictures to represent a lithium wire, according to the theory of metallic bonding.

37. Use the drawing you made in problem 36 to show how the lithium wire could be bent.

38. Use the drawing you made in problem 36 to show how an electron, from an exterior electric current, could move through the lithium wire.

Covalent and Ionic Bonds (Section 14-6)

39. Name the kind of bonding that occurs between (a) metal atoms and metal atoms (b) metal atoms and nonmetal atoms (c) nonmetal atoms and nonmetal atoms.

40. At ordinary room temperatures, sodium is a soft solid, chlorine is a gas, and sodium chloride is a hard solid. Explain these properties.

Predicting Melting and Boiling Points from Interparticle Forces (Section 14-7)

41. At low enough temperatures, all substances form liquids and solids. Why?

42. All of the elements that are gases at room temperatures are nonmetals. Suggest an explanation.

43. At room temperatures, two of the elements in group 7A are gases, but all of the elements in group 8A are gases. Suggest an explanation.

44. Predict whether oxygen or chlorine will have the higher boiling point, and explain your prediction.

45. Imagine that you're told that a compound is a gas at room temperatures, and that it is either hydrogen bromide or potassium bromide. How would you decide between these possibilities?

46. Imagine that you're shown one bottle containing a white crystalline solid and another containing a clear, colorless liquid. If the choices for the formulas for the solid and the liquid are C_6H_{14} and $C_{30}H_{62}$, how will you decide which is which?

47. Which of the compounds whose formulas follow would you expect to have the higher boiling point? Explain.

```
      H  H              H  H
      |  |              |  |
(a) H—C—C—H      (b) H—C—N:
      |  |              |  |
      H  H              H  H
```

48. Would you expect that two ammonium ions, NH_4^+, could form hydrogen bonds with one another? Explain.

49. Which of the interparticle forces is responsible for the fact that most of the elements are solids at room temperatures?

50. It can be estimated that, if water molecules didn't form hydrogen bonds, the boiling point of water would be about −100°C. Suggest some ways in which our planet would be different if water molecules didn't form hydrogen bonds.

15-1
In an aqueous solution there are attractive forces between the water molecules and the solute particles.

15-2
If the attractive forces between water molecules and solute particles are weak, only a small amount of the solute will dissolve.

15-3
The molarity of a solution is the number of moles of solute that are present in exactly one liter of the solution.

15-4
In pure water and in any aqueous solution there is a reaction between water molecules: $2\ H_2O(\ell) \rightleftarrows H_3O^+(aq) + OH^-(aq)$.

15-5
Strong and weak acids are compounds that produce hydronium ion, H_3O^+, in aqueous solutions.

15-6
Strong and weak bases are compounds that produce hydroxide ion, OH^-, in aqueous solutions.

15-7
The concentration of an acidic or basic solution can be expressed in molarity or in pH units: $pH = -\log[H_3O^+]$.

15-8
Acids and bases can be defined in other ways.

15-9
The most soluble ionic compounds consist of large ions with single charges.

Inside Chemistry:
How do soaps and detergents work?

Chapter Summary

Problems

Many ions of transition metals are colored in aqueous solution. From left to right, these aqueous solutions contain Cu^{2+}, Fe^{3+}, Zn^{2+}, Cr^{3+}, Co^{2+}, Ni^{2+}, and Mn^{2+}. *(Marna G. Clarke)*

15

Introduction to Aqueous Solutions

A solution, as we saw in Section 3-1, is a homogeneous mixture of two or more substances. An **aqueous solution** is a solution in which a substance, or more than one substance, is dissolved in water. In an aqueous solution water is designated as the **solvent,** and each dissolved substance is a **solute.**†

We know from everyday experience that aqueous solutions are uniquely important on our planet: Most of the surface of the earth is covered by aqueous solutions, and most of the mass of human beings and other animals consists of aqueous solutions. Because aqueous solutions are so widespread and so important, scientists have studied them intensively for many years and have learned much about them.

In this chapter we'll look at some of the fundamental properties of aqueous solutions. We'll give special attention to three classes of aqueous solutions that are of great importance in chemistry: solutions of ionic compounds, solutions of **acids,** and solutions of **bases.**◆

◆Before continuing with this chapter, you may want to review the nomenclature of covalent and ionic compounds in Sections 9-1 and 9-2.

15-1 In an aqueous solution there are attractive forces between the water molecules and the solute particles

When a substance dissolves in water, its particles—atoms, molecules, or ions—become attached to water molecules by some of the kinds of attractive forces described in Chapter 14. The particular kinds of interparticle forces that are present in an aqueous solution will depend on the nature of its solute particles.

In all aqueous solutions there are weak London forces between the water molecules and the solute particles. Water molecules and solute particles are packed close together in aqueous solutions. Because of this close packing, the electrons in water molecules and in adjacent solute particles move synchronously, and their synchronous movement creates London forces, as described in Section 14-2. The strength of London forces between particles increases with increasing numbers of electrons in the particles; because a water molecule contains only ten electrons, it can form only weak London forces. For this reason, London forces between water molecules and solute particles are always weak.

Exercise 15.1◆

Which would you expect to form stronger London forces with water molecules, helium atoms or neon atoms? Why?

◆Answers to Exercises are in Appendix 1.

† **Chemistry Insight**

There are more general definitions of solvent and solute that can be applied to all solutions: The **solvent** is the component of a solution that has retained its physical state during the solution process and that is present in the largest amount; any other component of the solution is a **solute.** In a mixture of 60% nitrogen gas and 40% oxygen gas, for example, nitrogen is the solvent, and in a solution made by dissolving 1 g of solid sodium chloride in 100 mL of liquid water, water is the solvent.

Because water molecules are dipoles, they form dipole–dipole interactions with solute molecules that are dipoles. For example, dipole–dipole interactions occur between water molecules, $\overset{\delta+}{H}-\overset{\delta-}{\ddot{\underset{\cdot\cdot}{O}}}-\overset{\delta+}{H}$, and molecules of carbon monoxide, $:\overset{\delta+}{C}\equiv\overset{\delta-}{O}:$. Opposite charges on adjacent dipoles attract one another, and like charges on adjacent dipoles repel one another. As a result, in an aqueous solution of carbon monoxide, the molecules of water and carbon monoxide are arranged in various ways with their opposite charges next to one another. These are two examples:

$$\overset{\delta-}{:\ddot{O}}(\overset{\delta+}{H})_2 \quad \cdots \quad :\overset{\delta+}{C}\equiv\overset{\delta-}{O}: \qquad\qquad (\overset{\delta+}{H})_2\ddot{O}:^{\delta-} \cdots :\overset{\delta+}{C}\equiv\overset{\delta-}{O}: \cdots :\ddot{O}:^{\delta-}(\overset{\delta+}{H})_2$$

Exercise 15.2

Would you expect water molecules to form dipole–dipole interactions with nitrogen molecules? Explain.

◆The features that must be present in a molecule to allow hydrogen bonding are described in Section 14-4.

◆As the line formula shows, the hydrogen peroxide molecule is bent.

Water molecules form hydrogen bonds with one another and with solute molecules that contain appropriate groups of atoms.◆ One example is the hydrogen peroxide molecule, $H-\ddot{\underset{\cdot\cdot}{O}}-\ddot{\underset{\cdot\cdot}{O}}-H$.◆ The following formulas show some ways in which water molecules and molecules of hydrogen peroxide can form hydrogen bonds with one another.

$$H_2\ddot{O}: \cdots H-\ddot{O}:-H \qquad\qquad H_2\ddot{O}:\cdots H-\ddot{O}-\ddot{O}-H\cdots:\ddot{O}H_2 \;\;(\ddot{O}\text{ atoms }\cdots H-\ddot{O}-H)$$

Exercise 15.3

A molecule of hydroxylamine, $H-\ddot{N}(H)-\ddot{\underset{\cdot\cdot}{O}}-H$, can form hydrogen bonds with water molecules in a variety of ways. Use line formulas to show one example of hydrogen bonding between a hydroxylamine molecule and a water molecule.

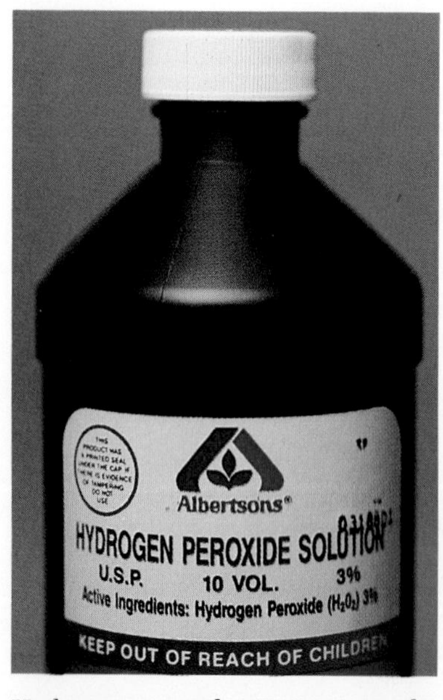

Hydrogen peroxide, H_2O_2, is used as a bleach and an antiseptic. *(David R. Frazier)*

With solute particles that are ions, water molecules form strong **ion–dipole interactions.** When an ionic compound dissolves in water, its ions become surrounded by water molecules. Because water molecules are dipoles, they arrange themselves around a cation with their oxygen (negative) ends pointing inward and around an anion with their hydrogen (positive) ends pointing inward, as shown here for sodium ion and chloride ion:

```
                                   ..
                                   O
                                  /..\
       H     H                   H    H
        \ .. /
  H      O       H               H       H
   \     ..     /               /         \
   :O:Na⁺:O:                 :O:    Cl⁻    :O:
   /     ..     \               \         /
  H      O       H               H       H
        /..\
       H    H                    H      H
                                  \ .. /
                                    O
                                    ..
```

The attractions between the opposite charges on an ion and a dipole are called **ion–dipole interactions.** Because an ion carries a greater charge than a dipole,◆ ion–dipole interactions are stronger attractive forces than dipole–dipole interactions.

◆The charges on ions are full charges, and the charges on dipoles are partial charges.

Ions with water molecules attached to them are said to be **hydrated.** The number of water molecules attached to an ion may vary. A typical sodium ion in aqueous solution, for example, may have from three to six water molecules attached to it.

Exercise 15.4

Use line formulas to show (a) four water molecules around a calcium ion (b) three water molecules around a fluoride ion.

15-2 If the attractive forces between water molecules and solute particles are weak, only a small amount of the solute will dissolve

Several factors determine the amount of a solute that will dissolve in a given quantity of water. One important factor is the strength of the attractive forces between solute particles and water molecules: If the forces are weak, only a small amount of the solute will dissolve.

For this reason, if the only attractive forces that are formed between water molecules and the particles of a solute are London forces, the solute will not be very soluble. In an aqueous solution of helium, for example, the only attractive forces between water molecules and helium atoms are weak London forces, so helium is only slightly soluble in water: At ordinary room temperatures, only 1.93 mg of helium will dissolve in 1.00 L of water.◆

◆Solubilities vary with temperature. In this book we'll assume that solutions are at ordinary room temperatures, unless otherwise stated.

Exercise 15.5

Which gas would you expect to be more soluble in water, helium or neon? Why?

In general, dipole–dipole interactions are weak attractive forces, so if the only forces between water molecules and solute molecules are dipole–dipole interactions and London forces, the solute will not be very soluble. For example, only 28.8 mg of carbon monoxide will dissolve in 1.00 L of water.

Exercise 15.6

A molecule of carbon monoxide, CO, and a molecule of nitrogen, N_2, have the same number of electrons. Which gas would you expect to be more soluble in water? Why?

If the molecules of a compound can form hydrogen bonds with water molecules, the compound will be much more soluble in water than would otherwise be expected. Hydrogen bonds, as we saw in Section 14-7, are moderately strong intermolecular forces, and they can increase the solubility of a solute substantially. For example, hydrogen, H_2, and oxygen, O_2, whose molecules cannot form hydrogen bonds, are only slightly soluble in water, but hydrogen peroxide, H_2O_2, whose molecules form hydrogen bonds with water molecules, is infinitely soluble in water. ◆

◆ Hydrogen peroxide (which is a liquid) and water form solutions in all proportions.

Exercise 15.7

Methane,

```
     H
     |
  H—C—H
     |
     H
```

and methanol,

```
     H
     |   ..
  H—C—O—H
     |   ..
     H
```

have very different solubilities in water: One is infinitely soluble and the other is only slightly soluble. Predict which compound has the greater solubility, and explain your choice.

Because ion–dipole interactions are strong interparticle forces, some ionic compounds are very soluble in water. For example, 357 g of sodium chloride will dissolve in 1.00 L of water. Not all ionic compounds are highly soluble, however, because factors other than the strength of ion–dipole interactions influence their solubility. Section 15.9 provides a detailed discussion of the factors that influence the solubility of ionic compounds.

15-3 The molarity of a solution is the number of moles of solute that are present in exactly one liter of the solution

The **concentration** of an aqueous solution is a measure of the amount of solute that is present in a given amount of the solution. For example, if each of two aqueous solutions has a volume of 100 mL, and if the first solution contains

1.00 g of dissolved sodium chloride and the second contains 2.00 g, then the second solution is twice as concentrated as the first. We'll use the concept of concentration often in discussing aqueous solutions.

Concentrations can be expressed in a variety of ways. The most general description classifies a solution as unsaturated or saturated: A solution containing less than the maximum amount of a solute it can hold is said to be **unsaturated;** a solution containing the maximum is said to be **saturated.**

The most common way of expressing chemical concentrations is in terms of molarity. The **molarity** of a solution is the number of moles of solute that are present in exactly one liter of the solution; molarity is moles of solute per liter of solution. The definition of molarity can be expressed as an equation:

$$M = \frac{\text{mol}}{\text{L}}$$

M is the molarity of a solution.
mol is the number of moles of solute in the solution.
L is the volume of the solution in liters.

This equation can be solved for any one of its three variables, if values for the other two are known.

Molarities are commonly expressed in two different ways. For example, suppose we have a solution that has been made by combining 1.00 mol (58.5 g) of sodium chloride with enough water to make 1.00 L of solution. For this solution,

$$M = \frac{\text{mol}}{\text{L}} = \frac{1.00 \text{ mol}}{1.00 \text{ L}} = 1.00 \text{ mol/L}$$

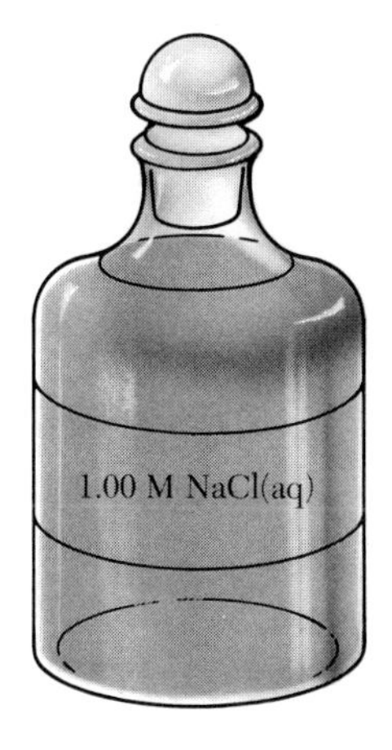

Because M = mol/L, we can express the concentration as 1.00 M. If we have this solution in a bottle and we want to write a label for the bottle to describe its contents, we can write "1.00 M NaCl(aq)." This label would be read "1.00 molar aqueous sodium chloride." The label could also be written "[NaCl(aq)] = 1.00 M" and would be read the same way.◆ Brackets, [], always designate the molar concentration of a substance; that is, they designate its concentration in moles per liter.

◆In writing a label for a bottle, the designation (aq) would be included to show that the solvent was water and not some other liquid. In writing answers to problems in this chapter, we can omit the designation (aq) because all of the solutions are understood to be aqueous solutions.

The following Examples and Exercises illustrate molarity calculations.

Example 15.1

Calculate the molarity of a solution made by combining 13.6 g of sodium chloride with enough water to make 250 mL of solution.

Solution

$$M = \frac{\text{mol}}{\text{L}}$$

Convert 13.6 g of NaCl to mol:

13.6 g = ? mol

g → mol

1 mol = 58.5 g for NaCl

g → mol

$$\left(\frac{13.6\ \cancel{g}}{}\right)\left(\frac{1\ \text{mol}}{58.5\ \cancel{g}}\right) = 0.232\ \text{mol}$$

Convert 250 mL to L:

250 mL = ? L

mL → L

1 L = 1000 mL

mL → L

$$\left(\frac{250\ \cancel{\text{mL}}}{}\right)\left(\frac{1\ \text{L}}{1000\ \cancel{\text{mL}}}\right) = 0.250\ \text{L}$$

Calculate molarity:

$$\text{M} = \frac{\text{mol}}{\text{L}} = \frac{0.232\ \text{mol}}{0.250\ \text{L}} = 0.928\ \text{mol/L}$$

[NaCl] = 0.928 M

Exercise 15.8

Calculate the molarity of a solution made by combining 1.07 g of potassium nitrate, KNO_3, with enough water to make 100 mL of solution.

Example 15.2

What mass, in grams, of lead(II) acetate will be needed to make 500 mL of 0.655 M $Pb(C_2H_3O_2)_2(aq)$?

Solution

Calculate moles of $Pb(C_2H_3O_2)_2$.

$$\text{M} = \frac{\text{mol}}{\text{L}} \qquad \text{mol} = (\text{M})(\text{L})$$

The molarity of this solution is 0.655 M or 0.655 mol/L, and its volume is 500 mL. Convert 500 mL to L:

mL → L

$$\left(\frac{500\ \cancel{\text{mL}}}{}\right)\left(\frac{1\ \text{L}}{1000\ \cancel{\text{mL}}}\right) = 0.500\ \text{L}$$

$$\text{mol} = (\text{M})(\text{L}) = \left(\frac{0.655\ \text{mol}}{1\ \cancel{\text{L}}}\right)\left(\frac{0.500\ \cancel{\text{L}}}{}\right) = 0.328\ \text{mol}$$

Convert 0.328 mol of $Pb(C_2H_3O_2)_2$ to g:

0.328 mol = ? g

mol → g

1 mol = 325 g for $Pb(C_2H_3O_2)_2$

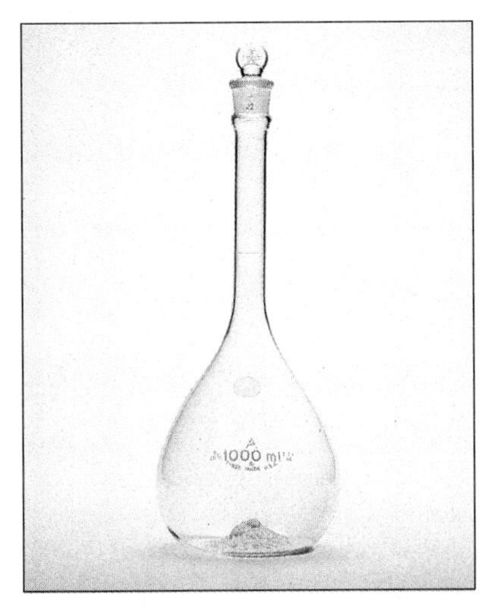

Figure 15.1
Using a volumetric flask to prepare 1000 mL of 0.100 M K_2CrO_4(aq). *(Marna G. Clarke)*

$$\text{mol} \rightarrow \text{g}$$

$$\left(\frac{0.328\ \cancel{\text{mol}}}{}\right)\left(\frac{325\ \text{g}}{1\ \cancel{\text{mol}}}\right) = 107\ \text{g}$$

This solution will be made by combining 107 g of $Pb(C_2H_3O_2)_2$ with enough water to make 500 mL of solution.

Exercise 15.9

(a) What volume of 1.25 M HCl(aq), in mL, will be needed to provide 0.414 mol of HCl? (b) What volume will be needed to provide 1.63 g of HCl?

Chemists work with aqueous solutions often, so they've found ways to prepare them quickly and accurately. One quick way to prepare an aqueous solution with a known concentration of solute is shown in Figure 15.1. A measured mass of a solute, in this example 19.4 g of the yellow solute potassium chromate, K_2CrO_4, is mixed with water in a **volumetric flask,** and more water is added until the level of the solution reaches a line etched into the neck of the flask; the flask has been made so that the amount of liquid that will fill it to the line is a specified volume, in this example 1000 mL.◆

◆Volumetric flasks are manufactured with volumes that are convenient for laboratory work, most commonly 1000 mL or less, and they're made to be accurate to about 0.01% of their volumes. For our calculations, we'll assume that, in a number specifying the volume of a volumetric flask, all of the digits are significant.

Exercise 15.10

Calculate the molarity of the solution described in the preceding paragraph.

It's often useful to prepare one aqueous solution from another by adding water; the process is called **dilution.** The volumes and molarities of the initial and final solutions have this relationship:

$V_1M_1 = V_2M_2$	V_1 and M_1 are the volume and molarity of the initial solution, and V_2 and M_2 are the volume and molarity of the final solution.

Example 15.3

A sample of 150 mL of 1.55 M NaCl(aq) was poured into a 500-mL volumetric flask, water was added until the level of liquid in the flask reached its etched line, and the contents of the flask were thoroughly mixed. Calculate the molarity of the solution in the flask.

Solution

$$V_1M_1 = V_2M_2$$

$$M_2 = \frac{V_1M_1}{V_2} = \frac{(150\ \cancel{mL})(1.55\ M)}{(500\ \cancel{mL})} = 0.465\ M$$

Exercise 15.11

(a) What volume of 6.00 M HCl(aq) should be diluted to make 200 mL of 1.50 M HCl(aq)? (b) In the dilution process described in part (a), how much water will be added to the original solution?

15-4 In pure water and in any aqueous solution there is a reaction between water molecules: $2\ H_2O(\ell) \rightleftarrows H_3O^+(aq) + OH^-(aq)$

In any sample of pure water and in any aqueous solution, water molecules are constantly colliding with one another. Some of these collisions result in a reaction in which two water molecules are converted to ions:

$$H_2\ddot{O}: + H{-}\ddot{O}{-}H \rightarrow H_2\ddot{O}\cdots H{-}\ddot{O}{-}H \rightarrow H_2\ddot{O}{-}H^+ + :\ddot{O}{-}H^-$$

This bond is breaking.

This bond is forming.

hydronium ion hydroxide ion

As the two water molecules collide, they momentarily form a cluster of atoms in which a hydrogen atom in one water molecule becomes attached to an unshared pair of electrons on the oxygen atom of the other water molecule. The cluster of atoms then breaks apart, and as it separates, the original bond to the hydrogen atom breaks and a new bond forms, producing a **hydronium ion,** H_3O^+, and a hydroxide ion, OH^-.◆ This reaction, which is referred to as the **autoionization of water,** is one of the most important reactions in chemistry.◆†

◆These ions are hydrated, as are all ions, in aqueous solution. There is evidence that, on the average, each hydronium ion and each hydroxide ion has three water molecules associated with it.

◆In Sections 15-5 through 15-8, we'll see that the autoionization of water is the basis for one of the most useful concepts in chemistry, the concept of acids and bases.

In a sample of pure water or in an aqueous solution the hydronium ions and hydroxide ions formed by the autoionization of water move about rapidly, and

† Chemistry Insight

It's important that you understand clearly how the autoionization reaction produces ions. If the process isn't clear to you from the description in the preceding text, try thinking of the reaction as occurring in two steps (even though actually it all occurs in one step). In the first step, imagine that one bond in a water molecule breaks, leaving the electrons in the bond on the oxygen atom:

$$H{-}\ddot{\underset{..}{O}}{-}H \rightarrow H^+ + :\ddot{\underset{..}{O}}{-}H^-$$

A hydrogen ion and a hydroxide ion form as a result of this reaction. In the second step, imagine that the hydrogen ion forms a covalent bond with an unshared pair of electrons on the oxygen atom of another water molecule:

$$H_2\ddot{O}: + H^+ \rightarrow H_2\ddot{O}{-}H^+$$

hydronium ion

The overall result of these two reactions is that two water molecules are converted to a hydronium ion, H_3O^+, and a hydroxide ion, OH^-.

they sometimes collide with one another. Some of these collisions result in a reaction that is the reverse of the original ionization:

$$H_2O{-}H^+ + :\ddot{\underset{..}{O}}{-}H^- \rightarrow H_2O\cdots H\cdots OH \rightarrow H_2\ddot{O}: + H{-}\ddot{O}{-}H$$

hydronium ion   hydroxide ion

This bond is forming.

This bond is breaking.

The autoionization of water is a **reversible reaction:** In water or any aqueous solution water molecules are constantly reacting to form hydronium ions and hydroxide ions, and at the same time hydronium ions and hydroxide ions are constantly reacting to form water molecules.

The reversible autoionization reaction is usually represented by this equation in molecular and condensed formulas:

$$2\ H_2O(\ell) \rightleftarrows H_3O^+(aq) + OH^-(aq)$$

In this equation the double arrow shows that the reaction is reversible.◆ The reaction that's shown from left to right in the equation for a reversible reaction is called the **forward reaction,** and the reaction that's shown from right to left is called the **reverse reaction.**

◆ Many chemical reactions are reversible; we'll see examples throughout the rest of this book. A double arrow in an equation always identifies a reversible reaction.

In water or any aqueous solution the forward and reverse reactions for the autoionization of water are occurring at the same rate. In each minute the number of water molecules converted to hydronium ions and hydroxide ions is the same as the number of hydronium ions and hydroxide ions converted to water molecules. Because of this balance, the autoionization reaction is said to be **at equilibrium.**◆

◆Equilibrium is a common and important characteristic of chemical reactions. We'll see many examples later in this book.

The autoionization reaction produces only a very small amount of hydronium ion and hydroxide ion: In pure water at each moment only about four water molecules in each billion have been converted to ions. It's useful to remember that the amounts of water and ions have this relationship:

$$\underset{\text{very large amount}}{2\ H_2O(\ell)} \rightleftarrows \underset{\text{very small amounts}}{H_3O^+(aq) + OH^-(aq)}$$

In pure water the concentration of hydronium ion and the concentration of hydroxide ion must be the same because, in the autoionization reaction, the ions are produced in equal numbers. The actual concentrations are $[H_3O^+] = [OH^-] = 1.00 \times 10^{-7}$ M.◆ In the following Sections we'll see that changes in these concentrations are characteristic of the solutions called acids and bases.

◆These concentrations vary slightly with temperature. In this book we'll assume that samples of water and aqueous solutions are at ordinary room temperatures.

15-5 Strong and weak acids are compounds that produce hydronium ion, H_3O^+, in aqueous solutions

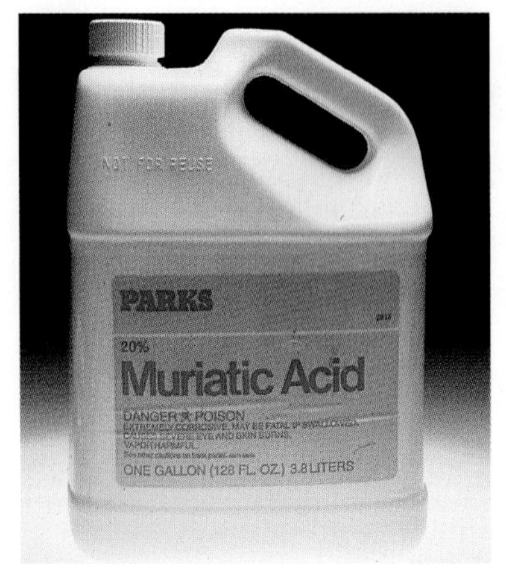

(Charles D. Winters)

Strong Acids

To introduce the class of substances called acids, we'll begin with the example of an acid commonly used in laboratory and industrial work, hydrochloric acid. Hydrochloric acid is a colorless aqueous solution made by dissolving hydrogen chloride gas, HCl(g), in water.◆

A molecule of hydrogen chloride is diatomic and contains a polar covalent bond, so the molecule is a dipole (Section 14-3): $\overset{\delta+}{H}—\overset{\delta-}{\ddot{C}l}:$. Because molecules of hydrogen chloride are dipoles, we'd expect them to be attracted to water molecules by dipole–dipole interactions, as well as by London forces. In Section 15-2, we saw that compounds whose molecules form only dipole–dipole interactions and London forces with water molecules are in general not very soluble in water. It's therefore surprising, at first, to find that huge amounts of hydrogen chloride will dissolve in water: At room temperatures, about 700 g of hydrogen chloride (compared with 357 g of sodium chloride) will dissolve in 1.00 L of water.

◆Hydrochloric acid, often sold under the name *muriatic acid*, has many uses; you may have seen it in the form of a masonry or tile cleaner. In 1989 the United States produced 2.6 million tons of hydrochloric acid, which indicates the industrial importance of this acid.

The explanation for this high solubility is that hydrogen chloride reacts with water as it dissolves. In this reaction, the polar covalent bond in the hydrogen chloride molecule is broken, forming two ions.

$$\underset{\text{water molecule}}{H_2\ddot{O}:} + \underset{\text{hydrogen chloride molecule}}{H—\ddot{C}l:} \rightarrow \underbrace{\underset{\text{hydronium ion}}{[H_2\ddot{O}—H]^+} + \underset{\text{chloride ion}}{:\ddot{C}l:^-}}_{\text{hydrochloric acid}}$$

The overall result is that the covalent compound hydrogen chloride is converted into an aqueous solution of hydronium ions and chloride ions, called hydrochloric acid; in hydrochloric acid, the hydronium ions and chloride ions are hydrated.

In molecular and condensed formulas the reaction of water with dissolved hydrogen chloride to form hydrochloric acid is

$$H_2O(\ell) + HCl(aq) \rightarrow H_3O^+(aq) + Cl^-(aq)$$

Because hydrogen chloride reacts with water to form hydronium ions, the concentration of hydronium ion in hydrochloric acid is greater than the concentration of hydronium ion in pure water. In pure water $[H_3O^+] = 1.00 \times 10^{-7}$ M, and in hydrochloric acid $[H_3O^+] > 1.00 \times 10^{-7}$ M.

An **acid** is a substance that forms hydronium ion, H_3O^+, in aqueous solution. In a solution of an acid the concentration of hydronium ion is greater than it is in pure water; that is, in a solution of an acid $[H_3O^+] > 1.00 \times 10^{-7}$ M.

As shown in Table 15.1, five other common compounds behave in the same way as hydrogen chloride. Hydrogen chloride and these five compounds are designated the six common strong acids.◆ It's important to remember the six common strong acids: HCl, HBr, HI, HNO_3, H_2SO_4, and $HClO_4$.

◆The reason for calling them *strong* acids will be explained shortly.

◆Nitric acid and sulfuric acid, two of the most commonly used industrial chemicals, are used to make explosives, fertilizers, dyes, paper, glue, and many other products. In 1989 the U.S. production of nitric acid was 8.0 million tons, and the production of sulfuric acid was more than 43 million tons.

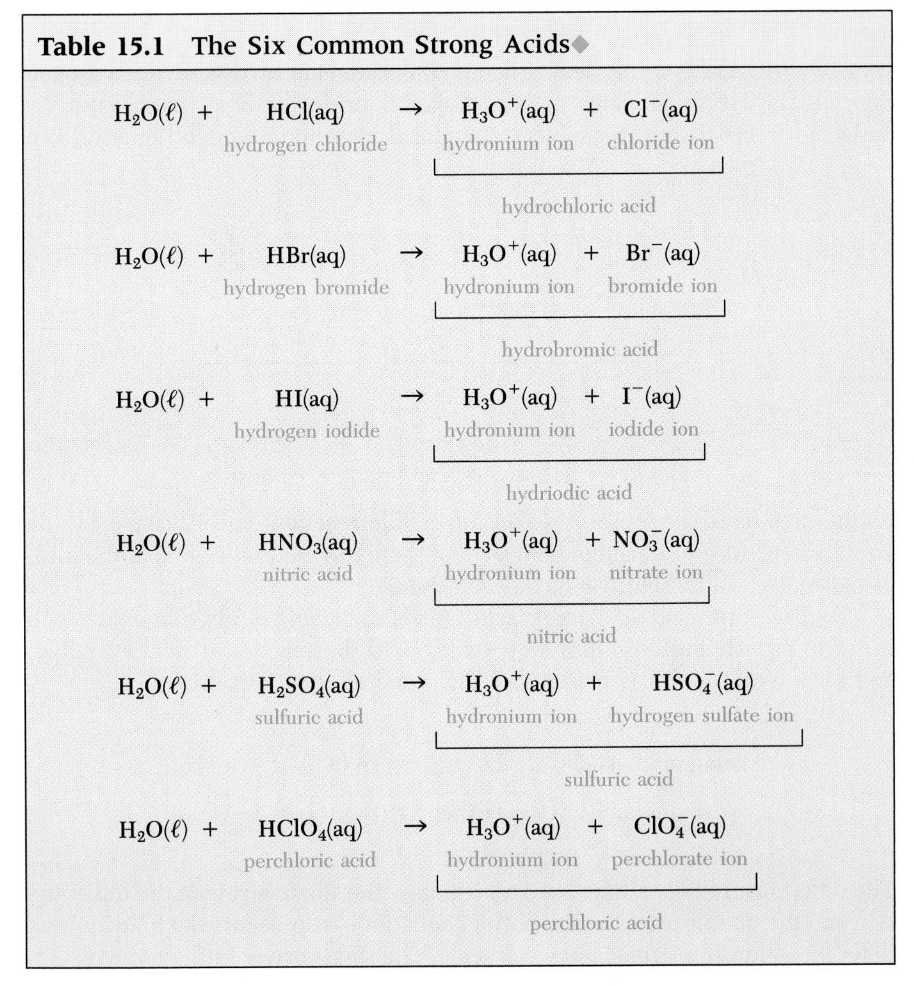

Table 15.1 The Six Common Strong Acids◆

$H_2O(\ell)$ + HCl(aq) → $H_3O^+(aq)$ + $Cl^-(aq)$
hydrogen chloride; hydronium ion; chloride ion — hydrochloric acid

$H_2O(\ell)$ + HBr(aq) → $H_3O^+(aq)$ + $Br^-(aq)$
hydrogen bromide; hydronium ion; bromide ion — hydrobromic acid

$H_2O(\ell)$ + HI(aq) → $H_3O^+(aq)$ + $I^-(aq)$
hydrogen iodide; hydronium ion; iodide ion — hydriodic acid

$H_2O(\ell)$ + $HNO_3(aq)$ → $H_3O^+(aq)$ + $NO_3^-(aq)$
nitric acid; hydronium ion; nitrate ion — nitric acid

$H_2O(\ell)$ + $H_2SO_4(aq)$ → $H_3O^+(aq)$ + $HSO_4^-(aq)$
sulfuric acid; hydronium ion; hydrogen sulfate ion — sulfuric acid

$H_2O(\ell)$ + $HClO_4(aq)$ → $H_3O^+(aq)$ + $ClO_4^-(aq)$
perchloric acid; hydronium ion; perchlorate ion — perchloric acid

As shown in Table 15.1, the word *acid* is used in two ways: to name the compound that dissolves to form hydronium ion, if the compound is a liquid (the pure compounds nitric acid, sulfuric acid, and perchloric acid are liquids under ordinary room conditions), and to name the aqueous solution of ions that's formed when the compound dissolves.

Exercise 15.12

The line formula for nitric acid is H—Ö—N(=Ö:)—Ö:. Use line formulas to write an equation for the reaction of nitric acid with water to form hydronium ion and nitrate ion.

Using HA to represent any strong acid, the general equation for the reaction of a strong acid with water is

$$H_2O(\ell) + HA(aq) \rightarrow H_3O^+(aq) + A^-(aq)$$

Weak Acids

Hydrofluoric acid is a colorless aqueous solution made by dissolving hydrogen fluoride gas, HF(g), in water.◆ As hydrogen fluoride gas dissolves in water, its molecules react with water molecules to form hydronium ion and fluoride ion:

◆ Hydrofluoric acid is used industrially to clean metals and ceramics and to etch glass.

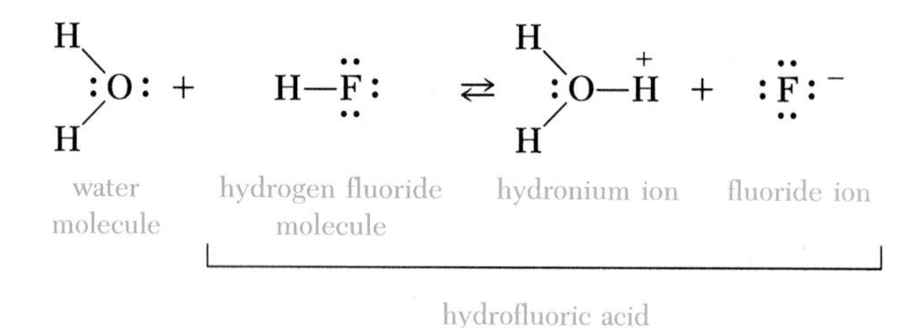

or

$$H_2O(\ell) + HF(aq) \rightleftharpoons H_3O^+(aq) + F^-(aq)$$

As the double arrow shows, this is a reversible reaction, and it establishes an equilibrium. At equilibrium about 97% of the hydrogen fluoride is in the form of molecules, and about 3% has formed ions.

Hydrofluoric acid is a *weak* acid. Acids are compounds that react with water to form hydronium ion; for a **strong acid** the reaction is not reversible, and for a **weak acid** it is reversible. The general equations are

$$\text{strong acid: } H_2O(\ell) + HA(aq) \rightarrow H_3O^+(aq) + A^-(aq)$$
$$\text{weak acid: } H_2O(\ell) + HA(aq) \rightleftharpoons H_3O^+(aq) + A^-(aq)$$

Glass etched with hydrofluoric acid. *(Kristen Brochmann, Fundamental Photographs, New York)*

The difference between these two equations—the single arrow in the top equation and the double arrow in the bottom equation—represents the fundamental difference between strong and weak acids. The single arrow in the top equation

† Chemistry Insight

To reinforce your understanding of this important concept, cover with your hand the reactants and the arrow in the equation $H_2O(\ell) + HCl(aq) \rightarrow H_3O^+(aq) + Cl^-(aq)$. The part of the equation that you can still see (the products) represents the solutes that are present in a solution made by dissolving hydrogen chloride in water.

shows that, as a strong acid dissolves in water, it is immediately and completely converted to ions. For example, the equation

$$H_2O(\ell) + HCl(aq) \rightarrow H_3O^+(aq) + Cl^-(aq)$$

shows that, as hydrogen chloride dissolves in water, it is immediately and completely converted to hydrated hydronium ions and hydrated chloride ions: In a solution made by dissolving hydrogen chloride, the only solutes are $H_3O^+(aq)$ and $Cl^-(aq)$.† The double arrow in the bottom equation on page 326 shows that, as a weak acid dissolves in water, it is immediately and partially converted to ions. For example, the equation

$$H_2O(\ell) + HF(aq) \rightleftarrows H_3O^+(aq) + F^-(aq)$$

shows that, as hydrogen fluoride dissolves in water, it is immediately and partially converted to hydrated hydronium ions and hydrated fluoride ions: In a solution made by dissolving hydrogen fluoride, the solutes are HF(aq), $H_3O^+(aq)$, and $F^-(aq)$, and the solute present in largest amount is HF(aq).††

It's important to be able to decide from the formula for an acid whether it will be strong or weak. We'll assume that, if an acid isn't one of the six common strong acids (HCl, HBr, HI, HNO_3, H_2SO_4, and $HClO_4$), it will be weak. Several weak acids are shown in Table 15.2 on page 328.

As shown in Table 15.2, the word *acid* is used in two ways: to name the compound that dissolves in water to form the acid solution, if the compound is a liquid (the pure compounds acetic acid and formic acid are liquids under ordinary room conditions), and to name the aqueous solution that's formed when the compound dissolves.◆

◆Nitrous acid, HNO_2, exists only in aqueous solutions, not as a pure compound. It's created in solution by the reaction of sodium nitrite with sulfuric acid:

$$2\ NaNO_2(aq) + H_2SO_4(aq) \rightarrow 2\ HNO_2(aq) + Na_2SO_4(aq)$$

Exercise 15.13

Decide whether each of the following acids is strong or weak, and write the appropriate equation for its ionization, using molecular and condensed formulas: (a) HCl (b) HF (c) HBr (d) HNO_2

†† Chemistry Insight

Cover with your hand the formula $H_2O(\ell)$ and the + sign on the left side of the equation $H_2O(\ell) + HF(aq) \rightleftarrows H_3O^+(aq) + F^-(aq)$. The part of the equation that you can still see represents the solutes that are present in a solution made by dissolving hydrogen fluoride in water.

Table 15.2 Examples of Weak Acids

$H_2O(\ell) + HF(aq) \rightleftharpoons H_3O^+(aq) + F^-(aq)$

hydrogen fluoride | hydronium ion | fluoride ion

hydrofluoric acid

$H_2O(\ell) + HC_2H_3O_2(aq) \rightleftharpoons H_3O^+(aq) + C_2H_3O_2^-(aq)$

acetic acid | hydronium ion | acetate ion

acetic acid

$H_2O(\ell) + HNO_2(aq) \rightleftharpoons H_3O^+(aq) + NO_2^-(aq)$

nitrous acid | hydronium ion | nitrite ion

nitrous acid

$H_2O(\ell) + HCN(aq) \rightleftharpoons H_3O^+(aq) + CN^-(aq)$

hydrogen cyanide | hydronium ion | cyanide ion

hydrocyanic acid

$H_2O(\ell) + HCHO_2(aq) \rightleftharpoons H_3O^+(aq) + CHO_2^-(aq)$

formic acid | hydronium ion | formate ion

formic acid

Polyprotic Acids

The molecules of some acids contain more than one hydrogen atom that can be transferred to a water molecule to form a hydronium ion. An example is sulfuric acid:

$$H_2O(\ell) + H_2SO_4(aq) \rightarrow H_3O^+(aq) + HSO_4^-(aq)$$

sulfuric acid | hydronium ion | hydrogen sulfate ion

$$H_2O(\ell) + HSO_4^-(aq) \rightleftharpoons H_3O^+(aq) + SO_4^{2-}(aq)$$

hydrogen sulfate ion | hydronium ion | sulfate ion

As these equations show, the transfer of the first hydrogen atom from a molecule of sulfuric acid is an irreversible reaction (H_2SO_4 is a strong acid), and the transfer of the second hydrogen atom is a reversible reaction (HSO_4^- is a weak acid).

A hydrogen atom that can be transferred from its molecule to a water molecule to form a hydronium ion is called an **ionizable hydrogen atom.** If a molecule contains some hydrogen atoms that are ionizable and some that are not, the symbols for those that are ionizable will be written first in its molecular formula. A molecule of acetic acid, for example, has one hydrogen atom that is ionizable and three that are not:

ionizable hydrogen atom

$$HC_2H_3O_2$$

nonionizable hydrogen atoms

The number of ionizable hydrogen atoms that are present in a molecule can be designated by the same prefixes that designate the numbers of atoms in the molecules of covalent compounds (Section 9-1). Because the hydrogen atom that's transferred leaves its electron behind, it can be thought of as forming a hydrogen ion, H^+, which becomes attached to a water molecule to form a hydronium ion, H_3O^+. Because a hydrogen ion is also a proton, a molecule that contains one ionizable hydrogen atom is said to be **monoprotic,** a molecule that contains two ionizable hydrogen atoms is said to be **diprotic,** and so forth. Any molecule that contains more than one ionizable hydrogen atom is said to be **polyprotic** (*poly-* means *many*). Examples of three common polyprotic acids are shown in Table 15.3.

Table 15.3 Three Polyprotic Acids

$H_2O(\ell)$	+	$H_2SO_3(aq)$ sulfurous acid	$\rightleftarrows H_3O^+(aq)$	+	$HSO_3^-(aq)$ hydrogen sulfite ion
$H_2O(\ell)$	+	$HSO_3^-(aq)$ hydrogen sulfite ion	$\rightleftarrows H_3O^+(aq)$	+	$SO_3^{2-}(aq)$ sulfite ion
$H_2O(\ell)$	+	$H_2S(aq)$ hydrogen sulfide	$\rightleftarrows H_3O^+(aq)$	+	$HS^-(aq)$ hydrogen sulfide ion
$H_2O(\ell)$	+	$HS^-(aq)$ hydrogen sulfide ion	$\rightleftarrows H_3O^+(aq)$	+	$S^{2-}(aq)$ sulfide ion
$H_2O(\ell)$	+	$H_3PO_4(aq)$ phosphoric acid	$\rightleftarrows H_3O^+(aq)$	+	$H_2PO_4^-(aq)$ dihydrogen phosphate ion
$H_2O(\ell)$	+	$H_2PO_4^-(aq)$ dihydrogen phosphate ion	$\rightleftarrows H_3O^+(aq)$	+	$HPO_4^{2-}(aq)$ hydrogen phosphate ion
$H_2O(\ell)$	+	$HPO_4^{2-}(aq)$ hydrogen phosphate ion	$\rightleftarrows H_3O^+(aq)$	+	$PO_4^{3-}(aq)$ phosphate ion

Citrus fruits contain sharp-tasting citric acid, $H_3C_6H_5O_7$. *(Marna G. Clarke)*

Exercise 15.14

Citric acid, $H_3C_6H_5O_7$, is one of the compounds responsible for the sharp taste of citrus fruits.

(a) How many ionizable hydrogen atoms are in a molecule of citric acid?
(b) Is the acid strong or weak?
(c) Write an equation in molecular and condensed formulas for the ionization of each ionizable hydrogen atom in a molecule of citric acid.

Abbreviated Equations for Acid Ionizations

We've seen that the ionization of an acid can be represented by an equation in line formulas or in molecular and condensed formulas; for example,

$$\mathrm{H_2\ddot{O}\!:} + \mathrm{H{-}\ddot{\underset{..}{Cl}}\!:} \rightarrow \mathrm{H_2\ddot{O}{-}\overset{+}{H}} + \mathrm{:\ddot{\underset{..}{Cl}}\!:^-}$$

or

$$H_2O(\ell) + HCl(aq) \rightarrow H_3O^+(aq) + Cl^-(aq)$$

An acid ionization can also be represented in a simplified form in which the hydrogen atom that leaves the molecule is shown as a hydrogen ion. Although equations of this kind give very limited information, they have the advantage of being very brief. The following equations show the ionizations of hydrogen chloride and acetic acid and the autoionization of water written in this simplified form:

$$HCl \rightarrow H^+ + Cl^-$$

$$HC_2H_3O_2 \rightleftarrows H^+ + C_2H_3O_2^-$$

$$H_2O \rightleftarrows H^+ + OH^-$$

Exercise 15.15

Using the abbreviated form, write an equation for the ionization of (a) nitric acid, HNO_3 (b) nitrous acid, HNO_2.

15-6 Strong and weak bases are compounds that produce hydroxide ion, OH^-, in aqueous solutions

◆ Sodium hydroxide is used industrially to make many kinds of products, including soaps, plastics, pharmaceuticals, rubber, rayon, and cellophane. In 1989 more than 11 million tons of sodium hydroxide was produced in the United States.

In many ways the concept of bases is similar to the concept of acids. To introduce the class of substances called bases, we'll begin with the example of a base that's commonly used in laboratory and industrial work, sodium hydroxide, NaOH.◆ Sodium hydroxide is a white solid (melting point 318°C), often sold in the form of pellets about the size of aspirin tablets.

Sodium hydroxide is an ionic compound and is very soluble in water: At ordinary room temperatures more than 1100 g of sodium hydroxide will dissolve in 1.00 L of water. As sodium hydroxide dissolves, its ions become hydrated:

$$NaOH(s) \rightarrow Na^+(aq) + OH^-(aq)$$

Because sodium hydroxide releases hydroxide ions into solution as it dissolves, the concentration of hydroxide ion in an aqueous solution of sodium hydroxide is greater than the concentration of hydroxide ion in pure water. In pure water $[OH^-] = 1.00 \times 10^{-7}$ M, and in aqueous sodium hydroxide $[OH^-] > 1.00 \times 10^{-7}$ M.

A **base** is a compound that produces hydroxide ion, OH^-, in aqueous solution. In a solution of a base the concentration of hydroxide ion is greater than it is in pure water; that is, in a solution of a base $[OH^-] > 1.00 \times 10^{-7}$ M.

Solid sodium hydroxide, NaOH(s), is a white ionic compound. *(Charles D. Winters)*

Hydroxide compounds are known for most of the metallic elements, but only the six compounds shown in Table 15.4 are classified as common strong bases. The other metal hydroxides either are rare, because the metal is scarce in nature (examples are CsOH and $Ra(OH)_2$), or are so poorly soluble in water that their aqueous solutions can be ignored for our purposes (examples are $Mg(OH)_2$, $Al(OH)_3$, and $Fe(OH)_3$).◆

◆The factors that determine the solubilities of ionic compounds are discussed in Section 15-9.

The word *base* is commonly used in two ways: to designate a compound that releases hydroxide ion as it dissolves and to designate the aqueous solution that's formed. For example, solid sodium hydroxide is commonly referred to as a base and so is aqueous sodium hydroxide.

There are also weak bases. The most common and most important example of a weak base is aqueous ammonium hydroxide, a colorless solution that's formed when ammonia gas, $NH_3(g)$, dissolves in water.◆ Ammonia is very soluble in water: Under ordinary room conditions, about 500 g of ammonia will dissolve in 1.00 L of water. As ammonia dissolves, it reacts with water:

◆Household ammonia is aqueous ammonium hydroxide. Ammonia gas is used as a fertilizer and in the production of many products, including explosives, plastics, and synthetic fibers. The United States produced more than 16 million tons of ammonia gas in 1989.

$$H{-}\overset{H}{\underset{H}{N}}: \; + \; H{-}\ddot{\underset{..}{O}}{-}H \; \rightleftarrows \; H{-}\overset{H}{\underset{H}{N}}{-}H^+ \; + \; :\ddot{\underset{..}{O}}{-}H^-$$

ammonia molecule   water molecule   ammonium ion   hydroxide ion

$$NH_3(aq) + H_2O(\ell) \rightleftarrows NH_4^+(aq) + OH^-(aq)$$

Table 15.4 The Six Common Strong Bases

Group 1A		
lithium hydroxide:	$LiOH(s)$	$\rightarrow Li^+(aq) + OH^-(aq)$
sodium hydroxide:	$NaOH(s)$	$\rightarrow Na^+(aq) + OH^-(aq)$
potassium hydroxide:	$KOH(s)$	$\rightarrow K^+(aq) + OH^-(aq)$
Group 2A		
calcium hydroxide:	$Ca(OH)_2(s)$	$\rightarrow Ca^{2+}(aq) + 2\ OH^-(aq)$
strontium hydroxide:	$Sr(OH)_2(s)$	$\rightarrow Sr^{2+}(aq) + 2\ OH^-(aq)$
barium hydroxide:	$Ba(OH)_2(s)$	$\rightarrow Ba^{2+}(aq) + 2\ OH^-(aq)$

(David R. Frazier Photolibrary)

Table 15.5 Examples of Weak Bases

ammonia:	$NH_3(aq) + H_2O(\ell) \rightleftarrows NH_4^+(aq) + OH^-(aq)$
aniline:	$C_6H_5NH_2(aq) + H_2O(\ell) \rightleftarrows C_6H_5NH_3^+(aq) + OH^-(aq)$
hydrazine:	$N_2H_4(aq) + H_2O(\ell) \rightleftarrows N_2H_5^+(aq) + OH^-(aq)$
methylamine:	$CH_3NH_2(aq) + H_2O(\ell) \rightleftarrows CH_3NH_3^+(aq) + OH^-(aq)$

The reaction is reversible, and it establishes an equilibrium. At equilibrium about 99.6% of the ammonia is in the form of molecules, and about 0.4% has formed ions.

All of the weak bases that we'll consider in this book are nitrogen compounds that undergo reversible reactions with water similar to the reaction shown for ammonia. Some examples are shown in Table 15.5. Ammonia is the only weak base whose formula you need to remember.

From the preceding descriptions, we can give general definitions of *strong base* and *weak base*. Bases are compounds that produce hydroxide ion in aqueous solution. **Strong bases** are certain soluble hydroxides of group 1A (LiOH, NaOH, and KOH) and group 2A ($Ca(OH)_2$, $Sr(OH)_2$, and $Ba(OH)_2$) that release hydroxide ion as they dissolve. **Weak bases** are certain compounds of nitrogen (ammonia, NH_3, is the most important) that undergo a reversible reaction with water to produce a cation and hydroxide ion.

Exercise 15.16

Decide which compound is a strong base: (a) $Zn(OH)_2$ (b) $Be(OH)_2$ (c) LiOH

Exercise 15.17

The compound pyridine, C_5H_5N, is a weak base. Following the examples in Table 15.5, write an equation for its reaction with water.

15-7 The concentration of an acidic or basic solution can be expressed in molarity or in pH units: $pH = -\log[H_3O^+]$

Aqueous solutions that are acids or bases have certain characteristic properties. Acidic solutions have a sharp smell and taste; a familiar example is vinegar, which is a dilute solution of acetic acid, $HC_2H_3O_2$. Basic solutions feel slippery and have an unpleasant brackish taste similar to that of salt water; soaps are weak bases, and these properties are familiar to you in soapy water.◆ Acids and bases also undergo certain characteristic reactions, as we'll see in Chapter 16.

◆It's not dangerous to feel, smell, or taste vinegar or soapy water because they're very dilute solutions of a weak acid and a weak base. In general, it's not safe to feel or taste chemicals, and they should be smelled only with caution.

Many of the properties of solutions of acids and bases depend on their concentrations. A more concentrated solution of an acid, for example, will have a sharper smell and taste, and a more concentrated solution of a base will have a stronger taste and feel more slippery. The concentration of an acid or a base can be expressed in molarity (Section 15-3), as shown in the following Examples and Exercises.

Example 15.4

What is the molarity of a solution made by dissolving 5.00 g of acetic acid in enough water to make 125 mL of solution?

Solution

$$M = \frac{\text{mol}}{\text{L}}$$

Convert 5.00 g of $HC_2H_3O_2$ to mol:

$$\text{g} \rightarrow \text{mol}$$

$$\left(\frac{5.00\ \cancel{\text{g}}}{}\right)\left(\frac{1\ \text{mol}}{60.0\ \cancel{\text{g}}}\right) = 0.0833\ \text{mol}$$

Convert 125 mL to L:

$$\text{mL} \rightarrow \text{L}$$

$$\left(\frac{125\ \cancel{\text{mL}}}{}\right)\left(\frac{1\ \text{L}}{1000\ \cancel{\text{mL}}}\right) = 0.125\ \text{L}$$

Calculate molarity:

$$M = \frac{\text{mol}}{\text{L}} = \frac{0.0833\ \text{mol}}{0.125\ \text{L}} = 0.666\ \text{mol/L}$$

$$[HC_2H_3O_2] = 0.666\ M$$

Vinegar is a dilute solution of acetic acid, $HC_2H_3O_2$. *(Leonard Lessin/Peter Arnold Inc.)*

Exercise 15.18

In college and university laboratories, it's common to find bottles labeled "dilute hydrochloric acid," containing solutions that are about 6 M HCl(aq). How many grams of solute are present in 30.0 mL of 6.00 M HCl(aq)?

Example 15.5

A solution was prepared by placing 4.50 g of potassium hydroxide in a 500-mL volumetric flask and filling the flask to its etched line with water. Calculate the molarity of hydroxide ion in the solution.

Solution

Calculate molarity of KOH:

$$M = \frac{\text{mol}}{\text{L}}$$

Convert 4.50 g of KOH to mol:

$$\text{g} \rightarrow \text{mol}$$

$$\left(\frac{4.50\ \cancel{\text{g}}}{}\right)\left(\frac{1\ \text{mol}}{56.1\ \cancel{\text{g}}}\right) = 0.0802\ \text{mol}$$

Convert 500 mL to L:

$$\text{mL} \rightarrow \text{L}$$

$$\left(\frac{500\ \text{mL}}{}\right)\left(\frac{1\ \text{L}}{1000\ \text{mL}}\right) = 0.500\ \text{L}$$

Calculate molarity of KOH:

$$M = \frac{\text{mol}}{\text{L}} = \frac{0.0802\ \text{mol}}{0.500\ \text{L}} = 0.160\ \text{mol/L}$$

$$[\text{KOH}] = 0.160\ \text{M}$$

Calculate molarity of OH^-:

Potassium hydroxide is a strong base that's completely dissociated into hydrated ions in aqueous solution:

$$\text{KOH(s)} \rightarrow \text{K}^+\text{(aq)} + \text{OH}^-\text{(aq)}$$

This equation shows that if 1 mol of KOH dissolves in water, it will form 1 mol of K^+(aq) and 1 mol of OH^-(aq). In a solution in which [KOH] = 0.160 M, it must be true that $[K^+]$ = 0.160 M and $[OH^-]$ = 0.160 M, so

$$[\text{OH}^-] = 0.160\ \text{M}$$

Exercise 15.19

Calculate the molarity of barium ion and the molarity of hydroxide ion in a solution made by dissolving 6.27 g of barium hydroxide in enough water to make 225 mL of solution.

In pure water and in any aqueous solution there is a fixed mathematical relationship between the molar concentrations of hydronium ion and hydroxide ion:

$$[\text{H}_3\text{O}^+][\text{OH}^-] = K_w$$

This equation is called the **ion–product equation** for water, and K_w is called the **ion–product constant** for water.◆ The numerical value of K_w is 1.00×10^{-14}.

◆ K_w is the product of two molarities. In pure water $[H_3O^+][OH^-] = (1.00 \times 10^{-7}\ M)^2 = 1.00 \times 10^{-14}$. The units on K_w, M^2, are customarily not shown.

The ion–product equation shows that, as the molarity of hydronium ion in an aqueous solution increases, the molarity of hydroxide ion decreases and vice versa. The equation can be used to calculate the molarity of hydroxide ion from the molarity of hydronium ion or vice versa.

Example 15.6

In a solution labeled [HBr(aq)] = 3.65×10^{-3} M, what is the molarity of hydroxide ion?

Solution

Hydrobromic acid is a strong acid, so it's completely converted to ions:

$$\text{H}_2\text{O}(\ell) + \text{HBr(aq)} \rightarrow \text{H}_3\text{O}^+\text{(aq)} + \text{Br}^-\text{(aq)}$$

In a solution in which $[HBr(aq)] = 3.65 \times 10^{-3}$ M, it will be true that $[H_3O^+] = 3.65 \times 10^{-3}$ M and $[Br^-] = 3.65 \times 10^{-3}$ M.

$$[H_3O^+][OH^-] = K_w$$

$$[OH^-] = \frac{K_w}{[H_3O^+]} = \frac{1.00 \times 10^{-14}}{3.65 \times 10^{-3}} = 2.74 \times 10^{-12}$$

$$[OH^-] = 2.74 \times 10^{-12}\ M$$

Exercise 15.20

In an aqueous solution of sodium hydroxide, the concentration of hydroxide ion was 7.44×10^{-2} M. Calculate $[H_3O^+]$.

On the basis of the ion–product equation, we can make these statements:

In pure water $[H_3O^+] = 1.00 \times 10^{-7}$ M and $[OH^-] = 1.00 \times 10^{-7}$ M.
In acidic solutions $[H_3O^+] > 1.00 \times 10^{-7}$ M and $[OH^-] < 1.00 \times 10^{-7}$ M.
In basic solutions $[H_3O^+] < 1.00 \times 10^{-7}$ M and $[OH^-] > 1.00 \times 10^{-7}$ M.◆

◆These relationships can be used to check the reasonableness of calculated values of $[H_3O^+]$ and $[OH^-]$ in aqueous solutions. For example, a calculated value of $[H_3O^+] = 4.25 \times 10^{-9}$ M for an acidic solution must be incorrect because in acidic solutions $[H_3O^+] > 1.00 \times 10^{-7}$ M.

We've seen that the terms *strong* and *weak* can be used to describe the extent to which an acid or a base is converted to hydrated ions in aqueous solution: A strong acid or base is one that's completely converted to hydrated ions by an irreversible reaction, and a weak acid or base is one that's only slightly converted to hydrated ions by a reversible reaction. As shown in Table 15.6, the terms *strong* and *weak* are also commonly used to describe, in an approximate way, the concentrations of acidic and basic solutions.

Because the concept of acids and bases is widely used, not only in chemistry but in other scientific and technological areas such as agriculture, biology,

Table 15.6 Descriptions of Aqueous Solutions of Acids and Bases

Solution		$[H_3O^+]$	$[OH^-]$
		1.0×10^{0} M	1.0×10^{-14} M
	strongly acidic	1.0×10^{-1} M	1.0×10^{-13} M
		1.0×10^{-2} M	1.0×10^{-12} M
Acid		1.0×10^{-3} M	1.0×10^{-11} M
		1.0×10^{-4} M	1.0×10^{-10} M
	weakly acidic	1.0×10^{-5} M	1.0×10^{-9} M
		1.0×10^{-6} M	1.0×10^{-8} M
Pure water	neutral	1.0×10^{-7} M	1.0×10^{-7} M
		1.0×10^{-8} M	1.0×10^{-6} M
	weakly basic	1.0×10^{-9} M	1.0×10^{-5} M
		1.0×10^{-10} M	1.0×10^{-4} M
Base		1.0×10^{-11} M	1.0×10^{-3} M
		1.0×10^{-12} M	1.0×10^{-2} M
	strongly basic	1.0×10^{-13} M	1.0×10^{-1} M
		1.0×10^{-14} M	1.0×10^{0} M

engineering, and medicine, it's useful to have a way to express the concentrations of acidic and basic solutions in a simpler numerical form than exponential notation. The **pH scale** expresses the concentrations of acidic and basic solutions on a scale from 0 to 14. The definition of pH is

$$pH = -\log[H_3O^+]$$

The quickest way to understand the significance of numbers on the pH scale is to see how these numbers relate to the corresponding molar concentrations of hydronium ion. These relationships are shown in Table 15.7.◆

◆The origin of the p in the symbol pH is in the German word *potenz*, power. You may find it useful to think of pH as standing for *power of hydronium ion*.

In Table 15.7 note these general characteristics of the pH scale:

In pure water, the pH is 7.◆
In an acidic solution, the pH is less than 7. The more strongly acidic the solution, the lower the pH.
In a basic solution, the pH is greater than 7. The more strongly basic the solution, the higher the pH.◆

◆The same is true for any neutral solution, that is, for a solution of any solute that doesn't produce H_3O^+ or OH^-. An aqueous solution of sodium chloride, for example, has a pH of 7.

◆These relationships can be used to check the reasonableness of calculated values of pH for aqueous solutions. For example, a calculated value of pH = 9.2 for an acidic solution must be incorrect because in acidic solutions pH < 7.

Exercise 15.21

Without looking at Table 15.7, identify each of the following solutions as being strongly or weakly acidic or strongly or weakly basic: (a) a solution with pH 8.00 (b) a solution with pH 1.00

Because the pH scale is logarithmic, a change of one pH unit corresponds to a tenfold change in the molar concentration of hydronium ion.◆ A solution with pH 3.00 has ten times the concentration of H_3O^+ as a solution with pH 4.00.

◆For a review of logarithms and antilogarithms, see Appendix 3.

Calculations of pH from $[H_3O^+]$ and vice versa are shown in the following Example and Exercise.

Table 15.7 The pH Scale

Solution		$[H_3O^+]$	pH	$[OH^-]$
Acid	strongly acidic	1.0×10^{0} M	0.00	1.0×10^{-14} M
		1.0×10^{-1} M	1.00	1.0×10^{-13} M
		1.0×10^{-2} M	2.00	1.0×10^{-12} M
		1.0×10^{-3} M	3.00	1.0×10^{-11} M
	weakly acidic	1.0×10^{-4} M	4.00	1.0×10^{-10} M
		1.0×10^{-5} M	5.00	1.0×10^{-9} M
		1.0×10^{-6} M	6.00	1.0×10^{-8} M
Pure water	neutral	1.0×10^{-7} M	7.00	1.0×10^{-7} M
Base	weakly basic	1.0×10^{-8} M	8.00	1.0×10^{-6} M
		1.0×10^{-9} M	9.00	1.0×10^{-5} M
		1.0×10^{-10} M	10.00	1.0×10^{-4} M
	strongly basic	1.0×10^{-11} M	11.00	1.0×10^{-3} M
		1.0×10^{-12} M	12.00	1.0×10^{-2} M
		1.0×10^{-13} M	13.00	1.0×10^{-1} M
		1.0×10^{-14} M	14.00	1.0×10^{0} M

Example 15.7

(a) Calculate the pH of a solution in which $[H_3O^+] = 2.6 \times 10^{-5}$ M.
(b) Calculate the pH of a solution in which $[OH^-] = 3.7 \times 10^{-9}$ M.
(c) Calculate the molarity of hydronium ion for a solution with pH 2.65.

Solution

(a) $pH = -\log[H_3O^+] = -\log(2.6 \times 10^{-5}) = 4.59$. On your calculator enter 2.6×10^{-5}, then press log, then press ±.

(b) Calculate $[H_3O^+]$:

$$K_w = [H_3O^+][OH^-] = 1.00 \times 10^{-14}$$

$$[H_3O^+] = \frac{K_w}{[OH^-]} = \frac{1.00 \times 10^{-14}}{3.7 \times 10^{-9}} = 2.7 \times 10^{-6}\text{ M}$$

Calculate pH:

$$pH = -\log[H_3O^+] = -\log(2.7 \times 10^{-6}) = 5.57$$

(c) To calculate $[H_3O^+]$ from pH, solve the equation that defines pH for $[H_3O^+]$:

$$pH = -\log[H_3O^+]$$

$$-pH = \log[H_3O^+]$$

$$\text{antilog}(-pH) = [H_3O^+]$$

$[H_3O^+] = \text{antilog}(-pH) = \text{antilog}(-2.65) = 2.2 \times 10^{-3}$ M. On your calculator enter 2.65, press ±, press 10^x (probably the second function on your log key).

Exercise 15.22

(a) Calculate the pH of a solution in which $[H_3O^+] = 8.4 \times 10^{-9}$ M.
(b) Calculate the pH of a solution in which $[OH^-] = 3.7 \times 10^{-3}$ M.
(c) Calculate the molarity of hydroxide ion in a solution with pH = 4.54.

pH meter. In laboratory work, pH is measured with indicators or pH meters. **Indicators** are compounds that change colors at certain H_3O^+ concentrations, and **pH meters** are electronic instruments that detect H_3O^+ concentrations and display them as digital pH readings. *(Chip Clark)*

15-8 Acids and bases can be defined in other ways

The definitions of *acid* and *base* that we've used are those that are most generally useful in understanding aqueous solutions: An acid is a compound that produces hydronium ion (H_3O^+) in aqueous solution, and a base is a compound that produces hydroxide ion (OH^-) in aqueous solution.

These definitions, together with two other definitions of acids and bases, can be found in Table 15.8 on page 338. As the table indicates, the definitions are usually identified by the names of the men who created them.

Each of the definitions of acids and bases shown in Table 15.8 focuses attention on a different feature of a reaction. For example, in the reaction

$$H_2\ddot{O}: + H{-}\ddot{\underset{..}{Cl}}: \rightarrow H_2\ddot{O}{-}\overset{+}{H} + :\ddot{\underset{..}{Cl}}:^-$$

Table 15.8 Three Definitions of Acids and Bases

Name	Definitions
Arrhenius Svante Arrhenius (1859–1927), Swedish chemist and physicist; Nobel Prize, 1903	Acid: compound that forms H_3O^+ in aqueous solution Base: compound that forms OH^- in aqueous solution Examples: $H_2O(\ell) + \underset{\text{acid}}{HCl(aq)} \rightarrow H_3O^+(aq) + Cl^-(aq)$ $\underset{\text{base}}{NaOH(s)} \rightarrow Na^+(aq) + OH^-(aq)$
Brønsted–Lowry Johannes N. Brønsted (1879–1947), Danish chemist, and Thomas M. Lowry (1874–1936), English chemist	Acid: proton (H^+) donor Base: proton (H^+) acceptor Examples: $\underset{\text{base}}{H_2\ddot{O}:} + \underset{\text{acid}}{H{-}\ddot{\underset{..}{Cl}}:} \rightarrow H_2\ddot{O}{-}\overset{+}{H} + :\ddot{\underset{..}{Cl}}:^-$
Lewis Gilbert N. Lewis (1875–1946), American chemist for whom Lewis symbols and formulas are also named	Acid: electron-pair acceptor Base: electron-pair donor Examples: $\underset{\text{base}}{H_2\ddot{O}:} + \underset{\text{acid}}{H{-}\ddot{\underset{..}{Cl}}:} \rightarrow H_2\ddot{O}{-}\overset{+}{H} + :\ddot{\underset{..}{Cl}}:^-$

◆In the original Arrhenius definition, an acid was a compound that produced H^+ in aqueous solution. We now understand that the species formed is actually H_3O^+.

the Arrhenius theory points to the formation of hydronium ion as the important feature of the reaction.◆ The Brønsted–Lowry theory emphasizes that a proton (hydrogen ion, H^+) is transferred from a hydrogen chloride molecule to a water molecule: The hydrogen chloride molecule donates a proton, and the water molecule accepts it.

$$\underset{\substack{\text{proton acceptor}\\ \text{base}}}{H_2\ddot{O}:} + \underset{\substack{\text{proton donor}\\ \text{acid}}}{H{-}\ddot{\underset{..}{Cl}}:} \rightarrow H_2\ddot{O}{-}\overset{+}{H} + :\ddot{\underset{..}{Cl}}:^-$$

The Lewis theory focuses on the fact that in this reaction an unshared electron pair on the oxygen atom in the water molecule forms a new bond: The water molecule donates the electron pair to create the new bond, and the hydrogen atom in the hydrogen chloride molecule accepts it.

$$\underset{\substack{\text{electron-pair donor}\\ \text{base}}}{H_2\ddot{O}:} + \underset{\substack{\text{electron-pair acceptor}\\ \text{acid}}}{H{-}\ddot{\underset{..}{Cl}}:} \rightarrow H_2\ddot{O}{-}\overset{+}{H} + :\ddot{\underset{..}{Cl}}:^-$$

Exercise 15.23

(a) In the forward (left-to-right) reaction shown in this equation

$$\begin{array}{c} H \\ | \\ H{-}\ddot{N}{:} \\ | \\ H \end{array} + \begin{array}{c} \ddot{O} \\ H \diagup \quad \diagdown H \end{array} \rightleftharpoons \begin{array}{c} H \\ | \\ H{-}N{-}H^{+} \\ | \\ H \end{array} + {:}\ddot{\underset{\cdot\cdot}{O}}{-}H^{-}$$

identify the Arrhenius base. (b) Identify the Brønsted–Lowry acid and base. (c) Identify the Lewis acid and base. Use arrows and labels, as shown in the equations in the preceding text, to justify your answers.

15-9 The most soluble ionic compounds consist of large ions with single charges

The example of sodium chloride illustrates the process by which ionic compounds dissolve in water (see Figure 15.2). A crystal of sodium chloride consists of alternating sodium ions and chloride ions, and when the crystal is dropped into water, each ion on its surface becomes surrounded by water molecules. Ion–dipole interactions form between the water molecules and the ions on the surface of the crystal, and these attractive forces draw the ions out of the lattice and into solution.

The process by which sodium chloride dissolves can be represented by this equation:

$$NaCl(s) \rightarrow Na^{+}(aq) + Cl^{-}(aq)$$

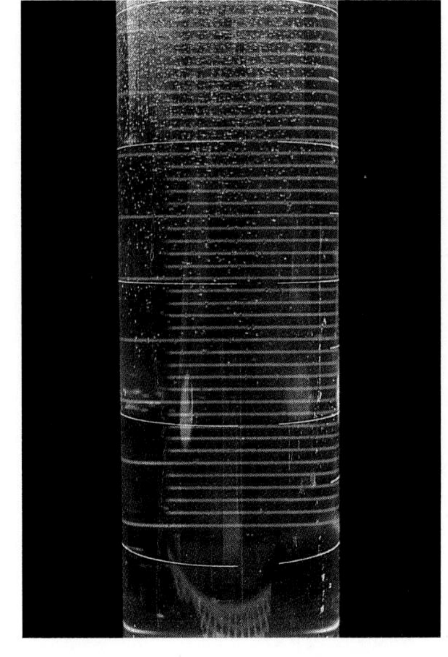

Crystals of sodium chloride dissolve as they sink through a column of water:

$$NaCl(s) \rightarrow Na^{+}(aq) + Cl^{-}(aq)$$

(Charles D. Winters)

Exercise 15.24

Write an equation, similar to the preceding one, for the process by which barium nitrate, $Ba(NO_3)_2$, dissolves in water.

Figure 15.2

Sodium chloride dissolving in water. (a) When a crystal of sodium chloride is dropped into water, ions on the surface of the crystal are surrounded by water molecules. The resulting ion–dipole interactions pull the crystal apart. (b) In solution, each ion is hydrated, that is, it's surrounded by water molecules to which it's attracted by ion–dipole interactions.

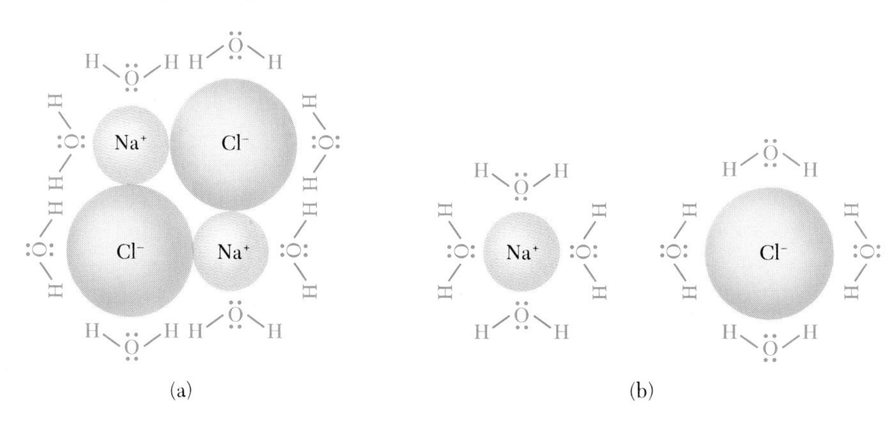

There is an upper limit to the amount of an ionic compound that will dissolve in a given amount of water, and the limit varies widely from one ionic compound to another. For example, 100 mL of water will dissolve a maximum of 35.7 g of sodium chloride, 200 g of ammonium nitrate (NH_4NO_3), or 1×10^{-9} g of aluminum phosphate ($AlPO_4$).◆

◆These solubilities are at 25°C, a common room temperature.

Ionic compounds in which the ions carry single charges are, in general, more soluble than compounds in which the ions carry multiple charges. In the preceding examples sodium chloride (Na^+ and Cl^-) and ammonium nitrate (NH_4^+ and NO_3^-) are more soluble than aluminum phosphate (Al^{3+} and PO_4^{3-}).

Exercise 15.25

In each pair, predict which compound will be more soluble: (a) KNO_3 or $BaSO_4$ (b) Al_2O_3 or $Al(C_2H_3O_2)_3$

Ionic compounds with larger ions are, in general, more soluble than compounds with smaller ions. In the preceding examples, ammonium nitrate, which consists of the large polyatomic ions NH_4^+ and NO_3^-, is more soluble than sodium chloride, which consists of the smaller monatomic ions Na^+ and Cl^-.

Example 15.8

In each pair, predict which compound will be more soluble, and explain your choice: (a) LiCl or CsCl (b) NaCl or $MgCl_2$

Solution

(a) CsCl should be more soluble (it is). Cesium is below lithium in group 1A, so Cs^+ is larger than Li^+.

(b) NaCl should be more soluble (it is). The magnesium ion carries a 2+ charge, and the sodium ion carries a 1+ charge. Also, Mg^{2+} is smaller than Na^+: Magnesium is to the right of sodium in the same period, so a magnesium atom is smaller than a sodium atom, and the same is true for their cations (see Sections 8-1 and 8-2).

Exercise 15.26

In each pair predict which compound will be less soluble and explain your choice: (a) KF or KI (b) ZnS or $Zn(NO_3)_2$

The principles stating that an ionic compound with larger ions or with single charges on its ions will be more soluble than a compound with smaller ions or with multiple charges on its ions are often useful in predicting relative solubilities of ionic compounds. More detailed solubility rules can be used to make more accurate predictions, as described in Appendix 4.◆

◆Your instructor may have a preference as to whether you should use the general principles given here or the rules in Appendix 4.

Because solubility varies with temperature, supersaturated solutions of ionic compounds can be prepared. The solubility of sodium chloride, for example, is greater at higher temperatures. If a saturated solution of sodium chloride is prepared at a higher temperature and cooled carefully to a lower one, it will contain more solute at the lower temperature than could be dissolved in it at that temperature; such a solution is said to be **supersaturated.** A supersaturated solution is unstable: If a crystal of sodium chloride is added to a supersaturated solution of sodium chloride, the excess solute will solidify, and the solution that remains will be saturated.

The molarity of a solution of an ionic compound is calculated in the usual way. It's sometimes useful to think of a solution of an ionic compound as a solution of cations and a solution of anions, and to identify the molarity for each species. For example, for a solution labeled $[Ni(NO_3)_2(aq)] = 0.500$ M, the solution process can be represented by the equation

$$Ni(NO_3)_2(s) \rightarrow Ni^{2+}(aq) + 2\ NO_3^-(aq)$$

This equation shows that 1 mole of $Ni(NO_3)_2(s)$ dissolves to form 1 mole of $Ni^{2+}(aq)$ and 2 moles of $NO_3^-(aq)$. In a solution for which $[Ni(NO_3)_2(aq)] =$ 0.500 M, the concentrations of the ions will be $[Ni^{2+}] = 0.500$ M and $[NO_3^-] =$ 2×0.500 M $= 1.00$ M.

Exercise 15.27

In a solution for which $[(NH_4)_3PO_4] = 0.0272$ M, what is $[NH_4^+]$?

Inside Chemistry

How do soaps and detergents work?

Taking a shower without soap isn't very effective for getting clean because skin and hair have a thin layer of natural oil that holds dirt particles and prevents water from washing them away. The water forms a layer over the oil layer, as shown in the following drawing.

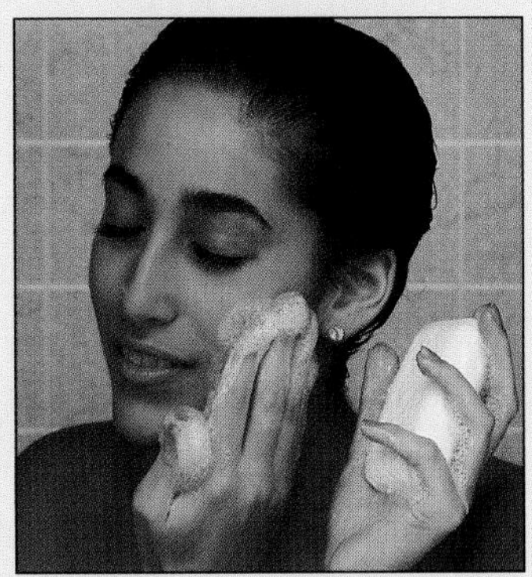
(Roy Morsch/The Stock Market)

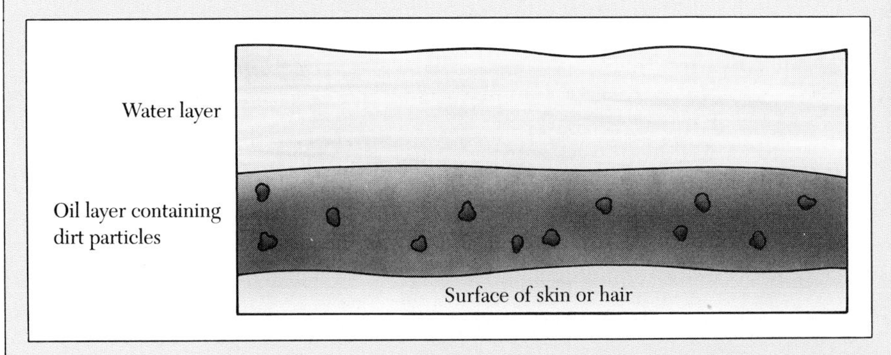

The oil on skin and hair isn't soluble in water because the oil molecules form only weak intermolecular forces with water molecules. This is the formula for a typical oil molecule:

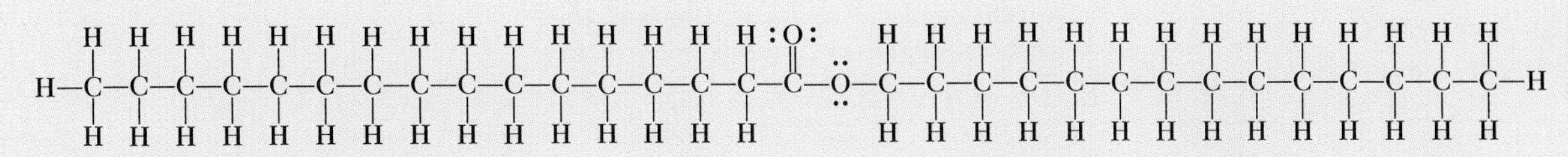

Dipole–dipole interactions with water molecules can occur at the oxygen atoms, but along most of the length of the oil molecule, only weak London forces with water molecules are possible.

A soap or detergent is an ionic compound with a very large polyatomic anion. These are typical examples:

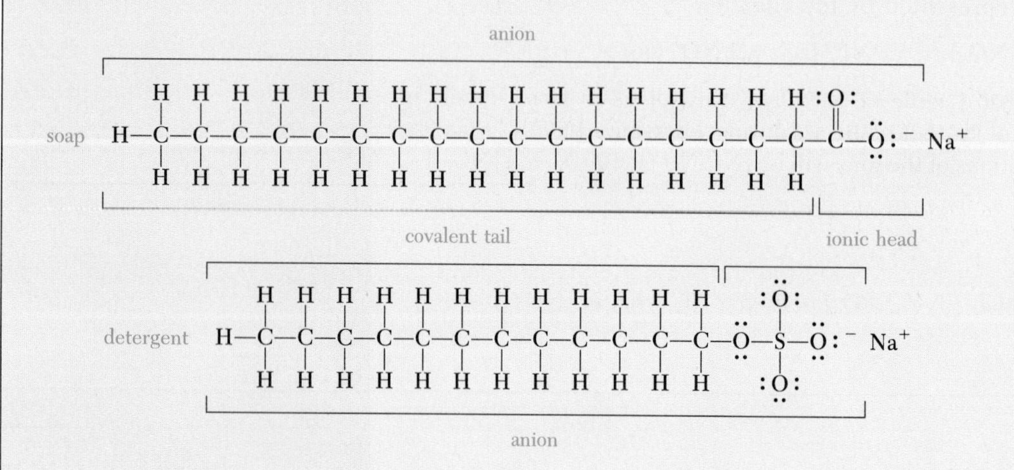

As these formulas show, each anion in a soap or detergent has a long covalent tail and a small ionic head.

When you wash with soap or detergent, the long covalent tails of the soap or detergent anions dissolve in the oil layer on your skin or hair by forming London forces; because the oil molecules and the anions are large and contain many electrons, the London forces between them are strong. The ionic heads of the anions dissolve in the water layer by forming ion–dipole attractions. Because the heads of the anions are soluble in the water layer and their tails are soluble in the oil layer, the anions arrange themselves so that they extend across the oil–water interface, as shown in the following drawing.

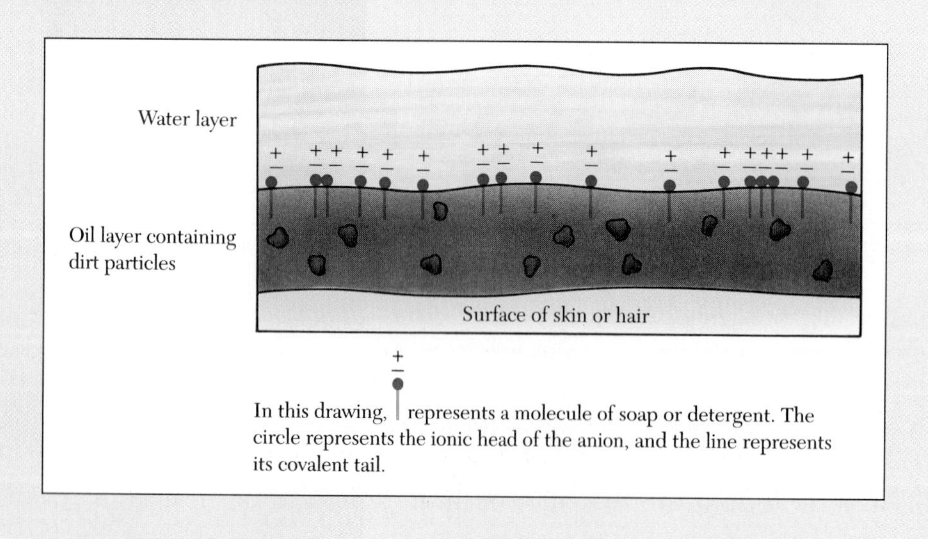

In this drawing, [symbol] represents a molecule of soap or detergent. The circle represents the ionic head of the anion, and the line represents its covalent tail.

In effect, the anions of the soap or detergent stitch the oil and water layers together. As a result, the oil layer, carrying dirt particles with it, becomes dispersed in the water layer and can be rinsed away, as shown in the following drawing.

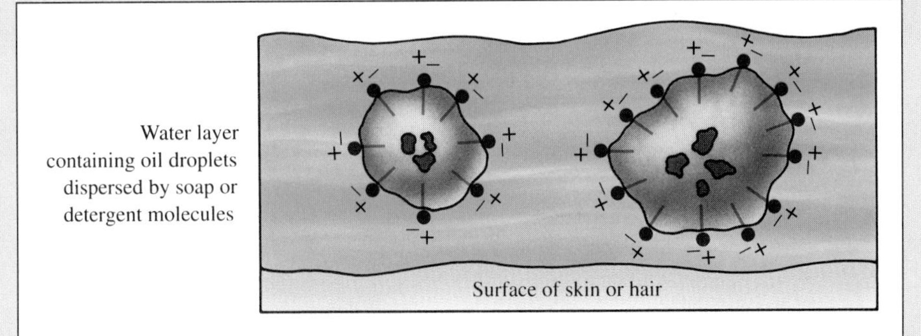

Chapter Summary: Introduction to Aqueous Solutions

Section	Subject	Summary	Check When Learned
	In an aqueous solution: solvent	Water.	
	solute	Substance dissolved in water.	☐
15-1	Kind of solute particles that will form the following kinds of interparticle forces with water molecules: London forces	All solute particles. The London forces are weak because each water molecule contains only ten electrons.	
	dipole–dipole interactions	Solute particles that are dipoles.	
	hydrogen bonds	Solute molecules that contain N, O, or F atoms with unshared electron pairs.	
	ion–dipole interactions	Solute particles that are ions.	☐
15-1	Hydrated ion	An ion in aqueous solution with water molecules attached to it by ion–dipole interactions. The number of water molecules may vary, with a typical number in the range from three to six.	☐

Section	Subject	Summary	Check When Learned
15-2	Relationship between solubility and strength of forces between water molecules and solute particles	If the attractive forces between water molecules and solute particles are weak, the solute will be only slightly soluble.	☐
15-3	Concentration	Amount of solute in a given amount of solution.	☐
15-3	Unsaturated and saturated	A solution is said to be unsaturated if it contains less solute than the maximum it can hold and is said to be saturated if it contains the maximum.	☐
15-3	Molarity	Moles of solute per liter of solution.	☐
15-3	Equation for molarity	$M = \frac{mol}{L}$. M is molarity, mol is moles of solute, and L is liters of solution.	☐
15-3	Meaning of 1.00 M NaCl(aq) [NaCl(aq)] = 1.00 M	1.00 molar aqueous sodium chloride. 1.00 molar aqueous sodium chloride.	☐
15-3	Meaning of brackets, [], for example in [NaCl]	Brackets always designate the molarity of the species shown in the brackets; e.g., [NaCl] means concentration of NaCl in moles per liter.	☐
15-3	Volumetric flask	A flask that's made so that, when it's filled with a liquid to a line etched on its neck, the volume of the liquid will be a specified volume, e.g., 1000 mL.	☐
15-3	Dilution	Preparation of one aqueous solution from another by adding water.	☐
15-3	In dilution, equation for relationship between initial volume and molarity and final volume and molarity	$V_1M_1 = V_2M_2$. V_1 and M_1 are the volume and molarity of the initial solution, and V_2 and M_2 are the volume and molarity of the final solution.	☐
15-4	Formulas for hydronium ion:		
	line formula	H, H bonded to :O—H with + charge ($[H_2\ddot{O}H]^+$)	
	condensed formula	H_3O^+	☐

Section	Subject	Summary	Check When Learned
15-4	Equations for autoionization of water:		
	in line formulas	$\mathrm{H_2}\!:\!\ddot{\mathrm{O}}\!: + \mathrm{H}\!-\!\ddot{\mathrm{O}}\!-\!\mathrm{H} \rightleftarrows \mathrm{H_2}\!:\!\mathrm{O}\!-\!\overset{+}{\mathrm{H}} + :\!\ddot{\mathrm{O}}\!-\!\overset{-}{\mathrm{H}}$	
	in molecular and condensed formulas	$2\ H_2O(\ell) \rightleftarrows H_3O^+(aq) + OH^-(aq)$	☐
15-4	Meaning of double arrow, $\rightleftarrows$	A double arrow in an equation shows that the reaction is reversible: The reaction occurs left-to-right as written, and it simultaneously occurs right-to-left.	☐
15-4	Forward reaction	In a reversible reaction, the left-to-right reaction.	☐
15-4	Reverse reaction	In a reversible reaction, the right-to-left reaction.	☐
15-4	Equilibrium	Condition of a reversible reaction in which the forward and reverse reactions are occurring at the same rate.	☐
15-4	In pure water $[H_3O^+] =$ $[OH^-] =$	1.00×10^{-7} M	☐
15-5	Acid	Substance that forms $H_3O^+(aq)$ in aqueous solution.	☐
15-5	Strong acid	Substance that forms $H_3O^+(aq)$ in aqueous solution by an irreversible reaction: $H_2O(\ell) + HA(aq) \rightarrow H_3O^+(aq) + A^-(aq)$	☐
15-5	The six strong acids	HCl, HBr, HI, HNO_3, H_2SO_4, $HClO_4$	☐
15-5	Weak acid	Substance that forms $H_3O^+(aq)$ in aqueous solution by a reversible reaction: $H_2O(\ell) + HA(aq) \rightleftarrows H_3O^+(aq) + A^-(aq)$	☐
15-5	Which acids are weak acids?	In this book we'll assume that all acids except the six strong acids shown in Table 15-1 are weak acids.	☐
15-5	Ionizable hydrogen atom	A hydrogen atom that can be transferred from its molecule to a water molecule to form a hydronium ion, e.g., the hydrogen atom in HCl.	☐
15-5	In a formula, the method for distinguishing ionizable from nonionizable hydrogen atoms	The symbols for the ionizable hydrogen atoms are written first in the formula. For example, $HC_2H_3O_2$ shows that the molecule contains one ionizable hydrogen atom and three nonionizable hydrogen atoms.	☐

Section	Subject	Summary	Check When Learned
15-5	Terms for designating the number of ionizable hydrogen atoms in one molecule of an acid	The prefixes are the same as those used in naming covalent compounds (Section 9-1): monoprotic, one ionizable hydrogen atom; diprotic, two ionizable hydrogen atoms; triprotic, three ionizable hydrogen atoms; and so forth.	☐
15-5	Polyprotic acid	Acid that contains more than one ionizable hydrogen atom per molecule.	☐
15-6	Base	Substance that produces $OH^-(aq)$ in aqueous solution.	☐
15-6	Strong base	Soluble metallic hydroxide of group 1A or 2A that releases $OH^-(aq)$ in aqueous solution.	☐
15-6	Six common strong bases	Group 1A: LiOH, NaOH, KOH. Group 2A: $Ca(OH)_2$, $Sr(OH)_2$, $Ba(OH)_2$.	☐
15-6	Weak base	Compound that reacts reversibly with water to form $OH^-(aq)$.	☐
15-6	Most important weak base	Ammonia, NH_3.	☐
15-6	Equation for reaction of ammonia with water	$NH_3(aq) + H_2O(\ell) \rightleftarrows NH_4^+(aq) + OH^-(aq)$	☐
15-7	Ion–product equation for water	$[H_3O^+][OH^-] = K_w = 1.00 \times 10^{-14}$	☐
15-7	Name for K_w	Ion–product constant for water.	☐
15-7	Molarities of hydronium ion and hydroxide ion in pure water in acidic solution in basic solution	 $[H_3O^+] = [OH^-] = 1.00 \times 10^{-7}$ M $[H_3O^+] > 1.00 \times 10^{-7}$ M and $[OH^-] < 1.00 \times 10^{-7}$ M $[H_3O^+] < 1.00 \times 10^{-7}$ M and $[OH^-] > 1.00 \times 10^{-7}$ M	☐
15-7	Definition of pH	$pH = -\log[H_3O^+]$	☐
15-7	Range of pH scale	0 to 14	☐
15-7	pH of pure water acidic solution basic solution	 7 below 7, down to 0 above 7, up to 14	☐
15-8	Definitions of acid and base: Arrhenius acid base	 Compound that forms H_3O^+ in aqueous solution. Compound that forms OH^- in aqueous solution.	

Section	Subject	Summary	Check When Learned
	Brønsted–Lowry		
	acid	Proton (H^+) donor.	
	base	Proton (H^+) acceptor.	
	Lewis		
	acid	Electron-pair acceptor.	
	base	Electron-pair donor.	☐
15-9	Equation in condensed symbols and formulas for process by which sodium chloride dissolves in water	$NaCl(s) \rightarrow Na^+(aq) + Cl^-(aq)$	☐
15-9	General principles for predicting solubilities of ionic compounds	Compounds in which the ions carry single charges are more soluble than compounds in which the ions carry multiple charges. Compounds with larger ions are more soluble than compounds with smaller ions.	☐
15-9	Supersaturated	A solution is said to be supersaturated if it contains more than the maximum amount of solute that it can ordinarily hold at a given temperature.	☐

Problems

Assume you can use the Periodic Table at the front of the book unless you're directed otherwise. Answers to odd-numbered Problems are in Appendix 1.

Interparticle Forces and Solubilities (Sections 15-1 and 15-2)

1. Which element would you expect to be more soluble in water, iodine or bromine? Why?
2. Which element would you expect to be more soluble in water, argon or krypton? Why?
3. Which element would you expect to be more soluble in water, hydrogen or oxygen? Why?
4. Which element would you expect to be more soluble in water, chlorine or argon? Why?
5. The line formula for ethanol (beverage alcohol) is

```
     H   H
     |   |   ..
  H—C—C—O—H
     |   |   ..
     H   H
```

   The ethanol molecule is a dipole. In an aqueous solution of ethanol, what interparticle forces would you predict are present between ethanol molecules and water molecules?

6. The line formula for methane (natural gas) is

```
      H
      |
   H—C—H
      |
      H
```

   The methane molecule is not a dipole. In an aqueous solution of methane, what interparticle forces would you predict are present between methane molecules and water molecules?

7. Ethanol (problem 5) is infinitely soluble in water. Suggest an explanation.
8. Methane (problem 6) is only slightly soluble in water. Suggest an explanation.

9. Under ordinary room conditions, one of the following compounds is a gas that's only slightly soluble in water, and the other is a liquid that's infinitely soluble in water. Predict which set of characteristics belongs to which compound, and explain your answer:

```
         H  H                    H  H
         |  |           ..       |  |     ..
(a) H—C—C—H      (b) H—O—C—C—O—H
         |  |           ..       |  |     ..
         H  H                    H  H
```

10. The line formulas for the following compounds are shown in Table 7.2. Predict whether each compound will be more than slightly soluble in water, and explain your answer: (a) sucrose (b) naphthalene

11. Cesium ion, Cs^+, is a very large ion. Use line formulas for water molecules to show how you would expect four water molecules to be arranged around a cesium ion in aqueous solution.

12. Iodide ion, I^-, is a very large ion. Use line formulas for water molecules to show how you would expect four water molecules to be arranged around an iodide ion in aqueous solution.

13. In the solution that's formed when hydrogen chloride dissolves in water, what interparticle forces are present between solute particles and water molecules?

14. In the solution that's formed when hydrogen fluoride dissolves in water, what interparticle forces are present between solute particles and water molecules?

Molarity (Sections 15-3, 15-7, and 15-9)

15. Calculate the molarity of a solution made by combining 3.00 g of potassium bromide with enough water to make 150 mL (3 significant digits) of solution.

16. Calculate the molarity of a solution made by combining 215 mg of zinc hydrogen sulfate with enough water to make 25.0 mL of solution.

17. How many milliliters of 3.00 M HBr(aq) will be needed to provide 0.250 mol of HBr?

18. How many milliliters of a solution labeled $[HNO_3(aq)] = 0.106$ M will be needed to provide 355 mg of HNO_3?

19. How many grams of barium nitrate will be needed to prepare 500 mL (3 significant digits) of a 0.155 M solution?

20. How many grams of tin(II) nitrite will be needed to prepare 125 mL of a 0.500 M solution?

21. Describe the process that would be used to prepare 100 mL of 0.750 M NaCl(aq), using a 100-mL volumetric flask.

22. Describe the process that would be used to prepare 250 mL of 1.00 M aqueous ammonium nitrate, using a 250-mL volumetric flask.

23. In a solution labeled $[HClO_4(aq)] = 0.250$ M, what is the concentration of perchlorate ion?

24. In a solution labeled 0.766 M $Sn(NO_3)_2(aq)$: (a) What is the concentration of nitrate ion? (b) What is the concentration of tin(II) ion?

25. How many milliliters of 6.00 M HI(aq) will be needed to prepare 75.0 mL of 2.00 M HI(aq)?

26. How many milliliters of 1.25 M calcium acetate will be needed to prepare 0.250 L of a solution in which the concentration of acetate ion is 0.650 M?

27. Describe the process that would be used to prepare 100 mL of 1.50 M HCl(aq) from 6.00 M HCl(aq), using a 100-mL volumetric flask.

28. How many milliliters of water should be added to 50.0 mL of 1.35 M aqueous ammonium sulfate to make a solution in which the concentration of ammonium ion is 0.550 M?

Autoionization of Water (Section 15-4)

29. Write an equation for the autoionization of water (a) in Lewis formulas and (b) in molecular and condensed formulas.

30. How many unshared electron pairs are (a) in a water molecule, (b) in a hydronium ion, and (c) in a hydroxide ion?

31. In the reaction between two water molecules to form a hydronium ion and a hydroxide ion: (a) How many bonds are broken? (b) How many bonds are formed?

32. In the reaction between a hydronium ion and a hydroxide ion to from two water molecules: (a) How many bonds are broken? (b) How many bonds are formed?

33. Write an equation in line formulas for this reaction:

$$H_3O^+ + H_2O \rightleftarrows H_2O + H_3O^+$$

34. Write an equation in line formulas for this reaction:

$$OH^- + H_2O \rightleftarrows H_2O + OH^-$$

35. In the liquid state hydrogen fluoride undergoes an autoionization reaction that's similar to that for water:

$$2\ HF \rightleftarrows H_2F^+ + F^-$$

Write the equation for this reaction, using Lewis symbols and line formulas.

36. In the liquid state ammonia undergoes an autoionization reaction that's similar to that for water:

$$2\ NH_3 \rightleftarrows NH_4^+ + NH_2^-$$

Write the equation for this reaction, using Lewis symbols and line formulas.

37. In your own words, explain the meaning of $\rightleftarrows$.

38. In your own words, explain the meaning of *equilibrium*.

Acids and Bases (Sections 15-5 and 15-6)

39. Explain the difference between hydrogen chloride and hydrochloric acid.

40. Hydrogen chloride is very soluble in water, but hydrogen and chlorine are not. Why?

41. Name the ions that are formed when each of the following compounds dissolves in water: (a) hydrogen iodide (b) perchloric acid (c) potassium hydroxide.

42. Which compound would you expect to be more soluble in water, hydrogen chloride or hydrogen iodide? Why?

43. Identify each of the following compounds as a strong or weak acid or base: (a) H_3PO_4 (b) N_2H_4 (c) $Ba(OH)_2$ (d) HNO_3.

44. Identify each of the following compounds as a strong or weak acid or base: (a) LiOH (b) NH_3 (c) KOH (d) $H_2C_2O_4$.

45. Phosphorous acid, H_2PHO_3, is a diprotic acid. Write equations for its two ionizations.

46. Sulfurous acid, H_2SO_3, is a diprotic acid. Write equations for its two ionizations.

47. Use Lewis symbols and formulas to write an equation for the reaction of hydrogen bromide with water.

48. Which strong acid is polyprotic?

49. The reaction of hydrogen chloride with water to form hydrochloric acid can be described by the abbreviated equation

$$HCl \rightarrow H^+ + Cl^-$$

Use this abbreviated form to write an equation for the reaction of each of the other five strong acids with water.

50. The reaction of acetic acid with water can be described by the abbreviated equation

$$HC_2H_3O_2 \rightleftarrows H^+ + C_2H_3O_2^-$$

Use this abbreviated form to write an equation for the reaction of each of the other weak acids shown in Table 15.2 with water.

51. Which of the following statements describe an acidic solution? (a) $[H_3O^+] = 1.00 \times 10^{-7}$ M (b) $[OH^-] = 1.00 \times 10^{-8}$ M (c) $[OH^-] < 1.00 \times 10^{-7}$ M

52. Which of the following statements describes a basic solution? (a) $[H_3O^+] > 1.00 \times 10^{-7}$ M (b) $[H_3O^+] = 1.00 \times 10^{-8}$ M (c) $[OH^-] = 1.00 \times 10^{-7}$ M

53. Which of the following statements describes an acidic solution? (a) $[H_3O^+] < [OH^-]$ (b) $[H_3O^+] > [OH^-]$ (c) $[H_3O^+] = [OH^-]$

54. Which of the following statements describes a basic solution? (a) $[H_3O^+] < [OH^-]$ (b) $[H_3O^+] > [OH^-]$ (c) $[H_3O^+] = [OH^-]$

55. What is $[OH^-]$ in a solution in which $[H_3O^+] = 1.00 \times 10^{-5}$ M?

56. What is $[H_3O^+]$ in a solution in which $[OH^-] = 1.00 \times 10^{-10}$ M?

57. What is $[OH^-]$ in a solution in which $[H_3O^+] = 7.48 \times 10^{-12}$ M?

58. What is $[H_3O^+]$ in a solution in which $[OH^-] = 3.77 \times 10^{-2}$ M?

pH (Section 15-7)

59. Which of the following pH values describes a solution that's weakly basic? (a) 1.50 (b) 7.00 (c) 8.00

60. Which of the following pH values describes a solution that's strongly acidic? (a) 9.00 (b) 6.00 (c) 1.00

61. What is the molarity of hydronium ion in a solution whose pH is 2.00?

62. What is the molarity of hydroxide ion in a solution whose pH is 3.00?

63. Calculate the pH of 0.500 M HI(aq).

64. Calculate the pH of 0.500 M NaOH(aq).

65. Calculate the pH of 0.288 M HCl(aq).

66. Calculate the pH of 0.200 M $Ba(OH)_2$(aq).

67. Calculate the pH of a solution made by dissolving 23.7 g of HBr in enough water to make 325 mL of solution.

68. Calculate the pH of a solution made by dissolving 3.49 g of KOH in enough water to make 252 mL of solution.

69. Calculate the pH of a solution made by dissolving 6.40 g of perchloric acid in enough water to make 0.250 L of solution.

70. Calculate the pH of a solution made by dissolving 4.20 g of calcium hydroxide in enough water to make 0.500 L of solution.

71. How many moles of HCl are present in 75.0 mL of a solution whose pH is 5.38?

72. How many moles of LiOH are present in 50.0 mL of a solution whose pH is 9.74?

73. How many grams of HNO_3 are present in 16.0 mL of a solution whose pH is 1.50?

74. How many grams of $Sr(OH)_2$ are present in 5.00 mL of a solution whose pH is 12.50?

75. How many milligrams of HBr are present in 5.00 mL of a solution whose pH is 4.00?

76. How many milligrams of calcium hydroxide are present in 5.00 mL of a solution whose pH is 9.00?

77. What is the pH of a solution in which $[OH^-] = 2.75 \times 10^{-4}$ M?

78. What is the molarity of hydroxide ion in a solution in which pH = 3.55?

79. A solution was prepared by combining 4.37 g of HNO_3 with enough water to make 325 mL of solution. (a) What is the molarity of hydronium ion in the solution? (b) What is the pH? (c) What is the molarity of hydroxide ion?

80. A solution was prepared by combining 52.9 mg of strontium hydroxide with enough water to make 2.00 mL of solution. (a) What is the molarity of hydroxide ion in the solution? (b) What is the molarity of hydronium ion? (c) What is the pH?

Definitions of Acids and Bases (Section 15-8)

81. In the autoionization of water, is water an Arrhenius acid or an Arrhenius base? Explain.

82. In the autoionization of water, is water a Brønsted–Lowry acid or a Brønsted–Lowry base? Explain.

83. The equation for the reaction of hydrogen fluoride with water is

$$HF(aq) + H_2O(\ell) \rightleftarrows H_3O^+(aq) + F^-(aq)$$

In the forward reaction identify the Brønsted–Lowry acid and base.

84. The equation for the reaction of hydrogen fluoride with water is

$$HF(aq) + H_2O(\ell) \rightleftarrows H_3O^+(aq) + F^-(aq)$$

In the reverse reaction, identify the Brønsted–Lowry acid and base.

85. The equation for the reaction of hydrogen fluoride with water is

$$HF(aq) + H_2O(\ell) \rightleftarrows H_3O^+(aq) + F^-(aq)$$

In the forward reaction, identify the Lewis acid and base.

86. The equation for the reaction of hydrogen fluoride with water is

$$HF(aq) + H_2O(\ell) \rightleftarrows H_3O^+(aq) + F^-(aq)$$

In the reverse reaction, identify the Lewis acid and base.

Solubility of Ionic Compounds (Section 15-9)

87. As a crystal of an ionic compound dissolves in water, the ions on its surface are surrounded by water molecules and drawn into solution. Which ions would you expect to be pulled more easily out of the lattice, those on a face of the crystal or those on an edge? Why?

88. As a crystal of an ionic compound dissolves in water, the ions on its surface are surrounded by water molecules and drawn into solution. Which ions would you expect to be pulled more easily out of the lattice, those on an edge of the crystal or those on a corner? Why?

89. Write an equation, using condensed formulas, for the process by which potassium acetate dissolves in water.

90. Write an equation, using condensed formulas, for the process by which iron(III) nitrate dissolves in water.

91. In each pair, predict which compound will be more soluble, and explain your choice. (a) $CaCl_2$ or $BaCl_2$ (b) $CaCl_2$ or $GaCl_3$ (c) KCl or KI

92. In each pair, predict which compound will be more soluble, and explain your choice. (a) LiF or CsI (b) $(NH_4)_2SO_4$ or $PbSO_4$ (c) $FePO_4$ or $NaClO_4$

93. All common ionic compounds in which the cation is ammonium ion are significantly soluble in water. Suggest an explanation.

94. All common ionic compounds in which the anion is acetate ion or nitrate ion are significantly soluble in water. Suggest an explanation.

95. In group 2A, the hydroxides of calcium, strontium, and barium are soluble enough to be included in the list of common strong bases, but the hydroxides of beryllium and magnesium are not. Suggest an explanation.

96. If you were going to choose an aluminum compound that would be significantly soluble in water, which would you choose, aluminum chloride or aluminum chlorate? Why?

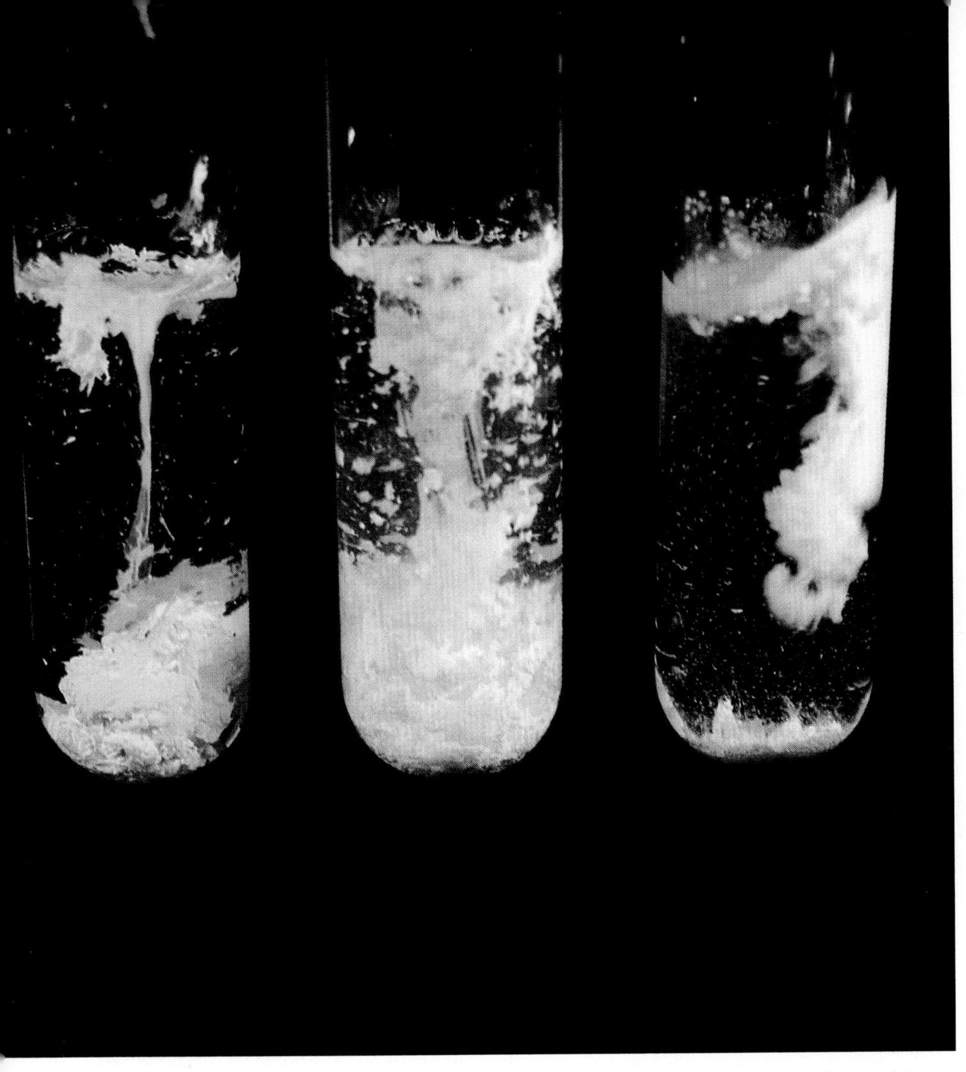

Reactions in aqueous solution forming, left to right, AgCl, AgBr, and AgI. *(Charles Steele)*

16
Reactions in Aqueous Solutions

Each moment, many millions of chemical reactions are taking place in the oceans, in the atmosphere, in the soil, and in living organisms, and most of these reactions occur in aqueous solutions. During the past two centuries, tens of thousands of chemists, biologists, agricultural scientists, soil scientists, ecologists, and other scientists have studied the reactions that occur in aqueous solutions and have learned much about them.

In this chapter we'll look briefly at five important kinds of reactions in aqueous solutions: reactions of metallic and nonmetallic oxides with water, reactions of acids with bases, reactions in which solid ionic compounds are formed, reactions in which gases are formed, and reactions in which one metal ion replaces another in solution. In studying these reactions, you'll be applying the basic concepts of chemistry that you learned throughout the earlier chapters of this book.

16-1 Some nonmetal oxides react with water to form solutions of acids, and some metal oxides react with water to form solutions of bases

Pure water exposed to the air slowly becomes slightly acidic. The change occurs because carbon dioxide from the air dissolves in and reacts with water to form a solution of the weak acid called carbonic acid:

$$2\,H_2O(\ell) + \underset{\text{carbon dioxide}}{CO_2(aq)} \rightleftarrows \underset{\text{hydronium ion}}{H_3O^+(aq)} + \underset{\text{hydrogen carbonate ion}}{HCO_3^-(aq)}$$

carbonic acid

Hydrogen carbonate ion is itself a weak acid:

$$H_2O(\ell) + \underset{\text{hydrogen carbonate ion}}{HCO_3^-(aq)} \rightleftarrows \underset{\text{hydronium ion}}{H_3O^+(aq)} + \underset{\text{carbonate ion}}{CO_3^{2-}(aq)}$$

Similar reactions occur with oxides of other nonmetals, and some of these reactions have important environmental consequences. For example, when coal that contains sulfur is burned, the sulfur (S_8) reacts with oxygen to form sulfur dioxide:

$$S_8(s) + 8\,O_2(g) \rightarrow 8\,SO_2(g)$$

In the atmosphere sulfur dioxide dissolves in and reacts with water to form a solution of the weak acid called sulfurous acid:

$$2\,H_2O(\ell) + \underset{\text{sulfur dioxide}}{SO_2(aq)} \rightleftarrows \underset{\text{hydronium ion}}{H_3O^+(aq)} + \underset{\text{hydrogen sulfite ion}}{HSO_3^-(aq)}$$

sulfurous acid

$$H_2O(\ell) + \underset{\text{hydrogen sulfite ion}}{HSO_3^-(aq)} \rightleftarrows \underset{\text{hydronium ion}}{H_3O^+(aq)} + \underset{\text{sulfite ion}}{SO_3^{2-}(aq)}$$

Exercise 16.1◆

In the atmosphere sulfur dioxide reacts with oxygen to form sulfur trioxide, and sulfur trioxide dissolves in and reacts with water to form sulfuric acid, H_2SO_4. Using molecular and condensed formulas, write an equation for each of these reactions.

◆Answers to Exercises are in Appendix 1.

Aqueous solutions of sulfurous acid and sulfuric acid that are formed in the atmosphere can fall to earth as **acid rain.** In North America and Europe acid rain has caused severe harm to forests and lakes and severe damage to stone buildings and statuary.

The oxides of some metals react with water to form basic solutions; this reaction is important for the oxides of the metals in groups 1A and 2A, except beryllium. Sodium oxide, for example, reacts with water to form aqueous sodium hydroxide:

$$Na_2O(s) + H_2O(\ell) \rightarrow 2\ Na^+(aq) + 2\ OH^-(aq)$$

Exercise 16.2

Using molecular and condensed formulas, write an equation for the reaction of calcium oxide with water to form aqueous calcium hydroxide.

Exercise 16.3

In the reaction of a metal oxide with water an oxide ion, $:\ddot{\underset{\cdot\cdot}{O}}:^{2-}$, reacts with a water molecule to form two hydroxide ions. Use Lewis symbols and line formulas to write an equation for this reaction.

The bronze statue and the yellow trees have been damaged by acid rain. *(Hans Wolf/The Image Bank; Runk/Schoenberger from Grant Heilman)*

16-2 Acids react with bases in a process called neutralization◆

◆As you read this Section, you'll need to be able to identify acids and bases as strong or weak from their formulas, as described in Sections 15-5 and 15-6.

Strong Acid and Strong Base

One of the most common and most important classes of chemical reactions is **neutralization,** the reaction of an acid with a base. As an example to introduce the concept of neutralization, we'll use the reaction of hydrochloric acid with aqueous sodium hydroxide.

As we saw in Chapter 15, hydrochloric acid is an aqueous solution that's formed when hydrogen chloride gas dissolves in and reacts with water,

$$H_2O(\ell) + HCl(aq) \rightarrow \underbrace{\underset{\text{hydronium ion}}{H_3O^+(aq)} + \underset{\text{chloride ion}}{Cl^-(aq)}}_{\text{hydrochloric acid}}$$

and aqueous sodium hydroxide is a solution that's formed when solid sodium hydroxide dissolves in water,

$$NaOH(s) \rightarrow \underbrace{\underset{\text{sodium ion}}{Na^+(aq)} + \underset{\text{hydroxide ion}}{OH^-(aq)}}_{\text{aqueous sodium hydroxide}}$$

When hydrochloric acid and aqueous sodium hydroxide are mixed, hydronium ions and hydroxide ions react with one another to form water molecules:

$$\underbrace{\boxed{H_3O^+}(aq) + Cl^-(aq)}_{\text{hydrochloric acid}} + \underbrace{Na^+(aq) + \boxed{OH^-}(aq)}_{\text{aqueous sodium hydroxide}} \rightarrow \underbrace{\boxed{2\ H_2O}(\ell)}_{\text{water}} + \underbrace{Na^+(aq) + Cl^-(aq)}_{\text{salt}}$$

The other products of the neutralization reaction are the hydrated sodium ion, $Na^+(aq)$, and the hydrated chloride ion, $Cl^-(aq)$; this combination of ions is referred to as a salt. A **salt** is any ionic compound that is not an acid, a base, or an oxide.◆

◆In everyday usage *salt* refers to one substance, table salt. In chemical usage *salt* refers to a class of compounds, and sodium chloride (table salt) is one compound in that class.

Exercise 16.4

Decide whether each of these compounds should be classified as a salt: (a) H_2SO_4 (b) Na_2SO_4 (c) Na_2O (d) SO_2

The preceding equation can be simplified by recognizing that two of the reactants, $Na^+(aq)$ and $Cl^-(aq)$, reappear unchanged in the products. That is, the sodium ion and the chloride ion, although they're present in the reaction mixture, do not undergo any change during the reaction; for this reason they're referred to as **spectator ions.** Because spectator ions don't participate in the reaction, we can take the symbols for them out of the equation, to make it simpler. An equation that includes spectator ions is called an **ionic equation,** and an equation that's formed by removing the spectator ions from an ionic equation is called a **net ionic equation.** For the neutralization reaction of hydrochloric acid with aqueous sodium hydroxide, these equations are

ionic: $H_3O^+(aq) + Cl^-(aq) + Na^+(aq) + OH^-(aq) \rightarrow 2\ H_2O(\ell) + Na^+(aq) + Cl^-(aq)$

net ionic: $H_3O^+(aq) + OH^-(aq) \rightarrow 2\ H_2O(\ell)$

Net ionic equations make it easy to see that all neutralization reactions between strong acids and strong bases are in fact the same reaction. For example, the equations for the reaction of nitric acid with aqueous potassium hydroxide are

ionic: $\underbrace{H_3O^+(aq) + NO_3^-(aq)}_{\text{nitric acid}} + \underbrace{K^+(aq) + OH^-(aq)}_{\text{aqueous potassium hydroxide}} \rightarrow \underbrace{2\ H_2O(\ell)}_{\text{water}} + \underbrace{K^+(aq) + NO_3^-(aq)}_{\text{salt}}$

net ionic: $H_3O^+(aq) + OH^-(aq) \rightarrow 2\ H_2O(\ell)$.

The net ionic equation for the reaction of a strong acid with a strong base is

$$H_3O^+(aq) + OH^-(aq) \rightarrow 2\ H_2O(\ell)$$

Example 16.1

Write ionic and net ionic equations for the reaction of hydrochloric acid with aqueous calcium hydroxide.

Solution

The equations for the formation of hydrochloric acid and aqueous calcium hydroxide are

$$H_2O(\ell) + HCl(aq) \rightarrow H_3O^+(aq) + Cl^-(aq)$$

$$Ca(OH)_2(s) \rightarrow Ca^{2+}(aq) + 2\ OH^-(aq)$$

In the reaction of hydrochloric acid with aqueous calcium hydroxide, the reactants are

$$H_3O^+(aq) + Cl^-(aq) + Ca^{2+}(aq) + 2\ OH^-(aq) \rightarrow$$

The products are water and the salt (the cation from the base and the anion from the acid). Two H_3O^+ ions are needed for two OH^- ions, and two Cl^- ions are needed for one Ca^{2+} ion. The balanced equation is

ionic: $2\ H_3O^+(aq) + 2\ Cl^-(aq) + Ca^{2+}(aq) + 2\ OH^-(aq) \rightarrow 4\ H_2O(\ell) + Ca^{2+}(aq) + 2\ Cl^-(aq)$

Removing the spectator ions gives

$$2\ H_3O^+(aq) + 2\ OH^-(aq) \rightarrow 4\ H_2O(\ell)$$

Dividing all of the coefficients by 2 gives

net ionic: $H_3O^+(aq) + OH^-(aq) \rightarrow 2\ H_2O(\ell)$

Exercise 16.5

Write ionic and net ionic equations for the neutralization of hydrobromic acid with aqueous barium hydroxide.

Weak Acid and Strong Base

In aqueous solution a weak acid exists almost entirely in the form of molecules, and only a very small portion of it is converted to ions (Section 15-5). In a solution of hydrofluoric acid, for example, the reversible ionization reaction is

$$H_2O(\ell) + HF(aq) \rightleftarrows H_3O^+(aq) + F^-(aq)$$

At equilibrium, 97% of the hydrogen fluoride exists as HF molecules, and only 3% has been converted to ions.

When a solution of hydrofluoric acid is mixed with a solution of sodium hydroxide, hydroxide ions from the base react with the most available form of the acid, the HF molecules. The equations are:

ionic: $HF(aq) + Na^+(aq) + OH^-(aq) \rightarrow H_2O(\ell) + Na^+(aq) + F^-(aq)$

net ionic: $HF(aq) + OH^-(aq) \rightarrow H_2O(\ell) + F^-(aq)$

Because a weak acid in aqueous solution exists predominantly in its molecular form, it should be shown in that form in ionic and net ionic equations. This is the general net ionic equation for the neutralization of a weak acid, HA, by a strong base:

$$HA(aq) + OH^-(aq) \rightarrow H_2O(\ell) + A^-(aq)$$

Exercise 16.6

Write ionic and net ionic equations for the neutralization of acetic acid, $HC_2H_3O_2(aq)$, with aqueous potassium hydroxide.

Strong Acid and Weak Base

The reasoning applied to weak acids can also be applied to weak bases. In aqueous solution a weak base exists almost entirely in the form of molecules, and only a very small portion of it is converted to ions (Section 15-6). In aqueous ammonia, for example, the reversible ionization reaction is

$$NH_3(aq) + H_2O(\ell) \rightleftarrows NH_4^+(aq) + OH^-(aq)$$

At equilibrium, more than 99% of the ammonia exists as NH_3 molecules, and less than 1% has been converted to ions.

When a solution of hydrochloric acid is mixed with aqueous ammonia, hydronium ions from the acid react with the most available form of the base, the NH_3 molecules. The equations are

ionic: $\underbrace{H_3O^+(aq) + Cl^-(aq)}_{\text{hydrochloric acid}} + \underbrace{NH_3(aq)}_{\text{aqueous ammonia}}$

$$\rightarrow \underbrace{H_2O(\ell)}_{\text{water}} + \underbrace{NH_4^+(aq) + Cl^-(aq)}_{\text{salt}}$$

net ionic: $H_3O^+(aq) + NH_3(aq) \rightarrow H_2O(\ell) + NH_4^+(aq)$

Because a weak base in aqueous solution exists predominantly in its molecular form, it should be shown in that form in ionic and net ionic equations. This is the general net ionic equation for the neutralization of a weak base, BN, by a strong acid:

$$H_3O^+(aq) + BN(aq) \rightarrow H_2O(\ell) + BNH^+(aq)$$

Exercise 16.7

Write ionic and net ionic equations for the neutralization of an aqueous solution of the weak base pyridine, C_5H_5N, by aqueous nitric acid.

Weak Acid and Weak Base

Weak acids and weak bases exist in aqueous solutions predominantly in their molecular forms, so they're shown as molecules in ionic and net ionic equations. These are the equations for the neutralization reaction of hydrofluoric acid with aqueous ammonia:

$$\text{ionic: } HF(aq) + NH_3(aq) \rightarrow NH_4^+(aq) + F^-(aq)$$

$$\text{net ionic: } HF(aq) + NH_3(aq) \rightarrow NH_4^+(aq) + F^-(aq)$$

Because none of the reactants or products is a spectator, the ionic and net ionic equations are identical.

This is the general equation for the neutralization of a weak acid, HA, by a weak base, BN:

$$HA(aq) + BN(aq) \rightarrow BNH^+(aq) + A^-(aq)$$

Exercise 16.8

Write ionic and net ionic equations for the reaction of the weak acid acetic acid, $HC_2H_3O_2(aq)$, with an aqueous solution of the weak base pyridine, $C_5H_5N(aq)$.

16-3 A neutralization reaction can be represented by a molecular, ionic, or net ionic equation

In the preceding Section we saw that a neutralization reaction can be represented by an ionic equation or by a net ionic equation. A neutralization reaction can also be represented by an equation in which all of the reactants and products are shown as condensed or molecular formulas; this kind of equation is called a **molecular equation.** These are the molecular, ionic, and net ionic equations for the reaction of hydrochloric acid with aqueous sodium hydroxide:

$$\text{molecular: } HCl(aq) + NaOH(aq) \rightarrow H_2O(\ell) + NaCl(aq)$$

$$\text{ionic: } H_3O^+(aq) + Cl^-(aq) + Na^+(aq) + OH^-(aq) \rightarrow 2\,H_2O(\ell) + Na^+(aq) + Cl^-(aq)$$

$$\text{net ionic: } H_3O^+(aq) + OH^-(aq) \rightarrow 2\,H_2O(\ell)$$

Each kind of equation—molecular, ionic, and net ionic—has advantages that make it especially useful for certain purposes.◆

◆We'll see later in this chapter that other kinds of reactions in aqueous solutions can also be represented by molecular, ionic, and net ionic equations.

A molecular equation is especially useful in laboratory work because it specifies, in a concise way, the solutions that should be used to carry out the reaction. The preceding molecular equation shows that this neutralization is carried out by combining two solutions: hydrochloric acid, HCl(aq), and aqueous sodium hydroxide, NaOH(aq).

An ionic equation is especially useful in understanding a reaction because it shows each of the reactants in the actual form that it takes in aqueous solution. For example, the preceding ionic equation shows that hydrochloric acid, sodium hydroxide, and sodium chloride exist as hydrated ions in aqueous solution and that water exists as molecules. The equation also shows that $Na^+(aq)$ and $Cl^-(aq)$ are spectator ions in this reaction.

A net ionic equation is especially useful in classifying a reaction because it shows only the ions or molecules that are breaking bonds or forming bonds during the reaction. The preceding net ionic equation shows that, in the neutralization of a strong acid with a strong base, the essential change is the reaction of a hydronium ion with a hydroxide ion to form two water molecules.

The following Examples and Exercises will give you practice in writing and interpreting molecular, ionic, and net ionic equations.

Example 16.2

Write molecular, ionic, and net ionic equations for the reaction of hydrofluoric acid with aqueous potassium hydroxide.

Solution

molecular: $HF(aq) + KOH(aq) \rightarrow H_2O(\ell) + KF(aq)$

ionic: $HF(aq) + K^+(aq) + OH^-(aq) \rightarrow H_2O(\ell) + K^+(aq) + F^-(aq)$
($HF(aq)$ is a weak acid and $KOH(aq)$ is a strong base.)

net ionic: $HF(aq) + OH^-(aq) \rightarrow H_2O(\ell) + F^-(aq)$
(In the ionic equation in this Example, $K^+(aq)$ is a spectator ion.)

Exercise 16.9

Write molecular, ionic, and net ionic equations for the reaction of hydriodic acid with aqueous ammonia.

Example 16.3

Identify the spectator ions in the reaction represented by this molecular equation:

$$H_2SO_3(aq) + 2\ NaOH(aq) \rightarrow 2\ H_2O(\ell) + Na_2SO_3(aq)$$

Solution

To identify the spectator ions in the reaction, write the ionic equation.

ionic: $H_2SO_3(aq) + 2\ Na^+(aq) + 2\ OH^-(aq)$
$\rightarrow 2\ H_2O(\ell) + 2\ Na^+(aq) + SO_3^{2-}(aq)$
($H_2SO_3(aq)$ is a weak acid and $NaOH(aq)$ is a strong base.)

The ion $Na^+(aq)$ appears in the reactants and the products, so it's a spectator ion.

Exercise 16.10

Identify the spectator ions in the reaction of nitric acid with aqueous ammonia.

16-4 In a precipitation reaction, an insoluble ionic compound is formed in an aqueous solution

Figure 16.1
Precipitation of cadmium sulfide. When a solution of cadmium nitrate is mixed with a solution of sodium sulfide, a solid precipitate of yellow-orange cadmium sulfide is formed. *(Chip Clarke)*

Figure 16.1 provides an example of one of the most dramatic kinds of chemical reactions, the formation of an insoluble ionic compound in an aqueous solution. We'll use the example of the reaction shown in Figure 16.1 to explain how such reactions occur.

The yellow-orange compound that appears in Figure 16.1 is cadmium sulfide, CdS. Its ions, Cd^{2+} and S^{2-}, are relatively small and carry multiple charges, so we'd predict, from the principles given in Section 15-9, that cadmium sulfide would not be significantly soluble in water. The prediction is correct: Only about 0.13 mg of cadmium sulfide will dissolve in 100 mL of water.

The reaction illustrated in Figure 16.1 was carried out by preparing and mixing two solutions, as shown in the following diagram:

Prepare aqueous cadmium nitrate:

$$Cd(NO_3)_2(s) \rightarrow \boxed{Cd^{2+}(aq) + 2\ NO_3^-(aq)}$$

Prepare aqueous sodium sulfide:

$$Na_2S(s) \rightarrow \boxed{2\ Na^+(aq) + S^{2-}(aq)}$$

Mix the two solutions.

$$Cd^{2+}(aq) + 2\ NO_3^-(aq) + 2\ Na^+(aq) + S^{2-}(aq) \rightarrow CdS(s) + 2\ Na^+(aq) + 2\ NO_3^-(aq)$$

Cadmium nitrate and sodium sulfide are very soluble in water (nitrate ion and sodium ion are large ions with single charges). By dissolving cadmium nitrate in water, a solution can be prepared with a high concentration of $Cd^{2+}(aq)$, and by dissolving sodium sulfide, a solution can be prepared with a high concentration of $S^{2-}(aq)$. When the two solutions are mixed, the concentration of cadmium sulfide in the mixture exceeds its maximum solubility. As a result, solid cadmium sulfide forms and slowly settles to the bottom of the container. The solid cadmium sulfide is called a **precipitate,** and its formation is called **precipitation.**

The preceding equation for the precipitation of cadmium sulfide is an ionic equation. These are the corresponding molecular and net ionic equations:

molecular: $Cd(NO_3)_2(aq) + Na_2S(aq) \rightarrow CdS(s) + 2\ NaNO_3(aq)$

net ionic: $Cd^{2+}(aq) + S^{2-}(aq) \rightarrow CdS(s)$

When aqueous lead(II) acetate is poured into aqueous potassium sulfate, insoluble lead(II) sulfate precipitates. *(Charles D. Winters)*

Example 16.4

A concentrated solution of lead(II) acetate is mixed with a concentrated solution of potassium sulfate. (a) Predict whether a precipitation reaction will occur. (b) If you predict a precipitation reaction, write molecular, ionic, and net ionic equations for it.

Solution

(a) The solutions are

$$Pb(C_2H_3O_2)_2(s) \rightarrow Pb^{2+}(aq) + 2\ C_2H_3O_2^-(aq)$$

$$K_2SO_4(s) \rightarrow 2\ K^+(aq) + SO_4^{2-}(aq)$$

When these solutions are mixed, these are the possible reactants:

$$Pb^{2+}(aq) + 2\ C_2H_3O_2^-(aq) + 2\ K^+(aq) + SO_4^{2-}(aq) \rightarrow$$

Because acetate ion and potassium ion are large ions with single charges, they should be soluble and remain in solution. Because lead(II) ion and sulfate ion carry multiple charges, lead(II) sulfate should be insoluble and form a precipitate, $PbSO_4(s)$.

(b) molecular: $Pb(C_2H_3O_2)_2(aq) + K_2SO_4(aq) \rightarrow PbSO_4(s) + 2\ KC_2H_3O_2(aq)$

ionic: $Pb^{2+}(aq) + 2\ C_2H_3O_2^-(aq) + 2\ K^+(aq) + SO_4^{2-}(aq) \rightarrow PbSO_4(s) + 2\ K^+(aq) + 2\ C_2H_3O_2^-(aq)$

net ionic: $Pb^{2+}(aq) + SO_4^{2-}(aq) \rightarrow PbSO_4(s)$

Exercise 16.11

A concentrated solution of barium nitrate is mixed with a concentrated solution of ammonium sulfide. (a) Predict whether a precipitation reaction will occur. (b) If you predict a precipitation reaction, write molecular, ionic, and net ionic equations for it.

16-5 If a reaction in an aqueous solution forms an insoluble gas, the gas will appear as bubbles in the solution

Figure 16.2
Formation of H_2 bubbles. When hydrochloric acid is poured over zinc, hydrogen gas is formed. Because hydrogen gas is nearly insoluble in water, it forms bubbles that rise through the reaction mixture. *(Charles Steele)*

Figure 16.2 shows a reaction that produces hydrogen gas in an aqueous solution. Because hydrogen is nearly insoluble in water, it appears as bubbles that rise rapidly through the reaction mixture.

Because reactions that form gas bubbles are easy to recognize, they're often used as experiments in introductory chemistry. In this Section we'll identify the most important kinds of reactions that produce insoluble gases. It's convenient to consider these reactions in three groups: reactions that produce hydrogen, reactions that produce carbon dioxide or sulfur dioxide, and reactions that produce hydrogen sulfide.

Reactions That Produce H_2

The reaction shown in Figure 16.2 was carried out by pouring hydrochloric acid over pieces of zinc. The equations are

molecular: $2\ HCl(aq) + Zn(s) \rightarrow H_2(g) + ZnCl_2(aq)$

ionic: $2\ H_3O^+(aq) + 2\ Cl^-(aq) + Zn(s) \rightarrow$
$H_2(g) + 2\ H_2O(\ell) + Zn^{2+}(aq) + 2\ Cl^-(aq)$

net ionic: $2\ H_3O^+(aq) + Zn(s) \rightarrow H_2(g) + 2\ H_2O(\ell) + Zn^{2+}(aq)$

This reaction is one member of a large group of reactions that show this pattern:

acid + metal → hydrogen + salt

Example 16.5

Write the molecular equation for the reaction of sulfuric acid with magnesium.

Solution

molecular: $H_2SO_4(aq) + Mg(s) \rightarrow H_2(g) + MgSO_4(aq)$

Exercise 16.12

Write the ionic equation for the reaction of hydrobromic acid with iron. Assume that the salt produced is iron(III) bromide.

Reaction of potassium with water. *(Charles Steele)*

Almost all metals react with strong acids to produce hydrogen. The highly reactive metals in groups 1A and 2A—the alkali metals and the alkaline-earth metals—also react with water to produce hydrogen. The pattern of these reactions is

water + group 1A or 2A metal → hydrogen + metal hydroxide

For example, these are the equations for the reaction of potassium with water:

molecular: $2\ H_2O(\ell) + 2\ K(s) \rightarrow H_2(g) + 2\ KOH(aq)$

ionic: $2\ H_2O(\ell) + 2\ K(s) \rightarrow H_2(g) + 2\ K^+(aq) + 2\ OH^-(aq)$

net ionic: $2\ H_2O(\ell) + 2\ K(s) \rightarrow H_2(g) + 2\ K^+(aq) + 2\ OH^-(aq)$

Example 16.6

The alkaline-earth metals, except beryllium, react with water to form hydrogen. Write the molecular equation for the reaction of calcium with water.

Solution

The reactants and products are

water + metal → hydrogen + metal hydroxide

$H_2O(\ell) + Ca(s) \rightarrow H_2(g) + Ca(OH)_2(aq)$

The balanced equation is

$2\ H_2O(\ell) + Ca(s) \rightarrow H_2(g) + Ca(OH)_2(aq)$

Reaction of calcium with water. *(Chip Clark)*

Exercise 16.13

Write the molecular equation for the reaction of lithium with water.

Reactions That Produce CO_2 or SO_2

Carbonate ion, CO_3^{2-}, and sulfite ion, SO_3^{2-}, react with acids to produce the gases carbon dioxide, CO_2, and sulfur dioxide, SO_2. The following molecular equations show examples of these reactions.

$$Na_2CO_3(aq) + 2\ HCl(aq) \rightarrow CO_2(g) + H_2O(\ell) + 2\ NaCl(aq)$$

$$Na_2SO_3(aq) + 2\ HCl(aq) \rightarrow SO_2(g) + H_2O(\ell) + 2\ NaCl(aq)$$

These reactions can be used as tests for carbonates and sulfites. Sulfur dioxide can be recognized by its sharp, unpleasant odor, similar to that of rotten eggs.

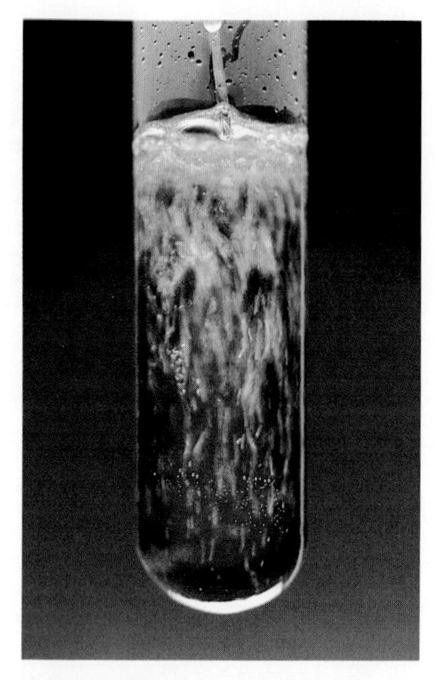

Figure 16.3
Formation of CO_2 bubbles. When hydrochloric acid is poured into aqueous sodium carbonate, bubbles of carbon dioxide are formed. *(Charles D. Winters)*

Example 16.7

Figure 16.3 shows the reaction of aqueous sodium carbonate with hydrochloric acid. From the molecular equation for this reaction,

$$Na_2CO_3(aq) + 2\ HCl(aq) \rightarrow CO_2(g) + H_2O(\ell) + 2\ NaCl(aq)$$

write the ionic and net ionic equations.

Solution

To create the ionic equation, show each dissolved compound in the form in which it occurs in aqueous solution:

$$2\ Na^+(aq) + CO_3^{2-}(aq) + 2\ H_3O^+(aq) + 2\ Cl^-(aq) \rightarrow CO_2(g) + 3\ H_2O(\ell) + 2\ Na^+(aq) + 2\ Cl^-(aq)$$

To create the net ionic equation, remove the spectator ions:

$$CO_3^{2-}(aq) + 2\ H_3O^+(aq) \rightarrow CO_2(g) + 3\ H_2O(\ell)$$

Exercise 16.14

Write molecular, ionic, and net ionic equations for the reaction of aqueous copper(II) sulfite will sulfuric acid.

Reactions That Produce H_2S

Sulfide ion reacts with acids to produce hydrogen sulfide gas; this is one example:

$$Na_2S(aq) + 2\ HCl(aq) \rightarrow H_2S(g) + 2\ NaCl(aq)$$

Hydrogen sulfide can be recognized by its rotten-egg odor, similar to that of sulfur dioxide but less sharp.

Exercise 16.15

Write molecular, ionic, and net ionic equations for the reaction of aqueous ammonium sulfide with hydrobromic acid.

16-6 In a replacement reaction, the ions of one metal replace the ions of another metal in an aqueous solution

In Figure 16.4 the container on the left shows a reaction mixture that has just been prepared by inserting one end of a bar of nickel metal into an aqueous solution of copper(II) sulfate; the blue color of the solution is caused by the $Cu^{2+}(aq)$ ions. The container on the right shows the changes that have occurred after the original reaction mixture has been allowed to stand: A coating of copper metal has formed on the nickel bar, and the solution has turned green because nickel metal has dissolved in it to form green $Ni^{2+}(aq)$ ions. The ionic equation for this reaction is

$$\underset{\text{nickel metal}}{Ni(s)} + \underset{\text{aqueous copper(II) sulfate}}{Cu^{2+}(aq) + SO_4^{2-}(aq)} \rightarrow \underset{\text{copper metal}}{Cu(s)} + \underset{\text{aqueous nickel sulfate}}{Ni^{2+}(aq) + SO_4^{2-}(aq)}$$

The preceding reaction, in which nickel ions replace copper ions in an aqueous solution, is an example of a replacement reaction.◆ In a replacement reaction, ions of one metal replace ions of another metal in an aqueous solution. The pattern of the reaction is:

◆Replacement reactions can also occur between reactants that are not in solution; these reactions were described in Section 9-6.

$$\text{metal A} + \begin{matrix}\text{aqueous solution}\\ \text{of salt of metal B}\end{matrix} \rightarrow \text{metal B} + \begin{matrix}\text{aqueous solution}\\ \text{of salt of metal A}\end{matrix}$$

Nickel ions will replace copper ions in aqueous solution, but not vice versa: If a bar of copper metal is immersed in aqueous nickel sulfate, no reaction occurs. The explanation for the replacement priority of one metal over another belongs in more advanced chemistry texts, but the priority can be predicted from the **activity series** shown on the right.◆ In this series the ions of each higher metal will replace the ions of each lower metal in an aqueous solution.

◆**Activity Series**

Li
K
Ba
Ca
Na
Mg
Al
Zn
Fe
Cd
Ni
Sn
Pb
Cu
Hg
Ag
Au

Example 16.8

Use the activity series to predict whether a reaction will occur if (a) a piece of copper is immersed in aqueous zinc nitrate (b) a piece of magnesium is immersed in aqueous silver nitrate. If you predict a reaction, write molecular, ionic, and net ionic equations for it.

(*Marna G. Clarke*)

Figure 16.4
Replacement of $Cu^{2+}(aq)$ by $Ni^{2+}(aq)$. In the container on the left, a bar of nickel metal has just been immersed in an aqueous solution of copper(II) sulfate. The blue color is caused by $Cu^{2+}(aq)$. The container on the right shows the changes that have occurred when the original reaction mixture has been allowed to stand. Copper metal has formed on the nickel bar, and nickel has entered the solution as green $Ni^{2+}(aq)$. The ionic equation is

$$\underset{\text{nickel metal}}{Ni(s)} + \underset{\text{aqueous copper(II) sulfate}}{Cu^{2+}(aq) + SO_4^{2-}(aq)} \rightarrow \underset{\text{copper metal}}{Cu(s)} + \underset{\text{aqueous nickel sulfate}}{Ni^{2+}(aq) + SO_4^{2-}(aq)}$$

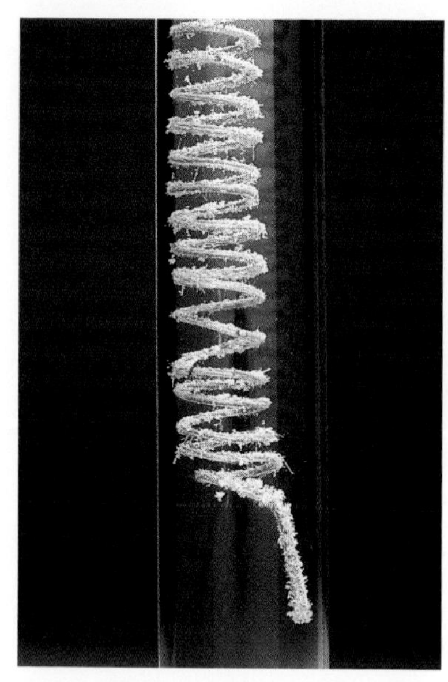

When a coil of magnesium metal is allowed to stand in aqueous silver nitrate, magnesium atoms enter the solution as magnesium ions, and silver ions leave the solution to form needles of silver metal on the surface of the magnesium. *(Charles D. Winters)*

Solution

(a) Copper is below zinc in the activity series, so no reaction should occur.

(b) Magnesium is above copper in the activity series, so a reaction should occur.

Begin by working out the ionic equation. The reactants are

$$\underset{\text{magnesium}}{Mg(s)} + \underset{\text{aqueous silver nitrate}}{Ag^{+}(aq) + NO_3^{-}(aq)} \rightarrow$$

In the replacement magnesium metal will form magnesium ions, and silver ions will form silver metal:

$$Mg(s) + Ag^{+}(aq) + NO_3^{-}(aq) \rightarrow Ag(s) + Mg^{2+}(aq) + NO_3^{-}(aq)$$

Balance the charges:

$$Mg(s) + 2\ Ag^{+}(aq) + 2\ NO_3^{-}(aq) \rightarrow 2\ Ag^{+}(aq) + Mg^{2+}(aq) + 2\ NO_3^{-}(aq)$$

From this ionic equation, write the molecular and net ionic equations:

molecular: $Mg(s) + 2\ AgNO_3(aq) \rightarrow 2\ Ag(s) + Mg(NO_3)_2(aq)$

ionic: $Mg(s) + 2\ Ag^{+}(aq) + 2\ NO_3^{-}(aq) \rightarrow 2\ Ag(s) + Mg^{2+}(aq) + 2\ NO_3^{-}(aq)$

net ionic: $Mg(s) + 2\ Ag^{+}(aq) \rightarrow 2\ Ag(s) + Mg^{2+}(aq)$

Exercise 16.16

Use the activity series to predict whether a reaction will occur if (a) a piece of cadmium is immersed in aqueous iron(III) nitrate (b) a piece of barium is immersed in aqueous aluminum acetate. If you predict a reaction, write molecular, ionic, and net ionic equations for it.

Replacement reactions provide simple examples of a large class of reactions called reduction–oxidation or redox reactions. In a **redox reaction,** electrons are transferred from one substance to another. The substance that loses electrons is said to be **oxidized,** and the substance that gains electrons is said to be **reduced.**◆ For example, from the net ionic equation

$$Mg(s) + 2\ Ag^{+}(aq) \rightarrow 2\ Ag(s) + Mg^{2+}(aq)$$

it's clear that in this reaction electrons are transferred from magnesium atoms to silver ions: Magnesium atoms are oxidized and silver ions are reduced.◆

◆Earlier in this book you've seen equations for other redox reactions, although they weren't identified in that way. For example, in the formation of any ionic compound from its elements, the metal is oxidized and the nonmetal is reduced. One example is

$$2\ Na\cdot + :\ddot{Cl}-\ddot{Cl}: \rightarrow 2\ Na^{+} + 2\ :\ddot{Cl}:^{-}$$

◆Chapter 17 discusses redox reactions in greater detail.

Exercise 16.17

In each net ionic equation you wrote in Exercise 16.16, identify the substance being oxidized and the substance being reduced.

16-7 For reactions in aqueous solutions, calculations of the quantities of reactants and products follow the usual steps in stoichiometry

Example 16.9 shows a typical stoichiometry problem of the kind we saw in Chapter 12. As shown in this Example, the steps in a stoichiometric calculation are

Translate given units into moles.	Translate moles of given reactant or product into moles of reactant or product to be found.	Translate moles into units required for answer.

Example 16.9

According to the equation

$$CH_4(g) + 2\ O_2(g) \rightarrow CO_2(g) + 2\ H_2O(g)$$

how many grams of oxygen will react with 5.00 g of methane?

Solution

$5.00\ g\ CH_4 = ?\ g\ O_2$

$$g\ CH_4 \rightarrow mol\ CH_4 \rightarrow mol\ O_2 \rightarrow g\ O_2$$

$$\left(\frac{5.00\ \cancel{g\ CH_4}}{}\right)\left(\frac{1\ \cancel{mol\ CH_4}}{16.0\ \cancel{g\ CH_4}}\right)\left(\frac{2\ \cancel{mol\ O_2}}{1\ \cancel{mol\ CH_4}}\right)\left(\frac{32.0\ g\ O_2}{1\ \cancel{mol\ O_2}}\right) = 20.0\ g\ O_2$$

Translate given units into moles. Translate moles of given reactant or product into moles of reactant or product to be found. Translate moles into units required for answer.

Stoichiometric calculations for reactions in aqueous solutions follow these same steps. For a reaction in solution the quantity of a reactant or product may be expressed as a volume of solution of a certain molarity, as shown in the following Examples and Exercises.

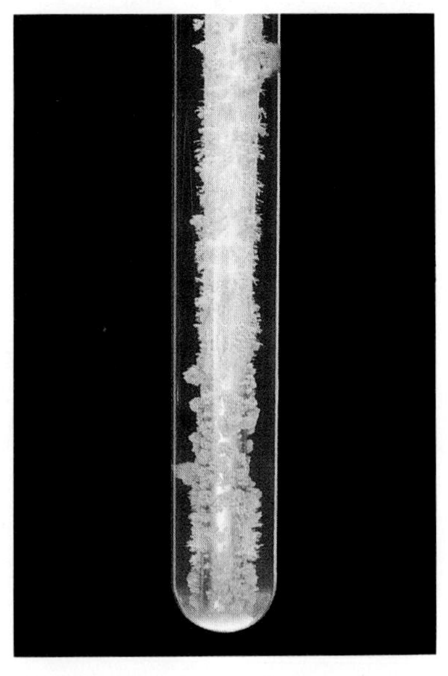

Aluminum is above silver in the activity series. When an aluminum rod is allowed to stand in aqueous silver nitrate, aluminum atoms enter the solution as aluminum ions, and silver ions leave the solution to form needles of silver metal on the surface of the aluminum. *(Charles D. Winters)*

Example 16.10

A replacement reaction can be carried out by immersing a piece of aluminum in an aqueous solution of copper(II) nitrate. The molecular equation for the reaction is

$$2\ Al(s) + 3\ Cu(NO_3)_2(aq) \rightarrow 3\ Cu(s) + 2\ Al(NO_3)_3(aq)$$

How many grams of aluminum will react with 16.0 mL of 0.500 M $Cu(NO_3)_2(aq)$?

Solution

16.0 mL $Cu(NO_3)_2$ = ? g Al

$$\text{mL } Cu(NO_3)_2 \rightarrow \text{mol } Cu(NO_3)_2 \rightarrow \text{mol Al}$$

$$\left(\frac{16.0 \text{ mL } Cu(NO_3)_2}{}\right)\left(\frac{0.500 \text{ mol } Cu(NO_3)_2}{1000 \text{ mL } Cu(NO_3)_2}\right)\left(\frac{2 \text{ mol Al}}{3 \text{ mol } Cu(NO_3)_2}\right)$$

solution is 0.500 M — from balanced equation

$$\rightarrow \text{g Al}$$

$$\left(\frac{27.0 \text{ g Al}}{1 \text{ mol Al}}\right) = 0.135 \text{ g Al}$$

Exercise 16.18

The molecular equation for the reaction of hydrochloric acid with solid sodium carbonate is

$$2\ HCl(aq) + Na_2CO_3(s) \rightarrow CO_2(g) + H_2O(\ell) + 2\ NaCl(aq)$$

How many mL of 1.25 M HCl(aq) will be needed to react with 0.746 g of Na_2CO_3?

Example 16.11

The molecular equation for the neutralization reaction between sulfuric acid and aqueous sodium hydroxide is

$$H_2SO_4(aq) + 2\ NaOH(aq) \rightarrow 2\ H_2O(\ell) + Na_2SO_4(aq)$$

How many mL of 1.30 M NaOH(aq) will be needed to react with 26.2 mL of 2.44 M H_2SO_4(aq)?

Solution

26.2 mL H_2SO_4 = ? mL NaOH

$$\text{mL } H_2SO_4 \rightarrow \text{mol } H_2SO_4 \rightarrow \text{mol NaOH} \rightarrow \text{mL NaOH}$$

$$\left(\frac{26.2 \text{ mL } H_2SO_4}{}\right)\left(\frac{2.44 \text{ mol } H_2SO_4}{1000 \text{ mL } H_2SO_4}\right)\left(\frac{2 \text{ mol NaOH}}{1 \text{ mol } H_2SO_4}\right)\left(\frac{1000 \text{ mL NaOH}}{1.30 \text{ mol NaOH}}\right) = 98.4 \text{ mL NaOH}$$

Exercise 16.19

According to the equation

$$Ni(NO_3)_2(aq) + (NH_4)_2S(aq) \rightarrow NiS(s) + 2\ NH_4NO_3(aq)$$

how many mL of 1.31 M $Ni(NO_3)_2$(aq) will be needed to react with 9.65 mL of 1.66 M $(NH_4)_2S$(aq)?

Example 16.12

The molecular equation for the neutralization reaction of nitric acid with aqueous barium hydroxide is

$$2\ HNO_3(aq) + Ba(OH)_2(aq) \rightarrow 2\ H_2O(\ell) + Ba(NO_3)_2(aq)$$

In a particular reaction 7.42 mL of 1.33 M HNO_3(aq) was needed to react with 19.2 mL of aqueous barium hydroxide. What was the molarity of the barium hydroxide solution?

Solution

Calculate the moles of $Ba(OH)_2$ in the $Ba(OH)_2$(aq) solution:

$$7.42\ mL\ HNO_3 = ?\ mol\ Ba(OH)_2$$

$$mL\ HNO_3 \rightarrow mol\ HNO_3 \rightarrow mol\ Ba(OH)_2$$

$$\left(\frac{7.42\ \cancel{mL\ HNO_3}}{}\right)\left(\frac{1.33\ \cancel{mol\ HNO_3}}{1000\ \cancel{mL\ HNO_3}}\right)\left(\frac{1\ mol\ Ba(OH)_2}{2\ \cancel{mol\ HNO_3}}\right)$$

$$= 4.93 \times 10^{-3}\ mol\ Ba(OH)_2$$

Convert 19.2 mL to L:

$$19.2\ mL = ?\ L$$

$$mL \rightarrow L$$

$$\left(\frac{19.2\ \cancel{mL}}{}\right)\left(\frac{1\ L}{1000\ \cancel{mL}}\right) = 0.0192\ L$$

Calculate molarity:

$$M = \frac{mol}{L} = \frac{4.93 \times 10^{-3}\ mol}{0.0192\ L} = 0.257\ mol/L$$

Solution is 0.257 M $Ba(OH)_2$(aq).

Exercise 16.20

The molecular equation for the precipitation reaction of aqueous lead(II) nitrate with aqueous sodium sulfide is

$$Pb(NO_3)_2(aq) + Na_2S(aq) \rightarrow PbS(s) + 2\ NaNO_3(aq)$$

In a particular reaction 5.37 mL of 0.848 M Na_2S(aq) was needed to react with 6.50 mL of $Pb(NO_3)_2$(aq). What was the molarity of the lead(II) nitrate solution?

Inside Chemistry How do antacids work?

Every day, millions of people take antacid medications to relieve stomach pain. The use of antacids is one of the most common applications of our understanding of the behavior of acids and bases.

Several substances in the human stomach carry out digestion, and one of them is hydrochloric acid. Under normal conditions a healthy human stomach produces about 650 mg of hydrogen chloride per day, dissolved in the stomach's fluids as hydrochloric acid. A stomach that's functioning abnormally may produce much more hydrogen chloride, as much as 2400 mg per day, and the excess hydrochloric acid can cause irritation of the stomach wall and lead to ulcers.

Antacids are compounds, usually hydroxides or carbonates, that destroy hydrochloric acid by reacting with it. Milk of Magnesia®, for example, is a suspension of particles of insoluble magnesium hydroxide, $Mg(OH)_2$. The neutralization reaction is

$$2\ HCl(aq) + Mg(OH)_2(s) \rightarrow MgCl_2(aq) + 2\ H_2O(\ell)$$

Tums® contains calcium carbonate, and the reaction with hydrochloric acid forms carbon dioxide gas:

$$2\ HCl(aq) + CaCO_3(s) \rightarrow CaCl_2(aq) + H_2O(\ell) + CO_2(g)$$

Both of these antacids contain carbonates. *(Charles D. Winters)*

Because hydroxides and carbonates are white and have a chalky taste, most antacid medications contain coloring and flavoring agents as well as their active ingredients.

Chapter Summary: Reactions in Aqueous Solutions

Section	Subject	Summary	Check When Learned
16-1	Water + nonmetal oxide →	Acid	☐
16-1	$2\ H_2O(\ell) + CO_2(aq) \rightarrow$	$H_3O^+(aq) + HCO_3^-(aq)$	☐
16-1	Water + metal oxide →	Base	☐
16-1	$H_2O(\ell) + Na_2O(s) \rightarrow$	$2\ Na^+(aq) + 2\ OH^-(aq)$	☐
16-2	Neutralization	Reaction of an acid with a base.	☐
16-2	Salt	An ionic compound that is not an acid, base, or oxide. Sodium chloride, NaCl, is one example.	☐

Section	Subject	Summary	Check When Learned
16-2	Spectator ion	In an ionic equation, an ion that appears in the reactants and in the products is a spectator ion. For example, in	☐
		$H_3O^+(aq) + Cl^-(aq) + Na^+(aq) + OH^-(aq) \rightarrow 2\ H_2O(\ell) + Na^+(aq) + Cl^-(aq)$	☐
		$Na^+(aq)$ and $Cl^-(aq)$ are spectator ions.	☐
16-2	Difference between ionic and net ionic equations	A net ionic equation is formed from the corresponding ionic equation by removing the spectator ions. For example, the net ionic equation that corresponds to the preceding ionic equation is	☐
		$H_3O^+(aq) + OH^-(aq) \rightarrow 2\ H_2O(\ell)$	☐
16-2	Net ionic equation for the reaction of		
	strong acid + strong base	$H_3O^+(aq) + OH^-(aq) \rightarrow 2\ H_2O(\ell)$	☐
	weak acid + strong base	$HA(aq) + OH^-(aq) \rightarrow H_2O(\ell) + A^-(aq)$	☐
	strong acid + weak base	$H_3O^+(aq) + BN(aq) \rightarrow H_2O(\ell) + BNH^+(aq)$	☐
	weak acid + weak base	$HA(aq) + BN(aq) \rightarrow A^-(aq) + BNH^+(aq)$	☐
16-3	Distinction among molecular, ionic, and net ionic equations	A molecular equation represents each reactant and product by a condensed or molecular formula; an ionic equation shows each reactant and product in the form in which it occurs in aqueous solution; and a net ionic equation omits spectator ions.	☐
16-3	For the reaction of hydrochloric acid with aqueous sodium hydroxide,		
	molecular equation:	$HCl(aq) + NaOH(aq) \rightarrow H_2O(\ell) + NaCl(aq)$	☐
	ionic equation:	$H_3O^+(aq) + Cl^-(aq) + Na^+(aq) + OH^-(aq) \rightarrow 2\ H_2O(\ell) + Na^+(aq) + Cl^-(aq)$	☐
	net ionic equation:	$H_3O^+(aq) + OH^-(aq) \rightarrow 2\ H_2O(\ell)$	☐
16-4	Precipitation reaction	Reaction in which an insoluble ionic compound is formed in an aqueous solution.	☐
16-4	Precipitate	The insoluble ionic compound that is formed in a precipitation reaction.	☐
16-5	Acid + metal →	Hydrogen + salt	☐
16-5	$HCl(aq) + Zn(s) \rightarrow$	The balanced molecular equation is $2\ HCl(aq) + Zn(s) \rightarrow H_2(g) + ZnCl_2(aq)$	☐

Section	Subject	Summary	Check When Learned
16-5	Water + group 1A metal →	Hydrogen + metal hydroxide	☐
16-5	$H_2O(\ell) + K(s) \rightarrow$	The balanced molecular equation is $2\ H_2O(\ell) + 2\ K(s) \rightarrow H_2(g) + 2\ KOH(aq)$	☐
16-5	Water + group 2A metal →	Hydrogen + metal hydroxide	☐
16-5	$H_2O(\ell) + Ca(s) \rightarrow$	The balanced molecular equation is $2\ H_2O(\ell) + Ca(s) \rightarrow H_2(g) + Ca(OH)_2(aq)$	☐
16-5	Gas that's produced from carbonate + acid	CO_2	☐
16-5	$Na_2CO_3(aq) + HCl(aq) \rightarrow$	The balanced molecular equation is $Na_2CO_3(aq) + 2\ HCl(aq) \rightarrow CO_2(g) + H_2O(\ell) + 2\ NaCl(aq)$	☐
16-5	Gas that's produced from sulfite + acid	SO_2	☐
16-5	$Na_2SO_3(aq) + HCl(aq) \rightarrow$	The balanced molecular equation is $Na_2SO_3(aq) + 2\ HCl(aq) \rightarrow SO_2(g) + H_2O(\ell) + 2\ NaCl(aq)$	☐
16-5	Gas that's produced from sulfide + acid	H_2S	☐
16-5	$Na_2S(aq) + HCl(aq) \rightarrow$	The balanced molecular equation is $Na_2S(aq) + 2\ HCl(aq) \rightarrow H_2S(g) + 2\ NaCl(aq)$	☐
16-6	Replacement reaction	Reaction in which the ions of one metal replace the ions of another metal in aqueous solution.	☐
16-6	General equation for replacement reaction	metal A + aqueous solution of salt of metal B → metal B + aqueous solution of salt of metal A	☐
16-6	Activity series	A list of symbols for metals that shows their replacement priority: The list is arranged so that the ions of a metal that's higher on the list will replace the ions of a metal that's lower, in aqueous solution.	☐
16-6	Reduction–oxidation (redox) reaction	Reaction in which electrons are transferred from one substance to another. One example is $Mg(s) + 2\ Ag^{+}(aq) \rightarrow Mg^{2+}(aq) + 2\ Ag(s)$	☐
16-6	Define: oxidized	In a redox reaction the substance that loses electrons is said to be oxidized. In the preceding example Mg is oxidized.	☐

Section	Subject	Summary			Check When Learned
	reduced	In a redox reaction the substance that gains electrons is said to be reduced. In the preceding example Ag^+ is reduced.			☐
16-7	Steps in stoichiometry:	Translate given units into moles.	Translate moles of given reactant or product into moles of reactant or product to be found.	Translate moles into units required for answer.	☐

Problems

Assume you can use the Periodic Table at the front of the book unless you're directed otherwise. Answers to odd-numbered Problems are in Appendix 1.

Reactions of Oxides with Water (Section 16-1)

1. Dinitrogen pentoxide is a gas that dissolves in and reacts with water to form aqueous nitric acid. Write a molecular equation for this reaction.
2. Cesium oxide is a solid that dissolves in and reacts with water to form aqueous cesium hydroxide. Write a molecular equation for this reaction.
3. Diphosphorus pentoxide is a solid that dissolves in and reacts with water to form aqueous phosphoric acid, H_3PO_4. Write a molecular equation for this reaction.
4. Strontium oxide is a solid that dissolves in and reacts with water to form aqueous strontium hydroxide. Write a molecular equation for this reaction.

Neutralization (Sections 16-2 and 16-3)

5. Complete and balance:

$$HBr(aq) + LiOH(aq) \rightarrow$$

6. Complete and balance:

$$HCl(aq) + Ba(OH)_2(aq) \rightarrow$$

7. Write a molecular equation for the reaction of hydriodic acid with aqueous potassium hydroxide.
8. Write a molecular equation for the reaction of perchloric acid with aqueous sodium hydroxide.
9. Write ionic and net ionic equations for the reaction described in problem 7.
10. Write ionic and net ionic equations for the reaction described in problem 8.
11. Write a molecular equation for the reaction of one mole of aqueous sulfuric acid with one mole of aqueous sodium hydroxide. The salt formed is sodium hydrogen sulfate.
12. Write a molecular equation for the reaction of one mole of aqueous sulfuric acid with two moles of aqueous sodium hydroxide. The salt formed is sodium sulfate.
13. Complete and balance:

$$HF(aq) + OH^-(aq) \rightarrow$$

14. Complete and balance:

$$HF(aq) + Ba(OH)_2(aq) \rightarrow$$

15. Write a molecular equation for the reaction of nitrous acid with aqueous sodium hydroxide, and name the salt that's produced.
16. Write a molecular equation for the reaction of two moles of nitrous acid with one mole of aqueous calcium hydroxide, and name the salt that's produced.

17. Write ionic and net ionic equations for the reaction described in problem 15.

18. Write ionic and net ionic equations for the reaction described in problem 16.

19. Write a molecular equation for the reaction of one mole of aqueous phosphoric acid, $H_3PO_4(aq)$, with (a) one mole of aqueous sodium hydroxide, (b) two moles of aqueous sodium hydroxide, (c) three moles of aqueous sodium hydroxide. In each reaction, name the salt that's produced.

20. Write a molecular equation for the reaction of aqueous phosphoric acid, $H_3PO_4(aq)$, with aqueous calcium hydroxide to produce aqueous calcium phosphate.

21. Write ionic and net ionic equations for the reaction described in part (b) of problem 19.

22. Write ionic and net ionic equations for the reaction described in problem 20.

23. Complete and balance:

$$HBr(aq) + NH_3(aq) \rightarrow$$

24. Complete and balance:

$$H_3O^+(aq) + NH_3(aq) \rightarrow$$

25. Write a molecular equation for the reaction of hydriodic acid with aqueous ammonia, and name the salt that's produced.

26. Write a molecular equation for the reaction of perchloric acid with aqueous ammonia, and name the salt that's produced.

27. Write ionic and net ionic equations for the reaction described in problem 25.

28. Write ionic and net ionic equations for the reaction described in problem 26.

29. Write a molecular equation for the reaction of hydrochloric acid with an aqueous solution of the weak base methylamine, $CH_3NH_2(aq)$.

30. Write a molecular equation for the reaction of aqueous perchloric acid with an aqueous solution of the weak base pyridine, $C_5H_5N(aq)$.

31. Complete and balance:

$$HC_2H_3O_2(aq) + NH_3(aq) \rightarrow$$

32. Complete and balance:

$$HCHO_2(aq) + NH_3(aq) \rightarrow$$

33. Write a molecular equation for the reaction of nitrous acid with aqueous ammonia, and name the salt that's produced.

34. Write a molecular equation for the reaction of hydrofluoric acid with an aqueous solution of the weak base methylamine, $CH_3NH_2(aq)$.

35. Write ionic and net ionic equations for the reaction described in problem 33.

36. Write a molecular equation for the neutralization reaction in which the salt produced is ammonium cyanide.

Precipitation (Section 16-4)

37. Decide whether a precipitation reaction should occur when concentrated solutions of the following reactants are mixed. If you predict a reaction, complete and balance the equation.
(a) $Al(NO_3)_3(aq) + Na_3PO_4(aq) \rightarrow$
(b) $Fe(C_2H_3O_2)_2(aq) + (NH_4)_2S(aq) \rightarrow$
(c) $Ba(NO_3)_2(aq) + KNO_3(aq) \rightarrow$

38. Decide whether a precipitation reaction should occur when concentrated solutions of the following reactants are mixed. If you predict a reaction, complete and balance the equation.
(a) $KNO_3(aq) + Na_2S(aq) \rightarrow$
(b) $NaC_2H_3O_2(aq) + Ca(NO_3)_2(aq) \rightarrow$
(c) $Ba(NO_3)_2(aq) + Na_2SO_4(aq) \rightarrow$

39. For each precipitation reaction you predicted in problem 37, write an ionic and a net ionic equation.

40. For each precipitation reaction you predicted in problem 38, write an ionic and a net ionic equation.

41. Write a balanced molecular equation for the precipitation reaction that occurs when a concentrated aqueous solution of sodium sulfide is mixed with a concentrated aqueous solution of mercury(II) acetate.

42. Write a balanced molecular equation for the precipitation reaction that occurs when a concentrated aqueous solution of sodium chromate, Na_2CrO_4, is mixed with a concentrated aqueous solution of lead(II) nitrate.

Gas Formation (Section 16-5)

43. Complete and balance:
(a) $HNO_3(aq) + Ni(s) \rightarrow$
(b) $HCl(aq) + Mg(s) \rightarrow$
(c) $H_2O(\ell) + Sr(s) \rightarrow$

44. Complete and balance:
(a) $HBr(aq) + Zn(s) \rightarrow$
(b) $HNO_3(aq) + Cd(s) \rightarrow$
(c) $H_2O(\ell) + Li(s) \rightarrow$

45. Complete and balance:
(a) $HCl(aq) + CaCO_3(s) \rightarrow$
(b) $HBr(aq) + FeS(s) \rightarrow$
(c) $HBr(aq) + CuSO_3(aq) \rightarrow$

46. Complete and balance:
(a) $HNO_3(aq) + FeCO_3(s) \rightarrow$
(b) $HI(aq) + K_2SO_3(aq) \rightarrow$
(c) $HCl(aq) + K_2S(s) \rightarrow$

47. Write ionic and net ionic equations for the molecular equation you wrote for part (c) of problem 45.

48. Write ionic and net ionic equations for the molecular equation you wrote for part (b) of problem 46.

49. Write molecular, ionic, and net ionic equations for the reaction of tin(II) sulfite with nitric acid.

50. Write molecular, ionic, and net ionic equations for the reaction of arsenic(III) sulfide with hydrobromic acid.

Replacement (Section 16-6)

51. Use the activity series in Section 16-6 to predict whether a reaction will occur between each pair of the following reactants. If you predict a reaction, complete and balance the equation.
(a) $Ag(s) + NaNO_3(aq) \rightarrow$
(b) $Cd(s) + Ni(NO_3)_2(aq) \rightarrow$
(c) $Mg(s) + Al(NO_3)_3(aq) \rightarrow$

52. Use the activity series in Section 16-6 to predict whether a reaction will occur between each pair of the following reactants. If you predict a reaction, complete and balance the equation.
(a) $Mg(s) + Zn(NO_3)_2(aq) \rightarrow$
(b) $Al(s) + Mg(NO_3)_2(aq) \rightarrow$
(c) $Zn(s) + Ni(NO_3)_2(aq) \rightarrow$

53. For each reaction you predicted would occur in problem 51, write ionic and net ionic equations.

54. For each reaction you predicted would occur in problem 52, write ionic and net ionic equations.

55. For each net ionic equation you wrote in problem 53, identify the substance being oxidized and the substance being reduced.

56. For each net ionic equation you wrote in problem 54, identify the substance being oxidized and the substance being reduced.

Solution Stoichiometry (Section 16-7)

57. According to the equation

$$Cu(s) + 2\ AgNO_3(aq) \rightarrow 2\ Ag(s) + Cu(NO_3)_2(aq)$$

how many grams of copper will react with 25.0 mL of 2.00 M $AgNO_3(aq)$?

58. According to the equation

$$2\ Al(s) + 3\ Hg(NO_3)_2(aq) \rightarrow 3\ Hg(\ell) + 2\ Al(NO_3)_3(aq)$$

how many grams of mercury will be produced by the reaction of 15.6 mL of 3.50 M $Hg(NO_3)_2(aq)$?

59. According to the equation

$$2\ Fe(NO_3)_3(aq) + 3\ Na_2S(aq) \rightarrow Fe_2S_3(s) + 6\ NaNO_3(aq)$$

how many milliliters of 6.00 M $Na_2S(aq)$ will be required to produce 1.00 g of $Fe_2S_3(s)$?

60. According to the equation

$$Pb(NO_3)_2(aq) + K_2CrO_4(aq) \rightarrow PbCrO_4(s) + 2\ KNO_3(aq)$$

how many milliliters of 2.00 M $K_2CrO_4(aq)$ will be required to react completely with 5.00 mL of 4.00 M $Pb(NO_3)_2(aq)$?

61. How many grams of hydrogen can be produced by the reaction of 20.0 mL of 6.00 M hydrochloric acid with magnesium?

62. How many milliliters of 3.00 M nitric acid will be needed to react with aqueous sodium carbonate to produce 0.500 g of carbon dioxide?

63. According to the equation

$$H_2SO_4(aq) + 2\ KOH(aq) \rightarrow 2\ H_2O(\ell) + K_2SO_4(aq)$$

how many milliliters of 1.65 M $H_2SO_4(aq)$ will be required to react with 13.7 mL of 2.00 M $KOH(aq)$?

64. According to the equation

$$2\ HNO_3(aq) + Ca(OH)_2(aq) \rightarrow 2\ H_2O(\ell) + Ca(NO_3)_2(aq)$$

how many milliliters of 0.0250 M $Ca(OH)_2(aq)$ will be required to react with 5.00 mL of 0.500 M $HNO_3(aq)$?

65. In a neutralization of aqueous ammonia with aqueous acetic acid, 14.3 mL of 1.75 M acetic acid was required to react with 10.0 mL of the ammonia solution. What was the molarity of the ammonia solution?

66. A reaction was carried out by adding solid potassium carbonate to hydrochloric acid: 3.55 g of potassium carbonate was needed to react with 12.5 mL of the acid. What was the molarity of the acid?

An electric current decomposes water into hydrogen (left) and oxygen. *(Charles D. Winters)*

17 A Closer Look at Equilibrium Reactions and Redox Reactions

EQUILIBRIUM REACTIONS

17-1
Equilibrium reactions are common in aqueous solutions and in gases.

17-2
In a reaction at equilibrium the rate of the forward reaction and the rate of the reverse reaction are equal.

17-3
If a reaction at equilibrium is disturbed, the reaction responds in a way that restores equilibrium.

17-4
A buffer is an equilibrium reaction that controls the pH of an aqueous solution.

17-5
From the balanced equation for an equilibrium reaction, we can write an important mathematical equation: the equilibrium-constant expression.

17-6
The numerical value of the equilibrium constant for a reaction can be calculated from the equilibrium concentrations of its reactants and products.

17-7
The equilibrium-constant expression for a reaction can be used to calculate the concentrations of its reactants and products at equilibrium.

REDOX REACTIONS AND ELECTROCHEMISTRY

17-8
In a redox reaction electrons are lost by one reactant and gained by another.

17-9
Redox reactions can be recognized by assigning oxidation numbers to atoms.

17-10
Equations for redox reactions in aqueous solutions are balanced by the half-reaction method.

17-11
Redox reactions are the source of electricity in dry cells and batteries.

17-12
Some redox reactions can be reversed by applying an electric current to their products.

Inside Chemistry:
How can iron be protected against corrosion?

Chapter Summary

Problems

Two of the most important classes of chemical reactions are equilibrium reactions, which were introduced in Chapter 15, and redox reactions, which were introduced in Chapter 16. This chapter discusses equilibrium and redox reactions in greater detail and describes some of their important applications.

Many of the reactions between reactants in aqueous solutions or between reactants that are gases are equilibrium reactions. This chapter describes the remarkable ability of equilibrium reactions to adjust to changing reaction conditions, and it describes a method for calculating the amounts of reactants and products that are present in a reaction at equilibrium.

Redox reactions—reactions in which electrons move from one reactant to another—are the basis of many important commercial products. This chapter describes methods for identifying redox reactions, and it describes several of their most important commercial applications, including the manufacture of dry cells and batteries.

EQUILIBRIUM REACTIONS

17-1 Equilibrium reactions are common in aqueous solutions and in gases

In Chapter 15 we saw examples of three kinds of equilibrium reactions that occur in aqueous solutions: the autoionization of water, the ionization of weak acids, and the ionization of weak bases.

autoionization of water

$$2\ H_2O(\ell) \rightleftarrows \underset{\text{hydronium ion}}{H_3O^+(aq)} + \underset{\text{hydroxide ion}}{OH^-(aq)}$$

ionization of a weak acid, acetic acid

$$H_2O(\ell) + HC_2H_3O_2(aq) \rightleftarrows \underset{\text{hydronium ion}}{H_3O^+} + \underset{\text{acetate ion}}{C_2H_3O_2^-(aq)}$$

ionization of a weak base, ammonia

$$NH_3(aq) + H_2O(\ell) \rightleftarrows \underset{\text{ammonium ion}}{NH_4^+(aq)} + \underset{\text{hydroxide ion}}{OH^-(aq)}$$

Equilibrium reactions also occur between solid ionic compounds and the ions in their saturated aqueous solutions; these reactions are called **solubility equilibria.** For example, if solid cadmium sulfide is in contact with a saturated aqueous solution of cadmium ions and sulfide ions, this solubility equilibrium occurs:

$$CdS(s) \rightleftarrows Cd^{2+}(aq) + S^{2-}(aq)$$

In the forward (left-to-right) reaction, cadmium ions and sulfide ions move out of the solid lattice and enter the solution as hydrated ions; in the reverse (right-to-left) reaction, hydrated ions leave the solution to occupy positions in the solid lattice.

Crystals of solid cadmium sulfide in contact with a saturated aqueous solution of cadmium ions and sulfide ions. In this system there is a continuous equilibrium reaction in which ions leave the solid state and enter the solution, and vice versa:

$$CdS(s) \rightleftarrows Cd^{2+}(aq) + S^{2-}(aq)$$

(Charles D. Winters)

Example 17.1◆

Write an equation for (a) the ionization of the weak acid benzoic acid, $HC_7H_5O_2$, (b) the solubility equilibrium of solid lead(II) iodide.

◆Answers to Exercises are in Appendix 1.

Solution

(a) $H_2O(\ell) + HC_7H_5O_2(aq) \rightleftarrows H_3O^+(aq) + C_7H_5O_2^-(aq)$
(b) $PbI_2(s) \rightleftarrows Pb^{2+}(aq) + 2\ I^-(aq)$

Exercise 17.1

Write an equation for (a) the ionization of the weak acid formic acid, $HCHO_2$, (b) the solubility equilibrium of silver sulfate.

Equilibrium reactions are also common between gases. One example is the reaction of nitrogen with hydrogen to form ammonia:

$$N_2(g) + 3\ H_2(g) \rightleftarrows 2\ NH_3(g)$$

Exercise 17.2

Write an equation for the equilibrium reaction of hydrogen gas with chlorine gas to form hydrogen chloride gas.

17-2 In a reaction at equilibrium the rate of the forward reaction and the rate of the reverse reaction are equal

In a reaction that is at equilibrium the forward reaction and the reverse reaction are occurring at the same rate. For example, imagine that the reaction

$$H_2O(\ell) + HC_2H_3O_2(aq) \rightleftarrows H_3O^+(aq) + C_2H_3O_2^-(aq)$$

is at equilibrium. If, in a split second, 100 water molecules react with 100 molecules of acetic acid, then in that same split second, 100 hydronium ions will react with 100 acetate ions.

An equilibrium reaction will automatically establish equal rates for its forward and reverse reactions. For example, imagine that the reaction

$$H_2(g) + I_2(g) \rightleftarrows 2\ HI(g)$$

is carried out by mixing hydrogen gas and iodine gas in a container. When the gases are first mixed, only the forward reaction can occur, because only hydrogen and iodine are present. As the forward reaction occurs, hydrogen iodide accumulates in the reaction mixture. As hydrogen iodide accumulates, the reverse reaction begins to occur, and as more hydrogen iodide accumulates, the reverse reaction goes faster and faster. Eventually, the rate of the reverse reaction equals the rate of the forward reaction and equilibrium is established.

An equilibrium can be established by starting with either the forward or the reverse reaction. For example, hydrogen iodide gas placed in a container will react to form hydrogen gas and iodine gas, and as hydrogen and iodine accumulate, they will begin to react with one another to form hydrogen iodide. Eventually, the equilibrium

$$H_2(g) + I_2(g) \rightleftarrows 2\ HI(g)$$

will be established.

17-3 If a reaction at equilibrium is disturbed, the reaction responds in a way that restores equilibrium

Equilibrium reactions have the remarkable capacity to accommodate to changing conditions. In this Section we'll see how reactions at equilibrium can accommodate to changes in concentrations of reactants, in temperature, and in pressure.

If a reaction is at equilibrium, an increase in the concentration of a reactant will increase the rates of the forward and reverse reactions. For example, imagine that the reaction

$$H_2(g) + I_2(g) \rightleftarrows 2\ HI(g)$$

is at equilibrium in a container; because the reaction is at equilibrium, the rate of the forward reaction and the rate of the reverse reaction are equal. Now imagine that the concentration of hydrogen gas is increased by adding more hydrogen to the container. Because more hydrogen is available to react, the rate of the forward reaction will increase, producing more hydrogen iodide. As more hydrogen iodide becomes available to react, the rate of the reverse reaction will also increase. Eventually, the rate of the forward reaction and the rate of the reverse reaction will again become equal, and equilibrium will be reestablished.

Exercise 17.3

If the reaction

$$N_2(g) + 3\ H_2(g) \rightleftarrows 2\ NH_3(g)$$

is at equilibrium and more ammonia is added, which reaction rate will increase first, the rate of the forward reaction or the rate of the reverse reaction?

The same reasoning can describe the effect of reducing the concentration of a reactant. Imagine that the reaction

$$H_2(g) + I_2(g) \rightleftarrows 2\ HI(g)$$

is at equilibrium in a container. Now imagine that the concentration of hydrogen gas is decreased by removing some hydrogen from the container. Because less hydrogen is available to react, the rate of the forward reaction will decrease, and less hydrogen iodide will be produced. As less hydrogen iodide becomes available, the rate of the reverse reaction will also decrease. Eventually the rates of the forward and reverse reactions will again become equal, and equilibrium will be reestablished.

A generalization known as **Le Châtelier's principle** describes the self-adjusting behavior of equilibrium reactions: If a reaction at equilibrium is disturbed, the reaction will respond in a way that will restore equilibrium.◆

◆Le Châtelier's principle is named for its creator, Henri-Louis Le Châtelier (1850–1936), French chemist.

A chemical reaction at equilibrium has a balance of reaction rates: The rate of the forward reaction equals the rate of the reverse reaction. There is also a balance of heat loss and gain: The forward reaction will either emit or absorb heat, and the reverse reaction will absorb or emit the same amount of heat. For example, in the equilibrium reaction

$$H_2(g) + I_2(g) \rightleftarrows 2\ HI(g)$$

the forward reaction emits heat and the reverse reaction absorbs the same amount of heat; that is, the forward reaction is exothermic and the reverse reaction is endothermic (Section 9-7):

$$H_2(g) + I_2(g) \rightleftarrows 2\ HI(g) + \text{heat}$$

If heat is added to a reaction at equilibrium, a new equilibrium will be established that increases the products of the endothermic reaction. For example, if heat is added to the preceding reaction, it will establish a new equilibrium in which more hydrogen and iodine, and less hydrogen iodide, will be present. The reverse is also true: If heat is removed from a reaction at equilibrium, it will establish a new equilibrium that increases the products of the exothermic reaction. For example, if heat is removed from the preceding reaction, it will establish a new equilibrium in which more hydrogen iodide, and less hydrogen and iodine, will be present.

The effect on an equilibrium reaction of adding or removing heat can be summarized in this way: If a reaction is at equilibrium and its temperature is raised (heat is added), the equilibrium will shift in the endothermic direction (the direction that absorbs heat); if its temperature is lowered (heat is removed), the equilibrium will shift in the exothermic direction (the direction that emits heat).

Example 17.2

The reaction

$$\text{heat} + 2\ OF_2(g) \rightleftarrows O_2(g) + 2\ F_2(g)$$

was at equilibrium in a container. If the container were cooled, would you expect the equilibrium to shift to the right or to the left?

Solution

Removing heat will shift the reaction in the exothermic direction, to the left.

Exercise 17.4

The reaction

$$N_2(g) + 3\ H_2(g) \rightleftarrows 2\ NH_3(g)$$

is exothermic from left to right. If this reaction were at equilibrium in a container and the container were heated, would you expect the equilibrium to shift in the direction of the forward or the reverse reaction?

Changes in pressure can disturb equilibrium reactions in which gases are reactants or products: A pressure increase will shift the equilibrium in the direction of fewer moles of gases, and a pressure decrease will shift the equilibrium in the direction of more moles of gases. The following Example illustrates this rule.

Example 17.3

(a) Will a pressure increase cause the equilibrium reaction

$$2\ OF_2 \rightleftarrows O_2(g) + 2\ F_2(g)$$

to shift right or left?

(b) Will a pressure decrease cause the equilibrium reaction

$$H_2(g) + I_2(g) \rightleftarrows 2\ HI(g)$$

to shift right or left?

(c) Will a pressure increase cause the equilibrium reaction

$$H_2O(\ell) \rightleftarrows H_2O(g)$$

to shift right or left?

Solution

(a) There are 2 moles of gas on the left and 3 moles of gas on the right, so a pressure increase will shift the equilibrium to the left. (b) There are 2 moles of gas on the left and 2 moles of gas on the right, so a pressure decrease will have no effect. (c) There are 0 moles of gas on the left and 1 mole of gas on the right, so a pressure increase will shift the reaction to the left.

Exercise 17.5

(a) Will a pressure increase cause the equilibrium reaction

$$CaCO_3(s) \rightleftarrows CaO(s) + CO_2(g)$$

to shift right or left? (b) Will a pressure decrease cause the equilibrium reaction

$$N_2(g) + 3\ H_2(g) \rightleftarrows 2\ NH_3(g)$$

to shift right or left?

17-4 A buffer is an equilibrium reaction that controls the pH of an aqueous solution◆

◆Before beginning this Section, you may want to review Section 16-2.

Aqueous solutions can be prepared in such a way that they will resist changes in pH. As an example, we'll consider a solution that's been prepared by dissolving hydrogen fluoride and sodium fluoride in water.

Hydrogen fluoride is a gas that dissolves in water to form a solution of hydrofluoric acid, a weak acid (Section 15-5):

$$HF(g) \rightarrow HF(aq)$$

$$\underset{}{H_2O(\ell)} + \underset{\text{dissolved hydrogen fluoride}}{HF(aq)} \rightleftarrows \underset{\text{hydronium ion}}{H_3O^+(aq)} + \underset{\text{fluoride ion}}{F^-(aq)}$$

Because hydrofluoric acid is a weak acid, most of the dissolved hydrogen fluoride remains in the molecular form, and only a small amount is converted to ions. The relative amounts are

$$\underset{\text{}}{H_2O(\ell)} + \underset{\text{large}}{HF(aq)} \rightleftarrows \underset{\text{small}}{H_3O^+(aq)} + \underset{\text{small}}{F^-(aq)}$$

Sodium fluoride is a soluble salt that dissolves in water to form hydrated sodium ions and hydrated fluoride ions:

$$NaF(s) \rightarrow Na^+(aq) + F^-(aq)$$

Adding sodium fluoride to a solution of hydrofluoric acid increases the amount of fluoride ion in the solution. As a result, the new relative amounts of reactants and products in the equilibrium are

$$H_2O(\ell) + \underset{\text{large}}{HF(aq)} \rightleftarrows \underset{\text{small}}{H_3O^+(aq)} + \underset{\text{large}}{F^-(aq)}$$

The equilibrium described by the preceding equation is a **buffer,** a reaction that controls the pH of a solution by reacting with an acid or base added to it. Suppose, for example, that acid is added to the solution just described. Most of the added hydronium ion will react with the large amount of fluoride ion in the solution, by the reverse reaction in the equilibrium. As a result, most of the added hydronium ion will be converted to $H_2O(\ell)$ and HF(aq) and will not be available in the solution. Because the buffer equilibrium reacts with added hydronium ion, the pH of the solution remains nearly unchanged when acid is added.

The buffer will also react with added base: Most of the added hydroxide ion will be neutralized by the large amount of molecular hydrogen fluoride present in the solution, by the reaction

$$OH^-(aq) + HF(aq) \rightarrow H_2O(\ell) + F^-(aq)$$

This reaction removes added hydroxide ion from the solution, and the pH remains nearly unchanged.

Example 17.4

A buffer solution was made by dissolving acetic acid and sodium acetate in water. (a) Write the equilibrium equation that describes the buffer. (b) Write the equation for the reaction of the buffer with acid. (c) Write the equation for the reaction of the buffer with base.

Solution

(a) $$H_2O(\ell) + \underset{\text{large}}{HC_2H_3O_2(aq)} \rightleftarrows \underset{\text{small}}{H_3O^+(aq)} + \underset{\text{large}}{C_2H_3O_2^-(aq)}$$

(b) $$H_3O^+(aq) + C_2H_3O_2^-(aq) \rightleftarrows H_2O(\ell) + HC_2H_3O_2(aq)$$

(c) $$OH^-(aq) + HC_2H_3O_2(aq) \rightarrow H_2O(\ell) + C_2H_3O_2^-(aq)$$

Exercise 17.6

A buffer was made by dissolving formic acid, $HCHO_2$, and sodium formate, $NaCHO_2$, in water. (a) Write the equilibrium equation that describes the buffer. (b) Write the equation for the reaction of the buffer with acid. (c) Write the equation for the reaction of the buffer with base.

Buffers are important in living organisms, to protect them against harmful changes in pH. In human blood, for example, the buffer

$$\underset{\text{large}}{2\ H_2O(\ell) + CO_2(aq)} \rightleftarrows \underset{\text{small}}{H_3O^+(aq)} + \underset{\text{large}}{HCO_3^-(aq)}$$

prevents biological reactions that produce acids or bases from causing harmful changes in blood pH.◆ The reactions of the buffer with acid and base are

$$H_3O^+(aq) + HCO_3^-(aq) \rightleftarrows 2\ H_2O(\ell) + CO_2(aq)$$

$$OH^-(aq) + CO_2(aq) \rightleftarrows HCO_3^-(aq)$$

◆Many biological reactions produce acids or bases. For example, during exercise muscles produce lactic acid, $HC_3H_5O_3$.

17-5 From the balanced chemical equation for an equilibrium reaction, we can write an important mathematical equation: the equilibrium-constant expression

For any equilibrium reaction it's possible to write a mathematical equation that describes an important relationship among the concentrations of the reactants and products. We saw one example in Section 15-7: For the autoionization of water

$$2\ H_2O(\ell) \rightleftarrows H_3O^+(aq) + OH^-(aq)$$

the mathematical equation, called the ion-product equation, is

$$[H_3O^+][OH^-] = K_W = 1.00 \times 10^{-14}$$

K_W is called the ion-product constant for water.

The mathematical equation that describes the relationship among the concentrations of reactants and products in an equilibrium reaction is called the **equilibrium-constant expression** for the reaction; the ion-product equation for water is one example of an equilibrium-constant expression. Figure 17.1 on page 382 describes the process for writing the equilibrium-constant expression for any equilibrium reaction from its balanced chemical equation.

In each equilibrium-constant expression, the symbol K is called the **equilibrium constant.** For each equilibrium reaction, K has a unique, fixed value at a given temperature. For the two examples shown in Figure 17.1

$$\frac{[H_3O^+][F^-]}{[HF]} = K = 6.8 \times 10^{-4} \qquad \text{and} \qquad [Ca^{2+}][F^-]^2 = K = 3.4 \times 10^{-11}$$

The symbol K is often given a subscript to show the kind of equilibrium reaction being described: K_W for the autoionization of water, K_a for the ionization of a weak acid, K_b for the ionization of a weak base, and K_{sp} for a solubility

Steps	Examples	
	$H_2O(\ell) + HF(aq) \rightleftarrows H_3O^+(aq) + F^-(aq)$	$CaF_2(s) \rightleftarrows Ca^{2+}(aq) + 2\ F^-(aq)$
Identify each reactant or product that is not a liquid or a solid, and write its symbol or formula in brackets to designate its concentration in moles per liter (molarity).	$[HF]$ $[H_3O^+]$ $[F^-]$	$[Ca^{2+}]$ $[F^-]$
For each symbol or formula in brackets add as its exponent the coefficient for that symbol or formula in the balanced chemical equation. An exponent of 1 doesn't need to be shown.	$[HF]$ $[H_3O^+]$ $[F^-]$	$[Ca^{2+}]$ $[F^-]^2$
Write a fraction, $\frac{\text{concentrations of products}}{\text{concentrations of reactants}}$, and set it equal to K.	$\frac{[H_3O^+][F^-]}{[HF]} = K$	$[Ca^{2+}][F^-]^2 = K$

Figure 17.1
Steps in writing an equilibrium-constant expression.

equilibrium. For a reaction in which the reactants and products are gases, K is usually not given a subscript. These are examples:

weak-acid equilibria

$$H_2O(\ell) + HF(aq) \rightleftarrows H_3O^+(aq) + F^-(aq) \qquad \frac{[H_3O^+][F^-]}{[HF]} = K_a$$

$$H_2O(\ell) + HC_2H_3O_2(aq) \rightleftarrows H_3O^+(aq) + C_2H_3O_2^-(aq) \qquad \frac{[H_3O^+][C_2H_3O_2^-]}{[HC_2H_3O_2]} = K_a$$

weak-base equilibria

$$NH_3(g) + H_2O(\ell) \rightleftarrows NH_4^+(aq) + OH^-(aq) \qquad \frac{[NH_4^+][OH^-]}{[NH_3]} = K_b$$

$$C_6H_5NH_2(aq) + H_2O(\ell) \rightleftarrows C_6H_5NH_3^+(aq) + OH^-(aq) \qquad \frac{[C_6H_5NH_3^+][OH^-]}{[C_6H_5NH_2]} = K_b$$

solubility equilibria

$$MgCO_3(s) \rightleftarrows Mg^{2+}(aq) + CO_3^{2-}(aq) \qquad [Mg^{2+}][CO_3^{2-}] = K_{sp}$$

$$Ag_2S(s) \rightleftarrows 2\ Ag^+(aq) + S^{2-}(aq) \qquad [Ag^+]^2[S^{2-}] = K_{sp}$$

gas equilibria

$$N_2(g) + 3\ H_2(g) \rightleftarrows 2\ NH_3(g) \qquad \frac{[NH_3]^2}{[N_2][H_2]^3} = K$$

$$N_2O_4(g) \rightleftarrows 2\ NO_2(g) \qquad \frac{[NO_2]^2}{[N_2O_4]} = K$$

Exercise 17.7

Write the equilibrium-constant expression for each of the following reactions:

(a) $H_2O(\ell) + HClO_2(aq) \rightleftarrows H_3O^+(aq) + ClO_2^-(aq)$

(b) $N_2H_4(aq) + H_2O(\ell) \rightleftarrows N_2H_5^+(aq) + OH^-(aq)$

(c) $Ca_3(PO_4)_2(s) \rightleftarrows 3\ Ca^{2+}(aq) + 2\ PO_4^{3-}(aq)$

(d) $2\ CO(g) + O_2(g) \rightleftarrows 2\ CO_2(g)$

17-6 The numerical value of the equilibrium constant for a reaction can be calculated from the equilibrium concentrations of its reactants and products

The numerical values of equilibrium constants are known for thousands of reactions; Table 17.1 shows a few examples.◆

◆Equilibrium constants vary with temperature; these values are at 25°C.

For an equilibrium reaction, the numerical value of the equilibrium constant can be calculated if the equilibrium concentrations of the reactants and products are known; the following Examples and Exercises show examples of these calculations.

Example 17.5

Barium fluoride is slightly soluble in water: 0.11 g of BaF_2 will dissolve in 100 mL (3 significant digits) of water. Calculate K_{sp} for BaF_2.

Solution

Write the chemical equation and the equilibrium-constant expression for the reaction.

$$BaF_2(s) \rightleftarrows Ba^{2+}(aq) + 2\ F^-(aq) \qquad [Ba^{2+}][F^-]^2 = K_{sp}$$

Table 17.1 Equilibrium Constants

Weak Acid	K_a	Weak Base	K_b
HF, hydrofluoric acid	6.8×10^{-4}	CH_3NH_2, methylamine	4.4×10^{-4}
HNO_2, nitrous acid	4.5×10^{-4}	NH_3, ammonia	1.8×10^{-5}
$HCHO_2$, formic acid	1.7×10^{-4}	N_2H_4, hydrazine	1.7×10^{-6}
$HC_2H_3O_2$, acetic acid	1.7×10^{-5}	$C_6H_5NH_2$, aniline	4.2×10^{-10}
HCN, hydrocyanic acid	4.9×10^{-10}		

Slightly Soluble Ionic Compound	K_{sp}
$MgCO_3$, magnesium carbonate	1.0×10^{-5}
PbI_2, lead(II) iodide	6.5×10^{-9}
CaF_2, calcium fluoride	3.4×10^{-11}
$Ca_3(PO_4)_2$, calcium phosphate	1×10^{-26}
Ag_2S, silver sulfide	6×10^{-50}

Calculate $[Ba^{2+}]$ and $[F^-]$:

In 100 mL of solution there are 0.11 g of BaF_2. Convert to moles of BaF_2:

$$\left(\frac{0.11\ \text{g BaF}_2}{}\right)\left(\frac{1\ \text{mol BaF}_2}{175\ \text{g BaF}_2}\right) = 6.3 \times 10^{-4}\ \text{mol BaF}_2$$

Calculate molarity of BaF_2:

$$M = \frac{\text{mol}}{\text{L}} = \frac{6.3 \times 10^{-4}\ \text{mol}}{0.100\ \text{L}} = 6.3 \times 10^{-3}\ \text{mol/L}$$

In this solution, $[BaF_2] = 6.3 \times 10^{-3}$ M. The preceding chemical equation shows that when 1 mol of BaF_2 dissolves, 1 mol of $Ba^{2+}(aq)$ and 2 mol of $F^-(aq)$ are formed. So, if $[BaF_2] = 6.3 \times 10^{-3}$ M, then $[Ba^{2+}] = 6.3 \times 10^{-3}$ M and $[F^-] = (2)(6.3 \times 10^{-3}\ \text{M}) = 1.3 \times 10^{-2}$ M.

Calculate K_{sp}:

$$[Ba^{2+}][F^-]^2 = K_{sp}$$
$$(6.3 \times 10^{-3}\ \text{M})(1.3 \times 10^{-2}\ \text{M})^2 = K_{sp} = 1.1 \times 10^{-6}$$

As shown in this Example and in Table 17.1, numerical values for equilibrium constants are customarily written without units.

Exercise 17.8

Calcium sulfate is slightly soluble in water: 0.13 g of $CaSO_4$ will dissolve in 200 mL (3 significant digits) of water. Calculate K_{sp} for $CaSO_4$.

Example 17.6

The reaction

$$CO(g) + 2\ H_2(g) \rightleftarrows CH_3OH(g)$$

was carried out in a container whose volume was 1.50 L. At equilibrium, there were 0.119 mol CO, 0.237 mol H_2, and 0.0313 mol CH_3OH in the container. Calculate K.

Solution

Write the equilibrium-constant expression for the reaction:

$$\frac{[CH_3OH]}{[CO][H_2]^2} = K$$

Calculate [CO], $[H_2]$, and $[CH_3OH]$:

$$\text{for [CO]: } M = \frac{\text{mol}}{\text{L}} = \frac{0.119\ \text{mol}}{1.50\ \text{L}} = 7.93 \times 10^{-2}\ \text{M}$$

$$\text{for }[H_2]\text{: } M = \frac{\text{mol}}{\text{L}} = \frac{0.237\ \text{mol}}{1.50\ \text{L}} = 0.158\ \text{M}$$

$$\text{for }[CH_3OH]\text{: } M = \frac{\text{mol}}{\text{L}} = \frac{0.0313\ \text{mol}}{1.50\ \text{L}} = 2.09 \times 10^{-2}\ \text{M}$$

Calculate K:

$$\frac{[CH_3OH]}{[CO][H_2]^2} = K$$

$$\frac{(2.09 \times 10^{-2}\ M)}{(7.93 \times 10^{-2}\ M)(0.158\ M)^2} = K = 10.6$$

Exercise 17.9

The reaction

$$PCl_3(g) + Cl_2(g) \rightleftarrows PCl_5(g)$$

was carried out in a 5.00-L container. At equilibrium there were 0.0185 mol PCl_3, 0.0870 mol Cl_2, and 0.0158 mol PCl_5 in the container. Calculate K.

17-7 The equilibrium-constant expression for a reaction can be used to calculate the concentrations of its reactants and products at equilibrium

For reactions that do not come to equilibrium, the concentrations of products that are formed can be determined by stoichiometry. For example, the reaction

$$H_2O(\ell) + HCl(aq) \rightarrow H_3O^+(aq) + Cl^-(aq)$$

is not an equilibrium reaction; hydrochloric acid is a strong acid. In a solution made by dissolving 0.75 mol of HCl in 1.0 L of water, the initial concentration of HCl(aq), before any reaction has occurred, will be 0.75 M:

	$H_2O(\ell)$ +	$HCl(aq) \rightarrow$	$H_3O^+(aq)$ +	$Cl^-(aq)$
initial		0.75 M	0 M	0 M

The reaction of HCl with H_2O occurs in a fraction of a second. When the reaction is over, all of the HCl has been converted to H_3O^+ and Cl^-. The final concentrations are◆

◆Because water is the solvent, its concentration is very large and remains nearly unchanged during the reaction.

	$H_2O(\ell)$ +	$HCl(aq) \rightarrow$	$H_3O^+(aq)$ +	$Cl^-(aq)$
initial		0.75 M	0 M	0 M
final		0 M	0.75 M	0.75 M

For equilibrium reactions, the concentrations of reactants and products that will be present at equilibrium can be calculated from the equilibrium-constant expression, as shown in the following Examples.

Example 17.7

A solution was prepared by dissolving 0.75 mol of hydrogen fluoride gas in enough water to make 1.00 L of solution. Calculate the concentrations of reactants and products, except water, at equilibrium.

◆The reaction occurs in a fraction of a second.

◆The solution would be found with the quadratic formula: For an equation of the form $ax^2 + bx + c = 0$, the solution is $x = \frac{-b \pm \sqrt{b^2 - 4ac}}{2a}$.

Solution

Hydrogen fluoride gas dissolves in water to form aqueous hydrogen fluoride:

$$HF(g) \rightarrow HF(aq)$$

The dissolved hydrogen fluoride reacts with water by an equilibrium process to form aqueous hydronium ion and aqueous fluoride ion:◆

$$H_2O(\ell) + HF(aq) \rightleftarrows H_3O^+(aq) + F^-(aq)$$

Before this reaction occurs, the concentration of dissolved hydrogen fluoride will be 0.75 M:

	$H_2O(\ell)$ +	$HF(aq)$ ⇄	$H_3O^+(aq)$ +	$F^-(aq)$
initial		0.75 M	0 M	0 M

After the reaction has reached equilibrium, some, but not all, of the HF(aq) will have been converted to $H_3O^+(aq)$ and $F^-(aq)$. If we designate the concentration of $H_3O^+(aq)$ at equilibrium as x, then the concentration of $F^-(aq)$ must also be x, because the chemical equation shows that $H_3O^+(aq)$ and $F^-(aq)$ are produced in equal molar quantities. The chemical equation also shows that, to produce x mol of $H_3O^+(aq)$, x mol of HF(aq) must have reacted, so the concentration of HF(aq) at equilibrium must be 0.75 M − x:

	$H_2O(\ell)$ +	$HF(aq)$ ⇄	$H_3O^+(aq)$ +	$F^-(aq)$
initial		0.75 M	0 M	0 M
equilibrium		0.75 M − x	x	x

These values for equilibrium concentrations can be substituted in the equilibrium-constant expression:

$$\frac{[H_3O^+][F^-]}{[HF]} = K_a \qquad \frac{(x)(x)}{(0.75 - x)} = 6.8 \times 10^{-4} \qquad \text{(from Table 17.1)}$$

Solving this equation for x would require complicated algebra.◆ The complicated algebra can be avoided if we assume that x will be a very small number; the assumption is warranted because K_a is a small number (6.8×10^{-4}), so only small amounts of $H_3O^+(aq)$ and $F^-(aq)$ will be present at equilibrium. If x is a very small number, than 0.75 − x will be nearly the same as 0.75; that is, $0.75 - x \approx 0.75$, where ≈ is read "equivalent to." Replacing 0.75 − x with 0.75 gives a much simpler equation:

$$\frac{(x)(x)}{(0.75)} = 6.8 \times 10^{-4}$$

$$x^2 = 5.1 \times 10^{-4}$$

To solve for x on your calculator, enter 5.1×10^{-4}, then press $\sqrt{x}$. The $\sqrt{x}$ key may be the 2nd function on the x^2 key.

$$x = 2.3 \times 10^{-2}$$

The calculated value of x is the concentration of $H_3O^+(aq)$ at equilibrium, so at equilibrium:

$$[H_3O^+] = x = 2.3 \times 10^{-2}\ M$$
$$[F^-] = x = 2.3 \times 10^{-2}\ M$$
$$[HF] = 0.75\ M - x = 0.73\ M$$

These numerical values show that, at equilibrium, only small amounts of $H_3O^+(aq)$ and $F^-(aq)$ are produced, and most of the dissolved HF remains in the form of HF(aq). The value for [HF] also shows that our assumption that $0.75 - x \approx 0.75$ is very nearly correct.

In solving any equilibrium problem in this book, you can make the kind of simplifying assumption shown in the preceding Example, to avoid complicated algebra.

Exercise 17.10

A solution was prepared by dissolving 0.80 mol of formic acid, $HCHO_2(\ell)$, in enough water to make 1.00 L of solution. Calculate the concentrations of reactants and products, except water, at equilibrium. For K_a, see Table 17.1.

Example 17.8

A solution was prepared by dissolving 4.04 g of the weak base methylamine, $CH_3NH_2(\ell)$ in enough water to make 200 mL (3 significant digits) of solution. For methylamine, $K_b = 4.4 \times 10^{-4}$. Calculate the concentrations of reactants and products, except water, at equilibrium. The reaction of dissolved methylamine with water is

$$CH_3NH_2(aq) + H_2O(\ell) \rightleftarrows CH_3NH_3^+(aq) + OH^-(aq)$$

Solution

Calculate the initial molarity of dissolved methylamine, before reaction with water has occurred:

$$\left(\frac{4.04\ g\ \cancel{CH_3NH_2}}{}\right)\left(\frac{1\ mol\ CH_3NH_2}{31.0\ g\ \cancel{CH_3NH_2}}\right) = 0.130\ mol\ CH_3NH_2$$

$$M = \frac{mol}{L} = \frac{0.130\ mol}{0.200\ L} = 0.650\ mol/L$$

The initial concentrations of reactants and products are

	$CH_3NH_2(aq)$	+ $H_2O(\ell)$ ⇄	$CH_3NH_3^+(aq)$	+ $OH^-(aq)$
initial	0.650 M		0 M	0 M

Designate the equilibrium concentration of one of the products as x. If we let x represent the equilibrium concentration of $CH_3NH_3^+(aq)$, then the equilibrium concentration of $OH^-(aq)$ must also be x, and the equilibrium concentration of $CH_3NH_2(aq)$ must be $0.650 - x$.

	$CH_3NH_2(aq)$	+ $H_2O(\ell)$ ⇄	$CH_3NH_3^+(aq)$	+ $OH^-(aq)$
initial	0.650 M		0 M	0 M
equilibrium	0.650 M − x		x	x

Substitute these values for the equilibrium concentrations in the equilibrium-constant expression.

$$\frac{[CH_3NH_3^+][OH^-]}{[CH_3NH_2]} = K_b \qquad \frac{(x)(x)}{(0.650 - x)} = 4.4 \times 10^{-4}$$

Assume that $0.650 - x \approx 0.650$ and solve for x.

$$\frac{x^2}{0.650} = 4.4 \times 10^{-4}$$

$$x = 1.7 \times 10^{-2}$$

The equilibrium concentrations are

$$[CH_3NH_3^+] = x = 1.7 \times 10^{-2}\ M$$

$$[OH^-] = x = 1.7 \times 10^{-2}\ M$$

$$[CH_3NH_2] = 0.650\ M - x = 0.633\ M$$

Exercise 17.11

A solution was prepared by dissolving 0.69 g of ammonia ($K_b = 1.8 \times 10^{-5}$) in enough water to make 155 mL of solution. Calculate the concentrations of reactants and products, except water, at equilibrium.

Example 17.9

The reaction

$$2\ HI(g) \rightleftarrows H_2(g) + I_2(g)$$

was carried out by placing 0.075 mol of hydrogen iodide in a 0.25-L container. Calculate the equilibrium concentrations of the reactant and products. At the temperature at which the reaction was carried out, $K = 2.0 \times 10^{-4}$.

Solution

Calculate the initial molarity of HI(g):

$$M = \frac{mol}{L} = \frac{0.075\ mol}{0.25\ L} = 0.30\ mol/L$$

	2 HI(g) ⇄	$H_2(g)$ +	$I_2(g)$
initial	0.30 M	0 M	0 M

Designate the equilibrium concentration of one of the products as x. If we let x represent the equilibrium concentration of H_2, then the equilibrium concentration of I_2 must also be x. According to the chemical equation, 2 mol HI produces 1 mol H_2 and 1 mol I_2, so 2x mol HI will be needed to produce x mol H_2 and x mol I_2; the equilibrium concentration of HI will be 0.30 M − 2x.

	2 HI(g) ⇄	$H_2(g)$ +	$I_2(g)$
initial	0.30 M	0 M	0 M
equilibrium	0.30 M − 2x	x	x

Substitute these values for the equilibrium concentrations in the equilibrium-constant expression:

$$\frac{[H_2][I_2]}{[HI]^2} = K \qquad \frac{(x)(x)}{(0.30 - 2x)^2} = 2.0 \times 10^{-4}$$

Assume that $0.30 - 2x \approx 0.30$ and solve for x:

$$\frac{x^2}{(0.30)^2} = 2.0 \times 10^{-4}$$

$$x = 4.2 \times 10^{-3}$$

The equilibrium values are

$$[H_2] = x = 4.2 \times 10^{-3}\ M$$

$$[I_2] = x = 4.2 \times 10^{-3}\ M$$

$$[HI] = 0.30\ M - 2x = 0.29\ M$$

Exercise 17.12

The reaction

$$PCl_5(g) \rightleftarrows PCl_3(g) + Cl_2(g)$$

was carried out by placing 0.17 mol of phosphorus pentachloride in a 0.300-L container. Calculate the concentrations of the reactant and products at equilibrium. At the temperature at which the reaction was carried out, $K = 3.0 \times 10^{-4}$.

Example 17.10

For PbI_2, $K_{sp} = 6.5 \times 10^{-9}$. Calculate the concentrations of $Pb^{2+}(aq)$ and $I^-(aq)$ in a saturated solution.

Solution

The equation for the solubility equilibrium is

$$PbI_2(s) \rightleftarrows Pb^{2+}(aq) + 2\ I^-(aq)$$

and the equilibrium-constant expression is

$$[Pb^{2+}][I^-]^2 = K_{sp}$$

Let x represent the molarity of one of the ions at equilibrium. If we let x represent the molarity of $Pb^{2+}(aq)$, then the molarity of $I^-(aq)$ will be 2x. Substitute these values in the equilibrium-constant expression and solve for x.

$$[Pb^{2+}][I^-]^2 = K_{sp}$$

$$(x)(2x)^2 = 6.5 \times 10^{-9}$$

$$4x^3 = 6.5 \times 10^{-9}$$

$$x^3 = 1.6 \times 10^{-9}$$

To solve for x on your calculator, enter 1.6×10^{-9}, press $x\sqrt{y}$; press 3; press =. The $x\sqrt{y}$ function is probably the 2nd function on the y^x key.

$$x = 1.2 \times 10^{-3}$$

The equilibrium values are

$$[Pb^{2+}] = x = 1.2 \times 10^{-3}\ M$$

$$[I^-] = 2x = (2)(1.2 \times 10^{-3}\ M) = 2.4 \times 10^{-3}\ M$$

Exercise 17.13

For $Zn(OH)_2$, $K_{sp} = 2.1 \times 10^{-16}$. Calculate the concentrations of $Zn^{2+}(aq)$ and $OH^-(aq)$ in a saturated solution.

Example 17.11

For CaF_2, $K_{sp} = 3.4 \times 10^{-11}$. How many grams of CaF_2 will be present in 75 mL of a saturated solution of CaF_2?

Solution

The equation for the solubility equilibrium is

$$CaF_2(s) \rightleftarrows Ca^{2+}(aq) + 2\ F^-(aq)$$

and the equilibrium-constant expression is

$$[Ca^{2+}][F^-]^2 = K_{sp}$$

Let x represent the molarity of $Ca^{2+}(aq)$ at equilibrium; the molarity of $F^-(aq)$ will be 2x.

$$[Ca^{2+}][F^-]^2 = K_{sp}$$

$$(x)(2x)^2 = 3.4 \times 10^{-11}$$

$$x = 2.0 \times 10^{-4}$$

The equilibrium concentrations are

$$[Ca^{2+}] = x = 2.0 \times 10^{-4}\ M$$

$$[F^-] = 2x = 4.0 \times 10^{-4}\ M$$

Calculate the number of grams of CaF_2 dissolved in 75 mL:

75 mL = ? g CaF_2

mL → mol $Ca^{2+}(aq)$ →

$$\left(\frac{75\ \cancel{mL}}{}\right)\left(\frac{2.0 \times 10^{-4}\ \cancel{mol\ Ca^{2+}(aq)}}{1000\ \cancel{mL}}\right)$$

↑

$[Ca^{2+}] = 2.0 \times 10^{-4}$ M

mol CaF_2 → g CaF_2

$$\left(\frac{1\ \cancel{mol\ CaF_2}}{1\ \cancel{mol\ Ca^{2+}(aq)}}\right)\left(\frac{78.1\ g\ CaF_2}{1\ \cancel{mol\ CaF_2}}\right) = 1.2 \times 10^{-3}\ g\ CaF_2$$

↑

For each mole of $Ca^{2+}(aq)$ in solution, one mole of CaF_2 must have dissolved.

Exercise 17.14

For $Fe(OH)_2$, $K_{sp} = 8.3 \times 10^{-16}$. How many milligrams of $Fe(OH)_2$ will be present in 235 mL of a saturated solution of $Fe(OH)_2$?

REDOX REACTIONS AND ELECTROCHEMISTRY

17-8 In a redox reaction electrons are lost by one reactant and gained by another

Metal atoms lose electrons easily and nonmetal atoms gain electrons easily, so a common reaction, as we saw in Chapter 8, is the transfer of electrons from metal atoms to nonmetal atoms to form ions. For example, sodium reacts with chlorine to form sodium chloride:

$$2\ \mathrm{Na}\cdot\ +\ :\ddot{\underset{\cdot\cdot}{\mathrm{Cl}}}-\ddot{\underset{\cdot\cdot}{\mathrm{Cl}}}:\ \rightarrow 2\ \mathrm{Na}^{+}\ +\ 2\ :\ddot{\underset{\cdot\cdot}{\mathrm{Cl}}}:^{-}$$

$$2\ \mathrm{Na(s)}\ +\ \mathrm{Cl_2(g)}\ \rightarrow\ 2\ \mathrm{NaCl(s)}$$

In this reaction each sodium atom loses one electron and each chlorine atom gains one electron.

A process in which an atom, ion, or molecule loses one or more electrons is called **oxidation,** and a process in which an atom, ion, or molecule gains one or more electrons is called **reduction.** In the reaction shown in the preceding equations, sodium is said to undergo oxidation and chlorine is said to undergo reduction; or we can say that sodium is **oxidized** and chlorine is **reduced.** The reaction is called a **reduction–oxidation** or **redox** reaction.

Reduction and oxidation always occur simultaneously because electrons are conserved; that is, electrons aren't created or destroyed, so an electron that is lost by one reactant must be gained by another reactant. The reactant that gains an electron is called the **oxidizing agent** (it causes the other reactant to be oxidized), and the reactant that loses an electron is called the **reducing agent** (it causes the other reactant to be reduced). In the preceding example sodium is oxidized and is the reducing agent; chlorine is reduced and is the oxidizing agent.

Exercise 17.15

For the reaction

$$2\ \mathrm{Mg}:\ +\ \cdot\ddot{\underset{\cdot\cdot}{\mathrm{O}}}-\ddot{\underset{\cdot\cdot}{\mathrm{O}}}\cdot\ \rightarrow 2\ \mathrm{Mg}^{2+}\ +\ 2\ :\ddot{\underset{\cdot\cdot}{\mathrm{O}}}:^{2-}$$

$$2\ \mathrm{Mg(s)}\ +\ \mathrm{O_2(g)}\ \rightarrow\ 2\ \mathrm{MgO(s)}$$

identify the substance that's being oxidized and the substance that's being reduced. Identify the oxidizing agent and the reducing agent.

Example 17.12

For the reaction

$$4\ \mathrm{K(s)} + \mathrm{O_2(g)} \rightarrow 2\ \mathrm{K_2O(s)}$$

identify the substance that's being oxidized and the substance that's being reduced. Identify the oxidizing agent and the reducing agent.

Solution

Because molecular and condensed formulas don't show electrons or ionic charges, it can be difficult to decide, from an equation written in molecular and condensed formulas, whether a redox reaction is occurring and, if it is, which reactant is gaining or losing electons. Rewriting the equation in Lewis symbols and line formulas shows the transfer of electrons:

$$4\ K\cdot + \cdot\ddot{\underset{..}{O}}—\ddot{\underset{..}{O}}\cdot \rightarrow 4\ K^+ + 2\ :\ddot{\underset{..}{O}}:^{2-}$$

Potassium loses electrons, is oxidized, and is the reducing agent; oxygen gains electrons, is reduced, and is the oxidizing agent.

17-9 Redox reactions can be recognized by assigning oxidation numbers to atoms

In a redox reaction electrons are transferred from one reactant to another. The reactant that loses electrons is oxidized and is the reducing agent, and the reactant that gains electrons is reduced and is the oxidizing agent.

In an equation written in Lewis symbols and line formulas, the transfer of electrons is easily recognized because the electrons are shown as dots. For example, in the equation

$$Ca: + :\ddot{\underset{..}{Br}}—\ddot{\underset{..}{Br}}: \rightarrow Ca^{2+} + 2\ :\ddot{\underset{..}{Br}}:^-$$

it's easy to see that 2 electrons are transferred from calcium to bromine. In this reaction calcium is oxidized and is the reducing agent, and bromine is reduced and is the oxidizing agent.

In an equation written in condensed or molecular formulas, a redox reaction is most easily recognized by assigning an **oxidation number** to each atom in the equation. In the symbol or formula for an element, the oxidation number for each atom is 0:

Write the oxidation number for each atom above the symbol for the atom. → $\overset{0}{Na}$ $\overset{0}{Cl_2}$

Write the sum of the oxidation numbers for all of the atoms of the same kind below the symbol for the atom. → $\underset{0}{Na}$ $\underset{0}{Cl_2}$

In the symbol for a monatomic ion, the oxidation number for the atom is the same as the charge on the ion:

$$\overset{1+}{\underset{1+}{Na^+}}\quad \overset{2+}{\underset{2+}{Mg^{2+}}}\quad \overset{1-}{\underset{1-}{Cl^-}}\quad \overset{2-}{\underset{2-}{O^{2-}}}$$

This rule also applies to formulas for ionic compounds:

$$\overset{1+1-}{KF}\quad \overset{2+\ 1-}{CaF_2}\quad \overset{3+\ \ 2-}{Al_2O_3}\quad \overset{1+\ \ 1-}{NaH}$$

The sum of the oxidation numbers in the formula for a compound is 0. → $\underset{1+1-}{KF}\quad \underset{2+\ 2-}{CaF_2}\quad \underset{6+\ \ 6-}{Al_2O_3}\quad \underset{1+\ \ 1-}{NaH}$

In the formula for a covalent compound of hydrogen, the oxidation number for each H atom is 1+; in the formula for a covalent compound of oxygen, the oxidation number for each O atom is 2−◆:

◆ Peroxides are exceptions: $\overset{1+\ 1-}{\underset{2+\ 2-}{H_2O_2}}$

$$\overset{1+\ 2-}{H_2O}$$

The sum of the oxidation numbers is 0. → $\underset{2+\ 2-}{H_2O}$

The oxidation numbers for other atoms are assigned from the formulas in which they occur by applying two principles: The sum of the oxidation numbers in a formula for a compound is 0; the sum of the oxidation numbers in the formula for a polyatomic ion is the same as the charge on the ion. The following Examples show the use of these principles to assign oxidation numbers.†

Example 17.13

Assign an oxidation number to each atom: (a) HCl (b) H_2S (c) SO_2 (d) $H_2C_2O_4$ (e) $Na_2S_2O_3$

Solution

(a) Write the known oxidation numbers:

1+?
HCl
1+?

The sum of all the oxidation numbers is 0, so

1+?
HCl
1+1− ← Sum is 0.

The oxidation number for the Cl atom must be 1−:

1+1−
HCl
1+1−

1+ ?	1+ ?	1+ 2−	? 2−	? 2−	4+2−
(b) H_2S	H_2S	H_2S	(c) SO_2	SO_2	SO_2
2+ ?	2+ 2−	2+ 2−	? 4−	4+4−	4+4−

1+ ? 2−	1+ ? 2−	1+ 3+ 2−	1+ ? 2−	1+ ? 2−	1+ 2+ 2−
(d) $H_2C_2O_4$	$H_2C_2O_4$	$H_2C_2O_4$	(e) $Na_2S_2O_3$	$Na_2S_2O_3$	$Na_2S_2O_3$
2+ ? 8−	2+ 6+ 8−	2+ 6+ 8−	2+ ? 6−	2+ 4+ 6−	2+ 4+ 6−

As parts (b), (c), and (e) show, the same atom—the sulfur atom, in these examples—can have different oxidation numbers in different formulas.

† Chemistry Insight

The oxidation number for an atom is the charge it would carry as an ion if all bonds were ionic, that is, if all bonding electrons were donated completely to the more electronegative atom. For example, the bonding in a water molecule is polar covalent:

δ−
O
H H
δ+ δ+

If we imagine that the bonds are completely ionic, we have

$$H^+ \; :\ddot{\underset{\cdot\cdot}{O}}:^{2-} \; H^+$$

The process of figuring out what charges ions would have if all bonds were ionic is slow, and it's usually quicker to work from the rules given in the text.

Example 17.14

Assign an oxidation number to each atom: (a) HS^- (b) CO_3^{2-} (c) $H_2PO_4^-$ (d) SO_4^{2-} (e) $C_2H_3O_2^-$

Solution

(a) Write the known oxidation numbers:

1+?
HS^-
1+?

The sum of the oxidation numbers is the same as the charge on the ion, 1−, so:

1+?
HS^-
1+2−

The oxidation number for the S atom must be 2−:

1+2−
HS^-
1+2−

? 2− ? 2− 4+2−
(b) CO_3^{2-} CO_3^{2-} CO_3^{2-}
? 6− 4+6− 4+6−

1+ ? 2− 1+ ? 2− 1+ 5+2−
(c) $H_2PO_4^-$ $H_2PO_4^-$ $H_2PO_4^-$
2+ ? 8− 2+ 5+8− 2+ 5+8−

? 2− ? 2− 6+2−
(d) SO_4^{2-} SO_4^{2-} SO_4^{2-}
? 8− 6+8− 6+8−

? 1+ 2− ? 1+ 2− 0 1+ 2−
(e) $C_2H_3O_2^-$ $C_2H_3O_2^-$ $C_2H_3O_2^-$
? 3+ 4− 0 3+ 4− 0 3+ 4−

Exercise 17.16

Assign an oxidation number to each atom: (a) Fe (b) O_2 (c) Fe_2O_3 (d) $HClO_3$ (e) $HClO_2$

An atom that's oxidized undergoes an increase in oxidation number, and an atom that's reduced undergoes a decrease in oxidation number; the reaction of sodium with chlorine is an example:

0 0 1+ 1−
$$2\ Na(s) + Cl_2(g) \rightarrow 2\ NaCl(s)$$
0 0 1+ 1−

In this reaction the oxidation number for each Na atom increases from 0 to 1+, and the oxidation number for each Cl atom decreases from 0 to 1−; sodium is oxidized and is the reducing agent; chlorine is reduced and is the oxidizing agent.

Example 17.15

Assign oxidation numbers to decide if each of the following equations represents a redox reaction. If it does, identify the oxidizing agent and the reducing agent:

(a) $2\ CO(g) + O_2(g) \rightarrow 2\ CO_2(g)$
(b) $HCl(aq) + NaOH(aq) \rightarrow H_2O(\ell) + NaCl(aq)$
(c) $Mg(s) + 2\ HNO_3(aq) \rightarrow Mg(NO_3)_2(aq) + H_2(g)$

Solution

2+2− 0 4+2−
(a) $2\ CO(g) + O_2(g) \rightarrow 2\ CO_2(g)$
2+2− 0 4+4−

The oxidation number for each C atom in CO increases from 2+ to 4+, and the oxidation number for each O atom in O_2 decreases from 0 to 2−, so this is a redox reaction. CO is oxidized and is the reducing agent; O_2 is reduced and is the oxidizing agent.

1+1− 1+ 2− 1+ 1+ 2− 1+ 1−
(b) $HCl(aq) + NaO\ H(aq) \rightarrow H_2O(\ell) + NaCl(aq)$
1+1− 1+ 2− 1+ 2+ 2− 1+ 1−

The oxidation number for each atom is the same in the reactants as it is in the products, so this is not a redox reaction.

0 1+ 5+ 2− 2+ 5+ 2− 0
(c) $Mg(s) + 2\ H\ N\ O_3(aq) \rightarrow Mg(NO_3)_2(aq) + H_2(g)$
0 1+ 5+ 6− 2+ 10+12− 0

The oxidation number for each Mg atom increases from 0 to 2+, and the oxidation number for each H atom in HNO_3 decreases from 1+ to 0, so this is a redox reaction. Mg is oxidized and is the reducing agent; HNO_3 is reduced and is the oxidizing agent.

Exercise 17.17

Assign oxidation numbers to decide if each of the following equations represents a redox reaction. If it does, identify the oxidizing agent and the reducing agent:

(a) $Fe_2O_3(s) + 2\ Al(s) \rightarrow 2\ Fe(s) + Al_2O_3(s)$
(b) $Na_2O(s) + H_2O(\ell) \rightarrow 2\ NaOH(aq)$
(c) $H_2(g) + I_2(g) \rightarrow 2\ HI(g)$

17-10 Equations for redox reactions in aqueous solutions are balanced by the half-reaction method

Table 17.2 on page 396 describes two oxidizing agents commonly used to carry out redox reactions in aqueous solutions. An important advantage of these compounds is that their aqueous solutions are colored, as shown in the table. As a redox reaction occurs, the color disappears, so it's easy to see that a reaction is taking place.

On the left, solid potassium permanganate, $KMnO_4(s)$, and an aqueous solution of potassium permanganate, $KMnO_4(aq)$; on the right, solid potassium dichromate, $K_2Cr_2O_7(s)$, and an aqueous solution of potassium dichromate, $K_2Cr_2O_7(aq)$. *(Charles D. Winters)*

Table 17.2 Two Common Oxidizing Agents

Oxidizing Agent	Color	Anion	Color of Anion in Aqueous Solution
$KMnO_4$ potassium permanganate	purple	MnO_4^- permanganate ion*	purple
$K_2Cr_2O_7$ potassium dichromate	orange	$Cr_2O_7^{2-}$ dichromate ion*	yellow

*A description of the bonding in MnO_4^- and $Cr_2O_7^{2-}$ requires more advanced theories of bonding than we've used in this text.

Exercise 17.18

Assign an oxidation number to each atom in MnO_4^- and $Cr_2O_7^{2-}$.

Equations for redox reactions in aqueous solution are more complicated than equations we've seen for other reactions. For example, this is the molecular equation for the reaction of potassium permanganate with iron(II) chloride in hydrochloric acid:

$$KMnO_4(aq) + 5\ FeCl_2(aq) + 8\ HCl(aq) \rightarrow KCl(aq) + MnCl_2(aq) + 5\ FeCl_3(aq) + 4\ H_2O(\ell).$$

The equations for redox reactions in aqueous solution can be simplified if we write them in the net ionic form; show H_3O^+ as H^+; and omit the designations (s), (ℓ), (g), and (aq). Throughout this Section we'll make these simplifications; with these changes the preceding equation becomes

$$MnO_4^- + 5\ Fe^{2+} + 8\ H^+ \rightarrow Mn^{2+} + 5\ Fe^{3+} + 4\ H_2O$$

Even in this simplified form, equations for redox reactions in aqueous solution would be difficult to balance without a systematic method. The method used to balance them, called the **half-reaction method,** is described in Example 17.16.◆ The strategy of the half-reaction method is to divide the net ionic equation into two parts—the half-reactions—balance each part separately, and then recombine them. To make the method easier to remember, an abbreviated outline of it follows the Example.

◆The half-reaction method is also called the ion-electron method.

Example 17.16

Permanganate ion reacts with chloride ion in aqueous acid to produce manganese(II) ion and chlorine. Write the balanced net ionic equation for the reaction.

Solution

1. **Write a skeleton equation for the reaction.**
   A skeleton equation for a redox reaction shows the reactants that undergo reduction and oxidation and their products, and shows whether the reaction occurs in acid or base:

$$MnO_4^- + Cl^- \rightarrow Mn^{2+} + Cl_2\ \text{(acid)}$$

2. Assign oxidation numbers and divide the skeleton equation into equations for two half-reactions.

$$\overset{7+2-}{\underset{7+8-}{MnO_4^-}} + \overset{1-}{\underset{1-}{Cl^-}} \rightarrow \overset{2+}{\underset{2+}{Mn^{2+}}} + \overset{0}{\underset{0}{Cl_2}}$$

The oxidation number for manganese decreases from 7+ to 2+, so it's being reduced, and the oxidation number for chlorine increases from 1− to 0, so it's being oxidized. One half-reaction shows the reduction and the other shows the oxidation:

$MnO_4^- \rightarrow Mn^{2+}$ $\qquad$ $Cl^- \rightarrow Cl_2$

3. Balance each half-reaction.

a. Insert coefficients to balance any atoms other than H or O:

$MnO_4^- \rightarrow Mn^{2+}$ $\qquad$ $2\ Cl^- \rightarrow Cl_2$

b. Balance O by inserting H_2O:

$MnO_4^- \rightarrow Mn^{2+} + 4\ H_2O$ $\qquad$ $2\ Cl^- \rightarrow Cl_2$

c. Balance H by inserting H^+

$MnO_4^- + 8\ H^+ \rightarrow Mn^{2+} + 4\ H_2O$ $\qquad$ $2\ Cl^- \rightarrow Cl_2$

d. Balance charge by inserting electrons, e^-.

In the equation on the left, the total charge on the left side of the equation is 1− + 8+ = 7+, and the total charge on the right side of the equation is 2+ + 0 = 2+. Adding 5 e^- to the left side of the equation gives a total charge of 2+ on each side. In the equation on the right, adding 2 e^- to the right side gives a total charge of 2− on each side:

$MnO_4^- + 8\ H^+ + 5\ e^- \rightarrow Mn^{2+} + 4\ H_2O$ $\qquad$ $2\ Cl^- \rightarrow Cl_2 + 2\ e^-$

4. Multiply the equation for each half-reaction by a number that will give the same number of electrons in each equation.

Multiplying the equation on the left by 2 and the equation on the right by 5 gives 10 electrons in each equation:

$2\ MnO_4^- + 16\ H^+ + 10\ e^- \rightarrow 2\ Mn^{2+} + 8\ H_2O$ $\qquad$ $10\ Cl^- \rightarrow 5\ Cl_2 + 10\ e^-$

5. Combine the balanced half-reactions.

Put all of the reactants together and all of the products together:

$$2\ MnO_4^- + 16\ H^+ + 10\ e^- + 10\ Cl^- \rightarrow 2\ Mn^{2+} + 8\ H_2O + 5\ Cl_2 + 10\ e^-$$

6. Eliminate the electrons (since there's the same number of them on each side) **and check the equation to be sure that it's balanced.**

$$2\ MnO_4^- + 16\ H^+ + 10\ Cl^- \rightarrow 2\ Mn^{2+} + 8\ H_2O + 5\ Cl_2$$

The equation contains the same number of atoms of each element on each side. The total charge on the left is 2− + 16+ + 10− = 4+ and the total charge on the right is 4+ + 0 + 0 = 4+, so the equation is balanced.

This is an abbreviated outline of the half-reaction method:

1. Write skeleton equation.
2. Assign oxidation numbers and divide skeleton into half-reactions.
3. Balance each half-reaction.
   a. Balance atoms other than H or O.
   b. Balance O with H_2O.
   c. Balance H with H^+.
   d. Balance charge with e^-.
4. Multiply half-reactions to make electrons equal.
5. Combine half-reactions.
6. Eliminate electrons and check for balance.†

Example 17.17

In acidic solution, sulfur dioxide reacts with bromine to produce sulfate ion and bromide ion. Write the balanced net ionic equation for the reaction.

Solution

$$SO_2 + Br_2 \rightarrow SO_4^{2-} + Br^- \text{ (acid)}$$

$$\overset{4+2-}{\underset{4+4-}{SO_2}} + \overset{0}{\underset{0}{Br_2}} \rightarrow \overset{6+2-}{\underset{6+8-}{SO_4^{2-}}} + \overset{1-}{\underset{1-}{Br^-}}$$

$SO_2 \rightarrow SO_4^{2-}$ $\quad$ $Br_2 \rightarrow Br^-$

---

† Chemistry Insight

Equations for redox reactions that don't take place in aqueous solution are usually simple enough to balance easily by comparing the number of atoms of each element in the products and in the reactants. For example, the unbalanced equation

$$Na(s) + Cl_2(g) \rightarrow NaCl(s)$$

is balanced by adding coefficients to show 2 formula units of NaCl on the right and 2 atoms of Na on the left:

$$2\ Na(s) + Cl_2(g) \rightarrow 2\ NaCl(s)$$

These equations can also be balanced by the half-reaction method:

$Na \rightarrow Na^+$ $\quad$ $Cl_2 \rightarrow Cl^-$

$Na \rightarrow Na^+$ $\quad$ $Cl_2 \rightarrow 2\ Cl^-$

$2\ Na \rightarrow 2\ Na^+ + 2\ e^-$ $\quad$ $Cl_2 + 2\ e^- \rightarrow 2\ Cl^-$

$2\ Na + Cl_2 + 2\ e^- \rightarrow 2\ Na^+ + 2\ e^- + 2\ Cl^-$

$2\ Na(s) + Cl_2(g) \rightarrow 2\ NaCl(s)$

$SO_2 + 2\ H_2O \rightarrow SO_4^{2-}$ $Br_2 \rightarrow 2\ Br^-$

$SO_2 + 2\ H_2O \rightarrow SO_4^{2-} + 4\ H^+$ $Br_2 + 2\ e^- \rightarrow 2\ Br^-$

$SO_2 + 2\ H_2O \rightarrow SO_4^{2-} + 4\ H^+ + 2\ e^-$

$SO_2 + 2\ H_2O + Br_2 + 2\ e^- \rightarrow SO_4^{2-} + 4\ H^+ + 2\ e^- + 2\ Br^-$

$SO_2 + 2\ H_2O + Br_2 \rightarrow SO_4^{2-} + 4\ H^+ + 2\ Br^-$

Exercise 17.19

Metallic zinc reacts with nitrate ion in aqueous acid to produce zinc ion and ammonium ion. Write the balanced net ionic equation for this reaction.

For a redox reaction that occurs in basic solution, the net ionic equation is first balanced as if the solution were acidic and then adjusted to show that it's basic, as shown in the following Example.

Example 17.18

Permanganate ion reacts with sulfite ion in aqueous base to produce manganese(IV) oxide and sulfate ion. Write the balanced net ionic equation for the reaction.

Solution

$MnO_4^- + SO_3^{2-} \rightarrow MnO_2 + SO_4^{2-}$ (base)

$MnO_4^- \rightarrow MnO_2$ $SO_3^{2-} \rightarrow SO_4^{2-}$

$MnO_4^- \rightarrow MnO_2 + 2\ H_2O$ $SO_3^{2-} + H_2O \rightarrow SO_4^{2-}$

$MnO_4^- + 4\ H^+ \rightarrow MnO_2 + 2\ H_2O$ $SO_3^{2-} + H_2O \rightarrow SO_4^{2-} + 2\ H^+$

$MnO_4^- + 4\ H^+ + 3\ e^- \rightarrow MnO_2 + 2\ H_2O$ $SO_3^{2-} + H_2O \rightarrow SO_4^{2-} + 2\ H^+ + 2\ e^-$

$2\ MnO_4^- + 8\ H^+ + 6\ e^- \rightarrow 2\ MnO_2 + 4\ H_2O$ $3\ SO_3^{2-} + 3\ H_2O \rightarrow 3\ SO_4^{2-} + 6\ H^+ + 6\ e^-$

$2\ MnO_4^- + 8\ H^+ + 6\ e^- + 3\ SO_3^{2-} + 3\ H_2O \rightarrow 2\ MnO_2 + 4\ H_2O + 3\ SO_4^{2-} + 6\ H^+ + 6\ e^-$

$2\ MnO_4^- + 2\ H^+ + 3\ SO_3^{2-} \rightarrow 2\ MnO_2 + H_2O + 3\ SO_4^{2-}$

Up to this point, the equation has been balanced by the six steps previously shown for the half-reaction method, as if the reaction were occurring in acidic solution. To show that the reaction occurs in basic solution, a seventh step is needed.

7. **To each side of the equation add a number of OH^- equal to the number of H^+ in the equation. Where H^+ and OH^- appear on the same side of the equation, combine them to form H_2O, and simplify the equation to give its final form. Check to be sure it's balanced.**

   $2\ MnO_4^- + 2\ H^+ + 2\ OH^- + 3\ SO_3^{2-} \rightarrow 2\ MnO_2 + H_2O + 3\ SO_4^{2-} + 2\ OH^-$

   $2\ MnO_4^- + 2\ H_2O + 3\ SO_3^{2-} \rightarrow 2\ MnO_2 + H_2O + 3\ SO_4^{2-} + 2\ OH^-$

   $2\ MnO_4^- + H_2O + 3\ SO_3^{2-} \rightarrow 2\ MnO_2 + 3\ SO_4^{2-} + 2\ OH^-$

The equation contains the same number of atoms of each element on each side. The total charge on the left is 2− + 0 + 6− = 8− and the total charge on the right is 0 + 6− + 2− = 8−, so the equation is balanced.

Example 17.19

Balance this equation by the half-reaction method:

$$MnO_4^- + C_2O_4^{2-} \rightarrow MnO_2 + CO_2 \text{ (base)}$$

Solution

$$\overset{7+\ 2-}{\underset{7+\ 8-}{MnO_4^-}} + \overset{3+\ 2-}{\underset{6+\ 8-}{C_2O_4^{2-}}} \rightarrow \overset{4+\ 2-}{\underset{4+\ 4-}{MnO_2}} + \overset{4+2-}{\underset{4+4-}{CO_2}}$$

$MnO_4^- \rightarrow MnO_2$ $\quad$ $C_2O_4^{2-} \rightarrow CO_2$

$MnO_4^- \rightarrow MnO_2 + 2\ H_2O$ $\quad$ $C_2O_4^{2-} \rightarrow 2\ CO_2$

$MnO_4^- + 4\ H^+ \rightarrow MnO_2 + 2\ H_2O$ $\quad$ $C_2O_4^{2-} \rightarrow 2\ CO_2 + 2\ e^-$

$MnO_4^- + 4\ H^+ + 3\ e^- \rightarrow MnO_2 + 2\ H_2O$ $\quad$ $3\ C_2O_4^{2-} \rightarrow 6\ CO_2 + 6\ e^-$

$$2\ MnO_4^- + 8\ H^+ + 6\ e^- \rightarrow 2\ MnO_2 + 4\ H_2O$$

$$2\ MnO_4^- + 8\ H^+ + 6\ e^- + 3\ C_2O_4^{2-} \rightarrow 2\ MnO_2 + 4\ H_2O + 6\ CO_2 + 6\ e^-$$

$$2\ MnO_4^- + 8\ H^+ + 3\ C_2O_4^{2-} \rightarrow 2\ MnO_2 + 4\ H_2O + 6\ CO_2$$

$$2\ MnO_4^- + 8\ H^+ + 8\ OH^- + 3\ C_2O_4^{2-} \rightarrow 2\ MnO_2 + 4\ H_2O + 6\ CO_2 + 8\ OH^-$$

$$2\ MnO_4^- + 8\ H_2O + 3\ C_2O_4^{2-} \rightarrow 2\ MnO_2 + 4\ H_2O + 6\ CO_2 + 8\ OH^-$$

$$2\ MnO_4^- + 4\ H_2O + 3\ C_2O_4^{2-} \rightarrow 2\ MnO_2 + 6\ CO_2 + 8\ OH^-$$

Exercise 17.20

Permanganate ion reacts with iodide ion in aqueous base to produce manganese(IV) oxide and iodate ion, IO_3^-. Write the balanced net ionic equation for the reaction.

17-11 Redox reactions are the source of electricity in dry cells and batteries

A movement of electrons through a wire is an electric current (Section 14-5). In a redox reaction electrons move from one reactant to another, and a redox reaction can be arranged to occur in such a way that the electrons move through a wire, creating an electric current. A device that uses a redox reaction to create an electric current is called a **voltaic cell.**◆

◆The voltaic cell is named for the Italian physicist Alessandro Volta (1745–1827).

To understand how a voltaic cell works, it's useful to compare two different methods for carrying out the same redox reaction: one arrangement in which the electrons move directly between reactants and another arrangement in which the electrons move through a wire. As an example, we'll use the reaction of copper atoms with aqueous silver ions:◆

◆This redox reaction is also a replacement reaction (Section 16-6).

$$Cu(s) + 2\ Ag^+(aq) \rightarrow Cu^{2+}(aq) + 2\ Ag(s)$$

This reaction between copper atoms and silver ions can be carried out by simply immersing a piece of copper in an aqueous solution of a silver salt, as shown in Figure 17.2. When the reaction is carried out in this way, silver ions move through the solution and make direct contact with copper atoms on the surface of the copper metal, and electrons are transferred directly from copper atoms to silver ions.

The same reaction can be carried out so that electrons move from copper atoms to silver ions through a wire, as shown in Figure 17.3. In the apparatus shown in the figure, the oxidation of copper atoms to copper ions occurs in the beaker on the left, and the reduction of silver ions to silver atoms occurs in the beaker on the right; the two beakers are connected by a wire, which passes through a voltmeter, and the beakers are connected also by a glass tube, called a salt bridge, containing aqueous potassium nitrate, KNO_3. The combination of the beakers, wire, and salt bridge is a voltaic cell in which an electric current flows through the wire from left to right.

To see how the cell works, begin with the beaker on the left. In this beaker a piece of copper metal is immersed in 1 M $CuSO_4(aq)$. Imagine that one copper atom breaks away from the surface of the copper metal and enters the solution as a copper(II) ion, leaving its two electrons on the piece of copper metal. The two electrons move from the piece of copper metal through the wire to the piece of silver metal in the beaker on the right, which contains 1 M $AgNO_3(aq)$. In that beaker two silver ions from the solution move to the piece of

Figure 17.2
Redox reaction between copper atoms and aqueous silver ions. This reaction was carried out by dissolving silver nitrate, $AgNO_3$, in water and immersing a piece of copper metal, twisted into a spiral shape, in the solution. After a few minutes, needles of silver metal formed on the copper wire and the solution turned blue from dissolved copper(II) ion. The net-ionic equation for the reaction is

$$Cu(s) + 2\ Ag^+(aq) \rightarrow Cu^{2+}(aq) + 2\ Ag(s)$$

When the reaction is carried out in this way, electrons move directly from copper atoms to silver ions. *(J. Morgenthaler)*

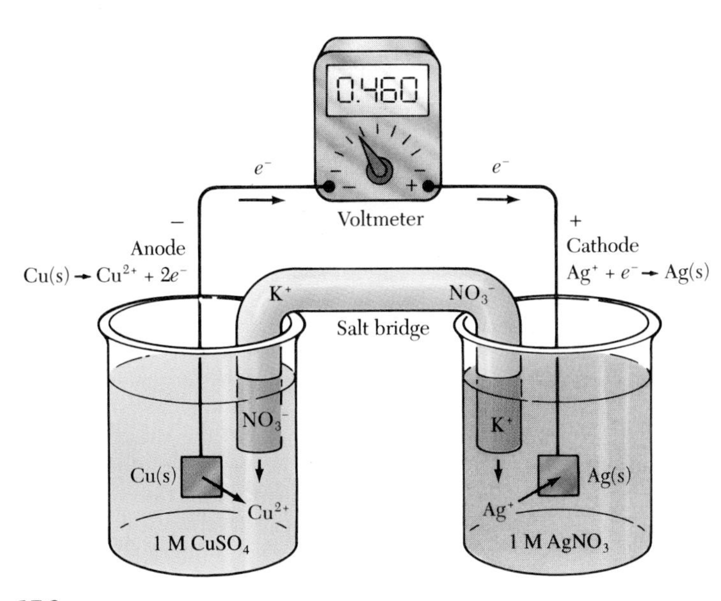

Figure 17.3
A voltaic cell. The cell uses the redox reaction

$$Cu(s) + 2\ Ag^+(aq) \rightarrow Cu^{2+}(aq) + 2\ Ag(s)$$

to generate an electric current. At the anode, Cu(s) atoms form $Cu^{2+}(aq)$ ions, and the electrons that leave the Cu atoms move through the wire to the cathode. At the cathode, the electrons combine with $Ag^+(aq)$ ions to form Ag(s) atoms. Ions from the salt bridge move into each beaker to maintain the same total charge on cations and anions. If you watched this cell operate, you'd see these changes: The piece of copper would get smaller as Cu atoms left it; the solution in the beaker on the left would turn darker blue as Cu^{2+} ions entered it; and the piece of silver would get larger as Ag atoms deposited on it.

silver metal, pick up the two electrons, and form silver atoms, which deposit on the surface of the silver metal. The function of the salt bridge is to maintain the same total charge on all cations and all anions in each beaker: As one Cu^{2+} ion enters the solution in the beaker on the left, two NO_3^- ions enter it from the salt bridge; as two Ag^+ ions leave the solution in the beaker on the right, two K^+ ions enter it from the salt bridge.

In a voltaic cell the pieces of metal on which the oxidation and reduction reactions take place are called **electrodes.** The electrode at which oxidation occurs is the **anode,** and the electrode at which reduction occurs is the **cathode.**

The cell described in Figure 17.3 generates 0.46 volt. The voltage generated by a voltaic cell depends on several factors, including the nature and concentrations of the reactants and the temperature. More advanced chemistry books describe methods for predicting cell voltages.

Dry cells and batteries are voltaic cells that have been designed to be portable. Four examples of commonly used voltaic cells are shown in Figures 17.4 through 17.6.

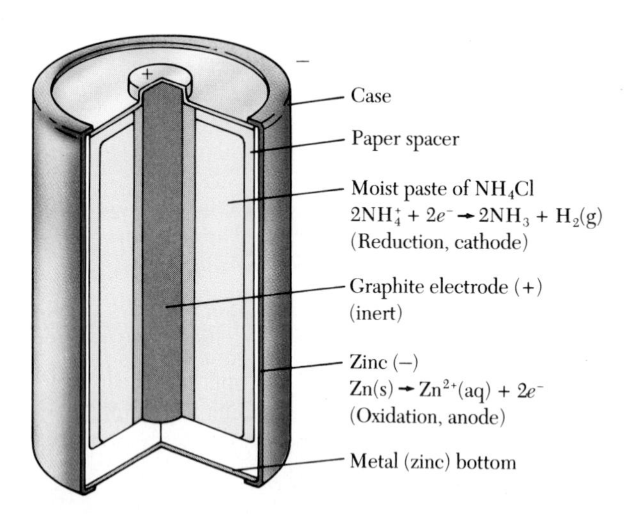

Figure 17.4
Common and alkaline batteries. When the anode and cathode are connected to one another, the cell generates an electric current. In this form of the voltaic cell the anode is a zinc can that also serves as the container for the battery.

In the common battery the anode reaction is the oxidation of zinc:

$$Zn(s) \rightarrow Zn^{2+}(aq) + 2\ e^-$$

The cathode is a graphite rod, immersed in a wet mixture or paste that contains ammonium chloride; the cathode reaction is the reduction of ammonium ion:

$$2\ NH_4^+(aq) + 2\ e^- \rightarrow 2\ NH_3(g) + H_2(g)$$

The overall reaction is the sum of the anode and cathode reactions:

$$Zn(s) + 2\ NH_4^+(aq) \rightarrow Zn^{2+}(aq) + 2\ NH_3(g) + H_2(g)$$

If the gases produced by this reaction, $NH_3(g)$ and $H_2(g)$, were allowed to accumulate, their increasing pressure would eventually burst the battery. To prevent the accumulation of $NH_3(g)$ and $H_2(g)$, other compounds are added to the paste to react with these gases, converting them into liquid and solid products.

In the alkaline battery potassium hydroxide is added to the paste between the electrodes, so the redox reactions occur in basic solution. The oxidation at the anode is

$$Zn(s) + 2\ OH^-(aq) \rightarrow Zn(OH)_2(s) + 2\ e^-$$

the reduction at the cathode is

$$2\ MnO_2(s) + 2\ H_2O(\ell) + 2\ e^- \rightarrow 2\ MnO(OH)(s) + 2\ OH^-(aq)$$

and the overall reaction is

$$Zn(s) + 2\ MnO_2(s) + 2\ H_2O(\ell) \rightarrow Zn(OH)_2(s) + 2\ MnO(OH)(s)$$

(Charles Steele)

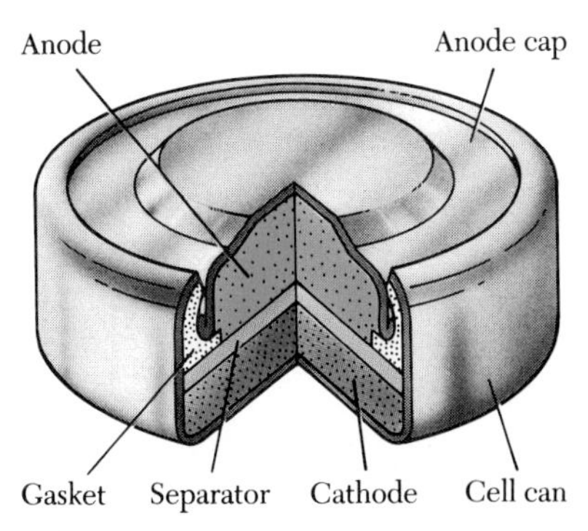

Mercury battery used in calculators, cameras, watches, and heart pacemakers. *(Photo courtesy Eveready Batteries)*

Figure 17.5
Mercury battery. The anode is zinc and the cathode is mercury(II) oxide; the electrodes are separated by a moist paste that contains potassium hydroxide. The oxidation at the anode is

$$Zn(s) + 2\ OH^-(aq) \rightarrow Zn(OH)_2(s) + 2\ e^-$$

the reduction at the cathode is

$$HgO(s) + H_2O(\ell) + 2\ e^- \rightarrow Hg(\ell) + 2\ OH^-(aq)$$

and the overall reaction is

$$Zn(s) + HgO(s) + H_2O(\ell) \rightarrow Zn(OH)_2(s) + Hg(\ell)$$

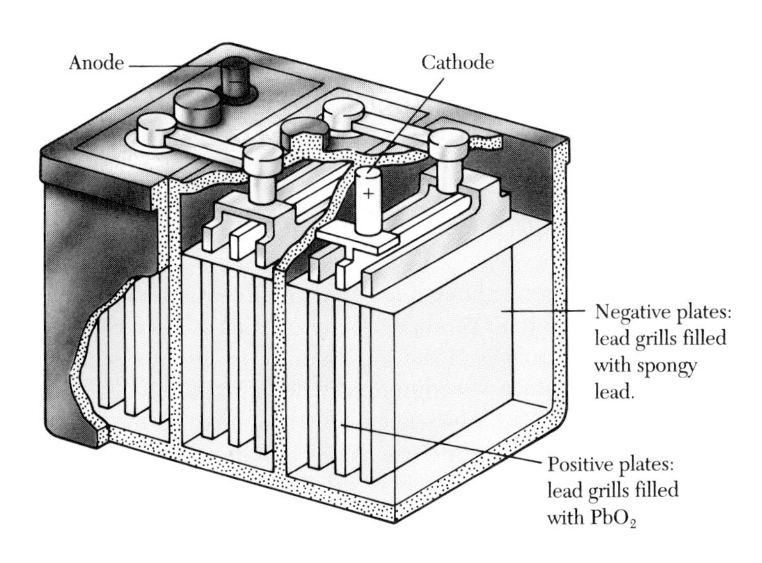

Lead storage battery. *(Tony Freeman/PhotoEdit)*

Figure 17.6
Lead storage battery. The battery consists of a series of cells with a combined voltage of about 12 V. The liquid in the battery is dilute sulfuric acid. The oxidation at the anode is

$$Pb(s) + SO_4^{2-}(aq) \rightarrow PbSO_4(s) + 2\ e^-$$

the reduction at the cathode is

$$PbO_2(s) + 4\ H_3O^+(aq) + SO_4^{2-}(aq) + 2\ e^- \rightarrow PbSO_4(s) + 6\ H_2O(\ell)$$

Because the battery can be recharged, it has a long useful lifetime under ordinary conditions.

Nicad batteries. *(Charles D. Winters)*

In addition to common and alkaline batteries nickel-cadmium (nicad) batteries now are in common use in calculators, photographic equipment, and other portable devices. In nicad batteries the oxidation at the anode is

$$Cd(s) + 2\ OH^-(aq) \rightarrow Cd(OH)_2(s) + 2\ e^-$$

the reduction at the cathode is

$$NiO_2(s) + 2\ H_2O(\ell) + 2\ e^- \rightarrow Ni(OH)_2(s) + 2\ OH^-(aq)$$

and the overall reaction is

$$Cd(s) + NiO_2(s) + 2\ H_2O(\ell) \rightarrow Cd(OH)_2(s) + Ni(OH)_2(s)$$

Nicad batteries have the important advantage over common and alkaline batteries of being rechargeable, so they have much longer useful lifetimes.

Batteries contain a variety of metals and metal compounds, such as $Cd(OH)_2$, Hg, HgO, $Ni(OH)_2$, Pb, $PbSO_4$, and Zn, that are toxic and that should not be released into the environment. Batteries should be disposed of in ways that do not contaminate the environment, and they should never be burned.

17-12 Some redox reactions can be reversed by applying an electric current to their products

Many redox reactions are not reversible under ordinary reaction conditions; that is, their products will not spontaneously re-form their reactants. For example, sodium reacts with chlorine to form sodium chloride

$$2\ Na(s) + Cl_2(g) \rightarrow 2\ NaCl(s)$$

but we know from everyday experience that sodium chloride does not spontaneously decompose into sodium and chlorine. Similarly, hydrogen and oxygen react spontaneously to form water

$$2\ H_2(g) + O_2(g) \rightarrow 2\ H_2O(\ell)$$

but we know from everyday experience that water does not spontaneously decompose into hydrogen and oxygen.

Some redox reactions that aren't spontaneously reversible can be reversed by applying an electric current to their products; the process is called **electrolysis.** Figure 17.7 shows the electrolysis of sodium chloride to produce sodium and chlorine. To carry out this process, sodium chloride is first melted so that its ions are free to move about.◆ An electric current is applied to the liquid sodium chloride by immersing two graphite electrodes in it and connecting the electrodes to a battery.◆ The battery acts as a pump for electrons, drawing them from the electrode on the right and forcing them onto the electrode on the left. At the left electrode—the cathode—sodium ions gain electrons and are reduced to sodium atoms; the sodium metal that's produced rises to the surface of the liquid sodium chloride.◆ At the right electrode—the anode—chloride ions lose electrons and are oxidized to chlorine atoms, which combine to form chlorine molecules; the chlorine gas that's produced rises to the surface of the liquid sodium chloride. The reduction at the cathode is

$$Na^+(\ell) + e^- \rightarrow Na(\ell)$$

◆As described in Chapter 14, the ions in a crystalline ionic solid are in fixed positions in their crystal lattice; when the solid is melted, the lattice breaks down and the ions can move throughout the body of the liquid.

◆Graphite is an appropriate choice for the electrodes because it conducts electricity and is inert—it doesn't react with sodium, chlorine, or sodium chloride. Other inert conductors could also be used.

◆Sodium chloride melts at 804°C and sodium melts at 98°C, so the sodium produced by the electrolysis shown in Figure 17.7 is a liquid.

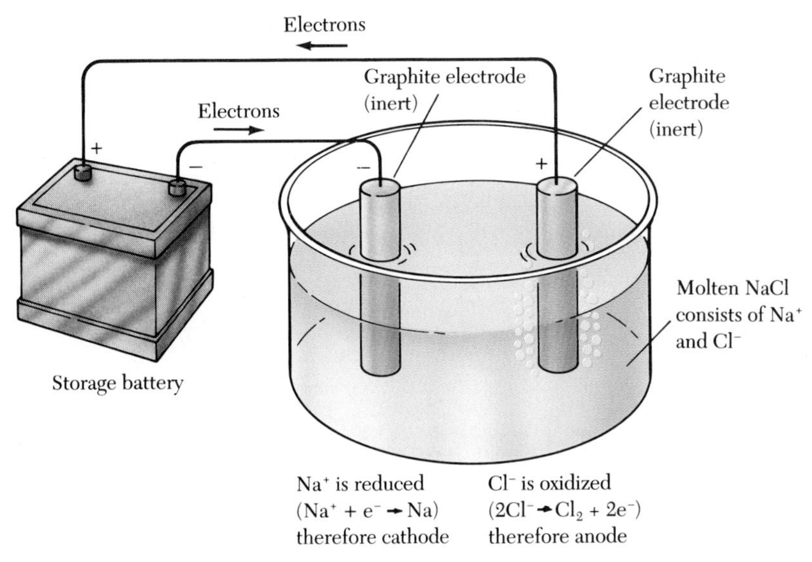

Figure 17.7
Electrolysis of sodium chloride. The sodium chloride is melted so that its ions can move to the electrodes. The battery pumps electrons from the anode to the cathode. At the cathode the liquid sodium metal that's formed rises to the surface of the liquid sodium chloride; at the anode the chlorine gas that's formed also rises to the surface of the liquid sodium chloride.

---

the oxidation at the anode is

$$2\ Cl^-(\ell) \rightarrow Cl_2(g) + 2\ e^-$$

and the overall reaction is

$$2\ Na^+(\ell) + 2\ Cl^-(\ell) \rightarrow 2\ Na(\ell) + Cl_2(g)$$

$$2\ NaCl(\ell) \rightarrow 2\ Na(\ell) + Cl_2(g)$$

The apparatus used to carry out an electrolysis is called an **electrolytic cell.** In a *voltaic* cell (Section 17-11) a spontaneous redox reaction generates an electric current; in an *electrolytic* cell an electric current from an exterior source causes an otherwise nonspontaneous redox reaction to occur. Voltaic cells and electrolytic cells are referred to by the general term **electrochemical cells,** and the branch of chemistry that describes electrochemical cells is called **electrochemistry.** In any electrochemical cell the electrode at which oxidation occurs is designated the anode, and the electrode at which reduction occurs is designated the cathode.

The electrolysis of water is shown in Figure 17.8. The reduction at the cathode, shown on the left in the photograph, is

$$2\ H_2O(\ell) + 2\ e^- \rightarrow H_2(g) + 2\ OH^-(aq)$$

the oxidation at the anode is

$$2\ H_2O(\ell) \rightarrow O_2(g) + 4\ H^+(aq) + 4\ e^-$$

and the overall reaction is

$$2\ H_2O(\ell) \rightarrow 2\ H_2(g) + O_2(g)$$

Figure 17.8
Electrolysis of water.
Cathode (left):

$$2\ H_2O(\ell) + 2\ e^- \rightarrow H_2(g) + 2\ OH^-(aq)$$

Anode (right):

$$2\ H_2O(\ell) \rightarrow O_2(g) + 4\ H^+(aq) + 4\ e^-$$

Overall:

$$2\ H_2O(\ell) \rightarrow 2\ H_2(g) + O_2(g)$$

As the photograph shows, the volume of hydrogen produced is twice the volume of oxygen produced. *(Charles D. Winters)*

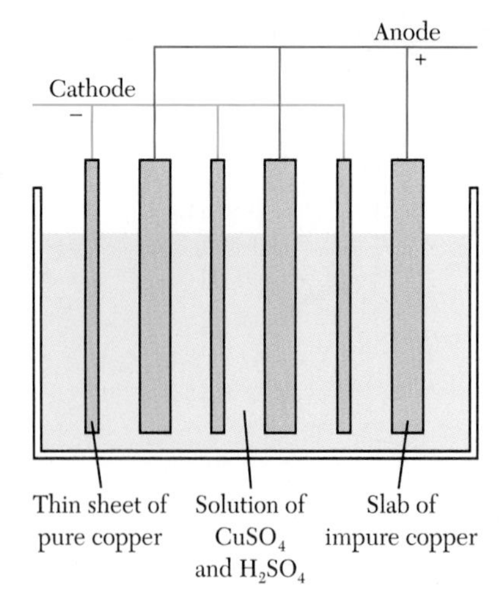

Figure 17.9
Electrorefining of copper. In these huge electrolytic cells large slabs of impure copper alternate with thin sheets of pure copper; the slabs and sheets are immersed in an acidic solution of copper(II) sulfate. The impure copper slabs are anodes, and the anode reaction is

$$Cu(s) \rightarrow Cu^{2+}(aq) + 2\ e^-$$

The copper(II) ions produced by this oxidation move through the solution and are reduced at the pure copper sheets, the cathodes:

$$Cu^{2+}(aq) + 2\ e^- \rightarrow Cu(s)$$

As the electrorefining cells run, the impure copper slabs shrink and the impurities from them fall to the bottom of the cell; at the same time, the pure copper sheets grow as more pure copper deposits on them. *(Sarco, Inc.)*

Among the important commercial applications of electrolysis is the production of aluminum, which was described in the Inside Chemistry essay at the end of Chapter 10. Another major commercial application is the electrolysis of sodium chloride to produce large quantities of sodium and chlorine.

Electrolysis is used to purify or refine some metals; the process is called **electrorefining.** The electrorefining of copper is shown in Figure 17.9; nickel, silver, and tin are also purified in this way.

Electrolysis is also used for **electroplating,** the process of coating objects with a thin layer of metal for protection or decoration. Figure 17.10 describes electroplating with copper; chromium, silver, and zinc are electroplated by similar methods.◆

◆ Electroplating with zinc is called **galvanization.**

Tableware being plated with silver at the cathode of an electrolytic cell. *(Reed & Barton Silversmiths)*

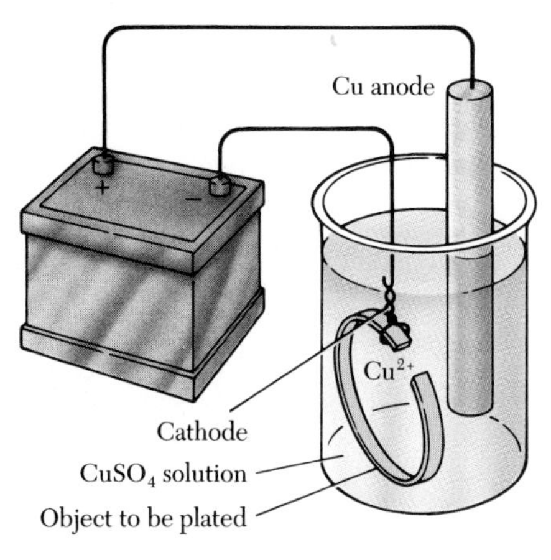

Baby shoes electroplated with copper. *(Charles Steele)*

Figure 17.10
Electroplating with copper. At the copper anode, copper atoms are oxidized and move into solution as copper(II) ions:

$$Cu(s) \rightarrow Cu^{2+}(aq) + 2\ e^-$$

At the cathode, copper(II) ions from the solution are reduced to copper atoms, which coat the object to be plated:

$$Cu^{2+}(aq) + 2\ e^- \rightarrow Cu(s)$$

Inside Chemistry — How can iron be protected against corrosion?

Iron and its alloys, especially the alloys called steel, are used in enormous quantities in manufacturing underground storage tanks, underground pipelines, the hulls of ships, and girders to support buildings and bridges. Iron is popular for these applications because it's cheap, strong, and easy to work with, but it has an important disadvantage: Iron corrodes easily.

The corrosion or rusting of iron is a redox reaction in which iron reacts with oxygen and water:

$$4\ Fe(s) + 3\ O_2(g) + 6\ H_2O(\ell) \rightarrow \underset{\text{iron(III) oxide trihydrate (rust)}}{2\ Fe_2O_3 \cdot 3\ H_2O}$$

In this reaction electrons move from iron atoms to oxygen atoms; iron is oxidized and oxygen is reduced.

Large iron or steel structures can be protected against corrosion by painting. The coating of paint prevents the corrosion reaction by preventing oxygen and water from coming into contact with the iron.

Large iron or steel objects can also be protected against corrosion by connecting them to pieces of magnesium metal. In this ingenious method, called **cathodic protection,** the pieces of magnesium are oxidized instead of the iron objects.

Cathodic protection depends on the fact that electrons will move from magnesium atoms to iron(III) ions; that is, the reaction

$$3\ Mg(s) + 2\ Fe^{3+}(aq) \rightarrow 3\ Mg^{2+}(aq) + 2\ Fe(s)$$

is spontaneous. For a piece of iron to corrode, its Fe

atoms must lose electrons and be oxidized to Fe^{3+} ions. If the piece of iron is connected to a piece of magnesium, electrons from Mg atoms, by the above reaction, will immediately replace the electrons lost by an Fe atom: The magnesium is oxidized and the iron is protected. In this protective reaction reduction occurs on the iron object, which is therefore the cathode. Figure 17.11 shows three examples of cathodic protection.

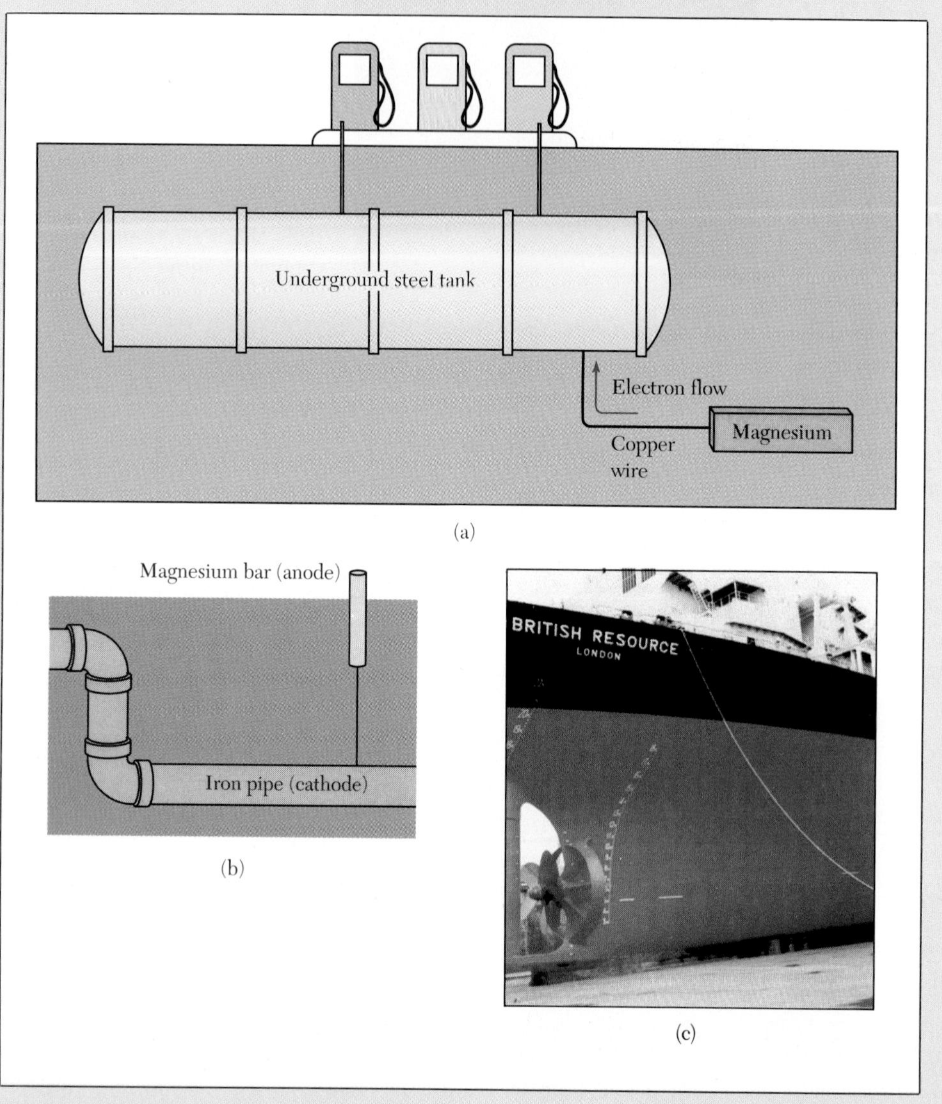

Figure 17.11
Cathodic protection. (a) An underground storage tank made of steel can be protected against corrosion by connecting the tank to a piece of magnesium: The magnesium is oxidized instead of the iron in the steel. Because the leakage of gasoline from underground tanks is a significant environmental hazard, federal law requires providing such tanks with cathodic protection. (b) An underground iron pipe is cathodically protected by a magnesium bar. The bar extends above the ground so that it can easily be replaced when it's almost completely oxidized. (c) The huge hull of this ship is cathodically protected by strips of titanium, which appear in this photo as horizontal yellow lines on the hull, along the propeller axis. *(Johnson-Matthey)*

Chapter Summary: A Closer Look at Equilibrium Reactions and Redox Reactions

Section	Subject	Summary	Check When Learned
17-1	Solubility equilibrium	Equilibrium reaction between a solid ionic compound and its ions in aqueous solution, for example, $CdS(s) \rightleftarrows Cd^{2}(aq) + S^{2-}(aq)$.	☐
17-2	Relationship between rates of forward and reverse reactions at equilibrium	The rates are equal.	☐
17-3	Effect on an equilibrium of		
	Adding reactant	The rates of the forward and reverse reactions increase, and equilibrium is reestablished.	☐
	Removing reactant	The rates of the forward and reverse reactions decrease, and equilibrium is reestablished.	☐
17-3	Le Châtelier's principle	If a reaction at equilibrium is disturbed, the reaction will respond in a way that will restore equilibrium.	☐
17-3	Effect on an equilibrium of		
	Raising the temperature	The equilibrium shifts in the endothermic direction (the products of the endothermic reaction increase).	☐
	Lowering the temperature	The equilibrium shifts in the exothermic direction (the products of the exothermic reaction increase).	☐
17-3	Effect on an equilibrium of		
	Increasing the pressure	The equilibrium shifts in the direction of fewer moles of gases.	☐
	Decreasing the pressure	The equilibrium shifts in the direction of more moles of gases.	☐
17-4	Buffer	A reaction that controls the pH of an aqueous solution by reacting with acid or base added to it.	☐
17-5	Equilibrium-constant expression	For an equilibrium reaction, the mathematical equation that shows the relationship among the molar concentrations of reactants and products at equilibrium.	☐
17-5	Equilibrium-constant expression for		
	Autoionization of water	$2\ H_2O(\ell) \rightleftarrows H_3O^+(aq) + OH^-(aq)$ $[H_3O^+][OH^-] = K_W = 1.00 \times 10^{-14}$	
	Ionization of HF as a weak acid	$H_2O(\ell) + HF(aq) \rightleftarrows H_3O^+(aq) + F^-(aq)$ $\frac{[H_3O^+][F^-]}{[HF]} = K_a$	

Section	Subject	Summary	Check When Learned
	Ionization of NH_3 as a weak base	$NH_3(g) + H_2O(\ell) \rightleftarrows NH_4^+(aq) + OH^-(aq)$ $\frac{[NH_4^+][OH^-]}{[NH_3]} = K_b$	
	Solubility equilibrium of CaF_2	$CaF_2(s) \rightleftarrows Ca^{2+}(aq) + 2\ F^-(aq)$ $[Ca^{2+}][F^-]^2 = K_{sp}$	
	Equilibrium reaction of $H_2(g)$ with $I_2(g)$ to produce HI(g)	$H_2(g) + I_2(g) \rightleftarrows 2\ HI(g)$ $\frac{[HI]^2}{[H_2][I_2]} = K$	☐
17-8	Oxidation	Loss of electrons, e.g., $Na \rightarrow Na^+ + e^-$	☐
17-8	Reduction	Gain of electrons, e.g., $Cl_2 + 2\ e^- \rightarrow 2\ Cl^-$	☐
17-8	Redox reaction	Reaction in which one reactant is oxidized and another is reduced, e.g., $2\ Na(s) + Cl_2(g) \rightarrow 2\ NaCl(s)$.	☐
17-8	Oxidizing agent	The reactant that's reduced, e.g., Cl_2 in the preceding reaction.	☐
17-8	Reducing agent	The reactant that's oxidized, e.g., Na in the preceding reaction.	☐
17-9	Oxidation numbers	Numbers assigned by rule to atoms, as a basis for identifying redox reactions.	☐
17-9	Rules for assigning oxidation numbers	In the symbol or formula for an element, the oxidation number for each atom is 0. In the symbol for a monatomic ion or in the formula for a compound that consists of monatomic ions, the oxidation number for each atom is the same as the charge on the ion. In the formula for a covalent compound of hydrogen, the oxidation number for each H atom is 1+. In the formula for a covalent compound of oxygen, the oxidation number for each O atom is 2− (in a peroxide, 1−). Oxidation numbers for other atoms are assigned from the formulas in which they occur by these principles: (1) The sum of the oxidation numbers in a formula for a compound is 0. (2) The sum of the oxidation numbers in the formula for a polyatomic ion is the same as the charge on the ion.	☐
17-9	Definition, in terms of oxidation number, of Oxidation Reduction	Increase in oxidation number. Decrease in oxidation number.	☐ ☐
17-10	Half-reaction	For a redox reaction, either the oxidation or the reduction reaction. For example, for $2\ Na(s) + Cl_2(g) \rightarrow 2\ NaCl(s)$ the oxidation half-reaction is $Na \rightarrow Na^+ + e^-$ and the reduction half-reaction is $Cl_2 + 2\ e^- \rightarrow 2\ Cl^-$.	☐

Section	Subject	Summary	Check When Learned
17-10	Skeleton equation	For a redox reaction, an abbreviated, unbalanced equation that shows only the reactants being oxidized and reduced and their products; if the reaction occurs in aqueous solution, the skeleton equation includes a statement that the reaction takes place in acid or base. This is an example: $MnO_4^-(aq) + Fe^{2+}(aq) \rightarrow Mn^{2+}(aq) + Fe^{3+}(aq)$ (acid)	☐
17-10	Half-reaction method for balancing redox equations	1. Write skeleton equation. 2. Assign oxidation numbers and divide skeleton into half-reactions. 3. Balance each half-reaction. a. Balance atoms other than H or O. b. Balance O with H_2O. c. Balance H with H^+. d. Balance charge with e^-. 4. Multiply half-reactions to make electrons equal. 5. Combine half-reactions. 6. Eliminate electrons and check for balance. 7. If the reaction occurs in basic solution, to each side of the equation add a number of OH^- equal to the number of H^+ in the equation. Where H^+ and OH^- appear on the same side of the equation, combine them to form H_2O, and simplify the equation to give its final form. Check to be sure it's balanced.	☐
17-11	Voltaic cell	Device that uses a redox reaction to create a flow of electrons through a wire—an electric current.	☐
17-11	Electrode	In a voltaic cell, a conducting surface at which oxidation or reduction occurs.	☐
17-11	Anode	The electrode at which oxidation occurs.	☐
17-11	Cathode	The electrode at which reduction occurs.	☐
17-11	Function of the salt bridge in a voltaic cell	Provides ions to the solutions in the cathode and anode compartments to maintain the same total charge on all cations and all anions in each compartment.	☐
17-11	Dry cell or battery	A voltaic cell designed to be portable.	☐
17-12	Electrolysis	Reversing a redox reaction by applying an electric current to its products.	☐
17-12	Electrolytic cell	Apparatus used to carry out electrolysis. The electrode at which oxidation occurs is the anode, and the electrode at which reduction occurs is the cathode.	☐
17-12	Electrochemical cell	General term for voltaic cell and electrolytic cell.	☐
17-12	Electrochemistry	The branch of chemistry that describes electrochemical cells.	☐
17-12	Electrorefining	Purification of metals by electrolysis.	☐
17-12	Electroplating	Coating objects with metals by electrolysis.	☐

Problems

Assume you can use the Periodic Table at the front of the book unless you're directed otherwise. Answers to odd-numbered Problems are in Appendix 1.

Types of Equilibrium Reactions (Section 17-1)

1. Identify the reaction represented by each of the following equations as a weak-acid equilibrium, a weak-base equilibrium, or a solubility equilibrium:
(a) $CH_3NH_2(aq) + H_2O(\ell) \rightleftarrows CH_3NH_3^+(aq) + OH^-(aq)$
(b) $BaSO_4(s) \rightleftarrows Ba^{2+}(aq) + SO_4^{2-}(aq)$
(c) $H_2O(\ell) + HNO_2(aq) \rightleftarrows H_3O^+(aq) + NO_2^-(aq)$

2. Identify the reaction represented by each of the following equations as a weak-acid equilibrium, a weak-base equilibrium, or a solubility equilibrium:
(a) $H_2O(\ell) + H_2SO_3(aq) \rightleftarrows H_3O^+(aq) + HSO_3^-(aq)$
(b) $N_2H_4(aq) + H_2O(\ell) \rightleftarrows N_2H_5^+(aq) + OH^-(aq)$
(c) $CaC_2O_4(s) \rightleftarrows Ca^{2+}(aq) + C_2O_4^{2-}(aq)$

3. Identify the reaction represented by each of the following equations as a weak-acid equilibrium, a weak-base equilibrium, or a solubility equilibrium:
(a) $Fe_2S_3(s) \rightleftarrows 2\ Fe^{2+}(aq) + 3\ S^{2-}(aq)$
(b) $Ag_2CO_3(s) \rightleftarrows 2\ Ag^+(aq) + CO_3^{2-}(aq)$
(c) $H_2O(\ell) + HPO_4^{2-}(aq) \rightleftarrows H_3O^+(aq) + PO_4^{3-}(aq)$

4. Identify the reaction represented by each of the following equations as a weak-acid equilibrium, a weak-base equilibrium, or a solubility equilibrium:
(a) $CaCO_3(s) \rightleftarrows Ca^{2+}(aq) + CO_3^{2-}(aq)$
(b) $C_5H_5N(aq) + H_2O(\ell) \rightleftarrows C_5H_5NH^+(aq) + OH^-(aq)$
(c) $H_2O(\ell) + HSO_4^-(aq) \rightleftarrows H_3O^+(aq) + SO_4^{2-}(aq)$

5. Write an equation for (a) the solubility equilibrium of barium sulfide (b) the weak-acid equilibrium of hydrogen cyanide (c) the weak-base equilibrium of dimethylamine, $(CH_3)_2NH$.

6. Write an equation for (a) the solubility equilibrium of tin(II) carbonate (b) the weak-acid equilibrium of hydrogen sulfite ion (c) the weak-base equilibrium of ethylamine, $C_2H_5NH_2$.

7. Write an equation for the solubility equilibrium of (a) calcium sulfide (b) magnesium hydroxide (c) copper(I) sulfide.

8. Write an equation for the solubility equilibrium of (a) aluminum hydroxide (b) aluminum sulfide (c) aluminum phosphate.

9. Dinitrogen tetroxide gas decomposes by an equilibrium reaction to form nitrogen dioxide gas. Write an equation for the reaction.

10. Bromine gas reacts with chlorine gas by an equilibrium reaction to form bromine monochloride gas. Write an equation for the reaction.

11. Sulfur dioxide gas reacts with oxygen gas by an equilibrium reaction to form sulfur trioxide gas. Write an equation for the reaction.

12. Phosphorus pentachloride gas decomposes by an equilibrium reaction to form phosphorus trichloride gas and chlorine gas. Write an equation for the reaction.

Rates of Forward and Reverse Reactions (Section 17-2)

13. The reaction

$$2\ NH_3(g) \rightleftarrows N_2(g) + 3\ H_2(g)$$

was at equilibrium in a container. In a certain time interval 4 mol of NH_3 decomposed into N_2 and H_2. In that same interval, how many moles of hydrogen reacted with nitrogen to form NH_3?

14. The reaction

$$NH_3(aq) + H_2O(\ell) \rightleftarrows NH_4^+(aq) + OH^-(aq)$$

was at equilibrium. In a certain time interval 0.5 mol of NH_3 reacted to form NH_4^+ and OH^-. In that same interval, how many moles of NH_4^+ and how many moles of OH^- reacted with one another to form NH_3 and H_2O?

Le Châtelier's Principle (Section 17-3)

15. The reaction

$$2\ NH_3(g) \rightleftarrows N_2(g) + 3\ H_2(g)$$

was at equilibrium in a container. If some of the hydrogen gas is removed from the container, how will the rates of the forward and reverse reactions change?

16. The reaction

$$H_2O(\ell) + HF(aq) \rightleftarrows H_3O^+(aq) + F^-(aq)$$

was at equilibrium. If some of the fluoride ion is removed, how will the rates of the forward and reverse reactions change?

17. The reaction

$$NH_3(aq) + H_2O(\ell) \rightleftarrows NH_4^+(aq) + OH^-(aq)$$

was at equilibrium. If sodium hydroxide is added, how will the rates of the forward and reverse reactions change?

18. The reaction

$$BaF_2(s) \rightleftarrows Ba^{2+}(aq) + 2\ F^-(aq)$$

was at equilibrium. If solid barium fluoride is added, how will the rates of the forward and reverse reactions change?

19. The reaction

$$2\ CO(g) + O_2(g) \rightleftarrows 2\ CO_2(g)$$

is exothermic in the forward direction. If this reaction were at equilibrium in a container and the container were heated, would the concentration of carbon monoxide increase or decrease?

20. The reaction

$$2\ HF(g) \rightleftarrows H_2(g) + F_2(g)$$

is endothermic in the forward direction. If this reaction were at equilibrium in a container and the container were cooled, would the concentrations of hydrogen and fluorine increase?

21. The reaction

$$H_2O(\ell) + HF(aq) \rightleftarrows H_3O^+(aq) + F^-(aq)$$

is endothermic in the reverse direction. Will raising the temperature increase or decrease the amount of hydrogen fluoride that's ionized?

22. The reaction

$$NH_3(g) + H_2O(\ell) \rightleftarrows NH_4^+(aq) + OH^-(aq)$$

is exothermic in the forward direction. Will cooling this reaction, at equilibrium, increase the concentration of ammonium ion?

23. If the reaction

$$PBr_3(g) + Br_2(g) \rightleftarrows PBr_5(g)$$

is at equilibrium, will increasing the pressure increase or decrease the concentration of phosphorus tribromide?

24. If the reaction

$$FeS(s) \rightleftarrows Fe^{2+}(aq) + S^{2-}(aq)$$

is at equilibrium, will decreasing the pressure increase or decrease the concentration of sulfide ion?

25. If the reaction

$$CO(g) + 3\ H_2(g) \rightleftarrows CH_4(g) + H_2O(g)$$

is at equilibrium, will decreasing the pressure increase or decrease the concentration of carbon monoxide?

26. If the reaction

$$COCl_2(g) \rightleftarrows CO(g) + Cl_2(g)$$

is at equilibrium, will increasing the pressure increase or decrease the concentration of carbon monoxide?

Buffers (Section 17-4)

27. A buffer was made by preparing a solution of the weak acid nitrous acid, $HNO_2(aq)$, and adding sodium nitrite, $NaNO_2$, to it. (a) Write the equilibrium equation that describes the buffer. (b) Write the equation for the reaction of the buffer with acid. (c) Write the equation for the reaction of the buffer with base.

28. A buffer was made by dissolving hydrogen cyanide in water and adding sodium cyanide to the solution. (a) Write the equilibrium equation that describes the buffer. (b) Write the equation for the reaction of the buffer with acid. (c) Write the equation for the reaction of the buffer with base.

29. For this buffer equilibrium:

$$H_2O(\ell) + \underset{\text{large}}{H_2PO_4^-(aq)} \rightleftarrows \underset{\text{small}}{H_3O^+(aq)} + \underset{\text{large}}{HPO_4^{2-}(aq)}$$

(a) Write the equation for the reaction of the buffer with acid. (b) Write the equation for the reaction of the buffer with base.

30. For this buffer equilibrium:

$$H_2O(\ell) + \underset{\text{large}}{HSO_4^-(aq)} \rightleftarrows \underset{\text{small}}{H_3O^+(aq)} + \underset{\text{large}}{SO_4^{2-}(aq)}$$

(a) Write the equation for the reaction of the buffer with acid. (b) Write the equation for the reaction of the buffer with base.

Writing Equilibrium-Constant Expressions (Section 17-5)

31. Write the equilibrium-constant expression for each reaction:
(a) $O_2(g) + 2\ F_2(g) \rightleftarrows 2\ OF_2(g)$
(b) $PbI_2(s) \rightleftarrows Pb^{2+}(aq) + 2\ I^-(aq)$
(c) $H_2O(\ell) + HNO_2(aq) \rightleftarrows H_3O^+(aq) + NO_2^-(aq)$

32. Write the equilibrium-constant expression for each reaction:
(a) $CH_3NH_2(aq) + H_2O(\ell) \rightleftarrows CH_3NH_3^+(aq) + OH^-(aq)$
(b) $NH_3(g) + 3\ I_2(g) \rightleftarrows NI_3(g) + 3\ HI(g)$
(c) $MgS(s) \rightleftarrows Mg^{2+}(aq) + S^{2-}(aq)$

33. Write the equilibrium-constant expression for each reaction:
(a) $Ba_3(PO_4)_2(s) \rightleftarrows 3\ Ba^{2+}(aq) + 2\ PO_4^{3-}(aq)$
(b) $PCl_3(g) + Cl_2(g) \rightleftarrows PCl_5(g)$
(c) $PCl_5(g) \rightleftarrows PCl_3(g) + Cl_2(g)$

34. Write the equilibrium-constant expression for each reaction:
(a) $H_2O(\ell) + H_2PO_4^-(aq) \rightleftarrows H_3O^+(aq) + HPO_4^{2-}(aq)$
(b) $Fe_2S_3(s) \rightleftarrows 2\ Fe^{3+}(aq) + 3\ S^{2-}(aq)$
(c) $H_2O(\ell) \rightleftarrows H_2O(g)$

35. Write the equilibrium-constant expression for (a) the solubility equilibrium of barium carbonate (b) the ionization of hydrogen carbonate ion.

36. Write the equilibrium-constant expression for (a) the solubility equilibrium of copper(II) phosphate (b) the ionization of the weak base dimethylamine, $(CH_3)_2NH(aq)$.

Calculations with Equilibrium-Constant Expressions (Sections 17-6 and 17-7)

37. Cadmium sulfide is very slightly soluble in water: 2×10^{-12} g of CdS will dissolve in 125 mL of water. Calculate K_{sp} for CdS.

38. Magnesium carbonate is slightly soluble in water: 0.074 g of $MgCO_3$ will dissolve in 275 mL of water. Calculate K_{sp} for $MgCO_3$.

39. In a saturated solution of zinc hydroxide the concentration of zinc ion is 7.2×10^{-9} M. Calculate K_{sp} for zinc hydroxide. Hint: From the concentration of the zinc ion calculate the concentration of hydroxide ion in the solution.

40. In a saturated solution of silver sulfide the concentration of silver ion is 4.9×10^{-17} M. Calculate K_{sp} for silver sulfide. Hint: From the concentration of silver ion calculate the concentration of sulfide ion in the solution.

41. The pH of a saturated solution of magnesium hydroxide is 10.50; calculate K_{sp} for magnesium hydroxide. Hint: First calculate $[H_3O^+]$, then calculate $[OH^-]$, then calculate the molarity of magnesium in the solution.

42. In a saturated solution of lead(II) chloride the concentration of chloride ion is 3.2×10^{-2} M. Calculate K_{sp} for lead(II) chloride.

43. A solution was prepared by dissolving 2.3 g of benzoic acid, $HC_7H_5O_2$, in enough water to make 125 mL of solution; K_a for benzoic acid is 6.3×10^{-5}. Calculate the concentrations of reactants and products, except water, at equilibrium.

44. A solution was prepared by dissolving 1.6 g of cyanic acid, HCNO, in enough water to make 75.0 mL of solution; K_a for cyanic acid is 3.5×10^{-4}. Calculate the concentrations of reactants and products, except water, at equilibrium.

45. A solution was prepared by dissolving 0.30 g of propionic acid, $HC_3H_5O_2$, in enough water to make 50.0 mL of solution; K_a for propionic acid is 1.3×10^{-5}. Calculate the concentrations of reactants and products, except water, at equilibrium.

46. A solution was prepared by dissolving 5.3 g of pyruvic acid, $HC_3H_3O_3$, in enough water to make 80.0 mL of solution; K_a for pyruvic acid is 1.3×10^{-2}. Calculate the concentrations of reactants and products, except water, at equilibrium.

47. A solution was prepared by dissolving 3.6 g of hydrogen fluoride in enough water to make 200 mL (3 significant digits) of solution; K_a for hydrofluoric acid is 6.8×10^{-4}. Calculate the pH of the solution. Hint: Calculate $[H_3O^+]$, then pH.

48. A solution was prepared by dissolving 0.30 g of acetic acid in enough water to make 50.0 mL of solution; K_a for acetic acid is 1.7×10^{-5}. Calculate the pH of the solution. Hint: Calculate $[H_3O^+]$, then pH.

49. A solution was made by dissolving 0.13 g of ammonia in enough water to make 75 mL of solution; K_b for ammonia is 1.8×10^{-5}. Calculate the pH of the solution. Hint: Calculate $[OH^-]$, then $[H_3O^+]$, then pH.

50. A solution was made by dissolving 0.064 g of hydrazine, N_2H_4, in enough water to make 100 mL (3 significant digits) of solution; K_b for hydrazine is 1.7×10^{-6}. Calculate the pH of the solution. Hint: Calculate $[OH^-]$, then $[H_3O^+]$, then pH.

51. For iron(II) sulfide, $K_{sp} = 6 \times 10^{-18}$. Calculate the molarity of sulfide ion in a saturated solution of iron(II) sulfide.

52. For barium fluoride, $K_{sp} = 1.0 \times 10^{-6}$. Calculate the molarity of fluoride ion in a saturated solution of barium fluoride.

53. For copper(II) hydroxide, $K_{sp} = 2.6 \times 10^{-19}$. How many milligrams of copper(II) hydroxide are present in 135 mL of a saturated solution?

54. For calcium sulfate, $K_{sp} = 2.4 \times 10^{-5}$. How many grams of calcium sulfate are present in 200 mL (3 significant digits) of a saturated solution?

55. The reaction

$$2\ HF(g) \rightleftarrows H_2(g) + F_2(g)$$

was carried out by placing 0.15 mol of HF in a 5.0-L container. Calculate the equilibrium concentrations of the reactant and products. At the temperature at which the reaction was carried out, $K = 1.0 \times 10^{-95}$.

56. The reaction

$$N_2(g) + O_2(g) \rightleftarrows 2\ NO(g)$$

was carried out. At equilibrium the concentration of N_2 was 0.023 M and the concentration of O_2 was 0.031 M. What was the concentration of NO? At the temperature at which the reaction was carried out, K = 0.0025.

57. The reaction

$$N_2(g) + O_2(g) \rightleftarrows 2\ NO(g)$$

was carried out by placing 3.0 mol each of N_2 and O_2 in a 1.00-L container. Calculate the equilibrium concentrations of the reactants and product. At the temperature at which the reaction was carried out, K = 0.0025.

58. The reaction

$$PCl_5(g) \rightleftarrows PCl_3(g) + Cl_2(g)$$

was carried out by placing 0.90 mol of PCl_5 in a 2.0-L container. Calculate the equilibrium concentrations of the reactant and products. At the temperature at which the reaction was carried out, $K = 2.0 \times 10^{-4}$.

Assigning Oxidation Numbers (Section 17-9)

59. Assign an oxidation number to each atom in each symbol or formula: (a) Mg (b) Mg^{2+} (c) O_2.

60. Assign an oxidation number to each atom in each symbol or formula: (a) Cl_2 (b) Cl^- (c) K.

61. Assign an oxidation number to each atom in each symbol or formula: (a) NH_3 (b) NO_2 (c) NH_4^+.

62. Assign an oxidation number to each atom in each symbol or formula: (a) CH_4 (b) CO (c) CO_2.

63. Assign an oxidation number to each atom in each symbol or formula: (a) Li (b) N_2 (c) Li_3N.

64. Assign an oxidation number to each atom in each formula: (a) P_4 (b) K_2O (c) P_2O_5.

65. Assign an oxidation number to each atom in each formula: (a) MnO_2 (b) $Mn(NO_3)_2$ (c) $Ca(MnO_4)_2$.

66. Assign an oxidation number to each atom in each formula: (a) CH_2O (b) $C_6H_{12}O_6$ (c) CaC_2O_4.

67. Assign oxidation numbers to decide if each of the following equations represents a redox reaction. If it does, identify the oxidizing agent and the reducing agent:
(a) $Ca(s) + Cl_2(g) \rightarrow CaCl_2(s)$
(b) $2\ Ca(s) + O_2(g) \rightarrow 2\ CaO(s)$
(c) $CaCO_3(s) \rightarrow CaO(s) + CO_2(g)$

68. Assign oxidation numbers to decide if each of the following equations represents a redox reaction. If it does, identify the oxidizing agent and the reducing agent:
(a) $Na(s) + 2\ H_2O(\ell) \rightarrow 2\ NaOH(aq) + H_2(g)$
(b) $2\ Na(s) + H_2(g) \rightarrow 2\ NaH(s)$
(c) $Zn(s) + 2\ HNO_3(aq) \rightarrow Zn(NO_3)_2(aq) + H_2(g)$

69. Assign oxidation numbers to decide if each of the following equations represents a redox reaction. If it does, identify the oxidizing agent and the reducing agent:
(a) $MgO(s) + 2\ HCl(aq) \rightarrow MgCl_2(aq) + H_2O(\ell)$
(b) $P_2O_5(s) + 3\ H_2O(\ell) \rightarrow 2\ H_3PO_4(aq)$
(c) $2\ (NH_4)_2S(aq) + Fe(NO_3)_2(aq) \rightarrow 2\ NH_4NO_3(aq) + FeS(s)$

70. Assign oxidation numbers to decide if each of the following equations represents a redox reaction. If it does, identify the oxidizing agent and the reducing agent:
(a) $2\ KClO_3(s) \rightarrow 2\ KCl(s) + 3\ O_2(g)$
(b) $2\ Al(s) + 3\ F_2(g) \rightarrow 2\ AlF_3(s)$
(c) $MgSO_3(s) + 2\ HCl(aq) \rightarrow MgCl_2(aq) + SO_2(g) + H_2O(\ell)$

Balancing Redox Equations (Section 17-10)

71. Balance each equation by the half-reaction method:
(a) $Cr_2O_7^{2-} + C_2O_4^{2-} \rightarrow Cr^{3+} + CO_2$ (acid)
(b) $MnO_2 + HNO_2 \rightarrow Mn^{2+} + NO_3^-$ (acid)
(c) $Mn^{2+} + BiO_3^- \rightarrow MnO_4^- + Bi^{3+}$ (acid)

72. Balance each equation by the half-reaction method:
(a) $As + ClO_3^- \rightarrow H_3AsO_3 + HClO$ (acid)
(b) $HNO_2 + Cr_2O_7^{2-} \rightarrow Cr^{3+} + NO_3^-$ (acid)
(c) $Cr_2O_7^{2-} + I^- \rightarrow Cr^{3+} + IO_3^-$ (acid)

73. Balance each equation by the half-reaction method:
(a) $Mn^{2+} + H_2O_2 \rightarrow MnO_2 + H_2O$ (base)
(b) $Mn^{2+} + ClO_3^- \rightarrow MnO_2 + ClO_2$ (base)
(c) $Cl_2 \rightarrow Cl^- + ClO_3^-$ (base)

74. Balance each equation by the half-reaction method:
(a) $MnO_4^- + NO_2^- \rightarrow MnO_2 + NO_3^-$ (base)
(b) $MnO_4^- + NO_2 \rightarrow MnO_2 + NO_3^-$ (base)
(c) $Cr(OH)_4^- + H_2O_2 \rightarrow CrO_4^{2-} + H_2O$ (base)

75. Balance each equation by the half-reaction method:
(a) $Cr_2O_7^{2-} + Fe^{2+} \rightarrow Cr^{3+} + Fe^{3+}$ (acid)
(b) $Pb(OH)_4^{2-} + ClO^- \rightarrow PbO_2 + Cl^-$ (base)
(c) $NO_3^- + Cu \rightarrow NO + Cu^{2+}$ (acid)

76. Balance each equation by the half-reaction method:
(a) $MnO_4^- + Br^- \rightarrow MnO_2 + BrO_3^-$ (base)
(b) $H_2S + NO_3^- \rightarrow NO_2 + S_8$ (acid)
(c) $Zn + NO_3^- \rightarrow NH_3 + Zn(OH)_4^{2-}$ (base)

77. Balance each equation by the half-reaction method:
(a) $S^{2-} + I_2 \rightarrow SO_4^{2-} + I^-$ (base)
(b) $MnO_4^- + SO_2 \rightarrow Mn^{2+} + SO_4^{2-}$ (acid)
(c) $CN^- + MnO_4^- \rightarrow CNO^- + MnO_2$ (base)

78. Balance each equation by the half-reaction method:
(a) $Hg_2^{2+} + H_2S \rightarrow Hg + S_8$ (acid)
(b) $Al + NO_3^- \rightarrow Al(OH)_4^- + NH_3$ (base)
(c) $Cr_2O_7^{2-} + H_2O_2 \rightarrow Cr^{3+} + O_2$ (acid)

79. Balance each equation by the half-reaction method:
(a) $Cr_2O_7^{2-} + Cl^- \rightarrow Cr^{3+} + Cl_2$ (acid)
(b) $S_8 + NO_3^- \rightarrow SO_2 + NO$ (acid)
(c) $Cr(OH)_4^- + ClO^- \rightarrow CrO_4^{2-} + Cl^-$ (base)

80. Balance each equation by the half-reaction method:
(a) $H_2O_2 + MnO_4^- \rightarrow O_2 + MnO_2$ (base)
(b) $Zn + NO_3^- \rightarrow Zn^{2+} + N_2$ (acid)
(c) $Co^{2+} + H_2O_2 \rightarrow Co(OH)_3 + H_2O$ (base)

Voltaic Cells (Section 17-11)

81. Following the example of Figure 17.3, sketch a voltaic cell in which the anode is a piece of zinc immersed in aqueous zinc nitrate and the cathode is a piece of copper immersed in aqueous copper(II) nitrate. The anode reaction is the oxidation of zinc atoms to zinc ions, and the cathode reaction is the reduction of copper(II) ions to copper atoms.

82. Following the example of Figure 17.3, sketch a voltaic cell in which the anode is a piece of cadmium immersed in aqueous cadmium nitrate and the cathode is a piece of silver immersed in aqueous silver nitrate. The anode reaction is oxidation of cadmium atoms to cadmium ions, and the cathode reaction is the reduction of silver ions to silver atoms.

83. Following the example of Figure 17.3, sketch a voltaic cell in which the anode is a piece of cadmium immersed in aqueous cadmium nitrate and the cathode is a piece of lead immersed in aqueous lead(II) nitrate. The anode reaction is the oxidation of cadmium ions, and the cathode reaction is the reduction of lead(II) ions to lead atoms.

84. Following the example of Figure 17.3, sketch a voltaic cell in which the anode is a piece of iron immersed in aqueous iron(II) nitrate and the cathode is a piece of silver immersed in aqueous silver nitrate. The anode reaction is the oxidation of iron atoms to iron(II) ions, and the cathode reaction is the reduction of silver ions to silver atoms.

Electrolysis (Section 17-12)

85. In the electrolysis of molten calcium chloride, chlorine gas is produced at the anode and liquid calcium is produced at the cathode. Write an equation for each of these reactions.

86. In the electrolysis of molten potassium bromide, liquid bromine is produced at the anode and liquid potassium is produced at the cathode. Write an equation for each of these reactions.

Nuclear chemistry provides our best methods for dating very old objects, as described in Section 18-8. These methods establish the age of this cave painting in Lascaux, France, at about 17,190 years. *(The New York Public Library)*

18 Nuclear Chemistry

◆ Before continuing with this chapter you may want to review the description of nuclear symbols in Section 4-5 and the description of nuclear equations in Section 4-8.

Nuclear reactions were briefly described in Section 4-8.◆ In these reactions the nuclei of atoms change by absorbing or emitting particles or energy; one example is the absorption of an alpha particle, $^{4}_{2}He$, by a $^{9}_{4}Be$ nucleus, with the formation of a $^{12}_{6}C$ nucleus and a neutron, $^{1}_{0}n$:

$$^{9}_{4}Be + ^{4}_{2}He \rightarrow ^{12}_{6}C + ^{1}_{0}n$$

Nuclear chemistry is the branch of chemistry that describes and explains nuclear reactions.

Nuclear chemistry is important as a science because it leads to an increasingly detailed understanding of the structure and behavior of atoms. In this chapter we'll see how the investigation of nuclear reactions has led to a variety of important scientific results, including the creation of new elements, the determination of the ages of ancient objects, and the explanation of the emission of energy from stars.

Nuclear chemistry is important as a technology because it provides sources of energy that can be applied to many human enterprises, including industrial production, warfare, and medicine. In this chapter we'll see how nuclear reactions have been applied to the generation of electricity, the manufacture of atomic weapons, and the treatment of cancer.

18-1 In a nuclear reaction an atomic nucleus emits a subatomic particle or energy

Nuclear reactions are fundamentally different from chemical reactions. In a chemical reaction the reacting atoms share, lose, or gain valence electrons—they make or break chemical bonds—and the nuclei of the atoms are unaffected. In a nuclear reaction no chemical bonds are formed or broken; the nuclei of the reacting atoms change by emitting particles or energy, and their valence electrons are unaffected.

The emission of particles or energy by nuclei in a nuclear reaction is called **radioactivity.** Table 18.1 describes the forms of radioactive emission and the symbols used to represent them.

From your earlier work in this book you're familiar with all of the forms of emission described in Table 18.1 except for the positron. A positron is a particle with the same mass as an electron but a positive charge. We sometimes express this relationship between the positron and the electron by calling the positron the *antiparticle* of the electron.

The following five types of nuclear reactions will be important for our purposes.

Alpha Emission

◆ In the equation for a nuclear reaction, the sum of the subscripts for the reactants must equal the sum of the subscripts for the products, and the same is true for the superscripts; see Section 4-8.

A nucleus that emits an alpha particle, $^{4}_{2}He$, loses two protons and two neutrons; as a result the atomic number of the nucleus decreases by 2 and its mass number decreases by 4. This is one example:◆

$$^{238}_{92}U \rightarrow ^{234}_{90}Th + ^{4}_{2}He$$

Table 18.1 Forms of Radioactive Emission

In each symbol the subscript shows charge and the superscript shows mass number: $^{1}_{0}n$ represents a particle (a neutron) whose charge is 0 and whose mass to the nearest whole number is 1 amu; $^{0}_{-1}e$ represents a particle (an electron) whose charge is 1− and whose mass to the nearest whole number is 0 amu.

Emission	Symbol	Alternative Symbol	Description
Neutron	$^{1}_{0}n$		Subatomic particle with mass 1.008665 amu and charge 0
Alpha particle	$^{4}_{2}He$	α	Cluster of two protons and two neutrons; a helium nucleus
Beta particle	$^{0}_{-1}e$	β^-	Subatomic particle with mass 0.0005846 amu and charge 1−; an electron
Gamma particle	$^{0}_{0}\gamma$		Photon of electromagnetic radiation with a very short wavelength, about 1×10^{-12} m
Positron	$^{0}_{1}e$	β^+	Subatomic particle with mass 0.0005846 amu and charge 1+; the antiparticle of the electron

Exercise 18.1◆

Write the equation for the emission of an alpha particle from $^{226}_{88}Ra$.

◆Answers to Exercises are in Appendix 1.

Beta Emission

In beta emission the *nucleus* of an atom emits an electron, $^{0}_{-1}e$ (a beta particle). An electron that emerges from an atomic nucleus can be thought of as being produced by the conversion of a neutron into a proton ($^{1}_{1}p$) and an electron:

$$^{1}_{0}n \rightarrow {}^{1}_{1}p + {}^{0}_{-1}e$$

neutron in nucleus; proton in nucleus; electron emitted from nucleus

By this conversion the emitting nucleus undergoes a net loss of one neutron and a net gain of one proton; its mass number remains the same, and its charge increases by one unit. This is one example:

$$^{99}_{43}Tc \rightarrow {}^{99}_{44}Ru + {}^{0}_{-1}e$$

$^{99}_{43}Tc$	$^{99}_{44}Ru$	$^{0}_{-1}e$
nucleus has:	nucleus has:	electron emitted from nucleus
43 protons	44 protons	
56 neutrons	55 neutrons	
99 (mass number)	99 (mass number)	

Exercise 18.2

Write the equation for the emission of a beta particle from $^{32}_{15}P$.

Gamma Emission

In Section 6-1 we saw that electrons that have been excited to higher energy levels emit energy in the form of electromagnetic radiation as they return to their ground-state energy levels. Nuclei can behave similarly. During radioactive emission a nucleus may gain energy and become excited; the nucleus can dissipate its excess energy by emitting gamma radiation, $^{0}_{0}\gamma$. Although gamma emission is a common form of radioactivity, equations for gamma emission will not be important in our discussion of nuclear reactions.

Electron Capture

In every atom the nucleus is surrounded by one or more electrons in various energy levels. Some atoms undergo a reaction called **electron capture,** in which the nucleus absorbs or captures an electron, $^{0}_{-1}e$, from the lowest energy level. This reaction is the reverse of the emission of an electron from a nucleus (beta emission). The net result of electron capture by a nucleus is the conversion of a proton to a neutron:

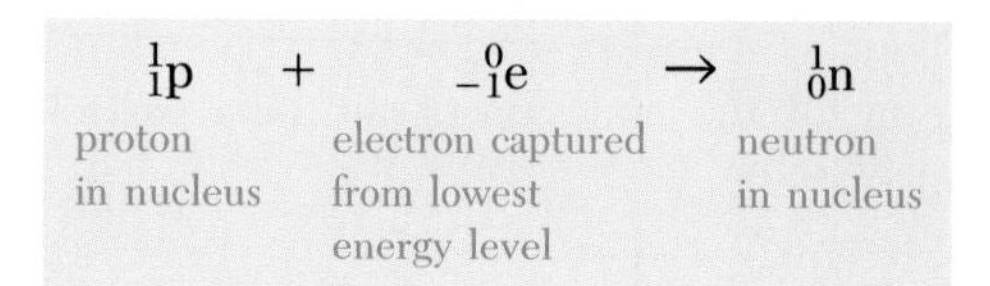

By this conversion the nucleus undergoes a net loss of one proton and a net gain of one neutron; its mass number remains the same and its charge decreases by one unit. This is one example:

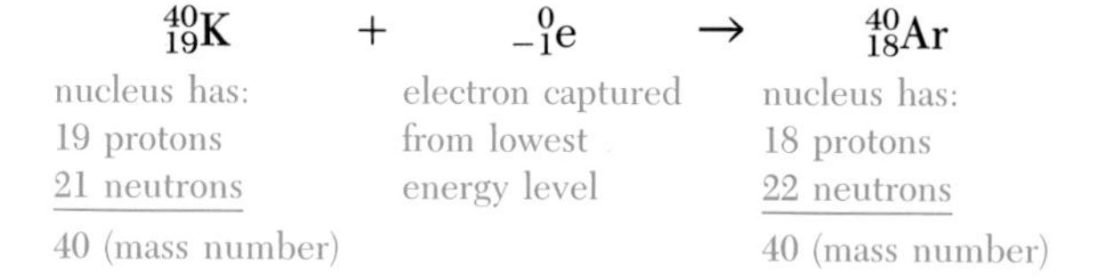

Exercise 18.3

Write the equation for electron capture by $^{25}_{13}Al$.

Positron Emission

A positron, $^{0}_{1}e$, emitted from a nucleus can be thought of as being produced by the conversion of a proton into a neutron:

$^{1}_{1}p$	$\rightarrow$	$^{1}_{0}n$	$+$	$^{0}_{1}e$
proton in nucleus		neutron in nucleus		positron emitted from nucleus

By this conversion the nucleus undergoes a net loss of one proton and a net gain of one neutron—the same net change caused by electron capture. The mass number of the nucleus remains the same and its charge decreases by one unit. This is one example:

$^{95}_{43}Tc$	$\rightarrow$	$^{95}_{42}Mo$	$+$	$^{0}_{1}e$
nucleus has:		nucleus has:		positron emitted
43 protons		42 protons		from nucleus
52 neutrons		53 neutrons		
95 (mass number)		95 (mass number)		

Exercise 18.4

Write the equation for positron emission by $^{25}_{13}Al$.

Because nuclei are surrounded by electrons, emitted positrons don't last long: When a positron collides with an electron, the two particles react to form gamma photons:

$$^{0}_{1}e + {}^{0}_{-1}e \rightarrow 2\ {}^{0}_{0}\gamma$$

18-2 You can predict whether a nucleus will undergo spontaneous emission

Some atomic nuclei are unstable—that is, they are spontaneously radioactive—and others are not. For example, the nucleus $^{210}_{84}Po$ is unstable; it spontaneously emits an alpha particle:

$$^{210}_{84}Po \rightarrow {}^{206}_{82}Pb + {}^{4}_{2}He$$

The product nucleus, $^{206}_{82}Pb$, is stable and does not undergo spontaneous emission.

To predict whether a nucleus will be stable or unstable, ask these questions:

Does the nucleus contain more than 83 protons? Any nucleus with more than 83 protons will be unstable; this rule takes precedence over other rules for nuclear stability. As shown in the preceding equation, the $^{210}_{84}Po$ nucleus is unstable.

Does the nucleus contain an odd number of protons or neutrons? An odd number of protons or neutrons increases the probability of nuclear instability. This rule isn't always true, but it's useful as a general guide, especially in deciding which of two nuclei is more likely to be unstable. For example, $^{40}_{19}K$ (19 protons and 21 neutrons) is unstable, but $^{39}_{19}K$ (19 protons and 20 neutrons) is stable.

Does the nucleus contain a *magic number* of protons or neutrons? These *magic numbers* of protons or neutrons tend to confer nuclear stability: 2,

20, 28, 50, 82; for neutrons 126 is also a magic number. In the examples already mentioned, $^{39}_{19}K$ (19 protons and 20 neutrons) is stable, and $^{40}_{19}K$ (19 protons and 21 neutrons) is unstable.

Exercise 18.5

Predict whether each of these nuclei will be stable, and explain your prediction: (a) $^{230}_{90}Th$ (b) $^{90}_{40}Zr$

18-3 From a graph of stable and unstable nuclei you can predict equations for spontaneous nuclear reactions

In a spontaneous nuclear reaction an unstable nucleus changes into a more stable one, and this change can be described on the graph shown in Figure 18.1. On this graph the horizontal and vertical axes designate, respectively, the number of protons and neutrons in nuclei. Each location on the graph corresponds to a nucleus with a certain number of protons and neutrons; for example, the location on the graph at which the two green lines meet corresponds to the nucleus $^{84}_{38}Sr$ (38 protons and 46 neutrons).

Exercise 18.6

On the graph in Figure 18.1, find the location that corresponds to (a) $^{40}_{19}K$ (b) $^{165}_{70}Yb$

The red dots in Figure 18.1 correspond to unstable nuclei, and the black dots correspond to stable nuclei. The band formed by the black dots is called the **band of stability.**

Exercise 18.7

Use Figure 18.1 to decide whether each of these nuclei is stable or unstable: (a) $^{40}_{20}Ca$ (b) $^{94}_{36}Kr$

In a spontaneous nuclear reaction an unstable nucleus becomes a more stable one. In terms of the graph, this means that the product (more stable) nucleus will lie closer to the band of stability than the reactant (unstable) nucleus; in some cases the product nucleus will be on the band of stability. For example, the unstable nucleus $^{47}_{20}Ca$, which is to the left of the band of stability, undergoes spontaneous beta emission to produce the more stable nucleus $^{47}_{21}Sc$, which is closer to the band of stability:

$$^{47}_{20}Ca \rightarrow ^{47}_{21}Sc + ^{0}_{-1}e$$

Exercise 18.8

Use Figure 18.1 to show that, in the following reaction, the product nucleus is closer to the band of stability than the reactant nucleus:

$$^{40}_{19}K + ^{0}_{-1}e \rightarrow ^{40}_{18}Ar$$

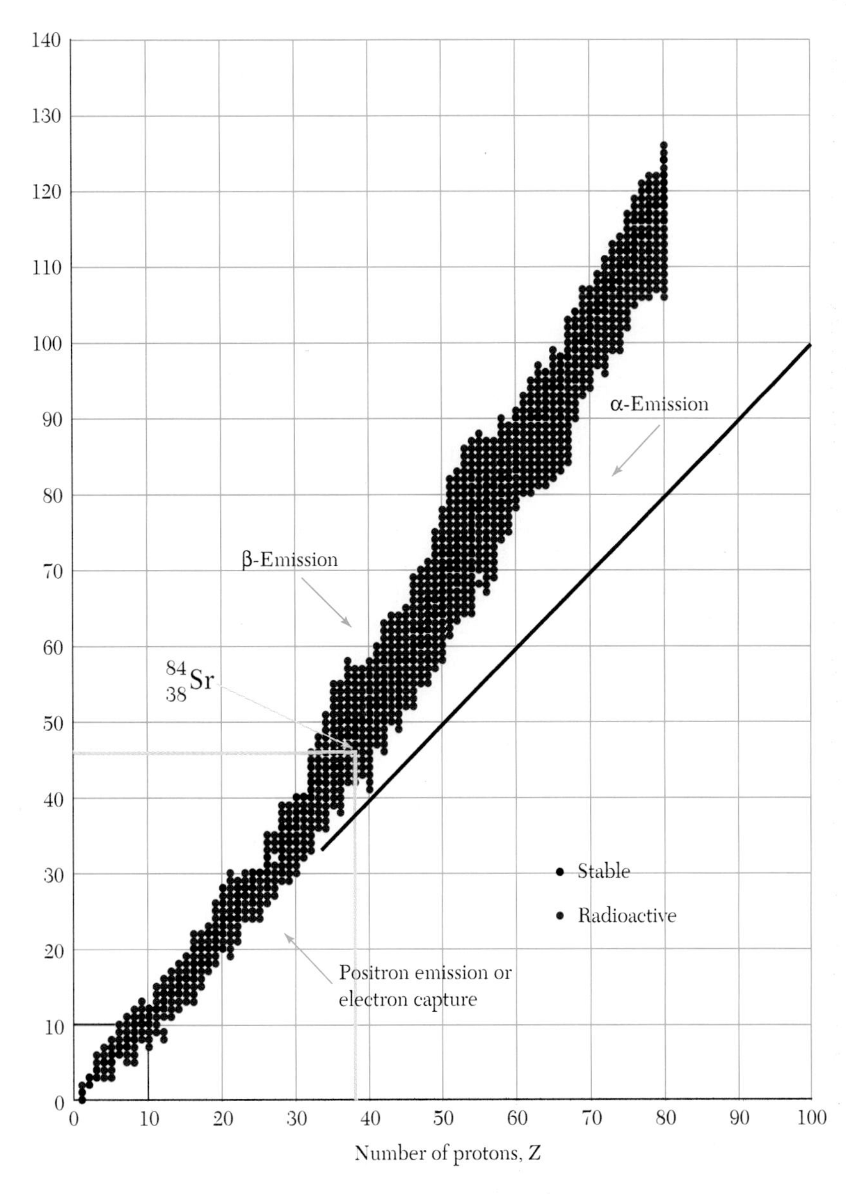

Figure 18.1
Stable and unstable nuclei. Each dot corresponds to a nucleus with a certain number of protons and neutrons. For example, the dot at the convergence of the green lines corresponds to the nucleus $^{84}_{38}$SR (38 protons and 46 neutrons). Black dots correspond to stable nuclei and red dots correspond to unstable nuclei. The black dots form a *band of stability*. Depending on its position with respect to the band of stability, an unstable nucleus can be predicted to react by alpha emission, beta emission, positron emission, or electron capture, as shown on the graph.

Figure 18.1 can be used to predict equations for spontaneous nuclear reactions. An unstable nucleus to the left of the band of stability will become more stable by increasing its number of protons (moving to the right on the graph) by beta emission. Similarly, an unstable nucleus to the upper right of the band will become more stable (moving left and downward) by alpha emission, and an unstable nucleus to the lower right of the band will become more stable (moving left) by positron emission or electron capture.◆

◆Predictions from these rules aren't always correct, but they're accurate often enough to be useful as general guides.

Example 18.1

Use Figure 18.1 to predict the equation for the spontaneous nuclear reaction, if any, of each of these nuclei: (a) $^{141}_{64}$Gd (b) $^{41}_{21}$Sc (c) $^{120}_{50}$Sn

Solution

(a) $^{141}_{64}Gd \rightarrow ^{137}_{62}Sm + ^{4}_{2}He$

(b) $^{41}_{21}Sc \rightarrow ^{41}_{20}Ca + ^{0}_{1}e$ *or* $^{41}_{21}Sc + ^{0}_{-1}e \rightarrow ^{41}_{20}Ca$

(c) This nucleus is stable, so no reaction occurs.

Exercise 18.9

Use Figure 18.1 to predict the equation for the spontaneous reaction, if any, of each of these nuclei: (a) $^{185}_{76}Os$ (b) $^{40}_{20}Ca$ (c) $^{95}_{37}Rb$

If the nucleus produced in a nuclear reaction is itself unstable—that is, if it is not in the band of stability—it will also undergo a spontaneous nuclear reaction, and this process will continue until a stable nucleus is produced. Each reaction can be predicted from the trends described in Figure 18.1.

Example 18.2

Use Figure 18.1 to predict the equation for the spontaneous reaction, if any, of each of the following nuclei. If the product nucleus is unstable, predict the equation for its reaction. (a) $^{128}_{52}Te$ (b) $^{59}_{29}Cu$

Solution

(a) $^{128}_{52}Te \rightarrow ^{128}_{53}I + ^{0}_{-1}e$ The product nucleus is unstable.

$^{128}_{53}I \rightarrow ^{128}_{54}Xe + ^{0}_{-1}e$

(b) $^{59}_{29}Cu \rightarrow ^{59}_{28}Ni + ^{0}_{1}e$ *or* $^{59}_{29}Cu + ^{0}_{-1}e \rightarrow ^{59}_{28}Ni$
The product nucleus is stable.

Exercise 18.10

Use Figure 18.1 to predict the equation for the spontaneous reaction, if any, of each of the following nuclei. If the product nucleus is unstable, predict the equation for its reaction. (a) $^{186}_{75}Re$ (b) $^{46}_{22}Ti$

18-4 By bombarding nuclei with subatomic particles, scientists have created new elements

The elements that occur naturally on earth are those with atomic numbers 1 through 92, but scientists have created elements with atomic numbers greater than 92 by bombarding atomic nuclei with subatomic particles. The elements that have been created by nuclear bombardment are called **transuranium elements** because they lie beyond uranium, element number 92, on the Periodic Table.

Scientists created the first transuranium element in 1940 at the University of California at Berkeley, by bombarding the nucleus $^{238}_{92}U$ with neutrons. The product nucleus, $^{239}_{92}U$, decays spontaneously by beta emission:

$$^{238}_{92}U + ^{1}_{0}n \rightarrow ^{239}_{92}U$$

$$^{239}_{92}U \rightarrow ^{239}_{93}Np + ^{0}_{-1}e$$

The new element was named *neptunium.*

Table 18.2 summarizes information about the transuranium elements.

Table 18.2 The Transuranium Elements

Atomic Number	Symbol	Name	Named After	Discovery
93	Np	Neptunium	the planet Neptune	U.S.A., 1940
94	Pu	Plutonium	the planet Pluto	U.S.A., 1940
95	Am	Americium	America	U.S.A., 1944
96	Cm	Curium	Marie Curie (1867–1935), Polish physicist and chemist in France,* and Pierre Curie (1859–1906), French physicist and chemist	U.S.A., 1944
97	Bk	Berkelium	Berkeley, California	U.S.A., 1950
98	Cf	Californium	California	U.S.A., 1950
99	Es	Einsteinium	Alfred Einstein (1879–1955), German physicist; American citizen from 1940	U.S.A., 1952
100	Fm	Fermium	Enrico Fermi (1901–1954), Italian physicist; in the U.S.A. after 1939	U.S.A., 1952
101	Md	Mendelevium	Dmitri Ivanovich Mendeleev (1834–1907), Russian chemist	U.S.A., 1955
102	No	Nobelium	Alfred Nobel (1833–1896), Swedish industrialist and philanthropist	U.S.A., 1958
103	Lr	Lawrencium	Ernest O. Lawrence (1901–1980), American physicist	U.S.A., 1961
104	Unq	Unnilquadium		U.S.A., 1969
105	Unp	Unnilpentium		U.S.A., 1970
106	Unh	Unnilhexium	See the Chemistry Insight box below.†	U.S.A. and U.S.S.R., 1974
107	Uns	Unnilseptium		U.S.S.R. and West Germany, 1981
108	Uno	Unniloctium		West Germany, 1984
109	Une	Unnilennium		West Germany, 1982

*Marie Curie won the Nobel Prize in physics with her husband in 1903 and the Nobel Prize in chemistry in 1911. She is one of only three persons who have won two Nobel Prizes in two different fields.

† Chemistry Insight

Newly discovered elements are given temporary names and three-letter symbols until their final names are agreed on. The temporary name for a new element is constructed from its atomic number: Each digit is assigned a root name, and these roots, plus the suffix *-ium*, are combined to form the name for the element. The roots are *nil* for 0, *un* for 1, *bi* for 2, *tri* for 3, *quad* for 4, *pent* for 5, *hex* for 6, *sept* for 7, *oct* for 8, and *enn* for 9.

Many of the transuranium elements have been produced only in small quantities. The West German scientists who prepared element 109 achieved the limit in producing minute quantities of an element: They reported the new element based on the detection of one atom.

A few isotopes of transuranium elements have important military or commercial uses. The isotope $^{239}_{94}Pu$ is used in atomic bombs and nuclear reactors, and $^{238}_{94}Pu$ is used as a power source in cardiac pacemakers, navigation buoys, and space satellites.◆ The isotope $^{241}_{95}Am$ may be in your home now, as the source of signals in a smoke detector.◆

◆The application of nuclear reactions in atomic bombs and nuclear reactors is described in Section 18-9.

◆The function of $^{241}_{95}Am$ in smoke detectors is described in Section 18-5.

Exercise 18.11

Are the transuranium elements metals, nonmetals, or metalloids? Are they representative or transition elements?

Exercise 18.12

Write the abbreviated electron configuration you predict for element 109.

Exercise 18.13

From the principles described in Section 18-2, would you predict that there are any stable isotopes of transuranium elements? Explain.

Exercise 18.14

Write equations for these reactions: (a) $^{238}_{94}Pu$ undergoing alpha emission (b) $^{239}_{94}Pu$ reacting with an alpha particle to produce a neutron and another transuranium element.

18-5 Alpha, beta, and gamma emissions can be measured because they cause atoms to lose electrons or to emit light

When a radioactive nucleus emits an alpha, beta, or gamma particle, the emitted particle usually collides in a fraction of a second with another atom; as a common result of this collision, electrons in the atom that's been struck are bumped to higher energy levels or are knocked completely off of the atom. Figure 18.2 shows these effects of radioactive emissions.

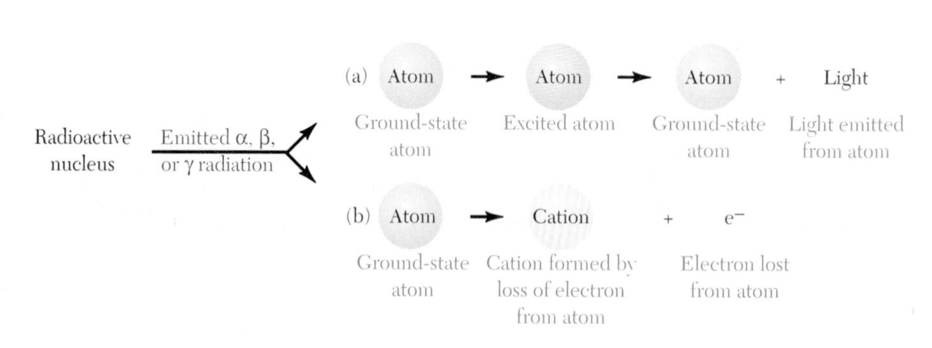

Figure 18.2
Collisions of radioactive emissions with atoms. When an alpha, beta, or gamma emission collides with an atom, it can bump electrons in the atom to higher energy levels or knock electrons off the atom.
(a) Collision of emitted radiation with a ground-state atom moves electrons to higher energy levels creating an excited atom; the electrons then spontaneously return to lower energy levels, losing energy in the form of light.
(b) Collision of emitted radiation with a ground-state atom knocks an electron completely off of the atom, creating a cation and a free electron.

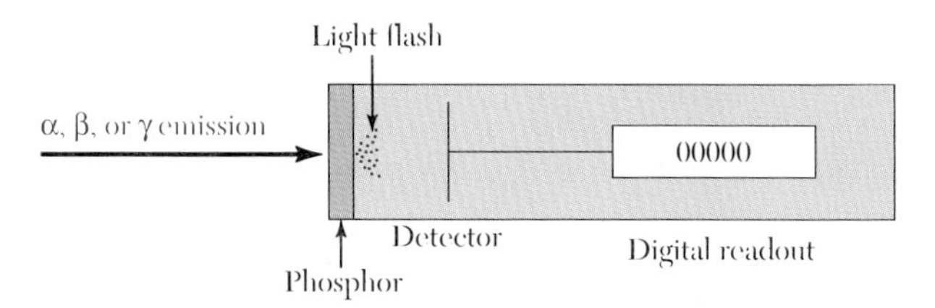

Figure 18.3
Features of a scintillation counter. When alpha, beta, or gamma emissions strike the phosphor, its atoms emit flashes of light by the process described in part (a) of Figure 18.2. These light flashes are counted by the detector, and the count is displayed on the digital readout.

As shown in Figure 18.2, collisions of radioactive emissions with atoms produce either light or cations and electrons.◆ These products of the collisions can be measured, and instruments that measure them are used to determine quantities of emitted radiation. The instruments that measure radioactive emissions in this way are classified as *scintillation counters* or *ionization detectors*.

◆Because alpha, beta, and gamma emissions can cause the formation of ions, they're sometimes referred to collectively as *ionizing radiation*.

Scintillation counters measure the light emitted by the process shown in part (a) of Figure 18.2.◆ As shown in Figure 18.3, a scintillation counter counts the light flashes produced by collisions of radioactive emissions with atoms and displays that count as a digital readout. The displayed number is a measure of the emitted radiation.

◆*Scintillation* means "emission of light in flashes."

Ionization detectors measure numbers of electrons produced by the ionization process shown in part (b) of Figure 18.2. The *Geiger counter*, shown diagrammatically in Figure 18.4, is a common form of ionization detector.◆ A Geiger counter measures the electrons produced by a radioactive emission and displays the measurement as a digital readout; the displayed number is a measure of the emitted radiation.

◆The Geiger counter is named for its inventor, Hans Geiger (1882–1947), a German physicist.

Scintillation counters and ionization detectors measure the quantities of alpha, beta, or gamma emissions that a radioactive substance produces in a given period of time; that is, these instruments measure the rate at which the unstable nuclei in a radioactive substance disintegrate. The unit used to describe rates of radioactive emission is the **curie (Ci)**.◆ The curie is defined as a rate of radioactivity in which 3.700×10^{10} unstable nuclei disintegrate in one second:

◆The curie is named for Marie Curie; see the footnote in Table 18.2.

$$1 \text{ Ci} = 3.700 \times 10^{10} \text{ disintegrations per second}$$

or

$$1 \text{ Ci} = \frac{3.700 \times 10^{10} \text{ disintegrations}}{1 \text{ second}}$$

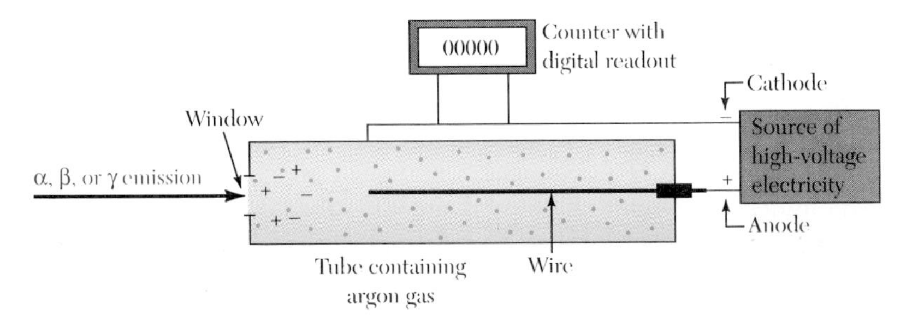

Figure 18.4
Features of a Geiger counter. An alpha, beta, or gamma emission that enters the tube collides with argon atoms, converting them into cations and free electrons. The cations are attracted to the walls of the tube, which are negatively charged, and the electrons are attracted to the positively charged wire down the center of the tube. Electrons that strike the wire create an electric current that's displayed on the counter as a digital readout.

Example 18.3

The rate of radioactivity in a sample of technetium was 2.3×10^{-2} Ci. Express this rate in disintegrations per second.

Solution

2.3×10^{-2} Ci = ? disintegrations per second

$$\text{Ci} \rightarrow \text{disintegrations per second}$$

$$\left(\frac{2.3 \times 10^{-2}\ \cancel{\text{Ci}}}{}\right)\left(\frac{3.700 \times 10^{10}\ \text{disintegrations per second}}{1\ \cancel{\text{Ci}}}\right)$$

$$= 8.51 \times 10^{8}\ \text{disintegrations per second}$$

Exercise 18.15

A Geiger counter measured 6.43×10^{5} disintegrations per second in a radioactive sample. Express this rate in curies.

The ability of alpha, beta, and gamma emissions to cause scintillation or ionization has important commercial uses. For example, the numbers and hands of a clock can be made to glow in the dark by painting them with a mixture of a phosphor and a radioactive substance: The scintillations of the phosphor appear as a faint, continuous light. Smoke detectors are an ingenious application of ionizing radiation. Inside a typical smoke detector, a tiny amount of $^{241}_{95}Am$ produces alpha and gamma emissions that ionize molecules of nitrogen and oxygen in the air, and the electrons produced in these ionizations are measured as an electric current. Smoke particles in the air interfere with the ionization process, reduce the current, and set off an alarm.

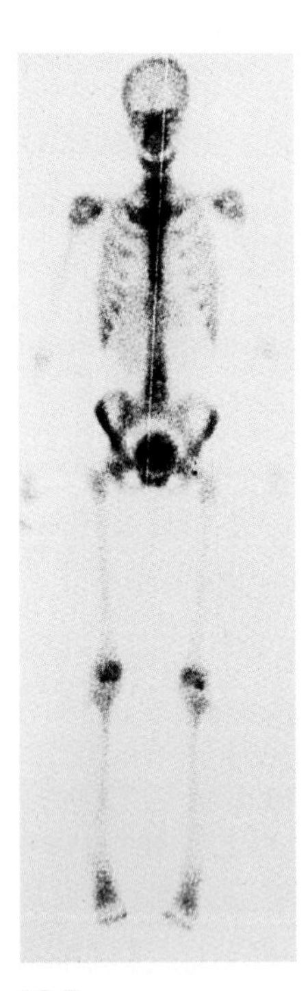

Figure 18.5
Skeletal imaging with radioactive $^{99}_{43}Tc$. The isotope $^{99}_{43}Tc$ concentrates in bones and emits gamma radiation. This image was made by injecting the subject with a solution of $^{99}_{43}Tc$ and measuring the gamma radiation from his body. *(Courtesy of DuPont Radiopharmaceuticals)*

Exercise 18.16

Write an equation for alpha emission by $^{241}_{95}Am$.

Exercise 18.17

The radioactivity in a smoke detector was rated at 1.0 microcurie. How many disintegrations per second were occurring in the detector?

18-6 Radioactive nuclei can be used to follow atoms in chemical reactions and to create images of tissues and organs in living human beings

The movements of radioactive nuclei can be followed because the emissions from the nuclei can be detected and measured. Following radioactive nuclei has made it possible for scientists to understand the steps in chemical reactions and to see the structures of living organisms more clearly than ever before.

All of the isotopes of an element, whether or not their nuclei are radioactive, will undergo the same chemical reactions, so radioactive isotopes—called **radioactive tracers**—can be used to follow atoms in chemical reactions. For example, imagine that oxygen molecules, $\cdot\ddot{\underset{\cdot\cdot}{O}}-\ddot{\underset{\cdot\cdot}{O}}\cdot$, are prepared in which one atom in each molecule is a radioactive isotope. We can designate the radioactive atom with an asterisk: $\cdot\overset{*}{\ddot{\underset{\cdot\cdot}{O}}}—\ddot{\underset{\cdot\cdot}{O}}\cdot$. If these oxygen molecules react with hydrogen molecules, 50% of the water molecules produced will be radioactive:

$$2\ H_2 + O_2 \rightarrow 2\ H_2O$$

$$2\ H—H + \cdot\overset{*}{\ddot{\underset{\cdot\cdot}{O}}}—\ddot{\underset{\cdot\cdot}{O}}\cdot \rightarrow H—\overset{*}{\ddot{\underset{\cdot\cdot}{O}}}—H + H—\ddot{\underset{\cdot\cdot}{O}}—H$$

If the water produced in this reaction is used in other reactions, the radioactive oxygen atoms can be followed in the compounds that are formed.

Exercise 18.18

Imagine that the water produced by the reaction shown in the preceding equation is used in this reaction with sulfur trioxide to produce sulfuric acid:

$$H_2O + SO_3 \rightarrow H_2SO_4$$

(a) What percentage of the sulfuric acid molecules will be radioactive? (b) In each molecule of sulfuric acid that's radioactive, how many of the oxygen atoms will be radioactive?

Radioactive tracers are especially useful in following the long and complex sequences of chemical reactions that occur in living organisms. For example, foods whose molecules contain radioactive isotopes can be used to follow the chemical reactions by which the foods are digested. Many important studies of the chemistry of living organisms have been carried out with radioactive tracers.

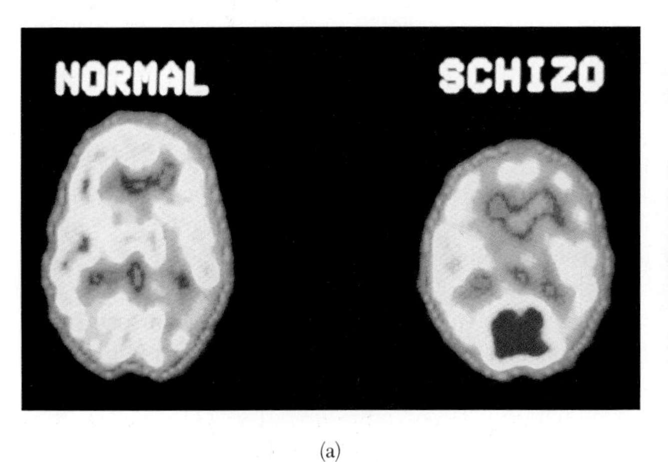

(a)

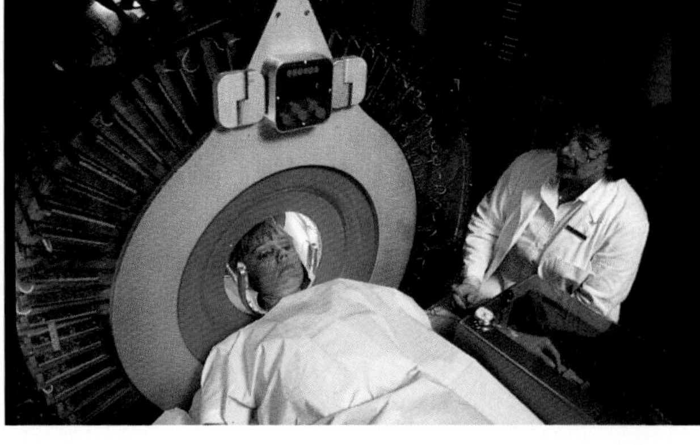

(b)

Figure 18.6
Brain imaging with radioactive $^{11}_{6}C$. The isotope $^{11}_{6}C$ can be used to prepare a radioactive form of the compound glucose, $C_6H_{12}O_6$; when glucose is injected into the bloodstream, it circulates throughout the body and is absorbed into many tissues. (a) These images were made by injecting normal and schizophrenic subjects with radioactive glucose and measuring gamma emissions from their brains. The images are artificially colored to identify differences in glucose concentration. (b) Many scintillation counters, arranged like spokes of a wheel, measure gamma radiation from the brain. The $^{11}_{6}C$ isotope emits positrons, which collide with electrons to produce gamma radiation, as described in Section 18-1:

$$^{11}_{6}C \rightarrow {}^{11}_{5}B + {}^{0}_{1}e$$

$$^{0}_{1}e + {}^{0}_{-1}e \rightarrow {}^{0}_{0}\gamma$$

The process of creating a brain image with positron-emitting isotopes is called Positron Emission Tomography (PET); the image shows a section or slice of the brain, and *tomo-* means *"slice."* The image is often referred to as a PET scan. *(NIH/Science Source/Photo Researchers, Inc.)*

Medical workers use radioactive tracers to create images of tissues and organs in living human beings; two examples are shown in Figures 18.5 and 18.6. The image in Figure 18.5 was created by detecting gamma emissions from $^{99}_{43}Tc$ injected into the subject's body; as the image shows, this isotope concentrates in the skeleton.◆ The images in Part (a) of Figure 18.6 were created by a similar process. Two subjects were injected with a solution of glucose (blood sugar, $C_6H_{12}O_6$) containing a radioactive carbon isotope, $^{11}_{6}C$, and the resulting gamma emissions from their brains were measured. The differences in the brain images correspond to differences in the normal and schizophrenic behavior of the subjects.

◆As described in Section 18-7, some radioactive emissions harm organisms. The tracers used to follow chemical reactions in organisms or in imaging studies of tissues and organs are selected to avoid harmful radiation.

18-7 Emissions from radioactive nuclei can harm human beings; the harm can be prevented by blocking the emissions with suitable shields

Some emissions from radioactive nuclei can destroy the structures of molecules by breaking their covalent bonds. These emissions are dangerous to organisms, including human beings, because the destruction of biological molecules damages tissues and disrupts essential biological processes.◆

◆Some molecules in reproductive cells contain the genetic information that will determine the characteristics of offspring; damage to these molecules can cause infertility or produce mutations.

The degree of potential danger to human beings from a radioactive source depends on the kinds and quantities of its emissions. In general, larger quantities of emissions and larger emitted particles are more damaging: Alpha emissions are more harmful than neutrons, and neutrons are more harmful than beta or gamma emissions.

The amount of biological damage that radiation can cause is expressed in units called **rems;** a single dose of 500 rems is usually fatal to a human being.◆ Because rems measure biological damage, they evaluate both quantity and kind of radiation: The same quantity of radiation will be evaluated at 50 rems for beta and gamma emissions, at 250 rems for neutrons, and at 500 rems for alpha emissions.

◆The name *rem* is the abbreviation for *roentgen equivalent in man;* the *roentgen* is a unit of radiation named for Wilhelm Konrad Roentgen (1845–1923, Nobel Prize 1901), the German physicist who discovered X-rays.

Radiation can be prevented from causing biological damage by blocking it with suitable shields. Alpha particles, the most damaging form of radiation, are also the most easily blocked; in human beings the skin is an effective shield, preventing alpha particles from reaching vulnerable internal organs.◆ The use of water as a radiation shield is shown in Figure 18.7.

◆Sources of alpha radiation that are not significantly dangerous as long as they're outside the body, because of shielding by the skin, can become very dangerous if they're swallowed or inhaled.

Emissions from radioactive nuclei are used to treat cancer in **radiation therapy.** In general, radiation is more damaging to cells that are growing more rapidly; cancer cells cause harm by growing too rapidly, so they're especially vulnerable to attack with radiation.

Radiation therapy must be carried out in such a way that cancer cells receive maximum radiation while normal cells receive minimum radiation. One way to focus radiation on cancer cells is shown in Figure 18.8. The radiation source is shielded so that only a narrow beam of radiation is directed onto the cancerous tissue. Another way to focus radiation on cancer cells is to introduce into the patient's body, orally or by injection, a radioactive isotope that will

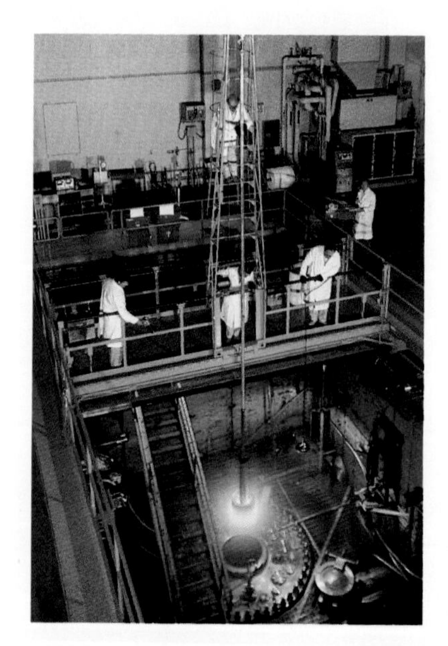

Figure 18.7
Water as a radiation shield. In this facility, at Oak Ridge National Laboratory in Tennessee, scientists carry out nuclear reactions to produce transuranium elements. Scintillations from emitted radiation cause the glow in the bottom of the pool; the water shields the workers from radiation. *(Oak Ridge National Laboratory.)*

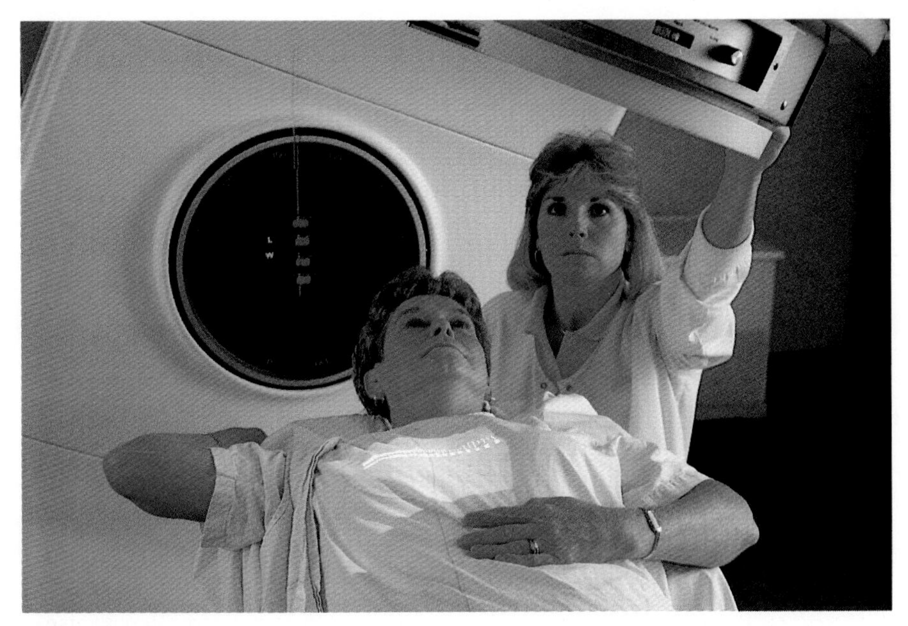

Figure 18.8
Radiation therapy. A technician adjusts a machine that will focus a beam of radiation on cancer cells in a patient's breast. *(©Joseph Nettis/Science Source/Photo Researchers, Inc.)*

concentrate in the cancerous tissue. In the treatment of thyroid cancer, for example, the isotope $^{131}_{53}I$, which emits beta and gamma radiation, is administered orally or intravenously as a solution of sodium iodide, and the radioactive iodide ions concentrate in the thyroid gland.

Exercise 18.19

Write an equation for beta emission by $^{131}_{53}I$.

18-8 Samples of some radioactive isotopes can be used as clocks to measure very long periods of time

Samples of radioactive isotopes are the best sources of information we have to establish the ages of objects that are hundreds, thousands, or millions of years old. In this Section we'll look briefly at the process of using information from radioactive samples to calculate the ages of very old objects.

A sample of a radioactive isotope has two characteristics that make it an ideal clock for measuring long periods of time: Its mechanism for keeping time—radioactive decay—can't be altered by anything that's done to the sample, and its rate of decay decreases continuously in a completely regular way.

A sample of a radioactive isotope is an unalterable clock because its rate of decay—its number of disintegrations per second—is a nuclear process that's unaffected by the physical or chemical condition of the sample. That is, the rate of decay is unaffected by the physical state of the sample (whether it's a solid, a liquid, or a gas) or by the form of chemical combination in which the radioactive isotopes occur. For this reason we can assume that the rate of decay of a radioactive sample has not been altered by any conditions the sample has been subjected to, even over millions of years.

A sample of a radioactive isotope is an unalterable clock, but it does run down: The rate at which any radioactive sample decays always steadily decreases over time. This decrease in the rate of decay, usually expressed as the *half-life* of the sample, is the basis for its use in measuring time.

To explain the concept of half-life, we'll use as our example the isotope $^{131}_{53}I$.◆ Imagine that we have a 1.000-g sample of pure $^{131}_{53}I$; every atom in the sample is $^{131}_{53}I$. We'll assume that this isotope decays by beta emission:

◆This isotope is used in treating thyroid cancer, as described in Section 18-7.

$$^{131}_{53}I \rightarrow {}^{131}_{54}Xe + {}^{0}_{-1}e$$

The half-life of $^{131}_{53}I$ is 8.07 days, and this means that after 8.07 days half of our 1.000-g sample will have decayed to $^{131}_{54}Xe$; at the end of this time we'll have 0.500 g of $^{131}_{53}I$ and 0.500 g of $^{131}_{54}Xe$. After another 8.07 days have passed, the quantity of $^{131}_{53}I$ in our sample will again be reduced by half; we'll then have 0.250 g of $^{131}_{53}I$ and 0.750 g of $^{131}_{54}Xe$. After each successive interval of 8.07 days, the mass of $^{131}_{53}I$ in our sample will again be reduced by half, as shown in Figure 18.9. The **half-life** of a radioactive isotope is the time in which half of a sample of the isotope will undergo decay.

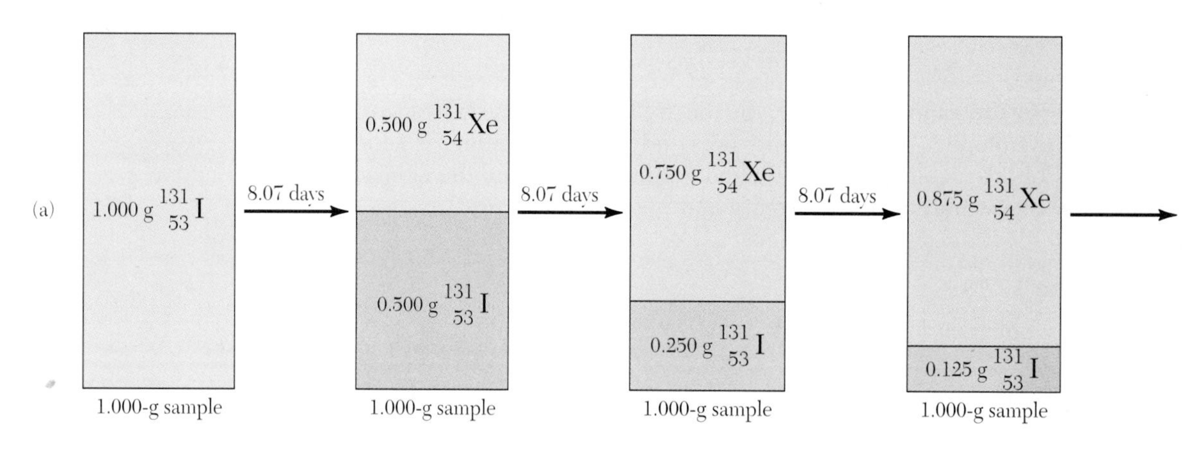

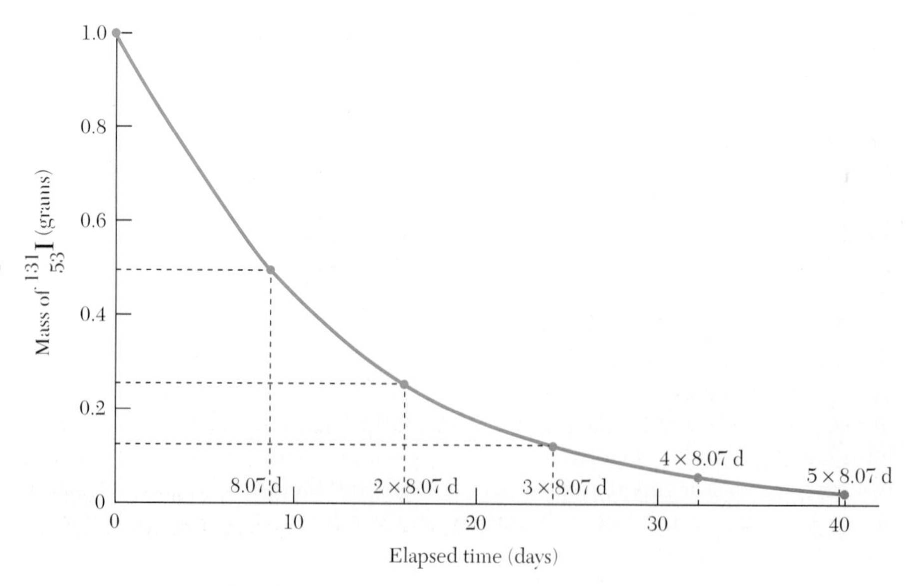

Figure 18.9
The isotope $^{131}_{53}I$ has a half-life of 8.07 days. The isotope decays by beta emission:

$$^{131}_{53}I \rightarrow {}^{131}_{54}Xe + {}^{0}_{-1}e$$

(a) In each successive period of 8.07 days the mass of $^{131}_{53}I$ in the sample decreases by half. (b) The graph shows the continuous decrease of the mass of $^{131}_{53}I$ with time.

Exercise 18.20

The gold isotope $^{198}_{79}Au$, used in the treatment of abdominal cancer, undergoes beta decay with a half-life of 2.69 days. (a) Write an equation for the beta decay of this isotope. (b) Imagine that you have a 3.000-g sample of pure $^{198}_{79}Au$. Draw and label a diagram like the one shown in Part (a) of Figure 18.9 to show the masses of $^{198}_{79}Au$ and $^{198}_{80}Hg$ in the sample after each of three successive intervals of 2.69 days each.

As the amount of radioactive isotope in a sample steadily decreases, the rate of radioactive emission from the sample also steadily decreases, because there are fewer and fewer nuclei to undergo decay. This steady decrease in the rate of radioactive decay is used to calculate the ages of radioactive samples. With this equation you can calculate the age of a radioactive sample:◆

◆Appendix 3 describes logarithms and antilogarithms.

$$t = 3.32t_{1/2} \log \frac{rate_i}{rate_t}$$

t is the age of the sample in time units, usually days or years
$t_{1/2}$ is the half-life of the radioactive isotope in the sample in time units, usually days or years
$rate_i$ is the initial rate of decay of the sample in disintegrations per second
$rate_t$ is the current rate of decay of the sample in disintegrations per second

For any radioactive sample $t_{1/2}$ and $rate_i$ will have fixed values. When the sample first begins to decay, $rate_i = rate_t$; from that time onward $rate_t$ steadily decreases, so $\frac{rate_i}{rate_t}$ steadily increases, $\log \frac{rate_i}{rate_t}$ increases, and t increases. In other words, the steady decrease in $rate_t$, the rate of radioactive decay of the sample, becomes, in this equation, a measure of time.

Example 18.4

The initial rate of decay for a sample of $^{131}_{53}I$ was 4.57×10^{15} disintegrations per second, and its rate of decay now is 1.23×10^{9} disintegrations per second; the half-life of $^{131}_{53}I$ is 8.07 days. How old is the sample?

Solution

$$t = 3.32t_{1/2} \log \frac{rate_i}{rate_t}$$

$$t = (3.32)(8.07\ d) \log \frac{(4.57 \times 10^{15}\ \cancel{\text{disintegrations s}^{-1}})}{(1.23 \times 10^{9}\ \cancel{\text{disintegrations s}^{-1}})} = 176\ d$$

The sample is 176 days old.

Exercise 18.21

The isotope $^{58}_{27}Co$, which is used to test for anemia, decays by electron capture and positron emission with a half-life of 71.9 days. A sample of this isotope had an initial rate of decay of 1.16×10^{15} disintegrations per second, and its rate now is 7.70×10^{14} disintegrations per second. Calculate the age of the sample.

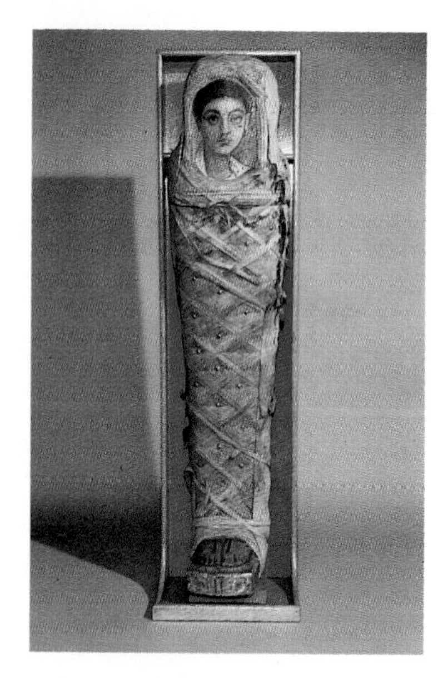

Carbon-14 dating establishes the age of this Egyptian mummy of a boy at about 1800 years. *(The Granger Collection)*

Most of the compounds that make up all plants and animals are compounds of carbon, and in every plant or animal some of the carbon atoms are the radioactive isotope $^{14}_{6}C$, a beta-emitter with a half-life of 5.73×10^3 years. This isotope is often used to establish the age of plant or animal remains, as described in the following Example and Exercise; the process is called **radiocarbon dating** or **carbon-14 dating.**

Example 18.5

The initial rate of radioactive decay of a newly formed sample containing $^{14}_{6}C$ is 918 disintegrations per second, and the rate of decay from a sample taken from the remains of a tree, in the form of ashes from a fire, is 681 disintegrations per second. Calculate the age of the remains.

Solution

$$t = 3.32 t_{1/2} \log \frac{\text{rate}_i}{\text{rate}_t}$$

$$t = (3.32)(5.73 \times 10^3 \text{ y}) \log \frac{(918 \cancel{\text{disintegrations s}^{-1}})}{(681 \cancel{\text{disintegrations s}^{-1}})} = 2.47 \times 10^3 \text{ y}$$

The remains are 2.47×10^3 years old.

Exercise 18.22

The initial rate of radioactive decay from a newly formed sample containing $^{14}_{6}C$ is 918 disintegrations per second; the half-life of $^{14}_{6}C$ is 5.73×10^3 y. A sample of wood from an ancient Egyptian tomb has a decay rate of 486 disintegrations per second. Calculate the age of the wood from the tomb.

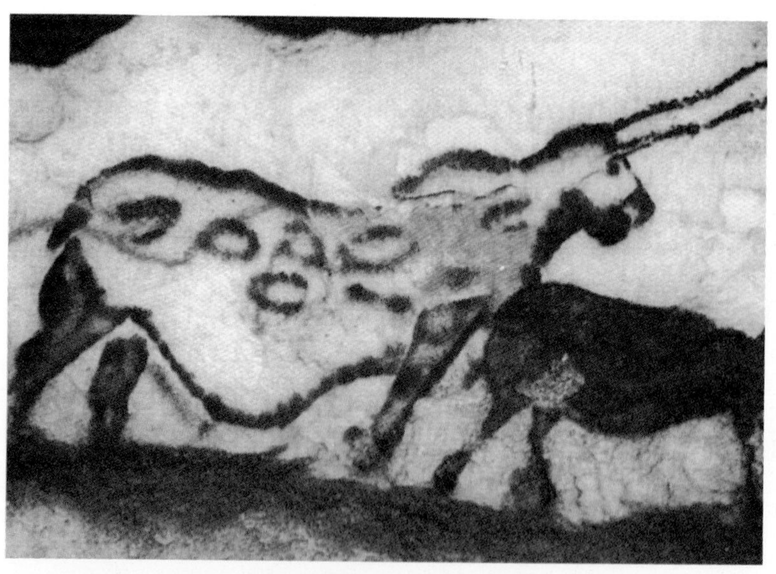

Carbon-14 dating establishes the age of this cave painting in Lascaux, France, at about 17,190 years. *(The New York Public Library)*

18-9 In nuclear fission heavy nuclei break up into lighter ones and release large amounts of energy

When some large atomic nuclei are struck with neutrons, the nuclei break into fragments and release large amounts of energy; this process, called **nuclear fission,** is the source of energy in nuclear power plants and in some nuclear weapons. One example of nuclear fission is the fragmentation of a $^{235}_{92}U$ nucleus:

$$^{235}_{92}U + ^{1}_{0}n \rightarrow ^{142}_{54}Xe + ^{90}_{38}Sr + 4\ ^{1}_{0}n$$

Exercise 18.23

The $^{235}_{92}U$ nucleus can undergo fission in more than one way. Write an equation for the reaction of a neutron with a $^{235}_{92}U$ nucleus to produce a $^{139}_{56}Ba$ nucleus, a $^{94}_{36}Kr$ nucleus, and the appropriate number of neutrons.

A nuclear fission reaction releases energy because the products of the reaction are more stable than the reactants; that is, the products of the reaction contain less energy than the reactants, and their difference in energy is the energy released during the reaction:◆

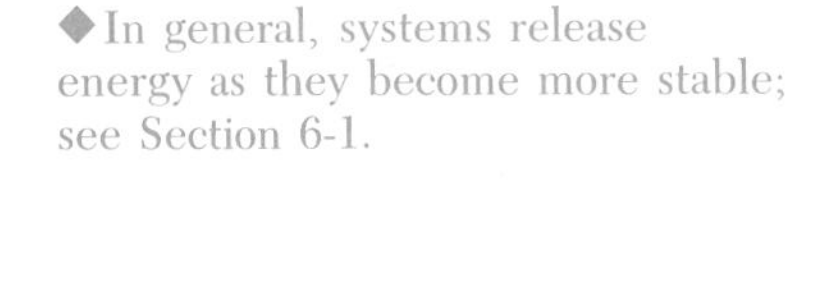
◆In general, systems release energy as they become more stable; see Section 6-1.

energy of reactants → energy of products + energy released

From this general description you can see that some nuclei are more stable than others. Figure 18.10 shows the relative stability of nuclei. As this graph shows, nuclei with mass numbers of about 50 are the most stable, while heavier and lighter nuclei are less stable.◆

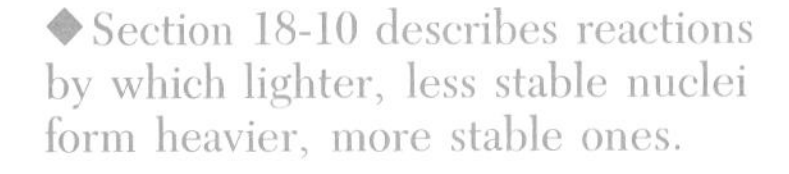
◆Section 18-10 describes reactions by which lighter, less stable nuclei form heavier, more stable ones.

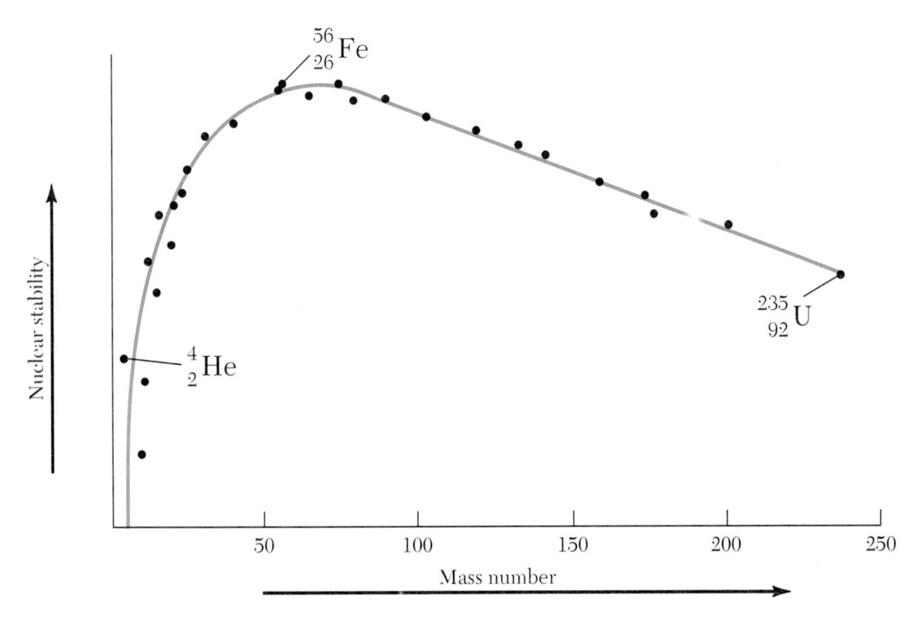

Figure 18.10
Relative stability of nuclei. Nuclei with mass numbers of about 50 are the most stable. Very light nuclei, such as $^{4}_{2}He$, and very heavy nuclei, such as $^{235}_{92}U$, are less stable.

In a fission reaction the mass of the products is slightly less than the mass of the reactants, and the mass that is lost appears as the energy released in the reaction. This astonishing transformation, the conversion of mass to energy, was predicted by Albert Einstein in 1905, more than 30 years before the first experiment in nuclear fission was carried out. Einstein's famous equation

$$E = mc^2$$

establishes a quantitative relationship among energy, mass, and the velocity of light. For our purposes the Einstein equation can be simplified to the form

$$1.00 \text{ g} = 8.97 \times 10^{10} \text{ kJ}$$

In a nuclear reaction, the disappearance of 1.00 g of matter will release 8.97×10^{10} kJ of energy.◆

◆This is enough energy to convert more than 43,000 tons of water at 100°C to steam.

In this nuclear power plant at Ventrop, West Germany, fission reactions are used to generate electrical energy. *(Michael Rosenfeld/The Image Bank)*

Exercise 18.24

In a nuclear reaction 0.274 g of mass was transformed into energy. Calculate the amount of energy produced.

Because fission reactions release enormous amounts of energy from small amounts of fuel, they have been used as the basis for nuclear weapons and nuclear power plants. These applications of nuclear fission depend on the creation of **chain reactions,** sequences of fission reactions in which each reaction causes several more. For example, when a $^{235}_{92}U$ nucleus is hit by a neutron, one of several reactions may occur:

$$^{235}_{92}U + ^{1}_{0}n \rightarrow ^{142}_{56}Ba + ^{92}_{36}Kr + 2\ ^{1}_{0}n$$

$$^{235}_{92}U + ^{1}_{0}n \rightarrow ^{89}_{37}Rb + ^{144}_{55}Cs + 3\ ^{1}_{0}n$$

$$^{235}_{92}U + ^{1}_{0}n \rightarrow ^{90}_{38}Sr + ^{143}_{54}Xe + 3\ ^{1}_{0}n$$

These reactions can be carried out so that the neutrons produced collide with other $^{235}_{92}U$ nuclei to create a chain reaction, as shown in Figure 18.11. In nuclear weapons the energy from chain reactions is released suddenly and destructively; in nuclear power plants the energy is released slowly and is used to generate electricity.

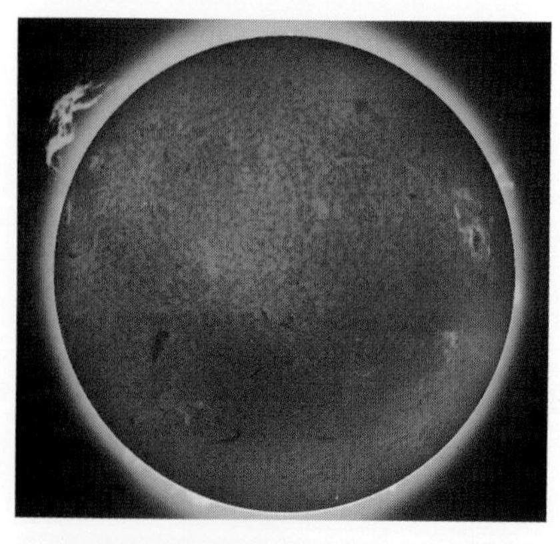

Fusion reactions are the source of the sun's energy.

18-10 In nuclear fusion light nuclei combine to form heavier ones and release large amounts of energy

Figure 18.10 on page 435 shows that nuclei with mass numbers of about 50 are the most stable, and that lighter nuclei are less stable. Because of this relative instability, some light nuclei will combine with one another to form heavier ones, with the release of large amounts of energy; the process is called **nuclear fusion.**

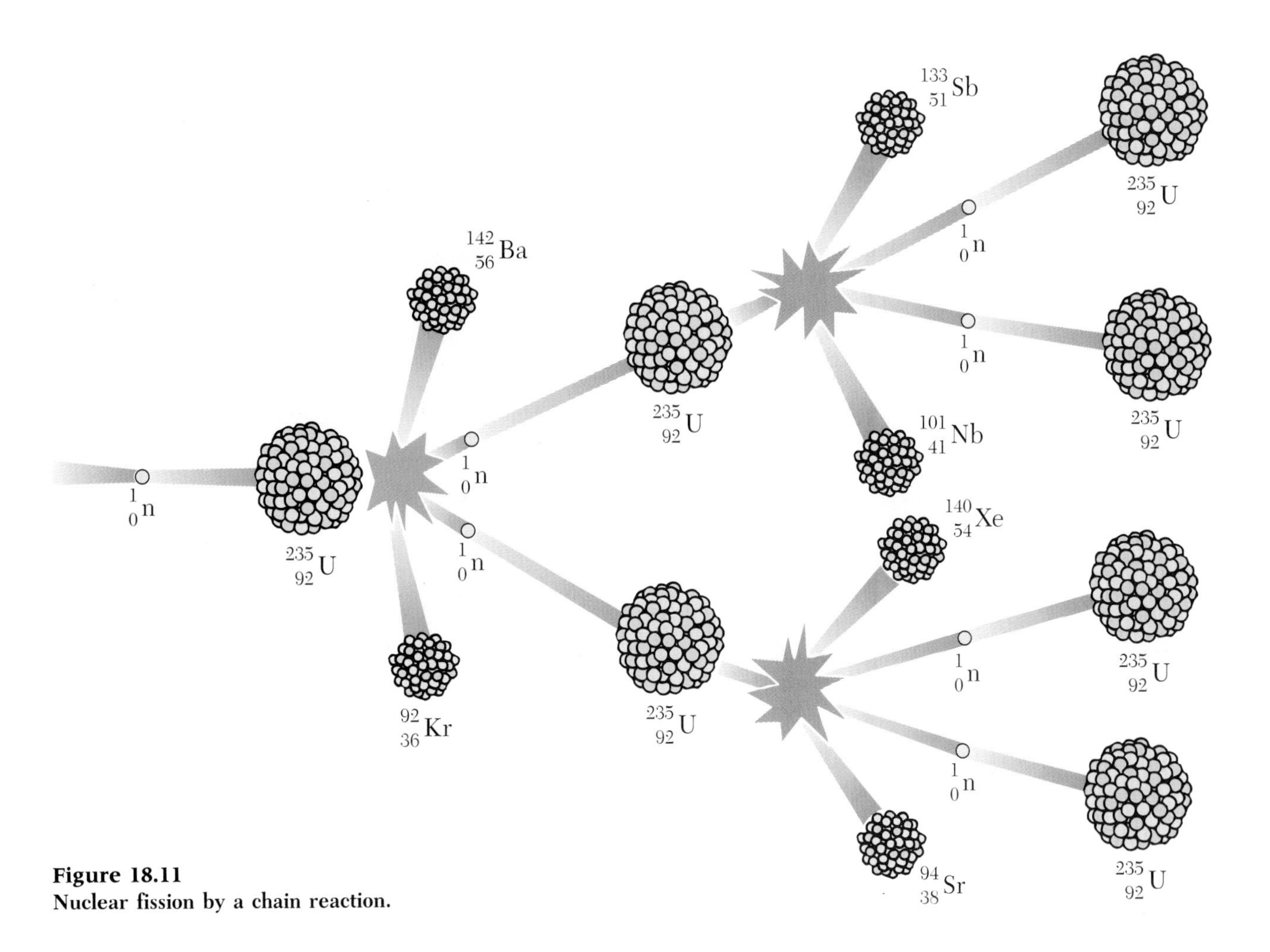

Figure 18.11
Nuclear fission by a chain reaction.

For nuclear fusion to occur, two atomic nuclei must be forced together. Two nuclei will resist being forced together because they both carry positive charges, and like charges repel. The repulsion between nuclei can be overcome by heating them to very high temperatures: at high enough temperatures the nuclei move so fast that they collide in spite of their mutual repulsion.◆

◆As the temperature of a sample of matter increases, its particles move faster; see the Inside Chemistry box at the end of Chapter 3.

In stars, temperatures are high enough to sustain nuclear fusion, and fusion reactions are believed to be the source of energy in stars, including the sun. One fusion reaction that apparently occurs in stars is the reaction of four hydrogen nuclei to form a helium nucleus, with positron emission:

$$4\ {}^{1}_{1}\text{H} \rightarrow {}^{4}_{2}\text{He} + 2\ {}^{0}_{1}\text{e}$$

This same fusion reaction has been carried out on earth, in hydrogen bombs; in a bomb, the high temperatures necessary for fusion are created for a moment by causing an explosion with nuclear fission.

The energy released in fusion, like that released in fission, comes from the transformation of matter into energy. Fusion has not yet been adapted to peaceful uses—for example, to the generation of electricity—because of the difficulty of sustaining the high temperatures it requires.

Inside Chemistry How old is the earth?

Primitive people imagine that the earth is an ageless foundation or center of the world, because it remains essentially unchanged in a human lifetime and because its existence extends back beyond human memory. One result of the creation of modern science, about 400 years ago, was the realization that the earth is not an ageless center of the world but a planet that must have been formed at a certain point in time. This realization made it possible to ask a new question: How old is the earth?

Until about forty years ago there was no reliable way to answer this question. We can now make reliable estimates of the earth's age by radioisotope dating.

One method for measuring the age of the earth depends on the radioactive decay of ${}^{238}_{92}U$, which occurs in rocks and undergoes a series of nuclear reactions to form the stable isotope ${}^{206}_{82}Pb$; the overall half-life for this series of reactions is 4.5×10^9 years. By measuring the masses of ${}^{238}_{92}U$ and ${}^{206}_{82}Pb$ in a rock sample, we can determine the age of the sample.

For example, one experimenter found the masses of these isotopes in a rock sample to be 0.621 mg of ${}^{238}_{92}U$ and 0.160 mg of ${}^{206}_{82}Pb$. To calculate the age of the earth from these data, first calculate the mass of the uranium isotope that must have decayed to produce this mass of the lead isotope:

$$0.160 \text{ mg } {}^{206}_{82}Pb = ? \text{ mg } {}^{238}_{92}U$$

$$\text{mg } {}^{206}_{82}Pb \rightarrow \text{mg } {}^{238}_{92}U$$

In this series of reactions one atom of ${}^{238}_{92}U$ produces one atom of ${}^{206}_{82}Pb$, so 1 mole of ${}^{238}_{92}U$ produces one mole of ${}^{206}_{82}Pb$, and the stoichiometric calculation is:

$$\text{mg } {}^{206}_{82}Pb \rightarrow \text{g } {}^{206}_{82}Pb \rightarrow \text{mol } {}^{206}_{82}Pb \rightarrow$$

$$\left(0.160 \cancel{\text{mg } {}^{206}_{82}Pb}\right)\left(\frac{1 \cancel{\text{g } {}^{206}_{82}Pb}}{1 \times 10^3 \cancel{\text{mg } {}^{206}_{82}Pb}}\right)\left(\frac{1 \cancel{\text{mol } {}^{206}_{82}Pb}}{206 \cancel{\text{g } {}^{206}_{82}Pb}}\right)$$

$$\text{mol } {}^{238}_{92}U \rightarrow \text{g } {}^{238}_{92}U \rightarrow \text{mg } {}^{238}_{92}U$$

$$\left(\frac{1 \cancel{\text{mol } {}^{238}_{92}U}}{1 \cancel{\text{mol } {}^{206}_{82}Pb}}\right)\left(\frac{238 \cancel{\text{g } {}^{238}_{92}U}}{1 \cancel{\text{mol } {}^{238}_{92}U}}\right)\left(\frac{1 \times 10^3 \text{ mg } {}^{238}_{92}U}{1 \cancel{\text{g } {}^{238}_{92}U}}\right)$$

$$= 0.185 \text{ mg } {}^{238}_{92}U$$

The total mass of ${}^{238}_{92}U$ that was originally in the sample, before the decay process began, was the mass of ${}^{238}_{92}U$ that's still present in the sample plus the mass of ${}^{238}_{92}U$ that's since been converted to ${}^{206}_{82}Pb$:

Total mass of ${}^{238}_{92}U$ in the sample before decay began

Earth from space. *(NASA)*

$= 0.621 \text{ mg } {}^{238}_{92}U$ (mass of ${}^{238}_{92}U$ in the sample now) $+$

$0.185 \text{ mg } {}^{238}_{92}U$ (mass of ${}^{238}_{92}U$ that's since decayed to form ${}^{206}_{82}Pb$) $= 0.806 \text{ mg } {}^{238}_{92}U$

In this sample the isotope that's undergoing decay is ${}^{238}_{92}U$, so the rate of decay at any given time will be directly proportional to the mass of ${}^{238}_{92}U$ in the sample. This means that

$$\frac{\text{rate}_i}{\text{rate}_t} = \frac{\text{mass of } {}^{238}_{92}U \text{ in the sample originally}}{\text{mass of } {}^{238}_{92}U \text{ in the sample now}} = \frac{0.806 \cancel{\text{mg}}}{0.621 \cancel{\text{mg}}}$$

From this ratio we can calculate the age of the sample:

$$t = 3.32 t_{1/2} \log \frac{\text{rate}_i}{\text{rate}_t}$$

$$t = (3.32)(4.5 \times 10^9 \text{ y}) \log \frac{0.806 \cancel{\text{mg}}}{0.621 \cancel{\text{mg}}} = 1.7 \times 10^9 \text{ y}$$

If we assume that the uranium in our rock was formed when the earth was formed, our calculation establishes the approximate age of the earth at 1.7 billion years.

Calculations based on the radioactive decay of several isotopes give the best estimate of the age of the earth as 4.6 billion years. Human beings have been on earth about three million years, about 0.075% of the time that has elapsed since the earth was formed.

Chapter Summary: Nuclear Chemistry

Section	Subject	Summary	Check When Learned
	Nuclear reaction	A reaction in which the nucleus of an atom changes by absorbing or emitting particles or energy.	☐
	Nuclear chemistry	The branch of chemistry that describes and explains nuclear reactions.	☐
18-1	Radioactivity	The emission of particles or energy by nuclei.	☐
18-1	Symbols for these forms of radioactive emission:		
	proton	${}^{1}_{1}p$	
	neutron	${}^{1}_{0}n$	
	alpha particle	${}^{4}_{2}He$ or α	
	beta particle	${}^{0}_{-1}e$ or β^-	
	gamma particle	${}^{0}_{0}\gamma$	
	positron	${}^{0}_{1}e$ or β^+	☐
18-1	Effect on the atomic number and mass number of a nucleus that emits:	Atomic Number / Mass Number	
	a proton	decreases by 1 / decreases by 1	
	an alpha particle	decreases by 2 / decreases by 4	
	a beta particle	increases by 1 / remains unchanged	
	a gamma particle	remains unchanged / remains unchanged	
	a positron	decreases by 1 / remains unchanged	☐
18-1	Equation for:		
	Conversion of a neutron to a proton by beta emission	${}^{1}_{0}n \rightarrow {}^{1}_{1}p + {}^{0}_{-1}e$	
	Conversion of a proton to a neutron by electron capture	${}^{1}_{1}p + {}^{0}_{-1}e \rightarrow {}^{1}_{0}n$	
	Conversion of a proton to a neutron by positron emission	${}^{1}_{1}p \rightarrow {}^{1}_{0}n + {}^{0}_{1}e$	
	Reaction of an electron with a positron	${}^{0}_{-1}e + {}^{0}_{1}e \rightarrow 2\ {}^{0}_{0}\gamma$	☐

Section	Subject	Summary	Check When Learned
18-2	Questions to ask in deciding whether a nucleus is stable	Does the nucleus contain: (a) more than 83 protons? (If it does, it's unstable.) (b) an odd number of protons or neutrons? (If it does, it's more likely to be unstable.) (c) a magic number of protons or neutrons? (If it does, it's more likely to be stable. The magic numbers are 2, 20, 28, 50, and 82; for neutrons 126 is also a magic number.)	☐
18-3	Band of stability	On a graph of number of neutrons (y axis) versus number of protons (x axis), the points that correspond to stable nuclei.	☐
18-3	Rules for using the band of stability to predict spontaneous nuclear reactions	An unstable nucleus: (a) to the left of the band will undergo beta emission. (b) to the upper right of the band will undergo alpha emission. (c) to the lower right of the band will undergo positron emission or electron capture.	☐
18-4	Transuranium elements	The elements beyond uranium, element 92, on the Periodic Table; all are artificially created.	☐
18-5	Scintillation counter	An instrument used to determine the quantity of alpha, beta, or gamma radiation emitted by a radioactive sample by measuring the number of scintillations caused by the radiation.	☐
18-5	Ionization detector	An instrument used to determine the quantity of alpha, beta, or gamma radiation emitted by a radioactive sample by measuring the number of electrons released in ionizations caused by the radiation.	☐
18-5	1 Curie (Ci)	3.700×10^{10} disintegrations per second	☐
18-6	Radioactive tracer	A radioactive atom in an element or compound, used to follow the changes that the element or compound undergoes in chemical reactions.	☐
18-7	General rules for relative potential harm from α, β, or γ emissions or neutrons	Larger particles are more harmful: α radiation is more harmful than neutron radiation, and neutron radiation is more harmful than β or γ radiation. Larger quantities of radiation are more damaging.	☐
18-7	rem	The unit used to describe biological damage by radiation; the abbreviation for *roentgen equivalent in man.*	☐
18-7	Radiation therapy	The use of emissions from radioactive nuclei to treat diseases, most commonly to treat cancer.	☐
18-8	Half-life($t_{1/2}$)	The time required for half of the radioactive isotopes in a sample to undergo decay; the half-life is a fixed quantity for a given radioactive isotope. For example, the half-life of $^{131}_{53}I$ (a beta-emitter) is 8.07 days, and this means that in each successive period of 8.07 days, half of the $^{131}_{53}I$ atoms in a sample will undergo decay.	☐

Section	Subject	Summary	Check When Learned
18-8	Equation for calculating the age of a radioactive sample	$t = 3.32t_{1/2} \log \frac{rate_i}{rate_t}$ t is the age of the sample in time units $t_{1/2}$ is the half-life of the radioactive isotope in the sample in time units $rate_i$ is the initial rate of decay of the sample in disintegrations per second $rate_t$ is the current rate of decay of the sample in disintegrations per second	☐
18-8	Radiocarbon dating or carbon-14 dating	The process of finding the age of a sample of animal or plant remains by measuring the rate of emission from ${}^{14}_{6}C$ (a beta-emitter) in the sample.	☐
18-9	Nuclear fission	A nuclear reaction in which one nucleus, hit by a neutron, splits into two or more nuclei and releases energy.	☐
18-9	General rule for the stability of isotopes, based on their mass numbers	In general, isotopes with mass numbers of about 50 are the most stable; isotopes with much larger or much smaller mass numbers are less stable, and they can become more stable by fission or fusion.	☐
18-9	Source of energy released in fission	Mass is converted to energy: 1.00 g = 8.97×10^{10} kJ.	☐
18-9	Chain reaction	A sequence of fission reactions in which each reaction releases neutrons that cause more reactions; the source of energy in nuclear power plants and some nuclear weapons.	☐
18-10	Nuclear fusion	A nuclear reaction in which two lighter nuclei combine to form a heavier one and release energy.	☐
18-10	Source of energy released in fusion	Mass is converted to energy: 1.00 g = 8.97×10^{10} kJ.	☐

Problems

Assume you can use the Periodic Table at the front of the book unless you're directed otherwise. Answers to odd-numbered problems are in Appendix 1.

Equations for Nuclear Reactions (Introduction and Section 18-1)

1. Define *nuclear reaction.*
2. In your own words, explain the difference between a chemical reaction and a nuclear reaction.
3. Explain the meaning of each number in ${}^{4}_{2}He$.
4. Explain the meaning of each number in ${}^{0}_{-1}e$.

5. Define *radioactivity*.

6. Describe the charge on each of these particles: (a) alpha (b) beta (c) gamma.

7. What is the mass number of (a) a proton (b) an alpha particle (c) a gamma particle?

8. What is the mass number of (a) a neutron (b) a beta particle (c) a positron?

9. Write an equation for electron capture by $^{288}_{91}Pa$.

10. Write an equation for electron capture by $^{160}_{68}Er$.

11. Write an equation for electron emission by $^{76}_{33}As$.

12. Write an equation for electron emission by $^{91}_{38}Sr$.

13. Write an equation for alpha emission by $^{238}_{94}Pu$.

14. Write an equation for alpha emission by $^{222}_{88}Ra$.

15. Write an equation for positron emission by $^{173}_{73}Ta$.

16. Write an equation for electron emission by $^{3}_{1}H$.

17. A $^{235}_{92}U$ nucleus spontaneously breaks up into a $^{231}_{90}Th$ nucleus and one other particle. Write an equation for this reaction.

18. A $^{230}_{90}Th$ nucleus spontaneously breaks up into an alpha particle and a radium nucleus with mass number 226. Write an equation for this reaction.

19. Write an equation for the reaction of a $^{27}_{13}Al$ nucleus with a neutron to form an alpha particle and a $^{24}_{11}Na$ nucleus.

20. Write an equation for positron emission by $^{120}_{53}I$.

21. Write an equation for the reaction of a neutron to form a proton and another subatomic particle.

22. Write an equation for the reaction of a proton with an electron to form a subatomic particle.

23. Write an equation for the reaction of a proton to form a positron and another subatomic particle.

24. Write an equation for the reaction of an electron with a positron.

25. The $^{127}_{55}Cs$ nucleus can undergo electron capture or positron emission. (a) Write equations for these reactions. (b) What do these reactions have in common?

26. The $^{119}_{52}Te$ nucleus can undergo electron capture or positron emission. (a) Write equations for these reactions. (b) What do these reactions have in common?

27. When a nucleus captures an electron, where does the electron come from?

28. When a nucleus emits an electron, where does the electron come from?

Predicting Nuclear Stability (Sections 18-2 and 18-3)

29. Without looking at Figure 18.1, decide whether each of these nuclei will be stable or unstable, and explain your answer: (a) $^{262}_{107}Uns$ (b) $^{184}_{77}Ir$ (c) $^{4}_{2}He$.

30. Without looking at Figure 18.1, decide whether each of these nuclei will be stable or unstable, and explain your answer: (a) $^{251}_{102}No$ (b) $^{44}_{21}Sc$ (c) $^{38}_{18}Ar$.

31. Without looking at Figure 18.1, decide which of these nuclei is more likely to be unstable, and explain your answer: $^{36}_{17}Cl$ or $^{37}_{17}Cl$.

32. Without looking at Figure 18.1, decide which of these nuclei is more likely to be stable, and explain your answer: $^{208}_{82}Pb$ or $^{209}_{82}Pb$.

33. Use Figure 18.1 to decide whether each of these nuclei is stable or unstable:
(a) $^{13}_{5}B$ (b) $^{183}_{78}Pt$ (c) $^{90}_{40}Zr$

34. Use Figure 18.1 to decide whether each of these nuclei is stable or unstable:
(a) $^{95}_{37}Rb$ (b) $^{136}_{58}Ce$ (c) $^{200}_{80}Hg$

35. Use Figure 18.1 to decide whether each of these nuclei is stable or unstable:
(a) $^{144}_{60}Nd$ (b) $^{7}_{3}Li$ (c) $^{43}_{22}Ti$

36. Use Figure 18.1 to decide whether each of these nuclei is stable or unstable:
(a) $^{132}_{51}Sb$ (b) $^{170}_{70}Yb$ (c) $^{120}_{50}Sn$

Predicting Equations for Spontaneous Nuclear Reactions (Section 18-3)

37. Use Table 18.1 to predict an equation for the first reaction in the radioactive decay of each unstable nucleus whose symbol is shown in problem 29.

38. Use Table 18.1 to predict an equation for the first reaction in the radioactive decay of each unstable nucleus whose symbol is shown in problem 30.

39. Use Table 18.1 to predict an equation for the first reaction in the radioactive decay of each unstable nucleus whose symbol is shown in problem 31.

40. Use Table 18.1 to predict an equation for the first reaction in the radioactive decay of each unstable nucleus whose symbol is shown in problem 32.

41. Use Table 18.1 to predict an equation for the first reaction in the radioactive decay of each unstable nucleus whose symbol is shown in problem 33.

42. Use Table 18.1 to predict an equation for the first reaction in the radioactive decay of each unstable nucleus whose symbol is shown in problem 34.

43. Use Table 18.1 to predict an equation for the first reaction in the radioactive decay of each unstable nucleus whose symbol is shown in problem 35.

44. Use Table 18.1 to predict an equation for the first reaction in the radioactive decay of each unstable nucleus whose symbol is shown in problem 36.

45. Use Figure 18.1 to predict the equation for the spontaneous reaction, if any, of each of the following nuclei. If the product nucleus is unstable, predict the equation for its reaction: (a) $^{81}_{40}Zr$ (b) $^{173}_{73}Ta$ (c) $^{70}_{30}Zn$

46. Use Figure 18.1 to predict the equation for the spontaneous reaction, if any, of each of the following nuclei. If the product nucleus is unstable, predict the equation for its reaction: (a) $^{60}_{30}Zn$ (b) $^{44}_{22}Ti$ (c) $^{150}_{65}Tb$

47. Use Figure 18.1 to predict the equation for the spontaneous reaction, if any, of each of the following nuclei. If the product nucleus is unstable, predict the equation for its reaction: (a) $^{108}_{48}Cd$ (b) $^{85}_{35}Br$ (c) $^{58}_{26}Fe$

48. Use Figure 18.1 to predict the equation for the spontaneous reaction, if any, of each of the following nuclei. If the product nucleus is unstable, predict the equation for its reaction: (a) $^{68}_{33}As$ (b) $^{121}_{49}In$ (c) $^{190}_{78}Pt$

Transuranium Elements (Section 18-4)

49. In your own words, define *transuranium element.*

50. Is unnilhexium a metal, nonmetal, or metalloid? Is it a transition element or a representative element?

51. If element number 110 is created, what will be its name and symbol?

52. If transuranium elements with successively higher atomic numbers continue to be created, what will be the atomic number of the first transuranium element that's not a metal? What will be its name and symbol?

53. Do you predict that any stable isotopes of transuranium elements will be found? Explain your answer.

54. Write the abbreviated electron configuration you predict for unnilpentium.

Scintillation Counters and Ionization Detectors (Section 18-5)

55. In scintillation counters, emissions from radioactive nuclei cause light flashes (scintillations). In your own words, explain this process.

56. In ionization detectors, emissions from radioactive nuclei cause ionizations. In your own words, explain this process.

57. Argon is commonly used as the substance that undergoes ionization in ionization detectors. What properties of argon make it a good choice for this use?

58. Would you expect helium atoms or argon atoms to be more easily ionized by collisions with radioactive emissions? Explain your answer.

59. The radioactivity in a sample of $^{198}_{77}Ir$ was measured as 6.3×10^{-5} Ci. Convert this value to disintegrations per second.

60. The radioactivity in a sample of $^{51}_{26}Fe$ was measured as 7.9×10^{-6} Ci. Convert this value to disintegrations per second.

61. The radioactivity in a sample of $^{204}_{84}Po$ was measured as 3.1×10^{-3} Ci. In this sample, how many nuclei disintegrate in one minute?

62. The radioactivity in a sample of $^{57}_{28}Ni$ was measured as 7.4×10^{-8} Ci. In this sample, how many nuclei disintegrate in 38 seconds?

Radioactive Tracers and Organ Imaging (Section 18-6)

63. Imagine that you have a sample of nitrogen in which one nitrogen atom in each molecule is a radioactive isotope. If this sample of nitrogen reacts with hydrogen to produce ammonia, what percentage of the ammonia molecules will contain a radioactive nitrogen atom?

64. Imagine that you have a sample of oxygen in which one oxygen atom in each molecule is a radioactive isotope. If this sample of oxygen reacts with carbon monoxide to produce carbon dioxide, what percentage of the carbon dioxide molecules will *not* contain a radioactive oxygen atom?

65. Imagine that you have a sample of hydrogen in which one hydrogen atom in each molecule is a radioactive isotope. If this sample of hydrogen reacts with nitrogen to produce ammonia, what percentage of the ammonia molecules will contain a radioactive hydrogen atom?

66. Imagine that you have a sample of chlorine in which one chlorine atom in each chlorine molecule is a radioactive isotope. If this sample of chlorine reacts with hydrogen to produce hydrogen chloride, what percentage of the hydrogen chloride molecules will contain a radioactive chlorine atom?

67. Suggest two reasons why radioactive isotopes that are alpha emitters would not be used to image organs in human beings.

68. In Positron Emission Tomography (PET) the images are actually created by gamma rays, not by positrons. Explain.

Biological Damage from Radiation (Section 18-7)

69. Which would you expect to be more dangerous, a radioactive sample that was emitting beta particles at a rate of 2.7×10^{-2} Ci or a sample that was emitting neutrons at the same rate? Why?

70. In your own words, explain the difference between curies and rems.

71. The isotope $^{203}_{80}Hg$ is a beta-emitter that's used to make brain images. Write the equation for beta emission by this isotope.

72. The isotope $^{137}_{55}Cs$ is a beta-emitter that's used to treat cancer. Write the equation for beta emission by this isotope.

73. The isotope $^{51}_{24}Cr$, which undergoes electron capture, is used to make spleen images. Write the equation for electron capture by this isotope.

74. The isotope $^{18}_{9}F$ is a positron emitter that's used to make bone images. Write the equation for positron emission by this isotope.

Half-Life and Dating with Radioactive Isotopes (Section 18-8)

75. The isotope $^{43}_{19}K$, used in heart imaging, undergoes beta decay with a half-life of 22.3 hours. (a) Write an equation for the beta decay of this isotope. (b) Imagine that you have a 1.000-g sample of pure $^{43}_{19}K$. Draw and label a diagram like the one shown in Part (a) of Figure 18.9 to show the masses of $^{43}_{19}K$ and $^{43}_{20}Ca$ in the sample after each of three successive intervals of 22.3 hours each.

76. The isotope $^{241}_{95}Am$, used to treat cancer, undergoes alpha emission with a half-life of 433 years. (a) Write an equation for the radioactive decay of this isotope. (b) Imagine that you have a 1.000-g sample of pure $^{241}_{95}Am$. Draw and label a diagram like the one shown in Part (a) of Figure 18.9 to show the masses of $^{241}_{95}Am$ and $^{237}_{93}Np$ in the sample after each of three successive intervals of 433 years each.

77. For the isotope described in problem 75, draw and label a graph like the one shown in Part (b) of Figure 18.9.

78. For the isotope described in problem 76, draw and label a graph like the one shown in Part (b) of Figure 18.9.

79. The initial rate of decay of a sample of $^{22}_{11}Na$, a positron emitter, was 2.65×10^{12} disintegrations per second, and its rate of decay now is 7.55×10^{7} disintegrations per second; the half-life of this isotope is 2.602 years. How old is the sample?

80. The initial rate of decay of a sample of $^{85}_{36}Kr$, a beta emitter, was 3.72×10^{11} disintegrations per second, and its rate of decay now is 8.44×10^{5} disintegrations per second; the half-life of this isotope is 10.72 years. How old is the sample?

81. The isotope $^{78}_{33}As$ decays by beta emission. A sample of this isotope had an initial rate of decay of 9.41×10^{12} disintegrations per second; its rate 4.000 hours later was 1.50×10^{12} disintegrations per second. Calculate the half-life of this isotope.

82. The isotope $^{142}_{61}Pm$ decays by positron emission. A sample of this isotope had an initial rate of decay of 6.19×10^{7} disintegrations per second; its rate 175 s later was 3.09×10^{6} disintegrations per second. Calculate the half-life of this isotope.

83. The isotope $^{194}_{76}Os$ undergoes beta emission with a half-life of 6.0 y. If the initial rate of decay of a sample of this isotope was 3.3×10^{-3} Ci, what will be its rate of decay after 1.0 y?

84. The isotope $^{232}_{92}U$ undergoes alpha emission with a half-life of 68.9 y. If the initial rate of decay of a sample of this isotope was 6.4×10^{-4} Ci, what will be its rate of decay after 50.0 y?

85. The isotope $^{52}_{25}Mn$ undergoes positron emission with a half-life of 5.591 d. A sample of this isotope had an initial rate of decay of 8.88×10^{9} disintegrations per second. Predict its rate of decay in curies after 127 hours.

86. The isotope $^{75}_{35}Br$ undergoes positron emission with a half-life of 1.62 h. A sample of this isotope had an initial rate of decay of 2.17×10^{-3} Ci. Predict its rate of decay after 2.0 days, in disintegrations per second.

87. The initial rate of decay from a newly formed sample containing $^{14}_{6}C$ is 918 disintegrations per second; the half-life of $^{14}_{6}C$ is 5.73×10^{3} y. A sample of wood from a carved figure in a Chinese temple shows a rate of 747 disintegrations per second. Calculate the age of the figure.

88. The initial rate of decay from a newly formed sample containing $^{14}_{6}C$ is 918 disintegrations per second; the half-life of $^{14}_{6}C$ is 5.73×10^{3} y. A sample of wood from a statue in a Japanese temple shows a rate of 846 disintegrations per second. Calculate the age of the statue.

89. The initial rate of decay from a newly formed sample containing $^{14}_{6}C$ is 918 disintegrations per second; the half-life of $^{14}_{6}C$ is 5.73×10^3 y. A sample of linen wrapping from an ancient scroll shows a rate of 721 disintegrations per second. Calculate the age of the wrapping.

90. The initial rate of decay from a newly formed sample containing $^{14}_{6}C$ is 918 disintegrations per second; the half-life of $^{14}_{6}C$ is 5.73×10^3 y. A sample of wood from an ancient chariot unearthed in Bulgaria shows a rate of 734 disintegrations per second. Calculate the age of the chariot.

Fission and Fusion (Sections 18-9 and 18-10)

91. In your own words, explain the difference between nuclear fission and nuclear fusion.

92. In your own words, explain the principle of a chain reaction.

93. What are the characteristics of fission and fusion reactions that have made them useful in creating weapons?

94. Fission reactions are used in nuclear power plants but fusion reactions are not. Why?

95. In a fission reaction 0.277 mg of matter was converted to energy. Calculate the quantity of energy produced.

96. In a fusion reaction 2.80×10^{13} J of energy was released. Calculate the mass of matter in milligrams that was converted to energy.

97. When an electron and a positron collide, they annihilate each other:

$$^{0}_{1}e + ^{0}_{-1}e \rightarrow 2\, ^{0}_{0}\gamma$$

An electron and a positron have the same mass, 5.846×10^{-4} amu; 1.00 g = 6.02×10^{23} amu. Calculate the amount of energy released when an electron and a positron collide.

98. The mass of a neutron is 1.01 amu, and 1.00 g = 6.02×10^{23} amu. How much energy would be created by the conversion of one neutron from matter to energy?

COVALENT BONDS FORMED BY CARBON ATOMS

19-1
A carbon atom in a molecule has four bonds.

19-2
Carbon atoms form single, double, or triple covalent bonds that are nonpolar or polar.

COMPOUNDS OF CARBON AND HYDROGEN

19-3
Carbon atoms can bond with one another to form chains or rings.

19-4
The name for a hydrocarbon is based on the number of carbon atoms in the longest chain in its molecule.

19-5
From the arrangement of the bonds in its molecule, a hydrocarbon is classified as *aromatic* or *aliphatic*.

19-6
In general, hydrocarbons whose molecules have more atoms have higher melting points and boiling points.

19-7
The combustion of hydrocarbons produces about two thirds of the energy used in the United States each year.

19-8
Hydrocarbons undergo two important kinds of reactions: substitution and addition.

VARIETIES OF ORGANIC COMPOUNDS

19-9
Organic compounds are classified by identifying characteristic groupings of atoms in their molecules.

19-10
Simple alcohols resemble water.

19-11
Simple amines resemble ammonia.

19-12
Simple organic acids are weak acids.

19-13
A polymer is an organic compound whose molecules are very long chains with repeating sequences of atoms.

Inside Chemistry:
What happens when TNT explodes?

Chapter Summary

Problems

These familiar objects and many others are made from polymers, described in Section 19-13. *(Charles D. Winters)*

19
Organic Chemistry

Organic compounds are covalent compounds whose molecules contain carbon atoms, and **organic chemistry** is the study of organic compounds.◆

◆Compounds of elements other than carbon are *inorganic compounds*, and the study of them is called *inorganic chemistry*.

Organic chemistry is an important part of science because organic compounds are widespread in nature, and they can be created synthetically in almost endless variety.◆ Of the ten million compounds whose formulas are now known, most are organic compounds.

◆We'll see in Chapter 20 that most of the compounds in biological systems are organic compounds.

Organic chemistry is an important part of technology because organic compounds are used to make an enormous range of commercial products; examples are paints, fuels, textiles, plastics, and pharmaceuticals.

In this chapter we'll see that the unique properties of carbon atoms allow them to form a nearly endless variety of molecules, and we'll look at some of the important commercial applications of organic compounds.

COVALENT BONDS FORMED BY CARBON ATOMS

Organic compounds are covalent compounds whose molecules contain carbon atoms. This part of the chapter describes the kinds of covalent bonds that carbon atoms form.

19-1 A carbon atom in a molecule has four bonds

The electron configuration for carbon is C^6: $1s^2 2s^2 2p^2$. This configuration shows that a carbon atom has four valence electrons ($2s^2 2p^2$); the Lewis symbol for a carbon atom shows these four valence electrons as dots: $\cdot\dot{\mathrm{C}}\colon$.◆

◆The relationship between the electron configuration for an atom and its Lewis symbol is described in Section 6-8.

Carbon atoms form covalent bonds. In forming covalent bonds, carbon atoms follow the octet rule: each valence electron in a carbon atom combines with one valence electron from another atom to form a covalent bond (a pair of shared electrons); the bonded carbon atom has four covalent bonds and a share in eight valence electrons.◆ For example, one carbon atom will bond with four hydrogen atoms:

◆There are rare exceptions—for example, carbon monoxide, $:\mathrm{C{\equiv}O}:$—but they won't be important for our purposes.

$$\cdot\dot{\mathrm{C}}\colon + 4\,\mathrm{H}\cdot \rightarrow \begin{array}{c} \mathrm{H} \\ \mathrm{H:\ddot{\underset{..}{C}}:H} \\ \mathrm{H} \end{array} \quad \text{or} \quad \begin{array}{c} \mathrm{H} \\ | \\ \mathrm{H{-}C{-}H} \\ | \\ \mathrm{H} \end{array}$$

Lewis formula line formula

The atoms in these formulas follow the octet rule: The carbon atom has a share in eight valence electrons, and each hydrogen atom has a share in two valence electrons.

The fundamental principle of organic chemistry is that a carbon atom in a molecule has four bonds. To make this rule easy to remember, we can write a *line symbol* for a carbon atom; in the **line symbol** for an atom, a line represents a valence electron the atom can share, that is, a line represents half of a covalent bond the atom can form. The line symbol for a carbon atom is $-\overset{|}{\underset{|}{\mathrm{C}}}-$.

We can use line symbols for other atoms as a convenient way to show the numbers of covalent bonds they'll form. A hydrogen atom, for example, has one valence electron and will form one covalent bond, so its line symbol is H—.◆ Following are the Lewis symbols and line symbols for the atoms whose bonding will be important in this chapter and in Chapter 20:

◆The number of bonds an atom can form is sometimes referred to as its *valence*. The valence of a hydrogen atom is 1, and the valence of a carbon atom is 4.

Lewis symbol: ·Ċ: H· ·N̈· ·Ö· ·S̈· :Ẍ·

Line symbol: —C— H— —N̈— —Ö— —S̈— :Ẍ—

In these symbols X stands for a halogen: F, Cl, Br, or I.

The following Example and Exercise show the use of line symbols to construct line formulas for simple molecules.

Example 19.1

Use line symbols to construct a line formula for each of the following molecules, and circle each bonded atom to show that it obeys the octet rule.
(a) HCl (b) Cl_2 (c) H_2O

Solution

(a) The atoms are H— and :C̈l—, and the molecule is H—C̈l: (hydrogen chloride). The line in a line symbol represents half of a covalent bond, so in constructing the formula for a molecule you join two lines to form one bond.

(H—C̈l:) The hydrogen atom has two valence electrons and the chlorine atom has eight.

(b) The atoms are :C̈l— and :C̈l—, and the molecule is :C̈l—C̈l: (chlorine).

(:C̈l—C̈l:) Each chlorine atom has eight valence electrons.

(c) The atoms are H—, H—, and —Ö—, and the molecule is H—Ö—H (water).

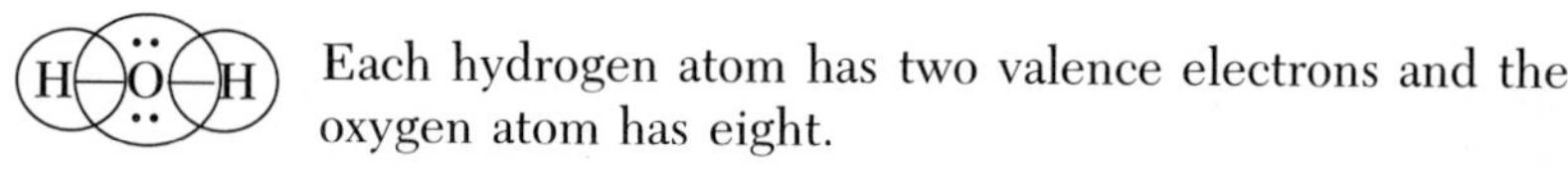

Each hydrogen atom has two valence electrons and the oxygen atom has eight.

Exercise 19.1◆

◆Answers to Exercises are in Appendix 1.

Use line symbols to construct a line formula for each of the following molecules, and circle each bonded atom to show that it obeys the octet rule.
(a) H_2S (b) CCl_4 (c) NH_3

19-2 Carbon atoms form single, double, or triple covalent bonds that are nonpolar or polar

A carbon atom in a molecule will be attached to other atoms by four covalent bonds. A carbon atom can be attached to another atom by a single, double, or triple covalent bond, depending on the number of bonds the other atom can form. For example, a hydrogen atom, H—, can form only one bond, so the attachment between a carbon atom and a hydrogen atom will be a single bond: —C—H. An oxygen atom, —Ö—, can form two bonds, so the attachment between a carbon atom and an oxygen atom may be a single bond or a double bond: —C—Ö— or —C=Ö. Table 19.1 shows the single, double, or triple bonds that can be formed between two carbon atoms or between a carbon atom and an atom of nitrogen, oxygen, sulfur, hydrogen, or a halogen. These are the common forms of bonding in the molecules of organic compounds.

Each of the bonds shown in Table 19.1 is nonpolar or polar, depending on the difference in electronegativity between the carbon atom and the other bonded atom; Figure 19.1 on page 450 shows the electronegativity values for the atoms that commonly bond with carbon. A single, double, or triple bond between two carbon atoms is nonpolar because the bonded atoms have the same electronegativity. A single, double, or triple bond between a carbon atom and an atom of another element is polar if the carbon atom and the other bonded atom have different electronegativities, and they usually do. Among the elements shown in Figure 19.1, sulfur and iodine have the same electronegativity as carbon, hydrogen has a lower electronegativity, and the other elements have higher electronegativities.

Example 19.2

Identify the bond on the right side of each carbon atom as single, double, or triple. Use Figure 19.1 to decide whether the bond is nonpolar or polar; if it's polar, use $\delta+$ and $\delta-$ to show its polarity. (a) H—C(H)(H)—H (b) H—C(H)=Ö (c) H—C(H)(H)—S̈—H

Table 19.1 Bonds Commonly Formed by Carbon Atoms
In this table, X stands for a halogen atom: F, Cl, Br, or I.

	Atom Bonded to Carbon					
Possible Forms of Bonding	—C— (four bonds)	—N̈— (three bonds)	—Ö—	—S̈—	:Ẍ—	H—
Single	—C—C—	—C—N̈—	—C—Ö—	—C—S̈—	—C—Ẍ:	—C—H
Double	—C=C—	—C=N̈—	—C=Ö	—C=S̈		
Triple	—C≡C—	—C≡N:				

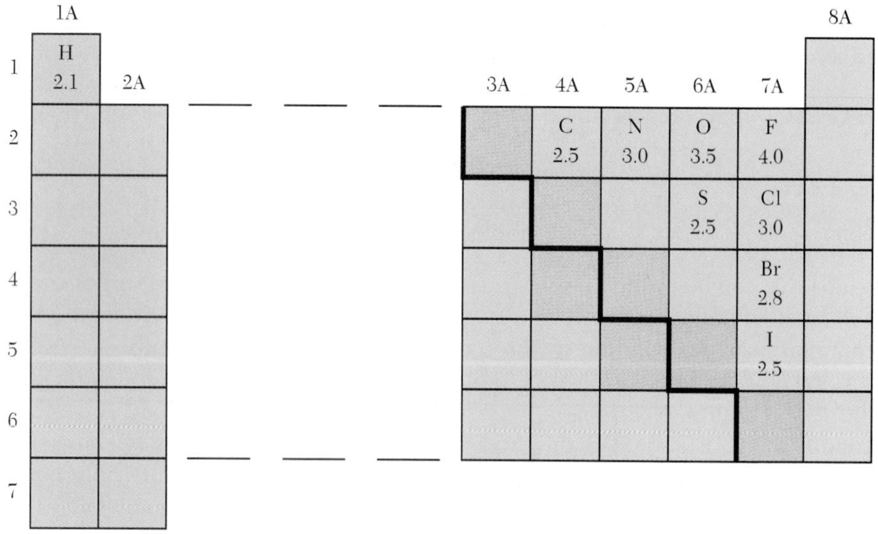

Figure 19.1
Electronegativity values for atoms that commonly bond with carbon. A bond between two carbon atoms is nonpolar because both of the bonded atoms have the same electronegativity. A bond between a carbon atom and a sulfur atom or an iodine atom is nonpolar because carbon, sulfur, and iodine have the same electronegativity. A bond between a carbon atom and an atom of hydrogen, nitrogen, oxygen, fluorine, chlorine, or bromine is polar because the two bonded atoms have different electronegativities.

Period	1A	4A	5A	6A	7A
1	H 2.1				
2		C 2.5	N 3.0	O 3.5	F 4.0
3				S 2.5	Cl 3.0
4					Br 2.8
5					I 2.5

Solution

(a) 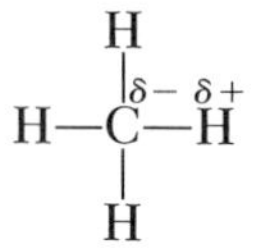 $H_3C^{\delta-}$—$H^{\delta+}$ (H—C—H with two more H on C) The bond is single and polar; C has a greater electronegativity than H.

(b) 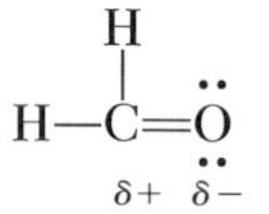 $H_2C^{\delta+}$=$\ddot{O}^{\delta-}$ (H—C(H)=O:) The bond is double and polar; O has a greater electronegativity than C.

(c) 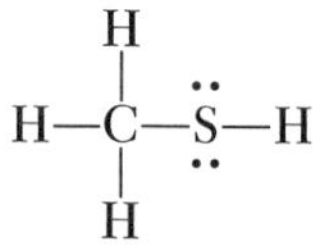 H_3C—$\ddot{S}$—H The bond is single and nonpolar; C and S have the same electronegativity.

Exercise 19.2

Identify the bond on the right side of each carbon atom as single, double, or triple. Use Figure 19.1 to decide whether the bond is nonpolar or polar; if it's polar, use $\delta+$ and $\delta-$ to show its polarity. (a) H—C≡N:

(b) 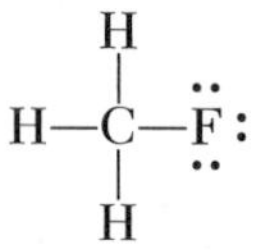 H_3C—$\ddot{F}$: (c) 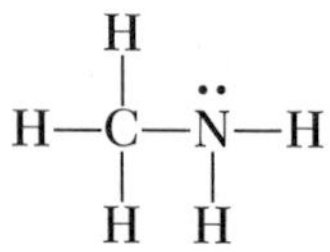 H_3C—$\ddot{N}$(H)—H

COMPOUNDS OF CARBON AND HYDROGEN

The simplest organic compounds are those whose molecules contain only carbon atoms and hydrogen atoms; these compounds are called **hydrocarbons.** This part of the chapter uses hydrocarbons as examples to introduce basic concepts of organic chemistry.

19-3 Carbon atoms can bond with one another to form chains or rings

Butane, C_4H_{10}, is used as a fuel in some cigarette lighters. Its condensed formula is $CH_3{-}CH_2{-}CH_2{-}CH_3$. *(Charles D. Winters)*

Carbon atoms can bond with one another to form chains that can be up to several thousand atoms long. In these chains the bonds between the carbon atoms can be single, double, or triple nonpolar covalent bonds. These are examples of hydrocarbon molecules that contain chains of four carbon atoms:

Line Formula	Condensed Formula
$\begin{array}{c} \quad H \quad H \quad H \quad H \\ \quad \vert \quad\; \vert \quad\; \vert \quad\; \vert \\ H{-}C{-}C{-}C{-}C{-}H \\ \quad \vert \quad\; \vert \quad\; \vert \quad\; \vert \\ \quad H \quad H \quad H \quad H \end{array}$	$CH_3{-}CH_2{-}CH_2{-}CH_3$
$\begin{array}{c} \quad H \quad H \quad H \quad H \\ \quad \vert \quad\; \vert \quad\; \vert \quad\; \vert \\ H{-}C{-}C{=}C{-}C{-}H \\ \quad \vert \qquad\qquad\; \vert \\ \quad H \qquad\qquad H \end{array}$	$CH_3{-}CH{=}CH{-}CH_3$
$\begin{array}{c} \qquad\qquad H \quad H \\ \qquad\qquad \vert \quad\; \vert \\ H{-}C{\equiv}C{-}C{-}C{-}H \\ \qquad\qquad \vert \quad\; \vert \\ \qquad\qquad H \quad H \end{array}$	$H{-}C{\equiv}C{-}CH_2{-}CH_3$

Exercise 19.3

Write the line formula and the condensed formula for the hydrocarbon molecule that has (a) three carbon atoms in a chain and only single bonds (b) four carbon atoms in a chain and a double bond on the first carbon atom (c) five carbon atoms in a chain and a triple bond between the third and fourth carbon atoms.

Chains of carbon atoms can be branched, as shown in these examples:

$$\begin{array}{c} CH_3 \\ \vert \\ CH_3{-}CH_2{-}CH{-}CH_2{-}CH_3 \end{array} \qquad \begin{array}{c} CH_3 \\ \vert \\ CH_2{=}C{-}CH_2{-}CH_3 \end{array} \qquad \begin{array}{c} CH_3{-}CH_2 \\ \vert \\ CH_3{-}C{\equiv}C{-}CH{-}CH_2{-}CH_3 \end{array}$$

Branches from carbon chains can themselves contain many carbon atoms; we'll consider only molecules with branches of one or two carbon atoms.

Exercise 19.4

Write the condensed formula for a hydrocarbon molecule that has (a) a four-carbon chain and a one-carbon branch off of the second carbon; (b) a six-carbon chain and a two-carbon branch off of the third carbon (c) a nine-carbon chain with a double bond between the third and fourth carbons, and a two-carbon branch off of the fifth carbon.

The end carbon atoms in a chain can bond with one another so that the chain forms a ring; compounds whose molecules are rings are referred to as **cyclic compounds.** These are examples:

$$\begin{array}{c} CH_2{-}CH_2 \\ \vert \qquad\quad \vert \\ CH_2{-}CH_2 \end{array} \qquad \begin{array}{c} CH \\ \diagup \qquad \diagdown\!\!\diagdown \\ CH_2 \qquad\quad CH \\ \diagdown \qquad \diagup \\ CH_2{-}CH_2 \end{array}$$

Exercise 19.5

Write the condensed formula for a hydrocarbon molecule that has (a) three carbon atoms in a ring (b) six carbon atoms in a ring and a double bond between two of them.

Because carbon atoms can bond to one another in a variety of ways, two or more different molecules can have the same numbers of carbon atoms and hydrogen atoms, for example,

$$CH_3{-}CH_2{-}CH_2{-}CH_3 \quad \text{and} \quad CH_3{-}\underset{}{\overset{\displaystyle CH_3\atop|}{C}H}{-}CH_3$$

The molecular formula for each of these compounds is C_4H_{10}.

Compounds whose molecules contain the same numbers of atoms of the same elements in different bonding arrangements are called **isomers,** and the compounds are said to show **isomerism.**

Exercise 19.6

Write formulas for the isomers of C_5H_{12}.

19-4 The name for a hydrocarbon is based on the number of carbon atoms in the longest chain in its molecule

Hydrocarbons whose molecules contain only single bonds are called **alkanes.** The system used to name alkanes is the basis for naming all organic compounds; in this system the suffix *-ane* identifies the compound as an alkane, and a prefix identifies the number of carbon atoms in the longest chain in the molecule. The formulas and names for the ten simplest alkanes are shown in Table 19.2.

A branch off of a carbon chain has a name that's formed from the name for the alkane with the same number of carbon atoms as the branch:

A petroleum refinery in Texas; refineries produce hydrocarbons from petroleum. *(Grant Heilman Photography)*

Table 19.2 Names for Alkanes

Number of Carbon Atoms in Chain	Formula	Name
1	CH_4	methane
2	$CH_3{-}CH_3$	ethane
3	$CH_3{-}CH_2{-}CH_3$	propane
4	$CH_3{-}(CH_2)_2{-}CH_3$	butane
5	$CH_3{-}(CH_2)_3{-}CH_3$	pentane
6	$CH_3{-}(CH_2)_4{-}CH_3$	hexane
7	$CH_3{-}(CH_2)_5{-}CH_3$	heptane
8	$CH_3{-}(CH_2)_6{-}CH_3$	octane
9	$CH_3{-}(CH_2)_7{-}CH_3$	nonane
10	$CH_3{-}(CH_2)_8{-}CH_3$	decane

one-carbon branch:	CH_3— methyl	named for	CH_4 methane
two-carbon branch:	CH_3—CH_2— ethyl	named for	CH_3—CH_3 ethane

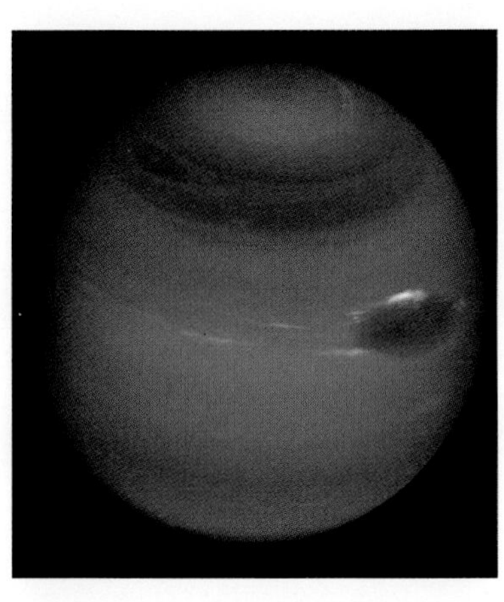

Methane gas, CH_4, is a major constituent of the atmosphere of the planet Jupiter. *(NASA)*

In a formula that shows a branched chain, number the carbon atoms from one end of the chain, and use these numbers to designate the position of the branch:

$$\underset{1}{CH_3}-\underset{2}{\overset{\overset{\displaystyle CH_3}{|}}{CH}}-\underset{3}{CH_2}-\underset{4}{CH_3}$$

2-methylbutane

Number the chain from the end that assigns the lower number to the branch:

$$\underset{8}{CH_3}-\underset{7}{CH_2}-\underset{6}{CH_2}-\underset{5}{CH_2}-\underset{4}{\overset{\overset{\displaystyle CH_3-CH_2}{|}}{CH}}-\underset{3}{CH_2}-\underset{2}{CH_2}-\underset{1}{CH_3}$$

4-ethyloctane

(If the chain were numbered left to right, the branch would be on carbon 5.)

Exercise 19.7

Name (a) $CH_3-CH_2-\overset{\overset{\displaystyle CH_3}{|}}{CH}-CH_2-CH_3$ (b) $CH_3-CH_2-\overset{\overset{\displaystyle CH_3}{|}}{CH}-CH_3$

Exercise 19.8

Write the formula for (a) 3-ethylhexane (b) 2-methylpentane

If a molecule has more than one kind of branch, name the branches in alphabetical order:

$$CH_3-\underset{\underset{\displaystyle CH_3}{|}}{CH}-\overset{\overset{\overset{\displaystyle CH_3}{|}}{\overset{\displaystyle CH_2}{|}}}{CH}-CH_2-CH_2-CH_3$$

3-ethyl-2-methylhexane

Describe two or more branches of the same kind with the prefixes used to describe the number of atoms in a molecule (Section 9-1): *di-, tri-, tetra-, penta-, hexa-, hepta-, octa-, nona-, deca-.*

$$CH_3-\overset{\overset{\displaystyle CH_3}{|}}{CH}-\overset{\overset{\displaystyle CH_3}{|}}{CH}-CH_3$$

2,3-dimethylbutane

Number the carbon chain from the end that assigns the lowest numbers to all of the branches:

$$\begin{array}{ccccccccccc}
 & & & & \mathrm{CH_3} & & & & & & \\
 & & & & | & & & & & & \\
 & & & & \mathrm{CH_2} & & & & \mathrm{CH_3} & & \\
 & & & & | & & & & | & & \\
\mathrm{CH_3} & - & \mathrm{CH_2} & - & \mathrm{CH} & - & \mathrm{CH_2} & - & \mathrm{C} & - & \mathrm{CH_3} \\
6 & & 5 & & 4 & & 3 & & |\,2 & & 1 \\
 & & & & & & & & \mathrm{CH_3} & &
\end{array}
\qquad
\begin{array}{ccccccccccc}
 & & & & \mathrm{CH_3} & & & & & & \\
 & & & & | & & & & & & \\
 & & & & \mathrm{CH_2} & & & & \mathrm{CH_3} & & \\
 & & & & | & & & & | & & \\
\mathrm{CH_3} & - & \mathrm{CH_2} & - & \mathrm{CH} & - & \mathrm{CH_2} & - & \mathrm{C} & - & \mathrm{CH_3} \\
1 & & 2 & & 3 & & 4 & & |\,5 & & 6 \\
 & & & & & & & & \mathrm{CH_3} & &
\end{array}$$

4-ethyl-2,2-dimethylhexane 3-ethyl-5,5-dimethylhexane

The sum of the numbers in the name on the left (8) is lower than the sum in the name on the right (13), so the name on the left is preferred.

Exercise 19.9

(a) Name

$$\begin{array}{ccccccccc}
 & & & & \mathrm{CH_3} & & & & \\
 & & & & | & & & & \\
 & & & & \mathrm{CH_2} & & & & \\
 & & & & | & & & & \\
\mathrm{CH_3} & - & \mathrm{CH_2} & - & \mathrm{C} & - & \mathrm{CH_2} & - & \mathrm{CH_3} \\
 & & & & | & & & & \\
 & & & & \mathrm{CH_2} & & & & \\
 & & & & | & & & & \\
 & & & & \mathrm{CH_3} & & & &
\end{array}$$

(b) Write the formula for 2,6-dimethylheptane.

The formula for an organic compound may not be written with all of the carbon atoms in its longest chain in a straight line. In naming a compound from its formula, be sure to find the longest carbon chain:

$$\begin{array}{ccccccccc}
 & & \mathrm{CH_3} & & & & & & \\
 & & | & & & & & & \\
 & & \mathrm{CH_2} & & & & & & \\
 & & | & & & & & & \\
\mathrm{CH_3} & - & \mathrm{CH} & - & \mathrm{CH_2} & - & \mathrm{CH_2} & - & \mathrm{CH_3}
\end{array}$$

3-methylhexane

Hydrocarbons that contain double bonds are called **alkenes,** and hydrocarbons that contain triple bonds are called **alkynes.** Name an alkene or an alkyne from the name for the corresponding alkane, and designate the location of the double or triple bond by numbering the atoms in the carbon chain:

$$\overset{1}{\mathrm{CH_3}}-\overset{2}{\mathrm{CH}}=\overset{3}{\mathrm{CH}}-\overset{4}{\mathrm{CH_2}}-\overset{5}{\mathrm{CH_3}} \qquad \overset{6}{\mathrm{CH_3}}-\overset{5}{\mathrm{CH_2}}-\overset{4}{\mathrm{CH_2}}-\overset{3}{\mathrm{C}}\equiv\overset{2}{\mathrm{C}}-\overset{1}{\mathrm{CH_3}}$$

2-pentene 2-hexyne

The suffix *-ene* or *-yne* designates an alkene or an alkyne. Number the chain in the direction that assigns the lower number to the first carbon atom attached to the double or triple bond.

Exercise 19.10

Name (a) $CH_2{=}CH{-}CH_2{-}CH_3$ (b) $CH_3{-}CH_2{-}C{\equiv}C{-}H$

Numbers assigned to designate the location of a double or triple bond can also be used to designate locations of branches:

$$\begin{array}{c} \quad CH_3 \ \ CH_3 \\ \quad | \qquad | \\ CH_3—CH—CH—CH_2—CH{=}CH_2 \end{array}$$

4,5-dimethyl-1-hexene

Some fruits, including oranges, spontaneously produce ethene gas, $CH_2{=}CH_2$, which causes fruits to ripen. The banana on the left was allowed to ripen without exposure to ethene; the banana in the center was exposed to ethene produced by the orange; and the banana on the right was exposed to ethene prepared by a chemical reaction. *(Runk/Shoenberger/Grant Heilman Photography)*

Exercise 19.11

Name $CH_3—C{\equiv}C—\overset{\displaystyle CH_3}{\overset{|}{C}}H—CH_3$

Molecules that contain several double or triple bonds are named by more complicated methods that we won't consider.

Cyclic compounds are named this way:

$$\begin{array}{c} CH_2—CH_2 \\ | \qquad\ | \\ CH_2—CH_2 \end{array} \qquad \begin{array}{c} CH_2 \\ \diagup \qquad \diagdown \\ CH_2 \qquad\quad CH_2 \\ \diagdown \qquad \diagup \\ CH_2—CH_2 \end{array}$$

cyclobutane cyclopentane

Exercise 19.12

Write the condensed formula for cyclopropane.

Some organic compounds have traditional names; for example, ethene, $CH_2{=}CH_2$, is also called ethylene, and ethyne, $H—C{\equiv}C—H$, is also called acetylene.

Acetylene gas, $H—C{\equiv}C—H$, is used as a fuel in welding torches. *(Bethlehem Steel)*

19-5 From the arrangement of the bonds in its molecule, a hydrocarbon is classified as *aromatic* or *aliphatic*

This arrangement of carbon atoms in a molecule:

$$\begin{array}{c} | \\ C \\ \diagup \quad \diagdown\!\!\diagdown \\ —C \qquad\quad C— \\ \| \qquad\qquad | \\ —C \qquad\quad C— \\ \diagdown \quad \diagup\!\!\diagup \\ C \\ | \end{array}$$

—six carbon atoms in a ring, with alternating single and double bonds—creates special and important properties. A compound whose molecule contains this arrangement is classified as an **aromatic compound.** The simplest aromatic hydrocarbon is benzene,

Benzene, C_6H_6, is a colorless, transparent liquid that boils at 80.1°C. *(Charles D. Winters)*

and the arrangement of carbon atoms that characterizes aromatic compounds is called a **benzene ring.**◆

Aliphatic compounds are those that aren't aromatic. All of the hydrocarbons whose formulas were shown in Sections 19-3 and 19-4 are aliphatic compounds.◆

The line formula for benzene describes the bonds in the benzene ring as alternating single and double bonds, but that description is misleading: All six bonds in a benzene ring are in fact the same, and they are neither single bonds nor double bonds. Because the bonds in a benzene ring aren't ordinary single or double bonds, they can't be accurately described by an ordinary line formula.

To understand the bonds in a benzene ring, picture the bonding electrons this way:

The electrons shown as dots could pair to form bonds in two ways:

or

or

In fact, the electrons shown as dots are shared equally among all of the carbon atoms in the ring, so that the bond between each adjacent pair of carbon atoms is neither single nor double.

◆When organic compounds were first being systematically investigated about a century ago, certain compounds with distinctive aromas were designated *aromatic compounds*, and the designation has been kept to identify compounds whose molecules contain benzene rings.

◆*Aliphatic* is from a Greek word that means "oily." Some of the aliphatic compounds that were first investigated were oily, and the designation has persisted.

Because the bonds in a benzene ring are neither single nor double, they can't easily be represented with Lewis or line formulas. One way to describe the bonds is by writing two line formulas that show the single and double bonds in exchanged positions:

```
          H                      H
          |                      |
  H       C       H      H       C       H
    \   /   \\   /         \   //  \   /
      C       C              C       C
      ||      |      ↔       |       ||
      C       C              C       C
    /   \   //   \         /   \\  /   \
  H       C       H      H       C       H
          |                      |
          H                      H
```

In this description of the benzene molecule the double-headed arrow has a complex meaning: It signals that neither of the two formulas alone is a good description of the molecule, and it suggests that a better description can be created by imagining a combination or hybrid of the two formulas.†

† Chemistry Insight

The double-headed arrow (↔) is *not* a double arrow (⇄), and it does *not* signify an equilibrium reaction. The double-headed arrow asks you to imagine superimposing the two formulas it joins, to make one formula. As an analogy, imagine a dog that's a cross between a German shepherd and a Doberman pinscher; the dog you imagine isn't a shepherd one moment and a pinscher the next, but a hybrid dog that blends features of both breeds.

Instead of imagining the benzene molecule as a combination of two formulas, we can write one formula that describes continuous bonding among all of the carbon atoms:

```
          H
          |
  H       C       H
    \   /   \   /
      C       C
      |  ( )  |
      C       C
    /   \   /   \
  H       C       H
          |
          H
```

We'll use this formula for benzene and similar formulas for other aromatic compounds.†

† Chemistry Insight

In more advanced chemistry textbooks benzene is often represented by the abbreviated formula ⌬. Each corner of the hexagon represents a carbon atom with a hydrogen atom attached to it. For our purposes the more complete formula shown in the text will be more useful.

The formulas for benzene are intended to show that its carbon atoms are joined to one another by single bonds and by a secondary form of bonding in which the bonding electrons move among all of the carbon atoms; the electrons that are shared in this secondary form of bonding are said to be **delocalized.**◆ Electron delocalization is also called **resonance.**

◆Section 14-5 describes electron delocalization in metals.

Compounds are known whose molecules consist of two or more benzene rings fused together; the simplest example is naphthalene:

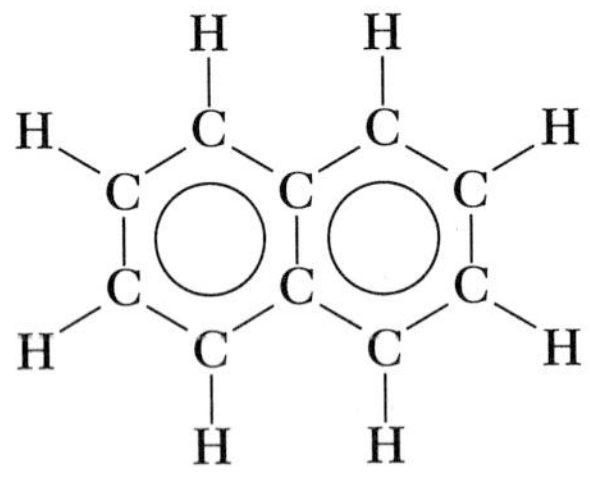

In our further work in this book we'll consider only aromatic compounds whose molecules contain one ring.

The hydrogen atoms on a benzene ring can be replaced by other atoms, for example:

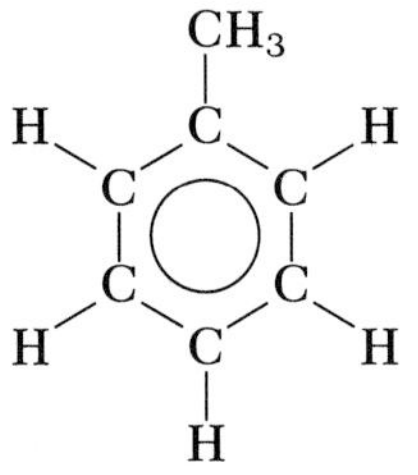

Methylbenzene

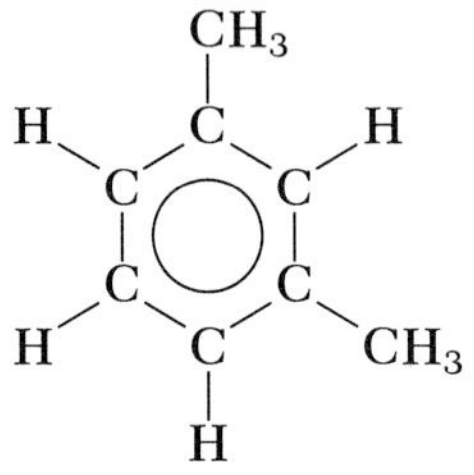

1,3-Dimethylbenzene

If two or more substituent groups are attached to the benzene ring, number the carbon atoms in the ring: Start at a carbon that has a group attached to it, and number in the direction that assigns the lowest numbers to the carbons with attached groups.

Exercise 19.13

Write the formula for (a) ethylbenzene (b) 1,4-diethylbenzene

19-6 In general, hydrocarbons whose molecules have more atoms have higher melting points and boiling points

The melting point of a solid or the boiling point of a liquid measures the strength of the attractive forces between its particles: The higher the melting point or boiling point, the stronger the interparticle forces.◆ In a solid or liquid alkane the particles are molecules, and the strongest attractive forces between the molecules are London forces.◆ Molecules with more electrons have stronger London forces, so alkanes whose molecules have more atoms—and therefore more electrons—have higher melting and boiling points.

◆The relationship between melting point or boiling point and the strength of interparticle forces is described in detail in Chapter 14.

◆London forces are caused by the synchronized movement of electrons in adjacent molecules, as described in Section 14-2.

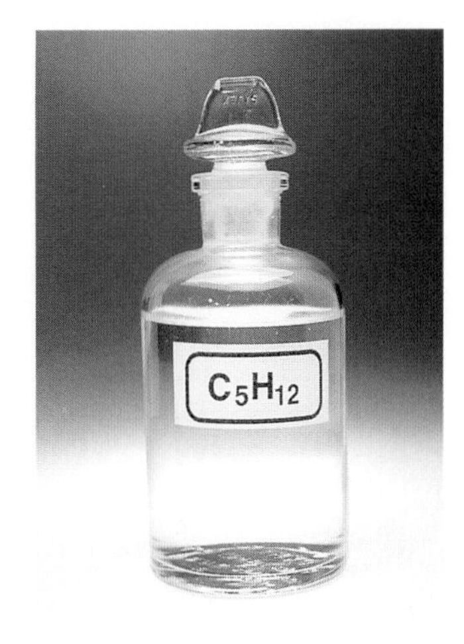

Pentane, C_5H_{12}, is a colorless, transparent liquid that boils at 36.1°C. Its condensed formula is $CH_3—CH_2—CH_2—CH_2—CH_3$. *(Charles D. Winters)*

Exercise 19.14

Under ordinary room conditions butane is a gas, and pentane and decane are liquids. What physical state would you predict for (a) octane (b) ethane? Explain your answers.

The general rule that compounds whose molecules contain more atoms have higher melting and boiling points also applies to alkenes, alkynes, and aromatic compounds; these are examples:

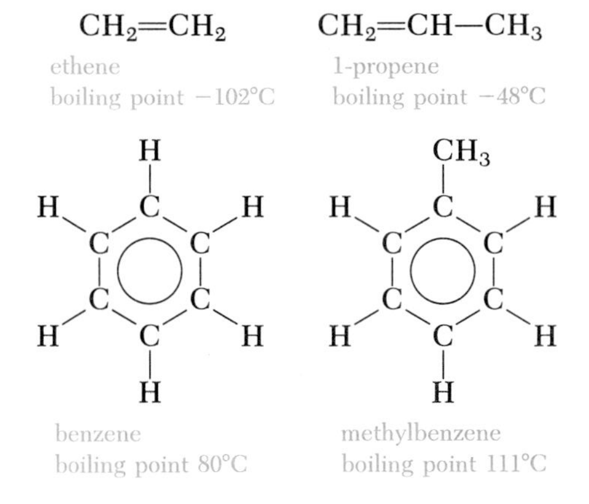

Exercise 19.15

Which compound would you expect to have the higher boiling point, methylbenzene or ethylbenzene?

Predictions of relative melting points or boiling points work well if the molecules being compared are members of the same series, that is, if both molecules are molecules of alkanes, alkenes, alkynes, or aromatic compounds; the predictions aren't reliable from one series to another, unless the molecules being compared contain very different numbers of atoms. For example, we can't make a reliable prediction of the relative boiling points of butane, $CH_3—CH_2—CH_2—CH_3$, and 2-butene, $CH_3—CH{=}CH—CH_3$, but it's safe to predict that octane, $CH_3—(CH_2)_6—CH_3$, will have a higher boiling point than ethene, $CH_2{=}CH_2$.

Exercise 19.16

Under ordinary room conditions, which compound is more likely to be a solid, propane or naphthalene? Explain your answer.

$CH_3—CH_2—CH_3$

propane

naphthalene

You can often predict correctly the relative melting points or boiling points of two isomeric hydrocarbons: The isomer whose molecule has the longer straight chain will have the higher melting point or boiling point. These are examples:

$$CH_3—CH_2—CH_2—CH_2—CH_3$$

pentane
boiling point 36°C

$$CH_3—\underset{}{\overset{CH_3}{\overset{|}{C}}}H—CH_2—CH_3$$

2-methylbutane
boiling point 28°C

An understanding of interparticle forces can also be used to explain the fact that hydrocarbons are insoluble in water.◆ The only attractive forces between hydrocarbon molecules and water molecules are weak London forces, so hydrocarbons don't dissolve in water.◆

◆You've probably seen examples of the insolubility of hydrocarbons: Lubricating oil and gasoline, which are mixtures of hydrocarbons, don't dissolve in water.

◆Sections 15-1 and 15-2 describe general principles for explaining water solubility in terms of intermolecular attractions.

19-7 The combustion of hydrocarbons produces about two thirds of the energy used in the United States each year

Hydrocarbons react with oxygen to produce carbon dioxide and water and release heat; the process is called **combustion.** For example, this is the equation for the combustion of ethane:

$$2\ CH_3—CH_3(g) + 5\ O_2(g) \rightarrow 2\ CO_2(g) + 6\ H_2O(\ell) + \text{heat}$$

The combustion reactions of hydrocarbons are the source of about two thirds of the energy used in the United States each year. Many kinds of hydrocarbons are burned to provide energy for many purposes. Methane, which occurs in underground deposits as natural gas, is used for generating electricity and for heating and cooking; propane, from petroleum, is used for heating and cooking in rural areas and as a fuel in some city buses. Complex mixtures of hydrocarbons, formulated from petroleum as gasolines and jet fuels, provide the energy that moves us, and moves the goods of commerce, from one place to another. Our heavy dependence on fuels from petroleum requires that we either maintain access to large supplies of petroleum or develop other fuels.

Oil and water don't mix because the only attractive forces between oil molecules and water molecules are weak London forces. *(©Kip Peticolas, Fundamental Photographs, New York)*

Gasoline is a mixture of hydrocarbons made from petroleum. Our enormous consumption of gasoline requires a huge distribution network to store and deliver it. *(Courtesy of Ashland Oil Company)*

Propane gas, $CH_3CH_2CH_3$, is used as a fuel in some rural areas. *(©Tom Hollyman/PhotoResearchers, Inc.)*

Exercise 19.17

About 20% of the energy used in the United States each year is produced from one chemical reaction: the combustion of methane. Write the equation for this reaction.

The worldwide combustion of hydrocarbons releases gigantic quantities of carbon dioxide, and the concentration of carbon dioxide in the atmosphere has increased by about 30% in the last century.◆ Carbon dioxide in the atmosphere traps heat on the surface of the earth, so the increase in atmospheric carbon dioxide may cause global warming.◆ Environmental scientists are intensively investigating the effects of increased atmospheric carbon dioxide, especially its potential for causing global warming.

◆The combustion of coal also produces carbon dioxide; coal is about 80% carbon and about 20% hydrocarbons. The combustion of coal provides about 25% of the energy used in the United States each year.

◆Carbon dioxide in the atmosphere, like the glass roof of a greenhouse, lets visible light in and traps heat, so global warming caused by carbon dioxide is called the *greenhouse effect*.

19-8 Hydrocarbons undergo two important kinds of reactions: substitution and addition

Hydrocarbons can undergo several kinds of reactions. For our purposes, two kinds will be important: *substitution reactions*, which are characteristic of alkanes and aromatic compounds, and *addition reactions*, which are characteristic of alkenes and alkynes.

Substitution Reactions of Alkanes and Aromatic Compounds

In a **substitution reaction** a hydrogen atom in a hydrocarbon molecule is replaced by another atom or group of atoms:

$$-\underset{|}{\overset{|}{C}}-H \rightarrow -\underset{|}{\overset{|}{C}}-Z$$

Substitution reactions are characteristic of alkanes and aromatic compounds.

Methane undergoes a substitution reaction with chlorine:

$$H-\underset{\underset{H}{|}}{\overset{\overset{H}{|}}{C}}-H + :\underset{\cdot\cdot}{\ddot{Cl}}-\underset{\cdot\cdot}{\ddot{Cl}}: \rightarrow H-\underset{\underset{H}{|}}{\overset{\overset{H}{|}}{C}}-\underset{\cdot\cdot}{\ddot{Cl}}: + H-\underset{\cdot\cdot}{\ddot{Cl}}:$$

If more molecules of chlorine are available, more hydrogen atoms will be replaced; for example:

$$H-\underset{\underset{H}{|}}{\overset{\overset{H}{|}}{C}}-H + 3\ :\underset{\cdot\cdot}{\ddot{Cl}}-\underset{\cdot\cdot}{\ddot{Cl}}: \rightarrow H-\underset{\underset{:\underset{\cdot\cdot}{Cl}:}{|}}{\overset{\overset{:\ddot{Cl}:}{|}}{C}}-\underset{\cdot\cdot}{\ddot{Cl}}: + 3\ H-\underset{\cdot\cdot}{\ddot{Cl}}:$$

Similar substitution reactions occur between methane or other alkanes and fluorine, chlorine, and bromine.

Exercise 19.18

Write an equation for the reaction of one molecule of methane with two molecules of bromine.

Substitution of different hydrogen atoms in a hydrocarbon molecule may produce isomers. For example, substitution on an end carbon atom in a propane molecule gives a different product than substitution on the central carbon atom:

$$CH_3{-}CH_2{-}CH_3 + Cl_2 \rightarrow CH_3{-}CH_2{-}CH_2{-}\ddot{\underset{..}{Cl}}: + H{-}\ddot{\underset{..}{Cl}}:$$

$$CH_3{-}CH_2{-}CH_3 + Cl_2 \rightarrow CH_3{-}\overset{\overset{:\ddot{Cl}:}{|}}{C}H{-}CH_3 + H{-}\ddot{\underset{..}{Cl}}:$$

Example 19.3

Write the formulas for all of the possible molecules that could be produced by substituting any two hydrogen atoms in a propane molecule with fluorine atoms.

Solution

$$CH_3{-}CH_2{-}\overset{\overset{:\ddot{F}:}{|}}{C}H{-}\ddot{\underset{..}{F}}: \qquad CH_3{-}\overset{\overset{:\ddot{F}:}{|}}{\underset{\underset{:\underset{..}{F}:}{|}}{C}}{-}CH_3$$

$$CH_3{-}\overset{\overset{:\ddot{F}:}{|}}{C}H{-}CH_2{-}\ddot{\underset{..}{F}}: \qquad :\ddot{F}{-}CH_2{-}CH_2{-}CH_2{-}\ddot{\underset{..}{F}}:$$

Exercise 19.19

Write the formulas for all of the possible molecules that could be produced by substituting any three hydrogen atoms in an ethane molecule with chlorine atoms.

Benzene undergoes a substitution reaction with chlorine:

$$C_6H_6 \text{ (benzene ring, each C bonded to H)} + :\ddot{\underset{..}{Cl}}{-}\ddot{\underset{..}{Cl}}: \xrightarrow{FeCl_3} C_6H_5{-}\ddot{\underset{..}{Cl}}: \text{ (benzene ring with one H replaced by } :\ddot{Cl}:) + H{-}\ddot{\underset{..}{Cl}}:$$

In this reaction iron(III) chloride is a catalyst. A **catalyst** is a substance that must be present with the reactants in order for a reaction to occur but that is unchanged at the end of the reaction. The presence of a catalyst is shown by writing the formula for the catalyst over the arrow in the equation for the reaction.[†] Benzene also undergoes substitution reactions with bromine, nitric acid, and sulfuric acid:

$$C_6H_6 + Br_2 \xrightarrow{FeBr_3} C_6H_5Br + HBr$$

$$C_6H_6 + HNO_3 \xrightarrow{H_2SO_4} C_6H_5NO_2 + H_2O$$

$$C_6H_6 + H_2SO_4 \rightarrow C_6H_5SO_3H + H_2O$$

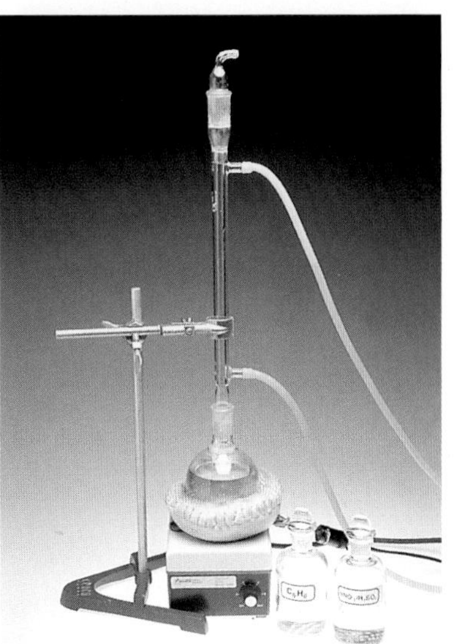

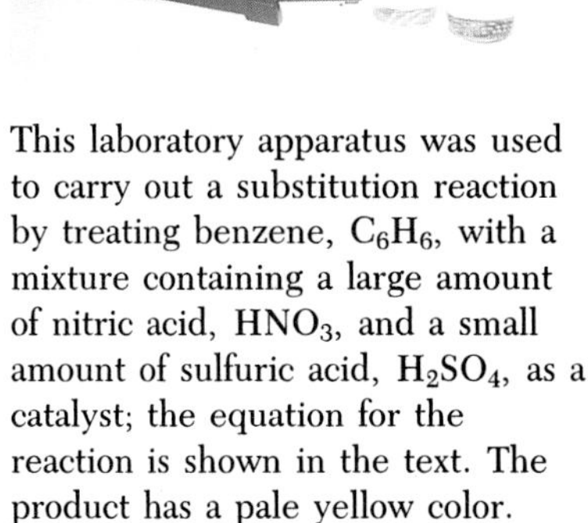

This laboratory apparatus was used to carry out a substitution reaction by treating benzene, C_6H_6, with a mixture containing a large amount of nitric acid, HNO_3, and a small amount of sulfuric acid, H_2SO_4, as a catalyst; the equation for the reaction is shown in the text. The product has a pale yellow color. *(Charles D. Winters)*

Other aromatic compounds undergo these and other substitution reactions to produce a wide variety of organic compounds.

Addition Reactions of Alkenes and Alkynes

In an **addition reaction** atoms add to a double or triple bond; addition reactions are characteristic of alkenes and alkynes, whose molecules contain double and

> [†] **Chemistry Insight**
>
> It would be more accurate to say that in the reaction of benzene with chlorine, iron(III) chloride must be present for the reaction to occur *at a reasonable rate;* without the catalyst the reaction would proceed too slowly to be worth carrying out. Catalysts increase reaction rates by interacting with reactants, and for many reactions, including the reaction of benzene with chlorine, these interactions are well understood. A detailed description of the function of catalysts belongs in more advanced chemistry books.

triple bonds. For example, a molecule of an alkene will react with a chlorine molecule to add chlorine atoms to the double bond:

$$CH_2{=}CH{-}CH_3 + :\overset{..}{\underset{..}{Cl}}{-}\overset{..}{\underset{..}{Cl}}: \rightarrow \begin{array}{ccc} CH_2 & - & CH{-}CH_3 \\ | & & | \\ :\underset{..}{Cl}: & & :\underset{..}{Cl}: \end{array}$$

A molecule of an alkyne will react with one or two chlorine molecules:

$$H{-}C{\equiv}C{-}CH_3 + :\overset{..}{\underset{..}{Cl}}{-}\overset{..}{\underset{..}{Cl}}: \rightarrow H{-}\begin{array}{ccc} C & = & C \\ | & & | \\ :\underset{..}{Cl}: & & :\underset{..}{Cl}: \end{array}{-}CH_3$$

$$H{-}C{\equiv}C{-}CH_3 + 2\ :\overset{..}{\underset{..}{Cl}}{-}\overset{..}{\underset{..}{Cl}}: \rightarrow H{-}\begin{array}{ccc} :\overset{..}{Cl}: & & :\overset{..}{Cl}: \\ | & & | \\ C & - & C \\ | & & | \\ :\underset{..}{Cl}: & & :\underset{..}{Cl}: \end{array}{-}CH_3$$

Similar addition reactions of alkenes and alkynes occur with hydrogen, bromine, and hydrogen bromide:

$$CH_2{=}CH{-}CH_3 + H{-}H \xrightarrow{Ni} CH_3{-}CH_2{-}CH_3$$

$$H{-}C{\equiv}C{-}CH_3 + :\overset{..}{\underset{..}{Br}}{-}\overset{..}{\underset{..}{Br}}: \longrightarrow \begin{array}{ccc} CH & = & C{-}CH_3 \\ | & & | \\ :\underset{..}{Br}: & & :\underset{..}{Br}: \end{array}$$

$$CH_2{=}CH{-}CH_3 + H{-}\overset{..}{\underset{..}{Br}}: \longrightarrow CH_3{-}\begin{array}{c} CH \\ | \\ :\underset{..}{Br}: \end{array}{-}CH_3$$

As this last example shows, in the addition of a hydrogen bromide molecule to a double or triple bond, the bromine atom becomes attached to the carbon atom that has fewer hydrogen atoms.

Exercise 19.20

Complete the equation for each reaction:

(a) $CH_2{=}CH_2 + H{-}H \xrightarrow{Ni}$

(b) $H{-}C{\equiv}C{-}H + :\overset{..}{\underset{..}{Br}}{-}\overset{..}{\underset{..}{Br}}: \longrightarrow$

(c) $CH_3{-}C{\equiv}C{-}H + H{-}\overset{..}{\underset{..}{Br}}: \longrightarrow$

VARIETIES OF ORGANIC COMPOUNDS

Millions of organic compounds are known whose molecules are combinations of carbon atoms with atoms of hydrogen, nitrogen, oxygen, sulfur, and the halogens. This part of the chapter describes simple examples of several varieties of these organic compounds.

19-9 Organic compounds are classified by identifying characteristic groupings of atoms in their molecules

The molecules of hydrocarbons—alkanes, alkenes, alkynes, and aromatic hydrocarbons—contain atoms of only two elements, hydrogen and carbon. Atoms of other elements can be attached to hydrocarbon molecules by addition or substitution reactions, as described in Section 19-8, and by a variety of other reactions. By these reactions millions of different compounds have been created.

The attachment of an atom of another element to a hydrocarbon molecule creates a characteristic grouping of atoms called a **functional group.** For example, the attachment of a chlorine atom creates the grouping

$$-\underset{|}{\overset{|}{C}}-\ddot{\underset{\cdot\cdot}{Cl}}:$$

Organic compounds can be classified by the functional groups in their molecules, as shown in Table 19.3 on page 466.

Each of the functional groups shown in Table 19.3 can be incorporated into many different hydrocarbon molecules, creating a nearly endless variety of organic compounds. In our further work in this chapter, we'll consider only the first three classes of compounds in Table 19.3: alcohols, amines, and carboxylic acids.

Exercise 19.21

Use Table 19.3 to identify the class of each compound:

$$\text{(a) } CH_3-\underset{\displaystyle CH_3}{\underset{|}{CH}}-\overset{\displaystyle :O:}{\overset{\|}{C}}-CH_3 \quad \text{(b) } CH_3-\underset{\displaystyle CH_3}{\underset{|}{CH}}-\overset{\displaystyle :O:}{\overset{\|}{C}}-H \quad \text{(c) } CH_3-\underset{\displaystyle CH_3}{\underset{|}{CH}}-\overset{\displaystyle :O:}{\overset{\|}{C}}-\ddot{\underset{\cdot\cdot}{O}}-H$$

Exercise 19.22

Use Table 19.3 to name each compound:

$$\text{(a) } CH_3-\ddot{\underset{\cdot\cdot}{O}}-H \quad \text{(b) } CH_3-\underset{\displaystyle H}{\underset{|}{\ddot{N}}}-H \quad \text{(c) } CH_3-CH_2-\overset{\displaystyle :O:}{\overset{\|}{C}}-\ddot{\underset{\cdot\cdot}{O}}-H$$

19-10 Simple alcohols resemble water

The characteristic functional group of **alcohols** is a hydroxyl group, $-\ddot{\underset{\cdot\cdot}{O}}-H$, attached to a carbon atom:

$$-\underset{|}{\overset{|}{C}}-\ddot{\underset{\cdot\cdot}{O}}-H$$

Table 19.3 Some Classes of Organic Compounds

Compound	Class of Compounds	Characteristic Functional Group
$CH_3-CH_2-\ddot{\underset{..}{O}}-H$ ethanol	alcohols	$-\overset{\vert}{\underset{\vert}{C}}-\ddot{\underset{..}{O}}-H$
$CH_3-CH_2-\ddot{\underset{\underset{H}{\vert}}{N}}-H$ ethanamine	amines	$-\overset{\vert}{\underset{\vert}{C}}-\ddot{\underset{\vert}{N}}-$
$CH_3-\overset{:O:}{\overset{\Vert}{C}}-\ddot{\underset{..}{O}}-H$ ethanoic acid	carboxylic acids	$-\overset{:O:}{\overset{\Vert}{C}}-\ddot{\underset{..}{O}}-H$
$CH_3-\overset{:O:}{\overset{\Vert}{C}}-H$ ethanal	aldehydes	$-\overset{:O:}{\overset{\Vert}{C}}-H$
$CH_3-CH_2-\ddot{\underset{..}{Cl}}:$ chloroethane	alkyl halides	$-\overset{\vert}{\underset{\vert}{C}}-\ddot{\underset{..}{X}}:$ X is F, Cl, Br, or I.
$CH_3-\overset{:O:}{\overset{\Vert}{C}}-CH_3$ 2-propanone	ketones	$-\overset{\vert}{\underset{\vert}{C}}-\overset{:O:}{\overset{\Vert}{C}}-\overset{\vert}{\underset{\vert}{C}}-$
$CH_3-C\equiv N:$ ethanenitrile	nitriles	$-\overset{\vert}{\underset{\vert}{C}}-C\equiv N:$
$CH_3-CH_2-\overset{:O:}{\overset{\Vert}{N}}-\ddot{\underset{..}{O}}:$ nitroethane	nitro compounds	$-\overset{\vert}{\underset{\vert}{C}}-\overset{:O:}{\overset{\Vert}{N}}-\ddot{\underset{..}{O}}:$
$CH_3-CH_2-\overset{:\ddot{O}:}{\underset{:\underset{..}{O}:}{S}}-\ddot{\underset{..}{O}}-H$ ethanesulfonic acid	sulfonic acids	$-\overset{\vert}{\underset{\vert}{C}}-\overset{:\ddot{O}:}{\underset{:\underset{..}{O}:}{S}}-\ddot{\underset{..}{O}}-H$
$CH_3-CH_2-\ddot{\underset{..}{S}}-H$ ethanethiol	thiols	$-\overset{\vert}{\underset{\vert}{C}}-\ddot{\underset{..}{S}}-H$

The system for naming simple alcohols is shown by these examples:

$CH_3—\ddot{\underset{..}{O}}—H$
methanol

$CH_3—CH_2—\ddot{\underset{..}{O}}—H$
ethanol

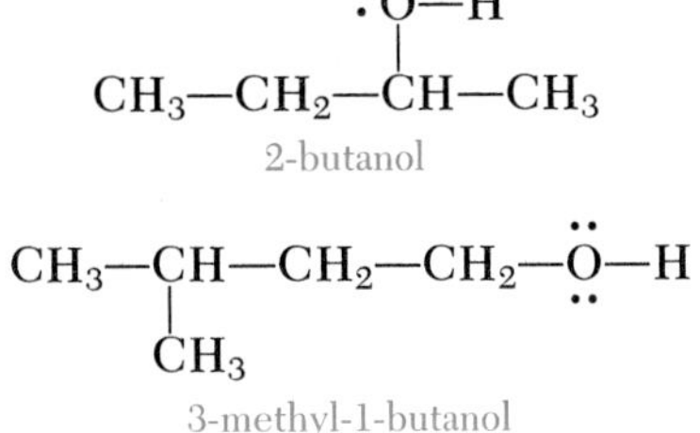

Name an alcohol from the corresponding alkane: Change the suffix of the alkane's name to *-ol*. If necessary, designate the location of the hydroxyl group by numbering the carbon chain from the end that assigns the lower number to the carbon atom attached to the hydroxyl group.

Methanol, CH_3OH, is a colorless, transparent liquid that boils at 64.7°C. *(Charles D. Winters)*

Exercise 19.23

Name: (a) $CH_3—CH_2—CH_2—\ddot{\underset{..}{O}}—H$ (b) $CH_3—\underset{CH_3}{\overset{:\ddot{O}—H}{C}}—CH_3$

Exercise 19.24

Write the formula for cyclopropanol.

The simple alcohols—those whose molecules contain only a few carbon atoms—resemble water. For example, methanol and ethanol are colorless liquids that boil at 65°C and 79°C, respectively; in each of these liquids the strongest intermolecular forces are hydrogen bonds (Section 14-4).

The solubility of the simple alcohols in water shows a general trend: Alcohols whose molecules have more carbon atoms are less soluble. For example, methanol and ethanol are infinitely soluble in water, 1-hexanol is slightly soluble, and 1-octanol is insoluble. This trend can be explained in terms of the interactions that occur between alcohol molecules and water molecules.◆ The hydroxyl group of an alcohol molecule is soluble in water because it forms hydrogen bonds with water molecules; these are examples of hydrogen bonds that can form between the hydroxyl group of a methanol molecule and water molecules:

◆Sections 15-1 and 15-2 describe general principles for explaining water solubility in terms of molecular interactions.

$$\begin{array}{c} H-\ddot{O}-H \\ \vdots \\ CH_3—\ddot{\underset{..}{O}}—H\cdots:\ddot{O}(H)_2 \end{array}$$

The hydrocarbon portion of an alcohol molecule is insoluble in water because it forms only weak London forces with water molecules.◆ Alcohols whose molecules have longer hydrocarbon portions are therefore less soluble.

◆We saw in Section 19-6 that hydrocarbons are insoluble in water.

Exercise 19.25

Use formulas to show hydrogen bonding between one ethanol molecule and three water molecules.

Many reactions of alcohols correspond to those of water. For example, water reacts with the alkali and alkaline-earth metals (Sections 10-1 and 10-2), and so do alcohols.◆

◆Alcohol molecules are bent at their oxygen atom as are water molecules, but the bent shapes of alcohol molecules aren't important for our purposes.

$$2\ Na\cdot + 2\ H{-}\ddot{\underset{\cdot\cdot}{O}}{-}H \rightarrow 2\ Na^+ + 2\ :\ddot{\underset{\cdot\cdot}{O}}{-}H^- + H{-}H$$

or

$$2\ Na + 2\ H_2O \rightarrow 2\ NaOH + H_2$$

$$2\ Na\cdot + 2\ CH_3{-}\ddot{\underset{\cdot\cdot}{O}}{-}H \rightarrow 2\ Na^+ + 2\ CH_3{-}\ddot{\underset{\cdot\cdot}{O}}:^- + H{-}H$$

or

$$2\ Na + 2\ CH_3OH \rightarrow 2\ NaOCH_3 + H_2$$

Calcium reacting with water. *(Chip Clark)*

Exercise 19.26

Calcium reacts with water:

$$Ca: + 2\ H{-}\ddot{\underset{\cdot\cdot}{O}}{-}H \rightarrow Ca^{2+} + 2\ :\ddot{\underset{\cdot\cdot}{O}}{-}H^- + H{-}H$$

or

$$Ca + 2\ H_2O \rightarrow Ca(OH)_2 + H_2$$

Write similar equations for the reaction of calcium with methanol.

When an acid dissolves in water, acid molecules transfer protons to water molecules (Section 15-5):

$$H_2\ddot{O}: + H{-}\ddot{\underset{\cdot\cdot}{Cl}}: \rightarrow H_2\ddot{O}{-}\overset{+}{H} + :\ddot{\underset{\cdot\cdot}{Cl}}:^-$$

A similar reaction occurs when an acid dissolves in an alcohol:

$$CH_3{-}CH_2{-}\ddot{\underset{\cdot\cdot}{O}}{-}H + H{-}\ddot{\underset{\cdot\cdot}{Cl}}: \rightarrow CH_3{-}CH_2{-}\underset{\cdot\cdot}{\overset{H}{\overset{|}{O}}}{-}\overset{+}{H} + :\ddot{\underset{\cdot\cdot}{Cl}}:^-$$

After an alcohol molecule accepts a proton from an acid molecule, a second reaction may occur. For example, the cation formed by adding a proton to ethanol, shown in the preceding equation, undergoes this reaction:

$$CH_3{-}CH_2{-}\underset{\cdot\cdot}{\overset{H}{\overset{|}{O}}}{-}\overset{+}{H} \rightarrow H^+ + CH_2{=}CH_2 + :\ddot{O}\begin{smallmatrix}H\\H\end{smallmatrix}$$

The curved arrows show movements of electron pairs: two bonds break, and one new bond forms.

$$H{-}\underset{H}{\overset{H}{C}}{-}\underset{H}{\overset{H}{C}}{-}\underset{\cdot\cdot}{\overset{H}{\overset{|}{O}}}{-}\overset{+}{H} \rightarrow H^+ + CH_2{=}CH_2 + :\ddot{O}\begin{smallmatrix}H\\H\end{smallmatrix}$$

Exercise 19.27

Write equations for the reactions of hydrogen chloride with 1-propanol to produce 1-propene.

Several simple alcohols are commercially important, as shown by the following examples.

$$CH_3-\ddot{O}-H$$

Methanol, also called methyl alcohol or wood alcohol, is used as a solvent in many industrial chemical processes. It's also used as an octane booster in gasolines and as an antifreeze agent in diesel oils.

$$CH_3-CH_2-\ddot{O}-H$$

Ethanol, also called ethyl alcohol, is the intoxicating agent in alcoholic beverages. It's also used as a solvent or reactant in industrial chemical reactions.

$$CH_3-CH(\ddot{O}H)-CH_3$$

2-Propanol, also called isopropyl alcohol or rubbing alcohol, is widely used as an industrial solvent and in such products as hand lotions and after-shave lotions.

$$H-\ddot{O}-CH_2-CH_2-\ddot{O}-H$$

This alcohol, commercially known as ethylene glycol, is used in large quantities as an automobile antifreeze. It's also used as an industrial solvent and as a raw material in the manufacture of plastics, as described in Section 19-13.

$$CH_2(\ddot{O}H)-CH(\ddot{O}H)-CH_2(\ddot{O}H)$$

Commercially known as glycerol or glycerin, this alcohol is used in large quantities as a sweetening agent in candies, as a raw material in the manufacture of nitroglycerin (Table 7.2), as a lubricant in cosmetics, as an antifrosting agent for windshields, and for many other purposes.

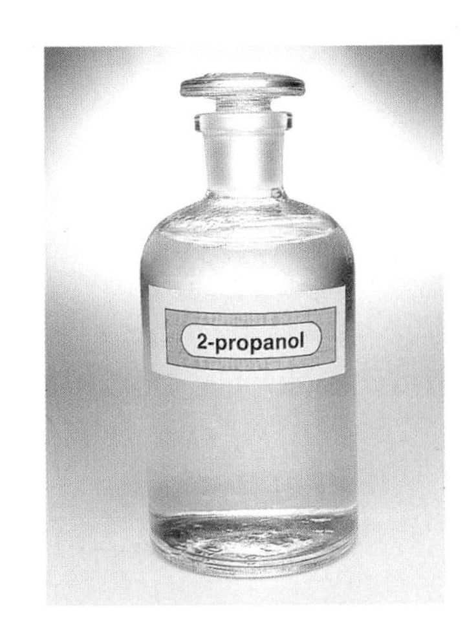

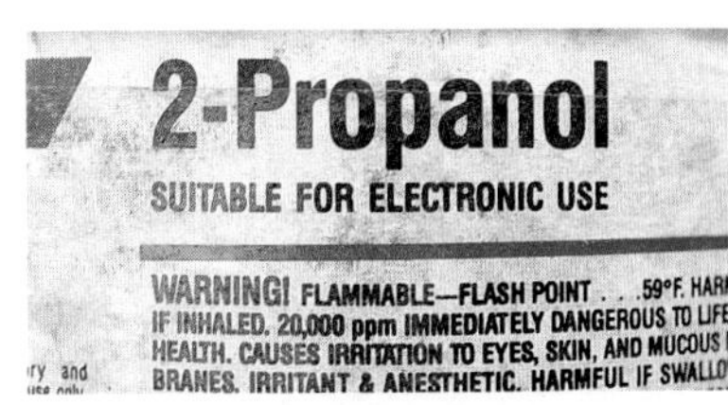

2-Propanol, $CH_3CH(OH)CH_3$, is a colorless, transparent liquid that boils at 82.5°C (top). Among the many industrial uses of 2-propanol is its use as a solvent to clean the components of electronic devices (bottom). *(Charles D. Winters)*

Because products such as rubbing alcohol and antifreeze contain alcohols, people sometimes drink them as substitutes for alcoholic beverages, with tragic results: Methanol, 2-propanol, and ethylene glycol are very poisonous.

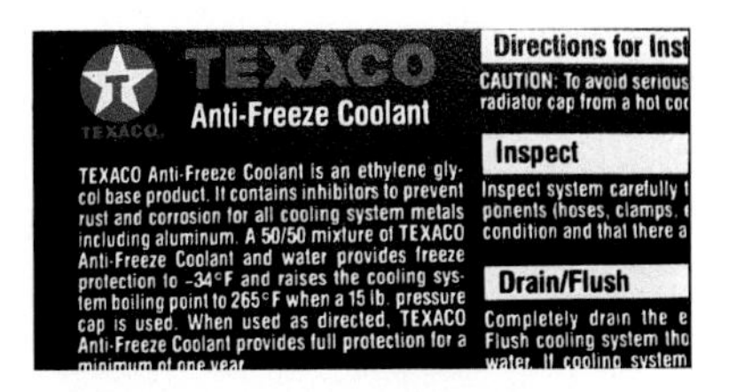

Ethylene glycol, $HOCH_2CH_2OH$, is used as an antifreeze agent in automobile radiators. *(Charles D. Winters)*

19-11 Simple amines resemble ammonia

The characteristic functional group of **amines** is a nitrogen atom attached to a carbon atom:

$$-\overset{|}{\underset{|}{C}}-\overset{..}{\underset{|}{N}}-$$

One, two, or three carbon atoms can be attached to the nitrogen atom; for example,

$$CH_3-\overset{..}{\underset{|}{N}}-H \quad (H) \qquad CH_3-\overset{..}{\underset{|}{N}}-CH_3 \quad (H) \qquad CH_3-\overset{..}{\underset{|}{N}}-CH_3 \quad (CH_3)$$

Glycerol, $HOCH_2CH(OH)CH_2OH$, is used as a sweetening agent in candies. *(Charles D. Winters)*

We'll limit our attention to amine molecules in which one carbon atom is bonded to the nitrogen atom; in these molecules, the group $-\ddot{N}(H)-H$ is called an **amino group.**

The system for naming simple amines is shown by these examples:

$$CH_3-\ddot{N}H-H$$

methanamine

$$CH_3-CH_2-\ddot{N}H-H$$

ethanamine

Name an amine from the corresponding alkane: Change the suffix of the alkane's name to *-amine*. If necessary, designate the location of the amino group by numbering the carbon chain from the end that assigns the lower number to the carbon atom attached to the amino group.

$$CH_3-CH_2-CH(:NH_2)-CH_3$$

2-butanamine

$$CH_3-CH(CH_3)-CH(:NH_2)-CH_3$$

3-methyl-2-butanamine

Exercise 19.28

Name $CH_3-CH(CH_3)-\ddot{N}H-H$

Under ordinary room conditions methanamine is a gas and the other simple amines are liquids. All of the simple amines have an ammonia-like odor.

The solubility of amines in water shows the same general trend as the solubility of alcohols, for similar reasons. In an amine molecule the amino group is soluble in water because it forms hydrogen bonds with water molecules, and the hydrocarbon portion is insoluble because it forms only London forces with water molecules. Amines whose molecules have larger hydrocarbon portions are therefore less soluble.

The simple amines resemble ammonia in being weak bases:◆ Amine molecules accept protons from water molecules by an equilibrium reaction. These are the reactions of ammonia and methanamine with water:

◆Ammonia is described as a weak base in Section 15-6.

$$H-\ddot{N}H-H(aq) + H_2\ddot{O}(\ell) \leftrightharpoons H-NH_2^{+}-H(aq) + :\ddot{O}-H^-(aq)$$

$$CH_3-\ddot{N}H-H(aq) + H_2\ddot{O}(\ell) \leftrightharpoons CH_3-NH_2^{+}-H(aq) + :\ddot{O}-H^-(aq)$$

The equilibrium for methanamine, like the equilibrium for ammonia, lies far to the left; at equilibrium most of the methanamine is in its molecular form, and only a small amount has reacted with water to form ions. The same is true for the other simple amines.

Exercise 19.29

Write an equation for the equilibrium reaction of ethanamine with water.

Because they are bases, simple amines react with acids. This is the molecular equation for the neutralization of aqueous methanamine with hydrochloric acid:◆

$$CH_3\text{—}\ddot{N}(H)\text{—}H(aq) + H\text{—}\ddot{\underset{\cdot\cdot}{Cl}}\!:(aq) \rightarrow CH_3\text{—}N(H)(H)\text{—}\overset{+}{H}(aq) + :\ddot{\underset{\cdot\cdot}{Cl}}\!:^{-}(aq)$$

◆The use of molecular equations to describe neutralization reactions is discussed in Section 16-3.

Exercise 19.30

Use line formulas to write a molecular equation for the neutralization of aqueous methanamine with nitric acid.

19-12 Simple organic acids are weak acids

The characteristic functional group of organic acids is the **carboxyl group:**

$$\text{—}\overset{\overset{:\ddot{O}:}{\|}}{C}\text{—}\ddot{\underset{\cdot\cdot}{O}}\text{—}H$$

Organic acids are often called **carboxylic acids.**

The system for naming simple organic acids is shown by these examples:

$$H\text{—}\overset{\overset{:O:}{\|}}{C}\text{—}\ddot{\underset{\cdot\cdot}{O}}\text{—}H$$

methanoic acid or formic acid

$$CH_3\text{—}\overset{\overset{:O:}{\|}}{C}\text{—}\ddot{\underset{\cdot\cdot}{O}}\text{—}H$$

ethanoic acid or acetic acid

$$CH_3\text{—}\underset{\underset{CH_3}{|}}{CH}\text{—}\overset{\overset{:O:}{\|}}{C}\text{—}\ddot{\underset{\cdot\cdot}{O}}\text{—}H$$

2-methylpropanoic acid

Name a simple acid from the corresponding alkane: Change the suffix of the alkane's name to *-oic* and add *acid*. Most simple acids have traditional names as well as systematic names. For example, methanoic acid and ethanoic acid are almost always called formic acid and acetic acid, respectively, so we'll use these names for them.◆

Acetic acid, $HC_2H_3O_2$, has many uses; for example, it's a component of solutions used to develop photographs. *(Charles D. Winters)*

◆Formic acid is named from the Latin word *formica*, which means "ant;" some ants inject formic acid as an irritant when they bite. Acetic acid, which occurs in vinegar, is named from the Latin word for vinegar, *acetum*.

Exercise 19.31

Name $CH_3{-}CH_2{-}CH_2{-}\overset{\overset{\Large :O:}{\|}}{C}{-}\ddot{\underset{\cdot\cdot}{O}}{-}H$

The simple organic acids are colorless liquids or white solids under ordinary room conditions. Several of them have characteristic unpleasant odors; butanoic acid, for example, smells like sweat, and octanoic acid smells like goats.

The solubility of acids in water shows the same general trend as the solubility of alcohols and amines, for similar reasons. In the molecule of a carboxylic acid the carboxyl group is soluble in water because it forms hydrogen bonds with water molecules, and the hydrocarbon portion is insoluble because it forms only London forces with water molecules. Acids whose molecules have longer hydrocarbon portions are therefore less soluble.

◆Weak acids, including acetic acid, are described in Section 15-5.

The simple organic acids are weak acids.◆ The equilibrium reaction of acetic acid with water is:

$$CH_3{-}\overset{\overset{:O:}{\|}}{C}{-}\ddot{\underset{\cdot\cdot}{O}}{-}H(aq) + H_2\ddot{O}:(\ell) \leftrightharpoons CH_3{-}\overset{\overset{:O:}{\|}}{C}{-}\ddot{\underset{\cdot\cdot}{O}}:^-(aq) + H_2\ddot{O}{-}H^+(aq)$$

At equilibrium most of the acetic acid is in its molecular form, and only a small amount has reacted with water to form ions. The same is true for the other simple organic acids.

Exercise 19.32

Write an equation for the equilibrium reaction of formic acid with water.

Carboxylic acids undergo neutralization reactions with bases. In these neutralizations, the essential reaction is the transfer of a proton from a carboxyl group to a hydroxide ion:

$$\underset{\text{carboxyl group of acid}}{{-}\overset{\overset{:O:}{\|}}{C}{-}\ddot{\underset{\cdot\cdot}{O}}{-}H} + \underset{\text{hydroxide ion of base}}{:\ddot{\underset{\cdot\cdot}{O}}{-}H^-} \rightarrow {-}\overset{\overset{:O:}{\|}}{C}{-}\ddot{\underset{\cdot\cdot}{O}}:^- + H{-}\ddot{\underset{\cdot\cdot}{O}}{-}H$$

◆The use of ionic equations to describe neutralization reactions is described in Section 16-3.

This is the ionic equation for the neutralization of aqueous acetic acid with aqueous sodium hydroxide:◆

$$CH_3{-}\overset{\overset{:O:}{\|}}{C}{-}\ddot{\underset{\cdot\cdot}{O}}{-}H(aq) + Na^+(aq) + :\ddot{\underset{\cdot\cdot}{O}}{-}H^-(aq)$$
$$\rightarrow CH_3{-}\overset{\overset{:O:}{\|}}{C}{-}\ddot{\underset{\cdot\cdot}{O}}:^-(aq) + Na^+(aq) + H{-}\ddot{\underset{\cdot\cdot}{O}}{-}H(\ell)$$

Exercise 19.33

Write an ionic equation for the neutralization of two moles of acetic acid with one mole of calcium hydroxide.

In the presence of acid catalysts, carboxylic acids react with alcohols to form the class of compounds called *esters* and with amines to form the class of compounds called *amides*. The following equations show examples of these reactions.

$$CH_3-\overset{\overset{:O:}{\|}}{C}-\ddot{\underset{..}{O}}-H + CH_3-\ddot{\underset{..}{O}}-H \xrightarrow{HCl} CH_3-\boxed{\overset{\overset{:O:}{\|}}{C}-\ddot{\underset{..}{O}}-}CH_3 + H-\ddot{\underset{..}{O}}-H$$

This functional group identifies an ester.

$$CH_3-\overset{\overset{:O:}{\|}}{C}-\ddot{\underset{..}{O}}-H + CH_3-\underset{\underset{H}{|}}{\ddot{N}}-H \xrightarrow{HCl} CH_3-\boxed{\overset{\overset{:O:}{\|}}{C}-\underset{\underset{H}{|}}{\ddot{N}}-}CH_3 + H-\ddot{\underset{..}{O}}-H$$

This functional group identifies an amide.

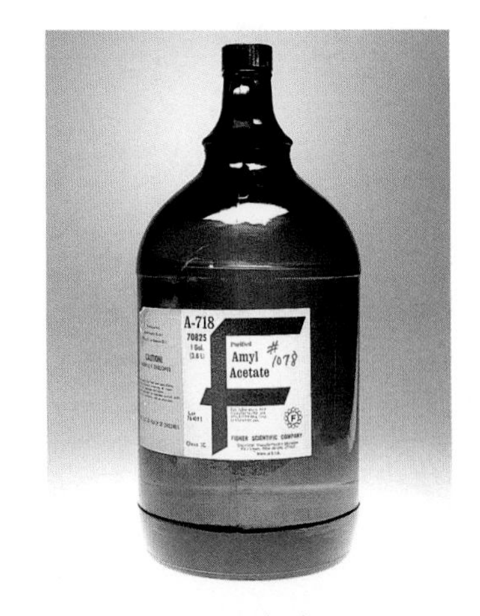

Amyl acetate—a colorless, transparent liquid that boils at 142°C—is an ester of acetic acid that's also called pear oil or banana oil because of its odor and taste. Its line formula is

$$CH_3-\overset{\overset{:O:}{\|}}{C}-\ddot{\underset{..}{O}}-CH_2-CH_2-\underset{\underset{CH_3}{|}}{CH}-CH_3$$

(Charles D. Winters)

Exercise 19.34

Complete the following equations.

(a) $CH_3-\overset{\overset{:O:}{\|}}{C}-\ddot{\underset{..}{O}}-H + CH_3-CH_2-\ddot{\underset{..}{O}}-H \xrightarrow{HCl}$

(b) $H-\overset{\overset{:O:}{\|}}{C}-\ddot{\underset{..}{O}}-H + H-\underset{\underset{H}{|}}{\ddot{N}}-H \xrightarrow{HCl}$

Esters and amides are important in *polymers*, as described in Section 19-13; they are also important in biochemical systems, as described in Chapter 20.

19-13 A polymer is an organic compound whose molecules are very long chains with repeating sequences of atoms

We saw in Section 19-12 that carboxylic acids react with amines to produce amides; one example is

$$CH_3-\overset{\overset{:O:}{\|}}{C}-\ddot{\underset{..}{O}}-H + CH_3-\underset{\underset{H}{|}}{\ddot{N}}-H \xrightarrow{HCl} CH_3-\overset{\overset{:O:}{\|}}{C}-\underset{\underset{H}{|}}{\ddot{N}}-CH_3 + H-\ddot{\underset{..}{O}}-H$$

Reactions of carboxylic acids with amines can be used to create very long chains of atoms. To create a long chain, a carboxylic acid whose molecule contains *two* carboxyl groups is mixed with an amine whose molecule contains *two* amino groups. For example:

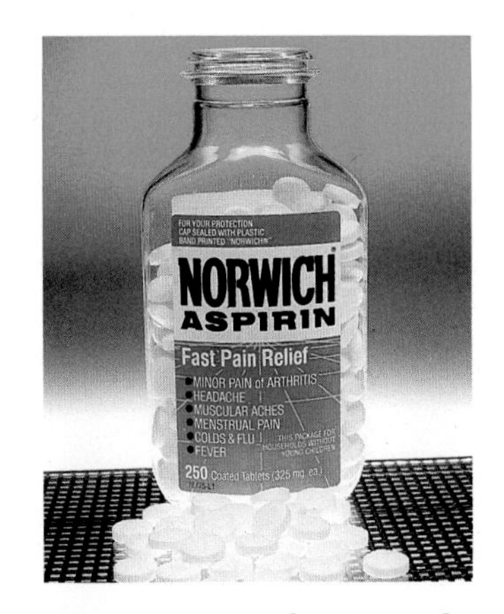

Aspirin is an ester of acetic acid. Its line formula is

[Structural formula of aspirin: benzene ring (C atoms with H substituents) bearing a $-\overset{\overset{:O:}{\|}}{C}-\ddot{\underset{..}{O}}-H$ group and, on the adjacent carbon, a $-\ddot{\underset{..}{O}}-\overset{\overset{:O:}{\|}}{C}-CH_3$ group]

(Charles D. Winters)

$$\mathrm{H{-}O{-}\overset{\overset{\displaystyle O}{\|}}{C}{-}(CH_2)_6{-}\overset{\overset{\displaystyle O}{\|}}{C}{-}O{-}H + H{-}\underset{\underset{\displaystyle H}{|}}{N}{-}(CH_2)_6{-}\underset{\underset{\displaystyle H}{|}}{N}{-}H}$$

$$\xrightarrow{\mathrm{HCl}} \mathrm{H{-}O{-}\overset{\overset{\displaystyle O}{\|}}{C}{-}(CH_2)_6{-}\overset{\overset{\displaystyle O}{\|}}{C}{-}\underset{\underset{\displaystyle H}{|}}{N}{-}(CH_2)_6{-}\underset{\underset{\displaystyle H}{|}}{N}{-}H + H{-}O{-}H}$$

The amide molecule produced in this reaction has a carboxyl group at one end, which can react with another amine molecule, and an amino group at the other end, which can react with another acid molecule. Successive reactions at each end of the molecule eventually produce a long chain that can be represented by this formula:

$$\mathrm{H{-}O{-}\overset{\overset{\displaystyle O}{\|}}{C}{-}(CH_2)_6{-}\overset{\overset{\displaystyle O}{\|}}{C}{-}\left[\underset{\underset{\displaystyle H}{|}}{N}{-}(CH_2)_6{-}\underset{\underset{\displaystyle H}{|}}{N}{-}\overset{\overset{\displaystyle O}{\|}}{C}{-}(CH_2)_6{-}\overset{\overset{\displaystyle O}{\|}}{C}\right]_n{-}\underset{\underset{\displaystyle H}{|}}{N}{-}(CH_2)_6{-}\underset{\underset{\displaystyle H}{|}}{N}{-}H}$$

In this formula the repeating sequence is shown in brackets, and n is some large number. The molecule represented by this formula is called a **polyamide** because it contains many amide linkages. A polyamide is one kind of polymer; a **polymer** is an organic compound whose molecules are very long chains that contain repeating sequences of atoms.◆ The polyamide whose formula is shown above is known commercially as nylon.

◆*Polymer* means "many parts."

A reaction between the two layers in the beaker produces nylon, which is wound onto the red glass rod. *(Charles D. Winters)*

In the reactions that produce polymers, the molecules that are formed typically have various lengths; the number (n) of repeating units in a polymer molecule is typically from several hundred to several thousand. Polymer molecules are therefore gigantic molecules with enormous molecular masses.

Names for many polymers or their abbreviations have become part of our everyday language: acrylic, alkyd, Dacron®, Formica®, nylon, Orlon®, polyester, polystyrene, polyurethane, PVC, rayon, Teflon®, vinyl, and so on. None of these terms existed sixty years ago. Their rapid spread into our everyday vocabulary is a measure of the rapid increase in the number of valuable uses we make of polymers.

Polymers have become so much a part of our everyday lives that it takes an effort of imagination to recognize how many uses we make of them. Look for a moment around your room. Are the walls painted? The paints are polymers. Is the floor carpeted? The fibers in the carpet are polymers. Are you working at a desk? The coating on its surface is a polymer. Do you wear glasses? The frames and lenses are probably polymers. Think for a moment of the objects you'll touch today—many of them, probably most of them, will be made of polymers: the case and buttons for your alarm clock, the containers for your milk and orange juice, the handle and bristles of your toothbrush, the bottle for your vitamin pills, your comb or brush, your pen, the keyboard of your typewriter or computer, at least some of your clothes, the steering wheel and upholstery of your car, the packages for your food, the tapes or disks for your stereo set, the case and buttons of your telephone, the cases and control knobs on your radio and television set, and probably the sheets you sleep between and the mattress you sleep on. The applications of polymers are a continuing revolution in the creation of consumer products.

The use of a polymer in creating consumer goods depends on its properties, and its properties depend on the structure of its molecules. For example, polyamides like the nylon polymer described earlier are easily formed into fibers that can be used to make textiles and ropes. Other kinds of polymers made from other kinds of reactants have properties that allow them to be formed into other kinds of shapes: films, foams, sheets, spheres, rods, tubes, and so forth.◆

Many thousands of polymers have been made and more are being created, so it's convenient to have a way to classify them. A polymer is commonly classified in two ways: by the kind of functional group that's repeated in the chain of its molecule—we've seen the example of a polyamide—and by the kind of reaction that produces the polymer. Two kinds of reactions are commonly used to produce polymers, *condensation polymerization* and *addition polymerization*.

These products are made from condensation polymers: the face shield from a polycarbonate, the twine from a polyamide, and the audio and video tapes from polyesters. *(Charles D. Winters)*

◆A *plastic* is a polymer that can be shaped with heat and pressure.

In the usual form of **condensation polymerization,** two reactants form a polymer and water. For example, the formation of a polyamide is a condensation polymerization that produces one water molecule for each amide linkage in the polyamide molecule:

$$(n+1)H-O-C(=O)-(CH_2)_6-C(=O)-O-H + (n+1)H-N(H)-(CH_2)_6-N(H)-H \xrightarrow{HCl}$$

$$H-O-C(=O)-(CH_2)_6-C(=O)-\left[N(H)-(CH_2)_6-N(H)-C(=O)-(CH_2)_6-C(=O)\right]_n-N(H)-(CH_2)_6-N(H)-H + n\,H-O-H$$

A polymer that's produced by condensation polymerization is classified as a **condensation polymer;** a polyamide is a condensation polymer.

Table 19.4 on pages 476 and 477 shows three examples of important condensation polymers. As shown in this table, a **monomer** is a reactant that forms a polymer.◆

◆*Monomer* means "one part."

Exercise 19.35

Use Table 19.4 to identify this polymer as a polyamide, polycarbonate, or polyester:

$$H-O-C(=O)-(CH_2)_4-C(=O)-\left[O-(CH_2)_4-O-C(=O)-(CH_2)_4-C(=O)\right]_n-O-(CH_2)_4-O-H$$

Exercise 19.36

For the polymer whose formula is shown in Exercise 19.35, write formulas for the corresponding monomers.

In **addition polymerization** monomer molecules with double bonds undergo successive addition reactions to form a chain; the polymer that's

Table 19.4 Three Important Condensation Polymers

Classification by Functional Group	Monomers	Polymer and Uses
Polyamide	$H{-}O{-}C({=}O){-}(CH_2)_6{-}C({=}O){-}O{-}H$ $H{-}N(H){-}(CH_2)_6{-}N(H){-}H$	$H{-}O{-}C({=}O){-}(CH_2)_6{-}C({=}O){-}[N(H){-}(CH_2)_6{-}N(H){-}C({=}O){-}(CH_2)_6{-}C({=}O)]_n{-}N(H){-}(CH_2)_6{-}N(H){-}H$ As synthetic fibers (nylon). Representative products are women's stockings and waterproof rope.
Polycarbonate	$H{-}O{-}C_6H_4{-}C(CH_3)_2{-}C_6H_4{-}O{-}H$ $Cl{-}C({=}O){-}Cl$	As clear plastic sheets (Lexan ®). Representative products are bulletproof glass and protective visors for helmets. $Cl{-}C({=}O){-}[O{-}C_6H_4{-}C(CH_3)_2{-}C_6H_4{-}O{-}C({=}O)]_n{-}O{-}C_6H_4{-}C(CH_3)_2{-}C_6H_4{-}O{-}H$

continued next page

Table 19.4 Three Important Condensation Polymers *(continued)*

Classification by Functional Group	Monomers	Polymer and Uses
Polyester	$H{-}O{-}C(=O){-}C_6H_4{-}C(=O){-}O{-}H$ $H{-}O{-}CH_2{-}CH_2{-}O{-}H$	$H{-}O{-}C(=O){-}C_6H_4{-}C(=O){-}[O{-}CH_2{-}CH_2{-}O{-}C(=O){-}C_6H_4{-}C(=O)]_n{-}O{-}CH_2{-}CH_2{-}O{-}H$ As fibers (Dacron ®); representative products are clothing and surgical patches for arteries. As films (Mylar ®); representative products are magnetic tapes for audio and video recording.

◆ Many addition polymerizations, including the polymerization of ethylene, require complex catalysts that we won't consider.

produced is classified as an **addition polymer.** The simplest addition polymer is polyethylene:◆

$$\text{n } CH_2{=}CH_2 \xrightarrow{\text{catalyst}} \left(\!-CH_2{-}CH_2-\!\right)_n$$

ethene, also called ethylene — growing chain of polyethylene

In the molecule of an addition polymer the ends of the chain are commonly occupied by atoms picked up at random from the environment; for example, a chain may end with a hydrogen atom or a hydroxyl group scavenged from a water molecule. For simplicity we'll show hydrogen atoms at the ends of these molecules, for example,

$$H\left(\!-CH_2{-}CH_2-\!\right)_nH$$

Polyethylene

Table 19.5 shows examples of important addition polymers.

Example 19.4

Use Table 19.5 to write the formula for Teflon®.

Solution

The formula for the monomer is $F_2C{=}CF_2$ (each F shown with three lone pairs: :F̤̈:)

so the formula for the addition polymer will be

$$H\left[\begin{array}{cc} F & F \\ | & | \\ -C- & -C- \\ | & | \\ F & F \end{array}\right]_n H$$

Exercise 19.37

Use Table 19.5 to write the formula for PVC.

These products are made from addition polymers: the packing material from polystyrene, the non-stick lining of the pan from a polymer of tetrafluoroethylene, the socks and sweater from polyacrylonitrile, the squeeze bottle from polyethylene, the paint from polyvinylacetate, the clear plastic sheet from a polymer of methyl methacrylate, and the plastic pipe from polyvinyl chloride. *(Charles D. Winters)*

Table 19.5 Important Addition Polymers*

Monomer	Common Name of Monomer	Name of Polymer (Trade Name)	Uses
$H_2C{=}CH_2$	Ethylene	Polyethylene (Polythene)	Squeeze bottles, bags, films, toys, electrical insulation
$H_2C{=}CH{-}CH_3$	Propylene	Polypropylene (Vectra, Herculon)	Bottles, films, indoor–outdoor carpets
$H_2C{=}CH{-}Cl$	Vinyl chloride	Polyvinyl chloride (PVC)	Floor tile, raincoats, pipe
$H_2C{=}CH{-}C{\equiv}N$	Acrylonitrile	Polyacrylonitrile (Orlon, Acrilan)	Rugs, fabrics
$H_2C{=}CH{-}C_6H_5$	Styrene	Polystyrene (Styrofoam, Styron)	Food and drink coolers, building insulation
$H_2C{=}CH{-}O{-}C({=}O){-}CH_3$	Vinyl acetate	Polyvinylacetate (PVA)	Latex paints, adhesives, textile coatings
$H_2C{=}C(CH_3){-}C({=}O){-}O{-}CH_3$	Methyl methacrylate	(Plexiglas, Lucite)	High-quality transparent objects, latex paints, contact lenses
$F_2C{=}CF_2$	Tetrafluoroethylene	(Teflon)	Gaskets, insulation, bearings, pan coatings

*From Joesten, et al., *World of Chemistry*. Philadelphia: Saunders College Publishing, 1991.

Inside Chemistry What happens when TNT explodes?

The explosive commonly called TNT is an aromatic compound whose molecule consists of a methyl group and three nitro groups attached to a benzene ring; the designation TNT is an abbreviation for *trinitrotoluene:*

benzene

methylbenzene or toluene

trinitrotoluene or TNT

TNT is produced commercially by a substitution reaction in which nitro groups are attached to a toluene molecule:

$$\text{toluene} + 3\,H{-}\ddot{O}{-}N(=O){-}\ddot{O}: \xrightarrow{H_2SO_4} \text{TNT} + 3\,H{-}\ddot{O}{-}H$$

The raw materials for this reaction—toluene, nitric acid, and sulfuric acid—are plentiful and inexpensive, so TNT is a relatively cheap explosive.

TNT, whose molecular formula is $C_7H_5N_3O_6$, is a yellow solid that explodes with terrific shattering force. When it explodes, TNT reacts with oxygen from the air to form a variety of products; for our purposes we can describe these products as nitrogen dioxide, carbon dioxide, and water:

$$C_7H_5N_3O_6(s) + 45\ O_2(g) \rightarrow 12\ NO_2(g) + 28\ CO_2(g) + 10\ H_2O(g)$$

The explosion that occurs when TNT reacts with oxygen results from two features of the reaction: The reaction occurs very fast, and it creates a large volume of gases from a small volume of solid. The increase in volume that occurs in a fraction of a second exerts devastating pressures on surrounding objects.

Explosive demolition of a building in Springfield, Illinois. *[(Barbara Van Cleve)/Tony Stone Worldwide]*

Chapter Summary: Organic Chemistry

Section	Subject	Summary	Check When Learned
	Organic compounds	Covalent compounds whose molecules contain carbon atoms.	☐
	Organic chemistry	Study of organic compounds.	☐
19-1	Electron configuration for carbon	C^6: $1s^2 2s^2 2p^2$	☐
19-1	Lewis symbol for carbon	$\cdot\dot{C}:$	☐
19-1	Number of covalent bonds formed by a carbon atom	4	☐
19-1	Meaning of a line symbol for an atom	In a line symbol, each line represents a valence electron the atom can share; that is, each line represents half of a covalent bond the atom can form.	☐
19-1	Line symbol for		
	carbon	—C— (with one line above and one below)	
	hydrogen	H—	
	nitrogen	—N— (two dots above, one line below)	
	oxygen	—O— (two dots above and below)	
	sulfur	—S— (two dots above and below)	
	fluorine	:F— (two dots above and below)	
	chlorine	:Cl— (two dots above and below)	
	bromine	:Br— (two dots above and below)	
	iodine	:I— (two dots above and below)	☐
19-2	Designate whether a carbon atom will form a single, double, or triple bond with an atom of:		
	carbon	single, double, or triple	
	hydrogen	single	
	nitrogen	single, double, or triple	

Section	Subject	Summary	Check When Learned
	oxygen sulfur fluorine chlorine bromine iodine	single or double single or double single single single single	☐
19-3	Hydrocarbon	Compound whose molecule contains only hydrogen atoms and carbon atoms.	☐
19-3	Cyclic compound	Compound whose molecule contains a ring of atoms, CH_2-CH_2 for example, CH_2-CH_2. (ring: CH_2-CH_2 over CH_2-CH_2, joined vertically)	☐
19-3	Isomers	Compounds whose molecules contain the same numbers of atoms of the same elements in different bonding arrangements; compounds with the same molecular formula but different line formulas.	☐
19-4	Alkanes	Hydrocarbons whose molecules contain only single bonds.	☐
19-4	Alkenes	Hydrocarbons whose molecules contain double bonds.	☐
19-4	Alkynes	Hydrocarbons whose molecules contain triple bonds.	☐
19-4	System for naming: Alkanes	Prefix designates number of carbon atoms in longest chain in molecule, and suffix is *-ane;* for example, $CH_3-CH_2-CH_2-CH_2-CH_3$ is pentane. If the chain is branched, use the suffix *-yl* for each branch, and number the chain from the end that assigns the lowest numbers to the branches; for example, the chain in CH_3-CH_2 $\quad$ CH_3 $CH_3-CH_2-CH-CH_2-C-CH_3$ CH_3 is numbered from the right end, and the compound is 4-ethyl-2,2-dimethylhexane.	
	Cycloalkanes	Add the prefix *cyclo-* to the name for the corresponding CH_2 alkane; for example, CH_2 $\quad$ CH_2 is cyclopentane. CH_2-CH_2	
	Alkenes	Change the suffix of the name for the corresponding alkane to *-ene;* if necessary, number the chain from the end that assigns the lower number to the first carbon atom attached to the double bond. For example, $CH_3-CH=CH-CH_2-CH_3$ is 2-pentene.	

Section	Subject	Summary	Check When Learned
	Alkynes	Change the suffix of the name for the corresponding alkane to *-yne;* if necessary, number the chain from the end that assigns the lower number to the first carbon atom attached to the triple bond. For example, $CH_3—CH_2—CH_2—C≡C—H$ is 1-pentyne.	□
19-5	Formula for benzene	H H C H C C C C H C H H	□
19-5	Meaning of the circle in the formula for benzene	The circle is intended to show that one electron from each carbon atom is shared among all of the carbon atoms; each adjacent pair of carbon atoms shares a single bond, shown by the line that joins them, and the additional bonding shown by the circle.	□
19-5	Resonance	Electron sharing by more than two atoms, as in the bonding represented by the circle in the formula for benzene; also called electron delocalization.	□
19-5	Aromatic compound	A compound whose molecule contains a benzene ring, C C C C C C	□
19-5	Aliphatic compound	A compound that isn't aromatic.	□
19-5	Meaning of ↔	The double-headed arrow is used to show resonance. For example, benzene can be represented as H H H C H H C H C C ↔ C C C C C C H C H H C H H H The double-headed arrow signals that the bonding in the benzene molecule should be imagined as a composite of the bonding in each formula joined by the arrow.	□

Section	Subject	Summary	Check When Learned
19-5	System for naming aromatic compounds	Name each group attached to the benzene ring; if necessary, number the carbon atoms in the ring to assign the lowest numbers to substituents. For example, [benzene ring structure with CH_3 at positions 1 and 3, H at the other positions] is 1,3-dimethylbenzene.	☐
19-6	General rule for relative melting points or boiling points of two hydrocarbons	Of two alkanes, the one with more atoms in its molecule will have stronger London forces and will therefore have the higher melting and boiling points; the same is true for alkenes, alkynes, and aromatic compounds. Of two isomers, the one with the longer straight chain will have the higher melting and boiling points.	☐
19-6	Solubility of hydrocarbons in water	Hydrocarbons are essentially insoluble in water because their molecules form only London forces with water molecules.	☐
19-7	Combustion of a hydrocarbon	Reaction of the hydrocarbon with oxygen to produce carbon dioxide, water, and heat, for example, $2\ CH_3{-}CH_3(g) + 5\ O_2(g) \rightarrow 2\ CO_2(g) + 6\ H_2O(\ell) + \text{heat}$	☐
19-8	Substitution reaction	A reaction in which a hydrogen atom in a hydrocarbon molecule is replaced by another atom or group of atoms: $-\overset{\vert}{\underset{\vert}{C}}-H \rightarrow -\overset{\vert}{\underset{\vert}{C}}-Z$	☐
19-8	Kinds of hydrocarbons that commonly undergo substitution reactions	Alkanes and aromatic compounds.	☐
19-8	Catalyst	A substance that must be present for a reaction to occur but that is unchanged at the end of the reaction. In the equation for the reaction, the formula for the catalyst is usually written over the arrow.	☐
19-8	Addition reaction	A reaction in which atoms add to a double or triple bond.	☐

Section	Subject	Summary	Check When Learned
19-8	Kinds of hydrocarbons that commonly undergo addition reactions	Alkenes and alkynes	☐
19-9	Functional group	A characteristic grouping of atoms attached to a hydrocarbon molecule.	☐
19-9	Class of compounds that corresponds to each of these functional groups:		
	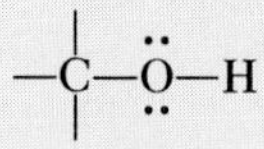 $-\overset{\vert}{\underset{\vert}{C}}-\ddot{\underset{\cdot\cdot}{O}}-H$	Alcohols.	
	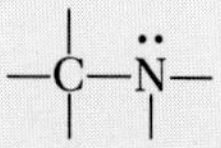 $-\overset{\vert}{\underset{\vert}{C}}-\overset{\cdot\cdot}{\underset{\vert}{N}}-$	Amines.	
	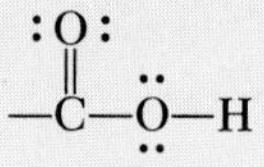 $-\overset{\overset{:O:}{\Vert}}{C}-\ddot{\underset{\cdot\cdot}{O}}-H$	Carboxylic acids.	☐
19-10	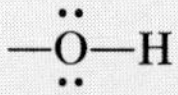 $-\ddot{\underset{\cdot\cdot}{O}}-H$	Hydroxyl group.	☐
19-10	System for naming alcohols	Change the suffix of the name for the corresponding alkane to *-ol;* if necessary, number the carbon chain from the end that assigns the lower number to the carbon atom that carries the hydroxyl group. For example, 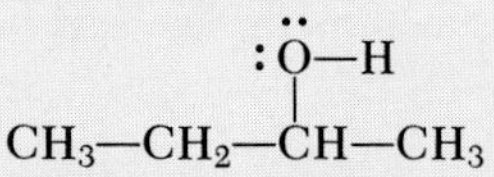 $CH_3-CH_2-\overset{\overset{:\ddot{O}-H}{\vert}}{C}H-CH_3$ is 2-butanol.	☐
19-10	Solubility of alcohols in water	Alcohols whose molecules have more carbon atoms are less soluble; methanol and ethanol are infinitely soluble.	☐
19-10	Reaction of methanol with:		
	sodium	$2\ Na + 2\ CH_3OH \rightarrow 2\ NaOCH_3 + H_2$	
	calcium	$Ca + 2\ CH_3OH \rightarrow Ca(OCH_3)_2 + H_2$	
	hydrogen chloride	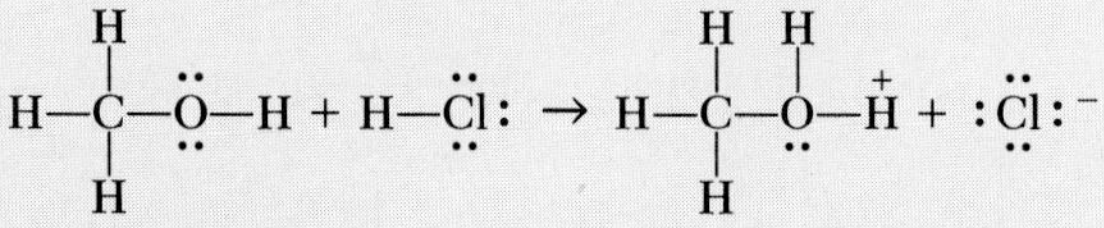 $H-\overset{\overset{H}{\vert}}{\underset{\underset{H}{\vert}}{C}}-\ddot{\underset{\cdot\cdot}{O}}-H + H-\ddot{\underset{\cdot\cdot}{Cl}}: \rightarrow H-\overset{\overset{H}{\vert}}{\underset{\underset{H}{\vert}}{C}}-\underset{\cdot\cdot}{\overset{\overset{H}{\vert}}{O}}-\overset{+}{H} + :\ddot{\underset{\cdot\cdot}{Cl}}:^-$	☐
19-11	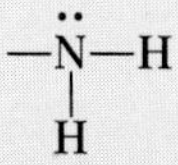 $-\overset{\cdot\cdot}{\underset{\underset{H}{\vert}}{N}}-H$	Amino group.	☐

Section	Subject	Summary	Check When Learned
19-11	System for naming amines	Change the suffix of the name for the corresponding alkane to *-amine;* if necessary, number the carbon chain from the end that assigns the lower number to the carbon atom that carries the amino group. For example, 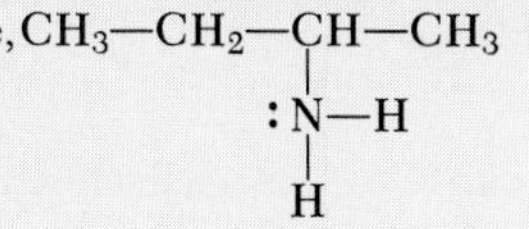 $CH_3—CH_2—CH(—\ddot{N}H_2)—CH_3$ is 2-butanamine.	☐
19-11	Solubility of amines in water	Amines whose molecules have more carbon atoms are less soluble.	☐
19-11	Reaction of methanamine with water	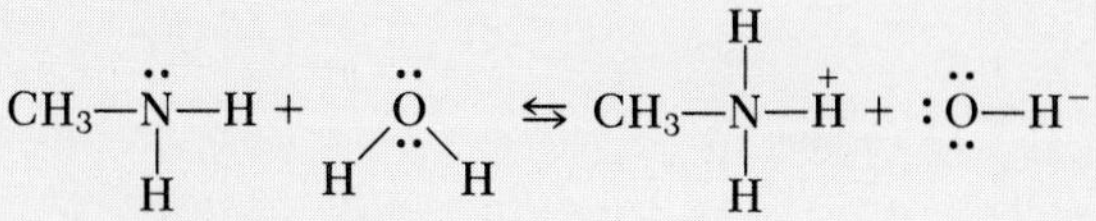 $CH_3—\ddot{N}H—H + H_2\ddot{O} \rightleftharpoons CH_3—NH_2—H^+ + :\ddot{O}—H^-$	
	hydrogen chloride	$CH_3—\ddot{N}H—H + H—\ddot{Cl}: \rightarrow CH_3—NH_2—H^+ + :\ddot{Cl}:^-$	☐
19-12	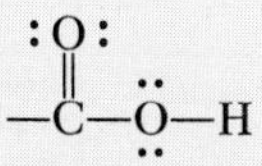 $—C(=\ddot{O})—\ddot{O}—H$	Carboxyl group.	☐
19-11	System for naming carboxylic acids	Change the suffix of the name for the corresponding alkane to *-oic* and add the word *acid.* Most simple acids have traditional as well as systematic names: $H—C(=\ddot{O})—\ddot{O}—H$ is named methanoic acid but is usually called formic acid.	☐
19-12	Solubility of carboxylic acids in water	Carboxylic acids whose molecules have more carbon atoms are less soluble.	☐
19-12	Reaction of methanoic acid with water	$H—C(=\ddot{O})—\ddot{O}—H(aq) + :\ddot{O}H_2(\ell) \rightleftharpoons H—C(=\ddot{O})—\ddot{O}:^-(aq) + H—\ddot{O}H_2^+$ (aq)	
	with aqueous hydroxide ion	$H—C(=\ddot{O})—\ddot{O}—H(aq) + :\ddot{O}—H^-(aq) \rightleftharpoons H—C(=\ddot{O})—\ddot{O}:^-(aq) + :\ddot{O}H_2(\ell)$	
	with methanol (HCl catalyst)	$H—C(=\ddot{O})—\ddot{O}—H + CH_3—\ddot{O}—H \xrightarrow{HCl} H—C(=\ddot{O})—\ddot{O}—CH_3 + H_2\ddot{O}$	

Section	Subject	Summary	Check When Learned
	with methanamine (HCl catalyst)	$H{-}C({=}O){-}O{-}H + CH_3{-}N(H){-}H \xrightarrow{HCl} H{-}C({=}O){-}N(H){-}CH_3 + H{-}O{-}H$	☐
19-12	Classify:		
	$H{-}C({=}O){-}O{-}CH_3$	Ester.	
	$H{-}C({=}O){-}N(H){-}CH_3$	Amide.	☐
19-13	Polymer	Organic compound whose molecules are very long chains that contain repeating sequences of atoms.	☐
19-13	Monomer	Reactant that forms a polymer.	☐
19-13	Condensation polymerization	Reaction that forms a polymer—for example, a polyamide—and produces water molecules.	☐
19-13	Condensation polymer	Polymer produced by condensation polymerization, for example, a polyamide.	☐
19-13	Addition polymerization	Reaction that forms a polymer—for example, polyethylene—by successive addition reactions of a monomer whose molecule contains a double bond.	☐
19-13	Addition polymer	Polymer produced by addition polymerization, for example, polyethylene.	☐

Problems

Assume you can use the Periodic Table at the front of the book unless you're directed otherwise. Answers to odd-numbered problems are in Appendix 1.

Lewis Symbols, Line Symbols, and Line Formulas (Section 19-1)

1. How many valence electrons does a carbon atom have?
2. In your own words state the octet rule specifically as it applies to the carbon atom.
3. Write the Lewis symbol and line symbol for (a) carbon (b) hydrogen (c) oxygen.
4. Write the Lewis symbol and line symbol for (a) nitrogen (b) sulfur (c) fluorine.

5. Write the line formula for the molecule formed by one atom of each of the following elements and the appropriate number of hydrogen atoms: (a) carbon (b) nitrogen (c) oxygen.

6. Write the line formula for the molecule formed by one atom of each of the following elements and the appropriate number of hydrogen atoms: (a) sulfur (b) chlorine (c) bromine.

7. Name each compound whose molecule is described in problem 5.

8. Name each compound whose molecule is described in problem 6.

9. Write the line formula for the compound whose molecule consists of one carbon atom and (a) three hydrogen atoms and one fluorine atom (b) two hydrogen atoms and two fluorine atoms (c) one hydrogen atom and three fluorine atoms.

10. Write the line formula for the compound whose molecule consists of one carbon atom and (a) two hydrogen atoms and two chlorine atoms (b) two chlorine atoms and two fluorine atoms (c) one hydrogen atom, one fluorine atom, one chlorine atom, and one bromine atom.

Carbon Atoms with Single, Double, or Triple Bonds That Are Nonpolar or Polar (Section 19-2)

11. Write the line formula for the compound whose molecule consists of two carbon atoms and (a) six hydrogen atoms (b) four hydrogen atoms (c) two hydrogen atoms.

12. Write the line formula for the compound whose molecule consists of two carbon atoms and (a) six chlorine atoms (b) four chlorine atoms (c) one fluorine atom and three chlorine atoms.

13. Write the line formula for the compound whose molecule consists of one carbon atom, one nitrogen atom, and (a) five hydrogen atoms (b) three hydrogen atoms (c) one hydrogen atom.

14. Write the line formula for the compound whose molecule consists of one carbon atom, one oxygen atom, and (a) four hydrogen atoms (b) two hydrogen atoms.

15. The molecular formula for phosgene, a poisonous gas used as a weapon in World War I, is $COCl_2$. Write its line formula.

16. Cyanogen, C_2N_2, is a highly poisonous gas. In a cyanogen molecule the carbon atoms are joined by a single bond. Write the line formula for the cyanogen molecule.

17. Use the symbols $\delta+$ and $\delta-$ and the information in Figure 19.1 to assign the polarity, if any, to the bond between the carbon atom and the atom bonded to it:

(a) $-\overset{|}{\underset{|}{\mathrm{C}}}-\overset{\cdot\cdot}{\underset{\cdot\cdot}{\mathrm{O}}}-$ (b) $-\overset{|}{\underset{|}{\mathrm{C}}}-\overset{\cdot\cdot}{\underset{|}{\mathrm{N}}}-$

18. Use the symbols $\delta+$ and $\delta-$ and the information in Figure 19.1 to assign the polarity, if any, to the bond between the carbon atom and the atom bonded to it:

(a) $-\overset{|}{\underset{|}{\mathrm{C}}}-\mathrm{H}$ (b) $-\overset{|}{\underset{|}{\mathrm{C}}}-\overset{|}{\underset{|}{\mathrm{C}}}-$

19. Use the symbols $\delta+$ and $\delta-$ and the information in Figure 19.1 to assign the polarity, if any, to the bond between the carbon atom and the atom bonded to it:

(a) $-\mathrm{C}{\equiv}\mathrm{C}-$ (b) $-\overset{|}{\underset{|}{\mathrm{C}}}-\overset{\cdot\cdot}{\underset{\cdot\cdot}{\mathrm{F}}}:$

20. Use the symbols $\delta+$ and $\delta-$ and the information in Figure 19.1 to assign the polarity, if any, to the bond between the carbon atom and the atom bonded to it:

(a) $-\overset{|}{\mathrm{C}}{=}\overset{\cdot\cdot}{\underset{\cdot\cdot}{\mathrm{O}}}$ (b) $-\overset{|}{\underset{|}{\mathrm{C}}}-\overset{\cdot\cdot}{\underset{\cdot\cdot}{\mathrm{I}}}:$

Chains and Rings of Carbon Atoms; Isomers (Section 19-3)

21. Write the line formula for the hydrocarbon molecule that has six carbon atoms in its chain and only single bonds.

22. Write the line formula for the hydrocarbon molecule that has five carbon atoms in its chain and a double bond between the second and third carbon atoms.

23. Write the line formula for the hydrocarbon molecule that has four carbon atoms in its chain and a triple bond between the two middle carbons.

24. Write the line formula for the hydrocarbon molecule that has six carbon atoms in its chain and two double bonds, one starting on the third carbon and the other starting on the fourth carbon.

25. Write the line formula and condensed formula for the hydrocarbon molecule that has a five-carbon chain, a double bond starting on the second carbon, and a one-carbon branch from the second carbon.

27. Write the line formula and condensed formula for the hydrocarbon molecule that consists of a five-carbon ring with only single bonds.

29. Write condensed formulas for the isomers of C_6H_{14}.

31. Write condensed formulas for the isomers of C_3H_6. One of them contains a ring.

26. Write the line formula and condensed formula for the hydrocarbon molecule that has a five-carbon chain, a double bond starting on the second carbon, and a two-carbon branch from the third carbon.

28. Write the line formula and condensed formula for the hydrocarbon molecule that consists of a six-carbon ring with one double bond.

30. How many different molecules have the molecular formula C_3H_8?

32. Write condensed formulas for the isomers of C_4H_8. One of them contains a ring.

Naming Aliphatic Hydrocarbons (Section 19-4)

33. In your own words, distinguish between *alkane* and *alkene*.

35. Name the straight-chain hydrocarbon with the molecular formula (a) CH_4 (b) C_3H_8 (c) C_5H_{12}.

37. Name:

$$\begin{array}{ccccccccc} & & & & & & CH_3 & & \\ & & & & & & | & & \\ CH_3 & - & CH_2 & - & CH_2 & - & CH & - & CH_3 \end{array}$$

39. Name:

$$\begin{array}{ccccccc} CH_3 & & & & & & \\ | & & & & & & \\ CH_2 & - & CH_2 & - & CH_2 & - & CH_2 \\ & & & & & & | \\ & & & & & & CH_3 \end{array}$$

41. Write the condensed formula for 3,3-dimethylpentane.

43. Write the condensed formula for 2-methyl-3-ethylpentane.

45. Write the condensed formula for cyclopropane.

47. Name:

$$CH_3-CH_2-CH{=}CH-CH_3$$

49. Name:

$$\begin{array}{ccccccccc} & & & & & & CH_3 & & \\ & & & & & & | & & \\ CH_3 & - & CH_2 & - & CH & = & C & - & CH_3 \end{array}$$

51. Write the condensed formula for 2,3-dimethyl-2-butene.

34. In your own words, distinguish between *alkene* and *alkyne*.

36. Name the straight-chain hydrocarbon with the molecular formula (a) C_2H_6 (b) C_4H_{10} (c) C_6H_{14}.

38. Name:

$$\begin{array}{ccccccccc} & & & & CH_3 & & & & \\ & & & & | & & & & \\ & & & & CH_2 & & & & \\ & & & & | & & & & \\ CH_3 & - & CH_2 & - & CH & - & CH_2 & - & CH_3 \end{array}$$

40. Name:

$$\begin{array}{ccccccccccc} & & CH_3 & & & & & & CH_3 & & \\ & & | & & & & & & | & & \\ & & CH_2 & & & & & & CH_2 & & \\ & & | & & & & & & | & & \\ CH_3 & - & CH & - & CH_2 & - & CH_2 & - & CH & - & CH_3 \end{array}$$

42. Write the condensed formula for 2,2,3,3-tetramethybutane.

44. Write the condensed formula for 4-ethyl-3,5-dimethylheptane.

46. Write the condensed formula for cycloheptane.

48. Name:

$$CH_3-CH_2-CH_2-C{\equiv}C-CH_2-CH_2-CH_3$$

50. Name:

$$\begin{array}{ccccccccccccccc} & & & & & & & & & & & & CH_3 & & \\ & & & & & & & & & & & & | & & \\ CH_3 & - & CH & - & CH_2 & - & C & \equiv & C & - & CH_2 & - & CH & - & CH_3 \\ & & | & & & & & & & & & & & & \\ & & CH_2 & & & & & & & & & & & & \\ & & | & & & & & & & & & & & & \\ & & CH_3 & & & & & & & & & & & & \end{array}$$

52. Write the condensed formula for 3,3-dimethyl-1-butyne.

Aromatic Hydrocarbons (Section 19-5)

53. In your own words, explain the meaning of *aromatic* in organic chemistry.

55. In your own words, explain what is meant by the statement that benzene shows resonance.

54. In your own words, explain the difference between *aliphatic* and *aromatic*.

56. In your own words, explain what is meant by the statement that benzene shows electron delocalization.

57. Explain the difference between ↔ and ⇄.

58. Explain the meaning of the circle in the formula

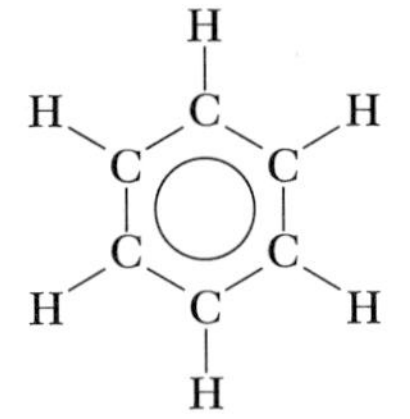

```
          H
          |
    H     C     H
     \  /   \  /
      C       C
      |  (O)  |
      C       C
     /  \   /  \
    H     C     H
          |
          H
```

59. Name:

```
          CH3
          |
   CH3    C     H
     \  /   \  /
      C       C
      |  (O)  |
      C       C
     /  \   /  \
    H     C     H
          |
          H
```

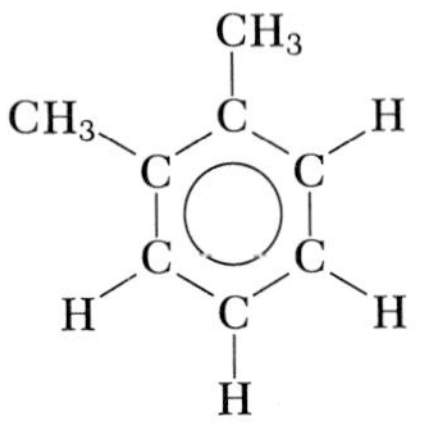

60. Write the formula for 1,2-diethylbenzene.

Predicting Relative Melting Points and Boiling Points of Hydrocarbons (Section 19-6)

61. Predict which compound will have the higher boiling point, ethane or hexane.

62. Predict which compound will have the higher melting point, octane or decane.

63. Predict which compound will have the lower melting point, pentane or 2,2-dimethylpropane.

64. Predict which compound will have the lower boiling point, cyclobutane or cyclohexane.

65. Under ordinary room conditions, which compound is more likely to be a gas, ethene or 1,4-dimethylbenzene? Explain your choice.

66. Under ordinary room conditions, which compound is more likely to be a gas, propane or decane? Explain your choice.

Reactions of Hydrocarbons (Sections 19-7 and 19-8)

67. Use condensed formulas to write an equation for the combustion of propane.

68. Use condensed formulas to write an equation for the combustion of ethyne (acetylene).

69. Classify each reaction as an addition or a substitution:

(a) $CH_2{=}CH_2 + :\ddot{Br}{-}\ddot{Br}: \rightarrow$ $CH_2{-}CH_2$ (each C bonded to $:\ddot{Br}:$)

(b) $CH_3{-}CH_3 + :\ddot{Br}{-}\ddot{Br}: \rightarrow$

$CH_3{-}CH_2{-}\ddot{Br}: + H{-}\ddot{Br}:$

70. Classify this reaction as an addition or a substitution:

```
          CH3
          |
    H     C     H
     \  /   \  /
      C       C
      |  (O)  |        + :Br—Br:
      C       C
     /  \   /  \
    H     C     H
          |
          H
```

```
                   CH3
                   |
             H     C     H
              \  /   \  /
  FeBr3        C       C
  ----->       |  (O)  |        + H—Br:
               C       C
              /  \   /  \
             H     C     H
                   |
                  :Br:
```

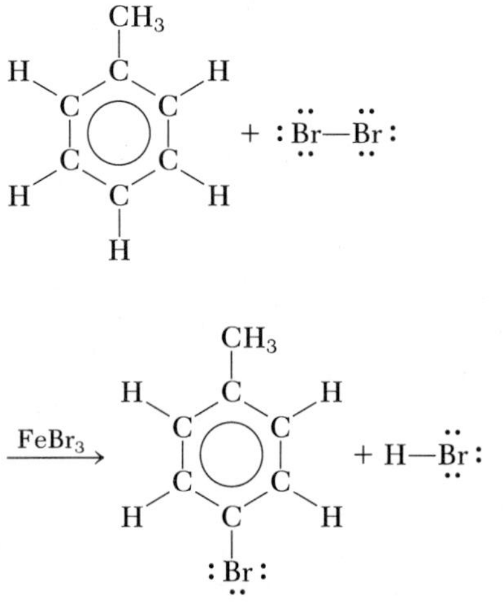

71. Complete this equation:

$$CH_3—CH{=}CH_2 + :\ddot{B}r—\ddot{B}r: \rightarrow$$

72. Complete this equation:

$$CH_3—C{\equiv}C—H + 2\ H—H \xrightarrow{Ni}$$

73. Use line formulas to write an equation for the reaction of one mole of methane with one mole of iodine.

74. Use line formulas to write an equation for the reaction of one mole of hydrogen iodide with one mole of ethyne (acetylene).

Classifying Compounds by Functional Groups (Section 19-9)

75. Use Table 19.3 to identify the class of each compound:
(a) $CH_3—\ddot{O}—H$ (b) $CH_3—\ddot{S}—H$

76. Use Table 19.3 to identify the class of each compound:

```
                          :O:
                           ‖
(a) CH3—N—H   (b) CH3—N—O:
        |
        H
```

77. Use Table 19.3 to circle and identify each functional group in Vitamin C:

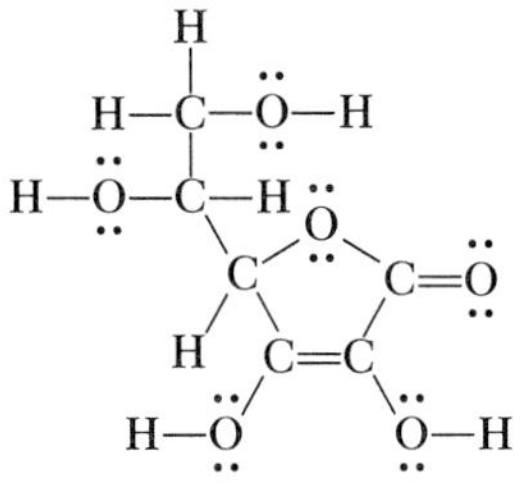

```
          H
          |
      H—C—O—H
          |
   H—O—C—H
          \    O
           C       C=O
          / \     /
         H   C=C
            /    \
        H—O       O—H
```

78. Methionine is one of the essential amino acids—compounds that form proteins and that must be present in the human diet for good health. Use Table 19.3 to circle and identify each functional group in methionine:

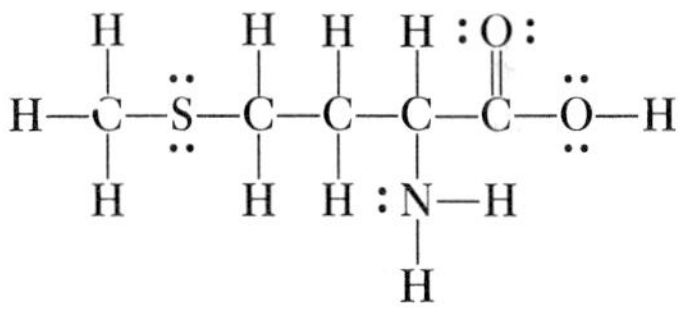

```
   H       H   H   H  :O:
   |       |   |   |   ‖
H—C—S—C—C—C—C—O—H
   |       |   |   |
   H       H   H  :N—H
                    |
                    H
```

Alcohols (Section 19-10)

79. Name:
(a) $CH_3—\underset{\displaystyle CH_3}{\underset{|}{C}}H—\ddot{O}—H$
(b) $CH_3—CH_2—CH_2—CH_2—CH_2—\ddot{O}—H$

80. Name:
(a) $H—\ddot{O}—CH_2—CH_3$
(b) $CH_3—\underset{\displaystyle CH_3}{\underset{|}{C}}H—CH_2—\ddot{O}—H$

81. Write the condensed formula for (a) 3-octanol (b) 2-methyl-1-propanol (c) cyclobutanol

82. Write the condensed formula for
(a) 3-methyl-2-butanol (b) 3-ethyl-3-pentanol
(c) cyclohexanol

83. Why does ethanol have a higher boiling point (79°C) than methanol (65°C)?

84. Which compound would you expect to have the higher boiling point, 1-propanol or 1-butanol? Explain your reasoning.

85. Ethylene glycol, whose formula is shown below, has a much higher boiling point (198°C) than ethanol (79°C). Suggest an explanation.

$$H—\ddot{O}—CH_2—CH_2—\ddot{O}—H$$

86. The formulas for ethylene glycol and glycerol are shown below. Which compound would you expect to have the lower boiling point? Explain your reasoning.

$$H—\ddot{O}—CH_2—CH_2—\ddot{O}—H$$

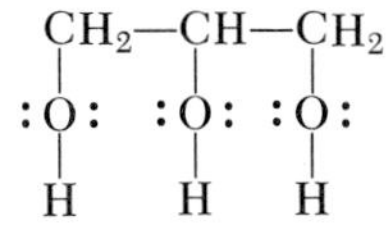

```
 CH2—CH—CH2
  |    |    |
 :O:  :O:  :O:
  |    |    |
  H    H    H
```

87. Would you expect glycerol, whose formula is shown below, to be significantly soluble in water? Explain your reasoning.

$$\begin{array}{ccccc} CH_2 & - & CH & - & CH_2 \\ | & & | & & | \\ :\ddot{O}: & & :\ddot{O}: & & :\ddot{O}: \\ | & & | & & | \\ H & & H & & H \end{array}$$

88. The formula for glucose (blood sugar) is shown below. Suggest an explanation for the fact that glucose is very soluble in water.

$$\begin{array}{ccccccc} & & H & & & & \\ & & | & & & & \\ & H- & C & -\ddot{O}-H & & & \\ & & | & & & & \\ & & C & \text{———} & \ddot{O} & & \\ H & / & | & & & \backslash & \ddot{O}-H \\ & C & H & & & C & \\ H-\ddot{O} & \backslash & :\ddot{O}-H & H & / & & H \\ & & C & \text{———} & C & & \\ & & | & & | & & \\ & & H & & :\ddot{O}: & & \\ & & & & | & & \\ & & & & H & & \end{array}$$

glucose

89. Use line formulas and Lewis symbols to write an equation for the reaction of ethanol with potassium.

90. Use line formulas and Lewis symbols to write an equation for the reaction of 2-propanol with lithium.

91. Write equations for the reactions of 1-butanol with hydrogen chloride to produce 1-butene.

92. Write equations for the reactions of cyclobutanol with hydrogen chloride to produce cyclobutene.

Amines (Section 19-11)

93. Name:

(a) $$\begin{array}{l} CH_3-CH_2-CH_2-\ddot{N}-H \\ \qquad\qquad\qquad\quad\ \ | \\ \qquad\qquad\qquad\quad\ \ H \end{array}$$

(b) $$\begin{array}{ccccc} & & CH_3 & & \\ & & | & & \\ CH_3 & - & C & - & \ddot{N}-H \\ & & | & & | \\ & & CH_3 & & H \end{array}$$

94. Name:

(a) $$\begin{array}{l} CH_3-CH_2-CH_2-CH_2-CH_2-\ddot{N}-H \\ \qquad\qquad\qquad\qquad\qquad\quad\ \ | \\ \qquad\qquad\qquad\qquad\qquad\quad\ \ H \end{array}$$

(b) $$\begin{array}{ccccc} & & CH_3 & & \\ & & | & & \\ & & CH_2 & & \\ & & | & & \\ CH_3-CH_2 & - & C & - & CH_3 \\ & & | & & \\ & & :N-H & & \\ & & | & & \\ & & H & & \end{array}$$

95. Write the condensed formula for (a) 3-heptanamine (b) 3-ethyl-3-pentanamine.

96. Write the condensed formula for
(a) 2,3-dimethyl-2-butanamine (b) cyclohexanamine.

97. What simple test could you use to distinguish between 1-propanol and 1-propanamine?

98. Would you expect ethanamine to be significantly soluble in ethanol? Explain.

99. Use line formulas to write an equation for the reaction of 2-propanamine with water.

100. Use line formulas to write an equation for the reaction of cyclopropanamine with water.

101. Use line formulas to write an equation for the reaction of 1-propanamine with hydrochloric acid.

102. Use line formulas to write an equation for the reaction of cyclopropanamine with hydrochloric acid.

Carboxylic Acids (Section 19-12)

103. Name:

(a) $$CH_3{-}CH_2{-}C({=}O){-}O{-}H$$

(b) $$CH_3{-}C(CH_3)_2{-}C({=}O){-}O{-}H$$

104. Name:

(a) $$H{-}O{-}C({=}O){-}CH_3$$

(b) $$CH_3{-}CH(CH_2{-}CH_3){-}C({=}O){-}O{-}H$$

105. Write the condensed formula for
    (a) 2,2,3-trimethylbutanoic acid
    (b) 3,3-diethylpentanoic acid

106. Write the condensed formula for
    (a) 4,4-dimethylpentanoic acid (b) heptanoic acid

107. Oxalic acid, whose formula is shown below, is a poisonous compound found in many plants. Would you expect oxalic acid to be significantly soluble in water? Explain.

$$H{-}O{-}C({=}O){-}C({=}O){-}O{-}H$$

oxalic acid

108. Citric acid, whose formula is shown below, is responsible for the characteristic sharp taste of citrus fruits. (a) What functional groups are present in a molecule of citric acid? (b) To what classes of organic compounds does citric acid belong?

$$H{-}O{-}C(CH_2{-}C({=}O){-}O{-}H)(CH_2{-}C({=}O){-}O{-}H){-}C({=}O){-}O{-}H$$

citric acid

109. Use condensed formulas to write an equation for the reaction of 2-methylpropanoic acid with water.

110. Oxalic acid, whose formula is shown in problem 107, is a weak acid and a diprotic acid. Use condensed formulas to write an equation for each of its reactions with water.

111. Use condensed formulas to write an equation for the reaction of 2,2-dimethylbutanoic acid with hydroxide ion.

112. Use condensed formulas to write an equation for the reaction of one mole of oxalic acid, whose formula is shown in problem 107, with two moles of hydroxide ion.

113. Use condensed formulas to write an equation for the reaction of one mole of oxalic acid, whose formula is shown in problem 107, with two moles of methanol in the presence of HCl as a catalyst.

114. Use condensed formulas to write an equation for the reaction of two moles of acetic acid with one mole of ethylene glycol, whose formula is shown below, in the presence of HCl as a catalyst.

$$H{-}O{-}CH_2{-}CH_2{-}O{-}H$$

ethylene glycol

115. Write the condensed formula for the ester that would be formed by 1-propanol and propanoic acid.

116. Write the condensed formula for the ester that would be formed by one molecule of citric acid, whose formula is shown in problem 108, and three moles of methanol.

117. Use condensed formulas to write an equation for the reaction of 2-methylpropanoic acid with methanamine in the presence of HCl as a catalyst.

118. Use condensed formulas to write an equation for the reaction of two moles of acetic acid with one mole of ethylenediamine, whose formula is shown below, in the presence of HCl as a catalyst.

$$H{-}N(H){-}CH_2{-}CH_2{-}N(H){-}H$$

ethylenediamine

119. Write the formula for the amide that would be formed by the reaction of 1-butanamine with butanoic acid.

120. Write the formula for the amide that would be formed by the reaction of one mole of oxalic acid, whose formula is shown in problem 107, with one mole of ethylenediamine, whose formula is shown in problem 118. Assume that each carboxyl group in the oxalic acid molecule reacts with one amino group in the ethylenediamine molecule; the product is a cyclic diamide.

Polymers (Section 19-13)

121. Write an equation for the reaction of oxalic acid with ethylenediamine to form a polyamide:

$$\mathrm{H{-}\ddot{\underset{..}{O}}{-}\overset{\overset{:\ddot{O}:}{\|}}{C}{-}\overset{\overset{:\ddot{O}:}{\|}}{C}{-}\ddot{\underset{..}{O}}{-}H}$$

oxalic acid

$$\mathrm{H{-}\underset{\underset{H}{|}}{\ddot{N}}{-}CH_2{-}CH_2{-}\underset{\underset{H}{|}}{\ddot{N}}{-}H}$$

ethylenediamine

122. Write an equation for the reaction of oxalic acid with ethylene glycol to form a polyester:

$$\mathrm{H{-}\ddot{\underset{..}{O}}{-}\overset{\overset{:\ddot{O}:}{\|}}{C}{-}\overset{\overset{:\ddot{O}:}{\|}}{C}{-}\ddot{\underset{..}{O}}{-}H}$$

oxalic acid

$$\mathrm{H{-}\ddot{\underset{..}{O}}{-}CH_2{-}CH_2{-}\ddot{\underset{..}{O}}{-}H}$$

ethylene glycol

123. Is the polymer whose formation is described in problem 121 an addition polymer or a condensation polymer?

124. Is the reaction described in problem 122 an addition polymerization or a condensation polymerization?

125. The monomer called styrene, whose formula is shown below, forms polystyrene, an addition polymer used to make insulating materials for beverage coolers and houses. Write a formula for polystyrene.

$$\mathrm{\overset{\overset{H}{|}}{\underset{\underset{H}{|}}{C}}{=}\overset{\overset{H}{|}}{C}{-}C_6H_5}$$

styrene

126. The monomer called acrylonitrile, whose formula is shown below, forms polyacrylonitrile, an addition polymer used under the commercial names Orlon® and Acrilan® to make rugs and fabrics. Write a formula for polyacrylonitrile.

$$\mathrm{\overset{\overset{H}{|}}{\underset{\underset{H}{|}}{C}}{=}\overset{\overset{H}{|}}{\underset{\underset{C\equiv N:}{|}}{C}}}$$

acrylonitrile

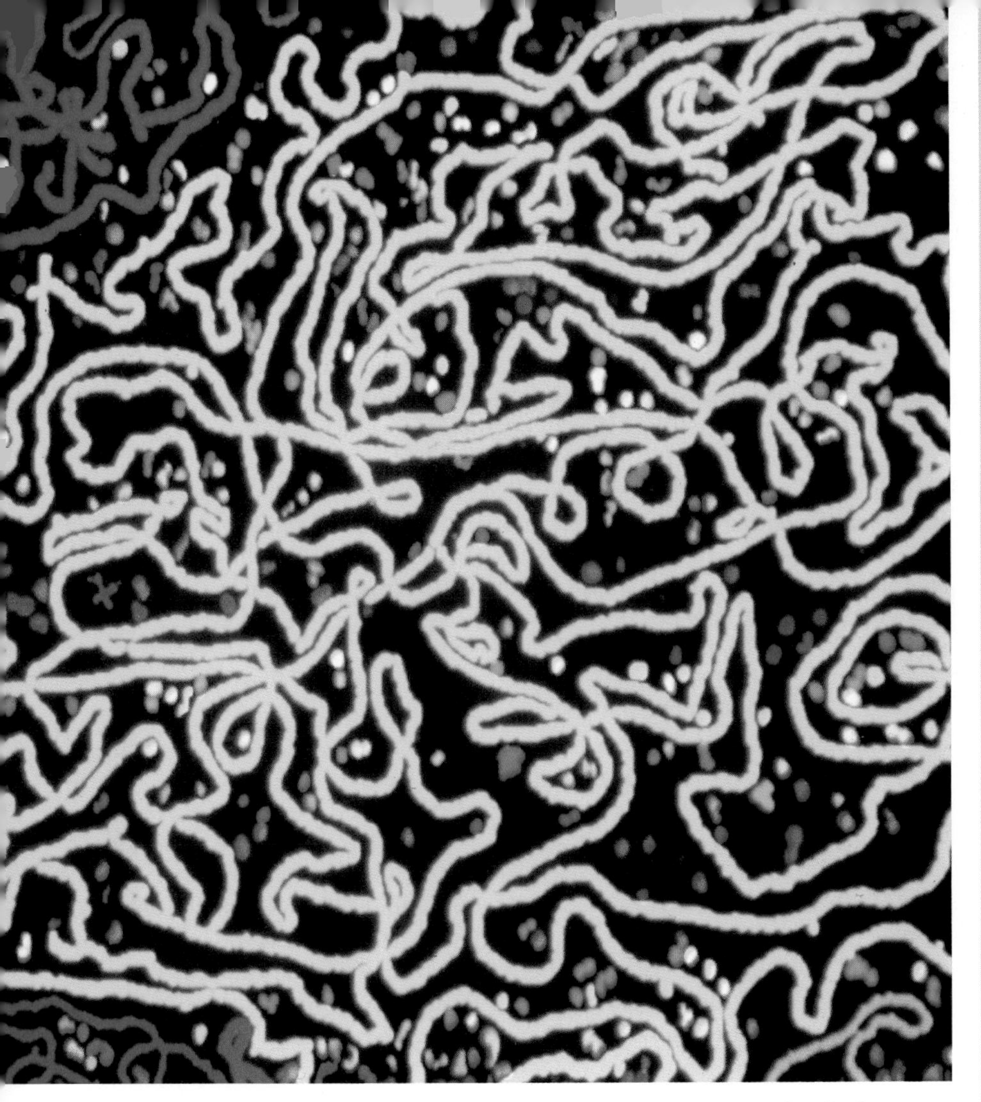

Human deoxyribonucleic acid (DNA). DNA controls the life processes in each cell of a human being, and it controls the transmission of life from one human generation to the next. In this photo the DNA is color-enhanced and greatly magnified. *(Dr. Thomas Broker/Phototake NYC)*

20
Biochemistry

Biochemistry is the branch of chemistry that describes and explains the chemical processes of living organisms. In our century biochemistry has created a revolution in our understanding of life processes by explaining them at their most fundamental level, the chemical level. As a result of this revolution, related fields such as biology, nutrition, and medicine have made advances that would have been nearly unimaginable a century ago.

In this chapter we'll look briefly at the achievements of biochemistry by considering four of the most important classes of biochemical compounds: carbohydrates, lipids, proteins, and nucleic acids. We'll see that each of these classes of compounds plays a unique and crucial role in sustaining life on earth.

The successes of biochemistry have been astonishing, but they are far from complete; many important questions about the nature of life remain unanswered. We'll complete our brief survey of biochemistry by looking at a few of these unanswered questions.

20-1 Carbohydrates are important in animals as sources of energy and in plants as structural materials

Carbohydrates are polyhydroxy aldehydes or ketones; that is, a molecule of a carbohydrate contains an aldehyde functional group or a ketone functional group and several alcohol functional groups.◆ These general characteristics of carbohydrate molecules are shown by the formulas for two important carbohydrates, glucose and fructose:

◆These functional groups are described in Table 19.3.

Fructose, whose line formula is shown in the text, is the intensely sweet carbohydrate that gives honey its sweet taste. *(Charles D. Winters)*

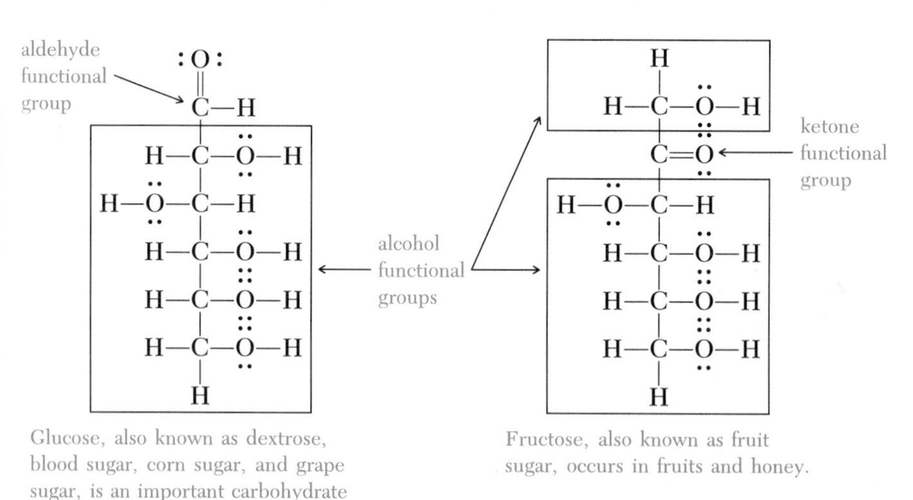

Glucose, also known as dextrose, blood sugar, corn sugar, and grape sugar, is an important carbohydrate in many species of plants and animals, including human beings.

Fructose, also known as fruit sugar, occurs in fruits and honey.

In their physical properties, glucose and fructose also show typical characteristics of carbohydrates: The compounds are white crystalline solids that are soluble in water and have a sweet taste.†

† Chemistry Insight

During the first chemical investigations of carbohydrates about a century ago, chemists coined the word *carbohydrates* because the molecular formulas for these compounds can be written in a way that suggests that they are hydrates (Section 8-11) of carbon. For example, glucose and fructose are isomers, and their common molecular formula, $C_6H_{12}O_6$, can be written $6(C \cdot H_2O)$.

Exercise 20.1

Suggest an explanation for the fact that glucose and fructose are significantly soluble in water.◆

◆Answers to Exercises are in Appendix 1.

Thousands of carbohydrates are known, and they have a variety of biological functions. Because glucose is widespread and important in plants and animals, we'll use it as our example in discussing carbohydrates.

The glucose molecule reacts with itself to form a ring:

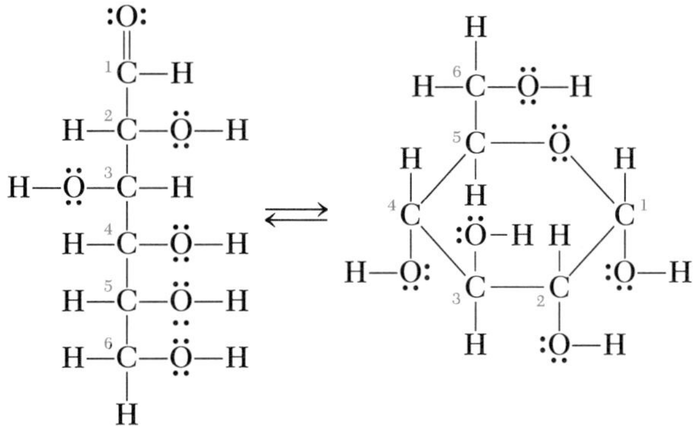

At equilibrium the ring form predominates, so the formula for glucose is usually written as a ring.◆

◆The chemical name for table sugar is *sucrose;* a sucrose molecule is a combination of a glucose molecule and a fructose molecule, each in a ring form. The formula for sucrose is shown in Table 7.2.

Glucose is the monomer for the polymer called **starch.** In a starch molecule the first carbon atom of one glucose unit is bonded through an oxygen atom to the fourth carbon atom of the next glucose unit:

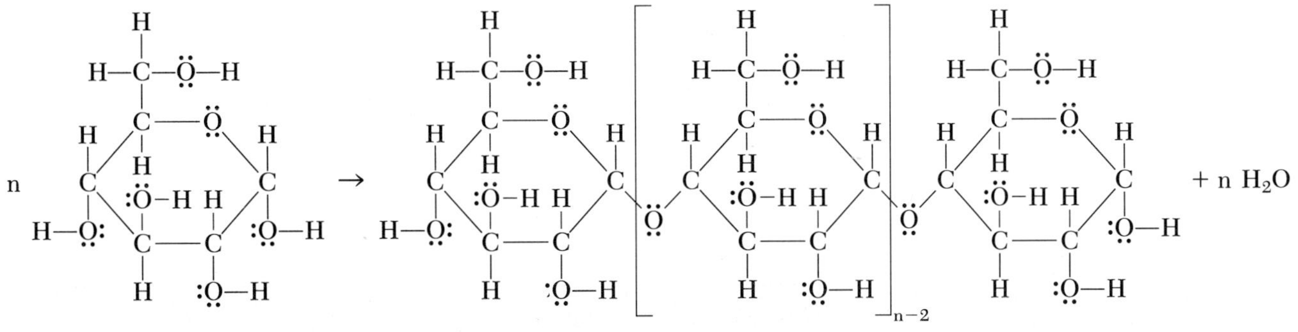

n glucose → starch $+ n\ H_2O$

glucose

starch

Some starch molecules are branched; a branch occurs at carbon 6 in a glucose unit, and each branch is itself a polyglucose chain.◆

◆The term *saccharide* (from a Greek word meaning "sugar") is sometimes used as a synonym for *carbohydrate;* glucose is a *monosaccharide*, and starch is a *polysaccharide*. Nutritionists describe glucose as a *simple carbohydrate* and starch as a *complex carbohydrate*.

Exercise 20.2

A molecule of the carbohydrate *maltose* consists of two glucose units bonded together in the same way that glucose units are bonded to one another in the chain of a starch molecule. (a) Write the line formula for maltose. (b) Write the molecular formula for maltose.

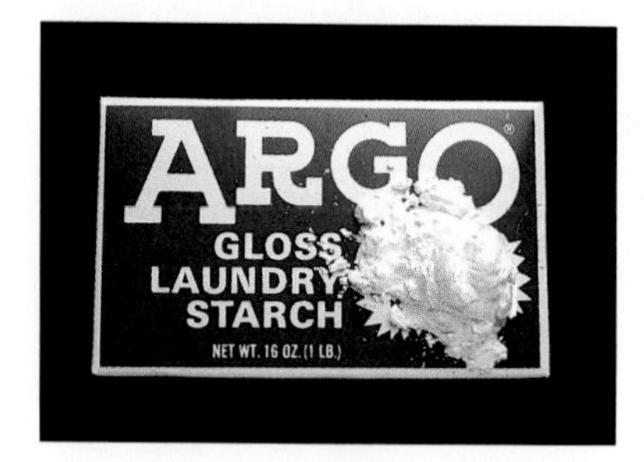

Starch, whose line formula is shown in the text, is a polymer of glucose. Glucose is a *simple carbohydrate*, and starch is a *complex carbohydrate*. *(Charles D. Winters)*

Exercise 20.3

From the examples you've seen so far, what suffix often appears in names for carbohydrates?

Every living thing continuously expends energy and must have energy supplied to it to sustain life. The ultimate source of energy for life on our planet is the sun, but very few organisms can make direct use of energy in the form of sunlight; the immediate source of energy for most organisms, including human beings, is a variety of organic compounds in which the sun's energy has been stored. Carbohydrates are among the most important compounds that store the sun's energy and make it available to living things.

The process by which energy in the form of sunlight becomes stored in organic compounds is called **photosynthesis.** The overall result of photosynthesis is the conversion of carbon dioxide and water into starch; in the process energy from sunlight is incorporated into the starch molecules:

$$CO_2 + H_2O + \text{Energy as sunlight} \rightarrow \text{Starch}$$

Chlorophyll, the green pigment in plants, is a catalyst for photosynthesis. *(Grant Heilman Photography)*

Photosynthesis occurs in two steps, the decomposition of water molecules and the formation of starch molecules.

Decomposition of Water Molecules

Chlorophyll, the green pigment in plants, catalyzes a reaction in which sunlight decomposes water molecules; the reaction produces oxygen, which is released into the atmosphere, and hydrogen atoms, which temporarily become attached in pairs to a large organic molecule. In this equation Y represents the large molecule:

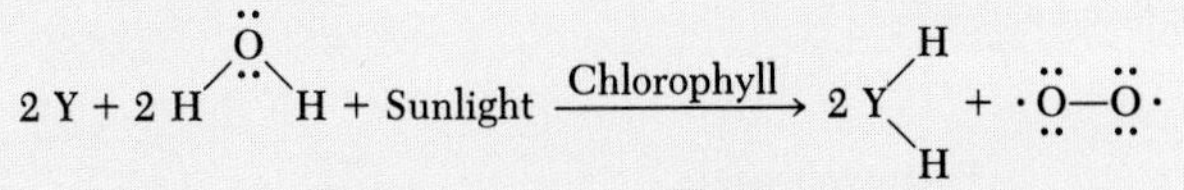

$$2\,\mathrm{Y} + 2\,\mathrm{H{-}\ddot{O}{-}H} + \text{Sunlight} \xrightarrow{\text{Chlorophyll}} 2\,\mathrm{Y}\begin{smallmatrix}\mathrm{H}\\ \mathrm{H}\end{smallmatrix} + \cdot\ddot{\mathrm{O}}{-}\ddot{\mathrm{O}}\cdot$$

Because this reaction requires light, it's called the **light reaction.**

Formation of Starch Molecules

The hydrogen atoms produced in the light reaction are transferred to carbon dioxide molecules, whose carbon atoms bond to one another to form glucose molecules. This equation summarizes several reactions:

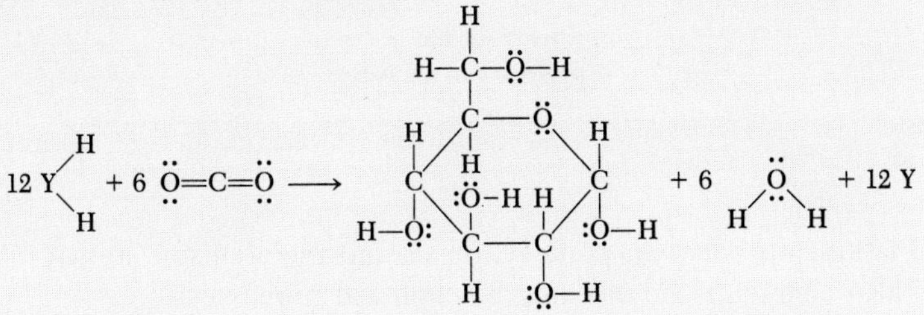

$$12\,\mathrm{Y}\begin{smallmatrix}\mathrm{H}\\ \mathrm{H}\end{smallmatrix} + 6\,\ddot{\mathrm{O}}{=}\mathrm{C}{=}\ddot{\mathrm{O}} \longrightarrow \begin{array}{c} \mathrm{H{-}\overset{H}{\underset{|}{C}}{-}\ddot{O}{-}H} \\ \mathrm{H\quad C{-}\ddot{O}\quad H} \\ \mathrm{C\quad \overset{H}{\ddot{O}}{-}H\;H\quad C} \\ \mathrm{H{-}\ddot{O}{:}\quad C{-}C\quad {:}\ddot{O}{-}H} \\ \mathrm{H\quad {:}\ddot{O}{-}H} \end{array} + 6\,\mathrm{H{-}\ddot{O}{-}H} + 12\,\mathrm{Y}$$

Because this second step in photosynthesis does not require light, it's called the **dark reaction.** The glucose molecules produced in the dark reaction polymerize to form starch molecules.

Figure 20.1
Photosynthesis

In its details, photosynthesis is a long and complex series of reactions that is only partially understood. Figure 20.1 summarizes the most important features of these reactions.

The digestive systems of animals use oxygen from the air to break down starch into carbon dioxide and water; as the starch molecules are broken down, they release the energy that was stored in them during photosynthesis. This process is the **metabolism** of starch. Figure 20.2 on page 500 summarizes the relationship between the photosynthesis of starch and its metabolism.

Exercise 20.4

In an early step in its metabolism, starch breaks down to form maltose (Exercise 20.2), which reacts with water to form glucose. Write an equation for the reaction of maltose with water to form glucose.

The photosynthesis of starch occurs in millions of plant species, and the metabolism of starch occurs in millions of animal species; in general, it's common for the same biochemical processes to be shared by many species. The

In overall result, metabolism reverses photosynthesis:

Photosynthesis in plants: $CO_2 + H_2O + \text{Energy} \rightarrow \text{Starch} + O_2$

Metabolism in animals: $\text{Starch} + O_2 \rightarrow CO_2 + H_2O + \text{Energy}$

The energy released in metabolism drives the many processes that maintain animal life—muscle contraction, nerve transmission, repair of tissue damage, and so forth.

Photosynthesis and metabolism form a cycle, sometimes referred to as the **carbon cycle:** the products of photosynthesis are the reactants for metabolism and vice versa, as shown in the following diagram.

Energy as sunlight

Photosynthesis
in plants

$CO_2 + H_2O$ Starch + O_2

Metabolism
in animals

Energy for life
processes in animals

Figure 20.2
Photosynthesis and metabolism of starch

cyclic relationship between photosynthesis and metabolism, shown in Figure 20.2, is also characteristic of many biochemical processes.

Carbohydrates perform one important biochemical function by storing energy from sunlight, and they perform another by providing **cellulose,** the primary structural material of plants. Like starch, cellulose is a polymer of glucose:

Magnified view of cellulose fibers in writing paper from wood pulp. *(Runk/Schoenberger/Grant Heilman Photography)*

Cellulose

As you'll see by comparing their formulas, starch and cellulose have slightly different bonding arrangements in their polymer chains. These slight differences in bonding create dramatic differences in biological function: Almost all animals can digest starch, but almost no animals can digest cellulose. Among biochemical compounds, it's common for a small change in bonding to make a large difference in biological function.

20-2 Lipids are important in animals as storage forms for energy and as structural materials

The class of biochemicals called **lipids** consists of thousands of compounds whose molecules contain a variety of functional groups; the common feature of lipids is the presence in their molecules of large hydrocarbon portions, as either chains or rings. These are typical examples:

$$CH_2-O-C(=O)-(CH_2)_{16}-CH_3$$
$$CH-O-C(=O)-(CH_2)_7-CH{=}CH-(CH_2)_7-CH_3$$
$$CH_2-O-C(=O)-(CH_2)_{14}-CH_3$$

A fat (a triacylglycerol)

Cholesterol

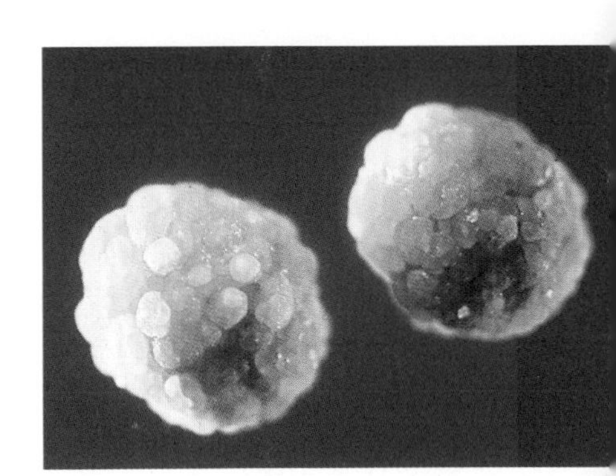

Human gallstones are almost pure cholesterol; each of these gallstones is about 0.5 cm in diameter. *(CBC/Phototake NYC)*

Because lipid molecules contain large hydrocarbon portions, lipids are insoluble in water and soluble in a variety of organic solvents.

In the following discussion of lipids, we'll use the two compounds whose formulas are shown above as our examples.

The first compound whose formula is shown above is a typical animal fat. The molecules of fats have the following features in common: (1) They are esters of the alcohol glycerol with carboxylic acids.◆ The fat whose formula is shown above is a triester of glycerol with three carboxylic acids:

◆Alcohols are described in Section 19-10; carboxylic acids and esters are described in Section 19-12.

$$
\begin{array}{ll}
CH_2-O-H & H-O-\overset{O}{\overset{\|}{C}}-(CH_2)_{16}-CH_3 \\
| & \\
CH-O-H & H-O-\overset{O}{\overset{\|}{C}}-(CH_2)_7-CH=CH-(CH_2)_7-CH_3 \\
| & \\
CH_2-O-H & H-O-\overset{O}{\overset{\|}{C}}-(CH_2)_{14}-CH_3 \\
\text{Glycerol} & \text{Carboxylic Acids}
\end{array}
$$

$$
\begin{array}{l}
CH_2-O-\overset{O}{\overset{\|}{C}}-(CH_2)_{16}-CH_3 \\
| \\
CH-O-\overset{O}{\overset{\|}{C}}-(CH_2)_7-CH=CH-(CH_2)_7-CH_3 \\
| \\
CH_2-O-\overset{O}{\overset{\|}{C}}-(CH_2)_{14}-CH_3 \\
\text{Fat}
\end{array}
$$

In biochemical terminology a triester of glycerol is called a **triacylglycerol.** (2) The carboxylic acid molecules that become parts of fat molecules typically have even numbers of carbon atoms, and they have long hydrocarbon chains—often about twenty carbon atoms; the chains are usually unbranched, and they may contain double bonds. Carboxylic acids whose molecules have these characteristics are called **fatty acids.**

Fats have several biological functions; for example, they insulate the internal organs of animals against temperature change and physical shock. Their most important function, however, is energy storage. In animals, carbohydrates are short-term energy stores, and fats are long-term energy stores. For example, when you first begin to exercise, carbohydrates are oxidized to provide the energy for muscle contraction; when your supply of carbohydrates is depleted, fats are oxidized to provide energy. Figure 20.3 describes the metabolism of a fat to provide energy.

Kangaroo rats. *(Tom McHugh/ PhotoResearchers, Inc.)*

Exercise 20.5

Complete and balance this equation for the oxidation of a fatty acid:

$$CH_3-(CH_2)_7-CH=CH-(CH_2)_7-\overset{O}{\overset{\|}{C}}-O-H + O_2 \rightarrow$$

One product of fat (and carbohydrate) metabolism is water. A few animal species use this **metabolic water** as their entire water supply. One example is the kangaroo rat, a desert animal that eats seeds rich in fat and can survive on a diet that includes no water.

Most of the energy stored in most animals is in the form of fat. A typical distribution of stored energy in an adult human male is 79% in fats, 20% in proteins, and 1% in carbohydrates.◆

◆ Proteins are discussed in Section 20-3.

The complete metabolism of a fat requires many reactions that are summarized here in two steps, hydrolysis and oxidation.

In the hydrolysis step, the fat reacts with water to form glycerol and fatty acids:

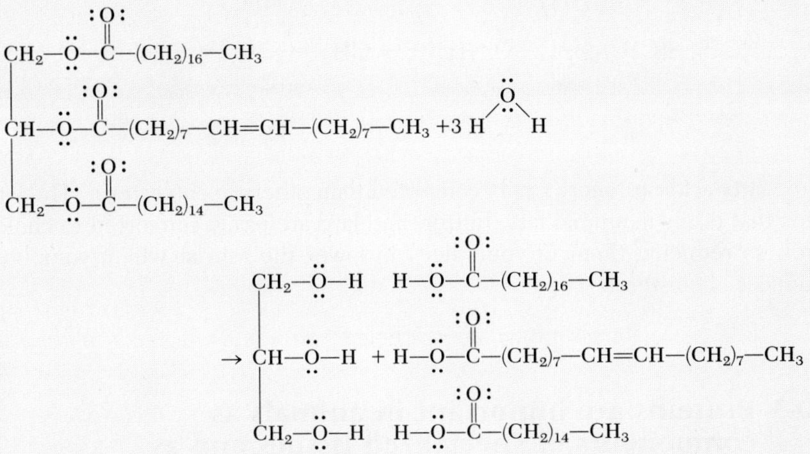

$$\begin{array}{l} CH_2{-}O{-}C({=}O){-}(CH_2)_{16}{-}CH_3 \\ | \\ CH{-}O{-}C({=}O){-}(CH_2)_7{-}CH{=}CH{-}(CH_2)_7{-}CH_3 + 3\,H{-}O{-}H \\ | \\ CH_2{-}O{-}C({=}O){-}(CH_2)_{14}{-}CH_3 \end{array}$$

$$\rightarrow \begin{array}{l} CH_2{-}O{-}H \quad H{-}O{-}C({=}O){-}(CH_2)_{16}{-}CH_3 \\ | \\ CH{-}O{-}H \; + H{-}O{-}C({=}O){-}(CH_2)_7{-}CH{=}CH{-}(CH_2)_7{-}CH_3 \\ | \\ CH_2{-}O{-}H \quad H{-}O{-}C({=}O){-}(CH_2)_{14}{-}CH_3 \end{array}$$

In the oxidation step, each fatty acid is oxidized, in a long series of reactions, to form carbon dioxide and water. These abbreviated, unbalanced equations outline the oxidation of one fatty acid:

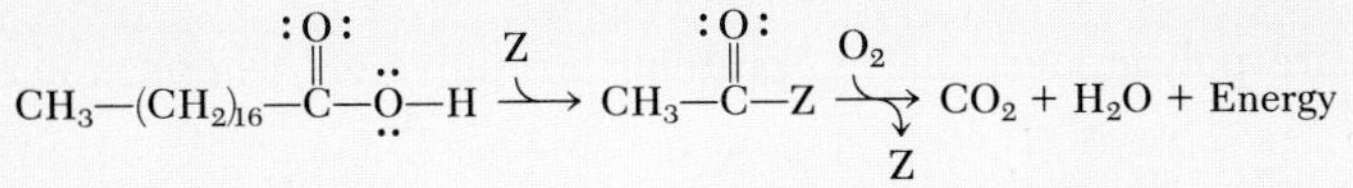

$$CH_3{-}(CH_2)_{16}{-}C({=}O){-}O{-}H \xrightarrow{Z} CH_3{-}C({=}O){-}Z \xrightarrow{O_2} CO_2 + H_2O + \text{Energy} \quad (+\,Z)$$

The first stage of the oxidation process cuts the fatty acid molecule into two-carbon fragments that are attached to a large organic molecule, represented here as Z; in the second stage the two-carbon fragments are oxidized to carbon dioxide and water. The balanced overall equation for the oxidation of the fatty acid shown above is

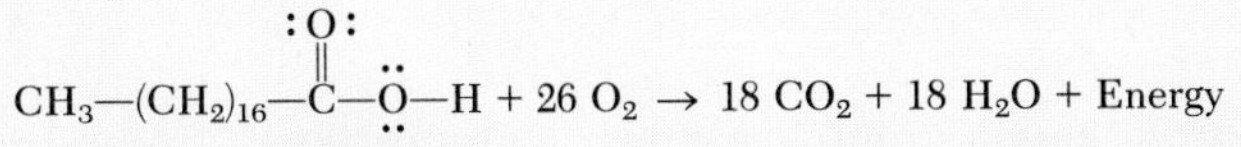

$$CH_3{-}(CH_2)_{16}{-}C({=}O){-}O{-}H + 26\,O_2 \rightarrow 18\,CO_2 + 18\,H_2O + \text{Energy}$$

Figure 20.3
Metabolism of a fat

Cholesterol, our second example of a lipid, has several biological functions in animals; for example, cholesterol is an important component of cell membranes, and it is converted by biochemical reactions into other important compounds, such as sex hormones.◆ Our discussion of cholesterol will focus on its role in disease.

◆The relationship of cholesterol to sex hormones is described in the *Inside Chemistry* essay at the end of Chapter 7.

Cholesterol from digested food travels in the bloodstream in particles called low-density lipoprotein (LDL) particles. The surfaces of many body cells contain LDL receptors that allow LDL particles to pass out of the bloodstream into the cells, where the particles are broken down.

If the bloodstream carries more LDL particles than the LDL receptors can accommodate, cholesterol particles, called plaques, may deposit in arteries; the arterial deposition of cholesterol is the disease called **atherosclerosis.** Arterial blockage by atherosclerosis can cause heart failure or stroke.◆

◆As a result of a rare genetic defect, the body cells of some persons have no LDL receptors. These persons suffer from atherosclerosis from birth, and they rarely live longer than 20 years.

Atherosclerosis cannot be prevented simply by reducing dietary cholesterol because the body produces cholesterol, primarily from fatty acids; an adult on a low-cholesterol diet typically synthesizes about 800 mg of cholesterol

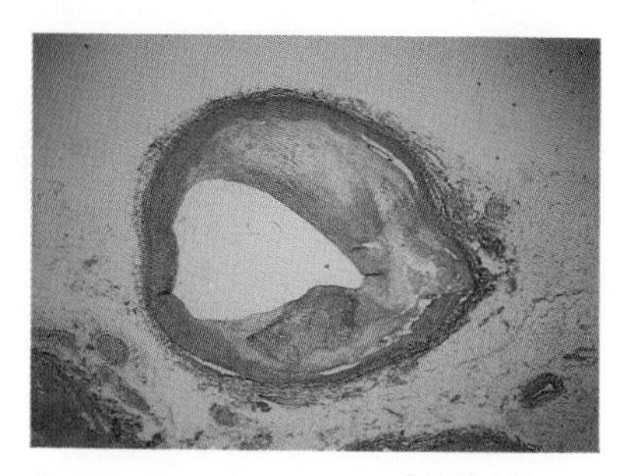

Human coronary artery blocked with cholesterol plaques. *(©Martin Rotker/Phototake NYC)*

per day. In this process, fatty acid molecules are cut into two-carbon fragments, as described in Figure 20.3, and these fragments are then assembled, in a long series of reactions, into cholesterol molecules:

$$\underset{\text{Fatty acid}}{CH_3{-}(CH_2)_{16}{-}\overset{\overset{:O:}{\|}}{C}{-}\ddot{\underset{\cdot\cdot}{O}}{-}H} \xrightarrow{Z} \underset{\text{Two-carbon fragment}}{CH_3{-}\overset{\overset{:O:}{\|}}{C}{-}Z} \rightarrow \text{Cholesterol}$$

Some fatty acids are more easily converted than others to cholesterol. The fatty acids that occur in animal fats, butter, and lard are easily converted to cholesterol, so reducing them in your diet can lower the rate at which your body produces cholesterol.

20-3 Proteins are important in animals as components of specialized tissues and as enzymes, the catalysts that control biochemical reactions

Proteins are polyamides; that is, proteins are polymers whose molecules are chains that contain many amide functional groups, $-\overset{\overset{:O:}{\|}}{C}-\ddot{N}-$. ◆ The monomers for proteins are **amino acids**; an amino acid is a compound whose molecule contains an amino group, $-\underset{\underset{H}{|}}{\ddot{N}}-H$, and a carboxyl group, $-\overset{\overset{:O:}{\|}}{C}-\ddot{\underset{\cdot\cdot}{O}}-H$.

◆ Polyamides are described in Section 19-13.

The general formula for amino acids is

$$H{-}\underset{\cdot\cdot}{\overset{\overset{H}{|}}{N}}{-}\underset{\underset{R}{|}}{CH}{-}\overset{\overset{:O:}{\|}}{C}{-}\ddot{\underset{\cdot\cdot}{O}}{-}H$$

in which R represents a group of atoms that differs from one amino acid to another. **Proteins** are polymers of amino acids.†

About twenty amino acids are commonly found in proteins; a representative group of eight is shown in Table 20.1. The metabolism of human beings can synthesize some amino acids but not others. Amino acids needed for good health that the body cannot produce, or cannot produce in the needed amounts, are called **essential amino acids;** they must be supplied by diet. The amino acids in Table 20.1 are the essential amino acids for human beings.

† Chemistry Insight

If you're wondering why an amino acid molecule doesn't react with itself, congratulations! It does. An amino acid molecule is both an acid, because of its carboxyl group, and a base, because of its amino group, so it reacts with itself to transfer a proton from the carboxyl group to the amino group:

$$H_2N{-}CH(R){-}C(=O){-}OH \rightleftarrows H_3N^+{-}CH(R){-}C(=O){-}O^-$$

At equilibrium, most of an amino acid is in the ionic form shown on the right. For our purposes, however, it will be simpler to write the formula for an amino acid in the un-ionized form shown on the left.

Table 20.1 Essential Amino Acids

Under ordinary room conditions all of the essential amino acids are white crystalline solids. Amino acids are often referred to by their three-letter abbreviations.

Name and Abbreviation	Formula
Isoleucine, Ile	$H_2N{-}CH(CH(CH_3)CH_2CH_3){-}C(=O){-}OH$
Leucine, Leu	$H_2N{-}CH(CH_2CH(CH_3)CH_3){-}C(=O){-}OH$
Lysine, Lys	$H_2N{-}CH((CH_2)_4NH_2){-}C(=O){-}OH$

(continued)

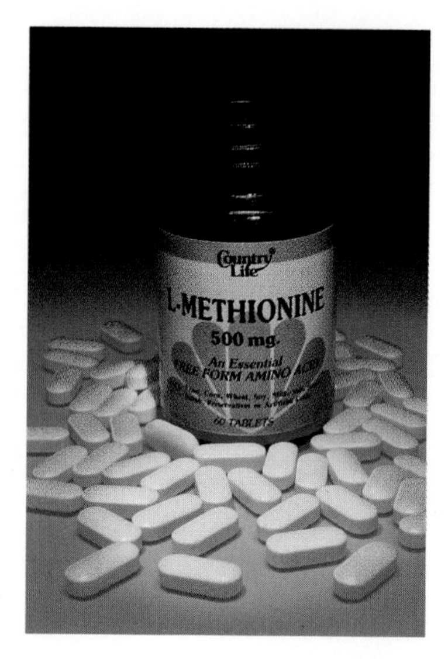

Methionine. *(Charles D. Winters)*

Table 20.1 Essential Amino Acids *(continued)*

Under ordinary room conditions all of the essential amino acids are white crystalline solids. Amino acids are often referred to by their three-letter abbreviations.

Name and Abbreviation	Formula
Methionine, Met	$H{-}\ddot{N}(H){-}CH[(CH_2)_2{-}\ddot{S}{-}CH_3]{-}C({=}\ddot{O}){-}\ddot{O}{-}H$
Phenylalanine, Phe	$H{-}\ddot{N}(H){-}CH(CH_2{-}C_6H_5){-}C({=}\ddot{O}){-}\ddot{O}{-}H$
Threonine, Thr	$H{-}\ddot{N}(H){-}CH[CH(\ddot{O}{-}H){-}CH_3]{-}C({=}\ddot{O}){-}\ddot{O}{-}H$
Tryptophan, Trp	$H{-}\ddot{N}(H){-}CH(CH_2{-}C_8H_5\ddot{N}H){-}C({=}\ddot{O}){-}\ddot{O}{-}H$ (indole ring)
Valine, Val	$H{-}\ddot{N}(H){-}CH[CH(CH_3){-}CH_3]{-}C({=}\ddot{O}){-}\ddot{O}{-}H$

The polymerization of three amino acids to form a simple polyamide is illustrated by this equation:

```
     H      :O:              H      :O:              H      :O:
     |       ||   ..         |       ||   ..         |       ||   ..
H—N—CH—C—O—H + H—N—CH—C—O—H + H—N—CH—C—O—H →
  ..         ..           ..         ..           ..         ..
     CH2                       CH2                      (CH2)4
H    |    H                    |                          |
  \  C  /                      CH—CH3                   H—N—H
   C   C                       |                          ..
   | ○ |                       CH3
   C   C
  /  C  \
H    |    H
     H
    Phe                        Leu                       Lys
```

```
        N-terminal        amide groups        C-terminal
        residue                               residue
     H     ↓      :O: H         :O: H       ↓      :O:                      ..
     |             ||  |          ||  |              ||   ..                 O
H—N—CH—C—N—CH—C—N——CH—C—O—H + 2 H   H
  ..                ..            ..                    ..                  ..
       CH2                CH2              (CH2)4
H      |      H           |                 |
  \    C    /             CH—CH3         H—N—H
    C     C               |                 ..
    |  ○  |               CH3
    C     C
  /    C    \
H      |      H
       H
                              Phe–Leu–Lys
```

The portion of an amino acid that's incorporated into a polymer chain is called an **amino acid residue;** in the chain, each residue is joined through an amide group to the next residue. As shown in the equation above, the molecule of an amino acid polymer has an N-terminal end with a free amino group and a C-terminal end with a free carboxyl group; by custom the formula for a chain is written with its N-terminal residue at the left.

Exercise 20.6

Write the formula for Thr–Val–Leu.

Protein molecules can contain hundreds or thousands of amino acid residues, and their structures can be very complex. ◆ The molecular chains of most proteins curl up on themselves to form complicated three-dimensional shapes that are essential to the protein's biological function.

◆ Chains of fewer than a few dozen amino acid residues are commonly called *peptides*, and the C–N bond in each amide group is called a *peptide bond*.

One important function of proteins is their formation of specialized tissues. Muscles, tendons, skin, hair, nails, and many other biological structures are made of proteins.

The shapes of protein molecules are essential to their functions. Egg white is nearly pure protein. The protein molecules in undisturbed egg white have their natural shapes, and the egg white is transparent (left). If the shapes of the molecules are changed—for example, by beating the egg white—the transparency is lost (right). *(Charles D. Winters)*

Proteins have a uniquely important function in controlling biochemical processes. In your body at this moment millions of biochemical reactions are taking place, and the rate of almost every one of these reactions is controlled by a specific protein acting as a catalyst.◆ These protein catalysts are called **enzymes.**

◆Catalysts are described in Section 19-8.

The control exerted by enzymes is crucial for an organism's survival, and the delicacy of their control is nearly unimaginable. In the ordinary process of walking across a room, for example, millions of biochemical reactions must occur in your muscles, tendons, visual system, skin, spinal cord, and brain, and each of them must occur at its appropriate rate. An enzyme controls each reaction. Biochemists are only beginning to understand the ways in which enzymes control biochemical reactions; research on enzymes is growing because of its importance for biochemistry and for medicine.

20-4 Nucleic acids control all of the life processes of each organism, and they control the transmission of life from one generation to the next

The substance called deoxyribonucleic acid (DNA) controls life on our planet. It controls the life of each organism by dictating the structures of all of the organism's proteins—including enzymes—and it controls the transmission of life from one generation to the next by carrying hereditary information from parent to offspring. What is DNA, and how does it exert its extraordinary control over life?

The Structure of DNA

DNA is a polymer whose molecules have a complex structure. In a molecule of DNA the repeating unit is a *nucleotide;* a **nucleotide** molecule is a combination of a phosphoric acid molecule, a base molecule, and a carbohydrate molecule.◆ These are the components of one nucleotide:

◆To remember the components of a nucleotide, think of ABC for acid, base, carbohydrate.

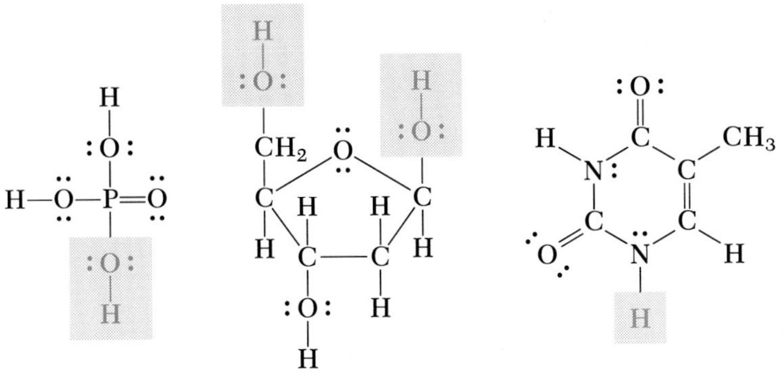

Phosphoric acid

Deoxyribose, a carbohydrate

Thymine, a base

The blue patches mark the sites at which these molecules combine to form the nucleotide.

This is the formula for the nucleotide assembled from these components:

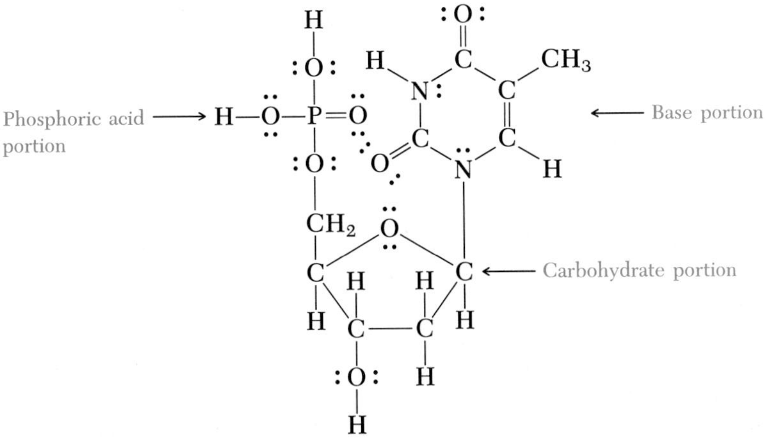

Phosphoric acid portion

Base portion

Carbohydrate portion

A nucleotide

In all DNA nucleotides the phosphoric acid portion and the carbohydrate portion are the same; that is, all DNA nucleotides have the generic formula

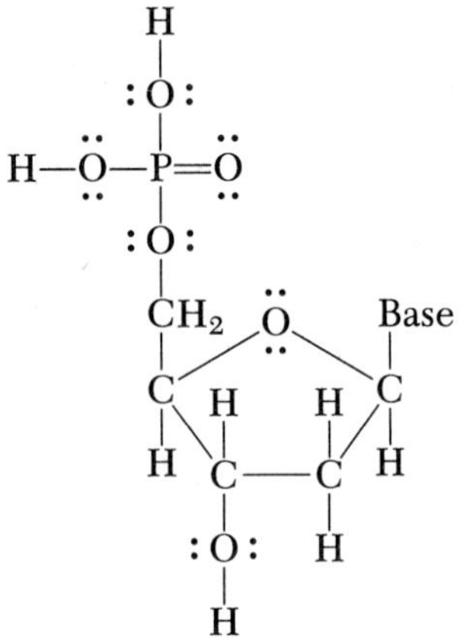

and they differ from one another only in the base portions of their molecules. Four bases commonly occur in DNA nucleotides:◆

◆These nitrogen compounds are called bases because, like the amines described in Section 19-11, all of them are weak bases.

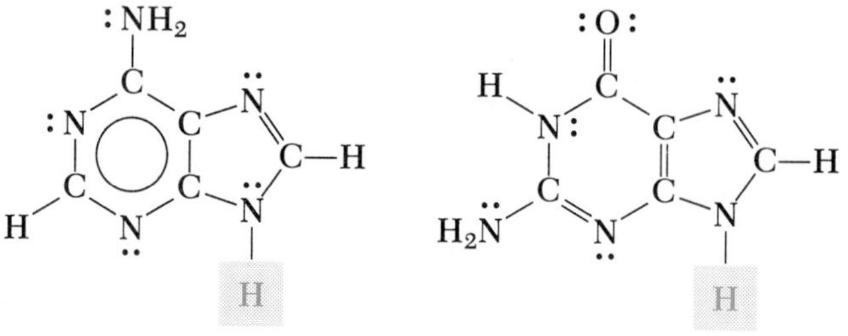

Adenine, A

Guanine, G

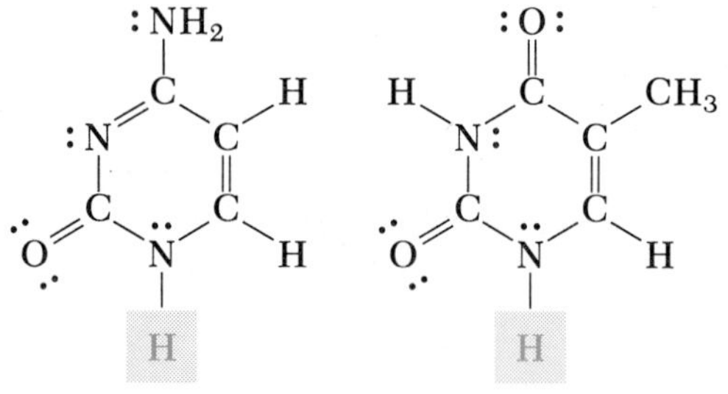

Cytosine, C

Thymine, T

The blue patch marks the site at which the base attaches to the carbohydrate in a nucleotide.

The bases are often referred to by their one-letter abbreviations, A, G, C, and T.

Exercise 20.7

Write the formula for the DNA nucleotide that contains guanine.

In abbreviated form the formulas for the four DNA nucleotides are:

```
         H                                  H
         |                                  |
        :O:                                :O:
         |                                  |
  H—O—P=O                            H—O—P=O
         |                                  |
        :O:                                :O:
         |                A                 |                G
      5' CH2   O          |               CH2   O           |
         |   /   \        |                |   /   \        |
      4' C          C 1'                   C  H    H  C
         |  \  H   / |                     |  \ |  | /  |
         H   C ——C   H                     H   C ——C    H
            3'  2' |                            |     |
           :O:   H                            :O:   H
            |                                  |
            H                                  H

         H                                  H
         |                                  |
        :O:                                :O:
         |                                  |
  H—O—P=O                            H—O—P=O
         |                                  |
        :O:                                :O:
         |                C                 |                T
        CH2   O           |               CH2   O           |
         |   /   \        |                |   /   \        |
         C  H    H  C                      C  H    H  C
         |  \ |  | /  |                    |  \ |  | /  |
         H   C ——C    H                    H   C ——C    H
             |     |                           |     |
           :O:   H                            :O:   H
            |                                  |
            H                                  H
```

As shown in the first formula above, the carbon atoms in the carbohydrate portion can be designated by numbering them.†

DNA is a polymer of the four nucleotides whose formulas are shown above. In the DNA molecule the phosphoric acid portion of each nucleotide is attached to the 3′ carbon atom in the next nucleotide; this formula shows three nucleotide units in a DNA chain:

† **Chemistry Insight**

Rules for numbering atoms in complicated molecules require that the numbers used in the carbohydrate portion of a nucleotide be 1′, 2′, 3′, 4′, and 5′, rather than 1, 2, 3, 4, and 5; the numbers are read "one prime," "two prime," and so forth. In the name *deoxyribonucleic acid, deoxy-* refers to the 2′ carbon atom: It's the only carbon atom in the carbohydrate ring that has no oxygen atom bonded to it.

(Chain continues)

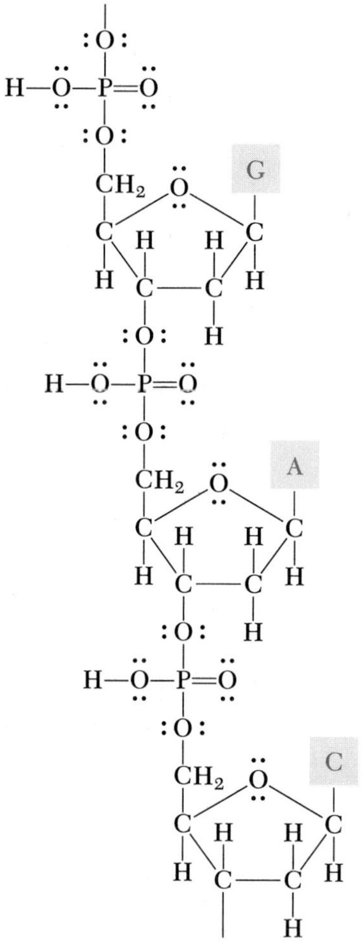

G

A

C

(Chain continues)

Under biological conditions the hydrogen atom on each phosphoric acid portion is ionized, so this formula more accurately represents the molecule:

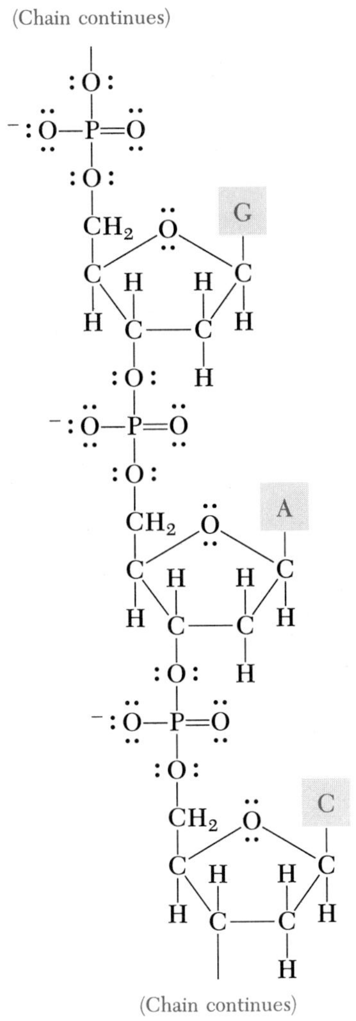

(Chain continues)

G

A

C

(Chain continues)

In its usual form, a molecule of DNA consists of two polynucleotide chains attached to one another by hydrogen bonds.◆ Hydrogen bonding occurs between these **complementary base pairs:**

◆The nature of hydrogen bonds is discussed in Section 14-4.

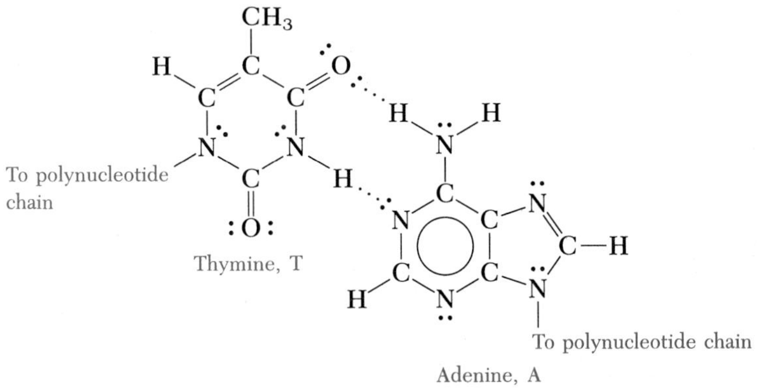

To polynucleotide chain

Thymine, T

To polynucleotide chain

Adenine, A

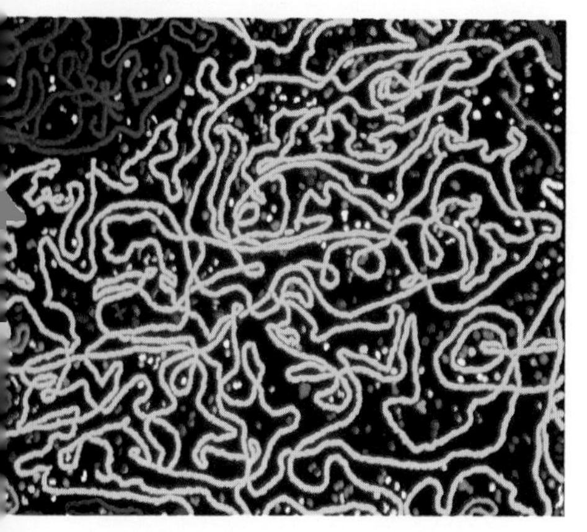

Human DNA. In this photo the DNA is color-enhanced and greatly magnified. *(Dr. Thomas Broker/ Phototake NYC)*

To polynucleotide chain

To polynucleotide chain

Cytosine, C

Guanine, G

Because hydrogen bonding is limited to the pairs A–T and G–C, the sequence of nucleotides in one chain of DNA must be complementary to the sequence in the chain attached to it. For example, if a sequence of bases in the nucleotides in one chain is ATGCTACGGATTCAA, then the complementary sequence in the other chain is TACGATGCCTAAGTT.

Exercise 20.8

Write the complementary base sequence for CGGATAACC.

The two complementary chains of a DNA molecule twist around one another, as shown in Figure 20.4; the shape of the molecule is described as a **double helix.**◆

◆The double helix structure of DNA was discovered in 1953 by Francis H. C. Crick (British, born 1916) and James D. Watson (American, born 1928); their discovery was based in part on work by Maurice H. F. Wilkin (British, born 1916). Crick, Watson, and Wilkin shared a Nobel Prize in 1963.

The Function of DNA

How does DNA control the life processes of organisms, and how does it pass hereditary information from each generation to the next?

DNA controls the life processes of an organism by dictating the structures of all of its proteins, that is, by dictating the sequence of amino acid residues in the chain of each of its protein molecules. The proteins include the enzymes that control almost all of the organism's biochemical reactions; by dictating the structures of an organism's enzymes, DNA controls its biochemical reactions. This is the overall pattern of control:

Sequence of nucleotides in DNA → Sequence of amino acid residues in enzymes → Enzymatic control of biochemical reactions

The sequence of bases in a polynucleotide chain of a DNA molecule corresponds to the sequence of amino acid residues in a protein chain: Each successive three-base sequence in DNA corresponds to a particular amino acid in a protein. Table 20.2 on page 516 shows the three-base sequences that correspond to the essential amino acids described in Table 20.1.

You can use the information in Table 20.2 to predict the sequence of amino acid residues in a protein molecule from the corresponding sequence of bases

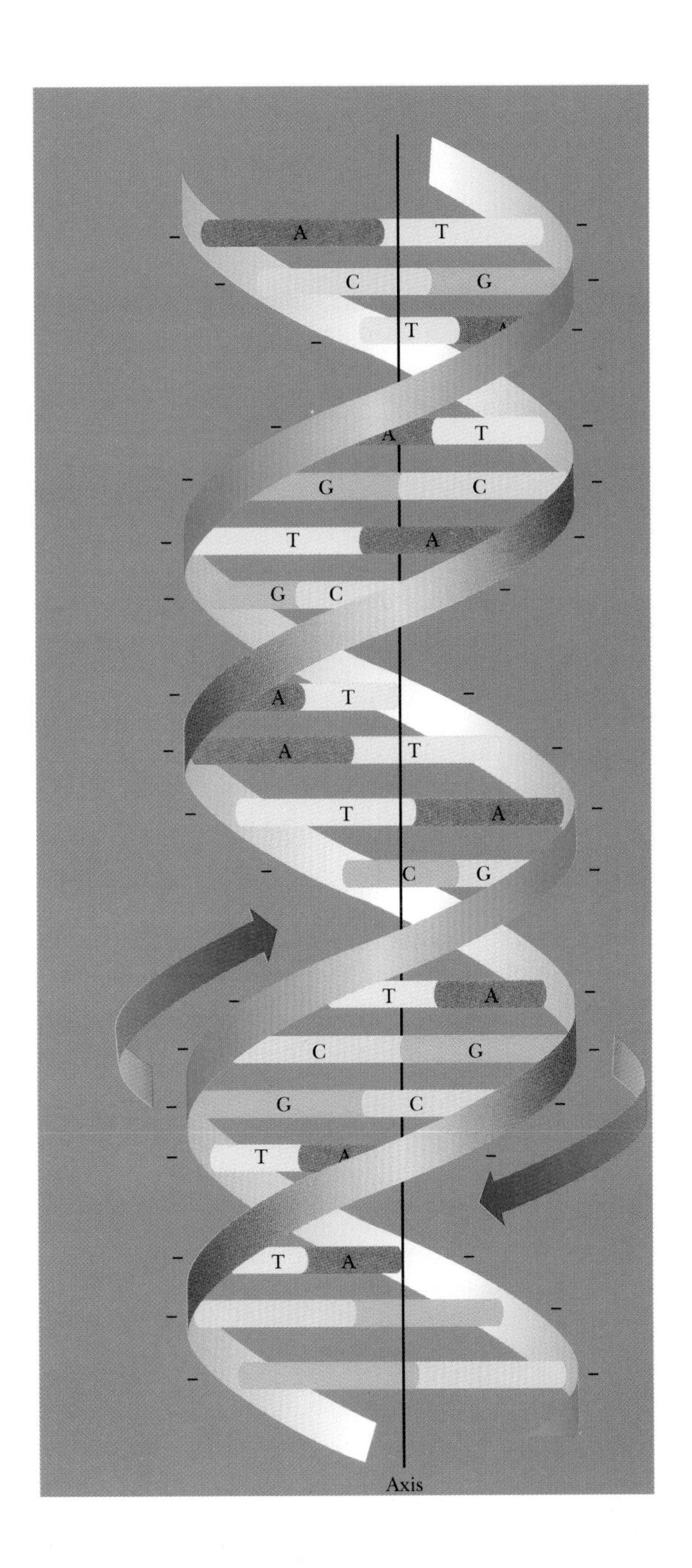

Figure 20.4
The double helix of DNA. Two strands of polynucleotides twist around one another, held together by hydrogen bonds between their complementary base pairs, A–T and G–C. The negative charges occur at the ionized phosphoric acid groups.

Table 20.2 Correspondence Between Three-Base Sequences in DNA and Essential Amino Acids

Amino Acid	Three-Base Sequence in DNA
Isoleucine, Ile	TAT
Leucine, Leu	GAC
Lysine, Lys	TTT
Methionine, Met	TAC
Phenylalanine, Phe	AAA
Threonine, Thr	TGT
Tryptophan, Trp	ACC
Valine, Val	CAG

in a DNA chain, and vice versa. For example, the base sequence CAGAAAGAC corresponds to the amino acid sequence Val–Phe–Leu.

Exercise 20.9

Use the information in Table 20.2 to write the base sequence in DNA that corresponds to this amino acid sequence in a protein: Trp–Phe–Ile–Thr.

◆ Biochemists around the world are cooperating with one another to identify the complete sequence of bases in the human genome, which contains about three billion base pairs.

In a DNA molecule all of the three-base sequences that dictate the amino acid sequence for a particular protein are the **gene** for that protein, and all of the genes for an organism are its **genome.**◆ A list of all possible three-base sequences (there are 64 of them) and their corresponding amino acids is called the **genetic code.**

DNA passes life from one generation to the next. In sexual or asexual reproduction each parent organism passes some or all of its DNA to its offspring. Because the DNA in an organism determines all of its characteristics, offspring resemble their parents.

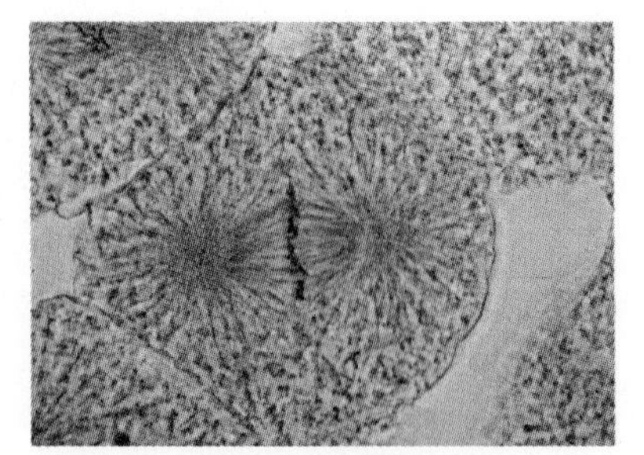

During cell division, chromosomes—rod-like structures that contain DNA—are divided and passed on to each new cell. *(Eric Gravel Phototake NYC)*

20-5 In living organisms, biochemical reactions are organized and controlled in a variety of ways

In your body in the next second, many millions of biochemical reactions will take place. To sustain your life and health, each of these reactions must be carefully controlled, and all of them must be organized in balance with one another. In all living things, organization and control of biochemical processes are essential for life and health.

In this Section we'll look briefly at a few of the many ways in which biochemical processes are organized and controlled.

Control by DNA Determination of Enzyme Structures

We saw in Section 20-4 that DNA controls biochemical reactions by determining enzyme structures. Certain sequences of nucleotides in a DNA molecule dictate corresponding sequences of amino acids in an enzyme, and the structure of the enzyme determines its role in the catalysis of a particular biochemical reaction. This pattern of DNA control occurs in all living organisms; it is one of the most general forms of biochemical control.

Control by Activation and Deactivation of Enzymes

Enzymes are extraordinarily effective catalysts: Most biochemical reactions do not occur at all in the absence of their enzymes, and they occur at very high rates when sufficient concentrations of their enzymes are present. The occurrence and rates of biochemical reactions are often regulated by control of the concentrations of their enzymes.

In one form of regulation, an enzyme is produced from a protein precursor—called a **zymogen**—when it is needed:

$$\underset{\text{Protein that does not act as a catalyst}}{\text{Zymogen}} \rightarrow \underset{\text{Protein that does act as a catalyst}}{\text{Enzyme}}$$

This form of regulation allows quick control over a biochemical reaction: While the zymogen is present, no reaction occurs, but as soon as the zymogen is converted to its enzyme, the reaction can occur at a very high rate.

Most biological processes are long sequences of biochemical reactions, each catalyzed by its own enzyme; in these sequences a product of one reaction becomes a reactant in the next:

$$\text{Initial Reactants} \xrightarrow{\text{Enzyme 1}} \text{Product 1} \xrightarrow{\text{Enzyme 2}} \text{Product 2} \dashrightarrow \text{Final Product}$$

In the form of control called **feedback inhibition**, the final product, as it accumulates, reacts with the first enzyme in the sequence, and reversibly converts it to an inactive form:

$$\text{Final Product} + \text{Active Enzyme 1} \rightleftarrows \text{Inactive Enzyme 1}$$

When enough of the final product is present, this equilibrium shifts to the right, following Le Châtelier's principle (Section 17-3), and the entire sequence of reactions shuts down. When the final product has been used up and more is needed, the equilibrium shifts left, and the sequence of reactions resumes.

Control by Locations of Reactions in Cells

Animals and plants that are large enough to be seen with the naked eye are composed of many millions of tiny units called **cells**; Figure 20.5 shows a typical

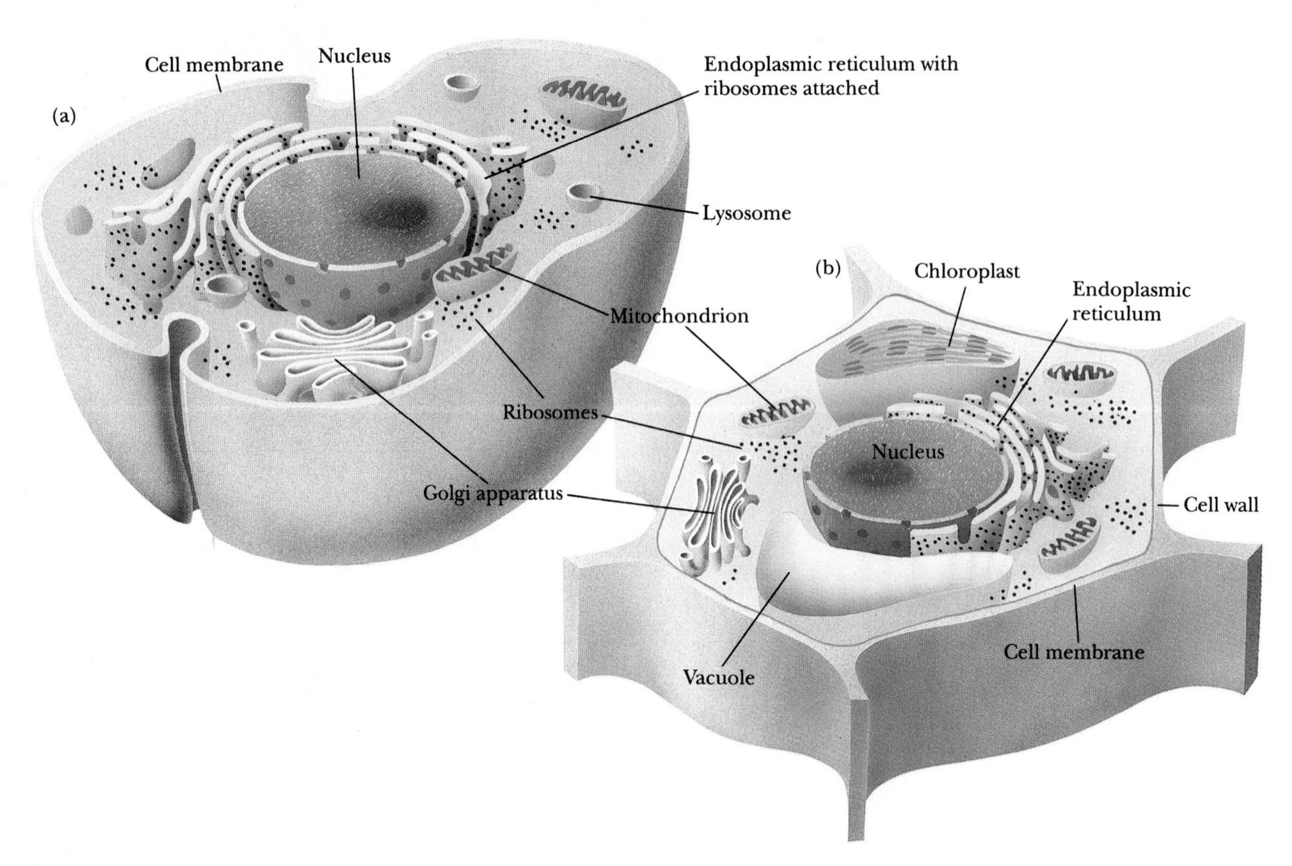

Figure 20.5
Typical animal and plant cells. Each part of an animal cell (a) or plant cell (b) provides a site at which certain biochemical processes occur. The nucleus contains the cell's DNA. The mitochondrion provides a site at which fatty acids and other molecules are oxidized; the ribosomes are the site of protein synthesis; and the Golgi apparatus moves proteins out of the cell. Lysosomes in animal cells store enzymes, and chloroplasts in plant cells are the site of photosynthesis.

animal cell and a typical plant cell. As shown in Figure 20.5, each cell consists of many parts arranged in a complex structure.

The parts of a cell control and direct the biochemical reactions that occur in it. For example, the cell membrane allows certain molecules to enter the cell but not others. A cell's DNA is in its nucleus, and protein synthesis occurs on its ribosomes;◆ certain molecules read the sequence of bases in DNA in the nucleus and carry that information to the ribosomes, where the base sequence is translated into the sequence of amino acids in a growing protein chain. Similarly, each of the other parts of a cell provides a site at which certain biochemical reactions occur and are controlled by their enzymes.

◆In the word *deoxyribonucleic*, the suffix *-nucleic* refers to the location of DNA in cell nuclei.

Control by Hormones

The biochemical control mechanisms we've considered so far operate within individual cells. There are also controls that operate between groups of cells. A **hormone** is a biochemical compound that's produced by one group of cells and that travels to other cells, where it causes and controls biochemical reactions.

Human beings have dozens of hormones with a wide variety of chemical structures and biological functions. One example is **thyroxine,** a hormone produced in the thyroid gland at the base of the neck. Thyroxine migrates in the bloodstream to cells throughout the body, where it regulates reactions that use oxygen as a reactant.

```
       ..              ..
      :I:    H        :I:    H
        \   /           \   /
         C—C             C—C                    :O:
  ..    /   \     ..    /   \                   ||   ..
H—O—C (   )  C—O—C (   )  C—CH2—CH—C—O—H
  ..    \   /     ..    \   /         |          ..
         C—C             C—C         :NH2
        /   \           /   \
      :I:    H        :I:    H
       ..              ..
```

Thyroxine

Iodized salt contributes to good health by providing iodine atoms that are incorporated into molecules of thyroxine, the hormone produced in the thyroid gland. The line formula for thyroxine is shown in the text. *(Charles D. Winters)*

Control by Reaction Cycles

From the broadest point of view, all life on earth consists of continuously repeating cycles of biochemical reactions, ultimately driven by energy from the sun. The carbon cycle, which was described in Figure 20.2, is one example.

Figure 20.6 shows another large and crucial reaction cycle, the **nitrogen cycle.** In a simplified description, these are the forms of nitrogen in the nitrogen cycle:

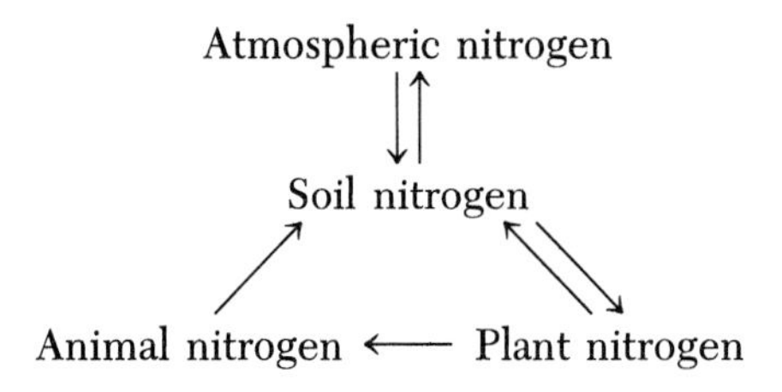

The atmosphere, which is about 80% elemental nitrogen, is an enormous reservoir of nitrogen for biochemical processes. The conversion of atmospheric nitrogen to soil nitrogen—called **nitrogen fixation**—occurs in three very different ways: by the action of lightning, by the action of soil bacteria, and by the action of human beings.

Lightning causes reactions between nitrogen and oxygen in the atmosphere to form nitrogen oxides that dissolve in atmospheric water and fall to earth in rain. In the soil these oxides eventually become dissolved nitrate ion, $NO_3^-(aq)$.

Some soil bacteria contain enzymes, called *nitrogenases*, that catalyze the conversion of atmospheric nitrogen to ammonia, NH_3, which dissolves in water to form aqueous ammonia and aqueous ammonium ion, $NH_4^+(aq)$. Other soil bacteria carry out conversions of ammonium ion first to nitrite ion, $NO_2^-(aq)$, and then to nitrate ion, $NO_3^-(aq)$.

Early in this century chemists and soil scientists created methods to carry out nitrogen fixation: Nitrogen from the atmosphere is converted either to ammonia or to nitrates and then injected in these forms into the soil. Eventually, the injected nitrogen is stored in the soil as $NO_3^-(aq)$.

Plants convert nitrate ion from soil into amino acids and proteins, and animals obtain their nitrogen by eating plants. Animal excretions return nitro-

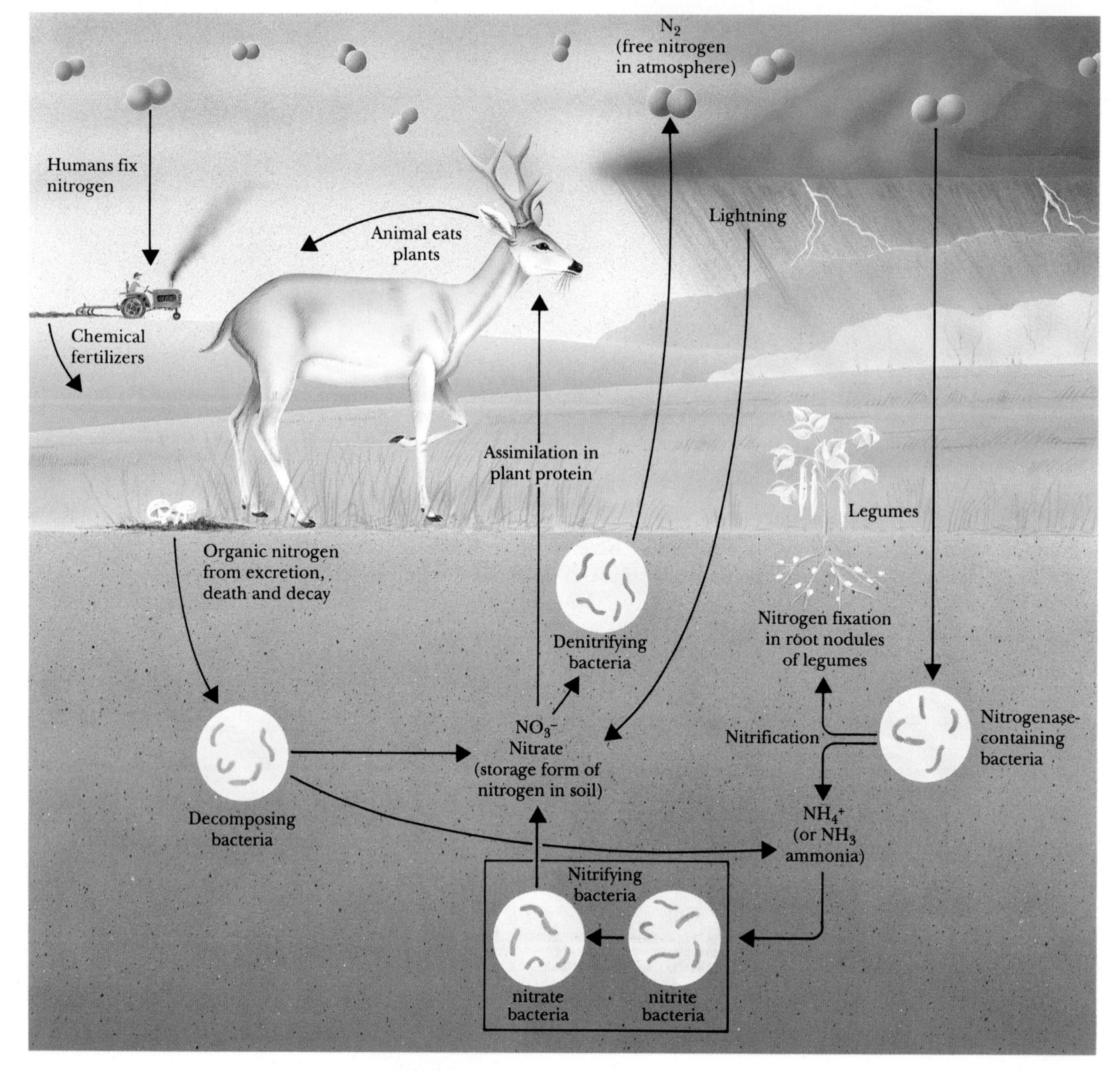

Figure 20.6
The nitrogen cycle. Atmospheric nitrogen, N_2, enters the soil by the actions of lightning, bacteria, and human beings; in the soil, nitrogen is stored as nitrate ion, NO_3^-. Growing plants make protein from the nitrate ion in soil, and animals get their nitrogen by eating plants; animal excretions and the decaying remains of animals and plants return nitrogen to the soil. The cycle is completed when soil bacteria convert nitrate ion to elemental nitrogen.

gen to the soil, and when plants and animals die, their decay returns nitrogen to the soil. The nitrogen cycle is completed by soil bacteria that convert $NO_3^-(aq)$ to $N_2(g)$, which rises into the atmosphere.

20-6 Many important questions in biochemistry remain unanswered

The body of knowledge that makes up modern chemistry has grown at an astonishing rate: Dalton created the atomic theory at the start of the nineteenth century, and by the end of that century the science of chemistry was developed fully enough to be applied to the study of the complex chemical systems of living organisms. In our century biochemistry has provided an increasingly sophisticated understanding of the molecules and reactions that are the basis of life. The growth of biochemistry is one of the greatest intellectual achievements of our time, and modern medicine is one application of that knowledge.

Although the scope of biochemical knowledge is now enormous, many important biochemical questions remain unanswered. These are a few examples:

How do enzymes work? As we've seen, almost all of the many millions of reactions that occur in living systems are catalyzed by enzymes. Thousands of enzymes have been identified, but only a handful are understood well enough to allow a complete explanation of their functions as catalysts.

What are the causes of diseases? Our understanding of diseases has increased enormously but is still very slight. For example, we do not know, at the biochemical level, the causes of the common cold, most cancers, or schizophrenia. Significant progress in understanding diseases will come about only as we learn more about the biochemistry of healthy organisms.

How do drugs work? Because we do not fully understand the biochemical basis of most diseases, we cannot understand how drugs relieve or cure them. The biochemical actions of most drugs are not fully understood; as a result, the development of new drugs is to a large extent a trial-and-error process that is slow and expensive.

What causes aging? Is aging a result of environmental damage to the organism, or is it an accumulation of errors by the organism's own biochemical processes? Or is it something else entirely? Because we do not know the answers to these questions, we do not know whether it is possible to slow or prevent the aging process.

How did life on earth begin? Did life on earth begin as spontaneous reactions of compounds dissolved in primordial oceans, or was it seeded from life elsewhere in the universe? What was the chemical nature of the first life forms, and how did they reproduce themselves? We do not know.

What is learning? As you read these words, what biochemical events in your brain make it possible for you to understand them? Why do you remember some experiences and not others? How can learning be made easier? These are fundamental questions about the biochemistry of human beings that no one can yet answer.

These and many other similarly important questions are under investigation by biochemists around the world. Our increasing ability to answer them will give us increasing control over life itself.

Inside Chemistry What is biotechnology?

Biotechnology is a new and rapidly growing industry that uses the discoveries of biochemistry to create products with commercial value. According to some predictions, the commercial successes of biotechnology will be even greater than those of computer technology. ◆

◆ In the fiscal year ending June 30, 1991—a year of recession for most of the U.S. economy—sales in the biotechnology industry increased by 38% over the previous year.

Biotechnology uses many biochemical discoveries, especially those that allow the manipulation of DNA. As one example, we'll consider the process called *cloning*.

Cloning is commonly carried out with relatively simple one-celled organisms called *bacteria*. For the purposes of cloning, bacteria have two important characteristics: They multiply rapidly, and their DNA can be artificially altered.

In bacterial DNA the ends of the double helix are joined together; that is, bacterial DNA is circular. By the use of certain enzymes, a bacterial DNA molecule can be opened:

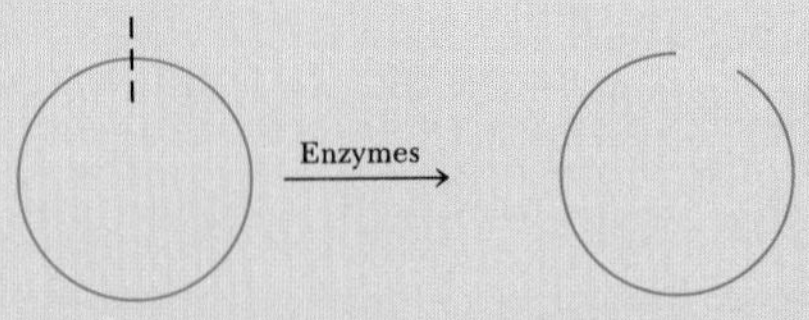

Similarly, enzymes can be used to selectively remove a section from the DNA molecule of an animal:

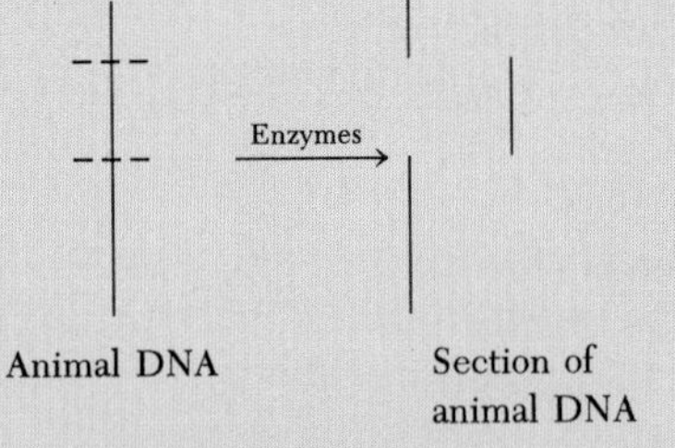

With other enzymes, the section of animal DNA can be spliced into the bacterial DNA to produce *recombinant bacterial DNA*:

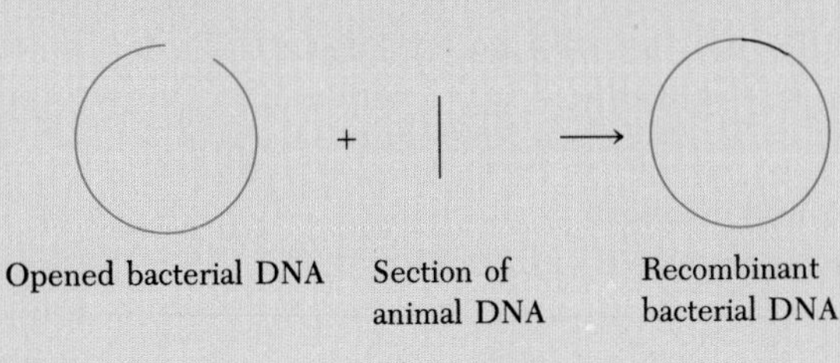

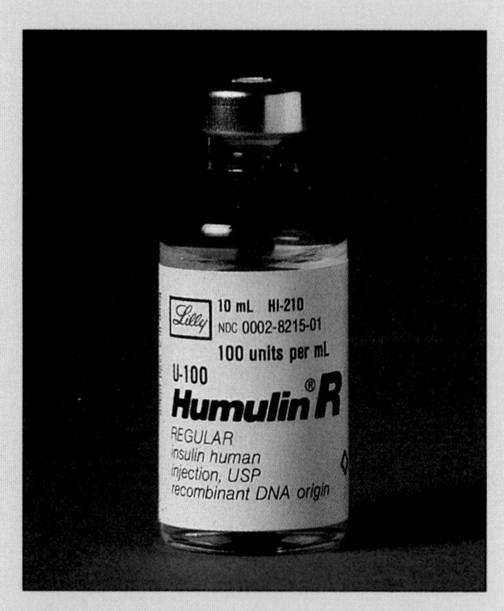

Human insulin produced by biotechnology. *(SIY/Science Source/ PhotoResearchers, Inc.)*

When bacteria that contain recombinant DNA reproduce, their offspring also contain recombinant DNA; cloning is the production of offspring that contain artificially modified DNA.

Cloning is commercially useful because the section of animal DNA that's inserted into bacterial DNA can be chosen to carry the genetic code for a particular protein; all of the bacteria with the recombinant DNA will then produce that protein. In this way bacteria can be used to make proteins that would otherwise be difficult and expensive to produce. One of the products now made in this way is the hormone insulin, used by diabetics to control their levels of blood sugar.

Chapter Summary: Biochemistry

Section	Subject	Summary	Check When Learned
	Biochemistry	Branch of chemistry that describes and explains chemical processes of living organisms.	☐
20-1	Carbohydrates	Polyhydroxy aldehydes or ketones: A carbohydrate molecule contains an aldehyde or ketone functional group and several alcohol functional groups.	☐
20-1	Open-chain and ring formulas for glucose	The two forms are in equilibrium, with the ring form predominating: 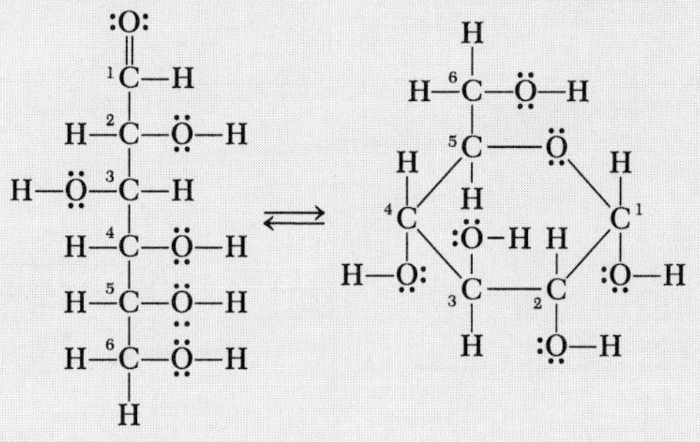 :O: ‖ ¹C—H H—²C—Ö—H H—Ö—³C—H H—⁴C—Ö—H H—⁵C—Ö—H H—⁶C—Ö—H H ⇌ (ring form)	☐
20-1	Starch	Polymer of glucose; in the polymer chain, the first carbon atom of one glucose unit is bonded through an oxygen atom to the fourth carbon atom of the next glucose unit: 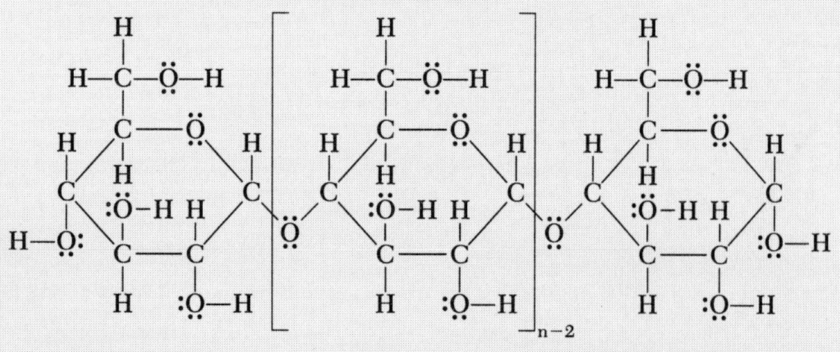 [structure, repeating unit]$_{n-2}$	☐
20-1	Photosynthesis	Process by which energy from sunlight is stored in carbohydrates by plants: $CO_2 + H_2O + \text{Energy as sunlight} \rightarrow \text{Starch}$	☐
20-1	Two steps in photosynthesis:	Y is a large organic molecule:	
	Light reaction	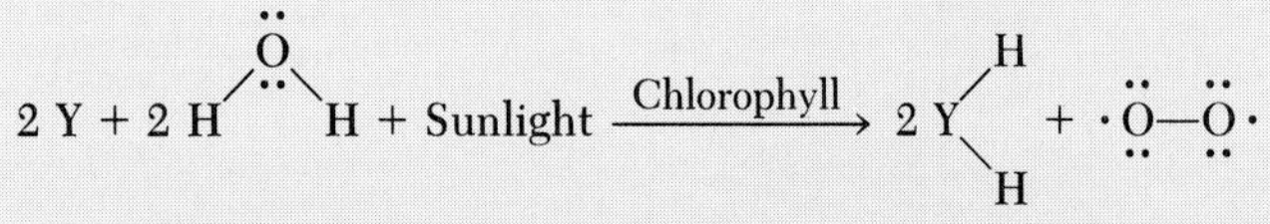 $2\ Y + 2\ H{-}\ddot{O}{-}H + \text{Sunlight} \xrightarrow{\text{Chlorophyll}} 2\ YH_2 + \cdot\ddot{O}{-}\ddot{O}\cdot$	

Section	Subject	Summary	Check When Learned
	Dark reaction	$12\,\mathrm{Y}\langle^{\mathrm{H}}_{\mathrm{H}} + 6\,\ddot{\underset{\cdot\cdot}{\mathrm{O}}}{=}\mathrm{C}{=}\ddot{\underset{\cdot\cdot}{\mathrm{O}}} \rightarrow \text{Glucose} + 6\,\mathrm{H}{-}\ddot{\underset{\cdot\cdot}{\mathrm{O}}}{-}\mathrm{H} + 12\,\mathrm{Y}$	☐
20-1	Metabolism	The process by which foods—for example, carbohydrates—are broken down by an organism.	☐
20-1	Metabolism of starch	$\text{Starch} + O_2 \rightarrow CO_2 + H_2O + \text{Energy}$	☐
20-1	Carbon cycle	Cyclic relationship between photosynthesis of starch in plants and metabolism of starch in animals: Energy as sunlight Photosynthesis in plants $CO_2 + H_2O$ → Starch + O_2 Metabolism in animals Energy for life processes in animals 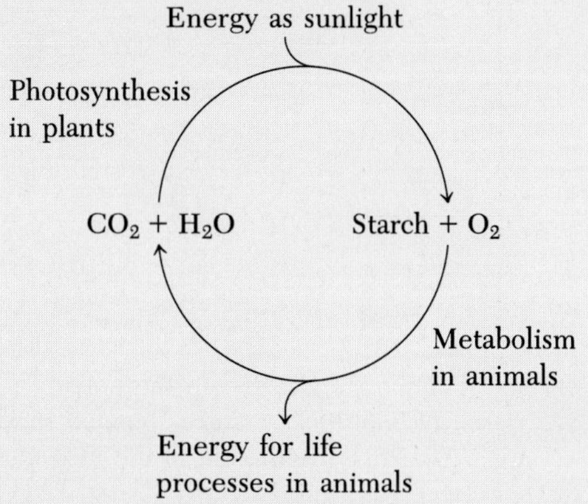	☐
20-1	Cellulose	Polymer of glucose; its bonding differs slightly but importantly from that in starch.	☐
20-1	Two important functions of carbohydrates	Energy storage (starch) and structural material (cellulose).	☐
20-2	Lipids	Biochemical compounds whose molecules contain large hydrocarbon portions; besides these hydrocarbon portions, the molecules of lipids contain a variety of functional groups.	☐
20-2	Fats	Triacylglycerols, esters of glycerol with fatty acids. Their generic formula is: 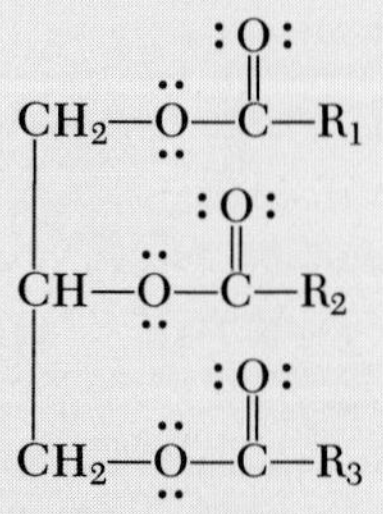 $CH_2{-}\ddot{\underset{\cdot\cdot}{O}}{-}C({=}\ddot{O}{:}){-}R_1$ $\vert$ $CH{-}\ddot{\underset{\cdot\cdot}{O}}{-}C({=}\ddot{O}{:}){-}R_2$ $\vert$ $CH_2{-}\ddot{\underset{\cdot\cdot}{O}}{-}C({=}\ddot{O}{:}){-}R_3$ where R_1, R_2, and R_3 represent different hydrocarbon chains.	☐

Section	Subject	Summary	Check When Learned
20-2	Fatty acids	Carboxylic acids found in fats. Their molecules typically have even numbers of carbon atoms and long, unbranched carbon chains (often about twenty carbon atoms); the chains may contain double bonds.	☐
20-2	Most important function of fats	Long-term energy storage in animals. (Carbohydrates are used for short-term energy storage.)	☐
20-2	Two steps in metabolism of fat:		
	Hydrolysis	$\begin{array}{l} CH_2{-}O{-}C({=}O){-}R_1 \\ \vert \\ CH{-}O{-}C({=}O){-}R_2 \\ \vert \\ CH_2{-}O{-}C({=}O){-}R_3 \end{array} + 3\,H{-}O{-}H \rightarrow \begin{array}{l} CH_2{-}O{-}H \\ \vert \\ CH{-}O{-}H \\ \vert \\ CH_2{-}O{-}H \end{array} + \begin{array}{l} H{-}O{-}C({=}O){-}R_1 \\ H{-}O{-}C({=}O){-}R_2 \\ H{-}O{-}C({=}O){-}R_3 \end{array}$	☐
	Oxidation	Z is a large organic molecule: $R{-}C({=}O){-}O{-}H \xrightarrow{Z} CH_3{-}C({=}O){-}Z \xrightarrow{O_2} CO_2 + H_2O + \text{Energy}$ (Z released in the second step)	
20-2	Metabolic water	Water formed in an organism as a product of metabolic reactions, for example, water produced in the metabolism of fats or carbohydrates.	☐
20-2	LDL particles	Low-density lipoprotein particles, the form in which cholesterol travels in the bloodstream.	☐
20-2	LDL receptors	Sites on body cells that allow LDL particles to enter the cells, where they can be broken down.	☐
20-2	Atherosclerosis	Disease in which cholesterol plaques deposit in arteries.	☐
20-3	Amino acids	Compounds whose formulas contain an amino group and a carboxyl group; the general formula for amino acids is $H{-}N(H){-}CH(R){-}C({=}O){-}O{-}H$ where R is a group of atoms that differs from one amino acid to another.	☐
20-3	Proteins	Polymers of amino acids. The general formula for a protein is $H{-}N(H){-}CH(R){-}C({=}O){-}\left[N(H){-}CH(R){-}C({=}O)\right]_n{-}N(H){-}CH(R){-}C({=}O){-}O{-}H$	☐

Section	Subject	Summary	Check When Learned
20-3	Amino-acid residue	Portion of an amino acid molecule incorporated into a protein molecule.	☐
20-3	Essential amino acids	Amino acids needed for good health that the body cannot produce, or cannot produce in the needed amounts, so they must be supplied in the diet.	☐
20-3	Three-letter abbreviations for Isoleucine Leucine Lysine Methionine Phenylalanine Threonine Tryptophan Valine	 Ile Leu Lys Met Phe Thr Trp Val	☐
20-3	Two important functions of proteins	Proteins form specialized tissues—for example, hair and nails—and enzymes.	☐
20-3	Enzymes	Proteins that function as catalysts in biochemical reactions.	☐
20-4	DNA	Deoxyribonucleic acid, a polymer of nucleotides.	☐
20-4	Nucleotide	Combination of phosphoric acid, carbohydrate, and base; the general formula for a nucleotide of DNA is 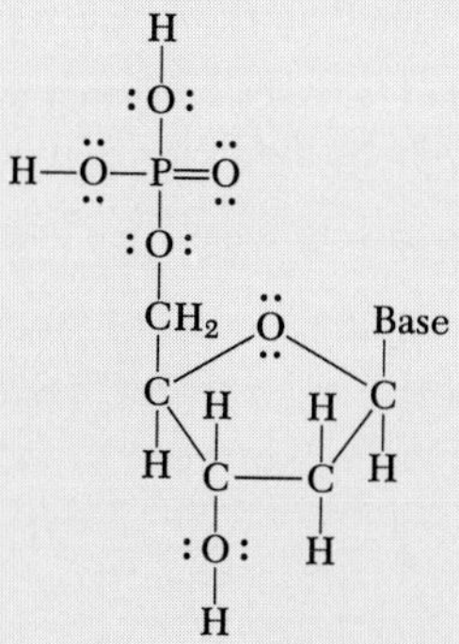 (structure below)	☐

```
              H
              |
             :O:
       ..     |  ..
     H—O—P=O
       ..     |  ..
             :O:
              |
             CH2  ..O..   Base
              |  /     \   |
              C  H    H  C
              | \ |   | / |
              H  C——C   H
                 |    |
               :O:    H
                 |
                 H
```

Section	Subject	Summary	Check When Learned
20-4	Four bases found in DNA	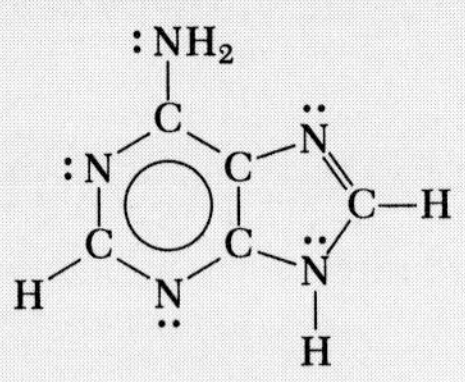 [structural formulas] Adenine, A; 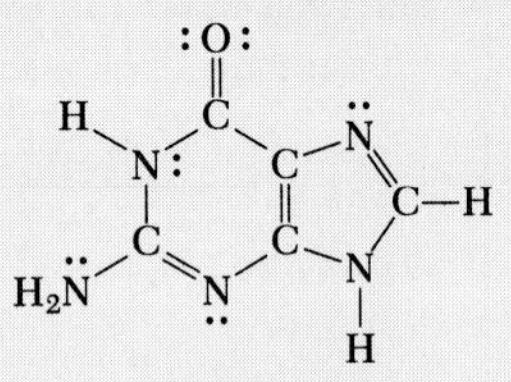 Guanine, G; 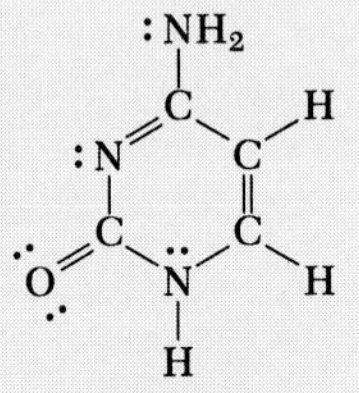 Cytosine, C; 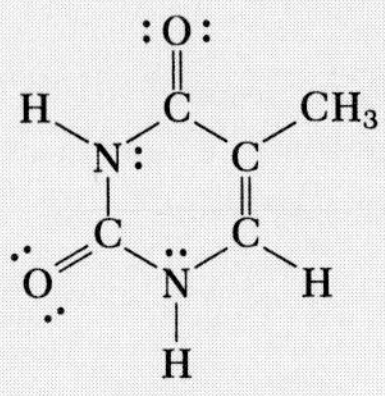 Thymine, T	☐
20-4	Bonding between nucleotides in DNA chain	The phosphoric acid portion of each nucleotide is attached to the 3′ carbon atom in the next nucleotide.	☐
20-4	Overall structure of DNA	Two polynucleotide chains twist around one another in a double helix; the chains are held together by hydrogen bonding between complementary base pairs.	☐
20-4	Hydrogen bonding between complementary base pairs	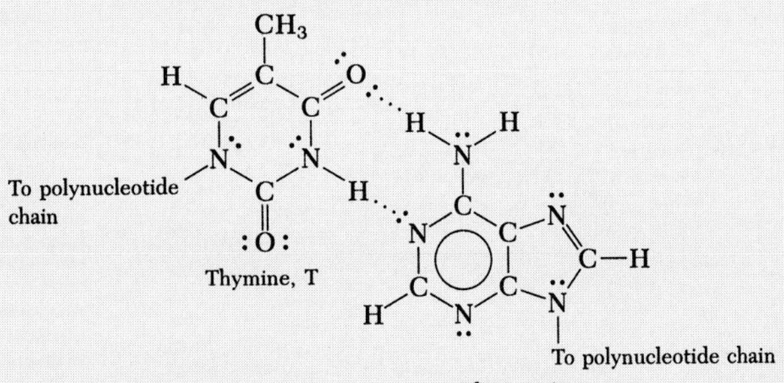 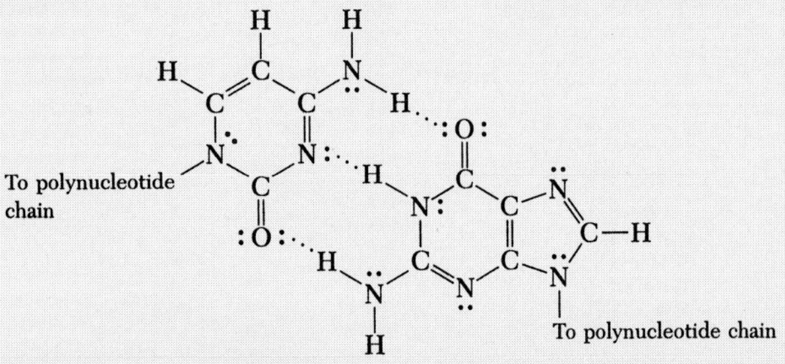 [structural formulas] To polynucleotide chain; Thymine, T; To polynucleotide chain; Adenine, A. To polynucleotide chain; To polynucleotide chain; Cytosine, C; Guanine, G	☐

Section	Subject	Summary	Check When Learned
20-4	Complementary sequence for TGTCAG	For each base, the complementary base is determined by the base pairing shown above: ACAGTC.	☐
20-4	Relationship between DNA structure and protein structure	Each three-base sequence in a DNA strand corresponds to a specific amino-acid residue in a protein chain.	☐
20-4	Genetic code	A list of all possible three-base sequences and their corresponding amino acids.	☐
20-4	Gene	In a DNA molecule, all of the three-base sequences that dictate the sequence of amino-acid residues in the molecule of a particular protein.	☐
20-4	Genome	All of the genes for an organism.	☐
20-5	Zymogen	A protein, not active as a catalyst, that can be converted to an enzyme.	☐
20-5	Feedback inhibition	Form of biochemical control in which the final product of a sequence of reactions reversibly inactivates the enzyme that catalyzes the first reaction in the sequence.	☐
20-5	Cell	Fundamental, microscopic biological unit. All animals and plants consist of one or more cells; a human being consists of many millions of cells.	☐
20-5	Hormone	Biochemical compound produced by one group of cells in an organism that travels to other cells, where it causes and controls biochemical reactions.	☐
20-5	Nitrogen cycle	Cyclic movement of nitrogen atoms, bonded in various ways, through the atmosphere, soil, and living organisms: 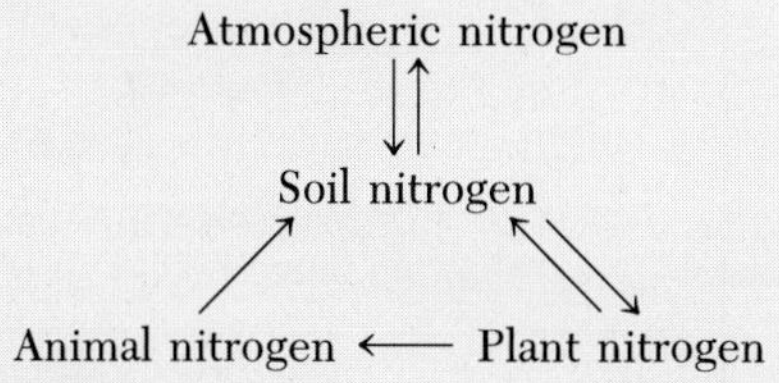 Atmospheric nitrogen ⇅ Soil nitrogen; Plant nitrogen ⇄ Soil nitrogen; Plant nitrogen → Animal nitrogen; Animal nitrogen → Soil nitrogen	☐
20-5	Nitrogen fixation	Conversion of atmospheric nitrogen (N_2) to soil nitrogen (ultimately NO_3^-).	☐

Problems

Assume you can use the Periodic Table at the front of the book unless you're directed otherwise. Answers to odd-numbered problems are in Appendix 1.

Carbohydrates (Section 20-1)

1. In your own words, define *carbohydrate*.

2. What three kinds of functional groups are present in the molecules of carbohydrates?

3. Which of the following compounds could be classified as a carbohydrate? Explain your answer.

   (a) $H{-}C({=}O){-}CH_2{-}CH_3$ (b) $H{-}C({=}O){-}CH(OH){-}CH_2{-}O{-}H$

4. Which of the following compounds could be classified as a carbohydrate? Explain your answer.

   (a) $CH_3{-}C({=}O){-}CH_3$ (b) $CH_2(OH){-}C({=}O){-}CH_2(OH)$

5. Write the line formula for a carbohydrate molecule with four carbon atoms, an aldehyde group on the first carbon atom, and a hydroxyl group on each of the other carbon atoms.

6. Write the line formula for a carbohydrate molecule with four carbon atoms, a ketone group on the second carbon atom, and a hydroxyl group on each of the other carbon atoms.

7. Which of the compounds whose formulas are shown in problem 3 would you expect to be more soluble in water? Explain.

8. Which of the compounds whose formulas are shown in problem 4 would you expect to be more soluble in water? Explain.

9. On the basis of your answer to problem 7, write an explanation, in one or two sentences, for the fact that many carbohydrates are significantly soluble in water.

10. On the basis of your everyday experience, which is more soluble in water, glucose or its polymers—starch and cellulose?

11. Glucose molecules react with themselves to establish an equilibrium between an open-chain form and a ring form. Use line formulas to write an equation for this equilibrium.

12. See the line formulas for glucose and fructose at the beginning of Section 20-1. (a) What are their molecular formulas? (b) What are their simplest formulas?

13. In your own words, describe the crucial role of starch in sustaining life on our planet.

14. In your own words, define *photosynthesis*.

15. In your own words, describe the role of water molecules in photosynthesis.

16. In your own words, describe the role of carbon dioxide molecules in photosynthesis.

17. In your own words, explain what is meant by the *metabolism* of starch.

18. What is the ultimate source of the energy released in the metabolism of starch?

19. In your own words, describe the carbon cycle.

20. In the carbon cycle, what is the molecular form of carbon in the atmosphere? Suggest two sources for these molecules.

Lipids (Section 20-2)

21. In terms of its molecular structure, what is a fat? What functional groups are present in all fat molecules?

22. Does the term *fat* refer to one compound or to more than one compound? Explain.

23. From everyday experience you know that animal fats are not significantly soluble in water. Explain that fact in terms of their molecular structure.

24. From everyday experience you know that carbohydrates are in general more soluble in water than fats. Explain that fact in terms of their molecular structures.

25. In your own words, define *fatty acid.*

26. Which of the following compounds fits the description of a fatty acid? Explain.

(a) C_6H_5—$COOH$ (benzene ring with H on five carbons and a —C(=O)—O—H group on the sixth)

(b) $CH_3{-}(CH_2)_{18}{-}C(=O){-}O{-}H$

27. Use the general formula for a fat shown below to write an equation for its hydrolysis.

$$\begin{array}{l} CH_2{-}O{-}C(=O){-}R_1 \\ | \\ CH{-}O{-}C(=O){-}R_2 \\ | \\ CH_2{-}O{-}C(=O){-}R_3 \end{array}$$

28. What two compounds are produced by the complete oxidation of a fatty acid?

29. Complete and balance this equation for the oxidation of a fatty acid:

$$CH_3{-}(CH_2)_{14}{-}C(=O){-}O{-}H + O_2 \rightarrow$$

30. What is the most important function of fats in animals?

31. Camels store large quantities of fats in their humps, and they can survive for long periods without water. Suggest a relationship between these two characteristics of camels.

32. In your own words, explain the difference between the function of fats and the function of carbohydrates in animals.

33. From its structural formula, would you expect cholesterol to be significantly soluble in water? Explain.

34. According to Table 19.3, to what class of organic compounds does cholesterol belong?

35. What are LDL receptors, and what is their role in atherosclerosis?

36. Is it true that eliminating cholesterol from your diet will eliminate cholesterol from your bloodstream? Explain.

Proteins (Section 20-3)

37. Using *R* to represent the group of atoms that varies from one amino acid to the next, write the general formula for amino acids.

38. The molecular formula for the amino acid called *alanine* is $C_3H_7NO_2$. Write its line formula.

39. Use the formula you wrote in problem 37 to explain that an amino acid is both an acid and a base.

40. Because an amino acid is both an acid and a base, molecules of amino acids react with themselves as described in the *Chemistry Insight* box on page 505. Write an equation for the reaction of a molecule of valine (Table 20.1) with itself.

41. What are essential amino acids?

42. Write the name that corresponds to each abbreviation: (a) Thr (b) Phe (c) Val.

43. Use Table 20.1 to write the line formula for the amino-acid polymer Leu–Val–Phe.

44. Use Table 20.1 to write the line formula for the amino-acid polymer Ile–Ile–Lys–Val.

45. Beginning with one molecule each of Leu, Val, and Phe, how many different polymer chains can be made?

46. Beginning with one molecule each of Leu, Val, Phe, and Lys, how many different polymer chains can be made?

47. What are enzymes?

48. Why are enzymes important in biochemical systems?

49. Does the name *amino acid* refer to one compound or to more than one compound? Explain.

50. Does the name *protein* refer to one compound or to more than one compound? Explain.

Nucleic Acids (Section 20-4)

51. What are the three components of a nucleotide?

52. Using the word *Base* to represent its base component, write the general formula for a DNA nucleotide.

53. Name the bases that commonly occur in DNA nucleotides.

54. How many different nucleotides commonly occur in DNA?

55. Explain the significance of each part of the word *deoxyribonucleic acid.*

56. Does *DNA* refer to one compound or to more than one compound? Explain.

57. What are *complementary base pairs?* Using one-letter abbreviations for the DNA bases, which pairs are complementary?

58. In the DNA double helix, what kinds of bonds exist between complementary base pairs?

59. Write the complementary base sequence for ATTAGGCAT.

60. If the base sequence in one strand of DNA is GGGGGG, what is the base sequence in the complementary strand?

61. In your own words, explain the relationship between the sequence of bases in a strand of DNA and the corresponding sequence of amino acids in a protein chain.

62. In a strand of DNA, how many bases in a sequence code for one amino acid?

63. Use Table 20.2 to write the base sequence that corresponds to this sequence of amino acids in a protein: Val–Lys–Lys–Phe.

64. Use Table 20.2 to write the base sequence that corresponds to this sequence of amino acids in a protein: Ile–Leu–Trp–Phe.

65. Use Table 20.2 to write the sequence of amino acids that corresponds to this base sequence in DNA: ACCTGTCAG.

66. Use Table 20.2 to write the sequence of amino acids that corresponds to this base sequence in DNA: GACAAAACC.

67. In your own words, explain how the DNA in an organism controls its biochemical reactions.

68. Would you expect larger, more complex organisms to have larger DNA molecules? Explain.

69. What is a gene? What is a genome?

70. What is the *genetic code?*

71. How does DNA function in heredity?

72. Explain what is meant by this statement: The genetic code is apparently the same in all organisms, but the DNA is not.

Biochemical Organization and Control (Section 20-5)

73. In your own words, explain the statement that DNA controls life on our planet.

74. How are enzymes important in the control of biochemical processes?

75. What is the relationship between a zymogen and its enzyme?

76. How does the relationship between a zymogen and its enzyme effect control over the reaction catalyzed by the enzyme?

77. What is feedback inhibition, and how does it exert control over biochemical processes?

78. What are hormones?

79. Identify the forms of nitrogen that correspond to the parts of this diagram of the nitrogen cycle:

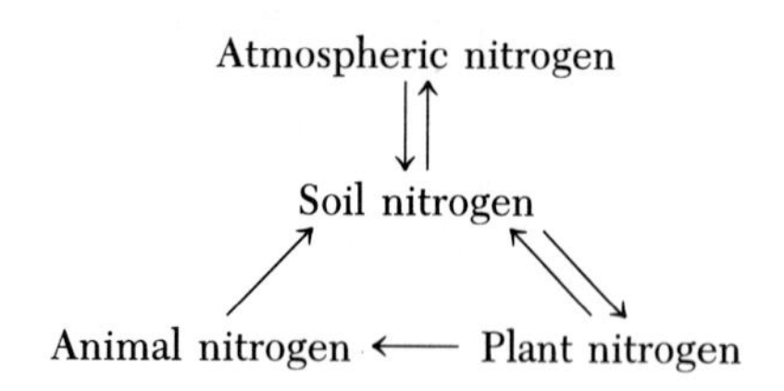

80. What is nitrogen fixation, and what is its importance for life on earth?

Answers to Exercises and Odd-Numbered Problems

Chapter 2—Mathematical Background

Answers to Exercises

2.1 (a) 7.27×10^2 (b) 1.96×10^2 (c) 2.32×10^6
2.2 (a) 310 (b) 52.3 (c) 101,200
2.3 (a) 1.75×10^{-2} (b) 5.55×10^{-4} (c) 8.56×10^{-3}
(d) 9.13×10^{-1} (e) 1.02

> Note: When writing decimal numbers less than 1, always put a zero to the left of the decimal as a place holder: Write "0.123," not ".123."

2.4 (a) 0.0066 (b) 0.00005 (c) 0.0226
2.5 (a) 7.34×10^3 (b) 1.678×10^{-4}
2.6 (a) 3.6×10^8 (b) 9.3 (c) 5.4×10^{-21}
(d) 1.6 (e) 7.2×10^{-11} (f) 2.4×10^{-1}
2.7 (a) 3 (b) 2 (c) 3 (d) 3 (e) 3
2.8 (a) 2 (b) 3 (c) 1 (d) 3
2.9 (a) 3.00×10^{-1} inches (b) 3.05 inches (c) 1.25×10^5 inches
2.10 (a) 2 (b) infinite (c) 3 (d) infinite (e) infinite
2.11 (a) 33.6 (b) 7.5
2.12 (a) 9.20×10^{-2} (b) 9.29×10^4
2.13 (a) 1 (b) 2 (c) 2 (d) 2
2.14 (a) 8.77×10^5 (b) 1.30×10^{-2} (c) 0.00334 (d) 3.56
2.15 (a) $x = 5$ (b) $x = 15$ (c) $x = 8$
2.16 (a) $y = n - m$ (b) $y = pq$ (c) $y = \frac{b}{a}$
2.17 (a) $P = \frac{nRT}{V}$ (b) $T = \frac{PV}{nR}$

Answers to Odd-Numbered Problems

2.1 (a) 8.375×10^3 (b) 3.68×10^{-4} (c) 1.07×10^{-1} (d) 5.0×10^1
2.3 (a) 71.3 (b) 0.0225 (c) 0.1274
2.5 (a) 5.23×10^4 (b) 4.95×10^{-4}
2.7 (a) 4.5×10^{-2} (b) 3.3×10^{-4} (c) 2.3×10^8
2.9 (a) 3 (b) 3 (c) 2 (d) 3 (e) 1

> Note: The abbreviation for inch is "in.", the only unit to have a period after it; the period is there to avoid confusion with the word "in".

2.11 (a) 3 (b) 4 (c) 2 (d) 1 (e) 4
2.13 (a) 2 (b) 3
2.15 Each counted item is a whole object. Each of these counted values has an infinite number of significant digits.
2.17 3.70×10^5 miles

Appendix 1

2.19 Each of these values is a defined value. All defined values have infinite significant digits.
2.21 (a) 7.54×10^4 (b) 0.0556 (c) 21.4 (d) 21.5
2.23 (a) 2.59×10^1 (b) -5.8 (c) 3.80×10^{-3} (d) 4.77×10^2 (e) 2.9×10^3
2.25 (a) 3.10×10^{-2} (b) 6.6×10^3 (c) 3.5 (d) 2×10^{-1} (e) 5×10^6
2.27 (a) $x = -3$ (b) $x = -5$ (c) $x = 8$
2.29 (a) $x = \frac{z}{y}$ (b) $z = xy$

Chapter 3—Measuring Matter and Energy

Answers to Exercises

3.1 (Other answers are possible.) Gases such as water vapor or air are invisible to our eyes yet consist of matter. Fire is visible to us because burning particles emit radiant energy, so even though fire is visible, fire itself has no mass.
3.2 For a short time, the small droplets of oil and of vinegar mix well enough to appear homogeneous; in fact, microscopic examination would show that this mixture is heterogeneous and not a homogeneous solution at all. (The term *dispersion* would be more correct.) After sitting, the small droplets of oil and of vinegar will have separated into layers and will require more shaking before being poured on salad.
3.3 Salt is decomposed into different substances, so it must be a compound.
3.4 We can say that the fact that sodium is not decomposed by an electric current *suggests*, but does not *prove*, that sodium is an element. There might be another technique for decomposing sodium that we are not aware of.
3.5 Inside the glass, sand is a solid phase and water is a liquid phase. The glass is another solid phase, and the air around the glass is a gas phase—for a total of four phases present. (To be precise, the air around the glass also contains some water vapor that has evaporated from the liquid water in the open glass. Since there is no visible boundary between the water vapor and the air, the water vapor does not constitute a separate phase. Water vapor in air is a homogeneous mixture, a solution.)
3.6 (a) μ = micro, 1×10^{-6} (b) c = centi, 1×10^{-2} (c) n = nano, 1×10^{-9} (d) k = kilo, 1×10^3 (e) d = deci, 1×10^{-1}
3.7 (a) m, milli (b) d, deci (c) k, kilo (d) c, centi (e) n, nano
3.8 (a) 1.06 qt = 1 L (b) 1 lb = 454 g (c) 1 in. = 2.54 cm
3.9 (a) 1 cg = 1×10^{-2} g (b) 1 mL = 1×10^{-3} L (c) 1 km = 1×10^3 m
3.10 3 ft^2
3.11 9 ft^3
3.12 4 dm^2
3.13 (a) 4 cm^3 (b) since 1 cm^3 = 1 mL, 4 cm^3 = 4 mL
3.14 74 g, rounded to 2 significant digits
3.15 1 K = −458°F
3.16 1 kcal = 4.184 kJ
3.17 (a) miles per hour (b) 1/h or h^{-1} (c) meters $second^{-1}$, m s^{-1} (d) 3.00×10^8 meters $second^{-1}$, or 3.00×10^8 m s^{-1}

3.18 at $\lambda = 400$ nm, $\nu = 7.50 \times 10^{14}\ s^{-1}$
at $\lambda = 500$ nm, $\nu = 6.00 \times 10^{14}\ s^{-1}$
at $\lambda = 600$ nm, $\nu = 5.00 \times 10^{14}\ s^{-1}$
at $\lambda = 700$ nm, $\nu = 4.29 \times 10^{14}\ s^{-1}$
at $\lambda = 800$ nm, $\nu = 3.75 \times 10^{14}\ s^{-1}$

3.19 Visible light is that portion of the electromagnetic spectrum detectable by our eyes, ranging in wavelength from about 400 nm to about 800 nm.

3.20 (a) 5.00 miles = 2.64×10^4 ft (b) 2.74×10^5 m = 2.74×10^2 km

3.21 (a) 0.158 mi^2 = 4.40×10^6 ft^2 (b) 1.00 g/cm^3 = 1.04 lb/pt

Answers to Odd-Numbered Problems

3.1 A compound is a single, pure substance, whereas a mixture is a combination of two or more substances.

3.3 (a) Tea is homogeneous—it is uniform throughout.
(b) Orange juice is heterogeneous—we can usually see pulp in it.
(c) Vinegar is homogeneous—it is uniform throughout.

3.5 If limestone were an element, it would not decompose into two different substances on heating. Limestone must be a compound.

3.7 Increasing the temperature of a piece of solid iron would change it to its liquid form (melting) and would eventually change it to its gaseous form (boiling).

3.9 All three states of matter are present: *solid* wood, *liquid* water, and *gaseous* air. The wood is a solid phase, the water is a liquid phase, and the air (and water vapor) is a gas phase.

3.11 (a) pico, 10^{-12} (b) mega, 10^6 (c) deci, 10^{-1}

3.13 (a) c, centi (b) k, kilo (c) n, nano

3.15 1 pound = 454 g

3.17 (a) 1 dL = 1×10^{-1} L (b) 1 km = 1×10^3 m

3.19 8 in.^{2}

3.21 8 m^3

3.23 22.3 cm^3, rounded to 3 significant digits

3.25 The density of osmium is 22.5 g/cm^3.

3.27 134 g, rounded to 3 significant digits

3.29 1.74 kg

3.31 The density of titanium is 4.50 g/cm^3.

3.33 1.26

3.35 1.47 mL, rounded to 3 significant digits

3.37 The lowest temperature on the Kelvin scale is 0 K, and on the Celsius scale, −273°C.

3.39 24°C

3.41 498 K

3.43 −150°F

3.45 1 cal = 4.184×10^{-3} kJ

3.47 λ, m (or some other length unit)

3.49 Wavelength and frequency are inversely proportional: As wavelength increases, frequency decreases, and vice versa. Red light has a longer wavelength than blue light and therefore has a lower frequency than blue light.

3.51 $c = \lambda\nu \qquad \nu = \dfrac{c}{\lambda}$

3.53 $3.00 \times 10^{18}\ s^{-1}$

3.55 3.00×10^{-6} m

3.57 300 in. (3 significant digits)

3.59 2.50×10^{13} pm
3.61 0.592 miles, rounded to 3 significant digits
3.63 46.9 kg
3.65 76.2 cm
3.67 1.00×10^{6} mL
3.69 1390 in.^{2}, rounded to 3 significant digits
3.71 3.00×10^{6} m^2
3.73 2.00×10^{15} cm^3
3.75 1080 J, rounded to 3 significant digits
3.77 8.75 kcal, rounded to 3 significant digits
3.79 6.71×10^{8} mi/h

Chapter 4—Atoms and Chemical Symbols

Answers to Exercises

4.1 The first of Dalton's postulates (the one defining the atom) is essentially the same as that proposed by the early Greek philosophers. The Greeks proposed nothing regarding the sizes of atoms or how atoms combine; these ideas can be credited to Dalton.

4.2 (a) potassium (b) sodium (c) copper (d) carbon
(e) calcium

4.3 Protons are particles, with a small positive electrical charge and a very small mass (but relatively large mass compared to an electron), that are constituents of all atoms and that can, under extreme conditions, be separated from their atoms.

4.4 For every atom to be electrically neutral, the number of positive charges must exactly equal the number of negative charges, so the number of protons must equal the number of electrons.

4.5

Name	Greek symbol	Description
beta	β	Stream of electrons moving at very high speeds.
gamma	γ	Electromagnetic radiation with very short wavelength.

4.6 An α particle has a charge of 2+, so it must contain two protons. The α particle has a mass four times greater than a proton, but it only has two protons. The other two mass units can be accounted for by two neutrons. Thus, an α particle contains two protons and two neutrons.

4.7 (a) Oxygen has atomic number 8, so it must have 8 protons.
(b) Potassium, atomic number 19, must have 19 protons and 19 electrons.

4.8 (a) An atom of $^{190}_{76}Os$ has 76 protons, 76 electrons, and 114 neutrons.
(b) $^{200}_{80}Hg$ (c) calcium

4.9 (a) $50\ p^+$ / $69\ n^0$ (nucleus) $50\ e^-$ (electrons) (b) magnesium (c) $^{64}_{29}Cu$

4.10 6.024×10^{23} amu in one g

4.11 (a) $^{14}_{7}N$ (b) approximate mass 14 amu

4.12 The atomic mass of hydrogen is 1.008 amu, just slightly more than 1. The isotope of hydrogen of mass 1, $^{1}_{1}H$, must be much more prevalent than $^{2}_{1}H$.

4.13 The heavier elements have many more neutrons than protons.

4.14 $^{239}_{94}Pu + ^{4}_{2}He \rightarrow ^{1}_{0}n + ^{242}_{96}Cm$
reactants products

Answers to Odd-Numbered Problems

4.1 The belief that matter consists of tiny, indivisible particles was a hypothesis because the Greeks did not—and could not—test it.

4.3 With the advent of modern nuclear physics in this century, the atom was shown to consist of smaller particles. Even though Dalton's notion of atoms is not rigorously correct, it serves as a useful tool to explain commonly observed behavior of substances.

4.5 Dalton's principle that atoms of different elements have different masses is true today, with one qualification. There are a few known cases in which an isotope of one element has the same mass number as an isotope of a different element; these isotopes of different elements would have the same mass when measured to the units place. The precise masses, carried to several decimal places, would give different values, however, because the mass of a proton and of a neutron are not exactly equal.

4.7 (a) H (b) He (c) Sn (d) Cu

4.9 (a) barium (b) boron (c) bromine (d) lead

4.11 When charged electrodes are placed above and below the stream of cathode rays, the rays bend away from the negatively charged electrode and toward the positively charged one. Since like charges repel and opposite charges attract, the cathode rays must carry a negative charge.

4.13 Although the magnitude of the charge is the same on the electron and the proton, the proton has 1837 times the mass of an electron.

4.15 An atom must have equal numbers of protons and electrons. An atom with 23 protons has 23 electrons.

4.17 An α particle contains two protons and two neutrons: it is the same as the nucleus of an atom of helium-4, that is, ${}^{4}_{2}He$ without electrons. An α ray is a stream of α particles.

4.19 β particle (electron) $<$ proton (1 amu) $<$ α particle (4 amu)

4.21 Rutherford's experiments on the deflection of particles by thin metal foil were persuasive evidence for a dense nucleus surrounded by large empty space, that is, the picture of the *nuclear atom.*

4.23 An atom's atomic number is the number of protons in its nucleus. An atom's mass number is the sum of the number of protons and neutrons in its nucleus.

4.25 The atomic number for an atom is identical to the number of protons it contains.

4.27 Statement (a) is true.

4.29 The mass number is the sum of the numbers of protons and neutrons. The mass of a proton is approximately 1 amu, and the mass of an electron is negligible, so the mass number gives the approximate mass of the atom.

4.31 An atom can have only whole numbers of protons and neutrons—fractional protons and neutrons don't exist. The sum of whole numbers can only be a whole number.

4.33 The symbol ${}^{39}_{19}K$ signifies an atom with 19 protons, 20 neutrons (to make mass number 39), and 19 electrons, which we have identified as potassium with the symbol K.

4.35 Element 53 is iodine, symbol I.

4.37 The standard for atomic mass is the atom of ${}^{12}_{6}C$. It is a convenient standard because its mass is similar in magnitude to the masses of other atoms. It would not be convenient to use the Empire State Building as our standard and have to express atomic masses as some minuscule fraction of our standard.

4.39 The units for atomic masses on the Periodic Table are *atomic mass units*, abbreviated amu.

4.41 Statement (b) is correct.

4.43 The atomic mass of boron would be 10 amu, that is, the average of the mass numbers 9, 10, and 11.

4.45 Nuclear fission is the process of splitting a larger nucleus into two or more smaller nuclei. Nuclear fusion is the opposite of fission: fusion is the process of combining two lighter nuclei into one heavier one.

4.47 A chain reaction, once initiated, continues for many cycles of the reaction. This usually happens by the products from the first reaction colliding with other atoms, generating more products that can in turn collide with more atoms, and so on.

4.49 Reactants: ${}^{235}_{92}U + {}^{1}_{0}n$

Products: ${}^{139}_{56}Ba + {}^{94}_{36}Kr + 3\ {}^{1}_{0}n$

The coefficient is the "3" before ${}^{1}_{0}n$ in the products.

4.51 ${}^{235}_{92}U \rightarrow {}^{231}_{90}Th + {}^{4}_{2}He$ (an α particle)

Chapter 5—Introduction to the Periodic Table

Answers to Exercises

5.1 (a) 80 (b) 11 (c) 50

5.2 (a) 7 (b) 12 (c) 19

5.3 (a) iron (b) fluorine (note that "u" comes before "o")

5.4 (a) in the nucleus: 5 protons, 6 neutrons; outside the nucleus: 5 electrons
(b) in the nucleus: 80 protons, 120 neutrons; outside the nucleus: 80 electrons

5.5 (a) metal (b) nonmetal (c) metalloid

5.6 (a) S (sulfur) and Se (selenium) are in the same group and have similar chemical properties. Sr (strontium) is in a different group and will have very different properties from the first two.
(b) Be (beryllium) and Ba (barium) are in the same group (alkaline-earth metals) and have similar chemical properties. B (boron) is in a different group and will have very different properties from the first two.

5.7 metals: sodium, magnesium, aluminum
metalloid: silicon
nonmetals: phosphorus, sulfur, chlorine, argon

5.8 Group 1A: hydrogen, lithium, sodium
Group 2A: beryllium, magnesium, calcium
Group 7A: fluorine, chlorine, bromine
Group 8A: helium, neon, argon

5.9 (a) K, He, S (b) barium, magnesium, silicon

5.10 neon = 10; nitrogen = 7; iron = 26; chlorine = 17

5.11 See the outline of the Periodic Table at the top of page 423; the transition metals are in yellow.

5.12 Group numbers of the representative elements are on the top of the Periodic Table. Period numbers are on the left. See the outline of the Periodic Table on the bottom of page 423.

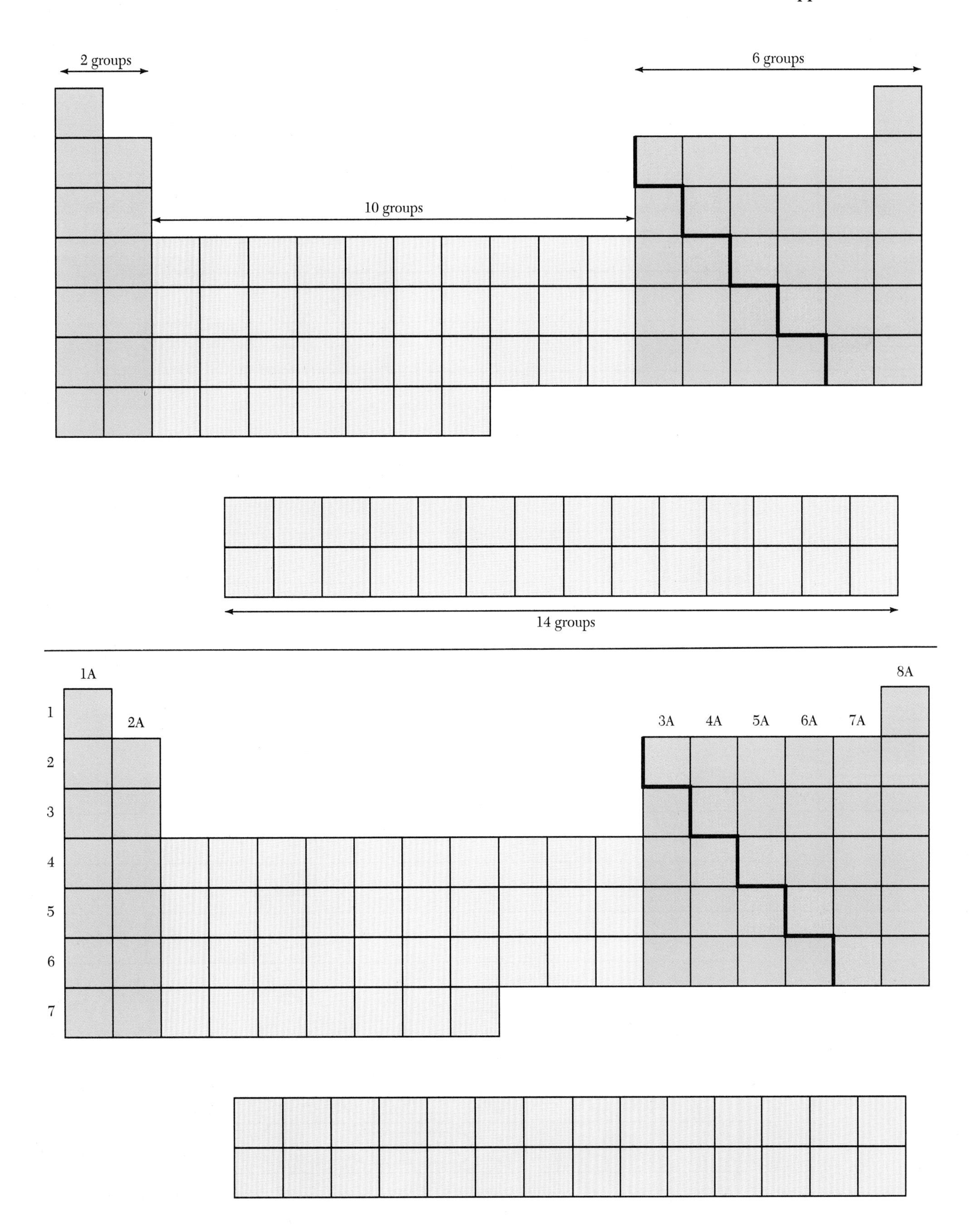

2 groups
6 groups
10 groups
14 groups
1A
8A
2A
3A
4A
5A
6A
7A
1
2
3
4
5
6
7

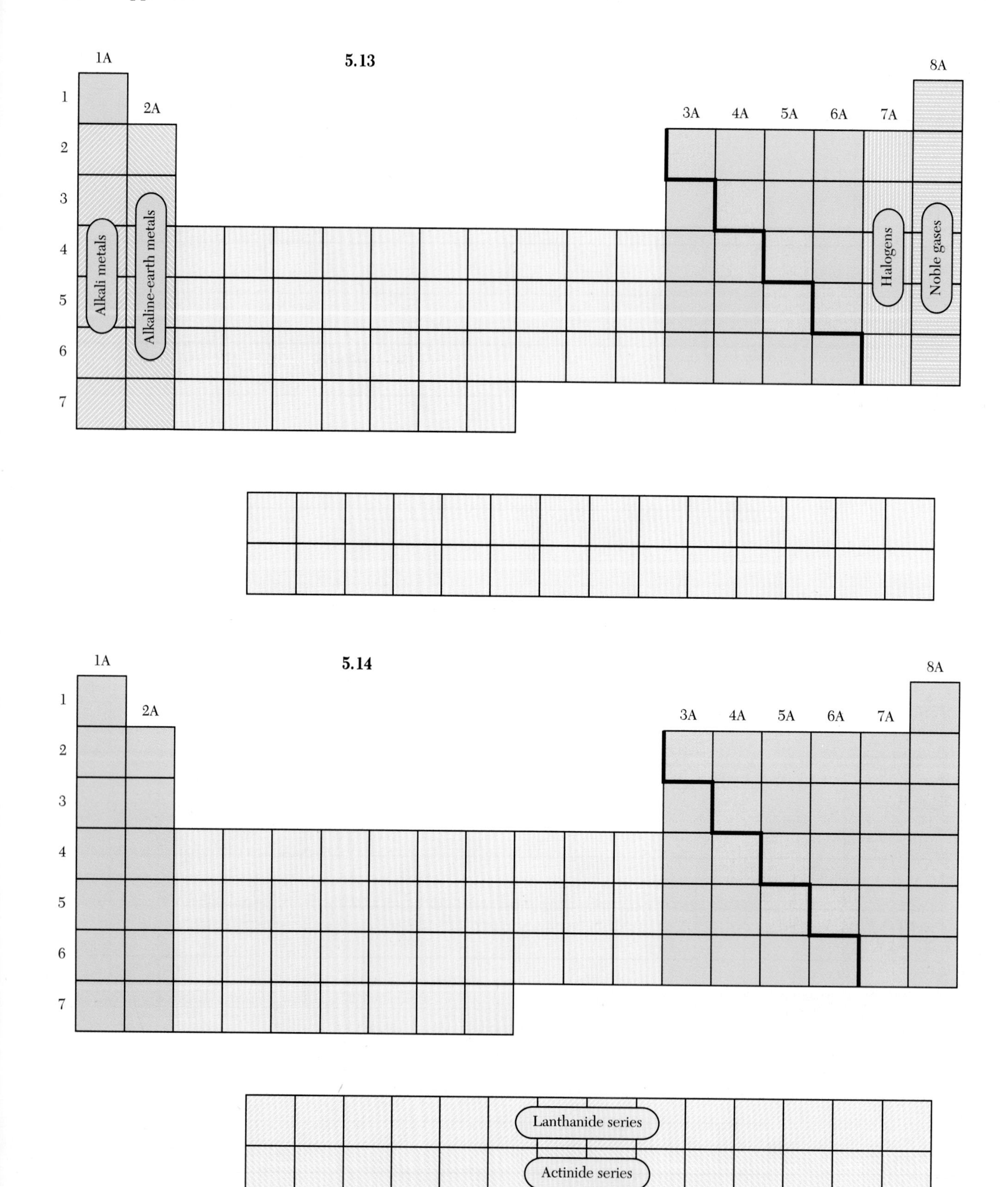
5.13
1A
2A
3A
4A
5A
6A
7A
8A
1
2
3
4
5
6
7
Alkali metals
Alkaline-earth metals
Halogens
Noble gases
5.14
1A
2A
3A
4A
5A
6A
7A
8A
1
2
3
4
5
6
7
Lanthanide series
Actinide series

5.15 Md and No, actinides; Eu, lanthanide
5.16

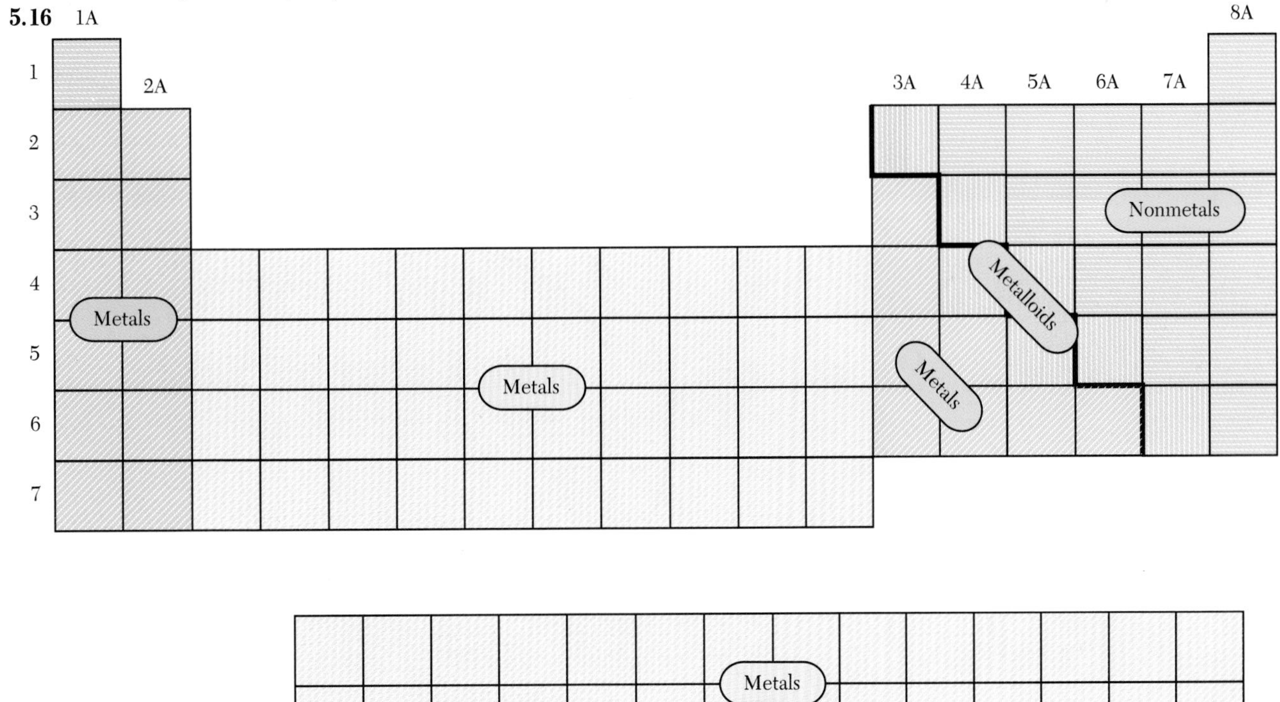

5.17 Phosphorus and silver are solids. Bromine is a liquid. Fluorine, nitrogen, and neon are gases.

Answers to Odd-Numbered Problems

5.1 (a) 13 (b) 13 (c) 14 (d) aluminum
5.3 (a) 40.1 (b) 12.0 (c) 23.0
5.5 The symbol $^{32}_{16}S$ refers to one atom of an element named sulfur, with 16 protons, 16 neutrons (to make mass number 32), and 16 electrons.
5.7 A metal is typically a shiny solid that conducts electricity and heat and is malleable (can be worked into different shapes). Copper exhibits all of these properties: A new penny is very shiny; copper wire conducts electricity and is easily bent; some brands of pots and pans have copper bottoms for efficient heat conduction.
5.9 boron
5.11 There are many more metals than metalloids and nonmetals combined.
5.13 The Periodic Table has seven rows or periods.
5.15 All of the elements in Group 8A are nonmetals. They are called the noble gases. As the name suggests, all of the 8A elements are gases.
5.17 The representative elements are the left two groups and the right six groups of the Periodic Table.
5.19 The lanthanide and actinide elements are displayed at the bottom of the Periodic Table, but they belong after lanthanum and actinium, respectively, beginning in Periods 6 and 7 of Group 3B.
5.21 The halogens are fluorine, chlorine, bromine, and iodine.
5.23 Scandium is the first of ten transition metals in Period 4. Yttrium is the first of ten transition metals in Period 5.

Chapter 6—Atomic Structure and the Periodic Table

Answers to Exercises

6.1 A *quantum* is a tiny, individual burst of energy. (*Quanta* is the plural form of *quantum.*) *Quantum theory* embodies the idea that a hot object emits energy in quanta rather than continuously. In energy emitted as light, each burst of energy, or quantum, is a *photon*.

6.2

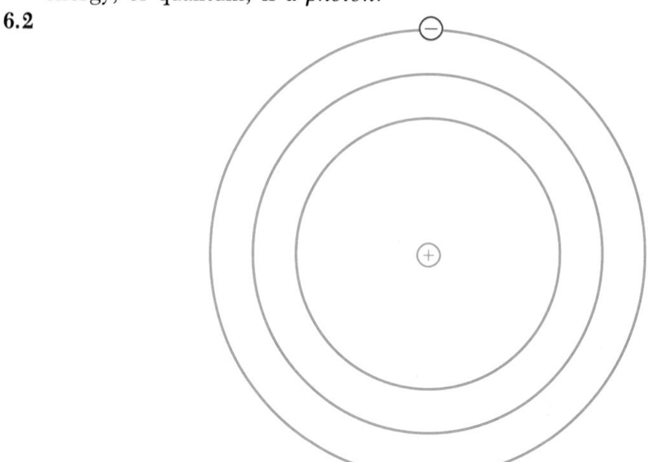

6.3 1.02×10^{-7} m

6.4 s = 1 orbital; p = 3 orbitals; d = 5 orbitals; f = 7 orbitals

6.5 energy level 1: s sublevel; energy level 2: s and p sublevels; energy level 3: s, p, and d sublevels; energy level 4: s, p, d, and f sublevels

6.6 (a) 1s (b) 3s (c) 5s (d) 2p (e) 3p (f) 3d
(g) 5d (h) 4f (i) 5f

6.7 (a) O^8: $1s^2 2s^2 2p^4$ (b) Zn^{30}: $1s^2 2s^2 2p^6 3s^2 3p^6 4s^2 3d^{10}$
(c) Am^{95}: $1s^2 2s^2 2p^6 3s^2 3p^6 4s^2 3d^{10} 4p^6 5s^2 4d^{10} 5p^6 6s^2 4f^{14} 5d^{10} 6p^6 7s^2 5f^7$

6.8 Ba^{56}: $[Xe]^{54} 6s^2$

6.9 (a) ground state (b) excited state

6.10 F^9: $1s^2 2s^2 2p^5$

F^9: $1s^2$ $2s^2$ $2p^5$

(↑↓) (↑↓) (↑↓)(↑↓)(↑)

Ne^{10}: $1s^2 2s^2 2p^6$

Ne^{10}: $1s^2$ $2s^2$ $2p^6$

(↑↓) (↑↓) (↑↓)(↑↓)(↑↓)

6.11 (a) Mn^{25}: $1s^2 2s^2 2p^6 3s^2 3p^6 4s^2 3d^5$

(b) $3d^5$: (↑)(↑)(↑)(↑)(↑)

(c) 20 paired electrons, 5 unpaired electrons

6.12 $4s^2 4p^3$

6.13 $5s^2 5p^2$

6.14 (a) ·S̈· (with a pair of dots above and below) (b) :K̈r: (with a pair of dots above and below) (c) K·

Answers to Odd-Numbered Problems

6.1 The solar-system model of the atom envisioned electrons orbiting the nucleus as planets orbit the sun.

6.3 Quantum theory is based on the principle that a hot object emits energy in individual bursts rather than continuously.

6.5 The Bohr model for the hydrogen atom proposed that an electron must be only in specific orbits around the nucleus, with the energy level of the orbit increasing with greater distance from the nucleus.
6.7 Energy level increases with greater distance from the nucleus, so energy level 3 must be closer to the nucleus than energy level 5.
6.9 The *ground state* of a hydrogen atom has its electron in the lowest possible energy level, the orbit closest to the nucleus. An *excited state* describes the electron in a higher energy level, farther from the nucleus.
6.11 (a) Light is emitted only when an electron moves from a higher energy level to a lower one. (b) 3.11×10^{-19} J
6.13 An electron in an atom can exist in different energy levels, that is, locations around the nucleus with a specific energy associated with them. Each energy level is made up of one or more sublevels. An electron in energy level 3, for example, might be in either the s, the p, or the d sublevel.
6.15 The s sublevel is present in every energy level.
6.17 The fourth energy level includes the s, p, d, and f sublevels.
6.19 The 2s orbital is spherical. (Note that the 1s orbital is not shown in the cross-section view; it would fit within the 2s orbital.)

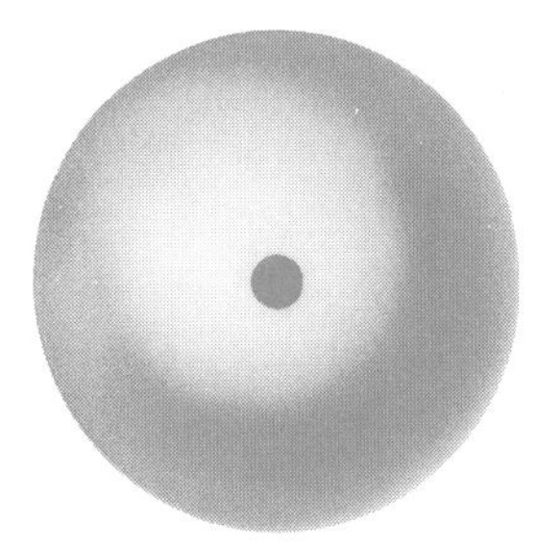

Outside view

Cross-section view

6.21 There is one orbital in an s sublevel and there are three orbitals in a p sublevel.
6.23 The maximum number of electrons for each sublevel: s = 2; p = 6; d = 10; f = 14.
6.25 Groups 1A (alkali metals) and 2A (alkaline-earth metals) are filling s sublevels.
6.27 4d
6.29 (a) 3s (b) 3d (c) 5d
6.31 An atom's *electron configuration* is the distribution of the atom's electrons in sublevels.
6.33 (a) F^{9}: $1s^2 2s^2 2p^5$
(b) I^{53}: $1s^2 2s^2 2p^6 3s^2 3p^6 4s^2 3d^{10} 4p^6 5s^2 4d^{10} 5p^5$
(c) K^{19}: $1s^2 2s^2 2p^6 3s^2 3p^6 4s^1$
(d) Zr^{40}: $1s^2 2s^2 2p^6 3s^2 3p^6 4s^2 3d^{10} 4p^6 5s^2 4d^2$
(e) Pr^{59}: $1s^2 2s^2 2p^6 3s^2 3p^6 4s^2 3d^{10} 4p^6 5s^2 4d^{10} 5p^6 6s^2 4f^3$
6.35 (a) N^{7}: $[He]^2 2s^2 2p^3$
(b) Bi^{83}: $[Xe]^{54} 6s^2 4f^{14} 5d^{10} 6p^3$
(c) Os^{76}: $[Xe]^{54} 6s^2 4f^{14} 5d^6$
(d) Mo^{42}: $[Kr]^{36} 5s^2 4d^4$
(e) Ho^{67}: $[Xe]^{54} 6s^2 4f^{11}$
6.37 cesium, Cs, element 55
6.39 (a) ground state (b) excited state

6.41 (a) H^1: $1s^1$ (b) Li^3: $1s^22s^1$

H^1: $1s^1$ — (↑) Li^3: $1s^2$ (↑↓) $2s^1$ (↑)

6.43 no unpaired electrons in argon
6.45 (a) $2s^22p^4$ (b) $4s^24p^4$ (c) $6s^26p^3$
6.47 (a) s^2p^1 (b) s^2p^2 (c) s^2p^4
6.49 (a) lithium (b) nitrogen (c) radium
6.51 (c) oxygen
6.53 (a) Li· (b) ·N̈· (c) Ra:
6.55 Na· Mg: ·Al: ·Si: ·P· ·S· :Cl· :Ar:
6.57 metal (For accuracy, helium should be included in this group; even though it is not a metal, helium has only s valence electrons.)
6.59 (a) Ca: (b) ·N̈· (c) :Br·

Chapter 7—Covalent Bonds

Answers to Exercises

7.1 (a)  (b)  (c) O:: C :: O (d) 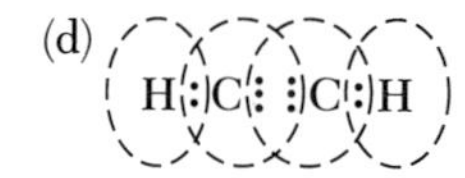

7.2 (a) all single bonds; all bonds are polar except the carbon-carbon bond

```
    H  H
H : C : C : O : H
    H  H
```

(b) all bonds are polar

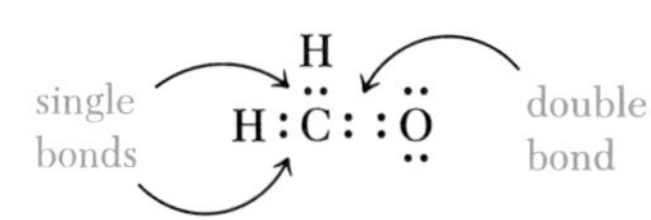

(c) both bonds are polar

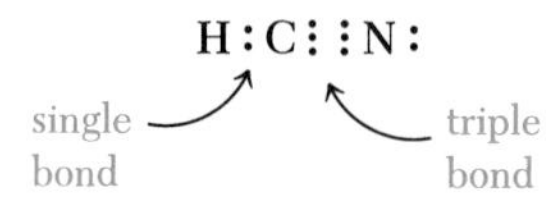

7.3 (a) hydrogen "octets" other octets

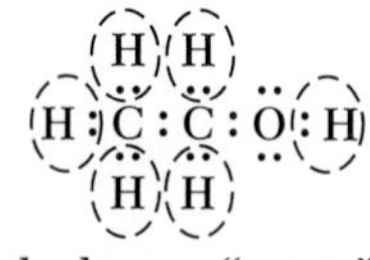

```
    H  H
H : C : C : O : H
    H  H
```

(b) hydrogen "octets" other octets

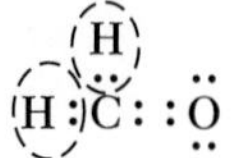

```
   H
H : C : : O
```

(c) hydrogen "octet"

H:C⋮⋮N:

other octets

7.4

(a)

```
     H  H        δ−
H : C : C : O : H  δ+
     H  H
```

(b)

```
     H
H : C : : O δ−
    δ+
```

(c) H : C⋮⋮N: (δ+ on C, δ− on N)

7.5

(a) H—C(H)(H)—C(H)(H)—O—H
C_2H_6O

(b) H—C(H)=O
CH_2O

(c) H—C≡N:
HCN

7.6

(a) δ+ H—Br: δ−

(b) δ− :F—O—F: δ− (δ+ on O)

(c) δ+ H—O—Cl(—O:)(—O:)—O: with δ− on each O and δ+ on Cl

7.7 44.0 amu, expressed to three significant digits

7.8 atomic mass = 126.9 amu; molecular mass = 254 amu, expressed to three significant digits

Answers to Odd-Numbered Problems

7.1 A *symbol* represents an atom, a *formula* represents a molecule, and an *equation* shows a chemical reaction of atoms and molecules.

7.3 A *coefficient* is a multiplier that tells how many times the symbol or formula appears in an equation. A *subscript* indicates how many atoms of a particular element are present in a molecule.

7.5 A *reaction* is the process of bond-breaking and bond-forming to make new compounds, whereas an *equation* is the written description of that process.

7.7 Two fluorine atoms form one fluorine molecule.

7.9 (a) An atom is a physical object made of protons, neutrons, and electrons; a symbol is the written representation of an atom. (b) A molecule is a chemical species composed of two or more atoms; a formula is the written representation of a molecule.

7.11 A covalent bond forms when the arrangement of shared electrons in shared orbitals is more stable (lower in energy) than the atoms by themselves.

7.13 s = single bond; d = double bond; t = triple bond

H—C≡C—C=C—H (with H atoms on the two central carbons); bonds labelled s, t, s, d, s, s

7.15 Representative answers are circled. You may have circled different atoms.

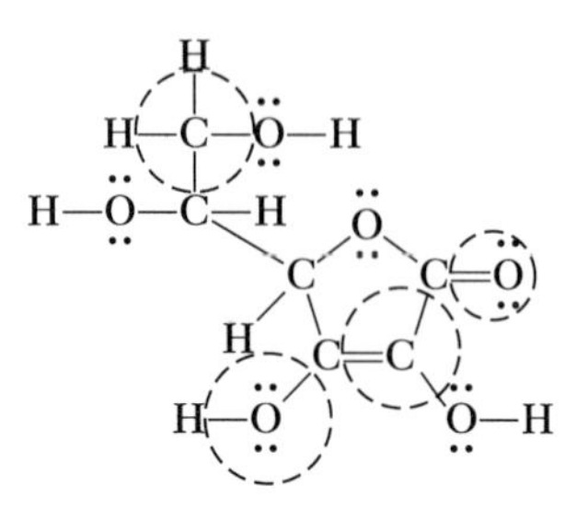

7.17 (a) O (b) O (c) I

7.19 A nonpolar covalent bond is a pair of electrons shared equally between two atoms with the same electronegativity, whereas a polar covalent bond is a pair of electrons shared unequally between two atoms of different electronegativity.

7.21

$^{\delta+}$H—$\ddot{\text{O}}^{\delta-}$—N$^{\delta+}$(=$\ddot{\text{O}}^{\delta-}$)—$\ddot{\text{O}}$:$^{\delta-}$

7.23

(a) :Br—Br:

This covalent bond is nonpolar.

(b) $^{\delta+}$:Br—F:$^{\delta-}$

(c) $^{\delta-}$:Cl—Si$^{\delta+}$—Cl:$^{\delta-}$ (with two more $^{\delta-}$:Cl: bonded to Si, above and below)

7.25

	nonpolar	polar
single	H—H	H—F:
double	H₂C=CH₂	O=C=O
triple	:N≡N:	:C≡O:

7.27

(a) :N≡N:

(b) H—N=N—H

(c) H₂N—NH₂

7.29 ZH_3

7.31 expressed to 3 significant digits: (a) 16.0 amu (b) 98.1 amu

Chapter 8—Ionic Bonds

Answers to Exercises

8.1 Arranged from smallest to largest atomic radius:
(a) krypton < arsenic < calcium (radius decreases left to right)
(b) nitrogen < carbon < silicon (nitrogen is to the right of carbon, and carbon is above silicon)

8.2 Arranged from smallest to largest nuclear charge:
fluorine < phosphorus < potassium

8.3 Arranged from smallest to largest shielding effect:
carbon < nitrogen < phosphorus

8.4 Arranged from largest to smallest ionization energy:
argon > chlorine > potassium

8.5 Arranged from largest to smallest electron affinity:
chlorine > bromine > selenium

8.6 (a) The metals lose electrons more easily. The difficulty of loss of electrons is measured by ionization energy. Ionization energy increases left to right across the Periodic Table, so elements on the left side of the Periodic Table (metals) lose their electrons more easily.
(b) The nonmetals add electrons more easily. The ease of adding electrons is measured by electron affinity. Electron affinity increases left to right across the Periodic Table, so the elements on the right side of the Periodic Table (nonmetals) add electrons more easily than the elements on the left side of the Periodic Table (metals).

8.7 $K\cdot \rightarrow K^+ + e^-$

8.8 $Ca: \rightarrow Ca^{2+} + 2\ e^-$

8.9 The Sr^{2+} cation has 38 protons and 36 electrons. The 2+ charge shows that it must have two fewer electrons than protons.

8.10 $\cdot Al: \rightarrow Al^{3+} + 3\ e^-$

8.11 $Fe^{26}: [Ar]^{18}(4s^2)\,3d^6$

8.12 $:\ddot{\underset{..}{I}}\cdot + e^- \rightarrow :\ddot{\underset{..}{I}}:^-$

8.13 $\cdot\ddot{\underset{..}{O}}\cdot + 2\ e^- \rightarrow :\ddot{\underset{..}{O}}:^{2-}$

8.14 The nitride ion has seven protons and ten electrons.

$\cdot\ddot{\underset{\cdot}{N}}\cdot + 3\ e^- \rightarrow :\ddot{\underset{..}{N}}:^{3-}$

8.15 (a) $Na\cdot + :\ddot{\underset{..}{Cl}}\cdot \rightarrow Na^+ + :\ddot{\underset{..}{Cl}}:^-$

(b) $\cdot Al: + \cdot\ddot{\underset{\cdot}{N}}\cdot \rightarrow Al^{3+} + :\ddot{\underset{..}{N}}:^{3-}$

(c) $Ca: + 2:\ddot{\underset{..}{Br}}: \rightarrow Ca^{2+} + 2:\ddot{\underset{..}{Br}}:^-$

8.16 The condensed formula is Na_3P.

$3\ Na\cdot + \cdot\ddot{\underset{\cdot}{P}}\cdot \rightarrow 3\ Na^+ + :\ddot{\underset{..}{P}}:^{3-}$

8.17 (a) $BaCl_2$ (b) Na_2O (c) SnS_2

8.18 The ions are Sr^{2+} and two I^-.

8.19 The bond between the alkali metal rubidium and the halogen bromine (nonmetal) will be ionic.

8.20 The bond between the two nonmetals phosphorus and chlorine will be covalent. It will be polar covalent because chlorine is more electronegative than phosphorus.

8.21
$$\text{H}-\ddot{\underset{..}{\text{O}}}-\overset{\overset{:\text{O}:}{\|}}{\text{C}}-\ddot{\underset{..}{\text{O}}}:^-$$

8.22 (a) $NaClO_3$ (b) K_2SO_3
8.23 (a) The ions are Na^+ and O^{2-}. (b) The ions are Na^+ and O_2^{2-}.
8.24 (a) Nitrogen and oxygen are nonmetals, so their bonds are covalent. The formula NO_2 represents a molecule.
(b) Iron is a metal, and NO_3^- is an anion. This compound is ionic, so $Fe(NO_3)_3$ represents a formula unit.
8.25 297 amu, rounded to 3 significant digits
8.26 95.0 amu, rounded to 3 significant digits

Answers to Odd-Numbered Problems

8.1 A covalent bond is a pair of electrons shared by two atoms, whereas an ionic bond is the attraction of a positively charged particle (a cation) and a negatively charged particle (an anion).
8.3 Atomic radius is the distance from the nucleus to the outer boundary of the atom's electrons in the outermost energy level.
8.5 (a) Kr (b) Ca (c) As
8.7 An atom's ionization energy is the amount of energy it takes to remove the atom's outermost electron.
8.9 (a) Kr (b) Ca (c) As
8.11 One explanation is that ionization energy increases left to right across the Periodic Table, so potassium will have a lower ionization energy than bromine. A better explanation includes these facts: first, potassium is larger than bromine, so its outermost electron is farther from the nucleus (and more easily removed) than is bromine's outermost electron. Second, potassium and bromine have the same number of inner electrons and have the same shielding effect on the outer electrons, but bromine has 16 more protons in the nucleus than potassium, so bromine's electrons will be held more tightly than potassium's.
8.13 Ionization energy decreases from top to bottom down the Periodic Table. As the nonmetals are higher in a group than the metals, the nonmetals will have higher ionization energies than the metals in the same group.
8.15 An atom's electron affinity is the amount of energy released when the atom gains an electron.
8.17 One explanation is that electron affinity increases left to right across the Periodic Table, so fluorine will have a larger electron affinity than lithium. A better explanation includes these facts: First, atomic radius decreases left to right across the Periodic Table, so the outermost electrons of fluorine are closer to the nucleus than the outermost electron of lithium; therefore, an additional electron can get closer to the fluorine nucleus than to the lithium nucleus. Second, the fluorine nucleus has 6 more protons than the lithium nucleus with no change in the shielding effect, so another electron will be much more strongly attracted to fluorine than to lithium.
8.19 An *ion* is a charged chemical species (an atom or group of atoms). A *cation* is a positively charged ion, and an *anion* is a negatively charged ion.
8.21 A potassium atom has 19 protons and 19 electrons. A potassium ion has a charge of 1+; it has 19 protons but only 18 electrons.
8.23 $Mg^{12}: [Ne]^{10}(3s^2)$
8.25 An atom of aluminum loses 3 electrons to form a cation with the electron configuration of the nearest noble gas, neon, an atom with a full octet.
$\cdot\text{Al}: \rightarrow Al^{3+} + 3\ e^-$

8.27 Removing one electron from a potassium atom forms a potassium ion with the same electron configuration as the nearest noble gas, argon. Removing another electron would be difficult because the stability of the filled octet would be lost.

8.29 A monatomic cation is always **smaller** than its atom, and a monatomic anion is always **larger** than its atom.

8.31 $:\ddot{\underset{\cdot\cdot}{I}}\cdot + e^- \rightarrow :\ddot{\underset{\cdot\cdot}{I}}:^-$

8.33 Ar

8.35 $3\ Mg: + 2\cdot\dot{\underset{\cdot}{N}}\cdot \rightarrow Mg_3\left[:\ddot{\underset{\cdot\cdot}{N}}:\right]_2$

8.37 (a) lead(IV) ion (b) mercury(II) ion (c) barium ion

8.39 (a) Na^+ (b) Hg_2^{2+} (c) Ni^{2+}

8.41

	H^-	S^{2-}	N^{3-}
Na^+	NaH	Na_2S	Na_3N
Sn^{2+}	SnH_2	SnS	Sn_3N_2
Sn^{4+}	SnH_4	SnS_2	Sn_3N_4

8.43 (a) CdO_2 (b) $AgClO$ (c) $NaC_2H_3O_2$

8.45 (a) $K^+ + Cl^-$ (b) $Cd^{2+} + S^{2-}$ (c) $Ca^{2+} + H^-$

8.47 A bond between two atoms is polar if the atoms have different electronegativity and nonpolar if the atoms have the same electronegativity. Two atoms of the same element must necessarily have identical electronegativity, so covalent bonds between atoms of the same element must be nonpolar.

8.49 The fluorine–bromine bond will be more polar than the fluorine–chlorine bond because the difference in electronegativity between fluorine and bromine is greater than the electronegativity difference between fluorine and chlorine. (Bromine is below chlorine on the Periodic Table, so bromine's electronegativity will be lower than chlorine's.)

8.51 (a) $\left[\begin{array}{c} H \\ | \\ H-N-H \\ | \\ H \end{array}\right]^+$ (b) $\left[\ddot{\underset{\cdot\cdot}{O}}=\ddot{N}-\ddot{\underset{\cdot\cdot}{O}}:\right]^-$

8.53 (a) $Zn(ClO_3)_2$ (b) $Pb(C_2H_3O_2)_2$ (c) $NaOH$

8.55 (a) Na_2S (b) Na_2SO_3 (c) Na_2SO_4

8.57 (a) $Ca^{2+} + SO_4^{2-}$ (b) $Al^{3+} + ClO_2^-$ (c) $NH_4^+ + CO_3^{2-}$

8.59 Molecules are combinations of nonmetals connected by covalent bonds: (a) H_2O (c) CO_2 (e) SO_2. A formula unit represents an ionic compound, a combination of a metal and a nonmetal: (b) Na_2O (d) PbO_2.

Chapter 9—Naming Compounds and Writing Equations

Answers to Exercises

9.1 (a) CS_2 (b) iodine monobromide (c) SF_6 (d) diphosphorus pentasulfide

9.2 (a) potassium iodide (b) $Ba(NO_3)_2$ (c) iron(II) bromide (d) Cu_3PO_4 (e) ammonium carbonate

9.3 (a) calcium perbromate (b) $Zn(IO_2)_2$

9.4 (a) barium hydrogen sulfate (b) $Ba(HSe)_2$
9.5 copper(II) sulfate pentahydrate
9.6 (a) This equation is not balanced. (b) This equation is balanced.
9.7

$$4\,H\cdot + 2\cdot\dot{C}: \rightarrow H{-}\underset{}{\overset{H}{\overset{|}{C}}}{=}\overset{H}{\overset{|}{C}}{-}H$$

9.8

$$2\,Na\cdot + \cdot\ddot{\underset{..}{S}}\cdot \rightarrow 2\,Na^{+} + :\ddot{\underset{..}{S}}:^{2-}$$

$$0 \;+\; 0 \;=\; 2(1+) \;+\; 2-$$

9.9 (a) This is a real reaction: Hydrogen exists as H—H.
(b) The reaction as written is imaginary because oxygen exists as O_2.

$$4\,K\cdot + \cdot\ddot{\underset{..}{O}}{-}\ddot{\underset{..}{O}}\cdot \rightarrow 4\,K^{+} + 2:\ddot{\underset{..}{O}}:^{2-}$$

9.10 $H_2 + F_2 \rightarrow 2\,HF$
9.11 $N_2 + O_2 \rightarrow 2\,NO$
9.12 $2\,H^+ + SO_4^{2-} \rightarrow H_2SO_4$
9.13

$$4\,Na\cdot + \cdot\ddot{\underset{..}{O}}{-}\ddot{\underset{..}{O}}\cdot \rightarrow 4\,Na^{+} + 2:\ddot{\underset{..}{O}}:^{2-}$$

9.14 $4\,Fe(s) + 3\,O_2(g) \rightarrow 2\,Fe_2O_3(s)$
9.15 $2\,H_2 + O_2 \rightarrow 2\,H_2O$ + heat
9.16 (a) combination and combustion (b) decomposition
9.17 (a) $2\,Mg(s) + O_2(g) \rightarrow 2\,MgO(s)$
magnesium oxygen magnesium oxide
combination
(b) $3\,Mg(s) + N_2(g) \rightarrow Mg_3N_2(s)$
magnesium nitrogen magnesium nitride
combination
9.18 $2\,Na(s) + Cl_2(g) \rightarrow 2\,NaCl(s)$
sodium chlorine sodium chloride
combination

Answers to Odd-Numbered Problems

9.1 (a) covalent (b) ionic (c) ionic
9.3 (a) tin(IV) hydride (b) tin(IV) oxide (c) methane
9.5 (a) aluminum sulfide (b) diphosphorus pentoxide (c) zinc iodide
9.7 (a) copper(II) nitrate (b) magnesium phosphate (c) aluminum nitride
9.9 (a) sodium acetate (b) calcium carbonate (c) calcium hydroxide
9.11 (a) CS_2 (b) $CdSO_4$ (c) CuI
9.13 (a) $Fe(CN)_2$ (b) H_2S (c) $Hg(HCO_3)_2$
9.15 (a) $Ba(ClO_3)_2$ (b) $Ba(BrO_3)_2$ (c) $Ba(IO_3)_2$
9.17 (a) LiOH (b) $NaClO_4$ (c) $(NH_4)_2S$
9.19 (a) $NiSO_3$ (b) NiS (c) $NiSO_4$
9.21

(a) $$3\,H\cdot + \cdot\ddot{\underset{\cdot}{N}}\cdot \rightarrow H{-}\underset{\underset{H}{|}}{\ddot{N}}{-}H$$

(b) $$2\,H\cdot + 2\cdot\ddot{\underset{\cdot}{N}}\cdot \rightarrow H{-}\ddot{N}{=}\ddot{N}{-}H$$

(c) $$4\,H\cdot + 2\cdot\ddot{\underset{\cdot}{N}}\cdot \rightarrow H{-}\underset{\underset{H}{|}}{\ddot{N}}{-}\underset{\underset{H}{|}}{\ddot{N}}{-}H$$

9.23 (a) $H_2 + I_2 \rightarrow 2\,HI$ (b) $3\,H_2 + N_2 \rightarrow 2\,NH_3$
(c) $2\,Li + H_2 \rightarrow 2\,LiH$

9.25 (a) $4\,Fe + 3\,O_2 \rightarrow 2\,Fe_2O_3$ (b) $2\,Zn + O_2 \rightarrow 2\,ZnO$
(c) $4\,K + O_2 \rightarrow 2\,K_2O$

9.27 (a) $2\,C_2H_2 + 5\,O_2 \rightarrow 4\,CO_2 + 2\,H_2O$ (b) $C_2H_4 + 3\,O_2 \rightarrow 2\,CO_2 + 2\,H_2O$
(c) $2\,C_2H_6 + 7\,O_2 \rightarrow 4\,CO_2 + 6\,H_2O$

9.29 (a) $Mg + Cl_2 \rightarrow MgCl_2$ (b) $3\,Ca + N_2 \rightarrow Ca_3N_2$
(c) $CH_4 + 4\,F_2 \rightarrow CF_4 + 4\,HF$

9.31 (a) $H^+ + CO_3^{2-} \rightarrow HCO_3^-$ (b) $H^+ + SO_3^{2-} \rightarrow HSO_3^-$
(c) $H^+ + S^{2-} \rightarrow HS^-$

9.33
(a) $2\,Na\cdot + :\underset{\cdot\cdot}{\ddot{Br}}-\underset{\cdot\cdot}{\ddot{Br}}: \rightarrow 2\,Na^+ + 2:\underset{\cdot\cdot}{\ddot{Br}}:^-$

(b) $Ca: + :\underset{\cdot\cdot}{\ddot{Cl}}-\underset{\cdot\cdot}{\ddot{Cl}}: \rightarrow Ca^{2+} + 2:\underset{\cdot\cdot}{\ddot{Cl}}:^-$

(c) $2\cdot Al: + 3:\underset{\cdot\cdot}{\ddot{F}}-\underset{\cdot\cdot}{\ddot{F}}: \rightarrow 2\,Al^{3+} + 6:\underset{\cdot\cdot}{\ddot{F}}:^-$

9.35 (a) $Sn^{2+} + 2\,ClO_3^- \rightarrow Sn(ClO_3)_2$ (b) $Cu^{2+} + CO_3^{2-} \rightarrow CuCO_3$
(c) $3\,Ni^{2+} + 2\,P^{3-} \rightarrow Ni_3P_2$

9.37 (a) $2\,K + Br_2 \rightarrow 2\,KBr$ (b) $Ba + Br_2 \rightarrow BaBr_2$
(c) $2\,Al + 3\,Br_2 \rightarrow 2\,AlBr_3$

9.39 (a) $S + O_2 \rightarrow SO_2$ (b) $2\,S + 3\,O_2 \rightarrow 2\,SO_3$ (c) $S + 3\,F_2 \rightarrow SF_6$

9.41 (a) This is an imaginary reaction. The real reaction:

$$\cdot\dot{C}: + 2:\underset{\cdot\cdot}{\ddot{F}}-\underset{\cdot\cdot}{\ddot{F}}: \rightarrow \begin{array}{c} :\ddot{F}: \\ | \\ :\underset{\cdot\cdot}{\ddot{F}}-C-\underset{\cdot\cdot}{\ddot{F}}: \\ | \\ :\underset{\cdot\cdot}{F}: \end{array}$$

(b) This is a real reaction.
(c) This is an imaginary reaction. The real reaction:

$4\,Na\cdot + \cdot\underset{\cdot\cdot}{\ddot{O}}-\underset{\cdot\cdot}{\ddot{O}}\cdot \rightarrow 4\,Na^+ + 2:\underset{\cdot\cdot}{\ddot{O}}:^{2-}$

9.43 (a) $3\,H_2 + N_2 \rightarrow 2\,NH_3$ (b) $H_2 + Br_2 \rightarrow 2\,HBr$
(c) $2\,CO + O_2 \rightarrow 2\,CO_2$

9.45 (a) $Hg(\ell) + F_2(g) \rightarrow HgF_2(s)$ + energy (b) $I_2(s) + Cl_2(g) \rightarrow 2\,ICl(s)$
(c) $2\,KClO_3(s)$ + heat $\rightarrow 2\,KCl(s) + O_2(g)$

9.47 All the reactions in Problem 27 are combustion reactions.

9.49 (a) combination (b) combination (c) combustion and combination

9.51 $2\,Na(s) + F_2(g) \rightarrow 2\,NaF(s)$ + heat
sodium fluorine sodium fluoride

9.53 $Zn(s) + Cl_2(g) \rightarrow ZnCl_2(s)$ + heat
zinc chlorine zinc chloride

9.55 $2\,H_2S(g) + 3\,O_2(g) \rightarrow 2\,SO_2(g) + 2\,H_2O(\ell)$

9.57 (a) $2\,M(s) + Cl_2(g) \rightarrow 2\,MCl(s)$
(b) $2\,M\cdot + :\underset{\cdot\cdot}{\ddot{Cl}}-\underset{\cdot\cdot}{\ddot{Cl}}: \rightarrow 2\,M^+ + 2:\underset{\cdot\cdot}{\ddot{Cl}}:^-$

9.59 (a) $4\,M(s) + O_2(g) \rightarrow 2\,M_2O(s)$
(b) $4\,M\cdot + \cdot\underset{\cdot\cdot}{\ddot{O}}-\underset{\cdot\cdot}{\ddot{O}}\cdot \rightarrow 4\,M^+ + 2:\underset{\cdot\cdot}{\ddot{O}}:^{2-}$

Chapter 10—The Periodic Table as a Guide to Formulas and Reactions

Answers to Exercises

10.1 (a) LiBr (b) $2\,Li + Br_2 \rightarrow 2\,LiBr$

10.2 $4\,K + O_2 \rightarrow 2\,K_2O$

10.3 $2\,Na(s) + 2\,H_2O(\ell) \rightarrow 2\,NaOH(aq) + H_2(g)$

10.4 $2\,Li(s) + 2\,H_2O(\ell) \rightarrow 2\,LiOH(aq) + H_2(g)$

10.5 (a) $2\ Na + H_2 \rightarrow 2\ NaH$ (b) KH (c) $2\ Li(s) + H_2(g) \rightarrow 2\ LiH(s)$
10.6 AsH_3
10.7 (a) CaF_2 (b) $Ca + F_2 \rightarrow CaF_2$
10.8 (a) MgO (b) $2\ Mg + O_2 \rightarrow 2\ MgO$ (c) BaO
10.9 $Sr(s) + 2\ H_2O(\ell) \rightarrow Sr(OH)_2(aq) + H_2(g)$
10.10 $2\ Al + 3\ F_2 \rightarrow 2\ AlF_3$
10.11 (a) Al_2O_3 (b) $4\ Al + 3\ O_2 \rightarrow 2\ Al_2O_3$
10.12 Ga_2O and Ga_2O_3
10.13 BCl_3 is boron trichloride (covalent); $AlCl_3$ is aluminum chloride; $TlCl_3$ is thallium(III) chloride; TlCl is thallium(I) chloride.
10.14 (a) $SiCl_4$ (b) $Si + 2\ Cl_2 \rightarrow SiCl_4$
10.15 SiH_4
10.16 $GeCl_4$
10.17 $Pb + 2\ F_2 \rightarrow PbF_4$
10.18 $2\ Sn + O_2 \rightarrow 2\ SnO$
10.19

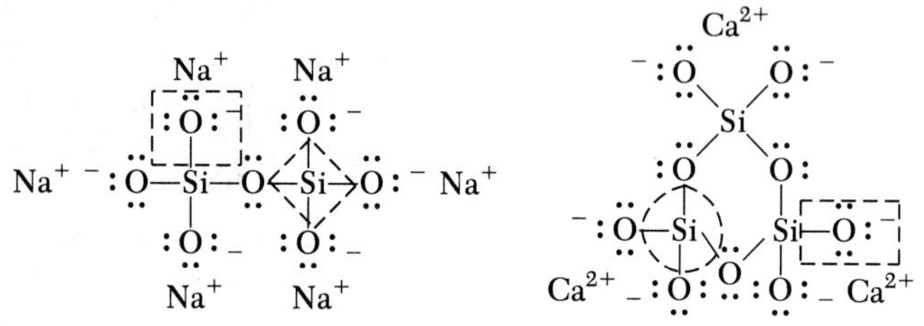

10.20 (a) PH_3 (b) K_3P (c) PI_3
10.21 PCl_3 is phosphorus trichloride, and PCl_5 is phosphorus pentachloride.
10.22 (a) $AsCl_3$ is arsenic trichloride, and SbI_3 is antimony triiodide.
(b) $Bi(NO_3)_3$ is bismuth(III) nitrate.
10.23 $Na_4P_2O_7$
10.24 (a) ZnO (b) $2\ Zn + O_2 \rightarrow 2\ ZnO$
10.25 (a) Na_2S (b) MgS (c) Al_2S_3
10.26 SCl_2
10.27 $S + O_2 \rightarrow SO_2$
10.28 H_2Se
10.29 (a) KBr (b) $SrBr_2$ (c) $AlBr_3$
10.30 $Ni + Cl_2 \rightarrow NiCl_2$
10.31 ClF is chlorine monofluoride; IBr is iodine monobromide; IF_5 is iodine pentafluoride.
10.32 $Cl_2 + F_2 \rightarrow 2\ ClF$
10.33 XeF_2 is xenon difluoride.
10.34 The perxenate ion has a charge of 4−.
10.35 Fluorine and oxygen are the two elements with the highest electronegativity and therefore have the greatest ability to pull electrons away from xenon atoms.
10.36 The outer or valence electrons of large atoms are farther from the nucleus and are more shielded by inner electrons, and therefore are held much less tightly than the outer electrons of small atoms. For this reason, the larger atoms Kr, Xe, and Rn share their outermost electrons in covalent bonds, and the smaller atoms He, Ne, and Ar do not.
10.37 (a) Element 47 is a transition metal. (b) It is filling the 4d sublevel.
(c) $[Kr]^{36}5s^24d^9$ (d) Element 47 is silver.
10.38 (a) zinc carbonate (b) iron(II) carbonate (c) $CuSO_3$
10.39 $Fe^{2+} + 6\ H_2O \rightarrow Fe(H_2O)_6^{2+}$
$Fe(H_2O)_6^{2+} + 2NO_3^- \rightarrow [Fe(H_2O)_6](NO_3)_2$

Answers to Odd-Numbered Problems

10.1 Hydrogen is a gas at room temperature and pressure. All the other 1A elements are solids.

10.3 s^1

10.5 cesium (francium has the same electronegativity but it is not a stable element)

10.7 (a) lithium iodide (b) potassium oxide (c) sodium sulfate

10.9 The formula for sodium chloride is NaCl. Both potassium and sodium are group 1A elements, and both fluorine and chlorine are group 7A elements, so the formula for potassium fluoride will be KF.

10.11 (a) $4\ Na + O_2 \rightarrow 2\ Na_2O$ (b) $2\ Na + 2\ H_2O \rightarrow 2\ NaOH + H_2$
(c) $6\ Na + N_2 \rightarrow 2\ Na_3N$

10.13 (a) $6\ Li + N_2 \rightarrow 2\ Li_3N$ (lithium nitride)
(b) $2\ Na + Br_2 \rightarrow 2\ NaBr$ (sodium bromide)
(c) $4\ K + O_2 \rightarrow 2\ K_2O$ (potassium oxide)

10.15 (a) 6A (b) Z is sulfur.

10.17 $2\ K + H_2 \rightarrow 2\ KH$ (potassium hydride)

10.19 s^2

10.21 (a) Mg^{2+} (b) Ca^{2+} (c) Sr^{2+}

10.23 (a) calcium iodide (b) magnesium cyanide (c) barium chlorite

10.25 (a) $CaBr_2$ (b) CaO (c) Ca_3N_2

10.27 $Ba(s) + 2\ H_2O(\ell) \rightarrow Ba(OH)_2(aq) + H_2(g)$

10.29 strontium sulfate

10.31 s^2p^1

10.33 (a) Al^{13}: $1s^22s^22p^63s^23p^1$
(b) By losing the outer three electrons, aluminum ion then has an electron configuration identical to the nearest noble gas. By losing three electrons, the atom becomes a 3+ ion.

10.35 Boron is a metalloid and has a higher electronegativity than aluminum; boron's electronegativity is closer to fluorine's electronegativity than is aluminum's. The difference in electronegativity between aluminum and fluorine is greater than the difference between boron and fluorine, so the aluminum-fluorine bond is more ionic than the boron–fluorine bond.

10.37 (a) boron triiodide (b) aluminum chloride (c) aluminum carbonate

10.39 (a) $AlBr_3$ (b) Al_2S_3 (c) AlN

10.41 The bonds in aluminum oxide would be more ionic than those in aluminum sulfide. Oxygen is more electronegative than sulfur, so the difference in electronegativity between aluminum and oxygen is larger than between aluminum and sulfur.

10.43 s^2p^2

10.45 (a) Pb^{82} : $[Xe]^{54}6s^24f^{14}5d^{10}6p^2$
(b) By losing just the two 6p electrons, a lead atom can become Pb^{2+} ion. By losing the two 6p electrons and the two 6s electrons, a lead atom can become Pb^{4+} ion.

10.47

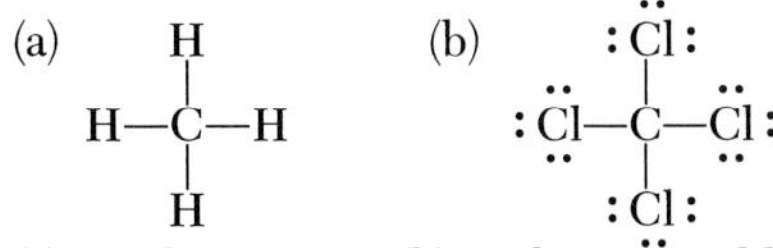

(b)

```
       ..
      :Cl:
  ..   |   ..
 :Cl—C—Cl:
  ..   |   ..
      :Cl:
       ..
```

10.49 (a) methane (b) carbon tetrachloride (c) tin(IV) iodide
(d) lead(II) oxide

10.51 (a) CS_2 (b) SnO (c) PbS_2

10.53 SnS, tin(II) sulfide; and SnS_2, tin(IV) sulfide

10.55 The bonds in tin(IV) fluoride would be more ionic than those in tin(IV) iodide. Fluorine is more electronegative than iodine, so the difference in electronegativity between tin and fluorine is greater than the difference between tin and iodine.

10.57 s^2p^3
10.59 (a) N^{3-} (b) P^{3-}
10.61 (a) ammonia (b) ammonium bromide
(c) arsenic trichloride (covalent)
10.63 Phosphine is PH_3; arsine is AsH_3; and stibine is SbH_3.
10.65 Nitrogen and chlorine are nonmetals, so they form a covalent compound. Bismuth and chlorine are a metal and a nonmetal, so they form an ionic compound.
10.67 s^2p^4
10.69 (a) O^8: $1s^22s^22p^4$ (b) $O^{2-} = 1s^22s^22p^6$
10.71 At room temperature and pressure, oxygen is a gas and sulfur is a solid.
10.73 Oxygen is more electronegative than sulfur. A polar covalent bond between oxygen and sulfur would find the oxygen partially negative.
10.75 (a) calcium oxide (b) potassium sulfide (c) hydrogen sulfide

10.77 (a) H—S̈—H (with two unshared pairs on S) (b) $[:\ddot{O}—\ddot{S}—\ddot{O}:]^{2-}$ with a third $:\ddot{O}:$ bonded below S

10.79 s^2p^5
10.81 (a) F^9: $1s^22s^22p^5$ (b) $F^- = 1s^22s^22p^6$
10.83 (a) hydrogen fluoride (b) sodium fluoride
(c) chlorine monofluoride
10.85 (a) $BaBr_2$ (b) barium bromide
10.87 $2\ K(s) + F_2(g) \rightarrow 2\ KF(s)$
10.89 s^2p^6
10.91 All of the elements in group 8A are gases.
10.93 In krypton difluoride the krypton atom is sharing two pairs of electrons, and it has three unshared electron pairs.
10.95

$:\ddot{O}—\ddot{Xe}—\ddot{O}:$ with $:\ddot{O}:$ bonded below Xe — xenon trioxide

10.97 NaH, MgH_2, AlH_3, SiH_4, PH_3, H_2S, HCl
10.99

SiH_4: Si bonded to four H (H—Si—H, with H above and below); PH_3: H—P̈—H with H below; H_2S: H—S̈—H with two unshared pairs; HCl: H—Cl with three unshared pairs on Cl

10.101 (a) Element 28 is a metal. (b) It is a transition element.
(c) It is a transition metal. (d) It is filling the 3d sublevel.
(e) $[Ar]^{18}4s^23d^8$ (f) Element 28 is nickel.
10.103 (a) $Cu_3(PO_4)_2$ (b) $Fe(HSO_3)_2$
10.105 $Cr^{3+} + 6\ NH_3 \rightarrow Cr(NH_3)_6^{3+}$
$Cr(NH_3)_6^{3+} + 3\ Br^- \rightarrow [Cr(NH_3)_6]Br_3$

Chapter 11—The Mole

Answers to Exercises

11.1 (a) 1 mol NH_3 = 17.0 g NH_3 (b) 1 mol $Ca(NO_3)_2$ = 164 g $Ca(NO_3)_2$
(c) 1 mol NO_2^- = 46.0 g NO_2^- (d) 1 mol O_2 = 32.0 g O_2
(e) 1 mol O atoms = 16.0 g O atoms (f) 1 mol Zn = 65.4 g Zn

11.2 (a) 171 g sucrose (b) 0.0107 mol sucrose
11.3 (a) 16.8 g $NaHCO_3$ (b) 219 g SF_6 (c) 55.0 g Cl^-
11.4 (a) 1 mol O_2 = 6.02×10^{23} O_2 molecules
(b) 1 mol O atoms = 6.02×10^{23} O atoms
(c) 1 mol CH_4 = 6.02×10^{23} CH_4 molecules
(d) 1 mol $FeCl_3$ = 6.02×10^{23} $FeCl_3$ formula units
(e) 1 mol O^{2-} = 6.02×10^{23} O^{2-} ions
11.5 (a) 2.99×10^{-17} g H_2O (b) 2.99×10^{-23} g H_2O
(c) 2.18×10^{23} K^+ ions
11.6 40.0% C; 6.73% H; 53.3% O
11.7 (a) KCO_2 (b) CH (c) $C_4H_5N_2O$ (d) NaCl
11.8 64.0 amu = simplest-formula mass
11.9 $C_6H_8O_6$
11.10 $C_8H_{10}N_4O_2$

Answers to Odd-Numbered Problems

11.1 (a) 52.0 g Cr (b) 17.3 g HNO_3 (c) 39.0 g O_2
11.3 (a) 14.0 g N_2 (b) 34.1 g H_2S (c) 213 g Na_3PO_4
11.5 (a) 0.0687 mol Mg (b) 0.00199 mol PbO (c) 0.475 mol O_2
11.7 (a) 0.507 mol Li (b) 0.00888 mol SO_3 (c) 0.00934 mol $CaSO_4$
11.9 (a) 1.66 mol Li (b) 5.71×10^{-4} mol N atoms (c) 0.0462 mol N_2
(d) 0.291 mol $SrCl_2$
11.11 (a) 0.444 mol Cr (b) 5.90 mol H_2O (c) 0.0208 mol NaCl
11.13 (a) 1.05 mol N_2 (b) 1.39 mol C (c) 10.1 mol CCl_4
11.15 (a) 0.0930 mol Fe (b) 0.235 mol CH_4 (c) 1.05×10^{-4} mol PCl_3
11.17 (a) 12.2 mol $Ca(NO_3)_2$ (b) 0.498 mol B (c) 0.0274 mol N_2O_5
(d) 12.3 mol $Zn(NO_3)_2$
11.19 (a) 4.96×10^{-4} mol Ni (b) 8.34 mol C (c) 0.0379 mol KBr
(d) 0.00262 mol CO
11.21 (a) 546 g SO_3 (b) 2.88×10^4 mg NaH_2PO_4 (c) 3.58×10^4 mol Cl_2
(d) 1.03×10^{24} formula units $MnSO_3$
11.23 (a) 2.99×10^3 g $Cu(NO_3)_2$ (b) 3.71×10^{16} atoms Ag
(c) 3.91×10^{18} molecules CCl_4 (d) 4.38×10^3 μg $KClO_3$
11.25 (a) 1.88×10^{22} O_2 molecules (b) 9.27×10^{-23} g per Fe atom
(c) 1.67×10^{21} molecules H_2O
11.27 63.2% C; 5.32% H; 31.6% O (These percentages sum to greater than 100% because of rounding.)
11.29 28.0% Ca; 49.7% Cl; 22.4% O
11.31 27.9% P; 72.3% S
11.33 41.7% Zn; 17.8% N; 40.8% O
11.35 78.2% Ag; 4.35% C; 17.4% O
11.37 (a) NaO (b) $NaCO_2$ (c) Na_2S
11.39 (a) molecular mass = 62.2 amu (b) SiH_3 (c) 31.1 amu
11.41 73.0 amu
11.43 $C_6H_{12}O_6$
11.45 $Li_2C_2O_4$; lithium oxalate
11.47 lead(IV) oxide
11.49 $C_{13}H_{18}O_2$
11.51 N_2H_4
11.53 C_6H_6

Chapter 12—Calculations Based on Chemical Equations: Stoichiometry

Answers to Exercises

12.1 1 mol CH_4 + 4 mol $Br_2 \rightarrow$ 1 mol CBr_4 + 4 mol HBr
12.2 102 g O_2
12.3 5.01 mg MgS
12.4 3.60 g Br_2
12.5 4.74 tons N_2; 1.02 tons H_2
12.6 Oxygen is the limiting reagent; 5.00 g O_2 would consume 1.62 g C_2H_2, leaving 8.4 g of acetylene.
12.7 2.35 mg AgF

Answers to Odd-Numbered Problems

12.1 1 mol H_2 + 1 mol $Cl_2 \rightarrow$ 2 mol HCl
12.3 2 mol SO_2 + 1 mol $O_2 \rightarrow$ 2 mol SO_3
12.5 Ca(s) + $Br_2(\ell) \rightarrow CaBr_2(s)$
1 mol Ca + 1 mol $Br_2 \rightarrow$ 1 mol $CaBr_2$
12.7 4 Al(s) + 3 $O_2(g) \rightarrow$ 2 $Al_2O_3(s)$
4 mol Al + 3 mol $O_2 \rightarrow$ 2 mol Al_2O_3
12.9 2 mol HgO = 1 mol O_2
12.11 1 mol Br_2 = 1 mol $C_2H_4Br_2$
12.13 1 mol O_2 = 2 mol BaO
12.15 2.00 mol O_2
12.17 0.500 mol $ZnCl_2$
12.19 0.417 mol NH_3
12.21 4.17 mol HgO
12.23 7.49 g SO_3
12.25 3.35 g BaO
12.27 2.64 g Ca
12.29 22.4 g $KClO_3$
12.31 15.9 g CaO
12.33 3.67 g Cl_2
12.35 1.13 g Ca
12.37 4.17×10^3 g Na_2SO_4
12.39 1.93 mg ZnO
12.41 2.39×10^3 mg PbO
12.43 701 g Al_2O_3
12.45 918 g HBr
12.47 3.71 g O_2
12.49 8.78×10^{27} molecules CO_2
12.51 0.664 g CH_4
12.53 2.00 mol BeO
12.55 0.667 mol $AlCl_3$
12.57 6.07 g NH_3
12.59 7.03 g SeO_2
12.61 6.09 g P_2O_5
12.63 61.4 g Mg_3N_2
12.65 2.80 g CaO

Chapter 13—Gases

Answers to Exercises

13.1 1.56×10^{-3} mol O_2; 0.0424 L; 298 K; 0.900 atm

13.2 (a) The pressure inside a tire equals the sum of the external atmospheric pressure plus the force of the stretched rubber of the tire. At the puncture, the pressure of the gas is higher than the external atmospheric pressure alone, so the gas particles escape through the puncture until the internal pressure equals the external atmospheric pressure. The tire loses its pressure; it goes flat.

(b) The force pushing in must equal the force pushing out. The internal pressure remains the same, but as the external pressure decreases at higher altitude, the internal pressure is higher than the external pressure. This difference pushes against the inside wall of the tire, making it expand until the force required to stretch the tire even more is exactly the force difference between the internal and external pressure.

13.3 0.233 mol He added

13.4 501 mL

13.5 1.22 atm

13.6 279 K

13.7 2.20 L Cl_2

13.8 104 L C_2H_2

13.9 8.88 L CO

13.10 426 K

13.11 807 torr

13.12 1.49 atm

13.13 2.92 L

13.14 26.8 ml O_2 required

13.15 3.90 g F_2

13.16 629 mg CO_2

Answers to Odd-Numbered Problems

13.1 (a) 0.345 g O_2 (b) 0.215 mol O_2 (c) 0.0238 mol O_2

13.3 (a) 50°C (b) 307 K (c) 136 K

13.5 0.0282 mol Cl_2; 0.561 L; 1.25 atm; 306 K

13.7 0.0549 mol NH_3; 2.55 L; 1.09 atm; 408 K

13.9 The height of the mercury column would increase. At lower elevations, the atmosphere is denser, with more particles of gas hitting a unit surface area. This is a pressure increase, reflected in the higher column of mercury.

13.11 As the particles of gas are forced into a smaller space, the collisions on the walls of the cylinder and face of the piston increase. The increased pressure on the face of the piston is felt as the resistance to pushing the piston farther down.

13.13 At the puncture, the particles of gas inside have only the external pressure opposing them, not the force from the expanded balloon. The balloon deflates as the particles of gas escape, eventually leaving the same pressure inside as outside the balloon. (The "pop" of a balloon is the balloon tearing, not the gas escaping.)

13.15 As the balloon rises, the external pressure decreases while the internal pressure remains constant. The difference in pressures must be equalized by stretching the balloon until the force of the stretched balloon equals the difference between the internal and external pressures.

13.17 Particles of a gas are in constant, random motion; that is, the particles are moving in all directions. To say that all oxygen molecules would move to one corner simultaneously would defy the logic of this principle of kinetic theory of gases. Common sense tells us that people in a room do not spontaneously suffocate.

13.19 The volume of methanol liquid (0.448 mL) is 0.126% the volume of methanol gas. The empty space in the gas is 99.9%.

13.21 The kinetic theory says that the number of particles in a gas sample and their motion are responsible for the volume, temperature, and pressure of the sample. At constant temperature and pressure, therefore, the volume of the sample must be directly proportional to the number (or moles) of gas particles—Avogadro's Law.

13.23 The kinetic theory says that the motion of gas particles influences volume, temperature, and pressure. As temperature increases, the average speed of the particles increases. To maintain constant pressure, the volume of the container must increase with the temperature increase.

13.25 The kinetic theory says that the motion of gas particles influences volume, temperature, and pressure. As temperature increases, the average speed of the particles increases. At constant volume, the particles of gas will hit the walls more frequently and with more force, which we measure as an increase in pressure.

13.27 4.69 L

13.29 2.01 L

13.31 0.218 g NH_3 was removed

13.33 489 mL

13.35 (a) $V_2 = \frac{V_1P_1n_2T_2}{P_2n_1T_1}$ (b) $n_2 = \frac{P_2V_2n_1T_1}{P_1V_1T_2}$

(c) $P_2 = \frac{V_1P_1n_2T_2}{V_2n_1T_1}$ (d) $T_2 = \frac{P_2V_2n_1T_1}{P_1V_1n_2}$

13.37 If $n_1 = n_2$ and $T_1 = T_2$, then:

$$\frac{P_1V_1}{n_1T_1} = \frac{P_2V_2}{n_2T_2} \Rightarrow P_1V_1 = P_2V_2 \Rightarrow \frac{V_1}{V_2} = \frac{P_2}{P_1}$$ Boyle's Law

13.39 If $n_1 = n_2$ and $V_1 = V_2$, then:

$$\frac{P_1V_1}{n_1T_1} = \frac{P_2V_2}{n_2T_2} \Rightarrow \frac{P_1}{T_1} = \frac{P_2}{T_2} \Rightarrow \frac{P_1}{P_2} = \frac{T_1}{T_2}$$

which is the mathematical law you wrote for Problem 24.

13.41 456 K

13.43 214 mL

13.45 An ideal gas is one that follows exactly the predictions of the kinetic theory and the gas laws.

13.47 6.24×10^4 torr-mL/mol-K

13.49 8.06 atm

13.51 0.151 mol CO_2

13.53 −222°C

13.55 2.35×10^{-3} g/mL

13.57 1.20×10^2 amu

13.59 225 mL PCl_5

13.61 82.5 mL O_2

13.63 51.5 L HCl

13.65 0.226 mol NF_3

13.67 0.234 mol Cl_2

13.69 7.11×10^3 mL O_2

Chapter 14—Liquids, Solids, and Changes of State

Answers to Exercises

14.1 Each O_2 molecule has 16 electrons and each N_2 molecule has 14, so oxygen has stronger London forces and therefore a higher boiling point. (The boiling point of O_2 is 90 K, whereas the boiling point of N_2 is 77 K.)

14.2 Each Br_2 molecule has 70 electrons and each Cl_2 has 34, so bromine has stronger London forces and therefore a higher boiling point. By coincidence, the boiling point of bromine happens to be above room temperature, making it a liquid at room conditions.

14.3 (a) O_2 is not a dipole—the atoms have identical electronegativity.
(b) NO is a dipole—the atoms have different electronegativity.
(c) Br_2 is not a dipole—the atoms have identical electronegativity.

14.4 (a) Only London forces are present in liquid H_2.
(b) In liquid HCl, both London forces (*always* present) and dipole–dipole interactions are present.
(c) Liquid HCl will have a higher boiling point than liquid H_2 for two reasons: (1) HCl has more electrons and therefore stronger London forces, and (2) HCl has dipole–dipole interactions, and H_2 does not.

14.5

```
H     H  ..
 \ .. /  O
  O  H /..\
  ..      H   H
```

14.6

```
   ..
   O       H
  /..\     |
 H    H  :Cl:
           ..
```

14.7

```
 H            ..           H
  \           O           /
  :O:···H  /..\  H···:O:
  /                      \
 H                        H
```

14.8

```
(a) H                      (b)
     \          ..              .        ..
     :O:···H—F:                  O—H···:F—H
     /          ..              /.       ..
    H                          H
```

14.9 Methane does not form hydrogen bonds. Methane does not have nitrogen, oxygen, or fluorine bonded to hydrogen, nor does methane have any unshared electron pairs.

14.10 (a) Methanol would form hydrogen bonds because it has a hydrogen bonded to an oxygen.
(b)

```
        H           H
        |    ..     |
     H—C—O—H···:O:
        |    ..     |
        H        H—C—H
                    |
                    H
```

14.11 Ammonia has N—H bonds and can form hydrogen bonds. Methane has neither nitrogen, oxygen, nor fluorine, so it cannot form hydrogen bonds. Ammonia melts at a higher temperature because of its stronger intermolecular forces.

14.12 The $3s^23p^1$ electrons are shared in metallic bonding.

14.13 Bending a wire is easy because the metal atoms are free to move with respect to one another. To pull the wire apart, it is necessary to overcome the very strong attractive forces between ions and the "sea" of electrons, a difficult process requiring much energy.

14.14 Silicon carbide should resemble the structure of diamond most closely. Diamond contains only atoms that form four single bonds to four other atoms. As silicon and carbon are both group 4A elements, the bonding of silicon should be similar to that of carbon, and the bonding network of silicon carbide should be similar to the pure carbon network in diamond.

14.15 (a) The Cl_2 molecule is covalent with London forces as its only interparticle forces. NaCl is an ionic compound with very strong interparticle forces. NaCl will have a much higher melting point than Cl_2.

(b) ClF will have a higher melting point than F_2. The ClF molecule has a dipole, so ClF molecules will be attracted by dipole–dipole interactions as well as by London forces. F_2 has no dipole and cannot have dipole–dipole interactions. Also, each ClF molecule has 26 electrons and each F_2 has 18, so ClF has stronger London forces than F_2.

(c) Neither H_2 nor O_2 has a dipole, so only London forces are at work here. Because each O_2 molecule has 16 electrons and each H_2 molecule has only 2, oxygen has stronger London forces and will melt at a higher temperature than hydrogen.

14.16 London forces alone would predict the order of increasing boiling points to be NH_3, PH_3, and AsH_3. Ammonia can form hydrogen bonds, however, increasing its boiling point, so the actual order is PH_3, AsH_3, and NH_3.

Answers to Odd-Numbered Problems

14.1 The process of a solid going directly to the gas phase without going through the liquid phase is called sublimation.

14.3 In the evaporation process, molecules of liquid occasionally overcome the attractive forces and leave the liquid phase, becoming a gas.

14.5 (a) The melting point is the temperature at which a solid changes into a liquid. The boiling point is the temperature at which a liquid turns into a gas.

(b) The melting point is the temperature at which the particles overcome the attractive force that keeps them in a rigid lattice. The boiling point is the temperature at which the particles overcome all remaining attractive forces, thereby separating themselves from all the other particles.

14.7 Solids are the most ordered state of matter, and gases are the most disordered.

14.9 The particles in a solid are in a rigid network and cannot move past one another. The particles in a liquid have greater freedom of motion and will flow to adopt the shape of the container.

14.11 In evaporation, some particles at the surface of the liquid have enough energy to leave the liquid phase and enter the gas phase; in boiling, particles in the body of the liquid have enough energy to form bubbles of gas.

14.13 Argon has more electrons than neon, so argon will have stronger London forces.

14.15 Bromine has more electrons than chlorine and therefore has stronger London forces. The melting (or freezing) point of bromine should be higher than that of chlorine.

14.17 The number of electrons a species has increases downward in the Periodic Table, thereby increasing the strength of the London forces, which in turn raises the melting and boiling points.

14.19 The electrons around a neon atom can be distributed unevenly, causing a dipole for only an instant. In the next instant they will be distributed in some other way. The hydrogen chloride molecule, however, has a permanent dipole: The chlorine is always more electronegative than hydrogen, so the chlorine always has a partial negative charge and the hydrogen always has a partial positive charge.

14.21 Each molecule of Br_2 has 70 electrons, and each molecule of HBr has 36 electrons. The London forces between molecules of Br_2 are strong because bromine has so many electrons. Even though H—Br has a permanent dipole, in this case the strong London forces in Br_2 are stronger than the weaker London forces plus dipole–dipole interactions of H—Br.

14.23

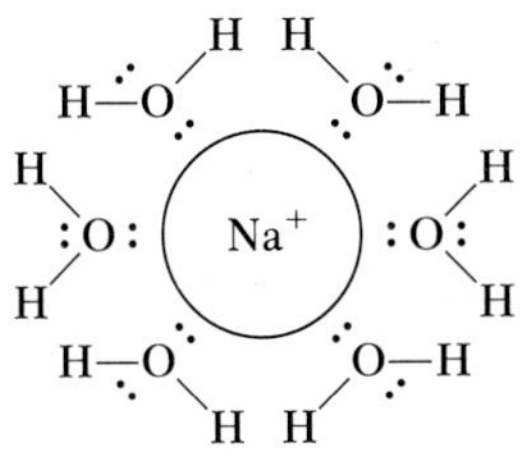

14.25

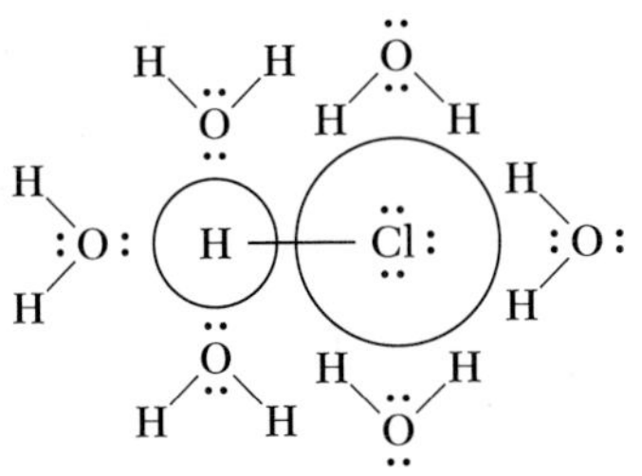

14.27 Many arrangements are possible, including these three:

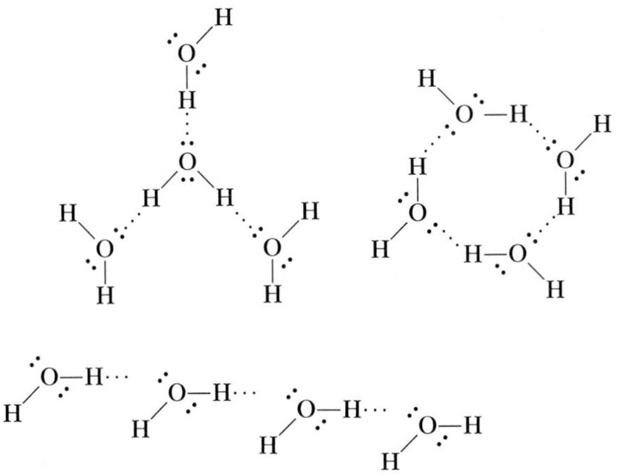

14.29

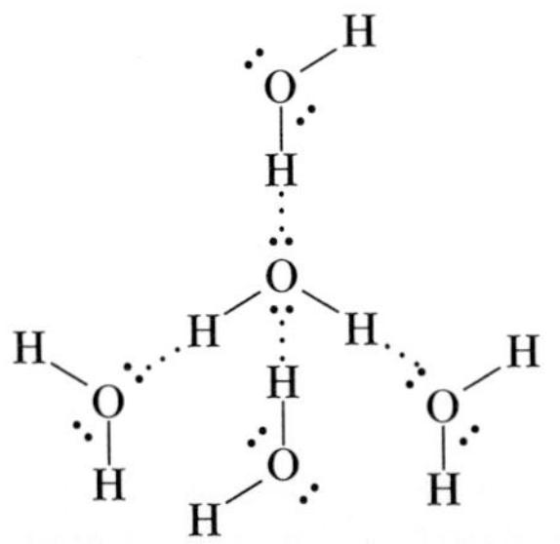

14.31 On London forces alone, we would predict that methane (10 electrons) would have a lower boiling point than silane (18 electrons); measurement shows this to be true. We would also predict that ammonia (10 electrons) would have a lower boiling point than phosphine (18 electrons); this prediction is not true. The hydrogen bonds that ammonia forms are stronger than London forces and increase ammonia's boiling point above that of phosphine.

14.33 Hydrogen, H_2, has only two electrons and therefore very weak London forces. Oxygen, O_2, has only London forces, but they are stronger than in H_2. Water, H_2O, has very strong hydrogen bonds, stronger than London forces. Hydrogen peroxide, H_2O_2, has greater London forces than water because hydrogen peroxide has the capacity for more hydrogen bonds than water as it has more unshared electron pairs.

14.35 Noble gases have only London forces. Alkali metals have metallic bonding, very much stronger than London forces. In fact, all the alkali metals have melting points higher than all the noble gases.

14.37 This sketch of a "bent" wire suggests that the atoms of lithium can slide past one another. Metallic bonding does not keep the atoms in fixed positions as we find in crystals.

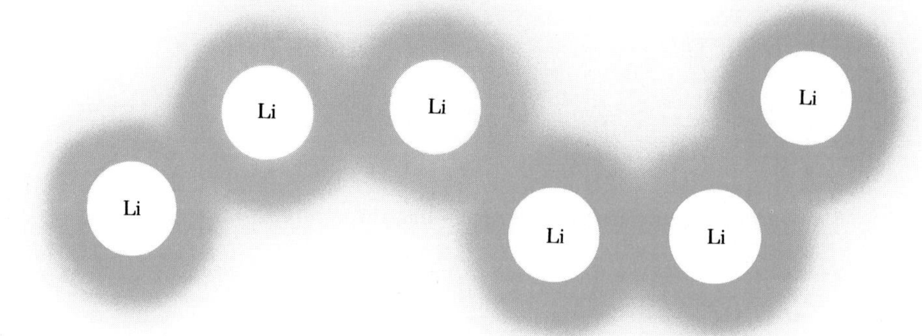

14.39 (a) metallic bonding (b) ionic bonding (c) covalent bonding

14.41 All substances have London forces, so if the temperature gets low enough, all substances will ultimately condense and freeze.

14.43 First, note that the noble gases are monatomic whereas the halogens form diatomic molecules, thereby doubling the number of electrons per molecule. All of the noble gases through xenon (54 electrons) are gases. The first two halogens, F_2 (18 electrons) and Cl_2 (34 electrons) are also gases. The next halogen, Br_2 (70 electrons), is a liquid; Br_2 has more electrons than any of the noble gases. The observation that all noble gases are gases is consistent with our understanding that the strength of London forces increases with increasing number of electrons.

14.45 You can decide by identifying their interparticle forces. Potassium bromide is a compound of a metal and a nonmetal, so its bonding will be ionic; the forces between its particles (ions) will be ionic bonds and London forces. Hydrogen bromide is a compound of two nonmetals, so its bonding will be polar covalent; the forces between its particles (molecules) will be dipole–dipole interactions and London forces. Ionic bonds are very strong forces, so almost all ionic compounds are solids at room conditions. It is likely that HBr is a gas.

14.47 Compound (b) has N—H bonds; it can form hydrogen bonds, which are much stronger than the London forces—the only interparticle force in compound (a). Compound (b) will have the higher boiling point.

14.49 The vast majority of the elements are solid metals. Metallic bonding is the important interparticle force.

Chapter 15—Introduction to Aqueous Solutions

Answers to Exercises

15.1 Each neon atom has 10 electrons and each helium atom has 2. Neon has stronger London forces with water than helium.

15.2 The nitrogen molecule, N_2, does not have a dipole so it cannot form dipole–dipole interactions with water or anything else.

15.3 Here are four possibilities:

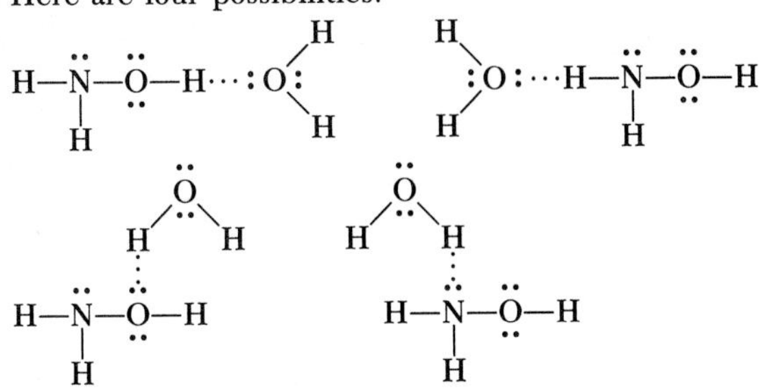

15.4

(a) Ca^{2+} (b) F^-

15.5 Each neon atom has 10 electrons and each helium atom has 2. Neon has stronger London forces with water than helium does, so neon will be more soluble in water than helium.

15.6 In water carbon monoxide would be more soluble than nitrogen. Carbon monoxide has a permanent dipole and would form dipole–dipole interactions with water. The N_2 molecule has no dipole, so weak London forces would be the only attractive force between nitrogen and water.

15.7 Methanol has much greater solubility in water than methane. Methanol has stronger attraction to water from London forces and from dipole–dipole interaction, but the most important force here is hydrogen bonding. Methanol can form hydrogen bonds with water, whereas methane cannot.

15.8 0.106 M $KNO_3(aq)$

15.9 (a) 331 mL of 1.25 M HCl(aq) (b) 0.0357 L of 1.25 M HCl(aq)

15.10 0.100 M $K_2CrO_4(aq)$

15.11 (a) 50.0 mL of 6.00 M HCl(aq) (b) 150 mL water needs to be added

15.12

$$H_2O + H{-}O{-}N(=O){-}O \rightarrow H_2O{-}H^+ + O{-}N(=O){-}O^-$$

15.13 *strong acids*

(a) $H_2O(\ell) + HCl(aq) \rightarrow H_3O^+(aq) + Cl^-(aq)$

(c) $H_2O(\ell) + HBr(aq) \rightarrow H_3O^+(aq) + Br^-(aq)$

weak acids

(a) $H_2O(\ell) + HF(aq) \rightleftarrows H_3O^+(aq) + F^-(aq)$

(c) $H_2O(\ell) + HNO_2(aq) \rightleftarrows H_3O^+(aq) + NO_2^-(aq)$

15.14 (a) Citric acid has three ionizable hydrogens; these are the hydrogens written first in the formula.

(b) Citric acid is not one of the six strong acids. It must be a weak acid.

(c) $H_2O(\ell) + H_3C_6H_5O_7(aq) \rightleftharpoons H_3O^+(aq) + H_2C_6H_5O_7^-(aq)$

$H_2O(\ell) + H_2C_6H_5O_7^-(aq) \rightleftharpoons H_3O^+(aq) + HC_6H_5O_7^{2-}(aq)$

$H_2O(\ell) + HC_6H_5O_7^{2-}(aq) \rightleftharpoons H_3O^+(aq) + C_6H_5O_7^{3-}(aq)$

15.15 (a) $HNO_3 \rightarrow H^+ + NO_3^-$

(b) $HNO_2 \rightleftharpoons H^+ + NO_2^-$

15.16 Only (c) LiOH is a strong base.

15.17 $C_5H_5N(aq) + H_2O(\ell) \rightleftharpoons C_5H_5NH^+(aq) + OH^-(aq)$

15.18 6.57 g HCl

15.19 $[Ba^{2+}] = 0.163$ M $[OH^-] = 0.326$ M

15.20 $[H_3O^+] = 1.34 \times 10^{-13}$ M

15.21 (a) pH 8.00 is weakly basic (b) pH 1.00 is strongly acidic

15.22 (a) pH = 8.08 (b) pH = 11.57 (c) $[OH^-] = 3.4 \times 10^{-10}$ M

15.23

(a)

$$\text{H}-\underset{\text{H}}{\overset{\text{H}}{\text{N}}}: + \text{H}-\ddot{\text{O}}-\text{H} \rightleftharpoons \text{H}-\underset{\text{H}}{\overset{\text{H}}{\text{N}}}-\text{H}^+ + :\ddot{\text{O}}-\text{H}^-$$

compound that
produces hydroxide ion
BASE

(b)

$$\text{H}-\underset{\text{H}}{\overset{\text{H}}{\text{N}}}: + \text{H}-\ddot{\text{O}}-\text{H} \rightleftharpoons \text{H}-\underset{\text{H}}{\overset{\text{H}}{\text{N}}}-\text{H}^+ + :\ddot{\text{O}}-\text{H}^-$$

proton acceptor
BASE

proton donor
ACID

(c)

$$\text{H}-\underset{\text{H}}{\overset{\text{H}}{\text{N}}}: + \text{H}-\ddot{\text{O}}-\text{H} \rightleftharpoons \text{H}-\underset{\text{H}}{\overset{\text{H}^+}{\text{N}}}-\text{H} + :\ddot{\text{O}}-\text{H}^-$$

electron-pair
donor
BASE

electron-pair
acceptor
ACID

15.24 $Ba(NO_3)_2(s) \rightarrow Ba^{2+}(aq) + 2\ NO_3^-(aq)$

15.25 (a) KNO_3 has singly charged ions and will be more soluble than $BaSO_4$.

(b) $Al(C_2H_3O_2)_3$ has singly charged acetate ions and will be more soluble than Al_2O_3.

15.26 (a) KF will be less soluble than KI. Fluoride ion is very much smaller than iodide.

(b) ZnS will be less soluble than $Zn(NO_3)_2$. Sulfide ion has a 2− charge, compared with nitrate with a 1− charge.

15.27 $[NH_4^+] = 0.0816$ M

Answers to Odd-Numbered Problems

15.1 Iodine would be more soluble in water than bromine. Each I_2 molecule has 106 electrons and can form stronger London forces with water than each Br_2 molecule, which has only 70 electrons.

15.3 Oxygen would be more soluble in water than hydrogen. Each O_2 molecule has 16 electrons and can form stronger London forces with water than each H_2 molecule, which has only 2 electrons.

15.5 There are three interparticle forces between ethanol molecules and water molecules: London forces (which are always present); dipole–dipole interactions, as both water and ethanol have δ^+ and δ^- charges; and hydrogen bonding, as both water and ethanol have O—H bonds.

15.7 The three types of interparticle forces discussed in Problem 5 make very strong attraction between water molecules and ethanol molecules, so that ethanol and water are mutally soluble in all proportions.

15.9 Compound (a), C_2H_6, is a gas and is only slightly soluble in water, because the only attractive forces it can have are weak London forces. Compound (b), $C_2H_6O_2$, can form London forces, dipole–dipole interactions, and hydrogen bonds; these strong attractive forces make it a liquid and soluble in all proportions with water. (Compound (b) is ethylene glycol, used as antifreeze in automobiles.)

15.11

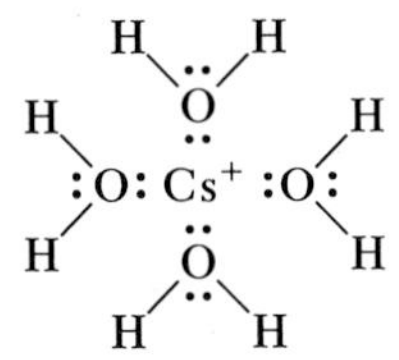

15.13 London forces are always present. Also, because both water and hydrogen chloride are polar, they will have dipole–dipole interactions.

15.15 [KBr] = 0.168 M

15.17 83.3 mL of 3.00 M HBr(aq) required

15.19 20.2 g $Ba(NO_3)_2$

15.21 Dissolve 4.39 g NaCl in enough water to make 100 mL of solution in the volumetric flask.

15.23 $[ClO_4^-] = 0.250$ M

15.25 25.0 mL of 6.00 M HI(aq) needed

15.27 Transfer 25.0 mL of 6.00 M HCl(aq) to a 100-mL volumetric flask. Add enough water to make 100 mL of solution.

15.29 (a) $H_2O: + H\text{—}O\text{—}H \rightleftharpoons H_2O\text{—}H^+ + :O\text{—}H^-$

(b) $2\ H_2O(\ell) \rightleftharpoons H_3O^+(aq) + OH^-(aq)$

15.31 (a) One O—H bond is broken. (b) One O—H bond is formed.

15.33

$H_2O\text{—}H^+ + :OH_2 \rightleftharpoons H_2O: + H\text{—}O(H)H^+$

15.35

$H\text{—}F: + H\text{—}F: \rightleftharpoons H\text{—}F\text{—}H^+ + :F:^-$

15.37 The double-arrow symbol means that a reaction is reversible, that is, that both the forward reaction (written left to right) and the reverse reaction (written right to left) can take place.

15.39 Hydrogen chloride is HCl molecules in a pure state; it happens to be a gas. Hydrochloric acid, written HCl(aq), is a solution of hydrogen chloride gas dissolved in water.

15.41 (a) hydronium ion and iodide ion (b) hydronium ion and perchlorate ion (c) potassium ion and hydroxide ion

15.43 (a) H_3PO_4 is a weak acid. (b) N_2H_4 is a weak base. (c) $Ba(OH)_2$ is a strong base. (d) HNO_3 is a strong acid.

15.45 $H_2O(\ell) + H_2PHO_3(aq) \rightleftarrows H_3O^+(aq) + HPHO_3^-(aq)$
$H_2O(\ell) + HPHO_3^-(aq) \rightleftarrows H_3O^+(aq) + PHO_3^{2-}(aq)$

15.47 $H_2O(\ell) + HBr(aq) \rightarrow H_3O^+(aq) + Br^-(aq)$

$$H_2\ddot{O}\!: + H\!-\!\ddot{Br}\!: \rightarrow H_2\ddot{O}\!-\!H^+ + :\ddot{Br}:^-$$

15.49 $HBr \rightarrow H^+ + Br^-$
$HI \rightarrow H^+ + I^-$
$HNO_3 \rightarrow H^+ + NO_3^-$
$H_2SO_4 \rightarrow H^+ + HSO_4^-$
$HClO_4 \rightarrow H^+ + ClO_4^-$

15.51 Expressions (b) and (c) describe acidic solutions.

15.53 Expression (b) describes an acidic solution.

15.55 $[OH^-] = 1.00 \times 10^{-9}$ M

15.57 $[OH^-] = 1.34 \times 10^{-3}$ M

15.59 Only (c) describes a solution that is weakly basic.

15.61 $[H_3O^+] = 1.0 \times 10^{-2}$ M

15.63 pH = 0.301

15.65 pH = 0.541

15.67 pH = 0.045

15.69 pH = 0.592

15.71 3.2×10^{-7} mol HCl

15.73 0.032 g HNO_3

15.75 0.040 mg HBr

15.77 pH = 10.439

15.79 (a) $[H_3O^+] = 0.213$ M (b) pH = 0.672
(c) $[OH^-] = 4.69 \times 10^{-14}$ M

15.81 In the autoionization of water:

$$\underset{1}{H_2\ddot{O}\!:} + \underset{2}{H\!-\!\ddot{O}\!-\!H} \rightleftarrows H_2\ddot{O}\!-\!H^+ + :\ddot{O}\!-\!H^-$$

water is both an Arrhenius acid and base. Water molecule **1** is the Arrhenius base because it forms hydroxide ion by taking H^+ from water molecule **2.** Water molecule **2** is the Arrhenius acid because it forms hydronium ion by giving H^+ to water molecule **1.**

15.83 HF is the Brønsted-Lowry acid as it donates H^+ to water. H_2O is the Brønsted-Lowry base as it accepts H^+ from HF.

15.85 HF is the Lewis acid as it accepts the pair of electrons that the oxygen in water shares. Water is the Lewis base as the oxygen of water shares an electron pair with a proton from HF.

15.87 The ions on an edge would be more easily surrounded by water molecules and therefore would leave the crystal more easily than ions on a crystal face.

15.89 $KC_2H_3O_2(s) \rightarrow K^+(aq) + C_2H_3O_2^-(aq)$

15.91 (a) $BaCl_2$ has a much larger cation and will be more soluble than $CaCl_2$.
(b) Gadolinium is to the right of calcium in the same period, so we can predict that Ga^{3+} is smaller than Ca^{2+}. Also, Ga^{3+} has a higher charge than Ca^{2+}. Both of these suggest that $CaCl_2$ would be more soluble than $GaCl_3$.
(c) KI has a much larger anion and will be more soluble than KCl.

15.93 Ammonium has a 1+ charge and is very large compared with monatomic cations. Both features make ammonium salts very soluble.

15.95 The larger the cation, the more water molecules can surround it, making it more soluble in water. Beryllium and magnesium are the smallest of the group 2A elements, and their hydroxides are not soluble enough to be considered strong bases.

Chapter 16—Reactions in Aqueous Solutions

Answers to Exercises

16.1 $2\ SO_2(g) + O_2(g) \rightarrow 2\ SO_3(g)$
$SO_3(g) \rightarrow SO_3(aq)$
$SO_3(aq) + H_2O(\ell) \rightarrow H_2SO_4(aq)$

16.2 $CaO(s) + H_2O(\ell) \rightarrow Ca^{2+}(aq) + 2\ OH^-(aq)$

16.3 $:\ddot{\underset{..}{O}}:^{2-} + H{-}\ddot{\underset{..}{O}}{-}H \rightarrow :\ddot{\underset{..}{O}}{-}H^- + :\ddot{\underset{..}{O}}{-}H^-$

oxide

16.4 (a) acid (b) salt (c) oxide (d) covalent compound

16.5 *ionic*
$2\ H_3O^+(aq) + 2\ Br^-(aq) + Ba^{2+}(aq) + 2\ OH^-(aq) \rightarrow$
$4\ H_2O(\ell) + Ba^{2+}(aq) + 2\ Br^-(aq)$
net ionic
$H_3O^+(aq) + OH^-(aq) \rightarrow 2\ H_2O(\ell)$

16.6 *ionic*
$HC_2H_3O_2(aq) + K^+(aq) + OH^-(aq) \rightarrow$
$H_2O(\ell) + K^+(aq) + C_2H_3O_2^-(aq)$
net ionic
$HC_2H_3O_2(aq) + OH^-(aq) \rightarrow H_2O(\ell) + C_2H_3O_2^-(aq)$

16.7 *ionic*
$H_3O^+(aq) + NO_3^-(aq) + C_5H_5N(aq) \rightarrow$
$H_2O(\ell) + NO_3^-(aq) + C_5H_5NH^+(aq)$
net ionic
$H_3O^+(aq) + C_5H_5N(aq) \rightarrow H_2O(\ell) + C_5H_5NH^+(aq)$

16.8 *ionic*
$HC_2H_3O_2(aq) + C_5H_5N(aq) \rightarrow C_5H_5NH^+(aq) + C_2H_3O_2^-(aq)$
net ionic
$HC_2H_3O_2(aq) + C_5H_5N(aq) \rightarrow C_5H_5NH^+(aq) + C_2H_3O_2^-(aq)$

16.9 *molecular*
$HI(aq) + NH_3(aq) \rightarrow NH_4I(aq)$
ionic
$H_3O^+(aq) + I^-(aq) + NH_3(aq) \rightarrow H_2O(\ell) + NH_4^+(aq) + I^-(aq)$
net ionic
$H_3O^+(aq) + NH_3(aq) \rightarrow H_2O(\ell) + NH_4^+(aq)$

16.10 *ionic*
$H_3O^+(aq) + NO_3^-(aq) + NH_3(aq) \rightarrow$
$H_2O(\ell) + NO_3^-(aq) + NH_4^+(aq)$
The nitrate ion, $NO_3^-(aq)$, is a spectator ion.

16.11 (a) Barium sulfide, BaS, will precipitate.
(b) *molecular*
$Ba(NO_3)_2(aq) + (NH_4)_2S(aq) \rightarrow BaS(s) + 2\ NH_4NO_3(aq)$
ionic
$Ba^{2+}(aq) + 2\ NO_3^-(aq) + 2\ NH_4^+(aq) + S^{2-}(aq) \rightarrow$
$BaS(s) + 2\ NH_4^+(aq) + 2\ NO_3^-(aq)$
net ionic
$Ba^{2+}(aq) + S^{2-}(aq) \rightarrow BaS(s)$

16.12 *ionic*
$6\ H_3O^+(aq) + 6\ Br^-(aq) + 2\ Fe(s) \rightarrow$
$3\ H_2(g) + 6\ H_2O(\ell) + 2\ Fe^{3+}(aq) + 6\ Br^-(aq)$
16.13 *molecular*
$2\ H_2O(\ell) + 2\ Li(s) \rightarrow H_2(g) + 2\ LiOH(aq)$
16.14 *molecular*
$CuSO_3(aq) + H_2SO_4(aq) \rightarrow SO_2(g) + H_2O(\ell) + CuSO_4(aq)$
ionic
$Cu^{2+}(aq) + SO_3^{2-}(aq) + 2\ H_3O^+(aq) + SO_4^{2-}(aq) \rightarrow$
$SO_2(g) + 3\ H_2O(\ell) + Cu^{2+}(aq) + SO_4^{2-}(aq)$
net ionic
$2\ H_3O^+(aq) + SO_3^{2-}(aq) \rightarrow SO_2(g) + 3\ H_2O(\ell)$
16.15 *molecular*
$(NH_4)_2S(aq) + 2\ HBr(aq) \rightarrow H_2S(g) + 2\ NH_4Br(aq)$
ionic
$2\ NH_4^+(aq) + S^{2-}(aq) + 2\ H_3O^+(aq) + 2\ Br^-(aq) \rightarrow$
$H_2S(g) + 2\ H_2O(\ell) + 2\ NH_4^+(aq) + 2\ Br^-(aq)$
net ionic
$2\ H_3O^+(aq) + S^{2-}(aq) \rightarrow H_2S(g) + 2\ H_2O(\ell)$
16.16 (a) no reaction
(b) *molecular*
$3\ Ba(s) + 2\ Al(C_2H_3O_2)_3(aq) \rightarrow 2\ Al(s) + 3\ Ba(C_2H_3O_2)_2(aq)$
ionic
$3\ Ba(s) + 2\ Al^{3+}(aq) + 6\ C_2H_3O_2^-(aq) \rightarrow$
$2\ Al(s) + 3\ Ba^{2+}(aq) + 6\ C_2H_3O_2^-(aq)$
net ionic
$3\ Ba(s) + 2\ Al^{3+}(aq) \rightarrow 2\ Al(s) + 3\ Ba^{2+}(aq)$
16.17 In Exercise 16.16(b), barium atoms are being oxidized and aluminum ions are being reduced.
16.18 11.3 mL of 1.25 M $HCl(aq)$
16.19 12.2 mL of 1.31 M $Ni(NO_3)_2(aq)$
16.20 0.700 M $Pb(NO_3)_2(aq)$

Answers to Odd-Numbered Problems

16.1 $N_2O_5(g)$ dissolves in water to form $N_2O_5(aq)$.
$N_2O_5(aq) + H_2O(\ell) \rightarrow 2\ HNO_3(aq)$
16.3 $P_2O_5(s)$ dissolves in water to form $P_2O_5(aq)$.
$P_2O_5(aq) + 3\ H_2O(\ell) \rightarrow 3\ H_3PO_4(aq)$
16.5 $HBr(aq) + LiOH(aq) \rightarrow H_2O(\ell) + LiBr(aq)$
16.7 $HI(aq) + KOH(aq) \rightarrow H_2O(\ell) + KI(aq)$
16.9 *ionic*
$H_3O^+(aq) + I^-(aq) + K^+(aq) + OH^-(aq) \rightarrow$
$2\ H_2O(\ell) + K^+(aq) + I^-(aq)$
net ionic
$H_3O^+(aq) + OH^-(aq) \rightarrow 2\ H_2O(\ell)$
16.11 $H_2SO_4(aq) + NaOH(aq) \rightarrow H_2O(\ell) + NaHSO_4(aq)$
16.13 $HF(aq) + OH^-(aq) \rightarrow H_2O(\ell) + F^-(aq)$
16.15 $HNO_2(aq) + NaOH(aq) \rightarrow H_2O(\ell) + NaNO_2(aq)$
sodium nitrite
16.17 *ionic*
$HNO_2(aq) + Na^+(aq) + OH^-(aq) \rightarrow H_2O(\ell) + Na^+(aq) + NO_2^-(aq)$
net ionic
$HNO_2(aq) + OH^-(aq) \rightarrow H_2O(\ell) + NO_2^-(aq)$

16.19 (a) $H_3PO_4(aq) + NaOH(aq) \rightarrow H_2O(\ell) + NaH_2PO_4(aq)$
sodium dihydrogen phosphate
(b) $H_3PO_4(aq) + 2\ NaOH(aq) \rightarrow 2\ H_2O(\ell) + Na_2HPO_4(aq)$
sodium hydrogen phosphate
(c) $H_3PO_4(aq) + 3\ NaOH(aq) \rightarrow 3\ H_2O(\ell) + Na_3PO_4(aq)$
sodium phosphate

16.21 *ionic*
$H_3PO_4(aq) + 2\ Na^+(aq) + 2\ OH^-(aq) \rightarrow$
$2\ H_2O(\ell) + 2\ Na^+(aq) + HPO_4^{2-}(aq)$
net ionic
$H_3PO_4(aq) + 2\ OH^-(aq) \rightarrow 2\ H_2O(\ell) + HPO_4^{2-}(aq)$
16.23 $HBr(aq) + NH_3(aq) \rightarrow NH_4Br(aq)$
16.25 $HI(aq) + NH_3(aq) \rightarrow NH_4I(aq)$
ammonium iodide

16.27 *ionic*
$H_3O^+(aq) + I^-(aq) + NH_3(aq) \rightarrow H_2O(\ell) + NH_4^+(aq) + I^-(aq)$
net ionic
$H_3O^+(aq) + NH_3(aq) \rightarrow H_2O(\ell) + NH_4^+(aq)$
16.29 $HCl(aq) + CH_3NH_2(aq) \rightarrow CH_3NH_3Cl(aq)$
16.31 $HC_2H_3O_2(aq) + NH_3(aq) \rightarrow NH_4C_2H_3O_2(aq)$
16.33 $HNO_2(aq) + NH_3(aq) \rightarrow NH_4NO_2(aq)$
ammonium nitrite

16.35 *ionic*
$HNO_2(aq) + NH_3(aq) \rightarrow NH_4^+(aq) + NO_2^-(aq)$
net ionic
$HNO_2(aq) + NH_3(aq) \rightarrow NH_4^+(aq) + NO_2^-(aq)$
16.37 (a) $Al(NO_3)_3(aq) + Na_3PO_4(aq) \rightarrow AlPO_4(s) + 3\ NaNO_3(aq)$
(b) $Fe(C_2H_3O_2)_2(aq) + (NH_4)_2S(aq) \rightarrow FeS(s) + 2\ NH_4C_2H_3O_2(aq)$
(c) no reaction
16.39 (a) *ionic*
$Al^{3+}(aq) + 3\ NO_3^-(aq) + 3\ Na^+(aq) + PO_4^{3-}(aq) \rightarrow$
$AlPO_4(s) + 3\ Na^+(aq) + 3\ NO_3^-(aq)$
net ionic
$Al^{3+}(aq) + PO_4^{3-}(aq) \rightarrow AlPO_4(s)$
(b) *ionic*
$Fe^{2+}(aq) + 2\ C_2H_3O_2^-(aq) + 2\ NH_4^+(aq) + S^{2-}(aq) \rightarrow$
$FeS(s) + 2\ NH_4^+(aq) + 2\ C_2H_3O_2^-(aq)$
net ionic
$Fe^{2+}(aq) + S^{2-}(aq) \rightarrow FeS(s)$
16.41 $Na_2S(aq) + Hg(C_2H_3O_2)_2(aq) \rightarrow HgS(s) + 2\ NaC_2H_3O_2(aq)$
16.43 (a) $2\ HNO_3(aq) + Ni(s) \rightarrow H_2(g) + Ni(NO_3)_2(aq)$
(b) $2\ HCl(aq) + Mg(s) \rightarrow H_2(g) + MgCl_2(aq)$
(c) $2\ H_2O(\ell) + Sr(s) \rightarrow H_2(g) + Sr(OH)_2(aq)$
16.45 (a) $2\ HCl(aq) + CaCO_3(s) \rightarrow CO_2(g) + H_2O(\ell) + CaCl_2(aq)$
(b) $2\ HBr(aq) + FeS(s) \rightarrow H_2S(g) + FeBr_2(aq)$
(c) $2\ HBr(aq) + CuSO_3(s) \rightarrow SO_2(g) + H_2O(\ell) + CuBr_2(aq)$
16.47 *ionic*
$2\ H_3O^+(aq) + 2\ Br^-(aq) + Cu^{2+}(aq) + SO_3^{2-}(aq) \rightarrow$
$SO_2(g) + 3\ H_2O(\ell) + Cu^{2+}(aq) + 2\ Br^-(aq)$
net ionic
$2\ H_3O^+(aq) + SO_3^{2-}(aq) \rightarrow SO_2(g) + 3\ H_2O(\ell)$

16.49 *molecular*
$SnSO_3(s) + 2\ HNO_3(aq) \rightarrow SO_2(g) + H_2O(\ell) + Sn(NO_3)_2(aq)$
ionic
$SnSO_3(s) + 2\ H_3O^+(aq) + 2\ NO_3^-(aq) \rightarrow$
$SO_2(g) + 3\ H_2O(\ell) + Sn^{2+}(aq) + 2\ NO_3^-(aq)$
net ionic
$SnSO_3(s) + 2\ H_3O^+(aq) \rightarrow SO_2(g) + 3\ H_2O(\ell) + Sn^{2+}(aq)$

16.51 (a) no reaction
(b) $Cd(s) + Ni(NO_3)_2(aq) \rightarrow Ni(s) + Cd(NO_3)_2(aq)$
(c) $3\ Mg(s) + 2\ Al(NO_3)_3(aq) \rightarrow 2\ Al(s) + 3\ Mg(NO_3)_2(aq)$

16.53 (b) *ionic*
$Cd(s) + Ni^{2+}(aq) + 2\ NO_3^-(aq) \rightarrow$
$Ni(s) + Cd^{2+}(aq) + 2\ NO_3^-(aq)$
net ionic
$Cd(s) + Ni^{2+}(aq) \rightarrow Ni(s) + Cd^{2+}(aq)$
(c) *ionic*
$3\ Mg(s) + 2\ Al^{3+}(aq) + 6\ NO_3^-(aq) \rightarrow$
$2\ Al(s) + 3\ Mg^{2+}(aq) + 6\ NO_3^-(aq)$
net ionic
$3\ Mg(s) + 2\ Al^{3+}(aq) \rightarrow 2\ Al(s) + 3\ Mg^{2+}(aq)$

16.55 (b) Cadmium atoms are being oxidized, and nickel ions are being reduced.
(c) Magnesium atoms are being oxidized, and aluminum ions are being reduced.

16.57 1.59 g Cu

16.59 2.40 mL of 6.00 M $Na_2S(aq)$

16.61 0.121 g $H_2(g)$

16.63 8.30 mL of 1.65 M $H_2SO_4(aq)$

16.65 2.50 M $NH_3(aq)$

Chapter 17—A Closer Look at Equilibrium Reactions and Redox Reactions

Answers to Exercises

17.1 (a) $H_2O(\ell) + HCHO_2(aq) \rightleftarrows H_3O^+(aq) + CHO_2^-(aq)$
(b) $Ag_2SO_4(s) \rightleftarrows 2\ Ag^+(aq) + SO_4^{2-}(aq)$

17.2 $H_2(g) + Cl_2(g) \rightleftarrows 2\ HCl(g)$

17.3 Increasing the ammonia will cause an initial increase in the rate of the reverse reaction.

17.4 An exothermic reaction generates heat as a product. Adding heat will shift the equilibrium in the reverse direction.

17.5 (a) There are no moles of gas on the left and one mole of gas on the right, so a pressure increase will shift the equilibrium to the left.
(b) There are four moles of gas on the left and two moles of gas on the right, so a pressure decrease will shift the equilibrium to the left.

17.6 (a) $H_2O(\ell) + \underset{\text{large}}{HCHO_2(aq)} \rightleftarrows \underset{\text{small}}{H_3O^+(aq)} + \underset{\text{large}}{CHO_2^-(aq)}$
(b) $H_3O^+(aq) + CHO_2^-(aq) \rightleftarrows H_2O(\ell) + HCHO_2(aq)$
(c) $OH^-(aq) + HCHO_2(aq) \rightarrow H_2O(\ell) + CHO_2^-(aq)$

17.7 (a) $\dfrac{[H_3O^+][ClO_2^-]}{[HClO_2]} = K_a$ (b) $\dfrac{[N_2H_5^+][OH^-]}{[N_2H_4]} = K_b$

(c) $[Ca^{2+}]^3[PO_4^{3-}]^2 = K_{sp}$ (d) $\dfrac{[CO_2]^2}{[CO]^2[O_2]} = K$

17.8 $K_{sp} = 2.3 \times 10^{-5}$
17.9 $K = 49.1$
17.10 $[H_3O^+] = 0.012$ M; $[CHO_2^-] = 0.012$ M; $[HCHO_2] = 0.79$ M
17.11 $[NH_4^+] = 2.2 \times 10^{-3}$ M; $[OH^-] = 2.2 \times 10^{-3}$ M; $[NH_3] = 0.26$ M
17.12 $[PCl_3] = 0.013$ M; $[Cl_2] = 0.013$ M; $[PCl_5] = 0.56$ M
17.13 $[Zn^{2+}] = 3.7 \times 10^{-6}$ M; $[OH^-] = 7.4 \times 10^{-6}$ M
17.14 0.12 mg $Fe(OH)_2$
17.15 Magnesium is the reducing agent and is being oxidized. Oxygen is the oxidizing agent and is being reduced.

	0	0	3+ 2−	1+ 5+ 2−	1+ 3+ 2−
17.16	(a) Fe	(b) O_2	(c) Fe_2O_3	(d) H Cl O_3	(e) H Cl O_2
	0	0	6+ 6−	1+ 5+ 6−	1+ 3+ 4−

17.17 (a)

3+ 2− 0 0 3+ 2−
$Fe_2O_3(s) + 2\ Al(s) \rightarrow 2\ Fe(s) + Al_2O_3(s)$
6+ 6− 0 0 6+ 6−

Fe_2O_3 is reduced and is the oxidizing agent. Al is oxidized and is the reducing agent.

(b)

1+ 2− 1+ 2− 1+ 2− 1+
$Na_2O(s) + H_2O(\ell) \rightarrow 2\ NaO\ H(s)$
2+ 2− 2+ 2− 1+ 2− 1+

The oxidation number for each atom is the same in the reactants as it is in the products, so this is not a redox reaction.

(c)

0 0 1+1−
$H_2(g) + I_2(g) \rightarrow 2\ H\,I(g)$
0 0 1+1−

I_2 is reduced and is the oxidizing agent. H_2 is oxidized and is the reducing agent.

	7+ 2−	6+ 2−
17.18	MnO_4^-	$Cr_2O_7^{2-}$
	7+ 8−	12+ 14−

17.19 $4\ Zn + NO_3^- + 10\ H^+ \rightarrow 4\ Zn^{2+} + NH_4^+ + 3\ H_2O$
17.20 $2\ MnO_4^- + I^- + H_2O \rightarrow 2\ MnO_2 + IO_3^- + 2\ OH^-$

Answers to Odd-Numbered Problems

17.1 (a) weak-base equilibrium (b) solubility equilibrium
(c) weak-acid equilibrium
17.3 (a) solubility equilibrium (b) solubility equilibrium
(c) weak-acid equilibrium
17.5 (a) $BaS(s) \rightleftarrows Ba^{2+}(aq) + S^{2-}(aq)$
(b) $H_2O(\ell) + HCN(aq) \rightleftarrows H_3O^+(aq) + CN^-(aq)$
(c) $(CH_3)_2NH(aq) + H_2O(\ell) \rightleftarrows (CH_3)_2NH_2^+(aq) + OH^-(aq)$
17.7 (a) $CaS(s) \rightleftarrows Ca^{2+}(aq) + S^{2-}(aq)$
(b) $Mg(OH)_2(s) \rightleftarrows Mg^{2+}(aq) + 2\ OH^-(aq)$
(c) $Cu_2S(s) \rightleftarrows 2\ Cu^+(aq) + S^{2-}(aq)$
17.9 $N_2O_4(g) \rightleftarrows 2\ NO_2(g)$
17.11 $2\ SO_2(g) + O_2(g) \rightleftarrows 2\ SO_3(g)$
17.13 Six moles of hydrogen reacted with two moles of nitrogen to form four moles of ammonia.

17.15 The rate of the reverse reaction will decrease, producing less NH_3, which will lead to a decrease in the rate of the forward reaction. Eventually, the rates will be identical.

17.17 The rate of the reverse reaction will increase, producing more NH_3 and H_2O, leading to an increase in the rate of the forward reaction. Eventually, the rates will be identical.

17.19 increase

17.21 decrease

17.23 decrease

17.25 increase

17.27 (a) $H_2O(\ell) + \underset{\text{large}}{HNO_2(aq)} \rightleftarrows \underset{\text{small}}{H_3O^+(aq)} + \underset{\text{large}}{NO_2^-(aq)}$

(b) $H_3O^+(aq) + NO_2^-(aq) \rightleftarrows H_2O(\ell) + HNO_2(aq)$

(c) $OH^-(aq) + HNO_2(aq) \rightarrow H_2O(\ell) + NO_2^-(aq)$

17.29 (a) $H_3O^+(aq) + HPO_4^{2-}(aq) \rightleftarrows H_2O(\ell) + H_2PO_4^-(aq)$

(b) $OH^-(aq) + HPO_4^{2-}(aq) \rightarrow H_2O(\ell) + PO_4^{3-}(aq)$

17.31 (a) $\dfrac{[OF_2]^2}{[O_2][F_2]^2} = K$ (b) $[Pb^{2+}][I^-]^2 = K_{sp}$ (c) $\dfrac{[H_3O^+][NO_2^-]}{[HNO_2]} = K_a$

17.33 (a) $[Ba^{2+}]^3[PO_4^{3-}]^2 = K_{sp}$ (b) $\dfrac{[PCl_5]}{[PCl_3][Cl_2]} = K$ (c) $\dfrac{[PCl_3][Cl_2]}{[PCl_5]} = K$

17.35 (a) $[Ba^{2+}][CO_3^{2-}] = K_{sp}$ (b) $\dfrac{[H_3O^+][CO_3^{2-}]}{[HCO_3^-]} = K_a$

17.37 $K_{sp} = 6 \times 10^{-27}$

17.39 $K_{sp} = 1.4 \times 10^{-24}$

17.41 $K_{sp} = 1.5 \times 10^{-11}$

17.43 $[H_3O^+] = 3.1 \times 10^{-3}$ M; $[C_7H_5O_2^-] = 3.1 \times 10^{-3}$ M; $[HC_7H_5O_2] = 0.15$ M

17.45 $[H_3O^+] = 1.0 \times 10^{-3}$ M; $[C_3H_5O_2^-] = 1.0 \times 10^{-3}$ M; $[HC_3H_5O_2] = 0.081$ M

17.47 pH = 1.60

17.49 pH = 11.11

17.51 $[S^{2-}] = 2 \times 10^{-9}$ M

17.53 5.3×10^{-3} mg $Cu(OH)_2$

17.55 $[H_2] = 9.5 \times 10^{-50}$ M; $[F_2] = 9.5 \times 10^{-50}$ M; [HF] = 0.030 M

17.57 [NO] = 0.15 M; $[N_2] = 2.9$ M; $[O_2] = 2.9$ M

17.59 (a) $\overset{0}{\underset{0}{Mg}}$ (b) $\overset{2+}{\underset{2+}{Mg^{2+}}}$ (c) $\overset{0}{\underset{0}{O_2}}$

17.61 (a) $\overset{3-}{\underset{3-}{N}}\,\overset{1+}{\underset{3+}{H_3}}$ (b) $\overset{4+}{\underset{4+}{N}}\,\overset{2-}{\underset{4-}{O_2}}$ (c) $\overset{3-}{\underset{3-}{N}}\,\overset{1+}{\underset{4+}{H_4^+}}$

17.63 (a) $\overset{0}{\underset{0}{Li}}$ (b) $\overset{0}{\underset{0}{N_2}}$ (c) $\overset{1+}{\underset{3+}{Li_3}}\overset{3-}{\underset{3-}{N}}$

17.65 (a) $\overset{4+}{\underset{4+}{Mn}}\overset{2-}{\underset{4-}{O_2}}$ (b) $\overset{2+}{\underset{2+}{Mn}}(\overset{5+}{\underset{10+}{N}}\,\overset{2-}{\underset{12-}{O_3)_2}}$ (c) $\overset{2+}{\underset{2+}{Ca}}(\overset{7+}{\underset{14+}{Mn}}\overset{2-}{\underset{16-}{O_4)_2}}$

17.67 (a) $\overset{0}{\underset{0}{Ca}}(s) + \overset{0}{\underset{0}{Cl_2}}(g) \rightarrow \overset{2+}{\underset{2+}{Ca}}\overset{1-}{\underset{2-}{Cl_2}}(s)$

Cl_2 is reduced and is the oxidizing agent. Ca is oxidized and is the reducing agent.

0 0 2+ 2−

(b) $2\ Ca(s) + O_2(g) \rightarrow 2\ CaO(s)$

0 0 2+ 2−

O_2 is reduced and is the oxidizing agent. Ca is oxidized and is the reducing agent.

2+ 4+ 2− 2+ 2− 4+ 2−

(c) $CaC\ O_3(s) \rightarrow CaO(s) + C\ O_2(g)$

2+ 4+ 6− 2+ 2− 4+ 4−

The oxidation number for each atom is the same in the reactants as it is in the products, so this is not a redox reaction.

17.69 None of these is a redox reaction.

2+ 2− 1+1− 2+ 1− 1+ 2−

(a) $MgO(s) + 2\ H\,Cl(aq) \rightarrow MgCl_2(aq) + H_2O(\ell)$

2+ 2− 1+1− 2+ 2− 2+ 2−

5+ 2− 1+ 2− 1+ 5+ 2−

(b) $P_2\ O_5(s) + 3\ H_2O(\ell) \rightarrow 2\ H_3P\ O_4(aq)$

10+10− 2+ 2− 3+ 5+ 8−

3− 1+ 2− 2+ 5+ 2− 3− 1+ 5+ 2− 2+ 2−

(c) $(N\ H_4)_2S(aq) + Fe(N\ O_3)_2(aq) \rightarrow 2\ N\ H_4N\ O_3(aq) + FeS(s)$

6− 8+ 2− 2+ 10+12− 3− 4+ 5+ 6− 2+ 2−

17.71 (a) $Cr_2O_7^{2-} + 3\ C_2O_4^{2-} + 14\ H^+ \rightarrow 2\ Cr^{3+} + 6\ CO_2 + 7\ H_2O$
(b) $MnO_2 + HNO_2 + H^+ \rightarrow Mn^{2+} + NO_3^- + H_2O$
(c) $2\ Mn^{2+} + 5\ BiO_3^- + 14\ H^+ \rightarrow 2\ MnO_4^- + 5\ Bi^{3+} + 7\ H_2O$

17.73 (a) $Mn^{2+} + H_2O_2 + 2\ OH^- \rightarrow MnO_2 + 2\ H_2O$
(b) $Mn^{2+} + 2\ ClO_3^- \rightarrow MnO_2 + 2\ ClO_2$
(c) $6\ Cl_2 + 12\ OH^- \rightarrow 10\ Cl^- + 2\ ClO_3^- + 6\ H_2O$

17.75 (a) $Cr_2O_7^{2-} + 6\ Fe^{2+} + 14\ H^+ \rightarrow 2\ Cr^{3+} + 6\ Fe^{3+} + 7\ H_2O$
(b) $Pb(OH)_4^{2-} + ClO^- \rightarrow PbO_2 + Cl^- + H_2O + 2\ OH^-$
(c) $2\ NO_3^- + 3\ Cu + 8\ H^+ \rightarrow 2\ NO + 3\ Cu^{2+} + 4\ H_2O$

17.77 (a) $S^{2-} + 4\ I_2 + 8\ OH^- \rightarrow SO_4^{2-} + 8\ I^- + 4\ H_2O$
(b) $2\ MnO_4^- + 5\ SO_2 + 2\ H_2O \rightarrow 2\ Mn^{2+} + 5\ SO_4^{2-} + 4\ H^+$
(c) $3\ CN^- + 2\ MnO_4^- + H_2O \rightarrow 3\ CNO^- + 2\ MnO_2 + 2\ OH^-$

17.79 (a) $Cr_2O_7^{2-} + 6\ Cl^- + 14\ H^+ \rightarrow 2\ Cr^{3+} + 3\ Cl_2 + 7\ H_2O$
(b) $3\ S_8 + 32\ NO_3^- + 32\ H^+ \rightarrow 24\ SO_2 + 32\ NO + 16\ H_2O$
(c) $2\ Cr(OH)_4^- + 3\ ClO^- + 2\ OH^- \rightarrow 2\ CrO_4^{2-} + 3\ Cl^- + 5\ H_2O$

17.81

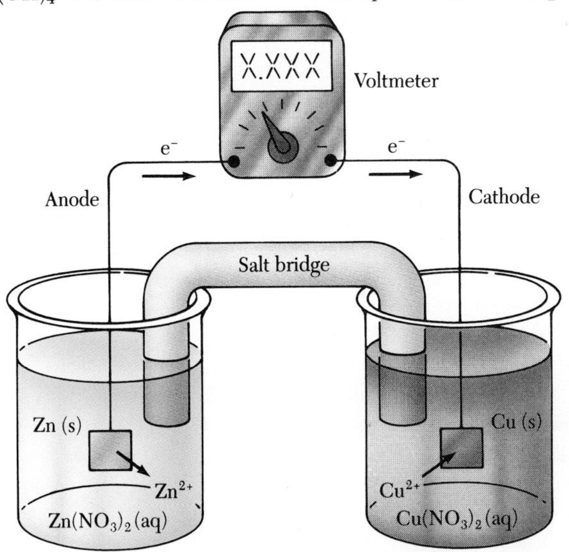

17.83

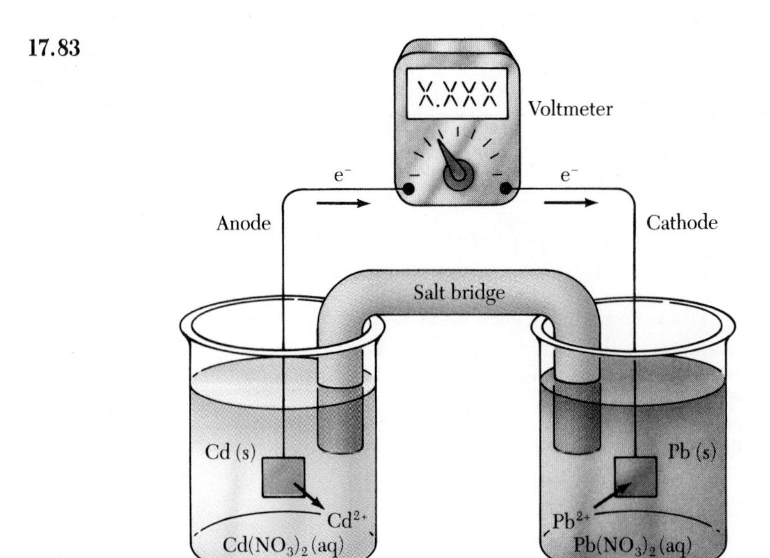

17.85 $Ca^{2+}(\ell) + 2\ e^- \rightarrow Ca(\ell)$
$2\ Cl^-(\ell) \rightarrow Cl_2(g) + 2\ e^-$

Chapter 18—Nuclear Chemistry

Answers to Exercises

18.1 ${}^{226}_{88}Ra \rightarrow {}^{222}_{86}Ru + {}^{4}_{2}He$
18.2 ${}^{32}_{15}P \rightarrow {}^{32}_{16}S + {}^{0}_{-1}e$
18.3 ${}^{25}_{13}Al + {}^{0}_{-1}e \rightarrow {}^{25}_{12}Mg$
18.4 ${}^{25}_{13}Al \rightarrow {}^{25}_{12}Mg + {}^{0}_{1}e$
18.5 (a) ${}^{230}_{90}Th$ will be unstable because it has more than 83 protons.
(b) ${}^{90}_{40}Zr$ will be stable because it has even numbers of protons and of neutrons, and it has a *magic number* of neutrons (50).
18.6 (a) ${}^{40}_{19}K$ is in the positron-emission or electron-capture group of nuclei.
(b) ${}^{165}_{70}Yb$ is in the α-emission group of nuclei.
18.7 (a) ${}^{40}_{20}Ca$ is a stable nucleus. (b) ${}^{94}_{36}Kr$ is not stable; it is a β-emitter.
18.8 ${}^{40}_{19}K$ is to the left of the band of stability. ${}^{40}_{18}Ar$ is closer to the band of stability.
18.9 (a) ${}^{185}_{76}Os \rightarrow {}^{181}_{74}W + {}^{4}_{2}He$ (b) ${}^{40}_{20}Ca$ is a stable nucleus.
(c) ${}^{95}_{37}Rb \rightarrow {}^{95}_{38}Sr + {}^{0}_{-1}e$
18.10 (a) ${}^{186}_{75}Re \rightarrow {}^{186}_{76}Os + {}^{0}_{-1}e$ The product nucleus is stable.
(b) ${}^{46}_{22}Ti$ is a stable nucleus.
18.11 The transuranium elements are metals and transition elements.
18.12 $[Rn]^{86}7s^2 5f^{14} 6d^7$
18.13 All of the transuranium elements have more than 83 protons, so none of the isotopes of the transuranium elements would be stable.
18.14 (a) ${}^{238}_{94}Pu \rightarrow {}^{234}_{92}U + {}^{4}_{2}He$ (b) ${}^{239}_{94}Pu + {}^{4}_{2}He \rightarrow {}^{242}_{96}Cm + {}^{1}_{0}n$
18.15 1.73×10^{-5} Ci
18.16 ${}^{241}_{95}Am \rightarrow {}^{237}_{93}Np + {}^{4}_{2}He$
18.17 3.7×10^4 disintegrations per second
18.18 (a) 50% of the H_2SO_4 molecules will be radioactive
(b) one radioactive oxygen per molecule of H_2SO_4
18.19 ${}^{131}_{53}I \rightarrow {}^{131}_{54}Xe + {}^{0}_{-1}e$
18.20 (a) ${}^{198}_{79}Au \rightarrow {}^{198}_{80}Hg + {}^{0}_{-1}e$

(b)

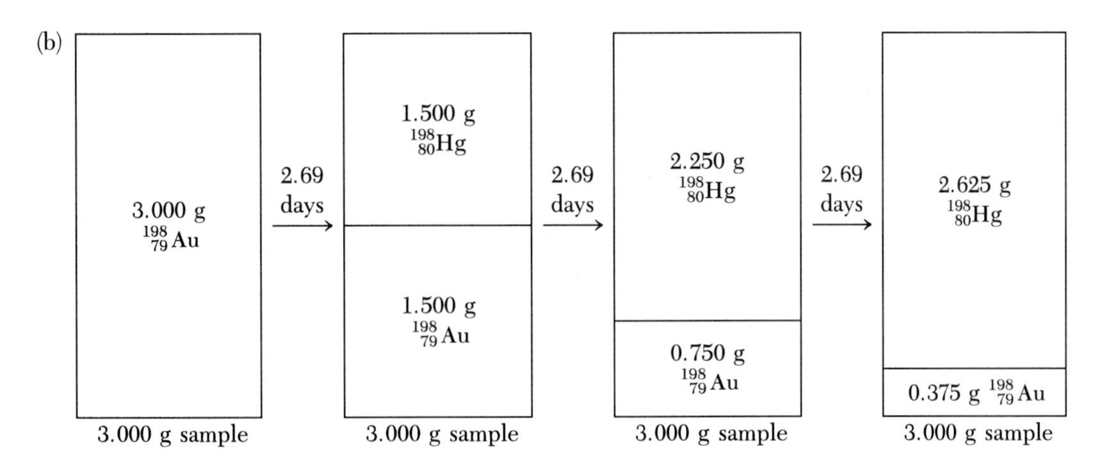

18.21 42.5 days
18.22 5.25×10^3 years
18.23 ${}^{235}_{92}U \rightarrow {}^{139}_{56}Ba + {}^{94}_{36}Kr + 2\ {}^{1}_{0}n$
18.24 2.46×10^{10} kJ

Answers to Odd-Numbered Problems

18.1 A *nuclear reaction* involves a change in the nucleus of an atom.
18.3 The "2" is the number of protons, the atomic number of the element. The "4" is the mass number, the sum of the protons and neutrons.
18.5 *Radioactivity* is the emission of particles or energy by nuclei in a nuclear reaction.
18.7 (a) 1 (b) 4 (c) 0
18.9 ${}^{288}_{91}Pa + {}^{0}_{-1}e \rightarrow {}^{288}_{90}Th$
18.11 ${}^{76}_{33}As \rightarrow {}^{76}_{34}Se + {}^{0}_{-1}e$
18.13 ${}^{238}_{94}Pu \rightarrow {}^{234}_{92}U + {}^{4}_{2}He$
18.15 ${}^{173}_{73}Ta \rightarrow {}^{173}_{72}Hf + {}^{0}_{1}e$
18.17 ${}^{235}_{92}U \rightarrow {}^{231}_{90}Th + {}^{4}_{2}He$
18.19 ${}^{27}_{13}Al + {}^{1}_{0}n \rightarrow {}^{24}_{11}Na + {}^{4}_{2}He$
18.21 ${}^{1}_{0}n \rightarrow {}^{1}_{1}p + {}^{0}_{-1}e$
18.23 ${}^{1}_{1}p \rightarrow {}^{1}_{0}n + {}^{0}_{1}e$
18.25 (a) ${}^{127}_{55}Cs + {}^{0}_{-1}e \rightarrow {}^{127}_{54}Xe$ *or* ${}^{127}_{55}Cs \rightarrow {}^{127}_{54}Xe + {}^{0}_{1}e$
(b) Both processes make the same isotope of xenon.
18.27 A captured electron comes from the lowest energy level in the atom, closest to the nucleus.
18.29 (a) unstable—more than 83 protons
(b) unstable—odd numbers of protons and of neutrons
(c) stable—even numbers of protons and of neutrons, including the magic number 2
18.31 ${}^{36}_{17}Cl$ has odd numbers of both protons and neutrons and is more likely to be unstable than ${}^{37}_{17}Cl$, which has a magic number of neutrons.
18.33 (a) unstable (b) unstable (c) stable
18.35 (a) stable (b) stable (c) unstable
18.37 (a) ${}^{262}_{107}Uns \rightarrow {}^{258}_{105}Unp + {}^{4}_{2}He$ (b) ${}^{184}_{77}Ir \rightarrow {}^{180}_{75}Re + {}^{4}_{2}He$
18.39 ${}^{36}_{17}Cl \rightarrow {}^{36}_{18}Ar + {}^{0}_{-1}e$
18.41 (a) ${}^{13}_{5}B \rightarrow {}^{13}_{6}C + {}^{0}_{-1}e$ (b) ${}^{183}_{78}Pt \rightarrow {}^{179}_{76}Os + {}^{4}_{2}He$
18.43 (c) ${}^{43}_{22}Ti \rightarrow {}^{43}_{21}Sc + {}^{0}_{1}e$ *or* ${}^{43}_{22}Ti + {}^{0}_{-1}e \rightarrow {}^{43}_{21}Sc$
18.45 (a) ${}^{81}_{40}Zr \rightarrow {}^{77}_{38}Sr + {}^{4}_{2}He$
${}^{77}_{38}Sr \rightarrow {}^{73}_{36}Kr + {}^{4}_{2}He$
(b) ${}^{173}_{73}Ta \rightarrow {}^{169}_{71}Lu + {}^{4}_{2}He$
${}^{169}_{71}Lu \rightarrow {}^{165}_{69}Tm + {}^{4}_{2}He$
(c) ${}^{70}_{30}Zn \rightarrow {}^{70}_{31}Ga + {}^{0}_{-1}e$
${}^{70}_{31}Ga \rightarrow {}^{70}_{32}Ge + {}^{0}_{-1}e$

18.47 (a) stable (b) ${}^{85}_{35}Br \rightarrow {}^{85}_{36}Kr + {}^{0}_{-1}e$ (c) stable
${}^{85}_{36}Kr \rightarrow {}^{85}_{37}Rb + {}^{0}_{-1}e$

18.49 A *transuranium element* is any element with atomic number greater than 92.

18.51 ununnilium, Uun

18.53 All of the transuranium elements have nuclei with more than 83 protons and will be unstable, regardless of the number of neutrons.

18.55 Radiation emitted from one atom can collide with a second atom, moving electrons in the second atom from a lower to a higher energy level. When the excited electrons return to lower energy levels light is emitted. We call these emissions of light *scintillations*.

18.57 Argon is a noble gas and is therefore unreactive and chemically stable. Argon is also denser than air and does not float away as helium or neon would.

18.59 2.3×10^6 disintegrations per second

18.61 6.9×10^9 disintegrations in one minute

18.63 50% of the ammonia molecules will contain a radioactive nitrogen atom.

18.65 87.5% of the ammonia molecules will contain at least one radioactive hydrogen atom.

18.67 Alpha particles are stopped by human tissue, so alpha emitters on the inside of a body could not be detected from the outside. Also, alpha particles can cause serious damage to tissue, so ingestion or inhalation of an alpha emitter would be hazardous to the person.

18.69 Neutrons are more damaging to tissue than beta particles (electrons), so at the same rate of emission, a sample emitting neutrons would be more dangerous than a sample emitting beta particles.

18.71 ${}^{203}_{80}Hg \rightarrow {}^{203}_{81}Tl + {}^{0}_{-1}e$

18.73 ${}^{51}_{24}Cr + {}^{0}_{-1}e \rightarrow {}^{51}_{23}V$

18.75 (a) ${}^{43}_{19}K \rightarrow {}^{43}_{20}Ca + {}^{0}_{-1}e$
(b)

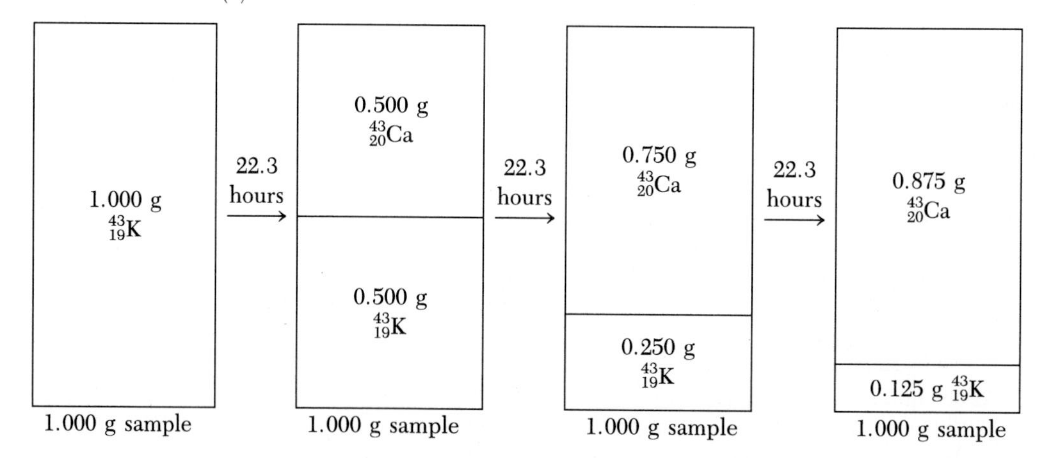

18.77

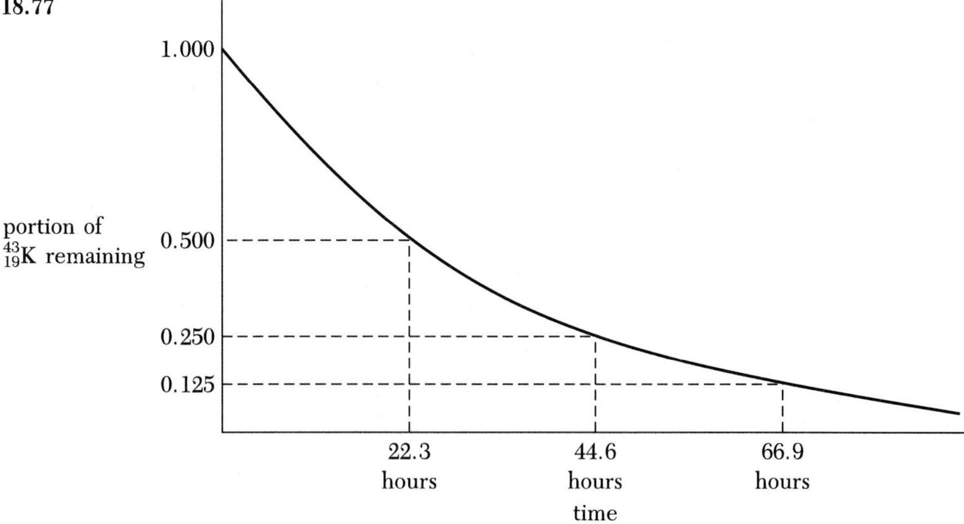

18.79 39.3 years
18.81 1.51 hour half-life
18.83 2.9×10^{-3} Ci
18.85 0.124 Ci
18.87 1.70×10^{3} years
18.89 2.00×10^{3} years
18.91 Fission is the splitting of large nuclei into smaller ones. Fusion is the combining of smaller nuclei into larger ones.
18.93 Both fission and fusion generate tremendous amounts of energy from small amounts of material.
18.95 2.48×10^{7} kJ
18.97 1.74×10^{-16} kJ

Chapter 19–Organic Chemistry

Answers to Exercises

19.1

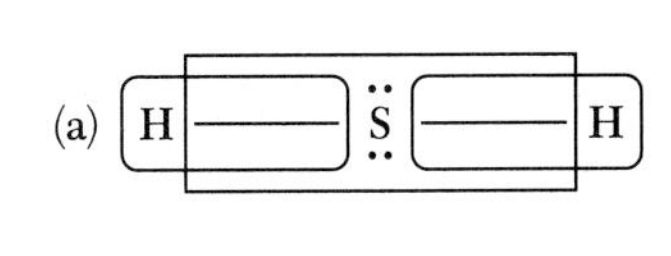

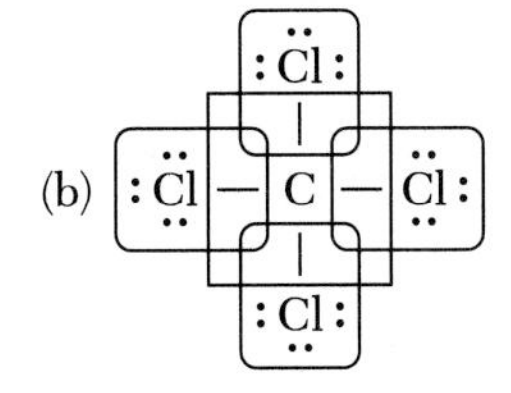

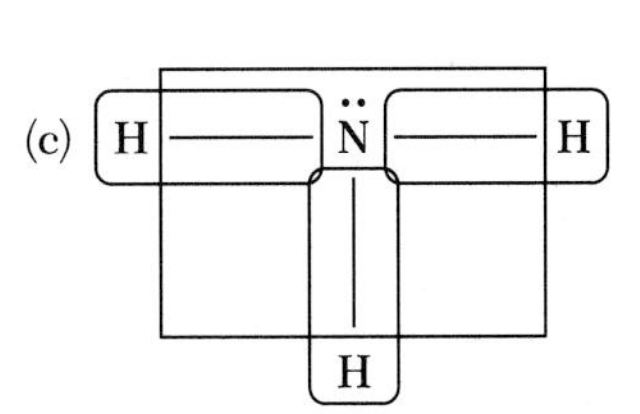

19.2

(a) $\overset{\delta+}{\text{H—C}}\equiv\overset{\delta-}{\text{N}}$: The bond is triple and polar; N has a greater electronegativity than C.

(b) H—$\overset{\delta+}{\text{C}}$—$\overset{\delta-}{\text{F}}$: (with H above and H below C) The bond is single and polar; F has a greater electronegativity than C.

(c)
```
   H   δ−
   | δ+ ··
H—C—N—H
   |  |
   H  H
```
The bond is single and polar; N has a greater electronegativity than C.

19.3

(a)
```
   H  H  H
   |  |  |
H—C—C—C—H
   |  |  |
   H  H  H
```
$CH_3—CH_2—CH_3$

(b)
```
   H  H  H  H
   |  |  |  |
H—C=C—C—C—H
         |  |
         H  H
```
$CH_2=CH—CH_2—CH_3$

(c)
```
   H  H        H
   |  |        |
H—C—C—C≡C—C—H
   |  |        |
   H  H        H
```
$CH_3—CH_2—C≡C—CH_3$

19.4

(a)
```
        CH3
        |
CH3—CH—CH2—CH3
```

(b)
```
        CH3—CH2
              |
CH3—CH2—CH—CH2—CH2—CH3
```

(c)
```
                     CH3—CH2
                           |
CH3—CH2—CH=CH—CH—CH2—CH2—CH2—CH3
```

19.5

(a)
```
    CH2
   /   \
CH2—CH2
```

(b)
```
       CH2
     /     \
 CH2        CH
  |         ||
 CH2        CH
     \     /
       CH2
```

19.6

$CH_3—CH_2—CH_2—CH_2—CH_3$

```
      CH3
      |
CH3—CH—CH2—CH3
```

```
      CH3
      |
CH3—C—CH3
      |
      CH3
```

19.7 (a) 3-methylpentane (b) 2-methylbutane

19.8

(a)
```
      CH3—CH2
            |
CH3—CH2—CH—CH2—CH2—CH3
```

(b)
```
      CH3
      |
CH3—CH—CH2—CH2—CH3
```

19.9 (a) 3,3-diethylpentane

(b)
```
      CH3                 CH3
      |                   |
CH3—CH—CH2—CH2—CH2—CH—CH3
```

19.10 (a) 1-butene (b) 1-butyne

19.11 4-methyl-2-pentyne

19.12
```
    CH2
   /   \
CH2—CH2
```

19.13

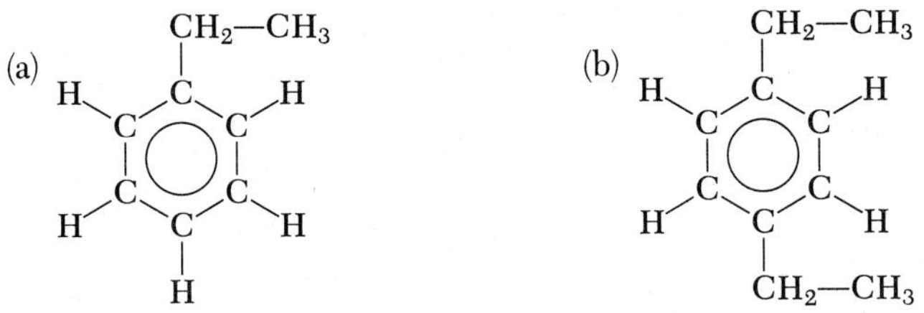

19.14 (a) Octane (8 carbons) is a liquid because its size is between pentane (5 carbons) and decane (10 carbons), which are both liquids.
(b) Ethane (2 carbons) is a gas because it is smaller than butane (4 carbons), which is also a gas.

19.15 Ethylbenzene (8 carbons) should have a higher boiling point than methylbenzene (7 carbons).

19.16 Naphthalene is more likely to be a solid since it has many more carbons and hydrogens, and therefore stronger London forces, than propane does. (Propane is more likely to be a gas since butane is a gas. See Exercise 19.14.)

19.17 $CH_4(g) + 2\ O_2(g) \rightarrow CO_2(g) + 2\ H_2O(\ell) + \text{heat}$

19.18

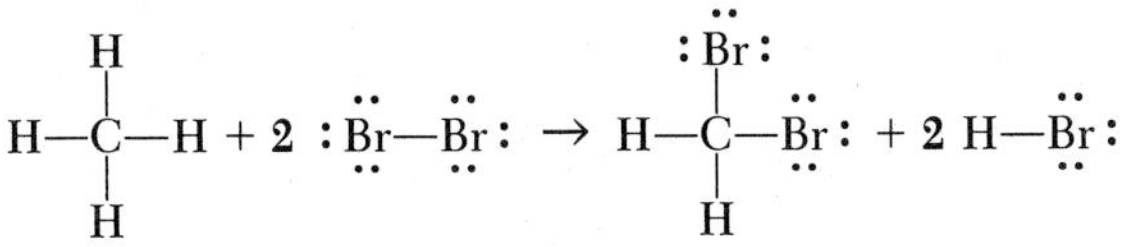

19.19

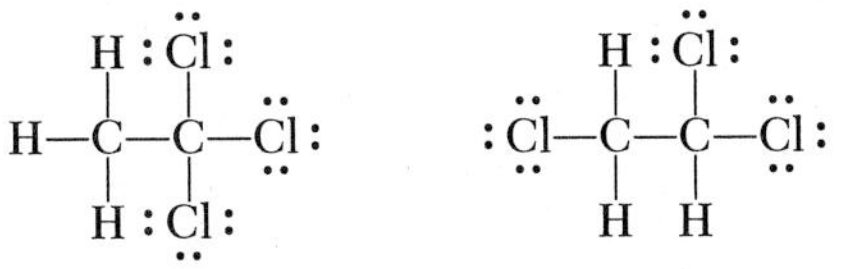

19.20

(a) $CH_3—CH_3$ (b) 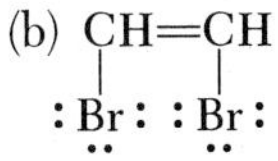 (c) 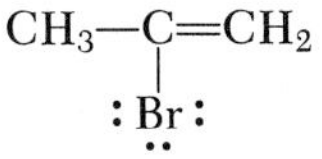

19.21 (a) ketone (b) aldehyde (c) carboxylic acid

19.22 (a) methanol (b) methanamine (c) propanoic acid

19.23 (a) 1-propanol (b) 2-methyl-2-propanol

19.24

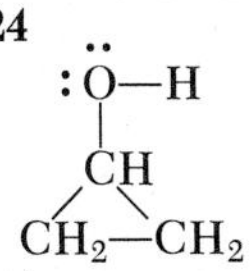

19.25

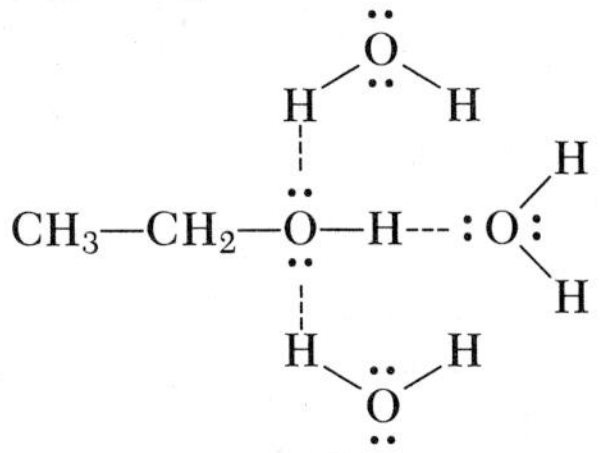

19.26 $Ca\!: + 2\ CH_3—\ddot{\underset{..}{O}}—H \rightarrow Ca^{2+} + 2\ CH_3—\ddot{\underset{..}{O}}\!:^{-} + H—H$

or $Ca + 2\ CH_3OH \rightarrow Ca(OCH_3)_2 + H_2$

19.27

$$\begin{array}{ccccccccccccccccccccccc}
 & & H & & H & & H & & & & & & & & & & H & & H & & H & & H \\
 & & | & & | & & | & & & & & & & & & & | & & | & & | & & | \\
H & - & C & - & C & - & C & - & \ddot{O} & - & H & + & H & - & \ddot{Cl}: & \rightarrow & H{-}C & - & C & - & C & - & \dot{O}^{+}{-}H + :\ddot{Cl}:^{-} \\
 & & | & & | & & | & & & & & & & & & & | & & | & & | & & \\
 & & H & & H & & H & & & & & & & & & & H & & H & & H & &
\end{array}$$

$$\begin{array}{ccccccccccccccccc}
 & & H & & H & & H & & H & & & & & H & & H & \quad H \\
 & & | & & | & & | & & | & & & & & | & & | & \ / \\
H & - & C & - & C & - & C & - & \ddot{O}^{+} & - & H & \rightarrow & H^{+} + H & - & C & - & C{=}C + :\ddot{O}: \\
 & & | & & | & & | & & & & & & & | & & | \quad | & \ \backslash \\
 & & H & & H & & H & & & & & & & H & & H \quad H & \quad H
\end{array}$$

19.28 2-propanamine

19.29

$$CH_3{-}CH_2{-}\underset{\underset{H}{|}}{\ddot{N}}{-}H\ (aq) + \underset{H\quad H}{\overset{\ddot{}}{\underset{\ddot{}}{O}}}\ (\ell) \rightleftarrows CH_3{-}CH_2{-}\overset{\overset{H}{|}}{\underset{\underset{H}{|}}{N}}{-}\overset{+}{H}\ (aq) + :\ddot{\underset{\cdot\cdot}{O}}{-}H^{-}\ (aq)$$

19.30

$$CH_3{-}\underset{\underset{H}{|}}{\ddot{N}}{-}H\ (aq) + H{-}\ddot{\underset{\cdot\cdot}{O}}{-}\overset{\overset{:O:}{\|}}{N}{-}\ddot{\underset{\cdot\cdot}{O}}:\ (aq) \rightarrow CH_3{-}\overset{\overset{H}{|}}{\underset{\underset{H}{|}}{N}}{-}\overset{+}{H}\ (aq) + :\ddot{\underset{\cdot\cdot}{O}}{-}\overset{\overset{:O:}{\|}}{N}{-}\ddot{\underset{\cdot\cdot}{O}}:^{-}\ (aq)$$

19.31 butanoic acid

19.32

$$H{-}\overset{\overset{:O:}{\|}}{C}{-}\ddot{\underset{\cdot\cdot}{O}}{-}H\ (aq) + :\overset{\ /H}{\underset{\ \backslash H}{O}}:\ (\ell) \rightleftarrows H{-}\overset{\overset{:O:}{\|}}{C}{-}\ddot{\underset{\cdot\cdot}{O}}:^{-}\ (aq) + H{-}\overset{\ /H}{\underset{\ \backslash H}{O}}:^{+}\ (aq)$$

19.33

$$2\ CH_3{-}\overset{\overset{:O:}{\|}}{C}{-}\ddot{\underset{\cdot\cdot}{O}}{-}H\ (aq) + Ca^{2+}\ (aq) + 2\ :\ddot{\underset{\cdot\cdot}{O}}{-}H^{-}\ (aq) \rightarrow 2\ CH_3{-}\overset{\overset{:O:}{\|}}{C}{-}\ddot{\underset{\cdot\cdot}{O}}:^{-}\ (aq) + Ca^{2+}\ (aq) + 2\ \underset{H\quad H}{\ddot{\underset{\cdot\cdot}{O}}}\ (\ell)$$

19.34

(a) $CH_3{-}\overset{\overset{:O:}{\|}}{C}{-}\ddot{\underset{\cdot\cdot}{O}}{-}CH_2{-}CH_3 + \underset{H\quad H}{\ddot{\underset{\cdot\cdot}{O}}}$

(b) $H{-}\overset{\overset{:O:}{\|}}{C}{-}\underset{\underset{H}{|}}{\ddot{N}}{-}H + \underset{H\quad H}{\ddot{\underset{\cdot\cdot}{O}}}$

19.35 polyester

19.36

$$H{-}\ddot{\underset{\cdot\cdot}{O}}{-}\overset{\overset{:O:}{\|}}{C}{-}(CH_2)_4{-}\overset{\overset{:O:}{\|}}{C}{-}\ddot{\underset{\cdot\cdot}{O}}{-}H + H{-}\ddot{\underset{\cdot\cdot}{O}}{-}(CH_2)_4{-}\ddot{\underset{\cdot\cdot}{O}}{-}H$$

19.37

$$H{-}\left[\begin{array}{ccc} H & & H \\ | & & | \\ C & - & C \\ | & & | \\ H & & :\underset{\cdot\cdot}{Cl}: \end{array}\right]_n{-}H$$

Answers to Odd-Numbered Problems

19.1 A carbon atom has four valence electrons.

19.3

(a) $\cdot\dot{C}:$ $\qquad -\overset{|}{\underset{|}{C}}-$

(b) H· H—

(c) ·Ö· —Ö—

19.5

(a) H—C(H)(H)—H (b) H—N̈(H)—H (c) H—Ö—H

19.7 (a) methane (b) ammonia (c) water

19.9

(a) H—C(H)(H)—F: (b) H—C(H)(F)—F: (c) H—C(F)(F)—F:

19.11

(a) H—C(H)(H)—C(H)(H)—H (b) H₂C=CH₂ (c) H—C≡C—H

19.13

(a) H—C(H)(H)—N̈(H)—H (b) H₂C=N̈—H (c) H—C≡N:

19.15

:Ö: ‖ :C̈l—C—C̈l:

19.17

(a) —C$^{\delta+}$—Ö$^{\delta-}$— (b) —C$^{\delta+}$—N̈$^{\delta-}$—

19.19

(a) —C≡C— nonpolar (b) —C$^{\delta+}$—F̈$^{\delta-}$:

19.21

H—C(H)(H)—C(H)(H)—C(H)(H)—C(H)(H)—C(H)(H)—C(H)(H)—H

19.23

H—C(H)(H)—C≡C—C(H)(H)—H

19.25

H—C(H)(H)—C(CH₃)=C(H)—C(H)(H)—C(H)(H)—H

CH_3—C(CH_3)=CH—CH_2—CH_3

19.27

$$\begin{array}{ccccc} & H & & H & \\ H & & C & & H \\ & C & & C & \\ H & & & & H \\ & C & - & C & \\ H & & & & H \\ & H & & H & \end{array} \qquad \begin{array}{ccc} & CH_2 & \\ CH_2 & & CH_2 \\ & CH_2-CH_2 & \end{array}$$

19.29

$$CH_3-CH_2-CH_2-CH_2-CH_2-CH_3 \qquad \begin{array}{c} \quad CH_3\;\; CH_3 \\ \quad | \quad\;\; | \\ CH_3-CH-CH-CH_3 \end{array}$$

$$\begin{array}{c} CH_3 \\ | \\ CH_3-CH-CH_2-CH_2-CH_3 \end{array} \qquad \begin{array}{c} CH_3 \\ | \\ CH_3-CH_2-CH-CH_2-CH_3 \end{array}$$

$$\begin{array}{c} CH_3 \\ | \\ CH_3-C-CH_2-CH_3 \\ | \\ CH_3 \end{array}$$

19.31

$$\begin{array}{c} CH_2 \\ / \;\; \backslash \\ CH_2-CH_2 \end{array} \qquad CH_2{=}CH-CH_3$$

19.33 An alkane is a hydrocarbon with only single bonds. An alkene is a hydrocarbon containing a carbon–carbon double bond.

19.35 (a) methane (b) propane (c) pentane

19.37 2-methylpentane

19.39 hexane

19.41

$$\begin{array}{c} CH_3 \\ | \\ CH_3-CH_2-C-CH_2-CH_3 \\ | \\ CH_3 \end{array}$$

19.43

$$\begin{array}{c} CH_3 \;\; CH_2-CH_3 \\ | \qquad | \\ CH_3-CH-CH-CH_2-CH_3 \end{array}$$

19.45

$$\begin{array}{c} CH_2 \\ / \;\; \backslash \\ CH_2-CH_2 \end{array}$$

19.47 2-pentene

19.49 2-methyl-2-pentene

19.51

$$\begin{array}{c} CH_3 \;\; CH_3 \\ | \qquad | \\ CH_3-C{=}C-CH_3 \end{array}$$

19.53 *Aromatic* refers to compounds containing rings of six carbons including alternating double and single bonds.

19.55 "Benzene shows resonance" means that benzene should be considered a composite or average of two different line formulas, each showing alternating single and double bonds between the carbon atoms.

19.57 Two arrows pointing in opposite directions indicate an equilibrium reaction with forward and reverse directions. The double-headed arrow with a single shaft has nothing to do with a reaction; it signifies that a compound should be considered as a composite or hybrid of the formulas related by the arrow.

19.59 1,2-dimethylbenzene

19.61 Hexane (six carbons) will have a higher boiling point than ethane (2 carbons).

19.63 2,2-Dimethylpropane (branched) will have a lower melting point than pentane (linear).

19.65 Ethene (2 carbons) is more likely to be a gas than is 1,4-dimethylbenzene (8 carbons) because ethene has many fewer atoms and therefore weaker London forces.

19.67 $CH_3—CH_2—CH_3\ (g) + \mathbf{5}\ O_2\ (g) \rightarrow \mathbf{3}\ CO_2\ (g) + \mathbf{4}\ H_2O\ (\ell)$ + heat

19.69 (a) addition (b) substitution

19.71

```
CH3—CH—CH2
     |   |
  :Br: :Br:
   ..   ..
```

19.73

```
     H                      H
     |      .. ..           |   ..          ..
H—C—H + :I—I:  →  H—C—I:  + H—I:
     |      .. ..           |   ..          ..
     H                      H
```

19.75 (a) alcohol (b) thiol

19.77

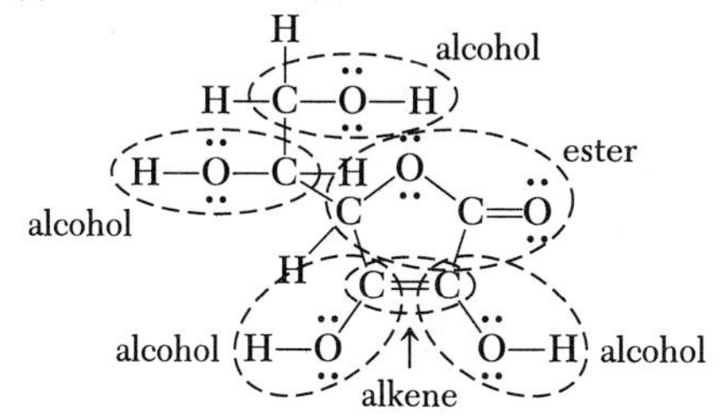

19.79 (a) 2-propanol (b) 1-pentanol

19.81

(a)

```
            ..
           :O—H
            |
CH3—CH2—CH—CH2—CH2—CH2—CH2—CH3
```

(b)

```
           ..
          :O—H
           |
CH3—CH—CH2
     |
    CH3
```

(c)

```
        ..
       :O—H
        |
CH2—CH
 |      |
CH2—CH2
```

19.83 Ethanol (C_2H_6O) has more atoms and therefore stronger London forces than methanol (CH_4O).

19.85 Ethylene glycol has two —OH groups per molecule and can form twice as many hydrogen bonds as each ethanol molecule. The more hydrogen bonds per molecule, the higher the boiling point.

19.87 The more hydrogen bonds a molecule can form, the more soluble it will be in water. Glycerol has three —OH groups, so it will have great solubility in water.

19.89 2 K· +2 CH₃—CH₂—Ö—H → 2 K⁺ + 2 CH₃—CH₂—Ö:⁻ + H—H

$2\ K\cdot + 2\ CH_3—CH_2—\ddot{\underset{..}{O}}—H \rightarrow 2\ K^+ + 2\ CH_3—CH_2—\ddot{\underset{..}{O}}:^- + H—H$

19.91

```
   H  H  H  H    ..              ..            H  H  H  H  H        ..
   |  |  |  |                                  |  |  |  |  |
H—C—C—C—C—O—H + H—Cl:  →  H—C—C—C—C—O⁺—H + :Cl:⁻
   |  |  |  |    ..             ..             |  |  |  |   ..      ..
   H  H  H  H                                  H  H  H  H

   H  H  H  H  H                           H  H     H          H
   |  |  |  |  |                           |  |     |         /
H—C—C—C—C—O⁺—H  →  H⁺ + H—C—C—C=C + :O:
   |  |  |  |  ..                          |  |  |  |          \
   H  H  H  H                              H  H  H  H           H
```

19.93 (a) 1-propanamine (b) 2-methyl-2-propanamine

19.95

(a)

$$\begin{array}{ccccccc} & & H—\ddot{N}—H & & & & \\ & & | & & & & \\ CH_3— & CH_2— & CH— & CH_2— & CH_2— & CH_2— & CH_3 \end{array}$$

(b)

$$\begin{array}{ccccc} & & H—\ddot{N}—H & & \\ & & | & & \\ CH_3— & CH_2— & C— & CH_2— & CH_3 \\ & & | & & \\ & & CH_2—CH_3 & & \end{array}$$

19.97 1-Propanamine has an odor like ammonia while 1-propanol does not. Chemically, 1-propanamine reacts with water to form a weakly basic solution whereas a solution of 1-propanol in water is neutral.

19.99

$$\begin{array}{ccc} H—\ddot{N}—H\ (aq) + & \ddot{\underset{\cdot\cdot}{O}} & \\ | & H \quad H & \\ CH_3—CH—CH_3 & & \end{array} (\ell) \leftrightarrows \begin{array}{c} H \\ | \\ H—N^{+}—H\ (aq) + :\ddot{\underset{\cdot\cdot}{O}}—H^{-}\ (aq) \\ | \\ CH_3—CH—CH_3 \end{array}$$

19.101

$$\begin{array}{ccc} & & H—\ddot{N}—H \\ & & | \\ CH_3— & CH_2— & CH_2 \end{array} (aq) + H—\ddot{\underset{\cdot\cdot}{Cl}}: \rightarrow \begin{array}{ccc} & & H \\ & & | \\ & & H—N^{+}—H \\ & & | \\ CH_3— & CH_2— & CH_2 \end{array} + :\ddot{\underset{\cdot\cdot}{Cl}}:^{-}$$

19.103 (a) propanoic acid (b) 2,2-dimethylpropanoic acid

19.105

(a)

$$\begin{array}{ccccc} & CH_3 & CH_3 & :O: & \\ & | & | & \| & \\ CH_3— & CH— & C— & C— & \ddot{\underset{\cdot\cdot}{O}}—H \\ & & | & & \\ & & CH_3 & & \end{array}$$

(b)

$$\begin{array}{cccccc} & CH_3— & CH_2 & & :O: & \\ & & | & & \| & \\ CH_3— & CH_2— & C— & CH_2— & C— & \ddot{\underset{\cdot\cdot}{O}}—H \\ & & | & & & \\ & CH_3— & CH_2 & & & \end{array}$$

19.107 Oxalic acid has four oxygen atoms and two hydrogen atoms allowing for hydrogen bonding with water. Oxalic acid has significant solubility in water.

19.109

$$\begin{array}{cccc} & CH_3 & :O: & \\ & | & \| & \\ CH_3— & CH— & C— & \ddot{\underset{\cdot\cdot}{O}}—H \end{array} (aq) + :\ddot{O}\begin{array}{l} ^{\diagup H} \\ _{\diagdown H} \end{array} (\ell) \leftrightarrows \begin{array}{cccc} & CH_3 & :O: & \\ & | & \| & \\ CH_3— & CH— & C— & \ddot{\underset{\cdot\cdot}{O}}:^{-} \end{array} (aq) + H—\ddot{O}\begin{array}{l} ^{\diagup H} \\ _{\diagdown H} \end{array}^{+} (aq)$$

19.111

$$\begin{array}{ccccc} & & CH_3 & :O: & \\ & & | & \| & \\ CH_3— & CH_2— & C— & C— & \ddot{\underset{\cdot\cdot}{O}}—H \\ & & | & & \\ & & CH_3 & & \end{array} (aq) + :\ddot{\underset{\cdot\cdot}{O}}—H^{-}\ (aq) \rightarrow \begin{array}{ccccc} & & CH_3 & :O: & \\ & & | & \| & \\ CH_3— & CH_2— & C— & C— & \ddot{\underset{\cdot\cdot}{O}}:^{-} \\ & & | & & \\ & & CH_3 & & \end{array} (aq) + \begin{array}{c} \ddot{\underset{\cdot\cdot}{O}} \\ H \quad H \end{array} (\ell)$$

19.113

$$\begin{array}{cccc} & :O: & :O: & \\ & \| & \| & \\ H—\ddot{\underset{\cdot\cdot}{O}}— & C— & C— & \ddot{\underset{\cdot\cdot}{O}}—H \end{array} + 2\ CH_3—\ddot{\underset{\cdot\cdot}{O}}—H \xrightarrow{HCl} \begin{array}{cccc} & :O: & :O: & \\ & \| & \| & \\ CH_3—\ddot{\underset{\cdot\cdot}{O}}— & C— & C— & \ddot{\underset{\cdot\cdot}{O}}—CH_3 \end{array} + 2 \begin{array}{c} \ddot{\underset{\cdot\cdot}{O}} \\ H \quad H \end{array}$$

19.115

$$\begin{array}{cccccc} & & :O: & & & \\ & & \| & & & \\ CH_3— & CH_2— & C— & \ddot{\underset{\cdot\cdot}{O}}— & CH_2— & CH_2—CH_3 \end{array}$$

19.117

$$CH_3\text{—}CH(CH_3)\text{—}C(\text{=}\ddot{O}\text{:})\text{—}\ddot{O}\text{—}H + CH_3\text{—}\ddot{N}(H)\text{—}H \xrightarrow{HCl} CH_3\text{—}CH(CH_3)\text{—}C(\text{=}\ddot{O}\text{:})\text{—}\ddot{N}(H)\text{—}CH_3 + H\text{—}\ddot{O}\text{—}H$$

19.119

$$CH_3\text{—}CH_2\text{—}CH_2\text{—}C(\text{=}\ddot{O}\text{:})\text{—}\ddot{N}(H)\text{—}CH_2\text{—}CH_2\text{—}CH_2\text{—}CH_3$$

19.121

$$(n + 1)\ H\text{—}\ddot{O}\text{—}C(\text{=}\ddot{O}\text{:})\text{—}C(\text{=}\ddot{O}\text{:})\text{—}\ddot{O}\text{—}H + (n + 1)\ H\text{—}\ddot{N}(H)\text{—}CH_2\text{—}CH_2\text{—}\ddot{N}(H)\text{—}H$$

$$\downarrow HCl$$

$$H\text{—}\ddot{O}\text{—}C(\text{=}\ddot{O}\text{:})\text{—}C(\text{=}\ddot{O}\text{:})\text{—}\left[\ddot{N}(H)\text{—}CH_2\text{—}CH_2\text{—}\ddot{N}(H)\text{—}C(\text{=}\ddot{O}\text{:})\text{—}C(\text{=}\ddot{O}\text{:})\right]_n\text{—}\ddot{N}(H)\text{—}CH_2\text{—}CH_2\text{—}\ddot{N}(H)\text{—}H + n\ H\text{—}\ddot{O}\text{—}H$$

19.123 a condensation polymer

19.125

H H
H— C—C —H
H
H C H
C C
C C
H C H
H
n

$$H\text{—}\left[CH(H)\text{—}CH(C_6H_5)\right]_n\text{—}H$$

Chapter 20—Biochemistry

Answers to Exercises

20.1 Glucose and fructose are significantly soluble in water because each molecule has several oxygen-containing functional groups that can form hydrogen bonds with water.

20.2 (a)

$CH_2\text{—}\ddot{O}\text{—}H$ $CH_2\text{—}\ddot{O}\text{—}H$
C —— $\ddot{O}$ C —— $\ddot{O}$
H H H H
H H
C :$\ddot{O}$—H H C C :$\ddot{O}$—H H C
H—$\ddot{O}$: C —— C $\ddot{O}$ C —— C :$\ddot{O}$—H
H H—$\ddot{O}$: H H—$\ddot{O}$:

maltose

(b) $C_{12}H_{22}O_{11}$

20.3 The suffix "-ose" is commonly used in the names of carbohydrates.

20.4

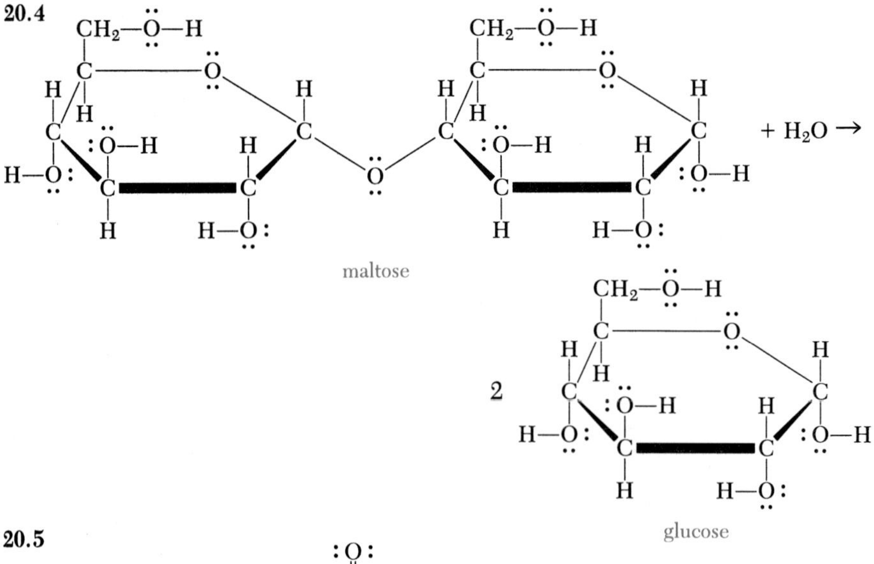

$+ H_2O \rightarrow$ 2

maltose

glucose

20.5

$$2\ CH_3\text{—}(CH_2)_7\text{—}CH\text{=}CH\text{—}(CH_2)_7\text{—}C(\text{=}O)\text{—}O\text{—}H + 51\ O_2 \rightarrow 36\ CO_2 + 34\ H_2O + \text{Energy}$$

20.6

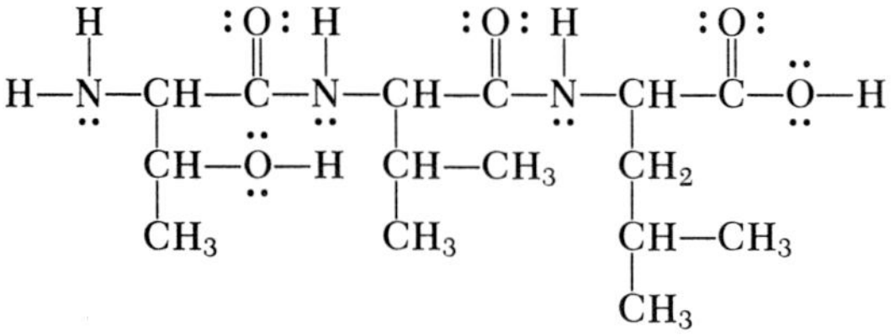

20.7

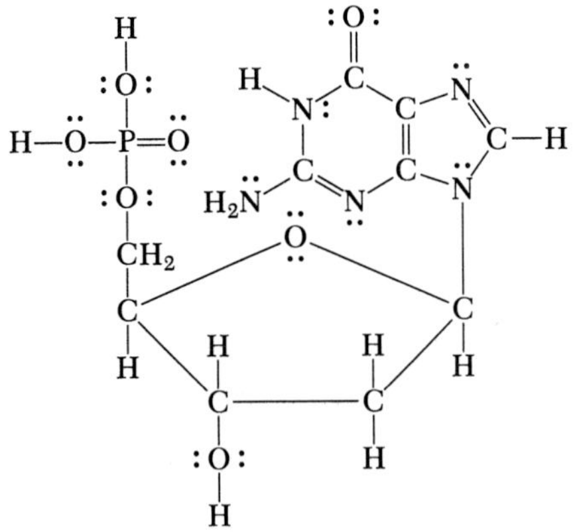

20.8 GCCTATTGG
20.9 ACCAAATATTGT

Answers to Odd-Numbered Problems

20.1 A *carbohydrate* is a compound of carbon, hydrogen, and oxygen, containing alcohol functional groups and either an aldehyde or a ketone functional group.

20.3 The compound in (a) is not a carbohydrate because it has no —OH groups. The compound in (b) is a carbohydrate because it is a dihydroxy aldehyde.

20.5

```
   :O:
    ‖
    C—H
    |  ..
 H—C—O—H
    |  ::
 H—C—O—H
    |  ::
 H—C—O—H
    |  ..
    H
```

20.7 The compound in (b) would be more soluble in water because it has three oxygens that can form hydrogen bonds with water. The compound in (a) has only one oxygen with which to form hydrogen bonds with water.

20.9 Many carbohydrates, even some very large carbohydrate molecules, are significantly soluble in water because they contain so many oxygens—almost as many oxygens as carbons. The more oxygens a compound contains, the more hydrogen bonds can be formed with water, and the greater their solubility.

20.11

```
      :O:
       ‖
       C—H                      ..
       |  ..                CH2—O—H
    H—C—O—H                |   ..
  ..   |  ..                C————O
H—O—C—H                H/|       ..      H
  ..   |  ..      ⇄      |/H              \   |
    H—C—O—H              C    ..            C
       |  ::        H—O: \ :O—H    H      / :O—H
    H—C—O—H          ..   C ▬▬▬▬▬▬ C       ..
       |  ::              |        |
    H—C—O—H              H      H—O:
       |  ..                       ..
       H
```

20.13 Not only does starch provide energy for animals, but starch also plays a crucial role in the carbon cycle. Plants synthesize starch from carbon dioxide, and animals consume and digest starch, producing carbon dioxide in the process. Without starch, the carbon cycle could not continue, and animal life as we know it could not survive.

20.15 There are three critical elements in photosynthesis: carbon, oxygen, and hydrogen. The carbon and oxygen come from carbon dioxide, and the hydrogen comes from water.

20.17 The *metabolism* of starch is the biochemical process of combining the starch molecule with oxygen, forming carbon dioxide and water—the reverse of photosynthesis. The energy released in the process sustains the organism.

20.19 The carbon cycle describes how plants and animals use carbon in complementary processes, recycling the limited amount of carbon on the earth.

20.21 A fat, also known as a triacylglycerol, is a combination of glycerol with long-chain carboxylic acids. All fats contain the ester functional group.

20.23 Fats include long chains of carbon and hydrogen. As these chains have neither oxygen nor nitrogen, water cannot form hydrogen bonds with the chains. Thus, fats cannot dissolve in water.

20.25 Fatty acids are carboxylic acids usually with an even number of carbons (up to about 20 carbons) in an unbranched chain.

20.27

$$\begin{array}{l} CH_2{-}O{-}C({=}O){-}R_1 \\ | \\ CH{-}O{-}C({=}O){-}R_2 \\ | \\ CH_2{-}O{-}C({=}O){-}R_3 \end{array} + 3\ H{-}O{-}H \rightarrow$$

$$\begin{array}{l} CH_2{-}O{-}H \\ | \\ CH{-}O{-}H \\ | \\ CH_2{-}O{-}H \end{array} + H{-}O{-}C({=}O){-}R_1 + H{-}O{-}C({=}O){-}R_2 + H{-}O{-}C({=}O){-}R_3$$

20.29

$$CH_3{-}(CH_2)_{14}{-}C({=}O){-}O{-}H + 23\ O_2 \rightarrow \mathbf{16}\ CO_2 + \mathbf{16}\ H_2O + \text{Energy}$$

20.31 Metabolism of fat produces water, so some of the water needs of the camel are satisfied by metabolism of the fat stored in its hump(s).

20.33 Cholesterol should be virtually insoluble in water. Cholesterol has 27 carbons and only one oxygen to form hydrogen bonds with water.

20.35 LDL receptors permit the breakdown of LDL particles containing cholesterol. With an insufficiency of LDL receptors, an animal will develop plaque deposits in the arteries, causing the disease atherosclerosis.

20.37

$$H{-}N(H){-}CH(R){-}C({=}O){-}O{-}H$$

20.39 The NH_2 group of an amino acid is weakly basic like ammonia. The carboxylic acid group is a weak acid. So amino acids have both acidic and basic characteristics.

20.41 Essential amino acids are those which humans cannot produce, or cannot produce in the amounts necessary for good health, so these amino acids must be present in the diet.

20.43

$$H{-}N(H){-}CH(CH_2{-}CH(CH_3){-}CH_3){-}C({=}O){-}N(H){-}CH(CH(CH_3){-}CH_3){-}C({=}O){-}N(H){-}CH(CH_2{-}C_6H_5){-}C({=}O){-}O{-}H$$

20.45 Six different polymer chains can be made from Leu, Val, and Phe.

20.47 Enzymes are protein molecules that serve as catalysts for biochemical reactions.

20.49 The term *amino acid* refers to a group of compounds, all of which contain an amino group and a carboxylic acid group.

20.51 A nucleotide is formed from a phosphoric acid molecule, a base molecule, and a carbohydrate molecule.
20.53 The four bases are adenine, guanine, cytosine, and thymine.
20.55 In *deoxyribonucleic acid, deoxyribo* refers to the carbohydrate portion, a ribose with one —OH missing at carbon-2′. *Nucleic* refers to the polymer of nucleotides. (Also, nucleic acids are present in the *nucleus* of the cell.) The *acid* refers to the acidic nature of the phosphoric acid part of DNA.
20.57 The *complementary base pairs* are those pairs that form hydrogen bonds between two chains of DNA. The complementary base pairs are A-T and C-G.
20.59 TAATCCGTA
20.61 The sequence of bases in DNA translates into a sequence of amino acids in a protein chain. Each amino acid in a protein corresponds to a three-base sequence in DNA.
20.63 CAGTTTTTTAAA
20.65 Trp-Thr-Val
20.67 DNA controls the synthesis of proteins, and proteins (enzymes) act as catalysts in virtually all biochemical reactions.
20.69 A *gene* is a series of three-base sequences that serve as the code for a complete protein. All of the genes for an organism are its *genome.*
20.71 The DNA double helix contains two complementary chains. In reproduction, all or part of a parent organism's DNA is passed to its offspring.
20.73 On our planet, all metabolic processes are controlled by enzymes, which in turn are determined by DNA. Thus, every organism's structure and function are a direct result of its DNA. Also, DNA transmits the characteristics of each organism to its offspring.
20.75 A zymogen is an inactive precursor to an enzyme. Through some transformation, a zymogen is activated into an enzyme.
20.77 Feedback inhibition is the process of the final product of a metabolic sequence inactivating the initial enzyme of the sequence. Another example of feedback inhibition is losing our appetite after eating a big meal.
20.79 Atmospheric nitrogen is N_2 gas. Soil nitrogen is nitrate ion. Plant and animal nitrogen are the amino acids, proteins, and other nitrogen-containing substances used in the metabolism of the organism.

Length, Mass, and Volume in the Metric System and the SI

The metric system was created in France during the French Revolution (1789–1798) and has become the system of weights and measures used by almost all countries. In 1960 the international General Conference on Weights and Measures adopted the SI, a selection of metric units to be used by all scientists.

The metric system and the SI have two kinds of units: **base units,** which are measured, and **derived units,** which are defined in terms of the base units.

The meter, the base unit for length, was originally described in 1795 as one ten-millionth of the length of the earth's quadrant from the pole to the equator, and that length was measured with the most accurate surveying instruments then available. In 1798 the length of the standard meter was recorded as the distance between two lines etched on a platinum bar, which was then carefully stored near Paris. In 1983 a new measurement was adopted for the meter: One meter is the distance light travels in a vacuum in 1/299,792,458 second. (Speed and other properties of light are discussed in Section 3-8.)

The kilogram is the base unit for mass. The standard for mass is a platinum–iridium cylinder whose mass is designated as exactly one kilogram, kept at Sèvres, France, near Paris.

Volume in the metric system and the SI is expressed in derived units: The standard is the cubic meter, a cube with an edge exactly one meter long. (The use of length units to express areas and volumes is discussed in Section 3-4). Chemists usually prefer the liter as a unit of volume; a liter is a cubic decimeter.

Three other SI base units are used in this book: the kelvin (Section 3-6), the second (Section 3-8), and the mole (Sections 11-1 and 11-2). The formal definitions for these units are more complicated than we'll need in this introductory course in chemistry, so we'll use the simpler definitions given in the text.

As we've seen in Section 3-3, the SI uses a system of prefixes to stand for powers of ten, and these prefixes are added to the names of units to designate the corresponding multiples of them. The complete set of SI prefixes and their symbols appears in the following table.

Appendix 2

SI Prefixes

Power of 10	Prefix	Abbreviation
1×10^{18}	exa	E
1×10^{15}	peta	P
1×10^{12}	tera	T
1×10^{9}	giga	G
1×10^{6}	mega	M
1×10^{3}	kilo	k
1×10^{2}	hecto	h
1×10	deka	da
1×10^{-1}	deci	d
1×10^{-2}	centi	c
1×10^{-3}	milli	m
1×10^{-6}	micro	μ (Greek letter mu, pronounced *mew*.)
1×10^{-9}	nano	n
1×10^{-12}	pico	p
1×10^{-15}	femto	f
1×10^{-18}	atto	a

Logarithms and Antilogarithms

The logarithm (log) to the base 10 of a number is the power to which 10 must be raised to equal the number; that is, if $n = 10^x$, then $\log n = x$. To use logs, you need to be able to calculate them and to understand their significance.

Learning to use logs is easy with a calculator. To find the log of a number, enter the number on your calculator and press the key labeled LOG; do that for the following numbers.

Number	Log	This means that
10	1	$10 = 10^1$
100	2	$100 = 10^2$
1000	3	$1000 = 10^3$
10 000	4	$10\,000 = 10^4$

A log doesn't have to be an integer (whole number). For example, you can see from the preceding table that for a number between 100 and 1000, the log will be between 2 and 3.

Numbers less than 1 have negative logs, for example:

Number	Log	This means that
0.1	−1	$0.1 = 10^{-1}$
0.01	−2	$0.01 = 10^{-2}$
0.001	−3	$0.001 = 10^{-3}$
0.0001	−4	$0.0001 = 10^{-4}$

Taking an antilogarithm (antilog) is the reverse of the process of taking a log: If $n = 10^x$, you find x by taking log n, and you find n by taking antilog x. To take the antilog of 2 on your calculator, enter 2 and press 10^x (it's probably the second function on the LOG key).

The number of significant digits in a number will be the same as the number of digits after the decimal in its log, as shown in the following table.

Number	Significant Digits	Log
3028	4	3.4812
0.00955	3	−2.020
7.62×10^{15}	3	15.882
1.00×10^{-14}	3	−14.000

Appendix 3

Exercise 1

(a) Calculate the logs of 121 and 954.

(b) Without using a calculator, identify the two integers between which the log of each of the following numbers will lie: 37; 55 122; 4. Confirm your answers with a calculator.

(c) Calculate the log of each of the following numbers: 1.57×10^2; 9.44×10^3; 9.44×10^5.
(d) Without using a calculator, identify the two integers between which the log of each of the following numbers will lie: 2.28×10^3; 3.84×10^7. Confirm your answers with a calculator.

(Answers: (a) 2.083 and 2.980 (b) 1 and 2; 4 and 5; 0 and 1 (c) 2.196; 3.975; 5.975 (d) 3 and 4; 7 and 8)

Exercise 2

Calculate the logs of 0.0246, 0.000737, 2.4×10^{-5}, 9.66×10^{-3}.

(Answers: −1.609, −3.133, −4.62, −2.015)

Exercise 3

(a) If the log of a number is 2.95, will the number be above or below 100? Above or below 1000? Closer to 100 or 1000? Take antilog 2.95 to confirm your answers.
(b) Find the antilogs of 6.35, 1.14, −2.27, −3.41.

(Answers: (a) Above 100; below 1000; closer to 1000; antilog is 8.9×10^2. (b) 2.2×10^6, 14, 5.4×10^{-3}, 3.9×10^{-4})

Exercise 4

(a) Calculate the logs of 1.00×10^{-7} and 2.020. (b) Calculate the antilogs of −3.165 and 23.780.

(Answers: (a) −7.000, 0.3054 (b) 6.84×10^{-4}, 6.03×10^{23})

Solubility Rules for Ionic Compounds

In the following tables a compound is described as soluble if more than 0.1 mole of it will dissolve in one liter of water.

Compounds That Contain These Ions Are Soluble	Except Compounds That Contain These Ions, Which Are Insoluble*
NH_4^+	No exceptions
Li^+, Na^+, K^+, Rb^+, Cs^+	No exceptions
NO_3^-	No exceptions
$C_2H_3O_2^-$	No exceptions
Cl^-	Ag^+, Hg_2^{2+}, Pb^{2+}
SO_4^{2-}	Sr^{2+}, Ba^{2+}, Pb^{2+}

*The listing "No exceptions" for an ion means that all of the compounds of the ion that we'll consider in this book are soluble. Some rarer compounds of these ions are insoluble.

Compounds That Contain These Ions Are Insoluble	Except Compounds That Contain These Ions, Which Are Soluble
CO_3^{2-}	NH_4^+, Li^+, Na^+, K^+, Rb^+, Cs^+
PO_4^{3-}	NH_4^+, Li^+, Na^+, K^+, Rb^+, Cs^+
S^{2-}	NH_4^+, Li^+, Na^+, K^+, Rb^+, Cs^+, Mg^{2+}, Ca^{2+}, Sr^{2+}, Ba^{2+}
OH^-	Li^+, Na^+, K^+, Rb^+, Cs^+, Sr^{2+}, Ba^{2+}

The rules in these tables can be used to decide whether a compound is soluble or insoluble, by looking at its formula. Many of these decisions become easy with a little practice. For example, if the formula for a compound shows that it contains sodium, you can immediately assume that the compound is soluble: NaCl, NaOH, Na_2SO_4, and Na_3PO_4 are all immediately recognizable as soluble compounds. Similarly for any compound that contains acetate ion: $Pb(C_2H_3O_2)_2$, $Zn(C_2H_3O_2)_2$, and $Fe(C_2H_3O_2)_3$ are all immediately recognizable as soluble compounds. Since the value in learning these rules lies in being able to apply them, the best way to master the tables is to apply the rules as you learn them. As you learn the tables, fill in the diagram on the next page, following the example shown for NH_4NO_3.

Appendix 4

	NO_3^-	$C_2H_3O_2^-$	Cl^-	SO_4^{2-}	CO_3^{2-}	PO_4^{3-}	S^{2-}	OH^-
NH_4^+	NH_4NO_3 soluble							
Li^+								
Na^+								
K^+								
Mg^{2+}								
Ca^{2+}								
Sr^{2+}								
Ba^{2+}								
Ag^+								
Hg_2^{2+}								
Pb^{2+}								
Al^{3+}								
Fe^{3+}								
Zn^{2+}								

Glossary

abbreviated electron configuration See *electron configuration.*

absolute scale See *temperature.*

acid In the Arrhenius theory, a substance that forms hydronium ion, $H_3O^+(aq)$, in aqueous solution; in the Brønsted-Lowry theory, a proton (H^+) donor; in the Lewis theory, an electron-pair acceptor (Sections 15-5 and 15-8). See also *strong acid, weak acid.*

actinide series Elements with atomic numbers 89 through 103, usually shown below the main body of the Periodic Table (Section 5-2). See also *transition metal* and *lanthanide series.*

activity series A list of symbols for metals that shows their replacement priority; the list is arranged so that the ions of a higher metal on the list will replace the ions of a lower metal, in aqueous solution (Section 16-6). See also *replacement reaction.*

addition reaction A reaction in which atoms add to a double or triple bond (Section 19-8).

alcohols The class of organic compounds whose characteristic functional group is the hydroxyl group, —Ö—H, attached to a carbon atom: —C—Ö—H (Section 19-10).

aliphatic compound See *aromatic compound.*

alkali metals In the Periodic Table, the elements in group 1A, except H: Li, Na, K, Rb, Cs, Fr (Section 5-2).

alkaline-earth metals In the Periodic Table, the elements in group 2A: Be, Mg, Ca, Sr, Ba, Ra (Section 5-2).

alkanes Hydrocarbons whose molecules contain only single bonds (Section 19-4).

alkenes Hydrocarbons whose molecules contain double bonds (Section 19-4).

alkynes Hydrocarbons whose molecules contain triple bonds (Section 19-4).

alpha (α) particles A form of radioactive emission that consists of helium nuclei, each containing two protons and two neutrons: ${}^{4}_{2}He$ (Sections 4-4 and 18-1). See also *radioactivity.*

amines The class of organic compounds whose characteristic functional group is a nitrogen atom attached to a carbon atom: —C—N̈— (Section 19-11).

amino acid residue See *proteins.*

amino acids Compounds whose molecules contain an amino group and a carboxyl group. The general formula for biochemically important amino acids is

H—N̈(H)—CH(R)—C(=Ö:)—Ö—H

where *R* is a group of atoms that differs from one amino acid to another (Section 20-3). See also *essential amino acids* and *proteins.*

analytical chemistry The branch of chemistry that undertakes the analysis of compounds to determine how their elements are combined (Section 3-1). See also *synthetic chemistry.*

anion See *ion.*

anode See *electrode.*

aqueous solution A solution in which the solvent is water (Introduction to Chapter 15). The symbol (aq) shows that a substance is dissolved in water; for example, NaCl(aq) represents an aqueous solution of sodium chloride (Section 10-1).

aromatic compound A compound whose molecule contains a benzene ring, [six-membered ring of C atoms with inscribed circle]. An organic compound that isn't classified as an aromatic compound is classified as an aliphatic compound (Section 19-5).

Arrhenius theory A theory of acids and bases that defines an acid as a compound that forms hydronium ion, $H_3O^+(aq)$, in aqueous solution and a base as a compound that forms hydroxide ion, $OH^-(aq)$, in aqueous solution. Named for Svante Arrhenius (1859–1927) (Section 15-8). See also *Brønsted-Lowry theory* and *Lewis theory*.

atherosclerosis A disease in which cholesterol plaques deposit in arteries (Section 20-2).

atmosphere (atm) A unit used to express pressures of gases; the normal pressure of the air at sea level. The atmosphere is used to define two other pressure units, the torr and the pascal (Pa): 1 atm = 760 torr = 101 kPa. The torr, also called the millimeter of mercury (mm of Hg), is named for Evangelista Torricelli (1608–1647); the pascal is named for Blaise Pascal (1623–1662) (Section 13-1).

atomic (covalent) radius An atomic radius measured as half the distance between the bonded nuclei of two atoms of the same element, usually measured in picometers (Section 8-1). See also *nucleus*.

atomic mass The average of the masses of an element's naturally occurring isotopes. In the Periodic Table at the front of the book, the atomic mass for each element is shown below its symbol. The units, not shown in the Periodic Table, are atomic mass units (Section 4-7).

atomic mass unit (amu) The unit of mass commonly used to express masses of protons, neutrons, electrons, atoms, ions, molecules, and formula units. The $^{12}_{6}C$ isotope is defined as having a mass of exactly 12 amu, so 1 amu is one-twelfth the mass of this isotope (Section 4-6).

atomic number The number of protons (also the number of electrons) in an atom, sometimes represented by the symbol Z. In the Periodic Table at the front of the book, the atomic number for each element is shown above its symbol (Section 4-5). See also *nuclear equation*.

autoionization of water An equilibrium reaction that occurs in pure water or in any aqueous solution in which water molecules form hydronium ion, $H_3O^+(aq)$, and hydroxide ion, $OH^-(aq)$ (Section 15-4):

$$2\ H_2O(\ell) \rightleftarrows H_3O^+(aq) + OH^-(aq)$$

Avogadro's Law A gas law that states that the volume of a gas sample is directly proportional to the number of moles of gas in the sample, at constant temperature and pressure: $V_1/V_2 = n_1/n_2$ at constant T and P. Named for Amadeo Avogadro (1776–1856) (Section 13-3). See also *Boyle's Law, Charles's Law, combined gas law,* and *ideal-gas equation*.

Avogadro's number The number of atoms, ions, formula units, or molecules in a mole: 6.02×10^{23}. Named for Amadeo Avogadro (1776–1856) (Section 11-2).

band of stability On a graph of number of neutrons (y axis) versus number of protons (x axis), the points that correspond to stable nuclei (Section 18-3).

base In the Arrhenius theory, a substance that produces hydroxide ion, $OH^-(aq)$, in aqueous solution; in the Brønsted-Lowry theory, a proton (H^+) acceptor; in the Lewis theory, an electron-pair donor (Sections 15-6 and 15-8). See also *strong base, weak base*.

base pairs See *complementary base pairs*.

battery See *dry cell*.

benzene ring See *aromatic compound*.

beta (β) particles A form of radioactive emission. Beta particles are electrons moving at very high speeds (Sections 4-4 and 18-1). See also *radioactivity*.

binary compound A compound that contains only two elements, for example HCl, H_2O, or Al_2O_3 (Section 9-1).

biochemistry The branch of chemistry that describes and explains the chemical processes of living organisms (Introduction to Chapter 20).

biotechnology The application of discoveries in biochemistry—especially discoveries of methods to manipulate DNA—to the creation of products with commercial value (*Inside Chemistry* essay at the end of Chapter 20).

boiling Formation of gas bubbles in the body of a liquid. Each pure liquid has a characteristic boiling point (temperature), which varies with pressure; if the pressure for a boiling point isn't specified, it's assumed to be one atmosphere (Section 14-1).

Boyle's Law A gas law that states that the volume of a gas sample is inversely proportional to its pressure, if the temperature and number of moles of gas are constant: $V_1/V_2 = P_2/P_1$ at constant n and T. Named for Robert Boyle (1627–1691) (Section 13-3). See also *Avogadro's Law, Charles's Law, combined gas law,* and *ideal-gas equation*.

Brønsted-Lowry theory A theory of acids and bases that defines an acid as a proton (H^+) donor and a base as a proton (H^+) acceptor. Named for Johannes N. Brønsted (1879–1947) and Thomas M. Lowry (1874–1936) (Section 15-8). See also *Arrhenius theory* and *Lewis theory*.

buffer A reaction that controls the pH of an aqueous solution by reacting with acid or base that is added to it (Section 17-4).

calorie (cal) A unit used to express quantities of heat energy. The calorie was originally defined as the quantity of heat needed to raise the temperature of 1 gram of water 1°C; it's now defined by the relationship 1 calorie = 4.184 joules (Section 3-7).

carbohydrates Polyhydroxy aldehydes or ketones important in biological systems as sources of energy and as structural materials (Section 20-1).

carbon cycle The cyclic relationship between photosynthesis and starch metabolism (Section 20-1).

carbon-14 dating See *radiocarbon dating*.

carboxylic acids The class of organic compounds whose characteristic functional group is the carboxyl group, $-\overset{\overset{\large :O:}{\|}}{C}-\ddot{\underset{\cdot\cdot}{O}}-H$ (Section 19-12).

catalyst A substance that must be present for a reaction to occur at a useful rate but that is unchanged at the end of the reaction (Section 19-8).

cathode See *electrode.*

cathode rays See *gas discharge tube.*

cation See *ion.*

cell The structural unit of plant and animal life (Section 20-5).

Celsius scale See *temperature.*

chain reaction A series of nuclear fission reactions in which each reaction causes several more, releasing huge amounts of energy (Sections 4-8 and 18-9). See also *nuclear fission.*

Charles's Law A gas law that states that the volume of a gas sample is directly proportional to its absolute temperature, if the pressure of the gas and the number of moles of gas are constant: $V_1/V_2 = T_1/T_2$ at constant n and P. Named for J.A.C. Charles (1746–1823) (Section 13-3). See also *Avogadro's Law, Boyle's Law, combined gas law, ideal-gas equation,* and *temperature.*

chemical bond A force that holds atoms together in a compound (Introductions to Chapters 7 and 8). See also *covalent bond, ionic bond,* and *octet rule.*

chemical change Change of composition (Section 1-2). The process by which a chemical change occurs is called a chemical reaction. See also *physical change.*

chemical equation A description of a chemical reaction, using symbols and formulas, for example,

$$2\ H_2 + O_2 \rightarrow 2\ H_2O$$

The substances that are present before the reaction (shown on the left in the equation) are the reactants, and the substances that are present after the reaction (shown on the right in the equation) are the products. The general overall statement of an equation is, "Reactants form products" (Section 7-2). See also *coefficient* and *molecular equation, ionic equation, net ionic equation.*

chemical reaction A process in which substances (reactants) interact with one another to form new and different substances (products); a process in which chemical bonds are broken or formed (Introduction to Chapter 7). See also *chemical equation.*

chemistry The experimental science that explains and predicts changes in the form and composition of matter (Section 1-2). The study of the ways in which elements interact (Section 3-1). See also *chemical reaction, analytical chemistry,* and *synthetic chemistry.*

coefficient A multiplier in a chemical equation or a nuclear equation that shows the number of times a symbol or formula appears in the equation, for example, either underlined numeral 2 in

$$\underline{2}\ H_2 + O_2 \rightarrow \underline{2}\ H_2O$$

If no coefficient is shown, the multiplier is assumed to be 1 (Sections 4-8 and 7-2).

combination reaction A chemical reaction in which two or more reactants form one product, for example

$$2\ H_2 + O_2 \rightarrow 2\ H_2O$$

(Section 9-8). See also *combustion reaction, decomposition reaction,* and *replacement reaction.*

combined gas law A gas law that combines Avogadro's Law, Boyle's Law, and Charles's Law: $P_1V_1/n_1T_1 = P_2V_2/n_2T_2$. In this equation P is pressure, V is volume, n is moles of gas, and T is temperature on the Kelvin scale (Section 13-5). See also *ideal-gas equation* and *temperature.*

combustion reaction A reaction in which a nonmetal or a compound of nonmetals reacts with oxygen to produce the oxides of the nonmetals and release heat (Sections 9-8 and 19-7). Examples are:

$$C(s) + O_2(g) \rightarrow CO_2(g)$$

$$CH_4(g) + 2\ O_2(g) \rightarrow CO_2(g) + 2\ H_2O(g)$$

complementary base pairs In the double helix of DNA, the pairs of bases that form hydrogen bonds with one another. Adenine forms hydrogen bonds with thymine, and guanine forms hydrogen bonds with cytosine (Section 20-4).

compound A pure substance that can be decomposed into simpler substances; a pure substance that consists either of atoms of different elements attached to one another by covalent bonds or of ions of different elements attached to one another by ionic bonds (Section 3-1, Introductions to Chapters 7 and 8). See also *covalent compound* and *ionic compound.*

concentration The amount of solute in a given amount of solution (Section 15-3). See also *molarity.*

condensation The formation of a liquid from a gas; the reverse of evaporation or of boiling (Section 14-1).

condensed formula A formula for an ionic compound that shows only the numbers of ions present in one formula unit, for example, $BaCl_2$. The subscript on each symbol shows the number of ions; no subscript means one ion (Section 8-6). See also *Lewis formula.*

coordination compound A compound in which an atom or cation of a transition metal forms covalent bonds with atoms or anions of nonmetals or with molecules (Section 10-9). See also *ligand.*

counted value A number that expresses a value found by counting objects that can't be divided, for example, 15 eggs. Because counted objects must be whole objects, a counted value has an infinite number of zeroes after its decimal point: 15 eggs means 15.(infinite zeroes) eggs (Section 2-3).

covalent bond Two electrons (one electron pair) shared between two atoms. In a single covalent bond, two electrons (one pair) are shared between two atoms; in a double covalent bond, four electrons (two pairs) are shared between two atoms; and in a triple covalent bond, six electrons (three pairs) are shared between two atoms (Sections

7-1 and 7-2). See also *nonpolar covalent bond, polar covalent bond, Lewis formula,* and *octet rule.*

covalent compound A compound whose fundamental units are molecules (Introduction to Chapter 7). See also *covalent bond, ionic compound,* and *molecular formula.*

covalent solid or network solid A solid in which all of the atoms are attached to one another by a network of covalent bonds. Covalent solids typically are hard and have high melting points; an example is diamond, C (Section 14-6).

curie (Ci) A unit used to designate rates of radioactive decay: 1 Ci = 3.700×10^{10} disintegrations per second (Section 18-5). Named for Marie Curie (1867–1935).

cyclic compound A compound whose molecule contains a ring of atoms, for example, cyclobutane,

$$\begin{array}{ccc} CH_2 & - & CH_2 \\ | & & | \\ CH_2 & - & CH_2 \end{array}$$

(Section 19-3).

decomposition reaction A chemical reaction in which one reactant forms two or more products; one example is

$$2\ H_2O \rightarrow 2\ H_2 + O_2$$

(Section 9-8). See also *combination reaction, combustion reaction,* and *replacement reaction.*

defined value A statement of an agreed relationship within a measurement system, for example, 1 foot = 12 inches or 1 kilogram = 1×10^3 grams. Defined relationships are exact, so there is an infinite number of significant digits on each side of the equation that states the definition: 1 foot = 12 inches means 1.(infinite zeroes) foot = 12.(infinite zeroes) inches (Section 2-4).

density Mass per unit volume; d = m/V. The usual SI units for the density of solids are g/cm^3 (Section 3-5).

deoxyribonucleic acid (DNA) See *nucleotide* and *nucleic acids.*

diatomic molecule A molecule consisting of two atoms, for example, H_2 or HCl (Section 7-3).

dilution Preparation of one aqueous solution from another, by adding water. In dilution, the equation for the relationship between initial volume and molarity and final volume and molarity is $V_1M_1 = V_2M_2$, in which V_1 and M_1 are the volume and molarity of the initial solution, and V_2 and M_2 are the volume and molarity of the final solution (Section 15-3).

dimensional analysis A systematic method for making calculations that involve converting one set of units into another, for example, converting pounds to grams or meters to centimeters (Section 3-9).

dipole An object that carries a negative charge at one end and a positive charge at the other end (Section 14-2). For example, in a molecule of hydrogen chloride, the electrons are displaced toward the more electronegative chlorine atom, so there is a partial negative charge on the chlorine atom and a partial positive charge on the hydrogen atom, and the molecule is a dipole: $\overset{\delta+}{H}-\overset{\delta-}{Cl}$. See also *dipole–dipole interactions* and *ion–dipole interactions.*

dipole–dipole interactions Attractions between opposite charges on adjacent dipoles (Section 14-3). For example, two molecules of hydrogen chloride will align themselves with their opposite charges next to one another:

$$\overset{\delta+}{H}-\overset{\delta-}{Cl}$$
$$\overset{\delta-}{Cl}-\overset{\delta+}{H}$$

double covalent bond See *covalent bond.*

dry cell A voltaic cell designed to be portable (Section 17-11). See *electrochemical cell.*

electric current A flow of electrons through a wire (Sections 14-5 and 17-11). See also *electrochemical cell.*

electrochemical cell An apparatus in which a spontaneous redox reaction creates an electric current (voltaic cell) or an apparatus in which an electric current causes a nonspontaneous redox reaction to occur (electrolytic cell) (Sections 17-11 and 17-12). See also *dry cell, electrolysis, electroplating,* and *electrorefining.*

electrochemistry The branch of chemistry that describes electrochemical cells (Section 17-12).

electrode In an electrochemical cell, a conducting surface at which oxidation or reduction occurs; the electrode at which oxidation occurs is the anode, and the electrode at which reduction occurs is the cathode (Section 17-11).

electrolysis The process of reversing a redox reaction by applying an electric current to its products; that is, the process of using an electric current to cause a nonspontaneous redox reaction to occur (Section 17-12). An apparatus used to carry out electrolysis is called an electrolytic cell; see *electrochemical cell.*

electromagnetic radiation Radiation consisting of waves, including X-rays, radio waves, microwaves, ultraviolet rays, and infrared rays, as well as visible light (Section 3-8). See also *frequency, velocity of electromagnetic radiation,* and *wavelength.*

electron A particle with a very small negative electrical charge, equal in magnitude but opposite in sign to that on the proton, and a very small mass (5.846×10^{-4} atomic mass units); a constituent of all atoms (Section 4-3). See also *gas discharge tube, neutron, nucleus,* and *beta particle.*

electron affinity The energy released when an electron is added to the outermost energy level of an atom. In general, electron affinity increases upward and to the right in the Periodic Table (Section 8-1).

electron configuration, abbreviated electron configuration The electron configuration for an atom is a description of the distribution of electrons in the sublevels of the atom, for example, Cl^{17}: $1s^2 2s^2 2p^6 3s^2 3p^5$. In an abbreviated electron configuration, only the sublevels in the period in which the element occurs are shown, and the filled sublevels from earlier periods are represented by the symbol for the preceding noble gas, for example, Cl^{17}: $[Ne]^{10} 3s^2 3p^5$ (Section 6-5).

electron delocalization See *resonance.*

electronegativity The ability of an atom to attract shared electrons toward itself. In general, electronegativity increases upward and to the right in the Periodic Table (Section 7-3). See also *nonpolar covalent bond* and *polar covalent bond.*

electroplating The process of coating objects with metals by electrolysis (Section 17-12).

electrorefining The process of purifying metals by electrolysis (Section 17-12).

element A pure substance that cannot be decomposed into simpler substances (Section 3-1). At the atomic level, an element is a substance that consists of atoms all of which have the same number of protons (the same atomic number) and the same number of electrons. See also *isotopes.*

endothermic, exothermic An endothermic reaction is a reaction that absorbs heat, and an exothermic reactions is a reaction that releases heat (Section 9-7). See also *chemical reaction* and *combustion.*

energy The capacity to move matter (Introduction to Chapter 3). Energy can occur in many forms; two of the most familiar are light and heat (Sections 3-6 through 3-8).

energy level In an atom, a spherical region, at a certain distance from the nucleus and centered on it, that one or more electrons could occupy. In theory, there are an infinite number of energy levels around a nucleus. According to the quantum theory, electrons can only be in these regions and not between them. An energy level consists of one or more sublevels, and a sublevel consists of one or more orbitals (Sections 6-1 and 6-2). See also *electron configuration, abbreviated electron configuration.*

enzymes Proteins that function as catalysts in biochemical reactions (Section 20-3). See also *catalyst* and *proteins.*

equation See *chemical equation.*

equilibrium The condition of a reversible reaction in which the forward reaction and the reverse reaction are occurring at the same rate (Section 15-4).

equilibrium-constant expression For an equilibrium reaction, the mathematical equation that shows the relationship among the molar concentrations of reactants and products at equilibrium; molar concentrations of liquids and solids don't appear in the equation (Section 17-5). For example, for the equilibrium reaction

$$H_2O(\ell) + HF(aq) \rightleftarrows H_3O^+(aq) + F^-(aq)$$

the equilibrium-constant expression is

$$\frac{[H_3O^+][F^-]}{[HF]} = K_a = 6.8 \times 10^{-4}$$

essential amino acids Amino acids needed for good health that the body cannot produce, or cannot produce in the needed amounts, so they must be supplied in the diet (Section 20-3). See also *amino acids.*

evaporation The movement of particles from the liquid state to the gas state, at the surface of a liquid (Section 14-1). See also *condensation.*

excited state The condition of an atom in which some of its electrons are not in the lowest available sublevels (Section 6-5). See also *ground state.*

exothermic See *endothermic, exothermic.*

exponent The power to which a number is raised; for example, the 3 in 10^3 (Section 2-1).

exponential notation, scientific notation Exponential notation is a system for writing numbers as powers of ten; for example, 325 can be written in exponential notation as 32.5×10^1, 3.25×10^2, 0.325×10^3, etc. In scientific notation, numbers are expressed in exponential notation with one nonzero digit to the left of the decimal; for example, 3.25×10^2 is written in scientific notation and 32.5×10^1 is not (Section 2-1).

factor A fraction that shows a relationship between two units; for example, the defined value 1 in. = 2.54 cm provides two factors, $\frac{1\text{ in.}}{2.54\text{ cm}}$ and $\frac{2.54\text{ cm}}{1\text{ in.}}$. Factors are used to make calculations by dimensional analysis (Section 3-9).

Fahrenheit scale See *temperature.*

fats Triacylglycerols, esters of glycerol with fatty acids; their generic formula is

$$\begin{array}{l} CH_2\text{—O—C(=O)—}R_1 \\ | \\ CH\text{—O—C(=O)—}R_2 \\ | \\ CH_2\text{—O—C(=O)—}R_3 \end{array}$$

where R_1, R_2, and R_3 represent different hydrocarbon chains (Section 20-2). Fats are a member of the class of biochemicals called lipids.

feedback inhibition A form of biochemical control in which the final product of a sequence of reactions reversibly inactivates the enzyme that catalyzes the first reaction in the sequence (Section 20-5).

fission See *nuclear fission.*

formula A collection of symbols that represents a formula unit (for example, NaCl), a molecule (for example, H_2O), or a polyatomic ion (for example, CO_3^{2-}) (Sections 7-2, 8-8, and 8-9).

formula mass The sum of the masses of the atoms or ions in a formula that represents a formula unit; the units are atomic mass units (Section 8-10).

formula unit For an ionic compound, a hypothetical unit that contains the number of atoms or ions of each element shown in the formula for the compound. For example, the formula for magnesium bromide is $MgBr_2$, and the formula unit for magnesium bromide is a hypothetical unit that consists of one magnesium ion and two bromide ions. For a polyatomic ion, the formula unit is one ion; for example, the formula unit for ammonium ion, NH_4^+, is one ammonium ion (Section 8-9). See also *formula mass.*

forward reaction The reaction shown from left to right in the chemical equation for a reversible reaction. For example, in the equation

$$3\ H_2(g) + N_2(g) \rightleftarrows 2\ NH_3(g)$$

the forward reaction is the reaction of hydrogen with nitrogen to form ammonia (Section 15-4). See also *reverse reaction.*

freezing, freezing point Freezing is the change of a substance from the liquid state to the solid state. The temperature at which freezing occurs is the freezing point; each pure solid has a characteristic freezing point. For a given substance, the freezing point and melting point are the same (Section 14-1). See also *melting, melting point.*

frequency (ν) In the description of electromagnetic radiation, the number of waves passing a point in one second; the units for frequency are s^{-1} or Hertz (Hz). As the wavelength of electromagnetic radiation increases, its frequency decreases (Section 3-8). See also *velocity of electromagnetic radiation.*

functional group A characteristic grouping of atoms attached to a hydrocarbon molecule (Section 19-9). For example, the characteristic functional group for carboxylic acids is the carboxyl group, $-\overset{\overset{\Large :O:}{\|}}{C}-\ddot{\underset{\cdot\cdot}{O}}-H$.

fusion See *nuclear fusion.*

gamma (γ) particles A form of radioactive emission that consists of electromagnetic radiation with very short wavelengths (Sections 4-4 and 18-1). See also *radioactivity.*

gas constant See *ideal-gas equation.*

gas discharge tube An apparatus used in early studies of atomic structure: an evacuated glass tube with electrodes sealed into it. When the electrodes are charged with high-voltage electricity, a glow—called cathode rays—appears between them. Cathode rays are now known to be a stream of electrons (Section 4-3).

gas laws Mathematical laws that describe the behavior of gases; the gas laws discussed in this book are Avogadro's Law, Boyle's Law, Charles's Law, the combined gas law, and the ideal-gas equation (Sections 13-3, 13-5, and 13-6).

Geiger counter See *ionization detector.*

gene In a DNA molecule, all of the three-base sequences that dictate the sequence of amino-acid residues in the molecule of a particular protein (Section 20-4). All of the genes for an organism are its genome. See also *genetic code.*

genetic code A listing of all possible three-base sequences in DNA (there are 64 of them) and their corresponding amino acids (Section 20-4). See also *gene.*

genome See *gene.*

gram In the metric system, the simplest unit for mass. The relationship to English units is 454 g = 1 lb (Section 3-3 and Appendix 2).

ground state The electron configuration of an atom in which all of its electrons are in the lowest available sublevels; the electron configuration for an atom that's predicted from the Periodic Table is the ground-state configuration (Section 6-5). See also *excited state.*

group A vertical column in the Periodic Table (Section 5-2). In general, elements in the same group have similar chemical properties, that is, they form compounds with similar formulas; for example, lithium and sodium are in group 1A, and the formulas for their compounds with oxygen are Li_2O and Na_2O. See also *period.*

half-life ($t_{1/2}$) The time required for half of the radioactive isotopes in a sample to undergo decay. The half-life is a fixed quantity for a given radioactive isotope (Section 18-8).

half-reaction For a redox reaction, either the oxidation or the reduction reaction (Section 17-10). For example, for

$$Mg(s) + 2\ Ag^+(aq) \rightarrow Mg^{2+}(aq) + 2\ Ag(s)$$

the oxidation half-reaction is

$$Mg(s) \rightarrow Mg^{2+}(aq) + 2\ e^-$$

and the reduction half-reaction is

$$Ag^+(aq) + e^- \rightarrow Ag(s)$$

halogens In the Periodic Table, the elements in group 7A: F, Cl, Br, I, At (Section 5-2).

Heisenberg uncertainty principle In quantum mechanics, a mathematical principle that states that the location of an electron in an atom can't be known with certainty. Named for Werner Heisenberg (1901–1976) (Section 6-2).

heterogeneous mixture, homogeneous mixture In a heterogeneous mixture, two or more substances can be seen; a homogeneous mixture has a uniform appearance, as if only one substance were present. Homogeneous mixtures that are gases or liquids are called solutions (Section 3-1).

homogeneous mixture See *heterogeneous mixture, homogeneous mixture.*

hormone A biochemical compound produced by one group of cells that travels to other cells in the organism, where it causes and controls biochemical reactions (Section 20-5).

hydrate A compound in which the formula unit includes one or more water molecules, for example, $CaCl_2 \cdot 2H_2O$ (Section 8-11).

hydrated ion An ion in aqueous solution with water molecules attached to it by ion–dipole interactions. The number of water molecules may vary, with a typical number in the range from three to six (Section 15-1). These are drawings of a hydrated sodium ion and a hydrated chloride ion:

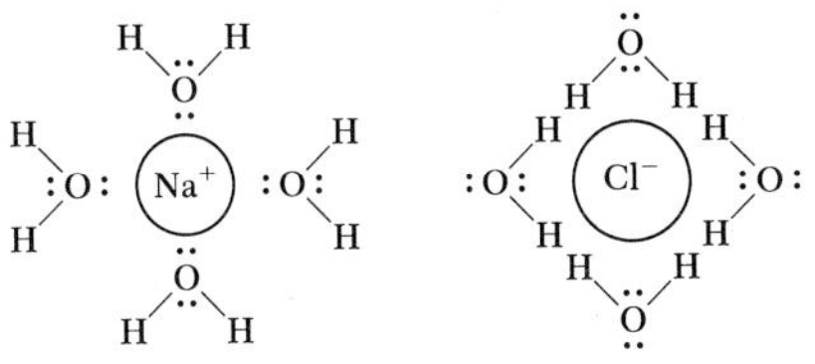

hydrocarbon A compound whose molecules contain only carbon atoms and hydrogen atoms (see Introduction to second section of Chapter 19).

hydrogen bond In a liquid or a solid, a bond that forms between a hydrogen atom in one molecule and a nitrogen, oxygen, or fluorine atom in an adjacent molecule. This example shows a hydrogen bond between two HF molecules:

$H—\ddot{\underset{..}{F}}: \cdots H—\ddot{\underset{..}{F}}:$ (Section 14-4).

hydronium ion

The cation $:O—H$ (with two H atoms, + charge) or H_3O^+; in the Arrhenius theory the hydronium ion is the characteristic cation of acids (Section 15-4). See also *autoionization of water.*

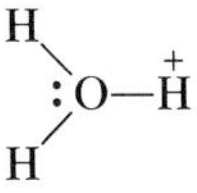

hypothesis A statement of a possible explanation for a law (Section 1-3). See also *theory.*

ideal gas A gas that behaves exactly as the kinetic theory and the gas laws predict (Section 13-6).

ideal-gas equation The most general gas law, PV = nRT, in which P is pressure in atmospheres; V is volume in liters; n is quantity of gas in moles; T is temperature in kelvins; and R is the gas constant, 0.0821 atm-L/mol-K (Section 13-6).

inner-transition element; inner-transition metal See *transition elements.*

interhalogen A compound formed between two halogens, for example, IBr (Section 10-7).

ion A charged particle formed when an atom or group of atoms gains or loses one or more electrons (Introduction to Chapter 8). An anion is a negatively charged ion formed when an atom or group of atoms gains one or more electrons, and a cation is a positively charged ion formed when an atom or group of atoms loses one or more electrons (Sections 8-2, 8-4, and 8-8). See also *monatomic ion* and *polyatomic ion.*

ionization detector An instrument used to determine the quantity of alpha, beta, or gamma radiation emitted by a radioactive sample by measuring the number of electrons released in ionizations caused by the radiation (Section 18-5). A Geiger counter, named for Hans Geiger (1882–1947), is a common form of ionization detector.

ion–dipole interaction The attractive force between an ion and a dipole. A cation and the negative end of a dipole attract one another because they have opposite charges, and similarly for an anion and the positive end of a dipole (Section 15-9). Ion–dipole interactions occur between ions and water molecules in aqueous solutions to form hydrated ions, for example,

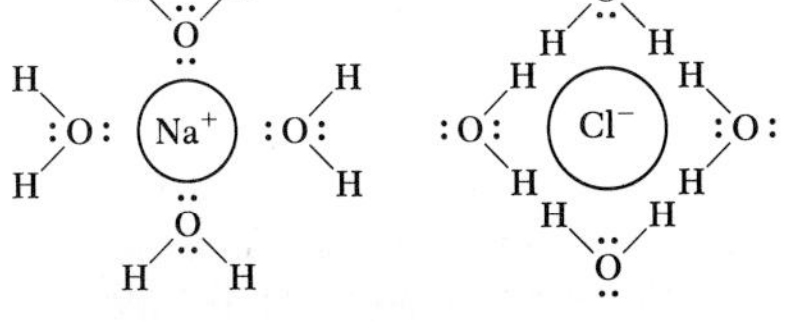

ionic bond The attractive force between oppositely charged ions (Introduction to Chapter 8). See also *chemical bond, ionic compound,* and *octet rule.*

ionic compound A compound whose fundamental particles are ions (Introduction to Chapter 8). In a sample of an ionic compound, the total charges on all of the cations must equal the total charges on all of the anions. See also *covalent compound, condensed formula, ionic bond,* and *formula unit.*

ionic equation See *molecular equation, ionic equation, net ionic equation.*

ionic solid A solid that consists of ions attached to one another by ionic bonds. Ionic solids typically are hard and have high melting points; an example is sodium chloride, NaCl (Section 14-6). See also *lattice.*

ionizable hydrogen atom A hydrogen atom that can be transferred, as a hydrogen ion, from its molecule to a water molecule to form a hydronium ion (Section 15-5). One example is the hydrogen atom in HCl:

$$H_2O(\ell) + HCl(aq) \rightarrow H_3O^+(aq) + Cl^-(aq)$$

ionization energy The energy required to remove an electron from the outermost energy level of an atom. In general, ionization energy increases upward and to the right in the Periodic Table (Section 8-1).

ion-product constant for water See *ion-product equation for water.*

ion-product equation for water The mathematical equation that describes the relationship between the concentrations of $H_3O^+(aq)$ and $OH^-(aq)$ in pure water or in any aqueous solution:

$$[H_3O^+][OH^-] = K_w = 1.00 \times 10^{-14}$$

In this equation, $[H_3O^+]$ is the molarity of hydronium ion and $[OH^-]$ is the molarity of hydroxide ion; K_w is called the ion-product constant for water (Section 15-7).

irreversible reaction See *reversible reaction, irreversible reaction.*

isomers Compounds whose molecules contain the same numbers of atoms of the same elements in different bonding arrangements; compounds with the same molecular formula but different line formulas (Section 19-3).

isotopes Atoms with the same number of protons (and electrons) and different numbers of neutrons; atoms with the same atomic number and different mass numbers. Examples are $^{16}_{8}O$ and $^{17}_{8}O$ (Section 4-6).

joule (J) A unit used to express quantities of heat energy; see also *calorie.* (The exact definition of the joule is based on SI units that aren't used in this book.) Named for J.P. Joule (1818–1889) (Section 3-7).

Kelvin scale See *temperature.*

kinetic theory of gases A theory that describes a gas sample as a collection of a large number of particles (atoms or molecules) in rapid, random motion. The volume of the sample is assumed to be much larger than the total volume of the particles; that is, most of the sample is assumed to be empty space between the particles (Section 13-2).

lanthanide series Elements with atomic numbers 57 through 71, usually shown below the main body of the Periodic Table (Section 5-2). See also *transition metal* and *actinide series.*

lattice The ordered arrangement of particles (atoms, ions, or molecules) in a solid (Section 14-1). See also *melting* and *freezing.*

law A statement that summarizes observations; see, for example, *Avogadro's Law, Boyle's Law, Charles's Law,* and *Periodic Law* (Section 1-3). See also *hypothesis* and *theory.*

LDL particles Low-density lipoprotein particles, the form in which cholesterol travels in the bloodstream (Section 20-2). See also *LDL receptors.*

LDL receptors Sites on body cells that allow LDL particles to enter the cells, where they can be broken down (Section 20-2). See also *LDL particles.*

Le Châtelier's principle A description of the adaptive behavior of equilibrium reactions: If a reaction at equilibrium is disturbed, the reaction will respond in a way that will restore equilibrium. Named for Henri-Louis Le Châtelier (1850–1936) (Section 17-3).

Lewis formula For a covalent compound, a combination of Lewis symbols that shows the bonding in a molecule and represents each valence electron by a dot, for example, $\ddot{\underset{\cdot\cdot}{O}}::C::\ddot{\underset{\cdot\cdot}{O}}$ (Section 7-4). For an ionic compound, the Lewis formula shows the numbers of ions, the charges on the ions, and the valence electrons, for example, $Na^{+}\left[:\ddot{\underset{\cdot\cdot}{Cl}}:\right]^{-}$. Named for Gilbert N. Lewis (1875–1946) (Section 8-6). See also *formula, condensed formula,* and *molecular formula.*

Lewis symbol A symbol for the atom of a representative element that shows valence electrons as dots (Section 6-8). Named for Gilbert N. Lewis (1875–1946). See also *Lewis formula.*

Lewis theory A theory of acids and bases that defines an acid as an electron-pair acceptor and a base as an electron-pair donor. Named for Gilbert N. Lewis (1875–1946) (Section 15-8). See also *Arrhenius theory* and *Brønsted-Lowry theory.*

ligand In a coordination compound, the species (atom or ion of a nonmetal or molecule) that bonds with the atom or cation of a transition metal (Section 10-9).

limiting reactant The reactant from which the least amount of product can be formed; the reactant that's used up first (Section 12-4). See also *chemical equation* and *chemical reaction.*

line formula A modified version of a Lewis formula in which each pair of electrons in a bond is represented by a line, for example, $\ddot{\underset{\cdot\cdot}{O}}{=}C{=}\ddot{\underset{\cdot\cdot}{O}}$ (Section 7-4). See also *molecular formula.*

line symbol A modified version of a Lewis symbol for an atom; in the line symbol a line represents each valence electron the atom can share (Section 19-1). For example, the Lewis symbol for a carbon atom is $\cdot\dot{C}:$ and the line symbol is $-\overset{|}{\underset{|}{C}}-$.

lipids Biochemical compounds whose molecules contain large hydrocarbon portions; besides these hydrocarbon portions, the molecules of lipids contain a variety of functional groups (Section 20-2). See also *fats.*

liter (L) In the metric system, the simplest unit for volume; a liter is a cubic decimeter. The relationship to English units is 1 L = 1.06 qt (Section 3-3 and Appendix 2).

London forces Attractions between opposite charges in instantaneous dipoles, caused by the synchronized movement of electrons in adjacent atoms, molecules, or ions (Section 14-2).

magic number A number of protons or neutrons that tends to confer nuclear stability: 2, 20, 28, 50, 82; for neutrons 126 is also a magic number (Section 18-2). See also *nuclear reaction.*

mass number The number of nucleons (protons and neutrons) in an atom, sometimes represented by the symbol A (Section 4-5). See also *nuclear equation.*

matter Anything that takes up space and has mass (Section 1-2). The most fundamental distinction in science is the distinction between matter and energy.

melting, melting point Melting is the change of a substance from the solid state to the liquid state; in melting, the lattice of the solid breaks down. The temperature at which melting occurs is the melting point; each pure substance has a characteristic melting point. For a given substance, the melting point and the freezing point are the same (Section 14-1). See also *freezing, freezing point.*

metabolic water Water formed in an organism as a product of metabolic reactions, for example, water produced in the metabolism of fats or carbohydrates (Section 20-2). See also *metabolism.*

metabolism The biochemical breakdown of a substance by an organism (Section 20-1 and 20-2). For example, the overall equation for the metabolism of starch is

$$\text{starch} + O_2 \rightarrow CO_2 + H_2O + \text{energy}$$

metal A solid, shiny substance that conducts heat and electricity well and that can be formed easily into complicated shapes such as wires, sheets, and tubes; the class of elements that lose electrons easily (Sections 5-1 and 8-2). See also *metalloid* and *nonmetal.*

metallic bonding Electron sharing (delocalization) throughout all of the atoms in a piece of metal (Section 14-5).

metalloid A substance with properties between those of a metal and those of a nonmetal: Metalloids conduct heat and electricity less well than metals and are less easily formed into complicated shapes such as sheets, wires, and tubes (Section 5-1).

meter In the metric system, the simplest unit for length (Section 3-3 and Appendix 2).

millimeter of mercury (mm of Hg) See *atmosphere.*

mixture A sample of matter that can be separated into two or more substances by physically separating its components from one another (Section 3-1). See also *heterogeneous mixture, homogeneous mixture.*

molar concentration See *molarity.*

molar volume of an ideal gas The volume, 22.4 liters, occupied by 1.00 mole of an ideal gas at standard temperature and pressure (STP) (Section 13-6).

molarity A unit of concentration: moles of solute per liter of solution. The equation for molarity is $M = \frac{\text{mol}}{\text{L}}$, in which M is molarity, mol is moles of solute, and L is liters of solution (Section 15-3).

mole (mol) A unit used to designate the quantity of a pure substance: 1 mol X = 6.02×10^{23} atoms, formula units, or molecules of X; also, 1 mol X = the atomic mass, formula mass, or molecular mass of X in grams (Sections 11-1 and 11-2).

molecular equation, ionic equation, net ionic equation A molecular equation is a chemical equation that represents each reactant or product by a condensed formula or a molecular formula. An ionic equation shows each reactant and product in the form in which it occurs in aqueous solution; in an ionic equation, an ion that appears in the reactants and in the products is called a spectator ion. A net ionic equation omits spectator ions (Section 16-3). These are examples:

molecular equation:

$$HCl(aq) + NaOH(aq) \rightarrow H_2O(\ell) + NaCl(aq)$$

ionic equation:

$$H_3O^+(aq) + Cl^-(aq) + Na^+(aq) + OH^-(aq) \rightarrow 2\ H_2O(\ell) + Na^+(aq) + Cl^-(aq)$$

net ionic equation:

$$H_3O^+(aq) + OH^-(aq) \rightarrow 2\ H_2O(\ell)$$

molecular formula A formula for a covalent compound that shows only the number of atoms of each element in one molecule, for example, CO_2. The subscript on each symbol shows the number of atoms; no subscript means one atom (Section 7-4). See also *Lewis formula, line formula,* and *simplest formula.*

molecular mass The mass of a molecule—the sum of the masses of its atoms—in atomic mass units (Section 7-7).

molecule An electrically neutral group of atoms attached to one another by covalent bonds; the fundamental unit of a covalent compound (Sections 7-2 and 8-9). See also *Lewis formula, line formula, molecular formula, simplest formula,* and *molecular mass.*

monatomic anion See *monatomic ion.*

monatomic cation See *monatomic ion.*

monatomic ion An ion formed from one atom. A monatomic anion is formed by adding one or more electrons to one atom; examples are F^-, O^{2-}, and N^{3-} (Sections 8-4, and 8-8). A monatomic cation is formed by removing one or more electrons from one atom; examples are Na^+, Ca^{2+}, and Al^{3+} (Sections 8-2, 8-3, and 8-5). See also *polyatomic ion.*

monomer See *polymer.*

net ionic equation See *molecular equation, ionic equation, net ionic equation.*

neutralization Reaction of an acid with a base (Section 16-2).

neutron A particle with no electrical charge and with a mass of 1.008665 atomic mass units, about the same mass as the proton. Neutrons are found in the nuclei of most atoms (Section 4-4). See also *electron, isotopes, mass number, nucleon,* and *nucleus.*

nitrogen cycle The cyclic movement of nitrogen atoms, bonded in various ways, through the atmosphere, soil, and living organisms (Section 20-5).

nitrogen fixation The conversion of atmospheric nitrogen (N_2) to soil nitrogen (ultimately NO_3^-) (Section 20-5).

noble gases In the Periodic Table, the elements in group 8A: He, Ne, Ar, Kr, Xe, Rn (Section 5-2).

nonmetal An element that's a poor conductor of heat and electricity and that can't be formed easily into complicated shapes; the class of elements that gain electrons easily

(Sections 5-1 and 8-4). See also *metal* and *metalloid.*

nonpolar covalent bond A covalent bond in which the electrons in the bond are shared equally between two atoms with the same electronegativity; for example, the bond in the hydrogen molecule, H—H, is a single nonpolar covalent bond, and the bond in the nitrogen molecule, :N≡N:, is a triple nonpolar covalent bond (Sections 7-2 and 7-3). See also *chemical bond, Lewis formula, line formula,* and *polar covalent bond.*

nuclear chemistry The branch of chemistry that describes and explains nuclear reactions (Introduction to Chapter 18). See also *nuclear equation, nuclear reaction, nucleus,* and *radioactivity.*

nuclear equation An equation that represents a nuclear reaction; for example,

$$^{238}_{90}U \rightarrow ^{234}_{90}Th + ^{4}_{2}He$$

represents fission of a uranium nucleus into a thorium nucleus and a helium nucleus (alpha particle). In each symbol, the atomic number of the nucleus is shown as a subscript and the mass number is shown as a superscript (Section 4-8). See also *chemical equation, chain reaction,* and *radioactivity.*

nuclear fission A nuclear reaction in which one nucleus, hit by a neutron, splits into two or more nuclei and releases energy (Sections 4-8 and 18-9).

nuclear fusion A nuclear reaction in which two lighter nuclei combine to form a heavier one and release energy (Sections 4-8 and 18-10).

nuclear reaction A reaction in which the nuclei of the reacting atoms change by emitting particles or energy. In a chemical reaction the reacting atoms share, lose, or gain valence electrons, and the nuclei of the atoms are unaffected; in a nuclear reaction the valence electrons of the atoms are unaffected (Section 18-1). See also *chemical reaction, nuclear chemistry,* and *nuclear equation.*

nucleic acids Polymers of nucleotides; in the polymer chain, the phosphoric acid portion of each nucleotide is attached to the 3′ carbon atom in the next nucleotide (Section 20-4). See also *nucleotide* and *polymer.*

nucleons The particles in the nucleus of an atom: protons and neutrons (Section 4-4). See also *atomic number, electron, mass number, nuclear equation,* and *nuclear reaction.*

nucleus The core of an atom. The nucleus is a cluster of the protons and neutrons in the atom; it therefore carries all of the positive charge and most of the mass of the atom. The electrons are outside of the nucleus at relatively very large distances from it (Section 4-4). See also *nuclear equation, nuclear reaction,* and *nucleons.*

nucleotide A compound whose molecule is a combination of a molecule of phosphoric acid, a molecule of a carbohydrate, and a molecule of a base; nucleic acids such as deoxyribonucleic acid (DNA) are polymers of nucleotides (Section 20-4). See also *nucleic acids.*

octet rule The rule that describes the ways in which atoms of representative elements form chemical bonds; the rule can be stated in several ways. For a covalent compound, the rule can be stated in these two forms: (1) A hydrogen atom that forms a covalent bond will share two electrons, and every other atom that forms a covalent bond will share enough electrons to give it a total of eight shared and unshared electrons in its valence shell. (2) An atom that forms a covalent bond will share enough electrons to give it the same number of valence electrons as a noble-gas atom: two valence electrons for a bonded hydrogen atom and eight valence electrons for any other bonded atom (Section 7-2). For an ionic compound: An atom of a metal will lose electrons until the cation that's formed has the same configuration of valence electrons as an atom of the preceding noble gas in the Periodic Table; an atom of a nonmetal will gain electrons until the anion that's formed has the same configuration of valence electrons as an atom of the next noble gas in the Periodic Table (Sections 8-2 and 8-4). See also *electron configuration, abbreviated electron configuration* and *ion.*

orbital In an atom a region with a certain shape in which there is a high (90%) probability that an electron could be found. An orbital is designated by the same letter—s, p, d, or f—as the sublevel in which it occurs. An s sublevel has one orbital, a p sublevel has three, a d sublevel has five, and an f sublevel has seven; each orbital can hold a maximum of two electrons (Section 6-2). See also *orbital diagram, energy level,* and *electron configuration.*

orbital diagram A description of the arrangement of the electrons in the orbitals of an atom. In the diagram each orbital is represented by a circle and each electron by an arrow; the arrows point up or down to show the two possible electron spins. Electrons entering a sublevel will occupy all of the orbitals with one electron each before pairing; paired electrons have opposite spins (Section 6-6). This is the orbital diagram for a nitrogen atom:

N^7: $1s^2$ $2s^2$ $2p^3$

(↑↓) (↑↓) (↑)(↑)(↑)

organic acids See *carboxylic acids.*

organic compounds See *organic chemistry.*

organic chemistry The study of covalent compounds whose molecules contain carbon atoms; these compounds are called organic compounds (Introduction to Chapter 19).

oxidation Loss of electrons; increase in oxidation number (Sections 16-6 and 17-8). See also *redox reaction.*

oxidation number A number assigned by rule to a bonded atom as a basis for identifying oxidation, reduction, and redox reactions (Section 17-9).

oxidizing agent See *redox reaction.*

partial charge, partial negative charge, partial positive charge See *polar covalent bond.*

pascal (Pa) See *atmosphere.*

percentage composition A statement of the percentage, by mass, of each element in a compound. For example, the percentage composition of sodium chloride, NaCl, is 39.3% Na and 60.7% Cl: In 100 g of sodium chloride, there are 39.3 g of sodium and 60.7 g of chlorine (Section 11-3).

period A horizontal row in the Periodic Table (Section 5-2). See also *group.*

periodic law The principle that the properties of the elements vary in a regular and recurring way with their atomic numbers. The periodic law is the basis for the organization of the modern Periodic Table (Introduction to Chapter 6).

periodic table An arrangement of the symbols for the elements in horizontal rows (periods) and vertical columns (groups) in order of increasing atomic number. Arranged in this way, the elements fall into three categories: metals, metalloids, and nonmetals. Elements in the same group have similar chemical properties, that is, they form compounds with similar formulas; for example, the compounds of the elements in group 6 with hydrogen are H_2O, H_2S, H_2Se, H_2Te, and H_2Po. The structure of the table also describes the electron configurations of the elements (Sections 5-1, 5-2, and 6-4). See also *periodic law.*

pH A scale designed to express the acidity or basicity of an aqueous solution in simple numbers: $pH = -\log [H_3O^+]$. The useful range of the pH scale is from 0 to 14. Pure water or a neutral aqueous solution has a pH of 7, acidic solutions have pHs below 7, and basic solutions have pHs above 7 (Section 15-7).

phase A region of matter separated by a visible boundary from the regions of matter next to it. For example, a layer of oil floating on a layer of water shows two phases, oil and water (Section 3-2).

photon The quantum of light energy (Section 6-1). See *quantum theory* and *electromagnetic radiation.*

photosynthesis The process by which energy in the form of sunlight becomes stored in organic compounds (Section 20-1). The overall process is:

$$CO_2 + H_2O + \underset{\text{sunlight}}{\text{energy as}} \rightarrow \text{starch}$$

physical change A change from one physical state to another; examples are melting, freezing, boiling, evaporation, and condensation (Section 1-2). See also *chemical change.*

Planck's constant The equation $E = h\nu$ describes the energy of a photon of electromagnetic radiation: E is the energy of the photon in joules; ν is the frequency of radiation in s^{-1}; and h is Planck's constant, 6.63×10^{-34} J · s. Named for Max Planck (1858–1947).

polar covalent bond A covalent bond in which the electrons in the bond are shared unequally between two atoms with different electronegativities. The electrons in the bond are displaced toward the more electronegative atom, so that atom has a partial negative charge ($\boldsymbol{\delta-}$) and the less electronegative atom has a partial positive charge ($\boldsymbol{\delta+}$) (Section 7-3). For example, the bond in a molecule of hydrogen chloride, $\overset{\delta+}{H}{-}\overset{\delta-}{\ddot{C}l}{:}$, is a single polar covalent bond, and each of the bonds in a molecule of carbon dioxide, $\overset{\delta-}{\ddot{O}}{=}\overset{\delta+}{C}{=}\overset{\delta-}{\ddot{O}}$, is a double polar covalent bond. See also *Lewis formula, line formula,* and *nonpolar covalent bond.*

polyatomic ion An ion that contains more than one atom; examples are H_3O^+, NH_4^+, OH^-, and NO_3^- (Section 8-8). See also *monatomic ion.*

polymer An organic compound whose molecules are very long chains with repeating sequences of atoms (Section 19-13). A monomer is a reactant that forms a polymer, for example:

$$n\ \underset{\text{Monomer (Ethylene)}}{CH_2{=}CH_2} \xrightarrow{\text{Catalyst}} \underset{\text{Polymer (Polyethylene)}}{{+}CH_2{-}CH_2{+}_n}$$

polyprotic acid An acid with more than one ionizable hydrogen atom per molecule (Section 15-5). For example, each molecule of sulfuric acid contains two ionizable hydrogen atoms:

$$H_2O(\ell) + H_2SO_4(aq) \rightarrow H_3O^+(aq) + HSO_4^-(aq)$$
$$H_2O(\ell) + HSO_4^-(aq) \rightleftarrows H_3O^+(aq) + SO_4^{2-}(aq)$$

positron A particle ($^{0}_{1}e$) with the same mass as an electron and a positive charge, emitted by some radioactive nuclei (Section 18-1). See also *electron* and *radioactivity.*

precipitate See *precipitation reaction.*

precipitation reaction A chemical reaction in which an insoluble ionic compound is formed in an aqueous solution; the insoluble compound appears as a solid precipitate (Section 16-4). One example is

$$Cd(NO_3)_2(aq) + Na_2S(aq) \rightarrow CdS(s) + 2\ NaNO_3(aq)$$

product See *chemical equation.*

proteins Polymers of amino acids (Section 20-3). The general formula for a protein is

$$H{-}N(H){-}CH(R){-}C({=}O){-}\left[N(H){-}CH(R){-}C({=}O)\right]_n{-}N(H){-}CH(R){-}C({=}O){-}O{-}H$$

The portion of an amino acid that appears in a protein chain is called an amino acid residue. See also *amino acids* and *polymers.*

proton A particle with a very small positive electrical charge, equal in magnitude but opposite in sign to that on the electron, and with a mass of 1.007276 atomic mass units (about 2000 times the mass of the electron). Protons are found in the nuclei of all atoms (Section 4-3). A hydrogen atom consists of one proton and one electron; when the electron is lost from a hydrogen atom, the ion that is formed consists of one proton. For this reason, the ion H^+ is sometimes referred to as a hydrogen ion and sometimes as a proton (Section 15-5). See also *isotopes, mass number, neutron,* and *nucleus.*

quantum mechanics A mathematical branch of physics that describes approximate locations for electrons in atoms (Section 6-2). See also *electron configuration, abbreviated electron configuration, energy level, orbital,* and *sublevel.*

quantum theory The theory that energy radiated from hot objects isn't emitted continuously, as it seems to be, but in many small bursts, called quanta (each burst is a quantum) (Section 6-1).

radiation therapy The use of emissions from radioactive nuclei to treat diseases, most commonly to treat cancer (Section 18-7). See also *radioactivity.*

radioactive tracer A radioactive atom in an element or compound, used to follow the changes that the element or compound undergoes in chemical reactions (Section 18-6). See also *radioactivity.*

radioactivity The emission of particles or energy by nuclei in a nuclear reaction (Sections 4-4 and 18-1). Natural radioactivity is spontaneous emission, and induced radioactivity is emission caused by bombardment with subatomic fragments. The fragments may be subatomic particles, for example, neutrons, or they may be nuclei, for example, alpha particles. See also *nuclear equation, nuclear reaction,* and *nucleus.*

radiocarbon dating The process of finding the age of a sample of animal or plant remains by measuring the rate of emission from $^{14}_{6}C$ (a beta-emitter) in the sample (Section 18-8).

reactant See *chemical equation.*

reaction See *chemical reaction* and *nuclear reaction.*

redox (reduction-oxidation) reaction A chemical reaction in which electrons are transferred from one reactant to another, for example,

$$Mg(s) + 2\ Ag^+(aq) \rightarrow Mg^{2+}(aq) + 2\ Ag(s)$$

The reactant that loses electrons (magnesium, in the preceding example) is said to be oxidized, and the reactant that gains electrons (silver ion, in the preceding example) is said to be reduced (Section 16-6). Oxidation is therefore defined as a loss of electrons or an increase in oxidation number, and reduction is defined as a gain of electrons or a decrease in oxidation number (Section 17-9). The reactant that's reduced is called the oxidizing agent, and the reactant that's oxidized is called the reducing agent (Section 17-8). See also *half-reaction, oxidation, reduction,* and *skeleton equation.*

reducing agent See *redox reaction.*

reduction Gain of electrons; decrease in oxidation number (Sections 16-6 and 17-8). See *redox reaction.*

rem A unit used to describe biological damage by radiation; the abbreviation for *roentgen equivalent in man* (Section 18-7). The roentgen is named for Wilhelm Konrad Roentgen (1845–1923).

replacement reaction A chemical reaction in which one element replaces another in a compound or in an aqueous solution (Sections 9-8 and 16-6). These are two examples:

$$Fe_2O_2 + 2\ Al \rightarrow Al_2O_3 + 2\ Fe$$

$$Mg(s) + 2\ Ag^+(aq) \rightarrow Mg^{2+}(aq) + 2\ Ag(s)$$

See also *activity series.*

representative elements In the Periodic Table, the elements in groups 1A through 8A. In their electron configurations, these elements are adding electrons to s and p sublevels (Sections 5-2 and 6-4). See also *transition elements.*

resonance Electron sharing across more than two adjacent atoms in a molecule; also called electron delocalization (Section 19-5).

reverse reaction The reaction shown from right to left in the chemical equation for a reversible reaction. For example, in the equation

$$3\ H_2(g) + N_2(g) \rightleftarrows 2\ NH_3(g)$$

the reverse reaction is the reaction of ammonia to form hydrogen and nitrogen (Section 15-4). See also *forward reaction.*

reversible reaction, irreversible reaction A reversible reaction is a chemical reaction in which the products can react with one another to re-form the reactants; in a chemical equation, a double arrow shows reversibility:

$$3\ H_2(g) + N_2(g) \rightleftarrows 2\ NH_3(g)$$

The products of an irreversible reaction cannot react with one another to re-form the reactants; in a chemical equation, a single arrow shows irreversibility:

$$HCl(aq) + NaOH(aq) \rightarrow H_2O(\ell) + NaCl(aq)$$

(Section 15-4). See also *forward reaction* and *reverse reaction.*

rounding off Deciding what the last significant digit in a calculated value will be (Section 2-5).

salt An ionic compound that's not an acid, base, or oxide; sodium chloride, NaCl, is one example (Section 16-2).

salt bridge In a voltaic cell, an aqueous solution of a salt that connects the cathode and anode compartments to maintain the same total charge on all cations and all anions in each compartment (Section 17-11). See *electrochemical cell.*

saturated, unsaturated, supersaturated An aqueous solution is said to be saturated if it contains the maximum amount of solute that can dissolve in the solvent at a given temperature and is said to be unsaturated if it contains less than the maximum (Section 15-3). By cooling a saturated solution, it's sometimes possible to prepare a supersaturated solution—one that contains more than the maximum amount of solute that could be dissolved in the solvent at the lower temperature (Section 15-9).

Schrödinger equation A mathematical equation used in quantum mechanics to calculate the probability of finding an electron in a certain region around a nucleus (Section 6-2). See also *energy level, orbital,* and *sublevel.*

scientific notation See *exponential notation, scientific notation.*

scintillation counter An instrument used to determine the quantity of alpha, beta, or gamma radiation emitted by a radioactive sample by measuring the number of scintillations caused by the radiation (Section 18-5).

shielding effect In an atom, the repulsion of electrons in outer energy levels by electrons in inner energy levels; because this repulsion has the effect of weakening the attraction between the outer electrons and the nucleus, it can be thought of as shielding the outer electrons from the nuclear charge (Section 8-1).

SI (le Système International d'Unités) A modified version of the metric system used by chemists (Section 3-3).

significant digits The digits in a measured value that show how precisely the measurement was made (Section 2-3).

simplest formula A formula that shows only the smallest whole-number relationships of atoms in a compound. If the molecular formula or condensed formula for a compound is known, the simplest formula can be found by dividing all of the subscripts in the molecular or condensed formula by the largest integer that will divide evenly into all of them; for example, for the molecular formula $C_6H_{12}O_6$, the simplest formula is CH_2O (Section 11-4).

single covalent bond See *covalent bond.*

skeleton equation For a redox reaction, an abbreviated, unbalanced equation that shows only the reactants being oxidized and reduced and their products; if the reaction occurs in aqueous solution, the skeleton equation includes a statement that the reaction takes place in acid or base (Section 17-10). This is an example:

$$MnO_4^- + Fe^{2+} \rightarrow Mn^{2+} + Fe^{3+} \quad \text{(acid)}$$

solubility equilibrium An equilibrium reaction between a solid ionic compound and its ions in saturated aqueous solution, for example,

$$CdS(s) \rightleftarrows Cd^{2+}(aq) + S^{2-}(aq)$$

(Section 17-1).

solute See *solution.*

solution A homogeneous mixture that's a liquid or a gas. (Solid homogeneous mixtures of metals are called alloys.) The solvent is the component of a solution that has retained its physical state in the solution process and that's present in largest amount; every other component is a solute (Section 3-1 and Introduction to Chapter 15). See also *aqueous solution* and *concentration.*

solvent See *solution.*

spdf notation The system of notation used to describe electron configurations, in which energy levels outward from the nucleus are represented by the integers 1, 2, 3, . . . ; sublevels are represented by the letters s, p, d, and f; and the number of electrons in a sublevel is shown as a superscript. One example is Cl^{17}: $1s^2 2s^2 2p^6 3s^2 3p^5$ (Section 6-5). See also *electron configuration, abbreviated electron configuration.*

specific gravity The density of a liquid divided by the density of water. In the SI, the specific gravity of a liquid has the same numerical value as its density and no units; for example, the density of mercury is 13.6 g/cm^3, and its specific gravity is 13.6 (Section 3-5).

spectator ion See *molecular equation, ionic equation, net ionic equation.*

standard temperature and pressure (STP) Conditions of temperature and pressure used as standards in the description of gas samples: 273 K and 1 atm (Section 13-6). See also *atmosphere.*

state One of the forms of matter described as solid, liquid, or gas. In a chemical equation, the state of a reactant or product can be shown by writing (s), (ℓ), or (g) after its symbol or formula (Section 9-6).

stoichiometry Calculations of masses of reactants or products in a chemical reaction by dimensional analysis, based on the chemical equation for the reaction (Introduction to Chapter 12).

STP See *standard temperature and pressure.*

strong acid, weak acid In the Arrhenius theory, a strong acid is a substance that forms hydronium ion in aqueous solution by an irreversible reaction:

$$H_2O(\ell) + HA(aq) \rightarrow H_3O^+(aq) + A^-(aq)$$

There are six common strong acids: HCl, HBr, HI, HNO_3, H_2SO_4, $HClO_4$. A weak acid is a substance that forms hydronium ion in aqueous solution by a reversible reaction:

$$H_2O(\ell) + HA(aq) \rightleftarrows H_3O^+(aq) + A^-(aq)$$

In this book, any acid except the six strong acids in the preceding list is assumed to be weak (Section 15-5). See also *reversible reaction, irreversible reaction.*

strong base, weak base In the Arrehnius theory, a strong base is a substance that forms hydroxide ion in aqueous solution by an irreversible reaction, for example,

$$NaOH(s) \rightarrow Na^+(aq) + OH^-(aq)$$

There are six common strong bases: LiOH, NaOH, KOH, $Ca(OH)_2$, $Sr(OH)_2$, and $Ba(OH)_2$. A weak base is a substance that forms hydroxide ion in aqueous solution by a reversible reaction; the most important example is ammonia:

$$NH_3(aq) + H_2O(\ell) \rightleftarrows NH_4^+(aq) + OH^-(aq)$$

(Section 15-6). See also *reversible reaction, irreversible reaction.*

sublevel A sublevel is a group of orbitals of the same kind. Sublevels are designated by the letters s, p, d, and f: An s sublevel can hold a maximum of two electrons, a p sublevel a maximum of six, a d sublevel a maximum of 10, and an f sublevel a maximum of 14 (Section 6-2). See also *electron configuration, abbreviated electron configuration; energy level; orbital;* and *spdf notation.*

sublimation Movement of particles from the solid state to the gas state (Section 14-1). See also *boiling, evaporation, freezing,* and *melting.*

substitution reaction A reaction in which a hydrogen atom in a hydrocarbon molecule is replaced by another atom or group of atoms:

$$-\overset{|}{\underset{|}{C}}-H \rightarrow -\overset{|}{\underset{|}{C}}-Z$$ (Section 19-80).

supersaturated See *saturated, unsaturated, supersaturated.*

symbol An abbreviation of the name for an element; examples are S for sulfur and Li for lithium. In its most common meaning, a symbol represents one atom of the element (Sections 4-2 and 4-5). See also *formula.*

synthetic chemistry The branch of chemistry that undertakes the preparation of compounds from elements or other compounds (Section 3-1). See also *analytical chemistry.*

temperature A measure of heat intensity. The three temperature scales that are most commonly used are the Fahrenheit scale (for Gabriel Daniel Fahrenheit, 1696–1736), the Celsius scale (for Anders Celsius, 1701–1744), and the Kelvin scale (for William Thomson, 1st Baron Kelvin, 1824–1907); for scientific work, the Celsius scale and the Kelvin scale (also called the absolute scale) are used. The equations for conversions among these scales are $K = C + 273$ and $(F - 32)/180 = C/100$ (Section 3-6).

theory A hypothesis that's trusted because it's been confirmed by tests; however, no scientific theory is ever trusted completely, because any theory may be disproved by further tests (Section 1-3).

torr See *atmosphere.*

transition elements, inner-transition elements; transition metals, inner-transition metals In the Periodic Table, the elements in the B groups. In their electron configurations, these elements are adding electrons to d and f sublevels (Sections 5-2 and 6-4). Used as specific terms, *transition elements* or *transition metals* refers to elements that are filling d sublevels, and *inner-transition elements* or *inner-transition metals* refers to elements that are filling f sublevels. Used as a general term, *transition elements* or *transition metals* refers to both of these specific classes (Section 10-9). See also *representative elements.*

transition metals See *transition elements.*

transmutation A nuclear reaction in which one nucleus is converted to another by bombarding it with subatomic particles, such as neutrons or alpha particles (Section 4-8). One example is

$${}^{235}_{92}U + {}^{1}_{0}n \rightarrow {}^{142}_{56}Ba + {}^{92}_{36}Kr + 2\,{}^{1}_{0}n$$

See also *nuclear equation* and *nuclear reaction.*

transuranium elements In the Periodic Table, the elements with atomic numbers higher than 92, the atomic number of uranium. All of these elements have been prepared synthetically, by nuclear reactions (Sections 4-8 and 18-4). See also *nuclear equation* and *nuclear reaction.*

triple covalent bond See *covalent bond.*

unit map In dimensional analysis, a diagram that uses arrows to show the unit conversions to be made in solving a problem. For example, for the unit conversions of kilograms to grams to milligrams, the unit map is kg → g → mg (Section 3-9).

unsaturated See *saturated, unsaturated, supersaturated.*

valence electrons In the atom of a representative element, the electrons in the s and p sublevels of the outermost energy level. For example, in F^9: $1s^22s^22p^5$, the valence electrons are the $2s^22p^5$ electrons. Because the outermost energy level contains the valence electrons, it's sometimes referred to as the valence shell (Section 6-7). See also *electron configuration; abbreviated electron configuration; Lewis formula; Lewis symbol; octet rule;* and *spdf notation.*

valence shell See *valence electrons.*

velocity of electromagnetic radiation (c) The speed with which electromagnetic radiation travels in a vacuum, 3.00×10^8 m/s (Section 3-8). See also *frequency* and *wavelength.*

voltaic cell See *electrochemical cell.*

volumetric flask A flask made so that, when it's filled with a liquid to a line etched on its neck, the volume of the liquid will be a specified volume, for example, 1000 mL. Volumetric flasks are often used to prepare aqueous solutions with accurately known concentrations (Section 15-3).

wavelength (λ) In the description of electromagnetic radiation, the distance from a point on one wave to the corresponding point on an adjacent wave; units often used for wavelength are meters or nanometers. As the wavelength of electromagnetic radiation increases, its frequency decreases (Section 3-8). See also *velocity of electromagnetic radiation.*

weak acid See *strong acid, weak acid.*

weak base See *strong base, weak base.*

zymogen A protein, not active as a catalyst, that can be converted to an enzyme (Section 20-5). See also *catalyst* and *enzyme.*

Index

Boldface page numbers indicate pages on which the index term is defined. Page numbers followed by f indicate figures; page numbers followed by t indicate tables.